Principles of Sustainable Energy

Mechanical Engineering Series

Frank Kreith, Series Editor

Alternative Fuels for Transportation
Edited by Arumugam S. Ramadhas

Computer Techniques in Vibration
Edited by Clarence W. de Silva

Distributed Generation: The Power Paradigm for the New Millennium
Edited by Anne-Marie Borbely and Jan F. Kreider

Elastic Waves in Composite Media and Structures: With Applications to Ultrasonic Nondestructive Evaluation
Subhendu K. Datta and Arvind H. Shah

Elastoplasticity Theory
Vlado A. Lubarda

Energy Audit of Building Systems: An Engineering Approach
Moncef Krarti

Energy Conversion
Edited by D. Yogi Goswami and Frank Kreith

Energy Management and Conservation Handbook
Edited by Frank Kreith and D. Yogi Goswami

The Finite Element Method Using MATLAB, Second Edition
Young W. Kwon and Hyochoong Bang

Fluid Power Circuits and Controls: Fundamentals and Applications
John S. Cundiff

Fundamentals of Environmental Discharge Modeling
Lorin R. Davis

Handbook of Energy Efficiency and Renewable Energy
Edited by Frank Kreith and D. Yogi Goswami

Heat Transfer in Single and Multiphase Systems
Greg F. Naterer

Heating and Cooling of Buildings: Design for Efficiency, Revised Second Edition
Jan F. Kreider, Peter S. Curtiss, and Ari Rabl

Intelligent Transportation Systems: Smart and Green Infrastructure Design, Second Edition
Sumit Ghosh and Tony S. Lee

Introduction to Biofuels
David M. Mousdale

Introduction to Precision Machine Design and Error Assessment
Edited by Samir Mekid

Introductory Finite Element Method
Chandrakant S. Desai and Tribikram Kundu

Machine Elements: Life and Design
Boris M. Klebanov, David M. Barlam, and Frederic E. Nystrom

Mathematical and Physical Modeling of Materials Processing Operations
Olusegun Johnson Ilegbusi, Manabu Iguchi, and Walter E. Wahnsiedler

Mechanics of Composite Materials
Autar K. Kaw

Mechanics of Fatigue
Vladimir V. Bolotin

Mechanism Design: Enumeration of Kinematic Structures According to Function
Lung-Wen Tsai

Mechatronic Systems: Devices, Design, Control, Operation and Monitoring
Edited by Clarence W. de Silva

The MEMS Handbook, Second Edition (3 volumes)
Edited by Mohamed Gad-el-Hak

MEMS: Introduction and Fundamentals
MEMS: Applications
MEMS: Design and Fabrication

Multiphase Flow Handbook
Edited by Clayton T. Crowe

Nanotechnology: Understanding Small Systems
Ben Rogers, Sumita Pennathur, and Jesse Adams

Nuclear Engineering Handbook
Edited by Kenneth D. Kok

Optomechatronics: Fusion of Optical and Mechatronic Engineering
Hyungsuck Cho

Practical Inverse Analysis in Engineering
David M. Trujillo and Henry R. Busby

Pressure Vessels: Design and Practice
Somnath Chattopadhyay

Principles of Solid Mechanics
Rowland Richards, Jr.

Principles of Sustainable Energy
Frank Kreith

Thermodynamics for Engineers
Kau-Fui Vincent Wong

Vibration Damping, Control, and Design
Edited by Clarence W. de Silva

Vibration and Shock Handbook
Edited by Clarence W. de Silva

Viscoelastic Solids
Roderic S. Lakes

Principles of Sustainable Energy

Frank Kreith
Lead Author

Jan F. Kreider
Co-Author

CRC Press
Taylor & Francis Group
Boca Raton London New York

CRC Press is an imprint of the
Taylor & Francis Group, an **informa** business

CRC Press
Taylor & Francis Group
6000 Broken Sound Parkway NW, Suite 300
Boca Raton, FL 33487-2742

Printed in the United States of America on acid-free paper
10 9 8 7 6 5 4 3

International Standard Book Number: 978-1-4398-1407-9 (Hardback)

Library of Congress Cataloging-in-Publication Data

Kreith, Frank.
Principles of sustainable energy / by Frank Kreith and Jan F. Kreider.
p. cm. -- (Mechanical engineering series ; 46)
Includes bibliographical references and index.
ISBN 978-1-4398-1407-9
1. Renewable energy sources. I. Kreider, Jan F., 1942- II. Title. III. Series.

TJ808.K74 2011
621.042--dc22 2010027386

Visit the Taylor & Francis Web site at
http://www.taylorandfrancis.com

and the CRC Press Web site at
http://www.crcpress.com

Contents

Foreword xvii
Preface xix
Contributing Authors xxiii

1. Introduction to Sustainable Energy 1
1.1 Global Sustainability 5
1.1.1 Sustainability Based on Water 5
1.1.2 Food Supply 9
1.1.3 Energy 10
1.2 Role of Energy Conservation 18
1.3 Nuclear Energy 26
1.3.1 Available Nuclear Resources 26
1.4 Hydrogen 29
1.5 Geothermal Energy 33
1.6 Global Status and Potential of Renewable Energy 36
1.6.1 Biomass and Hydroelectric Sources 36
1.6.2 Wind Power 39
1.6.3 Solar Energy 41
1.7 Summary of Renewable Energy Resources 43
1.8 Orders of Magnitude 46
Energy Units and Conversion Factors 54
Problems 57
Online Resources 60
References 61
Suggested Readings 63

2. Economics of Energy Generation and Conservation Systems 65
2.1 Time Value of Money 66
2.2 Inflation 70
2.3 Societal Costs 72
2.4 Total Life Cycle Costs 74
2.5 Internal Rate of Return 76
2.6 Levelized Cost of Energy 82
2.7 Example of a Typical Cash Flow Scenario 83
2.8 Energy Analysis Methodology 87
2.8.1 Process Chain Analysis 88
2.8.2 Input–Output (I/O) Analysis 89
2.9 EROI for a Wind Energy System 102
2.10 EROI for Nuclear Power 105

Problems 109
Appendix 2.A Capital Recovery Factor 112
References 115

3. Wind Energy 117
Contributing Author: Gary E. Pawlas

3.1 Wind Power in a Nutshell 117
3.2 Power and Energy 119
3.3 Fact or Fiction: Common Questions about Wind Turbines 122
3.3.1 Noise Issues 122
3.3.2 Lifetime Environmental Impact 123
3.3.3 Bird and Bat Kills 124
3.4 History of Wind Turbine Development: HAWTs and VAWTs 125
3.4.1 VAWT 128
3.4.2 HAWT 128
3.5 Introduction to Wind Turbine Performance 129
3.6 Aerodynamics 130
3.7 Wind Characteristics 134
3.7.1 Wind Generation 134
3.7.2 Distribution of Wind 135
3.7.3 Wind Speed Increasing with Height 137
3.7.4 Log Law Wind Speed Profile 138
3.7.5 Power Law Wind Speed Profile 140
3.7.6 Probability of Observing a Given Wind Speed 141
3.8 Turbine Performance 146
3.8.1 Control Schemes 156
3.9 Cost of Energy 158
3.9.1 Definitions 159
3.10 Wind Farms 163
3.11 Offshore Wind Energy 167
3.12 Additional Topics for Study 173
Acknowledgment 174
Problems 174
References 177

4. Capturing Solar Energy through Biomass 181
Contributing Authors: Robert C. Brown and Mark M. Wright

4.1 Introduction 181
4.2 Biomass Feedstocks 182
4.2.1 Introduction 182
4.2.2 Waste Materials 182
4.2.3 Energy Crops 183
4.2.4 Size of Resource Base 185

4.2.5 Important Properties of Biomass ... 187
4.2.6 Summary ... 191
Problems ... 191
4.3 Conversion of Biomass to Gaseous Fuels ... 192
4.3.1 Introduction ... 192
4.3.2 Anaerobic Digestion ... 192
4.3.3 Thermal Gasification ... 195
4.3.4 Summary ... 203
Problems ... 204
4.4 Transportation Fuels from Biomass ... 204
4.4.1 Introduction ... 204
4.4.2 Grain Ethanol ... 207
4.4.3 Cellulosic Ethanol ... 209
4.4.4 Other Fuels from Sugar Fermentation ... 210
4.4.5 Ethanol from Syngas ... 212
4.4.6 Ethanol from Bio-Oil ... 213
4.4.7 Methanol ... 213
4.4.8 Mixed Alcohols ... 214
4.4.9 Fischer–Tropsch Liquids ... 215
4.4.10 Fuels from Pyrolysis Oils ... 217
4.4.11 Fuels from Biocrude ... 219
4.4.12 Fuels from Lipids ... 219
4.4.13 Methylated Furans ... 223
4.4.14 Compressed Gases as Transportation Fuel ... 225
4.4.15 Well-to-Wheel Analysis of Transportation Fuels ... 226
4.4.16 Summary ... 228
Problems ... 229
4.5 Conversion of Biomass to Electricity ... 230
4.5.1 Introduction ... 230
4.5.2 Direct Combustion ... 230
4.5.3 Combustion Equipment ... 233
4.5.4 Summary ... 236
Problems ... 236
4.6 Conclusions ... 237
Additional Problems ... 237
References ... 240

5. Fundamentals of Solar Radiation ... 247
5.1 The Physics of the Sun and Its Energy Transport ... 247
5.2 Thermal Radiation Fundamentals ... 248
5.2.1 Black-Body Radiation ... 250
5.2.2 Radiation Function Tables ... 250
5.2.3 Intensity of Radiation and Shape Factor ... 252
5.2.4 Transmission of Radiation through a Medium ... 258

5.3 Sun–Earth Geometric Relationship ... 259
5.3.1 Solar Time and Angles ... 262
5.3.2 Sun-Path Diagram ... 269
5.3.3 The Shadow-Angle Protractor ... 271
5.4 Solar Radiation ... 275
5.4.1 The Extraterrestrial Solar Radiation ... 276
5.5 Estimation of Terrestrial Solar Radiation ... 280
5.5.1 Atmospheric Extinction of Solar Radiation ... 281
5.5.2 Solar Radiation on Clear Days ... 282
5.5.3 Solar Radiation on a Tilted Surface ... 283
5.5.4 Monthly Solar Radiation Estimation Models ... 287
5.6 Models Based on Long-Term Measured Horizontal Solar Radiation ... 290
5.6.1 Monthly Solar Radiation on Tilted Surfaces ... 290
5.6.2 Circumsolar or Anisotropic Diffuse Solar Radiation ... 295
5.6.3 Daily Solar Radiation on Tilted Surfaces ... 296
5.6.4 Hourly Solar Radiation on Tilted Surfaces ... 297
5.7 Measurement of Solar Radiation ... 299
5.7.1 Instruments for Measuring Solar Radiation and Sunshine ... 299
5.7.2 Detectors for Solar Radiation Instrumentation ... 303
5.7.3 Measurement of Sunshine Duration ... 303
5.7.4 Measurement of Spectral Solar Radiation ... 304
5.7.5 Solar Radiation Data and Web Sites ... 304
Problems ... 305
Appendix A.5: Calculation of True Solar Time ... 308
References ... 362

6. Photovoltaics ... 365
6.1 Semiconductors ... 367
6.1.1 *p–n* Junction ... 369
6.1.2 The Photovoltaic Effect ... 371
6.2 Analysis of Photovoltaic Cells ... 375
6.2.1 Efficiency of Solar Cells ... 380
6.2.2 Multijunction Solar Cells ... 381
6.2.3 Design of a Photovoltaic System ... 382
6.3 Manufacture of Solar Cells and Panels ... 384
6.3.1 Single Crystal and Polycrystalline Cells ... 384
6.3.2 Amorphous Silicon and Multijunction Thin-Film Fabrication ... 388
6.4 Design for Remote Photovoltaic Applications ... 389
6.4.1 Estimation of Loads and Load Profiles ... 390
6.4.2 Estimation of Available Solar Radiation ... 391
6.4.3 PV System Sizing ... 392
6.4.4 Water Pumping Applications ... 394

6.5 Thin-Film PV Technology 395
6.6 Multilayer PV Technology 400
Problems 406
References 410
Suggested Readings 411

7. Solar Heating and Cooling of Buildings 413
7.1 Radiative Properties and Characteristics of Materials 414
7.1.1 Selective Surfaces 419
7.1.2 Reflecting Surfaces 419
7.1.3 Transparent Materials 421
7.2 Flat-Plate Collectors 423
7.2.1 Liquid-Type Collectors 423
7.2.2 Air-Type Collectors 424
7.2.3 Glazings 425
7.2.3.1 Absorbers 426
7.2.4 Energy Balance for a Flat-Plate Collector 427
7.2.4.1 Collector Heat-Loss Conductance 428
7.2.5 Thermal Analysis of Flat-Plate Collector–Absorber Plate 433
7.2.6 Collector Efficiency Factor 437
7.2.7 Collector Heat-Removal Factor 438
7.2.8 Transient Effects 442
7.2.9 Air-Cooled Flat-Plate Collector 445
7.3 Evacuated Tube Collectors 446
7.3.1 Thermal Analysis of an Evacuated Tube Collector 448
7.4 Experimental Testing of Collectors 449
7.4.1 Testing Standards for Solar Thermal Collectors 452
7.4.1.1 Time Constant 452
7.4.1.2 Thermal Performance 453
7.4.1.3 Incidence Angle Modifier 453
7.5 Calculations of Heating and Hot Water Loads in Buildings 453
7.5.1 Calculation of Heat Loss 455
7.5.2 Internal Heat Sources in Buildings 458
7.5.3 The Degree-Day Method 458
7.5.4 Service Hot Water Load Calculation 462
7.6 Solar Water-Heating Systems 464
7.6.1 Natural Circulation Systems 464
7.6.2 Forced-Circulation Systems 467
7.6.3 Industrial Process Heat Systems 469
7.7 Liquid-Based Solar-Heating Systems for Buildings 469
7.7.1 Physical Configurations of Active Solar Heating Systems 470
7.7.2 Solar Collector Orientation 471
7.7.3 Fluid Flow Rates 471

7.7.4 Unglazed Transpired Wall System for Air Preheating 472
7.8 Methods of Modeling and Design of Solar Heating Systems 475
7.8.1 Design of a Liquid-Based Solar Heating System by *f*-Chart 475
7.8.2 The Utilizability Method 484
7.9 Solar Cooling 484
7.9.1 Cooling Requirements for Buildings 485
7.9.2 Vapor-Compression Cycle 489
7.9.3 Absorption Air-Conditioning 494
7.10 Solar Desiccant Dehumidification 505
7.10.1 Solid Desiccant Cooling System 505
7.10.2 Liquid Desiccant Cooling System 508
Problems 510
Appendix A 522
References 538

8. Solar Process Heat and Thermal Power 541
8.1 Historical Perspective 541
8.2 Solar Industrial Process Heat 544
8.2.1 SIPH for Textile Industries 549
8.2.2 SIPH System for Milk Processing 550
8.3 Parabolic Collectors 553
8.3.1 Compound Parabolic Concentrators 555
8.3.2 Single-Axis Tracking Parabolic Trough Collectors 559
8.4 Long-Term Performance of SIPH Systems 569
8.4.1 The Utilizability Method 570
8.4.2 Critical Solar Intensity Ratio *X* 571
8.4.3 Collection Period (Δt_c) 572
8.4.4 Empirical Expressions for Utilizability 573
8.4.5 Yearly Collector Performance 578
8.4.6 TRNSYS—Computer Simulation Program 581
8.5 Solar Thermal Power 581
8.5.1 Rankine Cycle 581
8.5.2 Components of a Rankine Power Plant 586
8.5.3 Choice of Working Fluid 588
8.6 Examples of Solar Thermal Power Plants 591
8.6.1 Parabolic Trough–Based Power Plant 591
8.6.2 Central Receiver Systems 596
8.7 Parabolic Dish Systems and Stirling Engines 605

8.7.1 Thermodynamics of a Stirling Cycle ... 606
8.7.2 Examples of Solar Stirling Power Systems ... 609
Problems ... 613
References ... 616

9. Passive Solar Heating, Cooling, and Daylighting ... 621
Contributing Author: Jeffrey H. Morehouse

9.1 Introduction ... 621
9.1.1 Distinction between a Passive System and Energy Conservation ... 621
9.2 Key Elements of Economic Consideration ... 622
9.2.1 Performance: Net Energy Savings ... 622
9.2.2 Cost: Over and Above "Normal" Construction ... 622
9.2.3 General System Application Status and Costs ... 622
9.3 Solar Thermosyphon Water Heating ... 623
9.3.1 Thermosyphon Concept ... 624
9.3.2 Thermo-Fluid System Design Considerations ... 625
9.4 Passive Solar Heating Design Fundamentals ... 627
9.4.1 Types of Passive Heating Systems ... 628
9.4.2 Fundamental Concepts for Passive Heating Design ... 630
9.5 Passive Design Approaches ... 630
9.5.1 The First Level: Generalized Methods ... 632
9.5.1.1 Load ... 632
9.5.1.2 Solar Savings Fraction ... 632
9.5.1.3 LCR ... 633
9.5.1.4 Storage ... 633
9.5.2 The Second Level: LCR Method ... 641
9.5.3 The Third Level: SLR Method ... 665
9.6 Passive Space Cooling Design Fundamentals ... 670
9.6.1 Solar Control ... 670
9.6.2 Natural Convection/Ventilation ... 670
9.6.3 Evaporative Cooling ... 673
9.6.4 Nocturnal and Radiative Cooling Systems ... 673
9.6.5 Earth Contact Cooling (or Heating) ... 675
9.6.5.1 Heat Transfer Analysis ... 676
9.6.5.2 Soil Temperatures and Properties ... 678
9.6.5.3 Generalized Results from Experiments ... 678
9.7 Daylighting Design Fundamentals ... 680
9.7.1 Lighting Terms and Units ... 680
9.7.2 Approach to Daylighting Design ... 680
9.7.3 Sun–Window Geometry ... 682
9.7.3.1 Solar Altitude Angle (α) ... 682

9.7.3.2 Sun–Window Azimuth Angle Difference (Φ) 682
9.7.4 Daylighting Design Methods 683
9.7.4.1 Lumen Method of Sidelighting (Vertical Windows) 683
9.7.4.2 Lumen Method of Skylighting 692
9.7.5 Daylighting Controls and Economics 697
Problems 700
Defining Terms 702
References 703
Suggested Readings 705
For Further Information 705

10. Energy Storage 707
10.1 Overview of Storage Technology 707
10.1.1 Applications 709
10.1.2 Technology Characterization 711
10.2 Mechanical Technologies 713
10.2.1 Pumped Hydroelectric Energy Storage 713
10.2.1.1 Turbines 715
10.2.2 Compressed Air Energy Storage 717
10.2.2.1 Round Trip Efficiency of CAES 719
10.2.2.2 Comparison between CAES and PHES 720
10.2.2.3 CAES Volumetric Energy Density 721
10.2.2.4 Existing CAES Plants 724
10.2.3 Flywheels 725
10.3 Direct Electrical Technologies 726
10.3.1 Ultracapacitors 726
10.3.2 Superconducting Magnetic Energy Storage 727
10.4 Fundamentals of Batteries and Fuel Cells 727
10.4.1 Principles of Battery Operation 729
10.4.2 Cell Physics 731
10.5 Rechargeable Batteries 732
10.5.1 Lead-Acid Batteries 733
10.5.2 Nickel Metal (Ni-Cd and Ni-MH) 734
10.5.3 Lithium Ion 736
10.5.4 Flow Batteries 737
10.6 Fuel Cells and Hydrogen 739
10.6.1 Principles of Fuel Cell Operation 739
10.6.2 Types of Fuel Cells 740
10.6.3 Generation of Hydrogen 742
10.6.4 Storage and Transport 743
10.6.5 Thermodynamics and Economics 744

10.7 Thermal Energy Storage 745
10.7.1 Sensible Heat 745
10.7.2 Phase Change Heat Storage 747
10.7.3 Thermochemical Storage 748
10.7.4 Applications 748
10.7.5 Thermal Storage for Concentrating Collector Systems 750
10.7.6 Overnight Storage for Buildings and Domestic Hot Water 752
10.7.7 Long-Term Performance of Solar Collector Systems with Storage 758
Problems 763
References 765
Battery Resources 767

11. Ocean Energy Conversion 769
11.1 Ocean Thermal Energy Conversion 769
11.1.1 Closed-Cycle Ocean Thermal Energy Conversion 771
11.1.2 Open-Cycle Ocean Thermal Energy Conversion 775
11.1.2.1 Direct Contact Evaporation and Condensation 778
11.1.3 Comparison of Open- and Closed-Cycle OTEC Systems 780
11.1.4 Cold-Water Pipe and Pumping Requirements 782
11.1.5 Economics 783
11.2 Tidal Energy 785
11.2.1 Introduction 785
11.2.2 Tidal Power 787
11.2.2.1 Economics of Tidal Power 791
11.3 Ocean Wave Energy 795
11.3.1 Deepwater Wave Power 795
11.3.2 Surface Wave Power 798
11.3.3 Wave Power Devices 801
11.3.3.1 Wave Capture Systems 801
11.3.3.2 The Oscillating Water Column 802
11.3.4 Wave Profile Devices 804
Problems 805
References 807

12. Transportation 809
12.1 Introduction 809
12.2 Alternative Fuels 810

12.3 Well-to-Wheel Analysis 812
12.4 Mass Transportation 813
12.5 Hybrid Electric Vehicles 815
12.6 Plug-In Hybrid Electric Vehicles 817
12.7 Advanced Ground Transportation with Biomass Fuel 821
12.7.1 Petroleum Requirement 823
12.7.2 Carbon Dioxide Emissions 823
12.8 Future All-Electric System 827
12.9 Hydrogen for Transportation 828
Problems 829
References 831

Index 833

Foreword

The book you are now holding, *Principles of Sustainable Energy*, arrives at an opportune moment. The growing worldwide realization of the need to leave behind our dependence on fossil fuels, and set course for a sustainable energy future, gives this volume abundant timeliness and relevancy. While support for clean energy alternatives has never been higher, the obstacles to achieving a true energy transformation are equally formidable. For those of us who have long known the value of clean energy technology, these are both exciting and challenging times.

This book examines the practical issues surrounding energy efficiency, and renewable energy concepts and systems. And a more appropriate author for such a work could not be found. Professor Frank Kreith has shown a lifelong devotion to exploring renewable technologies, and appropriately pushing them into mainstream use.

I first came to appreciate Prof. Kreith's depth of knowledge as an engineering student at Stanford. I benefited greatly from his authoritative text on heat transfer systems—it served as a foundation for much of my own early professional research. And later, Prof. Kreith's work as a pioneering solar energy engineer became even more pertinent when I was named director of the same Colorado laboratory where he had served as a senior research fellow.

To those who choose renewable energy as an academic field or career path, this text and the insights you gain from it will likely become fundamental to the achievements that await you. But, this text is not merely academic; it serves a valuable purpose in its clearheaded look at the energy picture from the ground up, and the environmental, economic, and sustainability benefits that renewable energy systems can provide.

Principles of Sustainable Energy combines the expertise earned from decades of practical and scholarly research with the most up-to-date analysis of the energy scene. For today, sustainable energy is no longer just a dream for the future; it is a growing field with incredible opportunity for anyone with the foresight to make its vision their own.

Dan E. Arvizu
President, Alliance for Sustainable Energy
Manager and operator of the National Renewable Energy Laboratory

Preface

This book is a combination of lecture notes for a course on sustainable energy that I introduced at the University of Colorado in 2006; contributions by experts in wind energy, passive solar design, and bioconversion; and updated material from the first edition of *Principles of Solar Engineering* by Frank Kreith and Jan F. Kreider, which was published in 1978, as well as from the second edition of *Principles of Solar Engineering* by D. Yogi Goswami, Frank Kreith, and Jan F. Kreider, which was published in 2000. The impetus for not merely revising the second edition was the recognition that transitioning from a fossil fuel–based economy to a sustainable energy system that utilizes renewable energy sources has now become inevitable and that this transition would be a challenge not only for engineering, but would also include economic and environmental considerations.

Furthermore, the two editions of *Principles of Solar Engineering* were essentially restricted to the technical aspects related to solar thermal energy technologies, and relegated wind power, photovoltaic energy conversion, ocean energy technologies, energy storage, and economics to a secondary consideration. This book has shifted the emphasis and can be used for engineering courses that have a broad perspective for planning a sustainable energy future, but is also suitable for courses that stress solar thermal energy technology.

The book is designed for courses at the senior or first-year graduate level in engineering and assumes that the reader has a basic understanding of thermodynamics and heat transfer, as well as knowledge of spreadsheets and computer skills. During the past decade, Web sites have been developed by the National Renewable Energy Laboratory (NREL) and other organizations that make it possible to obtain the performance of renewable energy.

Chapter 1 presents an overview of the physical limitations that the global availability of water, food, and energy place upon a sustainable population of our planet. It emphasizes the interrelation between the social, economic, environmental, and technical aspects of the sustainability challenge. It presents a summary of available resources, options of energy generation and conservation, and the need to consider the financial ramifications, as well as the energy return on energy investment of various technologies for future energy generation. This is particularly important during the transition from fossil to sustainable resources, because at the present time more than 80% of global energy generation comes from fossil fuels. Consequently, during the transition period when renewable energy systems have to be built, the construction of these systems will rely mostly on fossil fuel input, which inevitably poses a danger to the environment by causing pollution and global warming. It will also require that a certain percentage of our current fossil energy generation be set aside for the construction of renewable systems that will not generate any energy until they begin operation in the future. Therefore, it is of vital importance to utilize the most energy-efficient technologies to make the transition occur smoothly and without creating unnecessary economic, social, and environmental problems.

Chapter 2 covers the fundamentals of engineering economics with emphasis on the levelized cost of both energy conservation and generation. It also introduces a convenient method of estimating the energy return on energy investment (EROI) of technologies for conservation measures and generation of energy. Chapter 3 deals with wind energy, which is, at the present, not only one of the most advanced renewable technologies, but is also economically competitive with fossil fuels. Moreover, it has the highest return on energy investment of all renewable energy generation technologies. This is followed by Chapter 4 on biomass and biofuels, which are likely to play an important role, particularly in the future transportation sector. Chapter 5 deals with the fundamentals of solar radiation, followed by photovoltaic energy (PV) in Chapter 6. PV is currently experiencing an exponential growth in deployment, both for rooftop applications and for utility scale generation of electricity, and its cost is constantly decreasing. Traditional topics related to solar thermal heating and cooling, process heat, and power are presented in Chapters 7 and 8, followed by passive solar design methods and daylighting in Chapter 9. Part of Chapter 8 is devoted to technologies for concentrating solar radiation, because this may be an important facet of several future sustainable technologies.

It is well recognized that whereas fossil and nuclear technologies can provide uninterrupted base-load power, most of the renewable technologies are intermittent and will, therefore, require storage in order to eventually replace fossil-based energy generation. A major chapter (Chapter 10) is therefore devoted to energy storage, including hydrogen and fuel cells.

Chapter 11 deals with ocean energy. This is a form of sustainable energy, whose promise at the present time is uncertain because, despite the many ideas that have been proposed for conversion technologies such as wave and ocean thermal, no major plants have been built. The final chapter (Chapter 12) deals with ground transportation, which will become increasingly important as the availability of oil diminishes and its price escalates.

The book covers EROI and economics of nuclear power but not the details of nuclear fuel technology. I believe that the future of nuclear energy will depend largely on the economics of this technology. More than 100 nuclear power plants have operated safely for over 30 years, although no permanent safe disposal of the waste from this technology exists. But, in view of the fact that no nuclear power plant has been built in the United States during the past 30 years, the cost of future nuclear power plants and the long-term disposal of radioactive waste are uncertain. Hence, the level of deployment will have to wait the outcome of construction of new nuclear plants that will meet environmental and safety requirements to indicate what the true cost of nuclear power will be.

The preparation of this book has been aided by several contributing authors. Dr. Gary Pawlas wrote the chapter on wind energy. The chapter on biomass and biofuels has been contributed by Professor Robert Brown and Dr. Mark M. Wright, while I have updated Professor Jeffrey Morehouse's chapter on passive methods and daylighting from a previous version published in the *CRC Handbook of Energy Efficiency and Renewable Energy Technologies.*

Dr. Jan F. Kreider, my former doctoral student and long time professional collaborator did not participate with the writing of this book in any way, but gave me permission to use and/or adapt material from books that we have previously coauthored and proofread the final manuscript. To comply with previous contractual arrangements, I am happy to have Jan be named co-author on the title page in acknowledgment of his indirect contributions to this work. I also want to express my appreciation to my friend Dr. Yogi D. Goswami, who made major contributions to the second edition of *Principles of Solar Engineering*, but decided not to participate in the current project. I have received valuable help and support from students in the courses on sustainable energy that I have taught at the University of Colorado, Boulder, Colorado, and I would like to thank Professors Isaac Garaway and Michael Hannigan for assisting me in teaching some of these courses.

Finally, I want to specifically thank my editorial assistant, Bev Weiler, and my students, Terese Decker and Robert Threadgill. Without Bev, Terese, and Robert, who transcribed my dictations, this book could not have been completed. Robert also prepared the solutions manual for the book.

An effort has been made to simplify and streamline the parts of this book that have been adapted from the first and second editions of *Principles of Solar Engineering*. Moreover, technologies that have not found commercial applications during the past decade, such as solar ponds and solar thermal air collectors, have been deleted. In their place, I have put more detailed design procedures for parabolic trough and power tower concentrator systems, and information about high-efficiency multilayer PV cells with concentrators that are being developed for utility applications. And, it goes without saying that writing this book would not have been possible without the love and support from my wife, Marion.

The field of sustainable energy is vast and changing rapidly. A single book cannot cover all aspects of sustainable and renewable energy. I have therefore provided a list of Web sites that the reader will find useful in keeping up to date and delving in more detail into any of the topics covered in the book. I would like to receive comments from teachers and users of the book to correct errors and keep the information current in this exciting challenge of developing a sustainable energy future.

Frank Kreith

Editor's Note

After this book was typeset, the National Renewable Energy Laboratory announced that the Solar Advisory Model (SAM), which combines detailed performance models with typical financing (as described in Chapter 2) has been expanded to include all of the renewable energy technologies described in the book. SAM is the best available model to analyze the impact of physical system design on the overall economics (including the levelized cost of energy). The model is in the public domain and can be accessed at https://www.nrel.gov/analysis/sam/

Contributing Authors

Robert C. Brown
Department of Mechanical Engineering
Iowa State University
Ames, Iowa

Jeffrey H. Morehouse
Department of Mechanical Engineering
University of South Carolina
Columbia, South Carolina

Gary E. Pawlas
Department of Mechanical Engineering
University of Colorado
Boulder, Colorado

Mark M. Wright
Department of Mechanical Engineering
Iowa State University
Ames, Iowa

1

Introduction to Sustainable Energy

> It is evident that the fortunes of the world's human population, for better or for worse, are inextricably interrelated with the use that is made of energy resources.
>
> **—M. King Hubbert (1969)**

In 1983, the United Nations set up an independent body headed by Gro Harlem Brundtland, the former Prime Minister of Norway, entitled the World Commission on Environment and Development. The commission was asked to examine the critical, environmental, and developmental problems facing our planet, and to formulate realistic proposals to solve them. The proposed solution was to make sure that human progress can be sustained, but without bankrupting the resources of future generations. The outcome of this study was a book entitled *Our Common Future* published by Oxford University Press in 1987 [1]. It was reprinted six times that year and five times more in the subsequent 2 years. The jacket of the book reads "This is the most important document of the decade on the future of the world." It remains a valuable guide for sustainable development today.

Twenty years ago, the study concluded that worldwide efforts to guard and maintain human progress to meet human needs and to realize human ambition are unsustainable, both in developed and developing nations, because they use too many of our already overdrawn environmental resources to be affordable into the future without creating hazards for future generations. We are still continuing to borrow environmental capital from future generations without concern of repaying them. Our children will have to pay for our spendthrift ways while most of the decision makers of the day will be long dead before the planet feels the adverse effects of global warming, air pollution, and, most of all, depletion of land and fossil fuel resources.

The commission claimed that humanity has the ability to make development sustainable and provided a definition that can be a guide for this book: "Sustainable development should meet the needs of the present without compromising the ability of future generations to meet their own needs." The commission believed that sustainable global development requires that those who are more affluent adopt lifestyles that can be accommodated by the planet's ecological means, particularly their use of energy. Further, rapidly growing populations increase the pressure on resources. Sustainable development can, therefore, only be achieved if population size and growth are reduced to a point where they no longer exceed the productive potential of our ecosystems.

Painful choices will have to be made and, in the final analysis, sustainable development must rest on personal commitment and political will. The commission realized that future economic growth must be less energy intensive than growth in the past. Energy efficiency policy must be the primary vehicle of national energy strategies for sustainable development, but the commission also realized that energy efficiency can only buy time for the world to develop energy paths based on renewable sources and that this goal must be the foundation of any global energy strategy for the twenty-first century.

America has made two major energy transitions in its relatively short history. Before the Civil War, American industry depended on wood, wind, water, and animals for its heat, energy, and power. After the Civil War, these sources gave way to coal, which was more economical and technically more convenient for industrial process heat, home heating, and the development of railroad transportation in the country. Coal continued to supply the major part of U.S. energy needs between 1885 and the start of World War II in 1941. But changes in technology and the increased availability of oil and natural gas paved the way for the second energy transition, as shown in Figure 1.1. During the 1950s, the second transition from coal to oil and natural gas took place.

The second, as well as the first, transitions were from one or more relatively abundant type of fuel to another abundant type. Both resulted from

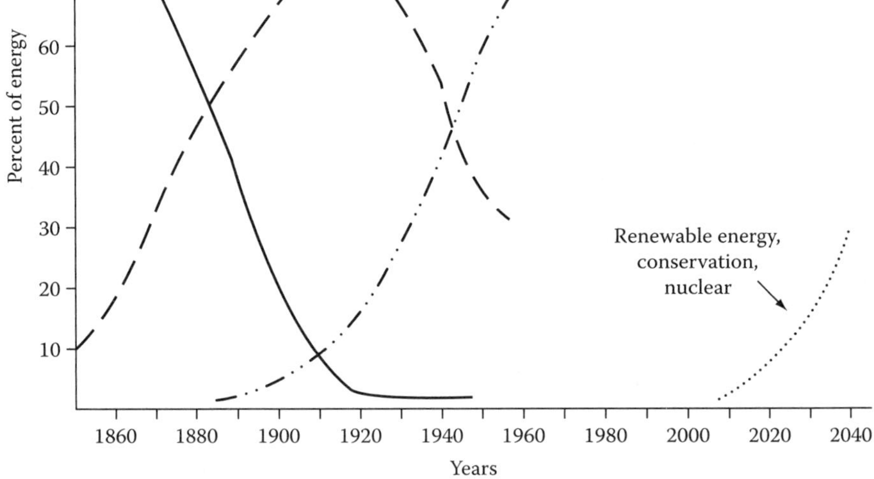

FIGURE 1.1
Schematic diagram of energy transitions.

technological advances, lower costs, greater versatility, and the ease of transporting the new energy sources. But neither of these transitions was the result of shortages or concerns about the environment. In fact, coal still remains today a vast energy resource and will continue to remain so for the foreseeable future.

The energy crisis facing America and the rest of the world today results from a divergence between the historically increasing energy demand of transportation, buildings, and industry, and the looming decrease in supplies of oil and gas with a concomitant increase in their costs and availability. To address this crisis, the world must make a new kind of energy transition, which is quite different from the previous two. It needs to be a transition from an infrastructure with abundant and cheap coal, oil, and gas, to one in which these resources will be reduced due to their decrease in availability and adverse effects on the climate, including global warming. In the 1960s, it was thought that nuclear energy would be the next major energy source of the future. However, the costs of nuclear energy turned out to be much higher than expected and there are still unresolved problems regarding the safe long-term deposition of nuclear waste and the safe dismantling of nuclear plants. In place of the hope for nuclear energy, various renewable energy sources have become the hope of the future. However, expectations from renewable sources are still facing an uphill battle due to their high initial investment cost and large area requirements. At this point in time, it does not appear that any single energy source will take the place of fossil fuels. Therefore, it is necessary to prepare for a condition in which many different energy sources will be used, and renewable sources will play an ever-increasing role. Thus, for the twenty-first century, we can expect that in addition to renewable energy sources, energy conservation and improvement in energy generation will become increasingly important. In the remainder of this book, we will look at these issues, as well as the economics and resource base of the various options from an engineering perspective.

Although this book deals mostly with the engineering challenges confronting our efforts to achieve sustainability, it is important to keep in mind that the development of energy sustainability is taking place in an era when at least four other relevant forces impinge on one another. The energy challenge involves also societal issues such as population explosion, resource depletion, and environmental degradation. As shown in Figure 1.2, all of these issues impinge upon the engineering design and economics of a sustainable energy future.

Although this book does not directly deal with these other issues, it is important to keep them in mind because, unless the political and social systems support the road map toward a sustainable energy future, social obstacles can impede the implementation of long-term technical solutions, which seem to be technically reasonable as well as economically acceptable. For the reader interested in these social issues, several books in the supplemental reading list at the end of the chapter will provide background information.

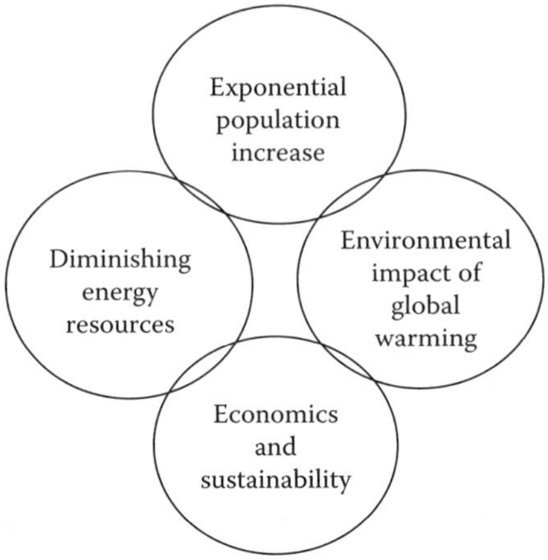

FIGURE 1.2
Energy and major global issues.

The energy choices made in the near future are among the most important of any choices in human history. While the questions may appear to be mostly technical and economic, they are actually much broader. Sustainability is really related to the basic assessment of major priorities in our society, as well as our attitude toward future generations. Renewable energy sources are not a panacea for our energy problems, but combined with an overall social and political strategy that includes energy conservation and a change in the future lifestyle for the most affluent in our society, the goal of a sustainable energy future appears feasible.

After the Industrial Revolution spread from Great Britain to other parts of the world at the end of the eighteenth century, world population started to increase rapidly. According to data from the United Nations [2], the world's population increased in the 118 years from 1804 to 1922 from 1 to 2 billion and doubled once again in the following 52 years reaching 4 billion in 1974. Although not really an engineering topic, the challenge of sustainable development is intimately linked to global population growth. Figure 1.3 shows historical data and growth projections for world populations from 1800 to 2050 according to the United Nations. The current world population is estimated to be 6.9 billion as of 2010. At the present rate of population increase (about 1.9% per year) the world population would reach 10 billion by the year 2020, but global limits on food production and water availability make this number of people unsustainable. This population trend poses enormous problems for future development. Fortunately, the rate of population growth has slowed appreciably in many developed countries, but not in Africa, the Middle East, and South America. In the United States, the population decreased for the first time in its history during 2009. But the United Nations predicts that the world population may reach 8 billion by 2028 unless changes in the social climate and availability of food and water reduce human production rates.

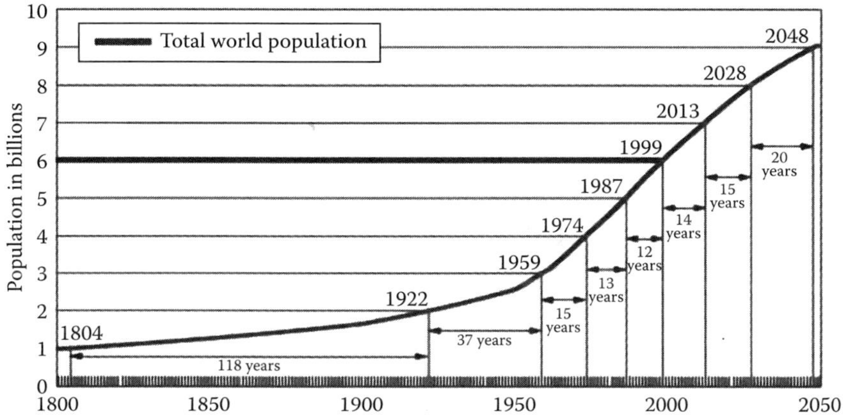

FIGURE 1.3
Annual additions and the annual growth rate of global population. (From Population Division of the Department of Economic and Social Affairs of the United Nations Secretariat, World Population Prospects: The 2008 Revision, http://esa.un.org/unpp.)

1.1 Global Sustainability

The three main ingredients for sustainable development are sufficiency in water, food, and energy. We will consider these three basic ingredients based on the carrying capacity of various regions.

1.1.1 Sustainability Based on Water

The World Bank [3] suggests that an indicator of water scarcity is an availability of less than 1000 m^3/person, while the Water Footprint Organization, in a 2008 study by Hoekstra and Chapagain [4], suggests that 1200 m^3/person is the minimum for a modest consumption footprint. We can easily calculate the carrying capacity at minimum needs for available freshwater, as well as for a modest consumption. Table 1.1 demonstrates the carrying capacities and deficits (i.e., the population excess) in millions based upon data for 1990. A negative number indicates a serious water problem that needs to be remedied by water imports or desalination to maintain the population. The last four columns show the potential capacity deficits in the years 1990 and 2025 for minimum and modest consumption. A negative sign indicates that there is a deficit between need and availability.

An inspection of Table 1.1 shows that there is already a deficit in water for many countries of the world. The population in those countries is already in excess of the human carrying capacity based on water. The last two columns and other projections show that the water deficit at the population growth rate predicted by the United Nations [5] would be acute in many parts of the

TABLE 1.1

Water Needs and Availability for Various Countries (All Figures in Millions)

Country	Available Freshwater (m^3/Capita)	Carrying Capacity (Minimum Needs)	Carrying Capacity (Modest Consumption)	Actual Population	Carrying Capacity Deficit 1990 (Minimum Needs)	Carrying Capacity Deficit 1990 (Modest Consumption)	Carrying Capacity Deficit 2025 (Minimum Needs)	Carrying Capacity Deficit 2025 (Modest Consumption)
Egypt	1123	57.0	49.0	52.0	6.0	−3.0	−36.0	−45.0
Ethiopia	2207	110.0	92.0	50.0	60.0	42.0	−20.0	−38.0
South Africa	1317	50.0	42.0	38.0	12.0	4.0	−23.0	−31.0
Israel	461	2.1	1.8	4.6	−2.5	−2.8	−5.9	−6.2
Jordan	327	1.3	1.1	4.0	−2.7	−2.9	−9.5	−9.7
Algeria	689	17.2	14.3	25.0	−7.8	−10.7	−34.8	−37.9
Kenya	636	15.2	12.5	24.0	−8.8	−11.5	−48.8	−51.5
Morocco	1117	27.9	23.3	25.0	2.9	−1.7	−19.1	−23.7
Suadi Arabia	306	4.6	3.8	15.0	−10.4	−11.2	−35.4	−36.2

world and we conclude that based upon the availability of water, the carrying capacity of the world would be exceeded in all but a very few countries. These data clearly indicate that control of world population is absolutely necessary in order to avoid water shortage.

This dilemma will be exacerbated as the need for energy increases. The situation has been recognized by the U.S. Department of Energy (DOE) in an extensive report, "Energy Demands on Water Resources," prepared by DOE [6] for the U.S. Congress in December 2006; it stated that

> Populations continue to grow and move to areas with already limited water supplies. The growing competition of water availability for energy production and electric-power generation has already been documented in many river basins. ... Available surface water supplies have not increased in 20 years, and groundwater tables and supplies are dropping at an alarming rate. New ecological water demands and changing climate could reduce available freshwater supplies even more.

Example 1.1

Water requirement for the condenser of a 1000 MW nuclear power plant.

Problem statement

Water at 60°F is to be used to condense steam from a nuclear power plant. Given a condensation pressure of 4 in Hg (abs.) the condensate temperature of the condensing steam is 125.4°F.* This medium-sized plant is designed to produce 1000 MW of electric power at an efficiency of 31.5%. If the temperature of the cooling water is not to exceed 110°F, estimate

(a) The rate of heat transfer in the condenser.
(b) The rate of flow of cooling water (gal/s).
(c) To eliminate the high cost of building long transmission lines and the additional associated losses, a central location is generally desired for any power plant. Discuss the implications of putting such power plants in the interior of the country near a population center. Consider environmental and other concerns.

Solution

(a) The rate at which the condenser rejects heat to the water must equal the rate at which the power plant produces waste heat. Our first step, then, is to calculate the rate of waste heat production. Note that only a fraction (31.5%) of the total energy input to the plant is converted to useful electric energy. The rest (68.5%) is waste heat. The rate of energy input is therefore

$$q_{\text{input}} = \frac{q_{\text{electric}}}{\eta} = \frac{1000\,\text{MW}}{0.315} = 3175\,\text{MW}$$

---

* This value can be found in the steam property tables of any standard thermodynamics textbooks or online at many engineering resource sites.

And the rate of waste heat production, which is equal to the rate of heat transfer in the condenser, is

$$q_{\text{waste}} = q_{\text{input}} - q_{\text{electric}} = 3175\,\text{MW} - 1000\,\text{MW}$$
$$= 2175\,\text{MW} \times \left(3.414 \times 10^6 \,\frac{\text{Btu/h}}{\text{MW}}\right) = 7.43 \times 10^9\,\text{Btu/h}$$

(b) The temperature increase of the cooling water in the condenser is obtained from a heat balance:

$$q_{\text{waste}} = \dot{m}c_p(T_{\text{out}} - T_{\text{in}})$$

With the specific heat of water (c_p) as 1.0 Btu/lbm°F, the mass flow rate of the cooling water is:

$$\dot{m} = \frac{q}{c_p(T_{\text{out}} - T_{\text{in}})} = \frac{7.43 \times 10^9\,\text{Btu/h}}{1.0\,\text{Btu/lbm}\,°\text{F} \times (110\,°\text{F} - 60\,°\text{F})} = 149 \times 10^6\,\text{lb/h}$$

Alternatively the volumetric flow rate in gal/s is

$$\dot{V} = \left(1 \times 10^6 \,\frac{\text{lb}}{\text{h}}\right) \times \left(0.464\,\frac{\text{gal}}{\text{lb}}\right) \times \left(\frac{1\,\text{h}}{3600\,\text{s}}\right) = 19{,}200\,\frac{\text{gal}}{\text{s}}$$

(c) Placing a power plant such as this one in the interior of the country near a population center presents a problem due to the amount of heat that must be rejected. To use local lakes or rivers would likely alter the water temperature too drastically to sustain the native aquatic life. Even if cooling were to be accomplished through a cooling tower by the evaporation of water into the atmosphere, the amount of water necessary may be overly taxing on the local water resources. There may also be concern over an accident, although nuclear power has an excellent safety.

To put the water use of thermal power plants into perspective, coal-fired power plants use between 110 and 300 gal/MWh, solar parabolic trough plants use between 760 and 920 gal/MWh, and nuclear plants use between 500 and 1000 gal/MWh. These estimates are based on plants that use superheated steam in a Rankine cycle for which the steam must be condensed into water in the condenser heat exchanger. In most instances, cooling water is not available directly from streams and lakes, and in order to obtain the low temperature water necessary for condensing the steam, cooling towers are used. In the cooling towers, waste heat is dissipated into the atmosphere as steam is evaporated into the air in a so-called wet cooled system. Although, air-cooled cooling towers in which the air is used directly as the heat sink could also be used, this approach reduces the efficiency of the plant and increases the cost of power. These technologies will be discussed further in Chapter 8. Because of the large amounts of water required to dissipate the heat from thermal power plants, wind energy and photovoltaic (PV) power generation are the preferred options in regions where water is in short supply.

1.1.2 Food Supply

The severity of the looming food crisis was brought into worldwide focus when Jacques Diouf, the Director of the U.N. Food and Agriculture Organization, corroborated the findings of U.N. food officials according to an Associated Press release (June 20, 2009) [7] that a record of 1 billion people are hungry, meaning that they have less than 1800 cal per day to eat. Diouf called these data "a rise in food insecurity," but other experts believe that it is the result of overpopulation, particularly in the world's most populous regions, Asia, Sub-Saharan Africa, and the Pacific where the percentage of hungry people has risen 11% during 2008.

It is obvious that in order to have sustainable development, the global food production must be adequate to feed the world's population. According to the data of the United Nations [2], the current global population is of the order of 6.6 billion, and will grow to a median of 7.8 billion in the year 2025 if current population increase is not reduced. It has been further estimated that the earth has 8.2 billion ac of potentially arable land. According to a study by the Millennium Institute [8], this acreage is divided according to the following levels of productivity:

- Highly productive—1.1 billion ac
- Total somewhat productive—2.2 billion ac
- Total slightly productive—4.9 billion ac
- Currently productive—3.6 billion ac
- Potentially productive—8.2 billion ac

The tabulation below demonstrates the "fair share" of productive acreage per person necessary to avoid starvation in the year 2007 and the year 2025.

	Year 2007 (ac)	Year 2025 (ac)
Highly productive	0.17	0.14
Somewhat productive	0.35	0.28
Slightly productive	0.78	0.63
Currently productive	0.57	0.46
Potentially productive if all acreage were fully utilized	1.30	1.05

There is an enormous difference in the way people are fed in different parts of the world. The per capita footprint in the United States is about 1.5 ac, but a modest per capita global footprint based upon diets with less meat and more grain is estimated to be about 1 ac per capita. If the rest of the world wanted to eat at the current U.S. level, the carrying capacity based on food needs is somewhere between 2.5 and 5.5 billion people, whereas, at a modest per capita global footprint, the global carrying capacity would be somewhere between 3.6 and 8.2 billion people. We are clearly approaching the population capacity limit based on food, and in many parts of the world people are undernourished or starving. Furthermore, soil depth and quality are diminishing as a result of urbanization and farm land is disappearing by expanding cities. It is therefore apparent that, even if the world diet would shift toward more grain and less meat, the global population carrying capacity predicted for 2025 would be in excess of the maximum available land. Moreover, it should be noted that current food production is heavily dependent on fertilizer, which is produced largely from natural gas. The availability of natural gas is bound to decrease as other demands by the increasing world population emerge. Thus, the affordability of natural gas and availability of fertilizer are bound to decrease. Once again, it is therefore apparent that based upon the availability of food, the global carrying capacity is being strained already and is soon likely to result in starvation and/or malnutrition of a large part of the world population. The religious dictum to "be fruitful and multiply" was based on conditions prevailing hundreds of years ago and is no longer reasonable. For humanity to survive, the rate of population growth must be limited to manageable proportions.

1.1.3 Energy

The energy supply required to provide a decent life for the people on earth is largely dependent on the efficiency of energy production, levels of energy conservation, and the standard of living the population expects. A study conducted some time ago [5] has identified a human standard of living based upon various criteria. Figure 1.4 is a plot of energy consumption per capita for various countries as a function of a composite indicator, developed by UNDP, to show countries' relative well-being in social as well as economic terms, called the Human Development Index (HDI).

It is apparent that the United States and Iceland have the highest energy consumption per capita, but their human satisfaction level is not much different from other developed countries such as France or Germany, which provide their populations with a good standard of living at a much lower energy consumption per capita. On the other hand, it can be seen that there are many countries in the poorer standard of living categories that have an energy consumption per capita less than 1000 kgoe per capita. The goal of this book is to examine the options available to provide energy for a sustainable future. To estimate sustainability based on energy, it is first necessary to select a reasonable standard of living in the country, which can then be used with a projection of its population to estimate the total energy requirements.

Figure 1.5 demonstrates the energy utilization and the contribution of various sources from 1949 to the year 2007. The enormous increase in energy utilization since the end of World War II in 1945 is the result of rebuilding parts of the world devastated by the war, improvement in the living standards of the middle class, and an exponential growth in world population. There is a direct coupling between the wealth of a country and its energy use. The data

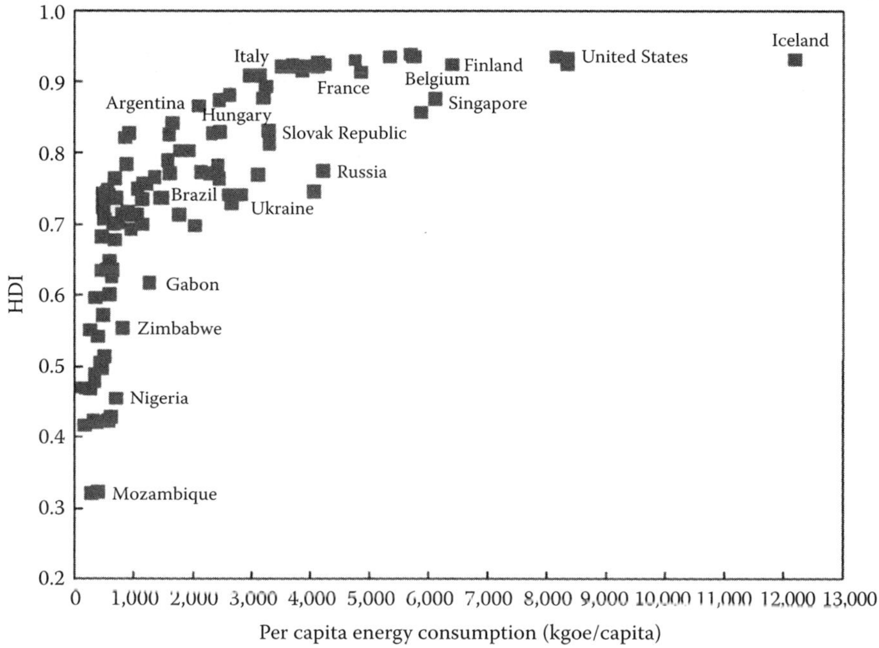

FIGURE 1.4
Relationship between HDI and per capita energy use, 1999–2000. (From Goldemberg, J. and Johansson, T.B. (eds.), *World Energy Assessment: Energy and the Challenge of Sustainability*, United Nations Development Programme, New York, 2004.)

in Figure 1.6 show a coupling between gross national product (GNP) and energy consumption per capita for several countries.

It is apparent that the energy efficiency of the United States and Canada is much lower than that of other developed countries such as France or Japan. Although the energy efficiency of the United States is increasing, it is still considerably below that of most other developed countries. The low energy efficiency of the United States has resulted in an enormous discrepancy between population and world energy consumption. Table 1.2 shows the percentage shares of world population, world gross domestic product (GDP) and energy consumption for some of the major countries of the world.

The United States, with less than 5% of the world's population, consumes almost a quarter of the total world energy and produces almost one-third of the world's GDP. In contrast, China with 20% of the world's population, contributes only 4% of the GDP, while using 11% of the world's energy consumption. However, this trend is rapidly changing. China is using an ever increasing percentage of world energy with an increase in the world's GDP, while at the same time contributing an ever increasing amount of waste products causing environmental degradation such as CO_2 generation and air pollution.

In order to get an idea of the scale of the proposed transition from current energy systems to renewable systems, one needs to examine the detailed

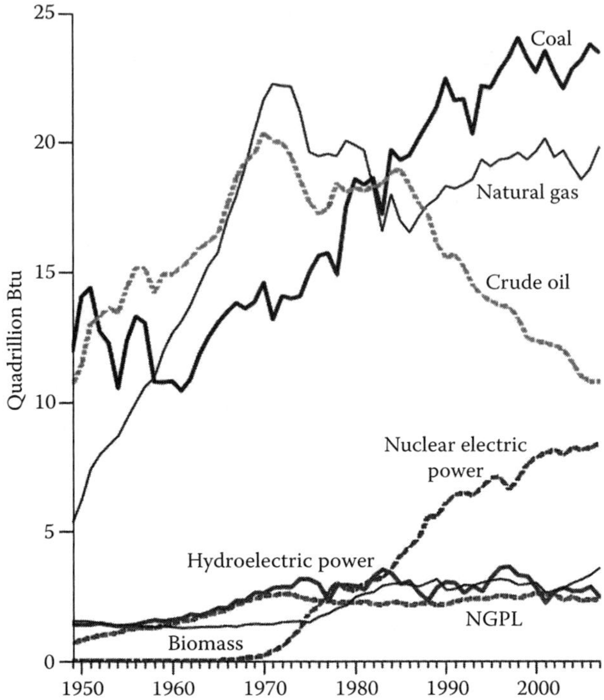

FIGURE 1.5
U.S. energy production by major source, 1949–2007. (From Energy Information Agency, *U.S. DOE Annual Energy Review*, Energy Information Agency, Washington, DC, 2008.)

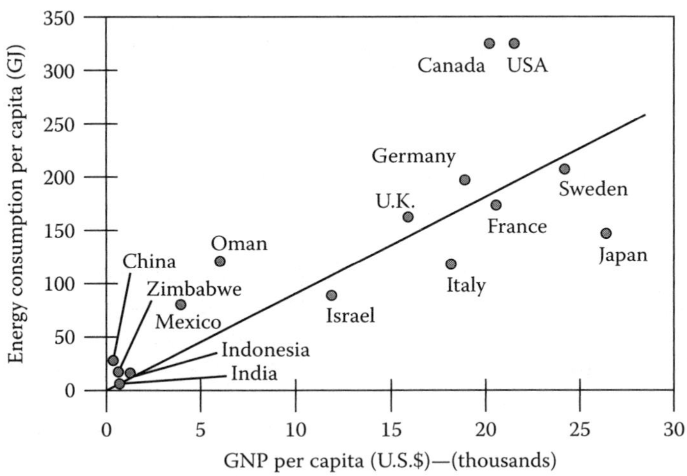

FIGURE 1.6
Comparison of energy use per capita per year versus GNP per capita for various countries. (Lines are provided only to show a general trend. 1 GJ = 10^9 J.). (From World Bank, *Development Data and Statistics*, 2003, www.worldbank.org/data (accessed May, 2007).)

TABLE 1.2

Percentage Shares of World Population, World GDP, and World Commercial Energy Consumption for Selected Countries

Country	% of World Population 2001[a]	% of World GDP 2002[b]	% of World Energy Consumption 2002[c]
United States	4.6	32	24
Japan	2.0	12	5
France	0.9	4	3
Germany	1.4	6	4
United Kingdom	1.0	5	2
China	20	4	11
India	17	2	4

[a] World Population 2001 was 6.2 billion. Country data from The Institute of Memetic Research. www.futuresedge.org/World_Population_Issues/

[b] World GDP 2002 was $32 trillion. Country data from World Development Indicators database, World Bank, July 2003. www.worldbank.org/data/

[c] World primary energy consumption 2002 was 9.4 billion tonnes of oil equivalent. Country data from BP. Statistical Review of World Energy 2003. www.bp.com/centres/energy/index.asp

pattern of energy supply and consumption. Figure 1.7 shows the U.S. energy production by major sources as a function of time. It is also useful to look at end use consumption by dividing it into four major sectors: transportation, residential, commercial, and industrial. These sectors consume, respectively, 29%, 21%, 18%, and 32% of total U.S. energy, as shown in Figure 1.7.

Total U.S. consumption in 2009 was about 102 quads, which is about one-fifth of the world's energy use, with fossil fuel accounting for about 85% of the total. The total contribution of renewable energy is only about 7%, with the majority derived from hydroelectric power generation. Figure 1.8 shows the percentage contribution of various renewable sources to the total. One can see that hydro contributes 36%, wood 32%, and biofuels 15%.The combined contribution of wind and solar energy is less than 6%.

The small contribution from solar and wind energy sources illustrates the dramatic change in the energy use pattern that will be necessary for a shift from fossil to renewable sources. An ancillary to that observation is the fact that the hardware necessary to begin producing energy from solar, biomass and wind sources will have to be built with an energy input from fossil fuels and nuclear power, which at present constitute the majority of our energy sources. Any future planning for the transition will therefore have to include the energy required to build the renewable and possibly also nuclear energy production facilities. This energy would be in addition to the normal needs of the world economy, which currently rely on fossil fuels almost entirely. Future planning will also have to deal with resistance by environmental groups to locating new large hydropower dams or wind-energy farms. Furthermore, availability of high-grade wind resources is

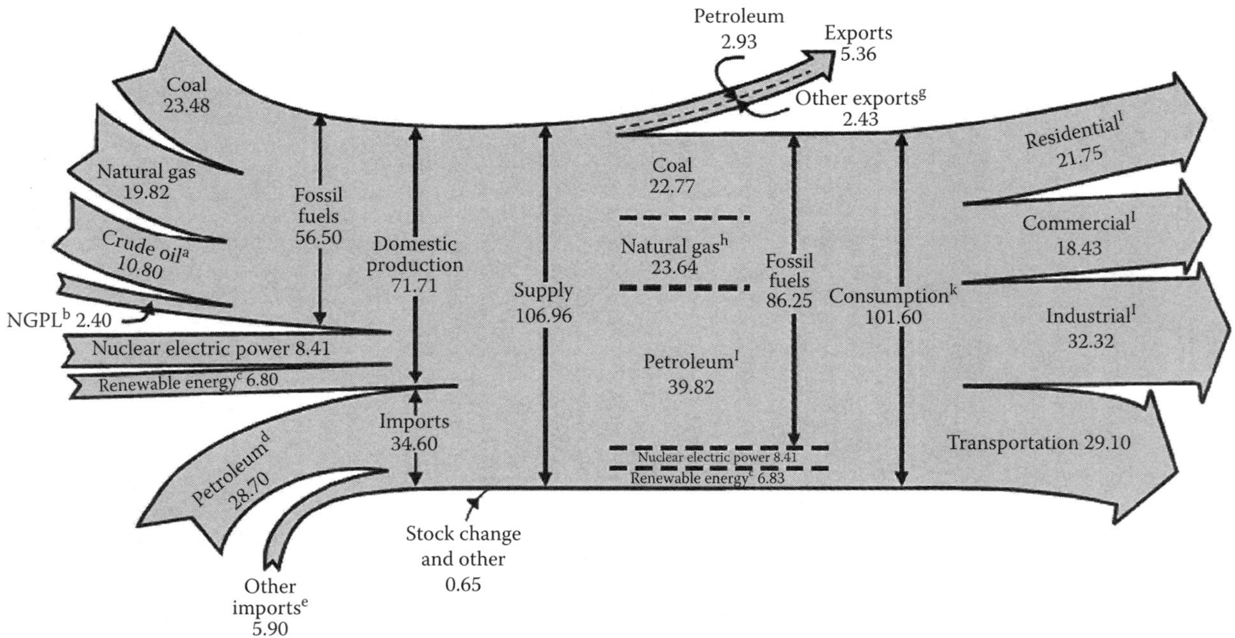

FIGURE 1.7

Energy flow 2007 (quadrillion Btu). (a) Includes lease condensate. (b) Natural gas plant liquids. (c) Conventional hydroelectric power, biomass, geothermal, solar/photovoltaic, and wind. (d) Crude oil and petroleum products. Includes imports into the Strategic Petroleum Reserve. (e) Natural gas, coal, coal coke, fuel ethanol, and electricity. (f) Adjustments, losses, and unaccounted for. (g) Coal, natural gas, coal coke and electricity. (h) Natural gas only; excludes supplemental gaseous fuels. (I) Petroleum products, including natural gas plant liquids, and crude oil burned as fuel. (J) Includes 0.03 quadrillion Btu of coal coke net imports. (k) Includes 0.11 quadrillion Btu of electricity net imports. (l) Primary consumption, electricity retail sales, and elctrical system energy losses, which are allocated to the end-use sectors in proportion to each sector's share of total electricity retail sales. *Notes:* Data are preliminary. Values are derived from source data prior to rounding for publication. Totals may not equal sum of components due to independent rounding. (From Energy Information Agency, *U.S. DOE Annual Energy Review*, Energy Information Agency, Washington, DC, 2008.)

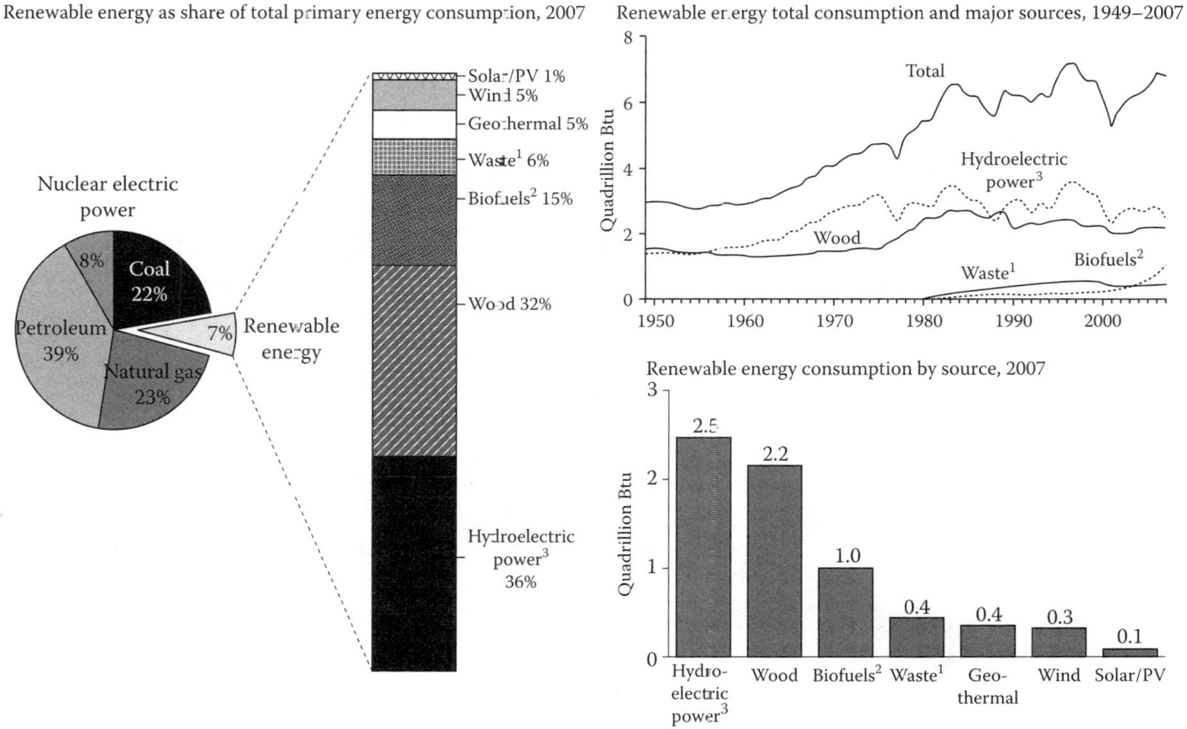

FIGURE 1.8
Renewable energy consumption by major sources. (1) Municipal solid waste from biogenic sources, landfill gas, sludge waste, agricultural byproducts, and other biomass. Through 2000, also includes non-renewable waste (municipal solid waste from non-biogenic sources, and tire-derived fuels). (2) Fuel ethanol and biodiesel consumption, plus losses and co-products from the production of fuel ethanol and biodiesel. (3) Conventional hydroelectric power. *Note:* Because vertical scales differ, graphs should not be compared. (From Energy Information Agency, *U.S. DOE Annual Energy Review,* Energy Information Agency, Washington, DC, 2008.)

limited and good wind resources are often located in sparsely inhabited regions that currently do not have viable transmission systems that can transport the power from the source to where it is needed in industrial parts of the country. The same applies, although to a lesser extent, to good solar radiation sources, many of which are located in parts of the country with limited water resources and some distance from industrial centers and large cities. This is one of the reasons why this book stresses the energy return on energy investment (EROI) concept.

It is well known that renewable energy technologies initially require a large investment in money, as well as in energy (currently mostly from fossil fuel sources), that must be repaid before the source becomes a useful net energy contributor. For a renewable system that does not require any fuel, such as PV, this concept is illustrated in Figure 1.9 where the solid line represents the energy input and the dotted line the energy output. After the construction of the system is completed and energy generation begins, it takes some time before the energy input is repaid and net energy delivery begins. The ratio of net energy delivered to energy input at the end of the project (ratio of dotted to solid lines) is the EROI, which will be considered further in Chapter 2. Therefore, when considering various options, we must ask "How much energy will this technology contribute during its life time compared to the initial energy investment?"

Another way to phrase this question is "What is the energy payback time?" Looking at Figure 1.9, construction begins at time zero, energy generation at time 1, energy input is repaid at time 2, and lifetime of PV system ceases at time 3.

Consider, for example, a thin-film PV system described by the National Renewable Energy Laboratory [10]. Thin-film PV modules use very little semiconductor material. The major energy costs for manufacturing are the substrate on which the thin films are deposited, the film-deposition process, and facility operation. NREL reported that it takes 120 kWh/m^2 to make frameless,

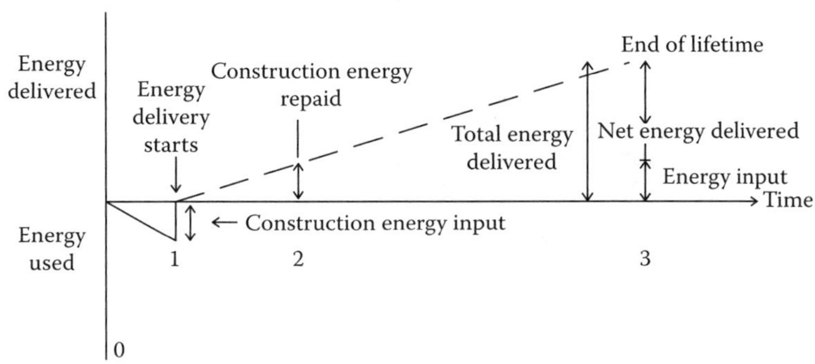

FIGURE 1.9
Schematic diagram of EROI for renewable technologies.

amorphous-silicon PV modules. Adding another 120 kWh/m^2 for a frame and support structure for a rooftop-mounted, grid-connected system, assuming 6% conversion efficiency (standard conditions) and 1700 kWh/m^2 per year of available sunlight energy, the energy payback is about 3 years for current thin-film systems. Deleting the frame, reducing use of aluminum in the support structure, assuming a conservative increase to 9% efficiency, and factoring in other improvements, the payback for thin-film PV could drop to just 1 year.

$CuInSe_2$ and CdTe modules are already being sold in the 9%–12% efficiency range, but their energy payback may be less, depending on design details, such as frames and mounting. Today, for an investment of 3 years worth of their energy output, rooftop PV systems can provide 30 years or more of clean energy. Support structures for ground-mounted systems, which might be more advantageous for utility generation, would add about another year to the payback period.

The next question we might ask is "How much CO_2 and pollution does PV avoid?" An average U.S. household uses 830 kWh of electricity per month. On average, producing 1000 kWh of electricity with solar power reduces emissions by nearly 8 lb of sulfur dioxide, 5 lb of nitrogen oxides, and more than 1400 lb of carbon dioxide. During its projected 30 years of clean energy production, a rooftop system with a 3 year energy payback and meeting about half of a middle-class household's electricity use would avoid conventional electrical plant emissions of more than 100 ton of carbon dioxide. From an energy-environmental perspective, PV is clearly a wise energy investment that affords impressive environmental benefits.

For conventional fossil energy systems such as a coal-fired or for nuclear power plants, the energy input from the fuel consumed must be subtracted from the energy delivered to consumers in the evaluation of the EROI. The same general approach can also be used to determine the EROI for liquid fuels, particularly the production of ethanol from corn.

The EROI for finding and producing oil in the United States has decreased from about 100 in the 1930s to about 30 in the 1970s and is somewhere between 11 and 18 today. This drop is a result of oil reservoirs being depleted and energy investment costs increasing as exploration and development are shifted to deeper levels or offshore [11]. A similar pattern of decline appears to occur in global petroleum production, but specific data are lacking. In contrast to the declining EROI for fossil fuels, the EROI for renewable energy technologies is improving. Exact figures are difficult to specify because many of the renewable technologies, such as wind or solar thermal, depend on the geographic location of the system.

From the perspective of energy sustainability, there are three general types of investments to consider. First, obtaining the energy; second, investment for maintenance of and replacing outdated existing infrastructure and technology; and third, discretionary expansion. This means that before we can think about building a new renewable energy generation infrastructure, we must be aware that we must invest in maintaining the existing energy

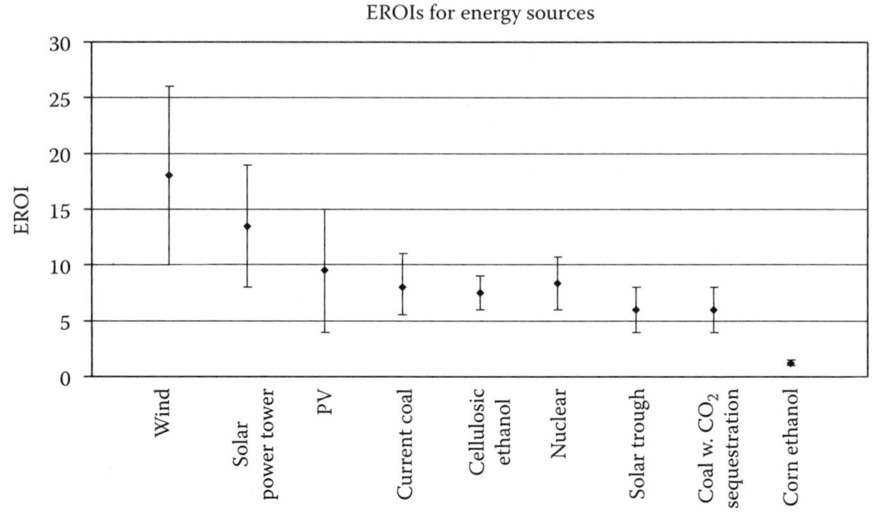

FIGURE 1.10
A comparison of the ratios of electric power or transportation fuel output to fossil fuel input for various technologies. (These estimates do not include social costs or the potential of recycling of the material in the generating facilities.)

system, in addition to the investment for building a new one. As the EROI of fossil fuel decreases, the amount of energy and dollars spent on maintaining the existing energy infrastructure will increase, while the remainder left for building the new renewable system is likely to decrease. With this in mind, an estimate of the range of the EROI for various energy generation technologies, including those for providing liquid fuels, are shown in Figure 1.10. These EROIs are based partly on existing systems, but those for the solar power tower, cellulosic ethanol and coal with CO_2 sequestration are based merely on analysis because no working, commercial full-sized systems exist as yet.

1.2 Role of Energy Conservation

Energy conservation can and must play an important role in any future energy strategy because it can reduce energy requirements and ameliorate adverse impacts on the environment rapidly and economically. Figure 1.11 shows that per capita energy consumption varies by as much as a factor of 3 between the United States and some European countries with almost the same level of HDI [5]. Even taking just the OECD European countries combined, the per capita energy consumption in the United States is twice as much. These data suggest that the per capita energy use of the United States could be reduced to the level of OECD Europe of 4.2 kW per capita by a combination of energy efficiency improvements and changes in the transportation infrastructure

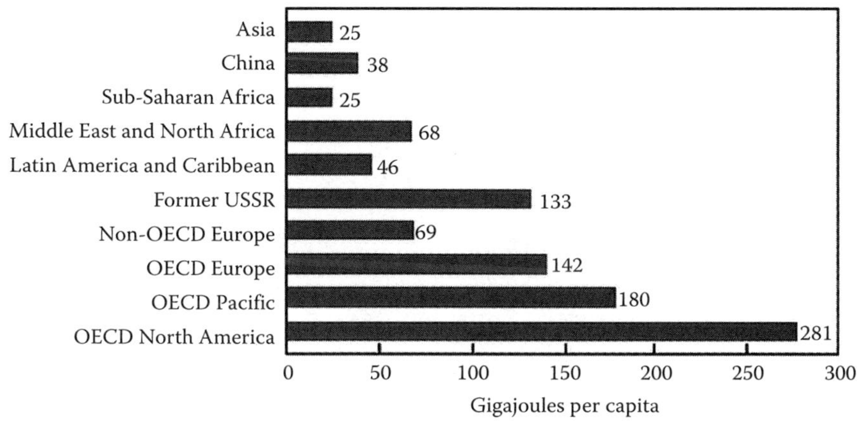

FIGURE 1.11
Per capita energy use by region (commercial and non-commercial) 2000. *Notes:* Asia excludes Middle East, China, and OECD countries; Middle East and North Africa comprises Algeria, Bahrain, Egypt, Iran, Iraq, Israel, Jordan, Kuwait, Lebanon, Libya, Morocco, Oman, Qatar, Saudi Arabia, Syria, Tunisia, United Arab Emirates and Yemen; Latin America and Caribbean excludes Mexico; OECD Pacific comprises Australia, Japan, Korea, and New Zealand; Former USSR comprises Armenia, Azerbaijan, Belarus, Estonia, Georgia, Kazakhstan, Kyrgyzstan, Latvia, Lithuania, Moldova, Russia, Tajikistan, Turkmenistan, Ukraine, and Uzbekistan; Non-OECD Europe consists of Albania, Bosnia and Herzegovina, Bulgaria, Croatia, Cyprus, Gibraltar, Macedonia, Malta, Romania, and Slovenia; OECD North America includes Mexico.

without significantly reducing the HDI. This is important because the United States currently uses about 25% of the energy production in the whole world.

Per capita energy use by region varies enormously, as shown in Figure 1.11. The present per capita energy consumption in the United States is 281 GJ, which is equivalent to about 9 kW per person, while the average for the whole world is only 2 kW per capita.

The Board of the Swiss Federal Institutes of Technology has developed a vision of a 2 kW per capita society by the middle of the century [5]. The vision is technically feasible. However, to achieve this vision will require a combination of increased R&D on energy efficiency and policies that encourage conservation and use of high efficiency systems. It will also require some structural changes in the transportation systems. According to the 2004 World Energy Assessment by UNDP, a reduction of 25%–35% in primary energy in the industrialized countries is achievable cost effectively in the next 20 years, without sacrificing the level of energy services. The report also concluded that similar reductions of up to 40% are cost effectively achievable in the transitional economies and more than 45% in developing economies. As a combined result of efficiency improvements and structural changes such as increased recycling, substitution of energy-intensive materials, etc., energy intensity could decline at a rate of 2.5% per year over the next 20 years [5].

The summary of the economic energy efficiency potentials in North America up to the year 2010 are shown in Table 1.3. It is apparent that the

TABLE 1.3

Economic Energy Efficiency Potentials in North America, 2010

Sector and Area	Economic Potential (%)		Energy Price Level Assumed	Base Year	Sources
	United States[a]	Canada			
Industry					
Iron and steel	4–8	29	United States:	United States:	United States: Interlab (1997),
Aluminum (primary)	2–4		scenario for price	1995	Brown and others (1998),
Cement	4–8		developments[b]		Romm (1999)
Glass production	4–8			Canada: 1990	
Refineries	4–8	23	Canada: price scenario		Canada: Jaccard and Willis (1996), Bailie and
Bulk chemicals	4–9	18	by province[c]		others (1998)
Pulp and paper	4–8	9			
Light manufacturing	10–18	7			
Mining	n.a.	9			
Industrial minerals	n.a.				
Residential	53				
Lighting	11–25		United States:	United States:	United States: Interlab (1997),
Space heating	16		scenario for price	1995	Brown and others (1998),
Space cooling	28–29		developments		OTA, (1992)
Water heating	10–33			Canada: 1990	
Appliances		13	Canada: Price scenario		Canada: Bailie and others (1998)
Overall					

Commercial and public					
Space heating	48		United States: scenario for price developments	United States: 1995	United States: Interlab (1997), Brown and others (1998)
Space cooling	48				
Lighting	25				
Water heating	10–20	9		Canada: 1990	Canada: Bailie and others (1998)
Refrigeration	31		Canada: price scenario		
Miscellaneous	10–33				
Overall	n.a.	9			
Transportation					
Passenger cars	11–17		United States: scenario for price developments	United States: 1997	United States: Interlab (1997), Brown and others (1998)
Freight trucks	8–9				
Railways	16–25				
Airplanes	6–11	3			
Overall	10–14		Canada: price scenario	Canada: 1990	Canada: Bailie and others (1998)

Source: Jochem, E., Energy end-use efficiency, in *World Energy Assessment: Energy and the Challenge of Sustainability*, United Nations Development Programme, New York, 2000.

[a] Industrial energy efficiency potentials in the United States reflect an estimated penetration potential under different conditions based on the Interlaboratory Working Group on Energy Efficient and Low-Carbon Technologies (1997). There are no separate estimates available for the economic potential. The economic potential under business-as-usual fuel price developments is estimated at 7% in energy-intensive industries and 16% in light industries.

[b] The Inter-Laboratory Working Group study (1997) used price scenarios for 1997–2010 to estimate the potential for energy efficiency improvement, based on the *Annual Energy Outlook 1997* scenario (EIA, 1996). The scenario assumes a 1.2% annual increase in oil prices from 1997 levels.

[c] For comparison; in 2010 light fuel oil prices are $6–8 a gigajoule at the 1999 exchange rate (Jaccard and Willis Energy Services, 1996).

greatest energy savings potential is in the transportation industry, followed by residential heating and cooling. The sources in the right-hand column refer to references in the United Nations study. In addition to the item cited in Table 1.3, it is believed that large energy savings are possible in office equipment, such as computers and communication. A similar estimate for the economic energy efficiency potential for Western Europe for the years 2010 and 2020 is presented in Table 1.4, where the resource references refer to the bibliography in [12]. Similar estimates for the energy-saving potential in Japan, Asia, and Latin America are presented in [12].

To illustrate the appropriate place of energy conservation in overall energy sustainability strategy, consider the heating fuel requirements for a typical residential structure, as shown in Figure 1.12.

Only relative quantities are shown in Figure 1.12 because actual fuel requirements will depend on extraneous factors such as climate and indoor comfort. Appropriate investments depend also on whether the analysis is for designing a new building or retrofitting an existing building. In the latter case, the initial thermal integrity of the building as well as the size and the architectural style need to be considered. When improving the thermal performance of a building, whether it be in retrofitting or initial design, conservation investments that yield the highest fuel saving (and hence, economic benefits) per dollar of investment should come before those that yield smaller savings. For example, it would not be appropriate to add storm windows, while neglecting to caulk leaks and add weather stripping beforehand because both of these steps yield a larger benefit with smaller investments. Furthermore, the installation of a solar heating system would not be appropriate until other less expensive improvements have been completed. As will be shown later, the size, application and location of the solar system would be important. A domestic solar hot water heater that can be used year-round would be far more beneficial than heating the building during the winter.

Note that the energy conservation investments in Figure 1.12 reduce the consumption of heating fuel, which would normally be coal or natural gas, and that the saved quantity of fuel is thus available for other needs. If it were not for the energy conservation investment, an energy producing investment would be needed to provide the needed fuel. In this sense, energy conservation can produce fuels and save dollars in much the same manner as energy production. Based on these criteria, it would not be wise to build a nuclear or solar power plant if the energy to be generated could be saved through less expensive conservation investment.

A similar situation exists relative to the consumption of electric energy with more energy-efficient appliances. Consider a refrigerator, which is normally the largest user of electricity in a home. Better insulation and other design changes could cut the average power consumption of a refrigerator considerably. For example, the replacement of an old refrigerator that uses 200 W with a newer one that uses 100 W results in the savings

TABLE 1.4

Economic Energy Efficiency Potentials in Western Europe, 2010 and 2020

Sector and Technological Area	Economic Potential (%)[a]		Energy Price Level Assumed	Base Year	Sources
	2010	2020			
Industry					
Iron and steel, coke ovens	9–15	13–20	1994	1995	Jochem and Bradke (1996), Ameling and others (1998)
Construction materials	5–10	8–15	1997	1997	
Glass production	10–15	15–25	1997	1997	ATLAS (1997)
Refineries	5–8	7–10	1995	1997	*Refining Processes* (1998)
Basic organic chemicals	5–10		1997	1996	Patel (1999), Brewer and Lopez (1998)
Pulp and paper		50	1996	1997	De Beer (1998)
Investment and consumer goods	10–20	15–25	1994	1995	Jochem and Bradke (1996), Böde and others (1999)
Food	10–15		1997	1997	Jochem and Bradke (1996)
Cogeneration in industry		10–20	1997	1997	ATLAS (1997), EC (1999)
Residential					
Existing buildings					
Boilers and burners	15–20	20–25	Today's prices	1997	ETSU (1994), Böde and others (1999)
Building envelopes	8–12	10–20	Today's prices	1995	Ziesing and others (1999)
New buildings		20–30	Today's prices	1995	Altner, Durr, Michelson (1995)
Electric appliances	20–30	35–45	1997	1997	GEA (1995), ECODROME (1999), Hennicke and other (1998), Boardman and others (1997)

(continued)

TABLE 1.4 (continued)

Economic Energy Efficiency Potentials in Western Europe, 2010 and 2020

Sector and Technological Area	Economic Potential (%)[a] 2010	2020	Energy Price Level Assumed	Base Year	Sources
Commercial public and agriculture					
Commercial buildings		30	8–13 cts/kWh	1995	Geiger and others (1999)
Electricity	10–20	20–37	4–10 cts/kWh	1997	ECODROME (1998)
Heat	10–25	15–25	Today's prices	1998	Zeising and others (1999)
Public buildings		30–40	7–15 cts/kWh	1992	Brechbühl (1992)
Agriculture and forestry		15–20	Today's prices		Neyer and Strebel (1996)
Horticulture		20–30	Today's prices		Arbeitsgemeinschaft (1992)
Decentralized cogeneration		20–30	Today's prices	1995	Ravel (1994)
Office equipment		40–50	1995	1995	Aebischer and others (1996), MACEBUR (1998), Hallenga and Kok (1998)
Transportation					
Cars	25		Today's prices	1995	IPSEP (1995)
Door-to-door integration	4			1995	Zeising and others (1999)
Modal split of freight transport		3[b]		1995	
Trains and railways		20	Today's prices	1999	Brunner and Gartner (1999)
Aircraft, logistics	15–20	25–30	Today's prices	1998	IPCC (1999a)

Source: Jochem, E., Energy end-use efficiency, in *World Energy Assessment: Energy and the Challenge of Sustainability*, United Nations Development Programme, New York, 2000.

[a] Assumes a constant structure or use of the sector or technology considered.

[b] Refers to the final energy use of the entire sector.

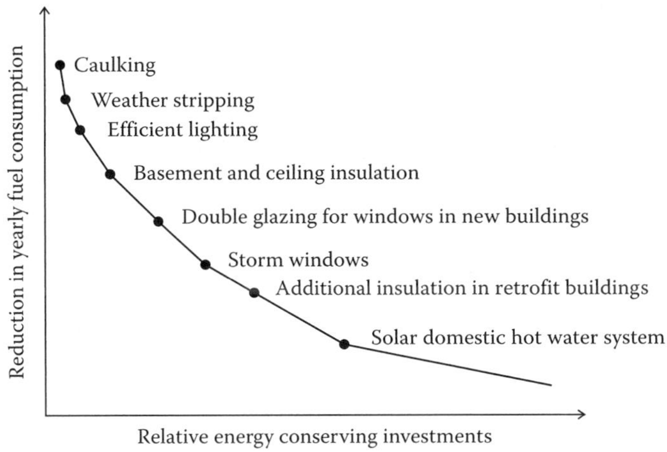

FIGURE 1.12
Relative energy savings of various conservation measures.

of 100 W of power and a savings of 880 kWhe/year, which (at $0.10/kWh) would be a savings of about $90 per year. Many other examples, such as improved air conditioning and energy-efficient lighting could be cited. Their energy as well as the dollar return on investment will be considered in later chapters.

Improving energy efficiency across all sectors of the economy should become a worldwide objective [5]. It should be noted, however, that free market price signals may not always be sufficient to increase energy efficiency. Hence, legislation on the state and/or national level for energy efficiency standards for equipment in the residential and commercial sector may be necessary. There is considerable debate whether incentives or mandates are the preferred way to improved energy efficiency. But some measures will be necessary because surveys indicate that consumers consistently rank initial purchase price high but energy use and operating costs quite low on the lists of attributes they consider when purchasing an appliance or construct a building. Incentives may be the preferred option, provided they induce decision makers to take appropriate conservation action. In the United States; however, national efficiency standards were imposed by Congressional legislation for a variety of residential and commercial appliances since 2004. Updated standards will take effect in the next few years for several more products. Outside the United States, over 30 countries have also adopted minimum energy performance standards for buildings and appliances. These measures have been economically attractive in the long term and can provide not only an appreciable reduction in energy consumption, but also reduce adverse environmental impacts.

1.3 Nuclear Energy

There is worldwide debate on the contribution that nuclear energy will make toward a sustainable energy future. Electricity production from nuclear energy began in 1954. That year, the first nuclear power plant generating heat from uranium was connected to the grid in the former USSR. Nuclear energy uses a carefully controlled fission reaction, which generates heat to produce steam or heat air that powers conventional turbines to generate electricity [13].

Nuclear power plants are similar to coal-fired or natural gas power plants, except for the energy source. In 2007, there were 439 reactor units in operations worldwide, contributing about 16% of the world's energy production. The majority of reactors are located in the United States (104 units), France (59 units), Japan (55 units), and Russia (31 units). In addition, there are at least 30 and possibly more units under construction, mainly in China. Figure 1.13 shows the nuclear reactors in operation in the various countries of the world in 2007. Two types of water-cooled reactors dominate the industry: pressurized water reactors (PWR) where the heated fluid from the reactor passes through an outside heat exchanger to generate steam for a turbine, and boiling water reactors where the steam is generated by surface boiling within the reactor vessel and sent directly to the turbine. Up-to-date details of the construction and operation of current and proposed nuclear power plants are presented in the *Nuclear Engineering Handbook* [13].

It is important to note that most reactors today are more than 20 years old (70%) and about 23% are more than 30 years old. As a result of the age of the reactors, within the next 10 or 20 years at least 150, or about 30%, will have to be shut down and replaced. The age distribution of operating reactors is shown in Figure 1.14.

Thus, at least 150 new reactors will have to be built before nuclear energy can make a net contribution to global needs. The cost of construction and the initial energy investment are enormous. A recent cost estimate for the construction of a new nuclear power plant in the United States was $7 per watt [14] and according to an estimate in [15], dismantling of the plant and safe storage of the waste may be of the same order of magnitude.

1.3.1 Available Nuclear Resources

Nuclear power uses uranium as the energy source. Available resources of uranium are divided into two groups: reasonably assured resources (RAR), and inferred resources (IR). RAR are the uranium in known deposits of sufficient size, grade, and configuration that it could be recovered within reasonable production costs with currently available mining and processing technology. IRs are inferred resources whose availability is uncertain.

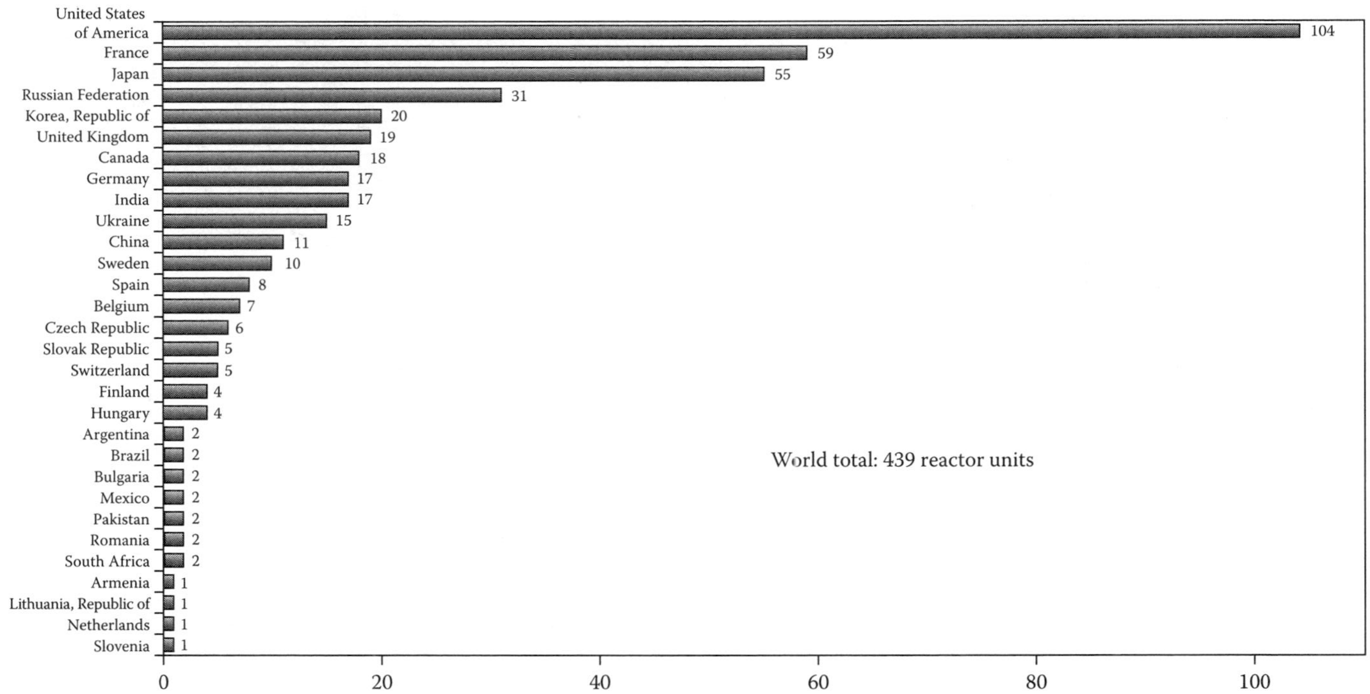

FIGURE 1.13
Number of reactors in operation worldwide (as of August 8, 2007). *Note:* Long-term shutdown units (5) are not counted. (From IAEA, *Power Reactor Information System (PRIS)*, Vienna, Austria, http://www.iaea.org/programmes/a2/)

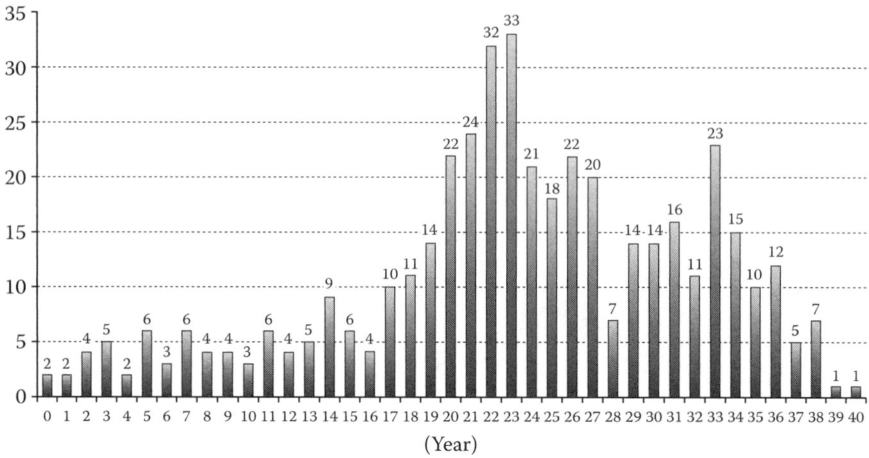

FIGURE 1.14
Number of operating reactors by age (as of June 26, 2007). *Note:* Age of a reactor is determined by its first grid connection. (From IAEA, *Power Reactor Information System (PRIS)*, Vienna, Austria, http://www.iaea.org/programmes/a2/)

The categories are further divided into various cost classes according to estimated extraction costs. The division of these classes into below \$50/kg, below \$80/kg, and below \$130/kg is widely accepted. According to some experts, RAR for less than \$80/kg uranium are called proven resources. It is estimated that of this type, about 2.6 million ton are available globally. When used in light water reactors, this corresponds to an energy equivalent of 28 billion tons of hard coal. Figure 1.15 shows the deposits of uranium and the various countries in which they are located. If mining costs up to \$130/kg are included, the global uranium sources are increased to 3.3 million tons. The total, irrespective of costs, may be as high as 15 million tons. The amount of energy that can be extracted from these uranium resources depends upon whether or not spent fuel from light water reactors can be reprocessed. Reprocessing all of the spent fuels from light water reactors and recycling the uranium and plutonium in mixed oxide (MOX) fuel to operate light water reactors with 30% MOX and 70% uranium would lead to a cumulative savings of some 600,000 ton of natural uranium in the next 40 years. However, current legislation in the United States forbids reprocessing.

Because of high cost and public safety concerns after the Chernobyl and Three Mile Island accidents no new nuclear power plants have been built in the United States since 1971. Thus, the future cost of nuclear power can only be estimated as shown in Chapter 2. The potential of nuclear power to make a viable contribution to a future energy scenario will largely depend on its cost, including the safe deposition and long-term storage of nuclear waste, which will not be known accurately until a new plant has been built and operated and a facility for long-term disposal of nuclear waste has been put into operation.

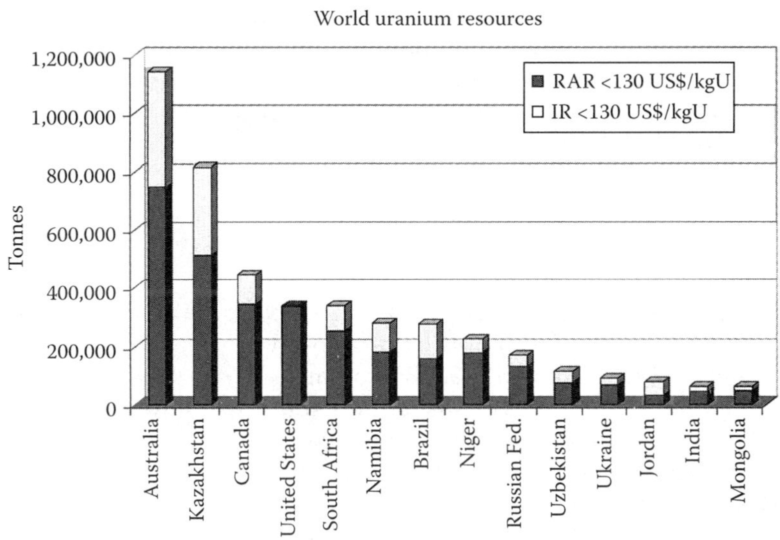

FIGURE 1.15
The deposits of uranium and the various countries in which they are located. (From *Nuclear Power Daily*, Progress Energy Florida signs contract for new, advanced-design nuclear plant, January 16, 2009, http://www.nuclearpowerdaily.com/reports/Progress_Energy_Florida_Signs_Contract_For_New_Advanced_Design_Nuclear_Plant_999.html (retrieved May 18, 2009).)

1.4 Hydrogen

Although the idea of a hydrogen economy has been promulgated by some energy analysts, as well as by ex-President George W. Bush, this is a somewhat misleading concept because hydrogen is an energy carrier but not an energy source. The concept of a hydrogen economy was proposed back in the 1870s as a fanciful speculation of Jules Verne's in his novel *The Mysterious Island* [16]. Hydrogen production was examined extensively in the 1970s by experts for the Institute of Nuclear Energy in Vienna and the Electric Power Research Institute [17,18]. The basic idea was to generate hydrogen by high-temperature nuclear reactions and then use the hydrogen to generate electricity, thereby replacing fossil fuels. The results of this study showed, however, that generating hydrogen with high-temperature thermal-nuclear methods was inferior in cost and efficiency to generating electricity from nuclear reactors and then producing hydrogen by electrolysis [18]. But the study also showed that using the electricity from the nuclear plants directly was preferable in cost and efficiency to the hydrogen path to generate electricity with a fuel cell.

Hydrogen is abundant on Earth, but only in chemically bound form. In order to use hydrogen as a fuel, it is necessary that it be available in unbound form. As a consequence of chemical reaction energies involved, a substantial energy input is needed to obtain unbound hydrogen. This energy input exceeds the energy released by the same hydrogen when used as a fuel. For example, to split water into hydrogen and oxygen according to the reaction

$$H_2O \rightarrow H_2 + \frac{1}{2}O_2$$

120 MJ/kg-hydrogen are needed (all gases at 25°C); while the reverse reaction of combining hydrogen and oxygen to give water (all gases at 25°C), ideally yields 120 MJ/kg-hydrogen. But, because no real process can be 100% efficient, more than 120 MJ/kg must be added to the first reaction, while less than 120 MJ/kg of useful energy can be recovered from the recombination. To evaluate the losses, it is, therefore, necessary to examine the "cradle-to-grave" energetics of hydrogen production processes quantitatively.

Figure 1.16 shows all the major pathways to produce hydrogen and to utilize it as an energy carrier. The top row of Figure 1.16 shows the primary energy sources: fossil fuels, nuclear materials, and renewable sources. The next three rows show the major processing steps for conversion of the primary energy into hydrogen. Below the hydrogen row are the two methods

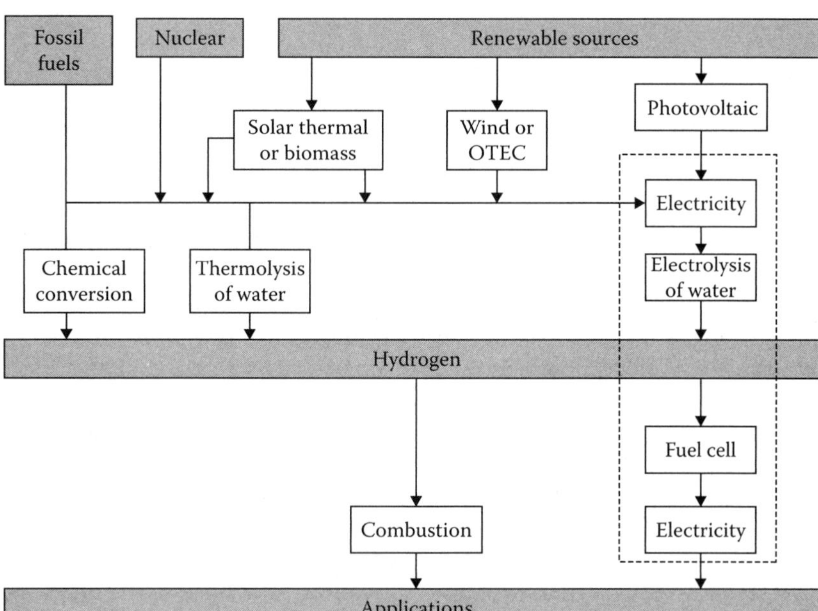

FIGURE 1.16
Pathways for hydrogen production and utilization. (From Kreith, F. and West, R.E., *J. Energ. Resour. Technol.*, 126, 249, 2004.)

of using hydrogen in energy applications: one is to combust hydrogen to produce heat for various applications, and the other is to generate electricity from hydrogen by means of a fuel cell.

Fossil fuels, nuclear energy, solar thermal (including OTEC), biomass, wind, and PVs can all be used to generate electricity. All of these, except PVs and wind, generate electricity by first producing heat, which is then converted to mechanical energy, which, in turn, is finally converted to electricity. PV cells generate electricity directly from solar radiation, while wind turbines directly generate mechanical energy and then electricity. In principle, some of the heat-producing technologies can also make hydrogen by thermolysis of water, i.e., heating of water to a sufficiently high temperature (greater than 3000 K) to break it into hydrogen and oxygen. However, practical engineering obstacles such as the cost and manufacture of materials to withstand these ultrahigh temperatures have relegated this process to future R&D rather than a viable near-term option.

In Figure 1.16, the dashed box isolates that portion of the pathway in which electricity is used to produce hydrogen by electrolysis of water and hydrogen subsequently is used to produce electricity via a fuel cell. These steps are common to all energy sources that produce hydrogen by electrolysis. The efficiency can be evaluated by examining the electrolysis and hydrogen utilization steps. These steps are shown in Figure 1.17 with estimated present and highly optimistic future efficiency for the electrolysis and fuel cell steps.

With present technology, it would take 2.9 kWh electricity input to produce 1 kWh of electricity output and with the most optimistic advanced efficiency, 1.9 kWh would only yield 1 kWh of output. The difference between input and output would be wasted because inefficiencies are involved in each of these sequential steps. Hence, the output of electricity via the hydrogen fuel cell path would cost approximately 2–3 times as much as the electricity input. Moreover, this cost ratio does not include the capital cost and nonelectrical

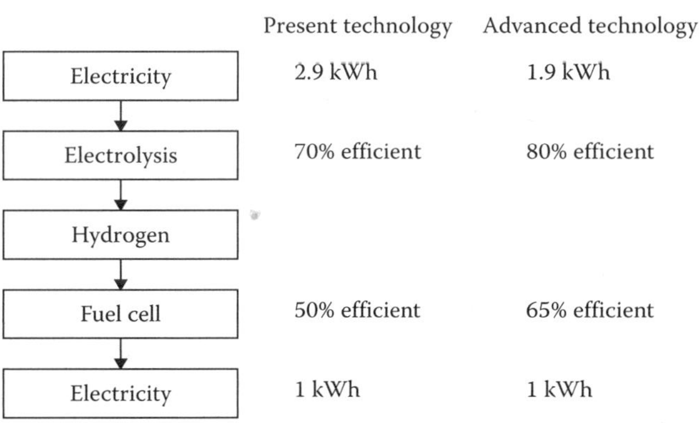

FIGURE 1.17
Producing electricity through hydrolysis.

operating cost of the electrolysis fuel cell and hydrogen storage equipment. Hence, it may be concluded [19] that the conversion of electricity to hydrogen by electrolysis and then the conversion of hydrogen to electricity via a fuel cell is inefficient and not a desirable basis for an economically and environmentally sound, sustainable energy policy.

Similarly, hydrogen produced by electrolysis could be used to produce heat by combustion. However, the efficiency of producing hydrogen by means of electrolysis is only 70%–80% and burning the hydrogen at an efficiency of 85% yields heat with an overall efficiency of about 60%–70%, while electricity can be converted to heat with essentially no losses. Hence, to use hydrogen made by electrolysis to produce heat is also inefficient and wasteful.

In summary, based upon a cradle-to-grave analysis, any currently available hydrogen production pathway irrespective of whether it uses fossil fuels, nuclear fuels, or renewable technologies as the primary energy source to generate electricity or heat is inefficient compared to using the electric power or heat from any of these sources directly. Hence, the idea of a hydrogen economy does not appear to be a viable concept with any currently available engineering process and equipment. This conclusion is further elucidated in [19] with appropriate additional references as a basis for the analysis.

There are processes described in the technical literature that can produce hydrogen from renewable or nuclear sources without using either electrolysis or thermolysis of water [20]. For example, biomass may be chemically converted to hydrogen by processes similar to those used with fossil fuels, or it may also be converted to hydrogen by biological conversion processes. Photochemical and photo-electrochemical reactions can produce hydrogen directly with solar radiation input. Thermochemical and hybrid thermochemical/electrochemical cycles use nuclear or solar thermal heat and electricity to drive chemical cycles that produce hydrogen from water. However, none of these ideas is anywhere close to commercialization, and they are topics for future R&D, not viable technologies for a sustainable energy policy [20,21].

A bird's eye view of the various energy technologies, including hydrogen fuel cells and nuclear energy, from an economic perspective is shown in Figure 1.18. In Figure 1.18, which was prepared by an independent investment company in 2009, the capital cost of various energy production facilities is plotted as a function of the consumer cost in cents/kWh with the time of construction required shown in the inside of the circle for each technology. It is apparent that the expected consumer cost from nuclear power is on par with various renewable technologies. The estimates shown in Figure 1.18 do not include social and environmental costs, storage facilities, or transmission lines. Furthermore, changes in the tax structure such as, for example, the imposition of a carbon tax, could substantially increase the consumer costs from some of the technologies shown and improve the competitiveness of nuclear energy.

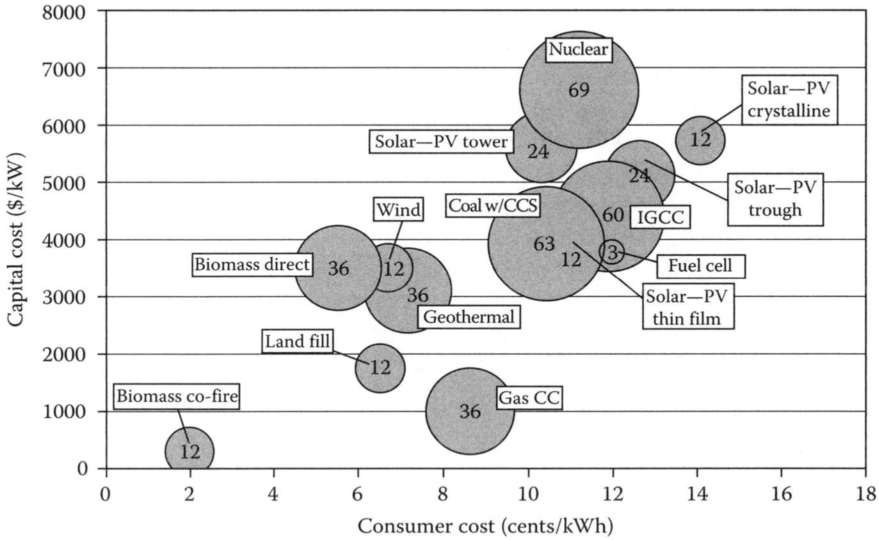

FIGURE 1.18
Consumer cost, capital cost, and construction times for various supply-side alternatives (circle size indicates construction time in months). (From Lazard, Ltd., *Energy Technology Assessment*, Lazard, Ltd., New York, pp. 11–13, 2009, available at: www.lazard.com. With permission.)

1.5 Geothermal Energy

Geothermal energy can use the heat in the interior of the earth for electric power generation, heating of buildings, or as a source of thermal energy for heat pumps. The majority of existing systems utilize only the thermal energy near the surface of the earth for direct heating and heat pump applications. The use of geothermal energy for ground-coupled heat pump technology uses the reservoir of constant temperature, shallow ground water and moist soil as the heat source during the winter and as the heat sink during summer cooling. The energy efficiency of geothermal heat pumps is 30% better than that of air-coupled heat pumps and at least 50% better than direct electric heating [22,23]. Application for heat pumps in individual buildings has been widely demonstrated and their use is entirely dictated by economic considerations [22].

Temperature within the earth increases with depth at an average of about 25°C per km. Spatial variations of the thermal energy within the earth give rise to concentrations near the surface that can be used directly. Heat is transferred from the deeper portions by conduction through rocks by the movement of hot, deep rock toward the surface and by deep circulation of water. Most high-temperature geothermal resources are associated with concentrations of heat caused by the movement of magma (melted rock) to near-surface position. Warm water near the surface at temperatures above 20°C can be used directly for various processes requiring low-temperature thermal energy as, for example, swimming pools, space heating, and domestic hot water. However, also industrial and agricultural processes are potential applications of geothermal use. In 1995, the United States was using over 500 TJ per year of energy from geothermal sources for direct use. Cities such as Boise (Idaho), Elko (Nevada), Clamon Falls (Oregon), and San Bernardino (California) have geothermal district heating systems where a number of commercial and residential buildings are connected to distributed pipelines circulating water at 54°C–90°C. Table 1.5 shows the U.S.-installed geothermal energy–generating capacity in MW, and Table 1.6, the geothermal energy for direct electric power generation by the 10 largest users worldwide.

TABLE 1.5

U.S.-Installed Geothermal Electrical Generating Capacity in MW_e

Rated State/Field	Plant Capacity	Type
California		
Casa Diablo	27	B
Coso	240	2F
East Mesa	37	2F
East Mesa	68.4	B
Honey Lake Valley	2.3	B
Salton Sea	440	2F
The Geysers	1797	S
Hawaii		
Puna	25	H
Nevada		
Beowawe	16	2F
Brady Hot Springs	21	2F
Desert Peak	8.7	2F
Dixie Valley	66	2F
Empire	3.6	B
Soda Lake	16.6	B
Steamboat	35.1	B
Steamboat	14.4	1F
Stillwater	13	B
Wabuska	1.2	B
Utah		
Roosevelt	20	1F
Cove Fort	2	B
Cove Fort	9	S

Note: S, natural dry steam; 1F, single flash; 2F, double flash; B, binary; H, hybrid flash and binary.

TABLE 1.6

Geothermal Energy for Direct Use by the 10 Largest Users Worldwide

Country	Flow Rate (kg/s)	Installed Power (MWt)	Energy Used (TJ/Year)
China	8,628	1915	16,981
France	2,889	599	7,350
Georgia	1,363	245	7,685
Hungary	1,714	340	5,861
Iceland	5,794	1443	21,158
Italy	1,612	307	3,629
Japan	1,670	319	6,942
New Zealand	353	264	6,614
Russia	1,240	210	2,422
United States	3,905	1874	13,890
Total	37,050	8664	112,441

Source: Freeston, D.H., Worldwide direct uses of geothermal energy, in *Proceedings of the World Geothermal Congress, 1995,* International Geothermal Association, Florence, Italy, 1995, pp. 15–26.

The U.S. Geological Survey has prepared an assessment of the geothermal energy potential in the United States [24]. The survey estimated the part of the identified accessible base that could be extracted and utilized in the reasonable future to be 23,000 MWe for 30 years. The undiscovered U.S. resource inferred from knowledge of earth science was estimated to be between 95,000 and 150,000 MWe for 30 years.

Recently, interest in enhanced geothermal systems (EGS, sometimes referred to as engineered geothermal systems) for primary energy recovery using heat mining technology designed to extract and utilize the earth's stored thermal energy at great depth has received a great deal of attention. A large-scale effort was mounted in 2006 and the report summarizing the results is available on the Internet at http://geothermal.inel.gov and www1.eere.energy.gov/geothermal/egs_technology.html. The EGS technology attempts to drill into the depth of the earth up to 1000 m or more and pump hot water through two interconnected pipes, which would then be used to raise steam in a heat exchanger to drive a conventional turbine. The success of the EGS efforts has so far only been marginal as described in the report referenced above. Although the study gave projections of potentially very large recoverable energy (estimated extractable portion of about 200,000 EJ, which is many times the annual consumption of the United States) no commercial system has yet been built and put into operation.

The concept of an EGS system is demonstrated in Figure 1.19 based upon the Ogachi project in Japan. Two injection wells were drilled to a depth of 1100 m, where a temperature of 240°C was reached. Water was then pumped down through one well into a fracture zone and up through the production well. However, problems were encountered in stimulating the water

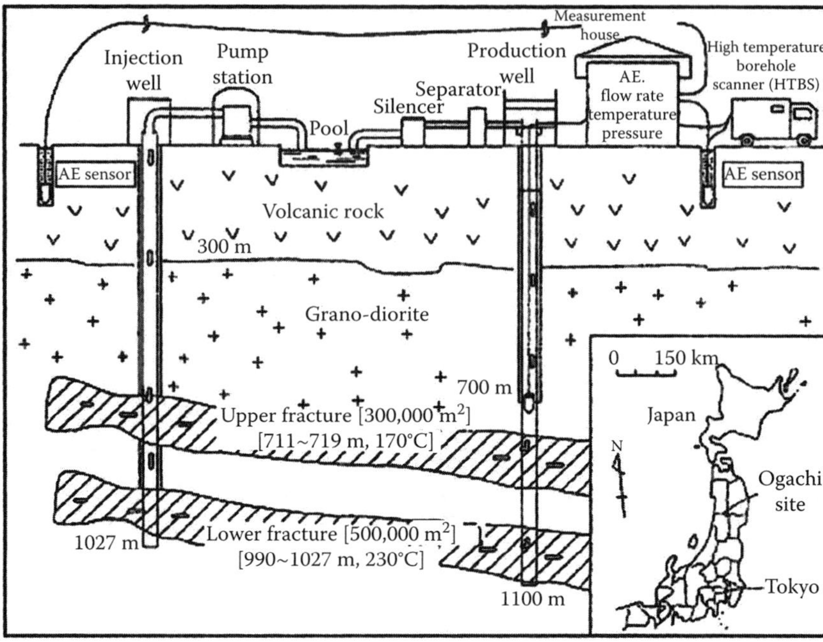

FIGURE 1.19
Schematic diagram of an enhanced geothermal system.

flow despite repeated efforts at improving the drilling and connecting of the reservoir. Geological studies are in progress to improve interconnections, which are necessary in engineered geothermal systems and the reader is referred to the assessment in [25] for further information.

1.6 Global Status and Potential of Renewable Energy

1.6.1 Biomass and Hydroelectric Sources

According to the data in Table 1.7, 13.3% of the world's total primary energy supply came from RE in 2003. However, almost 80% of this renewable energy supply was from biomass (Figure 1.20), and in developing countries it is mostly used by traditional open combustion for cooking, which is very inefficient and creates health problems. Because of their inefficient use, biomass resources presently supply only about 20% of what they potentially could, if converted by modern and more efficient technologies. As it stands, biomass provides only 11% of the world's total primary energy, which is much less than its real potential.

In 2003, shares of biomass and hydropower in the total primary energy mix of the world were about 11% and 2%, respectively. All of the other

TABLE 1.7

2003 Fuel Shares in World Total Primary Energy Supply

Source	Share (%)
Oil	34.4
Natural gas	21.2
Coal	24.4
Nuclear	6.5
Renewables	13.3

Source: Data from IEA, *World Energy Outlook*, IEA, Paris, France, 2004.

renewable sources, including solar thermal, solar PV, wind, geothermal and ocean combined, provided only about 0.5% of the total primary energy. During the same year, biomass combined with hydroelectric resources provided more than 50% of all the primary energy in Africa, 29.2% in Latin America, and 32.7% in Asia (Table 1.8).

Although theoretically harvestable biomass energy potential is of the order of 90 TW, the technical potential on a sustainable basis is of the order of 8–13 TW or 270 to 450 EJ/year [5]. This potential is 3–4 times the present electrical generation capacity of the world. It is estimated that by 2025, even the municipal solid waste (MSW) could generate up to 6 EJ/year.

The biggest future potential of biomass as an energy resource is its relatively straightforward transformation into transportation fuels. Biofuels have the potential to replace as much as 75% of the petroleum fuels in use for transportation in the United States today [26]. This is especially important in view of the expected decline of oil supplies worldwide. Biofuels for transportation will not require a lot of additional infrastructure development. Therefore, development of biofuels is being viewed very favorably by governments around the world. Biofuels, along with other transportation options such as electric vehicles will help diversify the fuel base for future transportation. Figure 1.21 shows that global ethanol production more than doubled between 2000 and 2005. Biodiesel production grew almost fourfold, although it started from a much smaller base. In 2005, the world ethanol

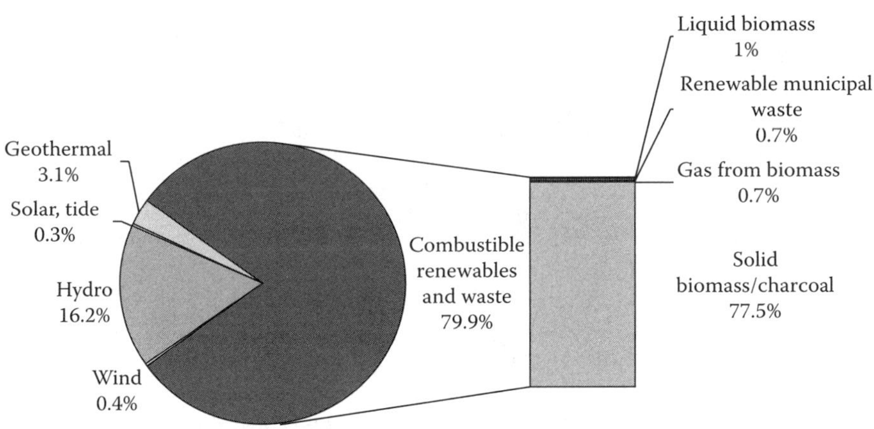

FIGURE 1.20
2003 Resource shares in world renewable energy supply. (Data from IEA, *World Energy Outlook*, IEA, Paris, France, 2004.)

TABLE 1.8
Share of Renewable Energy in 2003 Total Primary Energy Supply (TPES) on a Regional Basis

Region	TPES (Mtoe)	Renewables (Mtoe)	Share of Renewables in TPES (%)
Africa	558.9	279.9	50.1
Latin America	463.9	135.5	29.2
Asia	1,224.4	400	32.7
India	553.4	218	39.4
China	1,425.9	243.4	17.1
Non-OECD[a] Europe	103.5	9.7	9.4
Former USSR	961.7	27.5	2.9
Middle East	445.7	3.2	0.7
OECD	5,394.7	304.7	5.6
United States	2,280.8	95.3	4.2
World	10,578.7	1,403.7	13.3

Source: IEA, *Renewables Information*, IEA, Paris, France, 2005.
[a] Organisation for Economic Co-operation and Development.

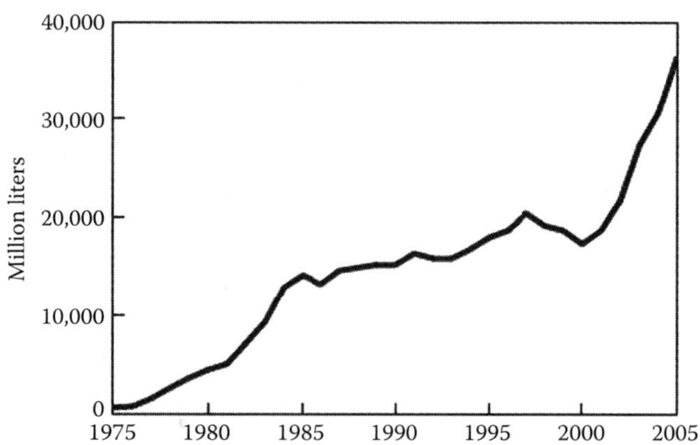

FIGURE 1.21
World ethanol production, 1975–2005. (From Worldwatch Institute, *Bio-Fuels for Transportation—Global Potential and Implications for Sustainable Energy in the 21st Century*, Worldwatch Institute, Washington, DC, 2006.)

production had reached about 36 billion liters per year while biodiesel production topped 3.5 billion liters during the same year. Table 1.9 shows the top five countries producing these fuels.

The cost of ethanol production in 2005 ranges from about €0.25 to about €1 per gasoline equivalent liter, as compared to the wholesale price of gasoline, which was between €0.4 and €0.6 per liter (Figure 1.22). Biodiesel costs, on the other hand, ranged between €0.20 and €0.65 per liter of diesel

TABLE 1.9

Top Five Ethanol and Biodiesel Producers in the World

	Ethanol Production (Million Liters/Year)		Biodiesel Production (Million Liters/Year)
Brazil	16,500	Germany	1,920
United States	16,230	France	511
China	2,000	United States	290
European Union	900	Italy	227
India	300	Austria	83

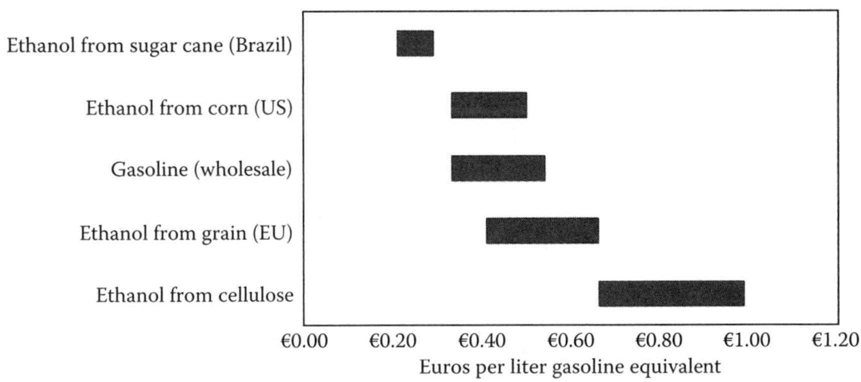

FIGURE 1.22
Cost ranges for ethanol and gasoline production, 2006. (From Worldwatch Institute, *Bio-Fuels for Transportation—Global Potential and Implications for Sustainable Energy in the 21st Century*, Worldwatch Institute, Washington, DC, 2006.)

equivalent (Figure 1.23). Figure 1.24 shows the feedstock used for these biofuels. An important consideration for biofuels is that the fuel not be produced at the expense of food while there are people going hungry in the world. This is of major concern when ethanol is produced from corn, but would not be of concern if biofuels were produced from cellulosic feed stock or municipal solid waste (MSW). According to a Worldwatch report [26], a city of 1 million people produces about 1800 ton of MSW and 1300 ton of organic waste every day, which, using present-day technology could produce enough fuel to meet the needs of 58,000 persons in the United States, 360,000 in France and nearly 2.6 million in China at current rates of per capita fuel use [26].

The total share of all renewable sources for electricity production in 2002 was about 17%, a vast majority (89%) of it being from hydroelectric power (Table 1.10).

1.6.2 Wind Power

Wind energy technology has progressed significantly over the last two decades. The technology has been vastly improved and capital costs have

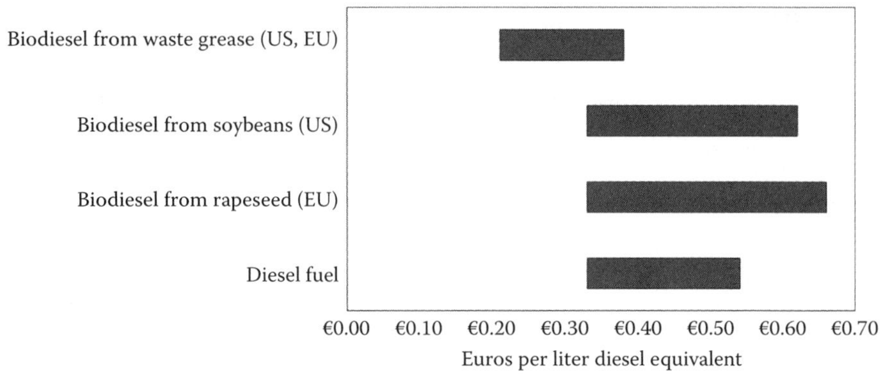

FIGURE 1.23
Cost ranges for biodiesel and diesel production, 2006. (From Worldwatch Institute, *Bio-Fuels for Transportation—Global Potential and Implications for Sustainable Energy in the 21st Century*, Worldwatch Institute, Washington, DC, 2006.)

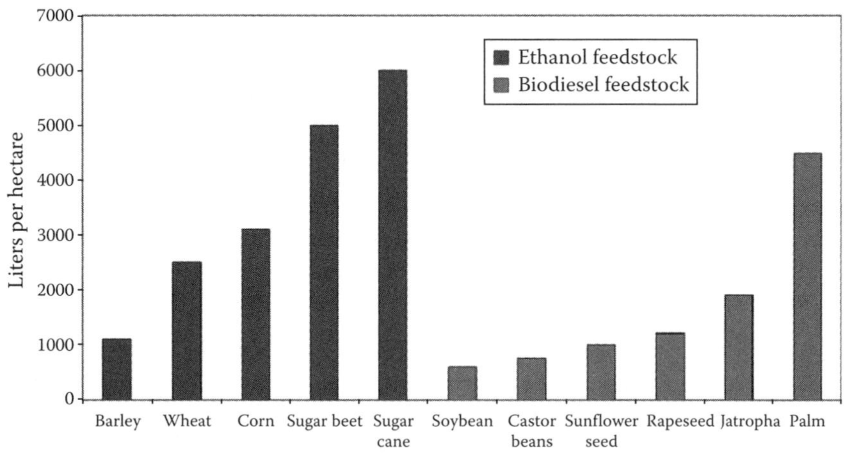

FIGURE 1.24
Biofuel yields of selected ethanol and biodiesel feedstock. (From Worldwatch Institute, *Bio-Fuels for Transportation—Global Potential and Implications for Sustainable Energy in the 21st Century*, Worldwatch Institute, Washington, DC, 2006.)

come down to as low as $1100 per kW. At this level of capital costs, wind power is already economical at locations with fairly good wind resources. Therefore, the average annual growth in worldwide wind energy capacity from 2000 to 2003 was over 30% (Figure 1.25) and it continued to grow at that rate in 2004 and 2005. The average growth in the United States over the same period was 37.7%. The total worldwide installed wind power capacity, which was 39 GW in 2003 (Figure 1.25), reached a level of 59 GW in 2005 [27]. The total theoretical potential for on-shore wind power for the world is around 55 TW with a practical potential of at least 2 TW [5], which is about two-thirds of the entire present worldwide generating capacity. The offshore wind energy potential is even larger.

TABLE 1.10

Electricity from Renewable Energy in 2002

	2002	
Energy Source	**TWh**	**(%)**
Hydropower	2610	89
Biomass	207	7
Wind	52	2
Geothermal	57	2
Solar	1	0
Tide/wave	1	0
Total	2927	100

Source: Data from IEA, *World Energy Outlook,* IEA, Paris, France, 2004.

1.6.3 Solar Energy

The amount of sunlight striking the earth's atmosphere continuously is 1.75×10^5 TW. Considering a 60% transmittance through the atmospheric cloud cover, 1.05×10^5 TW reaches the earth's surface continuously. If the irradiance on only 1% of the earth's surface could be converted into electric energy with a 10% efficiency, it would provide a resource base of 105 TW, while the total global energy needs for 2050 are projected to be about 25–30 TW. The present state of solar energy technologies is such that solar cell efficiencies have reached over 40% and solar thermal systems provide efficiencies of 40%–60%. With the present rate of technological development, these solar technologies will continue improving, thus bringing the costs down, especially with the economies of scale.

Solar PV panels have come down in cost from about \$30/W to about \$3/W in the last three decades. At \$3/W panel cost, the overall system cost is around \$6/W, which is still too high to compete with other resources for grid electricity. However, there are many off-grid applications where solar PV is

FIGURE 1.25
World wind energy installed capacity and growth rates. (Data from IEA, *World Energy Outlook,* IEA, Paris, France, 2004.)

TABLE 1.11

Growth in Photovoltaics Demand (2000–2003)

Region	Percentage Increase 2000–2003 (%)
OECD	32
OECD (Europe)	41.1
Germany	51.1

already cost effective. With net metering and governmental incentives, such as feed-in laws and other policies, even grid connected applications such as Building Integrated PV (BIPV) are becoming cost effective. As a result, the worldwide growth in PV production has averaged over 30% per year from 2000 to 2003, with Germany showing the maximum growth of over 51% (Table 1.11, Figure 1.26).

Solar thermal power using concentrating solar collectors was the first solar technology that demonstrated its grid power potential. A 354 MW solar thermal power plant has been operating continuously in California since 1988. Progress in solar thermal power stalled after that time because of poor government policy and lack of R&D. However, the last 5 years have seen a resurgence of interest in this technology and a number of solar thermal power plants around the world are under construction. The cost of power from these plants (which is so far in the range of 12–16 U.S. cents/kWh) has the potential to go down to 5 U.S. cents/kWh with scale-up and creation of a mass market. An advantage of solar thermal power is that thermal energy can be stored efficiently and fuels such as natural gas or biogas may be used as back-up to ensure continuous operation. If this technology is combined

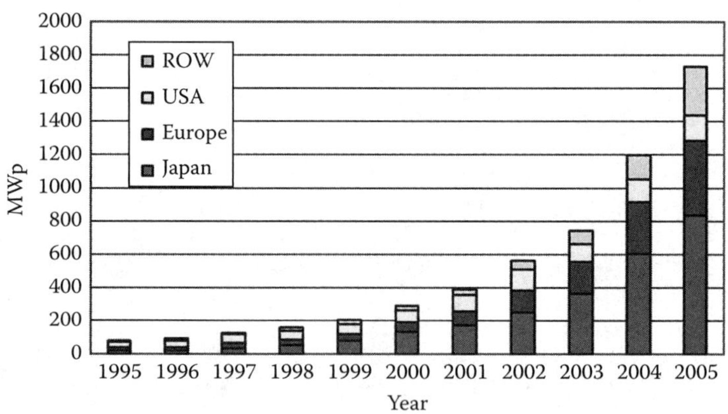

FIGURE 1.26
World solar PV production 1995–2005. (From Maycock, P., Annual review of the PV market, *PV News*, available at www.epia.org (accessed August 1, 2006).)

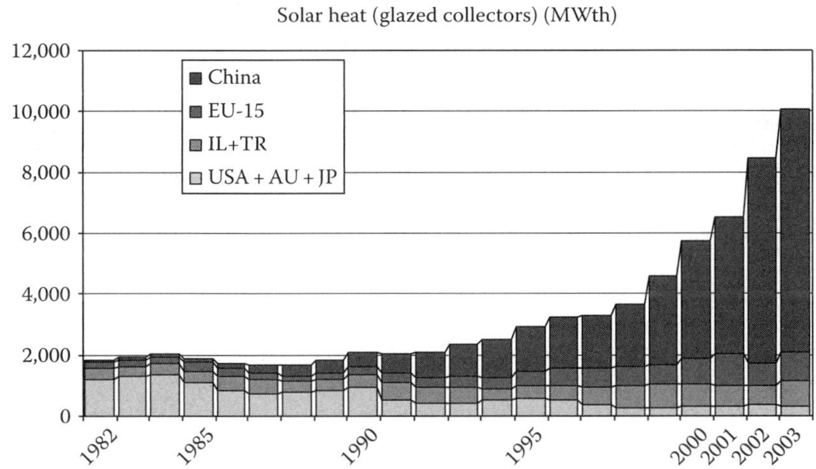

FIGURE 1.27
Deployment of solar heat (glazed) collectors, MWth. (From IEA, *World Energy Outlook*, IEA, Paris, France, 2004.)

with power plants operating on fossil fuels, it has the potential to extend the time frame of the existing fossil fuels.

Low-temperature solar thermal systems and applications have been well developed for quite some time. They are being actively installed wherever the policies favor their deployment. Figure 1.27 gives an idea of the rate of growth of solar thermal systems in the world. Just in 2003, over 10 MWth low-temperature solar collectors were deployed around the world, a vast majority of those being in China for domestic hot water.

1.7 Summary of Renewable Energy Resources

By definition, the term "reserves" does not apply to renewable resources. So we need to look at the annual potential of each resource. Table 1.12 summarizes the resource potential and the present costs and the potential future costs for renewable resources.

As in the case of other new technologies, it is expected that cost competitiveness of the renewable energy technologies will be achieved with R&D, scale-up, commercial experience, and mass production. The experience curves in Figure 1.28 show industry-wide cost reductions in the range of 10%–20% for each cumulative doubling of production for wind power, PV technologies, ethanol, and gas turbines [5]. Similar declines can be expected in solar thermal power and other renewable technologies. As seen from Figure 1.28, wind energy technologies have already achieved market maturity, and PV technologies are well on their way. Even though concentrating

TABLE 1.12

Potential and Status of Renewable Energy Technologies

Technology	Annual Potential	Operating Capacity 2005	Investment Costs US$ per kW	Current Energy Cost	Potential Future Energy Cost
Biomass energy	276–446 EJ Total				
Electricity	or 8–13 TW	~44 GW	500–6,000$/kWe	3–12 c/kWh	3–10 c/kWh
Heat	MSW ~ 6 EJ	~225 GWth	170–1,000$/kWth	1–6 c/kWh	1–5 c/kWh
Ethanol		~36 bln lit.	170–350$/kWth	25–75 c/L (ge)[a]	6–10 $/GJ
Biodiesel		~3.5 bln lit.	500–1,000$/kWth	25–85 c/L (de)[b]	10–15 $/GJ
Wind power	55 TW Theo.	59 GW	850–1,700 $/kWe	4–8 c/kWh	3–4 c/kWh
	2 TW practical				
Solar energy	>100 TW				
Photovoltaics		5.6 GW	5,000–10,000$/kWe	25–160 c/kWh	5–25 c/kWh
Thermal power		0.4 GW		12–34 c/kWh	4–20 c/kWh
Heat			2,500–6,000	2–25 c/kWh	2–10 c/kWh
			300–1,700		

Geothermal electricity	600,000 EJ useful resource base	9 GWe	800–3,000	2–10 c/kWh	1–8 c/kWh
Heat	5000 EJ economical in 40–50 years	11 GWth	200–2,000	0.5–5 c/kWh	0.5–5 c/kWh
Ocean energy					
Tidal	2.5 TW	0.3 GWe	1,700–2,500	8–15 c/kWh	8–15 c/kWh
Wave	2.0 TW		2,000–5,000	10–30 c/kWh	5–10 c/kWh
OTEC	228 TW		8,000–20,000	15–40 c/kWh	7–20 c/kWh
Hydroelectric	1.63 TW Theo.				
Large	0.92 TW Econ.	690 GW	1,000–3,500	2–10 c/kWh	2–10 c/kWh
Small		25 GW	700–8,000	2–12 c/kWh	2–10 c/kWh

Sources: Data from Goldemberg, J. and Johansson, T.B. (eds.), *World Energy Assessment: Energy and the Challenge of Sustainability*, United Nations Development Programme, New York, 2004; UNDP (2004) updated from other sources: Worldwatch Institute, *Bio-Fuels for Transportation—Global Potential and Implications for Sustainable Energy in the 21st Century*, Worldwatch Institute, Washington, DC, 2006; World Wind Energy Association, *Worldwide Wind Energy Bulletin*, World Wind Energy Association, Bonn, Germany, Available at www.wwindea.org (accessed October 12, 2006.); Photovoltoaic Barometer, EPIA (www.epia.org); World Geothermal Power Generation 2001–2005; GRC Bulletin; International Energy Annual, USDOE-EIA.

[a] ge—gasoline equivalent liter.

[b] de—diesel equivalent liter.

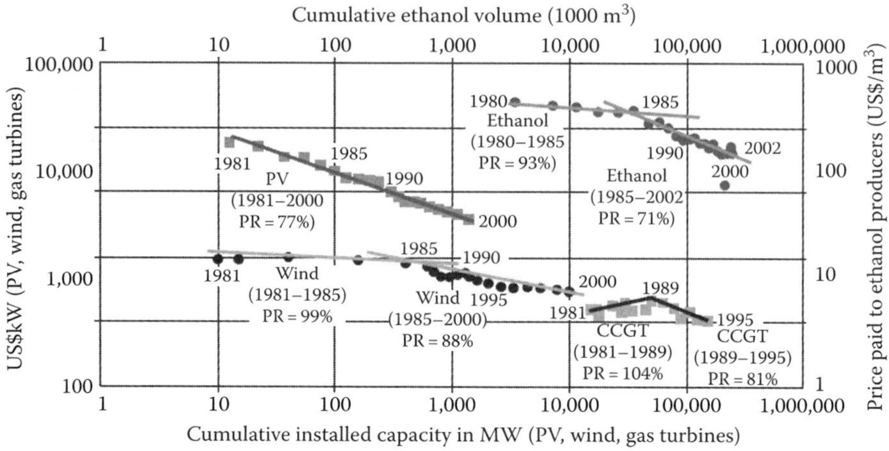

FIGURE 1.28
Experience curves for wind (From Neij, L. et al., The use of experience curves for assessing energy policy programs, Presented at EU/IEA Workshop, IEA, Paris, France, January 2003.), PV (From Parente, V. et al., *Prog. Photovolt. Rese. Appl.*, 10(8), 571, September 2002.), ethanol (From Goldemberg, J. et al., *Biomass Bioenergy*, 26(3), 301, March 2004.), and gas turbines. (From Claeson Colpier, U. and Cornland, D., *Energy Policy*, 30(4), 309, March 2002.). (Adapted from Goldemberg, J. and Johansson, T.B. (eds.), *World Energy Assessment*: *Energy and the Challenge of Sustainability*, United Nations Development Programme, New York.)

solar thermal power (CSP) is not shown in Figure 1.28, a GEF report estimates that CSP will achieve the cost target of about $0.05/kWh by the time it has an installed capacity of about 40 GW [28]. As a reference point, wind power achieved that capacity milestone in 2003.

1.8 Orders of Magnitude

In planning for a sustainable energy future, one encounters huge numbers such as the world greenhouse gas emission of 34 Gton of CO_2 equivalent per year or the world power consumption of 15 TW. Often, especially when thinking about energy policy, it is not possible to make detailed engineering calculations quickly enough. But, in many instances it is possible and very useful to make what is a called a "back-of-the-envelope" or order of magnitude estimates for various energy choices. This type of estimate often reveals the realism, or lack thereof, of making plans for obtaining energy under various scenarios. For such order of magnitude estimates, it is very useful to have some approximate data that can be used quickly and does not require detailed calculations or interpretations. In this connection, it is important to recognize that fossil fuel sources have a great energy density, while all of the renewable sources are very diffuse and therefore require large areas to deliver power

TABLE 1.13

Land or Water Area for Different Electric Power Generation Systems

Power per Unit Land or Water Area (W/m^2)	
Wind	2
Offshore wind	3
Tidal pools	3
Solar PV panels	5–20
Bioplantation	0.5
Hydroelectric	11
Ocean thermal	5
Concentrating solar power tower (desert)	15

needed in an industrial society. Table 1.13 shows the power generation per unit land or water area for various renewable energy technologies.

A broad perspective of the capital and levelized energy costs of renewables, as well as more traditional energy technologies has been prepared by Lazard a well-known financial advisory and asset management firm. The firm provides advice on mergers and acquisitions, as well as asset management services for large institutions and governments. The technology assessments shown below were prepared to advise its customers about investment in energy technologies. The details of the approach taken by Lazard are available in [29], but in subsequent chapters of the book, you will learn how to make such estimates on your own and how these estimates depend on assumptions such as the cost of money or the assumed life of the system. However, for the purpose of preliminary estimates, and particularly for making what is called a "back-of-the-envelope" calculation, the ranges of the cost figures provided by an impartial investment company will be useful. In Figure 1.29, the ranges of capital cost estimates (i.e., the cost of installing a given technology) are presented for conventional and alternative energy systems.

It is important to note that these estimates are based on current generation technologies for the conventional systems, which are considered mature. It is, therefore, not expected that the capital costs of these systems will substantially decrease in the future. However, this is not the case for the alternative energy systems, which are currently under intensive research, and for which new approaches may reduce capital costs. It should also be noted that the conventional systems, such as coal, nuclear, and gas, can operate continuously and can be located in many different localities, while most alternative energy systems operate only intermittently, may require extensive storage, and do not have a proven track record. Moreover, their performance is dependent on geographic locations, such as those with good wind resources.

Using the assumed capital costs in Figure 1.29, Lazard estimated the levelized cost of energy from conventional and alternative energy technologies, and are shown in Figure 1.30. The data in Figures and 1.29 and 1.30

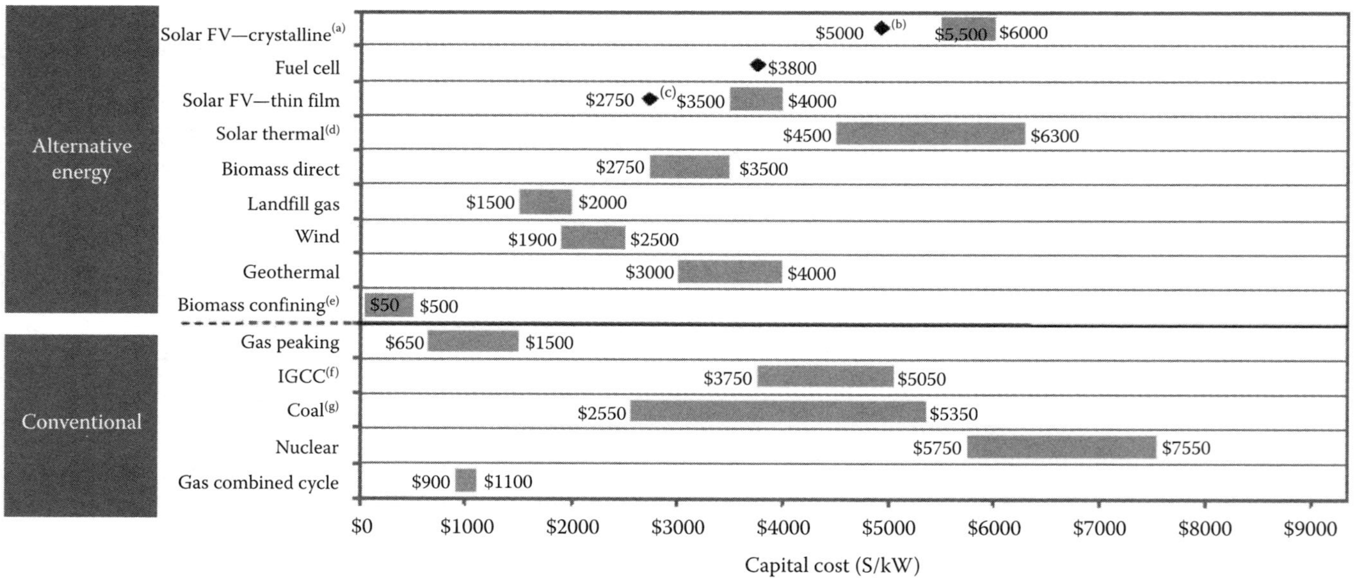

FIGURE 1.29
Capital costs for various electric power generation technologies in 2008. *Notes:* (a) Low and represents single-axis tracking crystalline. High end represents fixed installation. (b) Based on a leadign solar-crystalline company's guidance of 2010 total system cost of $5.00 per watt; company guidance for 2012 total system cost is $4.00 per watt. (c) Based on the leading thin-film company's guidance of 2010 total system cost of $2.75 per watt; company guidance for 2012 total system cost is $2.00 per watt. (d) Low end represents solar trough. High end represents solar tower. (e) Represents retrofit cost of coal plant. (f) High end incorporates 90% carbon capture and compression. (g) Based on advanced supercritical pulverized coal. High end incorporates 90% carbon capture and compression. (From Lazard, Ltd., *Energy Technology Assessment*, Lazard, Ltd., New York, 2009, available at: www.lazard.com; Palz, W. and Zibetta, H., *Int. J. Solar Energ.*, 10(3–4), 211, 1991.)

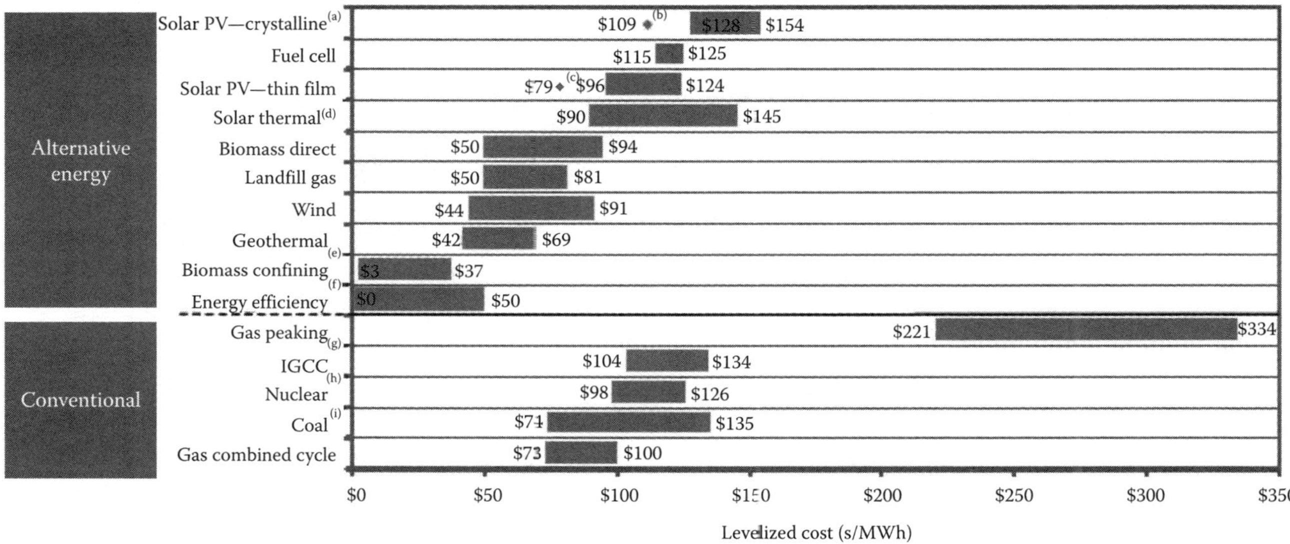

FIGURE 1.30
Levelized cost of energy comparison. *Notes:* Reflects production tax credit, investment tax credit, and accelerated asset depreciation as applicable. Assumes 2008 dollar, 60% debit as 7% import rate, 40% equity at 12% cost, 20-year economic life, 40% tax, and 5–20 year tax life. Assumes coal price of $2.50 per MMBtu and natural gas price of $8.00 per MMBtu. (a) Low end represents single-axis tracking crystalline. High end represents fixed installation. (b) and (c) Represents a leading solar crystalline company's targeted implined levelized cost of energy in 2010 assuming a total system cost of $5.00/watt. Company guidance for 2012 total system cost of $4.00/watt would imply a levelized cost of energy of $90/MWh. (d) Low end represents solar tower. High end represents solar trough. (e) Represents retrofit cost of coal plant. (f) Estimates per National Action Plan for Energy Efficiency; actual costs for various initiatives varies widely. (g) High end incorporates 90% carbon capture and compression. (h) Does not reflect potential economic impact of federal loan guarantees or other capture subsides. (i) Based on advanced supercritical pulverized coal. High end incorporates 90% carbon capture and compression. (From Lazard, Ltd., *Energy Technology Assessment*, Lazard, Ltd., New York, 2009, available at: www.lazard.com; Palz, W. and Zibetta, H., *Int. J Solar Energ.*, 10(3–4), 211, 1991.)

provide an overview of the future energy costs under reasonable assumptions made by the analysts. These estimates do not include social costs and, as a result, there are large uncertainties in what the real costs of energy in the future will be.

Integrated gas combined cycle (IGCC) is an emerging technology using coal to generate a syngas, which is then used to power a conventional gas turbine. In Figure 1.29, the high-cost figures are for IGCC and for coal with 90% carbon capture and disposal. However, these numbers are estimates. The technologies for carbon capture systems (CCS) have not been commercially demonstrated, and the general consensus is that CCS will not be available until 2025. Thomas D. Shop, the acting assistant secretary for fossil energy of the U.S. DOE, testified on April 16, 2007 before the Senate Energy and Natural Resources Committee, stating that CCS would be available in 2020–2025, but that "common, everyday employment" will not be seen until 2045. These projections assume success in several demonstration projects that have been proposed for 2010 and the immediate future thereafter. The high-end capital cost for solar thermal is for a power tower, which has not been fully commercialized and may be too high. The cost estimates for nuclear power do not include a facility for disposal of radioactive waste and the safe disposal of the plant after decommissioning. It should be noted that the levelized cost of electric power from a fuel cell assumes that hydrogen is generated from natural gas, which is not a renewable resource. The levelized cost is approximately three times higher if the hydrogen was produced by electrolysis from water and the required electricity would come from solar, fossil, or nuclear sources.

With alternative energy generation technologies becoming competitive with conventional generation technologies, it is important to be aware of such issues as location, dispatch characteristics such as base-load or intermittent, and contingencies such as carbon sequestration. The energy resources matrix shown in Table 1.14 summarizes the application potential of conventional and alternative energies from the perspective of a utility.

The following examples will illustrate the use of some of approximate gross numbers in technology assessment for policy decisions.

Example 1.2

Recently, the Arizona legislature passed a law requiring that by the year 2025, 15% of all the electric energy generation be from renewable sources. Assuming that the intended technology is to be silicon PV panels, calculate the collector area required and then estimate what percentage of the total land area of the state the solar system would occupy.

TABLE 1.14

Energy Resources: Matrix of Applications

		Levelized Cost of Energy	Carbon Neutral/ Rec Potential	State of Technology	Location			Dispatch			
					Customer Located	Central Station	Geography	Intermittent	Peaking	Load-Following	Base-Load
Alternative energy	Fuel cell	$115–125	?[a]	Emerging/ commercial	✓		Universal				✓
	Solar PV	$96–154	✓	Newly commercial	✓	✓	Universal	✓	✓		
	Solar thermal	$90–145	✓	Emerging		✓	Southwest	✓	✓	✓	
	Biomass direct	$50–94	✓	Mature		✓	Universal			✓	✓
	Wind	$44–91	✓	Mature		✓	Varies	✓			
	Geothermal	$42–69	✓	Commercial/ evolving		✓	Varies				✓
	Landfill gas	$50–81	✓	Mature		✓	Varies				✓
	Gas peaking	$221–334	×	Mature	✓	✓	Universal		✓		
	IGOC	$104–134	×[b]	Emerging[c]		✓	Colocated or mal				✓
Conventional	Gas combined cycle	$73–100	×	Mature	✓	✓	Universal			✓	✓
	Coal	$74–135	×[b]	Mature		✓	Colocated or mal				✓
	Nuclear	$98–126	✓	Mature/ emerging		✓	Colocated or mal				✓

Source: Lazard Ltd., *Energy Technology Assessment*, New York, 2009, available at: www.lazard.com

[a] Qualification for RPS *requirements* varies by location.

[b] Could be considered *carbon* neutral technology, assuming carbons capture and compression.

[c] Carbon capture and *compression* technologies are in emerging stage.

Solution

From the Arizona Energy Commission Web site, we find that in 2009, 30,800 GWh/year of electric energy were generated, and that expectations are that by the year 2025, 50,000 GWh/year will be required.

From the U.S. Solar Radiation Resource Map, found at the NREL Web site at http://rredc.nrel.gov/solar/old_data/nsrdb/redbook/atlas/Table.html, the solar radiation impinging on a flat plate facing south and inclined at the latitude, which is the most favorable orientation, is 5.5 kWh/m²/day, which is equivalent to about 2×10^3 kWh/m²/year. Assuming a PV efficiency of 12%, the amount of electric power delivered per year is 241 kWh/m². Hence, the collector area is

$$\frac{0.15\times10^{13}\ \text{kWh/year}}{240\ \text{kWh/year}} = 3.1\times10^{7}\ \text{m}^2$$

But in order to avoid shading one row of collector panels from another, they have to be spaced in such a manner as to require twice the land area, which means that approximately 6.2×10^7 m² of land are required.

Going again to the Internet, we find that the Arizona land area is 1.14×10^5 mi², which is equal to 29.5×10^{10} m². Hence, the area required to fulfill the mandate would cover 0.021% of the state of Arizona with PV collectors. However, from the NREL Web site, we can learn that a south-north tracking collector would receive twice the radiation that a stationary surface would and, therefore, the land area could be reduced to 0.01% of the state by using tracking instead of stationary collectors. We can estimate the cost of the installation from Figure 1.18, which indicates that the cost would be approximately $5.5/kW-hr. With a demand of 50,000 GWh/year, Arizona would need about 5.8×10^9 kW to satisfy this demand and that the installation would require an investment of about $31 billion to satisfy 15% of the electricity demand in the year 2025. Although the area required to meet the power demand is large, the collectors could be placed in desert where land is cheap. If the PV installations were combined with evaporative cooling systems to replace the electric load from conventional air conditioning, the electricity demand from fossil fuels could be substantially reduced.

Example 1.3

On July 17, 2008, former Vice President Al Gore gave a very popular talk (Transcript can be found at http://www.npr.org/templates/story/story.php?storyId=92638501) in which he proposes that in 10 years all our electric power generation be from carbon-free fuel. Alternatively, Senator McCain has suggested that nuclear power plants be installed in order to reduce the effects of global warming. But neither gave any specifics on how this transition might be accomplished or how much it would cost.

The following data is from the 2008 EIA—*Annual Energy Outlook* (http://www.eia.doe.gov/oiaf/forecasting.html):

Fossil based electricity generation in year 2020: 31.53 Quads (31.53×10^{15} Btu)
Total electricity generation in year 2020: 45.18 Quads (45.18×10^{15} Btu)

Assuming that we can save 8 Quads from this estimate by conservation and there is no replacement of fossil fuels in other sectors, such as transportation, estimate how many 2.5 MW wind turbines (average yearly capacity factor of 30%) would need to be built and installed to meet Mr. Gore's suggestion. Then estimate the total costs and land area of these wind turbines assuming that wind turbines cost about $1100/kW electric installed. Then estimate how many 1000 MW nuclear power plants with a 90% capacity factor would be required and what would be the capital cost if nuclear power installation cost $7000/kW electric.

Solution

Assuming conservation can reduce the 2020 electricity needs by 8 Quads, we are left with 23.53 Quads of electric energy that must be produced by clean sources.

(a) *Wind power*: Given a 30% annual capacity factor, we can calculate the yearly output of one wind turbine:

$$(0.30)\times(2.5\,\text{MW})\times\left(8760\,\frac{\text{h}}{\text{year}}\right)\times\left(3.4144\times10^{-9}\,\frac{\text{Quads}}{\text{MWh}}\right)$$

$$= 2.24\times10^{-5}\,\frac{\text{Quads}}{\text{year}}$$

Knowing that we need to produce a total of 23.53 Quads of electricity, we now calculate the number of wind turbines needed:

$$\frac{23.53\,\text{Quads}}{2.24\times10^{-5}\,\text{Quads/turbine}} = 1{,}050{,}000 \text{ turbines (about 1 million!)}$$

We can now determine the total land area required for wind farms in the United States in 2020. With about 50 ac of land area needed per turbine, we have

$$\left(50\,\frac{\text{ac}}{\text{turbine}}\right)\times1\times10^{6}\text{ turbines} = 50\,\text{million ac}$$

At $1100 per kW for wind turbines, the capital needed for the installation of these farms is

$$\left(\frac{\$1100}{\text{kW}}\right)\times\left(\frac{1000\,\text{kW}}{\text{MW}}\right)\times\left(\frac{2.5\,\text{MW}}{\text{turbine}}\right)\times(1\times10^{6}\text{ turbines}) = \$2.75\,\text{trillion}$$

(b) *Nuclear power*: The yearly electric energy produced by one 1000 MW nuclear power plant operating at 90% capacity is

$$(0.90)\times(1000\,\text{MW})\times\left(8760\ \frac{\text{h}}{\text{year}}\right)\times\left(3.4144\times10^{-9}\ \frac{\text{Quads}}{\text{MWh}}\right)$$

$$=0.0269\ \frac{\text{Quads}}{\text{year}}$$

Now we can calculate the number of nuclear plants needed to produce 23.53 Quads:

$$\frac{23.53\,\text{Quads}}{0.0269\,\text{Quads/plant}}=875\text{ plants}$$

At $7000 per kW for nuclear plants, we determine the capital needed for the construction of these plants:

$$\left(\frac{\$7000}{\text{kW}}\right)\times\left(\frac{1000\,\text{kW}}{\text{MW}}\right)\times\left(\frac{1000\,\text{MW}}{\text{plant}}\right)\times(875\,\text{plants})=\$6.13\,\text{trillion}$$

To provide some perspective, the total U.S. GNP is about $11 trillion/year.

Clearly, the cost of wind turbines is drastically lower than that of nuclear plants for this scenario. However, the choice cannot simply be made based on economic investment alone. Nuclear power has the advantage of being a more reliable and continuous source of power and is thus much better suited to meeting the needs of the electric grid. For wind to become a base-load energy provider some large-scale energy storage is required. Maybe a combination of both nuclear and wind should be considered.

Energy Units and Conversion Factors

In reading the literature pertaining to energy, one encounters an overwhelming number of units that are not usually used in conventional engineering calculations. To assist the reader in handling data from various sources that include unusual units such as barrels of oil equivalent, a conversion factor table is presented in Table 1.15.

TABLE 1.15
Energy Conversion Factors

	Btus	Quads	Calories	kWh	MWy
Btus	1	10^{-15}	252	2.93×10^{-4}	3.35×10^{-11}
Quads	10^{15}	1	2.52×10^{17}	2.93×10^{11}	3.35×10^{4}
Calories	3.97×10^{-3}	3.97×10^{-18}	1	1.16×10^{-6}	1.33×10^{-13}
kWh	3413	3.41×10^{-12}	8.60×10^{5}	1	1.14×10^{-7}
MWy	2.99×10^{10}	2.99×10^{-5}	7.53×10^{12}	8.76×10^{6}	1
bbls oil	5.50×10^{6}	5.50×10^{-9}	1.38×10^{9}	1612	1.84×10^{-4}
Tonnes oil	4.04×10^{7}	4.04×10^{-8}	1.02×10^{10}	1.18×10^{4}	1.35×10^{-3}
kg coal	2.78×10^{4}	2.78×10^{-11}	7×10^{6}	8.14	9.29×10^{-7}
Tonnes coal	2.78×10^{7}	2.78×10^{-8}	7×10^{9}	8139	9.29×10^{-4}
MCF gas	10^{6}	10^{-9}	2.52×10^{8}	293	3.35×10^{-5}
Joules	9.48×10^{-4}	9.48×10^{-19}	0.239	2.78×10^{-7}	3.17×10^{-14}
EJ	9.48×10^{14}	0.948	2.39×10^{17}	2.78×10^{11}	3.17×10^{4}

(continued)

TABLE 1.15 (continued)

Energy Conversion Factors

	Btus	Quads	Calories	kWh	MWy		
	bbls Oil Equivalent	Tonnes Oil Equivalent	kg Coal Equivalent	Tonnes Coal Equivalent	MCF Gas Equivalent	Joules	EJ
Btus	1.82×10^{-7}	2.48×10^{-8}	3.6×10^{-5}	3.6×10^{-8}	10^{-6}	1055	1.06×10^{-15}
Quads	1.82×10^{8}	2.48×10^{7}	3.6×10^{10}	3.6×10^{10}	10^{9}	1.06×10^{18}	1.06
Calories	7.21×10^{-10}	9.82×10^{-11}	1.43×10^{-7}	1.43×10^{-10}	3.97×10^{-9}	4.19	4.19×10^{-18}
kWh	6.20×10^{-4}	8.45×10^{-5}	0.123	1.23×10^{-4}	3.41×10^{-3}	3.6×10^{6}	3.6×10^{-12}
MWy	5435	740	1.08×10^{6}	1076	2.99×10^{4}	3.15×10^{13}	3.15×10^{-5}
bbls oil	1	0.136	198	0.198	5.50	5.80×10^{9}	5.80×10^{9}
Tonnes oil	7.35	1	1455	1.45	40.4	4.26×10^{10}	4.26×10^{-8}
kg coal	5.05×10^{-3}	6.88×10^{-4}	1	0.001	0.0278	2.93×10^{7}	2.93×10^{-11}
Tonnes coal	5.05	0.688	1000	1	27.8	2.93×10^{10}	2.93×10^{-8}
MCF gas	0.182	0.0248	36	0.036	1	1.06×10^{9}	1.06×10^{-9}
Joules	1.72×10^{-10}	2.35×10^{-11}	3.41×10^{-8}	3.41×10^{-11}	9.48×10^{-10}	1	10^{-18}
EJ	1.72×10^{8}	2.35×10^{7}	3.41×10^{10}	3.41×10^{7}	9.48×10^{8}	10^{18}	1

Notes: To convert from the first column units to other units, multiply by the factors shown in the appropriate row (e.g., 1 Btu = 252 cal). MWy = megawatt-year; bbls = barrels; tonnes = metric tons = 1000 kg = 2204.6 lb; MCF = thousand cubic feet; EJ = exajoule = 10^{18} J. Nominal calorific values assumed for coal, oil, and gas.

Problems

1.1 The hallways of a three-story building in Denver, Colorado use 600 lights per floor and each light is an incandescent bulb rated at 60 W.

a. Estimate the amount of energy that could be saved in August by turning the lights on when the sun sets, and turning them off when it becomes light.

b. Assuming that the power for the lights comes from a 30% efficient coal-fired power plant, estimate the amount of CO_2 that would be saved with this conservation measure. Assume that the chemical composition of the coal is $CH_{0.8}$ and that coal has a heating value of 29 MJ/kg.

c. Prepare a brief report, making recommendations for reasonable conservation measures.

1.2 On May 27, 2009, U.S. Senator Lamar Alexander proposed to build 100 new nuclear power plants in 20 years "for a rebirth of industrial America while we figure out renewable electricity." In his talk, he stated "This would double America's nuclear plants, which today produce 20% of all of our electricity and 70% of our pollution-free, carbon-free electricity."

Analyze the proposal of the senator from several perspectives:

a. Cost of 100 nuclear power plants.

b. Need to replace existing nuclear power plants to generate net power output.

c. Amount of water required for the condensers of the new power plants.

1.3 Two different fuels are being considered for a 1 MW (net output) heat engine, which can operate between the highest temperature produced during the burning of the fuels and an atmospheric temperature of 300 K. Fuel A burns at 2500 K, delivering 50 MJ/kg (heating value) and costs $2.00/kg. Fuel B burns at 1500 K, delivering 40 MJ/kg and costs $1.50/kg. Compare the fuel costs per hour of fuel A and fuel B, assuming that the heat engine operates

a. At Carnot efficiency

b. At 50% of Carnot efficiency

1.4 A solar energy collector produces a maximum temperature of 100°C. The collected energy is used in a cyclic heat engine that operates in a 5°C environment. Estimate the thermal efficiency if the engine operates at 40% of Carnot efficiency. How would the answer change if the collector were redesigned to focus the incoming solar radiation to enhance the maximum temperatures to 400°C.

1.5 Since meat is higher on the food chain than grains, the more calories you get from meat, the more agricultural land you use. Using the arable land area in your state and data from [30] (which concludes that 5.6 GJ metabolized energy is produced annually per hectare for meat and that 41.4 GJ of metabolized energy is produced annually per hectare for gain, estimate

a. How many vegetarians could you feed with land available in your state. Determine this as a number (percentage of 2011 population) and percentage of estimated population in 2030.
b. Repeat this estimate for the average American diet (the typical American gets two-thirds of his or her calories from meat and one-third from grain).

1.6 If 10% of your state would be covered with PV collectors, estimate the power produced in kWh/day per capita on average. If 10% of the people in your state had to drive 100 km/day to commute to work, how many kWh/day would that consume? Discuss the implications of these two estimates. How many collectors of 3 and 10 m would be required to produce the energy for commuting with an all-electric car with 60% efficient batteries?

1.7 Assuming that a heat pump has a coefficient of performance equal to 4, how much heat could this device deliver to the interior of a building with 1 kWh of electricity.

1.8 If there were a carbon tax of \$250/ton of CO_2 on fossil fuels, estimate the increase in the price of a barrel of oil.

1.9 In 2009, the administration of President Obama instituted a so-called "Cash for Clunkers" program. You are to evaluate the effectiveness of the program, which was intended to stimulate the economy and reduce the import of oil. According to the program announcement, an average of \$4000 per car in rebate was offered if a consumer brought in a "clunker" and exchanged it for a new car that would have a fuel efficiency equal to at least 10 mi/gal better than the car they traded. The total amount of money allocated to the program was \$3 billion.

a. What reduction in oil import per year, both in terms of barrels of oil and percentage of total import, was achieved by the program?
b. Assuming that imported oil cost \$55 per barrel, how long did it take for the savings in import to equal the investment in the program?
c. The energy required to build a new car, called the embedded energy in car construction, has been estimated to be 30,500 kWh/car by Argonne National Laboratory [31]. The embedded energy is mostly coal and nuclear, while the energy saved by the improved efficiency is mostly imported over the life of the car. Estimate how many years it would take to recoup the energy invested in the construction of the car by the savings in energy from the improved mileage of the new vehicle.
d. What reduction in CO_2 generation did the program achieve?

Discuss the effectiveness of the program from an energy perspective and what changes in the program you would recommend to make it more effective.

1.10 While completing your homework, imagine the study space you are using employs a space heater with a rating of 30 kW, a refrigerator (for snacks) that is rated for 5 kW, and a light fixture with two 100 W bulbs, all operating at a constant rate for the 1.5 h it takes you to finish.

a. What is the total power consumed at any given moment?
b. How much energy do you consume to complete the task?
c. Assume the utility company charges 6.5 cents per unit of energy in part b, how much did it cost you to complete the assignment?
d. If you were willing to pay the utility company $5, what would be the power rating of an additional appliance you could add to your study space?

1.11 In 2009, a group of environmentalists approached the Boulder City Council with a request to deny an application to continue operating the local coal-fired 200 MW power plant after 2010 in order to reduce CO_2 emissions. The group suggested that the coal-fired power plant be replaced by a nonpolluting solar plant.

To evaluate the feasibility of their request, a steel plate is placed in sunlight in Boulder at an angle equal to the latitude. The plate is treated with a coating that has an absorptivity for solar insolation of 0.9 and an emissivity in the long range spectral range of 0.2.

a. If the average insolation between 9 a.m. and 5 p.m. is equal to 400 W/m^2 and the convective heat transfer coefficient from the top surface of the plate is 8 W/m^{2}K, estimate the equilibrium temperature of the plate assuming the bottom is insulated and the effective sky temperature is 50 K.
b. If you could operate a heat engine between the plate temperature calculated in part (a) and the environment at 10°C, estimate the plate surface area required to generate power equal to the coal-fired power plant. Then comment on the result and compare it with a PV system of 12% efficiency.

1.12 A large nuclear power plant has a rated capacity of 1 GW electric. Its actual output is estimated to average about 80% of capacity due to the need for maintenance and imperfect match between supply and demand. Assuming that year-round average residential electricity needs are about 1 kW electric per housing unit and there are three residents per unit, estimate how many power plants would be needed to satisfy the demand of a metropolitan area with a population of 30 million.

1.13 The yearly average solar flux incident on a horizontal surface in the United States is about 180 W/m^2. The average total energy demand in the United States is approximately 100 quads (100×10^{18} J) per year. Assuming an average efficiency of PV cells of 14% and a biomass conversion efficiency around 1%, estimate the percentage of land required to provide the energy of the country by these two renewable technologies. (The total land area of the United States is about 9.2×10^6 km^2.)

Online Resources

The field of sustainable energy is changing so rapidly that a single book is not able to capture information in a real-time basis. The following Web sites will make it possible for the reader to access developments in the sustainable energy field and keep up to date. The list is not exhaustive, but will provide an entry into useful data.

www.technologyreview.com—MIT's publications on current technological advances in computing, internet, communications, energy, materials, biomedicine, and business.

www.nrel.gov—Publications from the National Renewable Energy Laboratory that focus on innovations for our nation's future energy. The following are some of the numerous tech subjects:

- Advanced vehicles and fuels
- Basic sciences
- Biomass
- Buildings
- Electric infrastructure systems
- Energy analysis
- Geothermal
- Hydrogen and fuel cells
- Solar
- Wind

http://www.eere.energy.gov—U.S. Department of Energy efficiency and renewable energy publications focusing on a wide range of programs from biomass, to solar energy, and building technologies.

http://cleanenergysector.com—Profits from the global energy transformation with clean and solar energy technologies.

http://sciencedaily.com contains current science advancements and articles of the latest science news from earth to space.

http://www.repp.org—Center for renewable energy and sustainable technology while building a renewable energy industry by breakdown of each state.

http://www.sciencedirect.com—Current science advancements in physical science and engineering.

http://www.skyfuel.com—Information from a leader in solar thermal technology.

http://www.rangefuels.com—Information from a Colorado company about cellulosic ethanol. It built a cellulosic ethanol plant in South Georgia.

http://www.lignol.ca—Cellulosic ethanol information.

http://www.awea.org—American wind energy association promotes wind energy and has the most up to date news on wind power.

http://www.renewableenergyworld.com—Global perspective on renewable energy.

http://www.theoildrum.com—Discussion about renewable energy from reputable sources.

http://www.aceee.org—Information and news about energy efficiency economy on a federal and a state level.

http://www.withouthotair.com—A textbook with careful analysis of the options for a sustainable energy system in Great Britain.

http://www.iea.org—The International Energy Agency provides useful information on global energy topics with important up-to-date data.

References

1. Brundtland, G.H. (1987) *Our Common Future,* World Commission on Environment and Development, Oxford University Press, New York.
2. United Nations. (1994) *World Population Prospects: 1994 Revision,* U.S. Census Bureau International Programs Center, International Database, New York.
3. World Bank (2003) *Development Data and Statistics,* www.worldbank.org/data (accessed May, 2007).
4. Hoekstra, A.Y. and Chapagain, A.K. (2008) *Globalization of Water: Sharing the Planet's Freshwater Resources,* Blackwell Publishing, Oxford, U.K.
5. Goldemberg, J. and Johansson, T.B. (eds.) (2004) *World Energy Assessment: Energy and the Challenge of Sustainability,* United Nations Development Programme, New York.
6. U.S. DOE (2006) Energy demands on water resources, Report to Congress, U.S. Department of Energy, Washington, DC.
7. Associated Press, U.N.: World's hungry now more than 1 billion, June 19, 2009, http://www.msnbc.msn.com/id/31449307/ (accessed June 20, 2009).
8. Barney, G.O., Blewett, J., and Barney, K.R. (1993) *Global 2000 Revisited: What Shall We Do?,* p. 11, The Millennium Institute, Alexandria, VA.
9. Energy Information Agency (2008) U.S. DOE *Annual Energy Review,* Energy Information Agency, Washington, DC.
10. National Renewable Energy Laboratory Solar Energy Technologies Program (2004) DOE/GO-102004-1847, January 2004.
11. Cleveland, C.J., Editor-in-Chief. (2004) *Encyclopedia of Energy* (six volumes), Elsevier Science, Oxford, U.K.
12. Jochem, E. (2000) Energy end-use efficiency, in: *World Energy Assessment: Energy and the Challenge of Sustainability,* United Nations Development Programme, New York.
13. Kok, K. (2009) *Nuclear Engineering Handbook,* CRC Press, Boca Raton, FL.
14. *Nuclear Power Daily* (2009) Progress Energy Florida signs contract for new, advanced-design nuclear plant, January 16, 2009, http://www.nuclearpowerdaily.com/reports/Progress_Energy_Florida_Signs_Contract_For_New_Advanced_Design_Nuclear_Plant_999.html (retrieved May 18, 2009).
15. Storm van Leeuwen, J.W. and Smith, P. (2008 updated), Nuclear power and the energy balance, at http://www.stormsmith.nl/ (accessed August 22, 2009).
16. Verne, J. (1888) *The Mysterious Island,* Atheneum, New York.
17. Shinnar, R. (2003) The hydrogen economy, fuel cells, and electric cars, *Technology in Society* 25(4), 453–476.
18. Shinnar, R., Shapira, D., and Zakai, A. (1981) Thermochemical and hybrid cycles for hydrogen production—A differential comparison with electrolysis, *IEC Process Design Development,* 20, 581.

19. Kreith, F. and West, R.E. (2004) Fallacies of a hydrogen economy: A critical analysis of hydrogen production and utilization, *Journal of Energy Resources Technology*, 126, 249–257.
20. National Research Council (2000) *Renewable Power Pathways—A Review of the U.S. Department of Energy's Renewable Energy Programs*, National Research Council, National Academics Press, Washington, DC.
21. National Research Council, National Academy of Engineering (2000) *The Hydrogen Economy: Opportunities, Costs, Barriers, and R&D Needs* Draft, The National Academies Press, Washington, D.C., see www.nap.edu
22. Renner, J.L. and Reed, M.J. (2008) Geothermal energy, in Goswami, D.Y. and Kreith, F. (eds.) *Energy Conversion*, CRC Press, Boca Raton, FL.
23. Kreith, F. and Goswami, Y. (2005) *CRC Handbook of Mechanical Engineering*, CRC Press, Boca Raton, FL.
24. Muffler, L.J.P. (ed.) (1979) Assessment of geothermal resources of the United States, U.S. Geological Survey, Circular 790, Washington, DC.
25. Tester, J.W., Blackwell, D., Petty, S., Richards, M., Moore, M., Anderson, B. et al. (2007) The future of geothermal energy, in: *Proceedings of the 32nd Workshop on Geothermal Reservoir Engineering*, Stanford University, Stanford, CA, Jan. 22–24, 2007. SGP-TR-183.
26. Worldwatch Institute (2006) *Bio-Fuels for Transportation—Global Potential and Implications for Sustainable Energy in the 21st Century*, Worldwatch Institute, Washington, DC.
27. World Wind Energy Association (2006) *Worldwide Wind Energy Bulletin*, World Wind Energy Association, Bonn, Germany. Available at www.wwindea.org (accessed October 12, 2006.).
28. Global Environmental Facility (2005) Assessment of the World Bank/GEF strategy for the market development of concentrating solar thermal power, GEF Report, GEF/c.25/Inf.11, GEF, Washington, DC.
29. Lazard, Ltd. (2009) *Energy Technology Assessment*, Lazard, Ltd., New York, Available at: www.lazard.com
30. Stehfest, E., Bouwman, L., van Vuuren, D.P., den Elzen, M.G.J., Eickhout, B., and Kabat, P. (July 2009) Climate benefits of changing diet, *Climactic Change* 95(1 and 2), 83–102.
31. Burnham et al. (2007), www.transportation.anl.gov/software/GREET/publications.html
32. IAEA, *Power Reactor Information System (PRIS)*, Vienna, Austria, http://www.iaea.org/programmes/a2/
33. IEA (2004) *World Energy Outlook*, IEA, Paris, France, 2004.
34. Maycock, P. (2006) Annual review of the PV market, *PV News*, available at www.epia.org (accessed August 1, 2006)
35. Neij L., Dannemand Andersen, P., Durstewitz, M., Helby, P., Hopper-Kilpper, M., and Morthorst, P.E. (2003) The use of experience curves for asssessing energy policy programmes. Presented at EU/IEA Workshop, January. IEA, Paris, France.
36. Claeson Colpier, U. and Cornland, D. (March 2002) The economics of the combined cycle gas turbine—An experience curve analysis, *Energy Policy*, 30(4), 309–316.

37. Parente, V., Goldemberg J., and Zilles, R. (September 2002) Comments on experience curves for PV modules, *Progress in Photovoltaics: Research and Applications*, 10(8), 571–574.
38. Goldemberg, J., Coelho, S.T., Nastari, P.M., and Lucon, O. (March 2004) Ethanol learning curve—The Brazilian experience, *Biomass and Bioenergy*, 26(3), 301–3047.
39. EIA (2008) *Annual Energy Outlook*, http://www.eia.doe.gov/oiaf/forecasting.html
40. EIA (1996) *Annual Energy Outlook 1997*, http://www.eia.doe.gov
41. M.K. Jaccard and Associates and Willis Energy Services Ltd. (1996) Industrial energy end-use analysis and conservation potential in six major industries in Canada, Report prepared for Natural Resources Canada, Chap. 5.
42. IEA (2005) *Renewables Information*, IEA, Paris, France.
43. Freeston, D.H. (1995) Worldwide direct uses of geothermal energy, in: *Proceedings of the World Geothermal Congress, 1995*, International Geothermal Association, Florence, Italy, pp. 15–26.

Suggested Readings

Boyle, G. (2004) *Renewable Energy—Power for a Sustainable Future*, Oxford University Press, Oxford, U.K.

Caputo, R. (2009) *Hitting the Wall—A Vision of a Secure Energy Future*, Morgan & Claypool Publishers, San Rafael, CA.

Goswami, D.Y., Kreith, F., and Kreider, J. (2000) *Principles of Solar Engineering*, 2nd edn., Taylor & Francis, Philadelphia.

Heinberg, R. (2005) *The Party's Over—Oil War and the Fate of Industrial Society*, New Society Publishers, Gabriola Island, Canada.

Kreith, F. and Kreider, J. (1978) *Principles of Solar Engineering*, McGraw-Hill, New York.

Kreider, J. and Kreith, F. (1981) *Solar Energy Handbook*, McGraw-Hill, New York.

Kreith, F. and West, R.E. (1996) *Handbook of Energy Efficiency*, CRC Press, Boca Raton, FL.

National Research Council (1982) *Energy in Transition 1985–2010: Final Report of the Committee on Nuclear and Alternative Energy Systems*, National Academies Press, Washington, DC.

Palz, W. and Zibetta, H. (1991) Energy payback time of photovoltaic modules, *International Journal of Solar Energy*, 10(3–4), 211–216.

Tester, J.W. (2005) *Sustainable Energy, Choosing among Options*, MIT Press, Cambridge, MA.

2

Economics of Energy Generation and Conservation Systems

> Anyone who believes exponential growth can go on forever in a finite world is either a madman or an economist.
>
> **—Kenneth Boulding (ca. 1980)**

This chapter presents a process for an economic assessment of energy systems at the simplest level able to give useful financial insight and enable a comparison between various energy options. No familiarity with microeconomics is presumed, and given the uncertainty in predicting a future economic environment, one could argue that added complexity may not be warranted. It should also be noted that the process presented here is equally applicable to renewable and nonrenewable energy systems, as well as to energy conservation projects.

Although the source of energy in renewable energy systems such as wind turbines and solar collectors is essentially free, there is a definite cost associated with energy generation by these systems [1]. In its simplest form, neglecting interest charges on capital and operating costs, one could calculate the cost of energy for a renewable or conservation system by assuming that the system has a life of t years and its initial cost of construction and installation is C_0 dollars. If, during its life, the system will on average receive every year an amount of energy Q, the unit cost of energy C_s is equal to the cost of installing the system divided by the total energy delivered during its lifetime, or

$$C_s = \frac{C_0}{Qt}. \tag{2.1}$$

For example, if a solar energy collector costs \$100 per m^2 of surface and has an expected life of 20 years, and is installed in a part of the country where the mean annual horizontal surface irradiance is 200 W/m^2 averaged over 24 h, the cost of solar energy will be equal to

$$C_s = \frac{100\ (\$/\text{m}^2) \times 1000\ \text{W/kW}}{200\ (\text{W/m}^2) \times 24\,(\text{h/day}) \times 365\ (\text{days/year}) \times 20\ \text{years}} = \$0.00285/\text{kWh} \tag{2.2}$$

It is obvious, however, that no collector will perform at 100% efficiency. Consequently, the collector efficiency enters the economics, because according to thermodynamic laws, only a fraction of the incident energy can be converted into a useful form such as heat. Assuming that the collector efficiency, η_c is 50%, the cost of solar energy will be twice that calculated above, or

$$C_s = \frac{C_0}{Qt\eta_c} = \$0.0057/\text{kWh} \tag{2.3}$$

Note that installing the very same collector in another part of the country where the mean irradiance is only 100 W/m^2 would again double the cost of the useful energy delivered to $0.0114/kWh, although the system cost would be the same. It is therefore apparent that the cost of solar energy depends on the place of deployment. The same approach can be used to evaluate the cost of energy saved by a conservation system, but the amount of energy produced, $Qt\eta_c$ in Equation 2.3, is replaced by the amount of energy saved during its lifetime.

An alternative method of evaluating the value of an energy generation or conservation system is to determine the time required for the system to payback the cost of its installation and operation. Consider, for instance, the payback period (PP) of a conservation system. The PP is equal to the cost of the capital equipment C_0 divided by the net value of yearly net energy savings, i.e., the difference between the energy saved and the energy used for operation and maintenance. The net value of the energy saved equals the amount of energy saved per year (kWh/year) times the unit cost of that energy (e.g., $/kWh). Once the yearly value of the saved energy and the cost of installing a conservation measure are known, the simple PP in years is

$$\text{PP} = \frac{\text{Cost of installation in \$}}{\text{Net value of energy saved per year in (\$/year)}} \tag{2.4}$$

This approach is acceptable for preliminary estimates if the PP is short, say less than 4 years. For a more precise estimate, the time value of money, the inflation rate, and the escalation in fuel costs must be considered.

2.1 Time Value of Money

If money is deposited in a bank, it earns interest. Similarly, when money is invested in a project or a business that is financed through a purchase of bonds or stock, a certain rate of return on the investment is expected. Conversely, if you borrow money from a bank, you have to pay interest (or dividends) as the cost of the money lent to you. These issues are known as the *time value of money*.

Since it is convenient to conduct economic analyses in a unit of fixed monetary value, a concept known as *present worth* is used. The present worth of a future cash flow or a future payment is the future value of that flow with the time value of the money removed. The opportunity for investing money at a given time with the expectation of a future return is known as the time value of money. For example, the future value X, of a sum of money P, invested today at an annual interest rate i is

$$X = P(1+i)^t \tag{2.5}$$

in which t is the future time in years or other time unit corresponding to the time basis of the interest rate i. This equation essentially states that an initial sum of money P appreciates by a multiplier factor, $(1 + i)$ each year, disregarding any general inflation.

Stated alternatively, a future sum of money X has a *present worth* P given by

$$P = \frac{X}{(1+i)^t} \tag{2.6}$$

This indicates that the present worth of a given amount of money in the future is discounted, in constant dollars, by a factor of $(1 + i)^{-t}$ for each year in the future. In this context, i is generally called the discount rate, but is also referred to as the cost of money or rate of return. Note that this approach assumes that i remains constant over the period of concern. In real life, however, it may well not be constant and constitutes a source of uncertainty.

The factor $(1+i)^{-t}$ is called the present worth factor [PWF(i, t)] and is given by

$$\text{PWF}(i,t) = \frac{1}{(1+i)^t} \tag{2.7}$$

To obtain the present value of a cash flow in the future at time t, it is multiplied by its PWF(i, t). Table 2.1 is a tabulation of PWFs for various interest rates, i, and number of years, n.

As an example, consider a discount rate of 5% and a cash inflow of \$1, 1 year from now. Using the above formula, the present worth is only \$0.95, approximately the present value of the cash inflow of \$1. Conversely, if \$0.95 was invested at 5% interest today, it would be worth approximately \$1 a year from now. The same approach can be used to evaluate cash flows at any future period from now.

The present worth concept is the key for a discounted cash flow analysis. Its use permits all calculations to be made in terms of *present discounted monies*. It demonstrates why revenues realized in the future are worth less than money earned at the present. For example, \$1 of revenue realized 40 years in the future at a discount rate i of 10%/year has a present worth of

TABLE 2.1

Present Worth Factors

	Interest Rate %													
Year	**0**	**1**	**2**	**3**	**4**	**5**	**6**	**7**	**8**	**9**	**10**	**11**	**12**	**13**
1	1.000	0.990	0.980	0.971	0.962	0.952	0.943	0.935	0.926	0.917	0.909	0.901	0.893	0.885
2	1.000	0.990	0.961	0.943	0.925	0.907	0.890	0.873	0.857	0.842	0.826	0.812	0.797	0.783
3	1.000	0.971	0.942	0.915	0.889	0.864	0.840	0.816	0.794	0.772	0.751	0.731	0.712	0.693
4	1.000	0.961	0.924	0.888	0.855	0.823	0.792	0.763	0.735	0.708	0.683	0.659	0.636	0.613
5	1.000	0.951	0.906	0.863	0.822	0.784	0.747	0.713	0.681	0.650	0.621	0.593	0.567	0.543
6	1.000	0.942	0.888	0.837	0.790	0.746	0.705	0.666	0.630	0.596	0.564	0.535	0.507	0.480
7	1.000	0.933	0.871	0.813	0.760	0.711	0.665	0.623	0.583	0.547	0.513	0.482	0.452	0.425
8	1.000	0.923	0.853	0.789	0.731	0.677	0.627	0.582	0.540	0.502	0.467	0.434	0.404	0.376
9	1.000	0.914	0.837	0.766	0.703	0.645	0.592	0.544	0.500	0.460	0.424	0.391	0.361	0.333
10	1.000	0.905	0.820	0.744	0.676	0.614	0.558	0.508	0.463	0.422	0.386	0.352	0.322	0.295
11	1.000	0.896	0.804	0.722	0.650	0.585	0.527	0.475	0.429	0.388	0.350	0.317	0.287	0.261
12	1.000	0.887	0.788	0.701	0.625	0.557	0.497	0.444	0.397	0.356	0.319	0.286	0.257	0.231
13	1.000	0.879	0.773	0.681	0.601	0.530	0.469	0.415	0.368	0.326	0.290	0.258	0.229	0.204
14	1.000	0.870	0.758	0.661	0.577	0.505	0.442	0.388	0.340	0.299	0.263	0.232	0.205	0.181
15	1.000	0.861	0.743	0.642	0.555	0.481	0.417	0.362	0.315	0.275	0.239	0.209	0.183	0.160
16	1.000	0.853	0.728	0.623	0.534	0.458	0.394	0.339	0.292	0.252	0.218	0.188	0.163	0.141
17	1.000	0.844	0.714	0.605	0.513	0.436	0.371	0.317	0.270	0.231	0.198	0.170	0.146	0.125
18	1.000	0.836	0.700	0.587	0.494	0.416	0.350	0.296	0.250	0.212	0.180	0.153	0.130	0.111
19	1.000	0.828	0.686	0.570	0.475	0.396	0.331	0.277	0.232	0.194	0.164	0.138	0.116	0.098
20	1.000	0.820	0.673	0.554	0.465	0.377	0.312	0.258	0.215	0.178	0.149	0.124	0.104	0.087

only \$0.0183 today. This explains why the far future has little influence on conventional business planning and why many businesses shy away from investments in renewable energy system or conservation measures that will not pay back until sometime in the future. The following example illustrates this concept.

Example 2.1

Assume that opportunity A generates benefits equal to \$100, \$150, and \$200 at the end of years 1, 2, and 3, respectively. Assume that opportunity B yields benefits of \$225 in year 2 and \$225 in year 3. Therefore over 3 years, both opportunities A and B yield benefits of \$450. However, the timing of the benefits received is different in each case. By using the present-worth technique, their two benefit flows can be viewed in terms of today's dollar value. Assume a 10% interest rate.

Solution

Step 1

Compute the PWF using Equation 2.7. With an interest rate of 10%, the calculation of PWF is as follows:

$$\text{Year 1} = \frac{1}{(1+0.10)^1} = \frac{1}{(1.10)^1} = 0.9091$$

$$\text{Year 2} = \frac{1}{(1+0.10)^2} = \frac{1}{(1.10)^2} = 0.8264$$

$$\text{Year 3} = \frac{1}{(1+0.10)^3} = \frac{1}{(1.10)^3} = 0.7513$$

Step 2

Compute the present worth of each opportunity benefit flow by multiplying the PWF by the annual benefit amount.

The results are shown in the table below.

		Annual Benefit (\$)		Present Worth (\$)	
Year	Present-Worth Factor	A	B	A	B
1	0.9091	100	0	90.91	0
2	0.8264	150	225	123.96	185.94
3	0.7513	200	225	150.26	169.04
Total	450	450	365.13	354.98	

Note that although the total benefits are the same, the present worth for case A is higher because the annual benefits occur sooner.

When future cash flows are fixed in size and regularly occur over a specific number of periods, the situation is known as an annuity. For an annuity, the present worth is given by Equation 2.8.

$$P = X\frac{(1+i)^n - 1}{i(1+i)^n} \tag{2.8}$$

For example, assume a cash flow of \$100/year at the end of each year for the next 5 years at a discount rate of 10%. According to Equation 2.8, the present value of the \$500 inflow at \$100/year is only worth, at present, \$379.

$$P = \$100 \times \frac{(1+0.10)^5 - 1}{0.10(1+0.10)^5} = \$379$$

2.2 Inflation

The cost and revenues of a project can be expressed either in "current dollars" or "constant dollars." Actual cash flows observed in the marketplace at a given time are called current dollar cash flows. They are the actual number of dollars required or acquired in the year the cost or the benefit is incurred. Current dollar cash flows will change over time because of inflation (or deflation). Constant dollar cash flows, $\bar{F}_n$, on the other hand, represent the number of dollars that would be required if the cost were paid in the base year, n. Cash flow expressed in current dollars in some year m, F_m, can be transferred into constant dollar cash flow in any year n by removing the effect of inflation, j, according to the formula below:

$$\bar{F}_n = \frac{F_m}{(1+j)^{m-n}} \tag{2.9}$$

This formula assumes a constant rate of inflation in the period between m to n years. The base year is defined as the year around which an analysis is structured. Most commonly, it is the initial period, which may be denoted by 0. In the base year, constant and current dollar cash flows are the same.

In practice, it is important to distinguish between real or effective and nominal discount rates. A real discount rate excludes inflation, whereas a nominal value includes inflation. When calculating a present value, it is important to remember that *the real discount rate must be used if cash flows are to be in constant dollars, whereas a nominal discount rate is to be used if the cash flow is in current dollars.*

If it is necessary to convert a stream of cash flows from current dollars to constant dollars or vice versa, it is useful to refer to an inflation rate index. There are a number of inflation rate indices, for example, the consumer price index (CPI) or the gross national product (GNP). The CPI measures the average change in prices of goods and services at the retail end. The GNP is the measure of output supplied by U.S. labor and property. Table 2.2 presents the changes in the GNP, as well as the consumer price indices from 1970 to 1993. Additional

TABLE 2.2

Selected Inflation Indices

Year	GDP Implicit Price Deflator (2000 = 100)[a]	GNP Implicit Price Deflator (2000 = 100)[a]	Consumer Price Indices (83–84 = 100)[b]	Producer Price Indices (82 = 100)[b]
1970	27.53	27.5	38.8	36.9
1971	28.91	28.9	40.5	38.1
1972	30.16	30.1	41.8	39.8
1973	31.85	31.8	44.4	45.0
1974	34.73	34.7	49.3	53.5
1975	37.99	38.0	53.8	58.4
1976	40.19	40.2	56.9	61.1
1977	42.74	42.7	60.6	64.9
1978	45.74	45.7	65.2	69.9
1979	49.54	49.5	72.6	78.7
1980	54.04	54.0	82.4	89.8
1981	59.12	59.1	90.9	98.0
1982	62.73	62.7	96.5	100.0
1983	65.19	65.2	99.6	101.3
1984	67.65	67.6	103.9	103.7
1985	69.71	69.7	107.6	103.2
1986	71.25	71.2	109.6	100.2
1987	73.19	73.2	113.6	102.8
1988	75.69	75.7	118.3	106.9
1989	78.55	78.5	124	112.2
1990	81.6	81.6	130.7	116.3
1991	84.4	84.4	136.2	116.5
1992	86.4	86.4	140.3	117.2
1993	88.4	88.4	144.5	118.9
1994	90.3	90.3	148.2	120.5
1995	92.1	92.1	152.4	124.8
1996	93.8	93.9	156.9	127.7
1997	95.4	95.4	160.5	127.6
1998	96.5	96.5	163	124.4
1999	97.9	97.9	166.6	125.5
2000	100.0	100.0	172.2	132.7
2001	102.4	102.4	177.1	134.2
2002	104.2	104.2	179.9	131.1
2003	106.4	106.4	184	138.1
2004	109.5	109.4	188.9	146.7
2005	113.0	113.0	195.3	157.4
2006	116.7	116.7	201.6	164.8
2007	119.8	119.8	207.3	172.7
2008	122.4	122.4	215.3	189.6

[a] Data from http://www.bea.gov/

[b] Data from http://www.bls.gov/

information and up-to-date data can be obtained from the U.S. Department of Commerce, which publishes the *Statistical Abstract of the United States* every year (available from http://www.ntis.gov/products).

The effect of inflation can be easily integrated into the simplified analysis presented above. The effect of inflation on the future value of an invested sum is to reduce the value by a factor $(1 + j)$ per year, where j is the inflation rate per year (e.g., dollars/dollars per year). Taking inflation into account, the future value of X from Equation 2.4 becomes

$$X = P\left(\frac{1+i}{1+j}\right)^t \tag{2.10}$$

Alternatively, the present worth of P under inflation is

$$P = \frac{X}{[(1+i)/(1+j)]^t} \tag{2.11}$$

Assuming that inflation is uniform during the period under consideration, an effective interest rate, i' can be defined to consider both interest and inflation effects. The future worth P then becomes

$$P = \frac{X}{(1+i')^t} = X * \text{PWF}(i', t) \tag{2.12}$$

where the real or effective discount rate, i' is

$$i' = \frac{1+i}{1+j} - 1 = \frac{i-j}{1+j} \tag{2.13}$$

For example, given an inflation rate of 6%, and a nominal discount rate of 10%, the real discount rate is 3.77%.

The PWF can then be modified to account for inflation by simply replacing the interest, i, with an effective interest, i'. For small inflation rates, j, the effective interest rate, i', is approximately the difference between the interest rate and inflation, or

$$i' = i - j \quad \text{for } j \tag{2.14}$$

2.3 Societal Costs

The use of conventional technologies, such as coal-fired power plants, to generate electricity creates environmental degradation and societal costs that are considerably higher than those from renewable technologies. These hidden costs are called *externalities* and are usually not included in conventional energy economics. Estimates of societal costs have been made by many

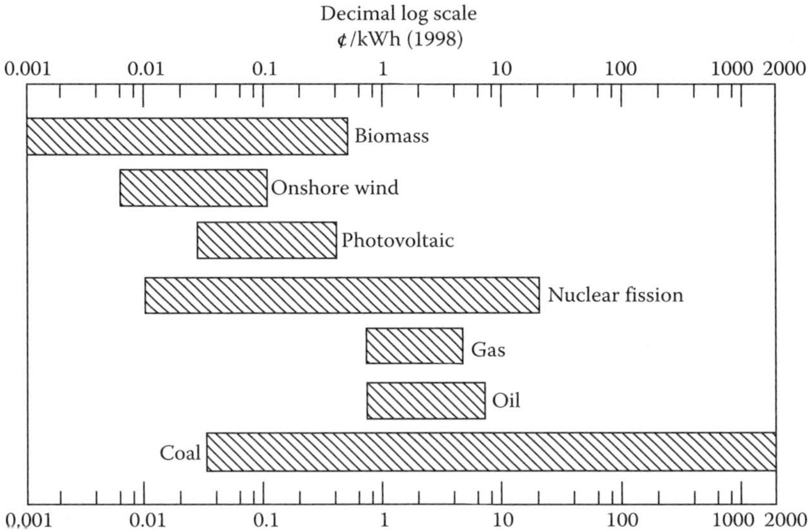

FIGURE 2.1
Externality estimates in ¢/kWh. (Reprinted from Stirling, A. Regulating the electricity supply industry by valuing environmental effects: How much is the Emperor wearing?, *Futures*, 24, 1024–1047. Copyright 1992, with permission from Elsevier.)

different studies. The results vary widely, because they depend on many factors that are difficult to pinpoint. For example, what is the appropriate discount rate for long-term effects? How does one assign a monetary value to items such as human health or life? How does one treat the effects of global warming?

The range of externality estimates, which does not include global warming, is shown in Figure 2.1 [2]. It is apparent that although renewable technologies have considerably less externality costs, the uncertainty in the actual value is substantial. A 1992 study by the American Solar Energy Society Economics of Solar Energy Technologies [3] estimated that the median externality cost of U.S. energy consumption is about $2/million Btu. For electricity, the estimate of societal cost is about $0.02/kWh, because typically 1 million Btu of heat is used to produce 100 kWh of electricity. Since 8 gal of gasoline contain about 1 million Btus, the societal cost of burning gasoline in automobiles is at least $0.25/gal. While it is difficult to precisely determine these externality costs, they are real, even if they do not appear on a utility bill. A methodology for evaluating the economic cost of the environmental impacts of energy generation is presented in [4].

The environmental effects of renewable energy technologies are small compared to the effects of fossil or nuclear technologies. The major societal cost of renewable technologies comes from building the equipment necessary for the future energy conversion. Once built, with the exception of biomass, renewable energy technologies do not emit polluting gases into the atmosphere. However, burning biomass produces less pollution than burning coal or natural gas, because biomass has a much smaller sulfur and ash content. Furthermore, burning biomass does not increase the concentration of CO_2 in the atmosphere, because an approximately equivalent amount of CO_2 is absorbed in growing the feedstock. The key goal for biomass utilization is either to grow the feedstock in a sustainable manner that does not impact food production, or to utilize waste products such as wood chips or industrial wastes.

2.4 Total Life Cycle Costs

A total life cycle cost (TLCC) analysis can evaluate differences in the costs and the timing of these costs between alternative projects. TLCC analysis should consider all significant dollar costs over the life of a project. These costs are then discounted to a base year value using present value analysis. Any revenue generated from the resale of an investment is also discounted to the base year and subtracted from present value costs. TLCC analysis includes many different costs and must use the appropriate form of the cash flows (current or constant dollars) so that the correct discount rate is applied. If the cash flows are not discounted with the appropriate rate, the analysis will be flawed. The formula for calculating the present value of TLCC is

$$\text{TLCC} = \sum_{n=1}^{N} \frac{C_n}{(1+i)^n} \tag{2.15}$$

where

C_n is the cost in period n, including finance charges, O&M costs, and energy costs

N is the analysis period

i is the annual discount rate [2]

Example 2.2

Assume that a nonprofit organization invests \$10,000 and estimates that operating and maintenance (O&M) costs will be \$1,300 in the first year and no salvage value. Assuming an inflation rate of 3%/year and a nominal discount rate of 12%, Table 2.3a demonstrates the TLCC evaluation, both in current dollars and in constant dollars. Note that the TLCC is the same, but the discount rate and cash flows are different. Since the major costs associated with most conservation and renewable energy technologies are the initial investment costs and O&M costs

TABLE 2.3a

Current and Constant Dollar TLCC Evaluation

Year	Investment	Discounted Investment	Current Dollar O&M Costs	Discounted O&M Costs (w/12% Nominal)	Constant Dollar O&M Costs	Discounted O&M Costs (w/8.74% Real)
0	$10,000	$10,000	$0	$0	$0	$0
1	0	0	$1,339	$1196	$1300	$1196
2	0	0	$1,379	$1100	$1300	$1100
3	0	0	$1,421	$1011	$1300	$1011
4	0	0	$1,463	$930	$1300	$930
5	0	0	$1,507	$855	$1300	$855
NPV		$10,000		$5091		$5091

TLCC = $15,091.

(including fuel), the formula for TLCC for residential, government, and nonprofit organizations simplifies to

$$\text{TLCC} = C_0 + \text{PVOM}$$

where
C_0 is the initial investment
PVOM is the present value of all O&M costs

or

$$\text{PVOM} = \sum_{n=1}^{N} \frac{(\text{O\&M})_n}{(1+i)^n} \tag{2.16}$$

In Table 2.3a, NPV is the net present value, which is defined in Equation 2.18.

If the investment is by a for-profit organization, taxes must be included in the calculations as shown below:

$$\text{TLCC} = C_0 - (T \times \text{PVDEP}) + \text{PVOM}(1-T) \tag{2.17}$$

where
T is the income tax rate
PVDEP is the present value of depreciation

Example 2.3

Suppose that the investor of Example 2.2 is a private company in the 34% federal income bracket, but pays no state income tax. The initial investment is $10,000, O&M is $1,300 per year in current dollars in year zero with an inflation rate of 3% and the nominal discount rate is 12%. The investor uses a "5 year double-declining depreciation schedule" as shown in column 4 and there is no salvage value. Table 2.3b shows the steps for evaluating the after-tax TLCC.

TABLE 2.3b

After-Tax TLCC Evaluation

Year	Investment	Discounted Investment	Depreciation	Discounted Depreciation	O&M Costs	Discounted O&M Costs
0	$10,000	$10,000			$0	$0
1	0	0	$4,000	$3,571	$1,339	$1,196
2	0	0	$2,400	$1,913	$1,379	$1,100
3	0	0	$1,440	$1,025	$1,421	$1,011
4	0	0	$1,080	$686	$1,463	$930
5	0	0	$1,080	$613	$1,507	$855
NPV		$10,000		$7,809		$5,091

TLCC = $10,000 – (0.34 × $7,809) + ($5,091 × 0.66) = $10,705.

2.5 Internal Rate of Return

The internal rate of return (IRR) analysis is a convenient procedure to compare a variety of investment activities. It is commonly used for accept/reject decisions by comparing the IRR with a minimal acceptable rate, often referred to as the hurdle rate. The IRR of an investment that has a series of future cash flows (F_0, F_1, …, F_n) is that rate of return on investment that sets the NPV of cash flows equal to zero. IRR implicitly assumes reinvestments of any return at the IRR rate, and is therefore not recommended for ranking projects. However, it has the advantage of giving an investor a quick comparison of the after-tax return on various financial instruments such as bonds, and providing the investor with a quick assessment of the project.

In equation form, IRR equals that rate of return for which

$$\text{NPV} = 0 = \sum_{n-0}^{N} \frac{F_n}{(1+D)^n} \tag{2.18}$$

where

NPV is the net present value of the capital investment

F_n is the cash flow received at time n

D is the rate of return that equates the present value of positive and negative cash flows when used as the discount rate.

Table 2.4 demonstrates the after-tax cash flow resulting from a project with the following assumptions (it is the same as in Example 2.3): An investment of $10,000 is made initially. The O&M cost in the first year is $1300 and it inflates at a rate of 3% per year thereafter. A 5 year double-declining

TABLE 2.4

After-Tax IRR

Year	Investment (a)	O&M Costs (b)	Revenues (c)	Depreciation (d)	Not Taxable Income (e) = (c − b − d)	Federal Income Tax (f) = e × tax	After Tax Cash Flow F_n(g) = (c − a − b − f)	$F_n/(1.61)^n$
0	$10,000	$0	$0	$0	N/A	N/A	−$10,000	
1	0	1339	10,300	4000	$4961	$1687	7,274	$4,516
2	0	1379	10,079	2400	6300	2142	6,558	2,530
3	0	1421	10,108	1440	7247	2464	6,223	1,490
4	0	1463	10,130	1080	7587	2580	6,087	900
5	0	1507	10,434	1080	7347	2668	6,259	580
								$10,010

depreciation schedule is used with a marginal federal income tax rate of 34%. There is no state income tax and no salvage value. The project produces 1000 units of energy in its first year of operation, 950 in the second, 925 in the third, and 900 in the fourth and fifth years. The energy is sold at $10/unit at year zero and inflates at 3% per year thereafter. The results of the IRR analysis calculations are shown in Table 2.4. The last column in the table demonstrates that if a return of 61% were earned from the after-tax cash flow each year, the total cash flow would equal $10,000, which is the initial investment. A trial-and-error method may be used to obtain the appropriate IRR. For this example, if the hurdle rate were higher than 61%, then an investment in another project would be appropriate, but if the hurdle rate is lower than 61%, the investment in this project represents a good opportunity.

Potential problems with IRR analyses are described in [2]. One of them is that IRR might overstate profitability, because it assumes that all interim proceeds are reinvested at a rate equal to the IRR. Hence, for projects of different scales and lives, it is possible for different ranking criteria such as NPV and IRR to produce conflicting results as a consequence of the differing reinvestment assumptions. The IRR method can, however, be modified, and this modified IRR analysis method can be found in reference [2].

PROBLEM

Estimate the payback time and the IRR for a 3 kW PV system installed in Boulder, Colorado (40°N latitude, 105°W longitude) in 2008 (see Figure 2.2). The cost of the system was $25,294.28. The electric utility serving the customer offered a rebate of $4.50 per Watt in order to meet the RE mandate in the Colorado constitution that requires that 20% of all electric power be from renewable energy by 2020. Moreover, the federal government offered a tax credit capped at $2000 for the 2008 tax year.

FIGURE 2.2
Kreith residence, 3 kW Sanyo high efficiency photovoltaic system. (See also Figure 6.1 in color insert.)

Use the following assumptions:

Price of electricity: \$0.152 per kWh
Average rate of power cost increase: 5% per year
Tilt: 20°
Azimuth: 180°
Discount rate: 6%
Inflation rate: 3%

To find the system's annual energy output, input the system parameters at the following Web site: http://rredc.nrel.gov/solar/codes_algs/PVWATTS/version1

Solution
To find the value of the annual savings, *A*, we first multiply the annual energy output by the price of electricity. Since the cost of electricity will increase 5% per year, we must project the cost of energy for the next 20 years. The future cost of energy, *X*, is defined in Equation 2.5, where *P* is the present cost of energy, *r* is the rate of increase and *n* is the year number.

$$X = P(1+r)^n$$

So, in the second year, $X = 0.152(1 + 0.05)^1 = \$0.16$. Then in the third year, as for each successive year, the value of *X* for the previous year becomes the value of *P*. So, in the 20th year, electricity will cost \$0.38 per kWh. Then, the electricity costs for each year are used to calculate annual savings by multiplying by 4207 kWh/year.

The future costs of energy for each of the next 20 years are shown below.

Year	**1**	**2**	**3**	**4**	**5**	**6**	**7**	**8**	**9**	**10**
Price/kWh	\$0.15	\$0.16	\$0.17	\$0.18	\$0.18	\$0.19	\$0.20	\$0.21	\$0.22	\$0.24
Year	**11**	**12**	**13**	**14**	**15**	**16**	**17**	**18**	**19**	**20**
Price/kWh	\$0.25	\$0.26	\$0.27	\$0.29	\$0.30	\$0.32	\$0.33	\$0.35	\$0.37	\$0.38

The system's annual energy output was found to be 4207 kWh per year from the given Web site. The annual savings are found by multiplying each year's respective cost of energy by the annual energy output. The results are shown in the table below.

Year	**1**	**2**	**3**	**4**	**5**	**6**	**7**
kWh	4207	4207	4207	4207	4207	4207	4207
Price/kWh	\$0.15	\$0.16	\$0.17	\$0.18	\$0.18	\$0.19	\$0.20
Annual savings	\$638.62	\$670.55	\$704.08	\$739.29	\$776.25	\$815.06	\$855.82

(continued)

(continued)

Year	8	9	10	11	12	13	14
kWh	4207	4207	4207	4207	4207	4207	4207
Price/kWh	$0.21	$0.22	$0.24	$0.25	$0.26	$0.27	$0.29
Annual savings	$898.61	$943.54	$990.71	$1,040.25	$1,092.26	$1,146.87	$1,204.22
Year	**15**	**16**	**17**	**18**	**19**	**20**	
kWh	4207	4207	4207	4207	4207	4207	
Price/kWh	$0.30	$0.32	$0.33	$0.35	$0.37	$0.38	
Annual savings	$1264.43	$1327.65	$1394.03	$1463.73	$1536.92	$1613.77	

The savings will be cumulative and must be adjusted for the effective interest rate. To find the cumulative savings, simply add all of the annual savings values. The cumulative savings is $21,116.67. The effective discount rate, denoted by i', is found in Equation 2.14,

$$i' = i - j$$

where
i is the discount rate
j is the inflation rate

Therefore, the effective discount rate is 3%. This value is used in Equation 2.6 to adjust the savings for the NPV of money

$$\text{NPV} = \frac{X}{(1+i')^{n-1}}.$$

After each year's savings are adjusted properly, they are added together to find the cumulative NPV savings. The cumulative NPV savings of each year is shown in the table below. After 20 years, the total cumulative NPV savings is $15,427.16.

Year	1	2	3	4	5
Annual savings	$638.62	$670.55	$704.08	$739.25	$776.25
Cumulative savings	$638.62	$1,309.18	$2,013.26	$2,752.54	$3,528.79
Adjusted annual savings	$638.62	$651.02	$663.66	$676.55	$689.69
Adjusted cumulative savings	$638.62	$1,289.65	$1,953.31	$2,629.86	$3,319.55
Year	**6**	**7**	**8**	**9**	**10**
Annual savings	$815.06	$855.82	$898.61	$943.54	$990.71
Cumulative savings	$4,343.86	$5,199.67	$6,098.28	$7,041.81	$8,032.53
Adjusted annual savings	$703.08	$716.73	$730.65	$744.84	$759.30
Adjusted cumulative savings	$4,022.63	$4,739.36	$5,470.01	$6,214.85	$6,974.15

(continued)

Year	**11**	**12**	**13**	**14**	**15**
Annual savings	$1040.25	$1,092.26	$1,146.87	$1,204.22	$1,264.43
Cumulative savings	$9072.78	$10,165.04	$11,311.91	$12,516.13	$13,780.56
Adjusted annual savings	$774.04	$789.07	$804.39	$820.01	$835.17
Adjusted cumulative savings	$7748.19	$8,537.26	$9,341.66	$10,161.167	$10,997.61
Year	**16**	**17**	**18**	**19**	**20**
Annual savings	$1,327.65	$1,394.03	$1,463.73	$1,536.92	$1,613.77
Cumulative savings	$15,108.21	$16,502.24	$17,965.98	$19,502.90	$21,116.67
Adjusted annual savings	$852.17	$868.72	$885.58	$902.78	$920.31
Adjusted cumulative savings	$11,849.77	$12,718.49	$13,604.07	$14,506.85	$15,427.16

After the deduction of the utility's rebate and the Federal tax credit, the final cost to the customer is $9794.28. After 20 years, the customer has made a profit of $5653.21. The PP is calculated by comparing the cost of the system to the customer and the value of the energy saved per year. From the table, we can see that the adjusted cumulative savings (the cost of the energy saved) equals the net cost of the system to the customer between the 13th and 14th year of operation. Therefore the payback time of the system is 13.5 years.

The IRR, found by using Equation 2.18, represents the summation of the discounted cash flows that will equal the initial investment. It is found by iterating on the discount rate until the summation of the yearly discounted cash flows equals zero. The results are shown in the table below, which shows the calculation for a discount rate of 9.08%, representing the IRR.

Table of Year-by-Year Calculations for IRR

Year	Net Investment $	O&M Cost $/Year	Annual Savings $/Year	Annual Adjusted Cash Flow[a] $/Year	Cash Flow Discounted at 9.08% $/Year
0	9794	0	−9794	−9794	−9794
1	0	0	671	692	634
2	0	0	705	733	616
3	0	0	740	777	599
4	0	0	777	824	582
5	0	0	816	873	565
6	0	0	857	925	549
7	0	0	900	981	534
8	0	0	945	1040	519
9	0	0	992	1102	504

(continued)

(continued)

Year	Net Investment $	O&M Cost $/Year	Annual Savings $/Year	Annual Adjusted Cash Flow[a] $/Year	Cash Flow Discounted at 9.08% $/Year
10	0	0	1042	1168	490
11	0	0	1094	1239	476
12	0	0	1148	1313	463
13	0	0	1206	1392	450
14	0	0	1266	1475	437
15	0	0	1329	1564	425
16	0	0	1396	1657	413
17	0	0	1539	1757	401
18	0	0	1539	1862	390
19	0	0	1616	1974	379
20	0	0	1697	2092	368

Note: Summation of discounted cash flows = Net present value = 0.
[a] Adjusted for inflation.

2.6 Levelized Cost of Energy

The levelized cost of energy (LCOE) is a procedure that is widely used for comparing different energy generation technologies. For example, LCOE could be used to compare the cost of energy from a renewable source with that from a nuclear generating unit.

The LCOE is defined as that cost that, if assigned to every unit of energy produced (or saved) by the system over the analysis period, will equal the TLCC when discounted back to the base year. The LCOE is given by the relation

$$\text{LCOE} = \frac{\text{TLCC}}{\sum_{n=1}^{N}\left(Q_n/(1+i)^n\right)} \tag{2.19}$$

The series $[(1 + i)^{-1} + (1 + i)^{-2} + \cdots]$ is equivalent to the ratio $[1 - (1 + i)^{-n}]/i$. The inverse of the above is called the capital recovery factor or CRF $(i, n) = i/[1 - (1 + i)^{-n}]$ and a tabulation of CRI for various values of i and n is shown in Appendix 2.A.

Note that, if the system output (Q_n), or the energy savings are constant over the time of analysis, the relation for LCOE simplifies to

$$\text{LCOE} = \left(\frac{\text{TLCC}}{Q}\right) \times \text{CRF} \tag{2.20}$$

where Q is the annual energy output or energy saving in the case of a conservation system. Using the data from the previous example, and assuming that the system produces 120,000 kWh per year, the capital recovery factor is 0.277 and, therefore

$$\text{LCOE} = (\$10{,}705/120{,}000) \times 0.277 = \$0.247/\text{kWh}.$$

The next example illustrates the LCOE estimate for a conservation measure.

Example 2.4

Calculate the TLCC and LCOE by replacing a 75 W incandescent light bulb by a 40 W fluorescent bulb. Assume that the incandescent bulb operates 6 h per night and needs to be replaced every year at a cost of \$1.00, while the fluorescent bulb lasts for 5 years with an initial cost of \$15. The cost of electricity over this period is \$0.08/kWh.

Solution

The yearly energy requirement for the incandescent lamp is 164 kWh, while the 5 year discounted O&M cost, i.e., the yearly cost of replacement, is

$$\$1\left[1+\left(\frac{1}{1.12}\right)+\left(\frac{1}{1.12}\right)^2+\left(\frac{1}{1.12}\right)^3+\left(\frac{1}{1.12}\right)^4\right]=\$4.03$$

The discounted electric power cost is \$47.35 if the electricity is paid at the end of each year. The 40 W bulb will require 87.6 kWh per year and, therefore

$$\text{TLCC} = \$15 + \$25.25 = \$40.25$$

Thus, the net savings is \$7.10, while the levelized cost of the saved energy is

$$\text{LCOE} = \frac{(\$15-\$4.03)}{76.4\ \text{kWh/year}} \times 0.277\ \text{year} = \$0.04/\text{kWh}$$

Note that LCOE for an energy conservation measure should be based on the incremental costs and savings for the system over the analysis period.

More extensive economic analyses, including taxes, incentives, depreciation, and salvage value, are presented in [2].

2.7 Example of a Typical Cash Flow Scenario

An example of the application of the present worth and levelization concept is shown in Table 2.5 for the generation of electricity by a nuclear light water reactor (LWR). The same general procedure can be applied for a side-by-side comparison with a renewable energy generation system, and the appropriate differences are noted in the table. The levelized unit cost of electricity produced in cents/kWh at the busbar (the entry to the electric transmission

TABLE 2.5

Lifetime-Levelized Busbar Cost of Electrical Energy from a Nuclear LWR*

e_b, the cost of electric power in cents per kilowatt hour (0.1 times mills per kilowatt hour) is the sum of

Capital-related costs:

$$\frac{100\varphi}{8766 \cdot L}\left(\frac{I}{K}\right)_{-c}\left[1+\frac{x+y}{2}\right]^{c}$$

Plus operating and maintenance (O&M) costs:

$$+\frac{100}{8766 \cdot L}\left(\frac{O}{K}\right)_{O}\left[1+\frac{yT}{2}\right]$$

Plus fuel costs:

$$+\begin{cases} \text{Nuclear} & \left[\dfrac{100}{24}\dfrac{F_o}{\eta B}\right]\left[1+\dfrac{yT}{2}\right] \\ \text{or} & \\ \text{fossil} & \left[\dfrac{0.0034 f_o}{\eta}\right]\left[1+\dfrac{yT}{2}\right] \end{cases}$$

where	Typical LWR Value
L = plant capacity factor: actual energy output + energy if always at 100% rated power	0.90
φ = annual fixed charge rate (i.e., effective "mortgage" rate) = $x/(1-\tau)$ where x is the discount rate, and τ is the tax fraction (0.4)	0.15/year
$X = (1-\tau)b \cdot r_b + (1-b)r_s$. in which b is the fraction of capital raised selling bonds (debt fraction), and r_b is the annualized rate of return on bonds, while r_s is the return on stock (equity)	0.09/year
$\left(\frac{I}{K}\right)_{-c}$ = overnight specific capital cost of plant, as of the start of construction, dollars per kilowatt: cost if it could be constructed instantaneously c years before startup in dollars without inflation or escalation	$6,335/kWe
y = annual rate of monetary inflation (or price escalation, if different)	0.04/year
c = time required to construct plant, years	5 years
T = prescribed useful life of plant, years	40 years
$\left(\frac{O}{K}\right)_{O}$ = specific operating and maintenance cost as of start of operation, dollars per kilowatt per year	$95/kWe year
η = plant thermodynamic efficiency, net kilowatts electricity produced per kilowatt of thermal energy consumed	0.33
F_O = net unit cost of nuclear fuel, first steady-state reload batch, dollars per kilogram of uranium; including financing and waste disposal charges, as of start of plant operation	$2,000/kg

TABLE 2.5 (continued)

Lifetime-Levelized Busbar Cost of Electrical Energy from a Nuclear LWR*

B = burnup of discharged nuclear fuel, megawatt days per metric ton	45,000
f_o = fossil fuel costs, at start of operation, cents per million British thermal units = (approximately) dollars per barrel times 16 for residual oil: dollars per ton times 4 for steam coal, cents per thousand standard cubic feet for natural gas, zero for solar or wind energy	

Thus, for a LWR nuclear power plant, using the representative values cited above

$$\text{Cap} + \text{O\&M} + \text{Fuel}$$

$$e_b = 12.8 + 2.2 + 1.0 = 16 \text{ cents/kWhr}_e$$

Source: Adapted from Tester, J.W., Drake, E.M., Golay, M.W. et al., *Sustainable Energy: Choosing Among Options*, MIT Press, Cambridge, MA, 2005, with updated cost figures. With permission from Pluto Press.

* Note that these costs represent only the cost of generating the electricity (i.e., excluding transmission and distribution). These costs are lifetime-average (i.e., "levelized") costs for a new plant starting operations today.

grid) is obtained by equating levelized revenue to levelized expenditures for capital costs, O&M costs, and fuel costs (if applicable). To add realism to this example, as proposed in [5], the cost of money is given by a weighted sum of specified returns on bonds and anticipated return on stocks. The carrying charge rate assumes that bond interest is tax-free, which may not be applicable in all cases. Furthermore, according to *Nuclear Power Daily*, Westinghouse Electric Company and The Shaw Group estimate the total cost for two 1105 MW generating nuclear units to be approximately $14 billion and therefore plant capital costs at time zero estimated are about $6335/kWh for the year 2009 [6]. This figure is computed as an overnight cost, that is, equivalent to hypothetical instantaneous construction, corrected for escalation and interest paid on borrowed funds over a construction period starting n years before operation begins. Additional details about this example can be found in [2].

The capacity factor, L, may vary among options. Typically, nuclear units must shut down approximately 1 month for every 2 years of service for refueling. Renewable options are constrained by the diurnal and intermittent nature of sun and wind. Wind turbine capacity factors are typically about 34% for favorable inland locations and up to 50% for offshore locations. Photovoltaic and solar thermal generation units are mostly on land and may have capacity factors somewhat lower. The evaluation of capacity factors for renewable energy systems will be treated in later chapters.

The useful life of a plant, T, is a very important factor, but somewhat difficult to specify. The useful life of a nuclear plant is somewhere between 30 and 40 years, whereas photovoltaic and solar thermal generation units may last as long as 40 or 50 years. Once the capital investment for a nuclear plant has been repaid, the cost of electricity produced from nuclear energy drops dramatically because repayment of the capital accounts for more than half of the levelized energy cost. Also note that societal costs have not been included in this example. Renewable options are generally capital intensive, and it becomes, therefore, very important to fully credit them with savings on externalities, which is often overlooked in conventional economic analyses.

The projection of energy cost is always subject to uncertainties resulting from tax incentives, technical innovations, market changes, inflation, and competition. For example, the cost of electricity from nuclear power was the topic of an international symposium of experts at the Keystone Center in June 2007. Their results, published as *Nuclear Power Joint Fact-Finding*, gave a summary of construction cost estimates and levelized energy cost presented in the tabulations below [7]. These estimates for low-cost and high-cost scenarios are shown in Tables 2.6 and 2.7. Clearly, the levelized cost

TABLE 2.6

Summary of Cost Assumptions

Main Assumptions (2007$)	Low Case	High Case
Overnight cost	$2950/kW	$2,950/kW
Plant life	40 years	30 years
Capital cost, including interest	$3600/kW	$4,000/kW
Capacity factor	90%	75%
Financial	8% debt, 12% equity, 50/50 ratio	8% debt, 15% equity, 50/50 ratio
Depreciation	15 year accelerated	15 year accelerated
Fixed O&M	$100/kW/year	$120/kW/year
Variable O&M	5 mil/kWh	5 mil/kWh
Fuel	1.2 cents/kWh	1.7 cents/kWh
Grid integration	$20/kW/year	$20/kW/year

Sources: Reprinted from *Nuclear Power Joint Fact-Finding*, The Keystone Center, Keystone, CO June 2007; Spreng, D.T., *Net Energy Analysis and the Energy Requirements of Energy Systems*, Praeger Publishing Company, New York, 1988.

TABLE 2.7

Summary of Levelized Cost (Cents/kWh)

Cost Category	Low Case	High Case
Capital costs	4.6	6.2
Fuel	1.3	1.7
Fixed O&M	1.9	2.7
Variable O&M	0.5	0.5
Total (levelized cents/kWh)	8.3	11.1

Sources: Reprinted from *Nuclear Power Joint Fact-Finding*, The Keystone Center, Keystone, CO June 2007; Spreng, D.T., *Net Energy Analysis and the Energy Requirements of Energy Systems*, Praeger Publishing Company, New York, 1988.

estimates for nuclear power from [7] or the example above, between $0.10 and $0.16 per kWh, are of the same order of magnitude as the projected costs from large solar or wind systems, as shown later in this book.

2.8 Energy Analysis Methodology

At this time, most of the world's energy production uses fossil resources and will continue to do so in the unavoidable period of transition from fossil to renewable resources. Since the supply of fossil fuels is limited, it is very important that the fossil-fueled energy product system be as energy efficient as possible and that the upfront energy investment for the renewable energy system be as small as possible. A net energy analysis (NEA) seeks to compare the amount of useful energy delivered by a technology to the total energy required to find, extract, process, deliver, and otherwise upgrade that energy to a useful form. The criterion by which one can evaluate the energy use efficiency is the energy return on energy investment (EROI) [8,9].

The EROI is the ratio of energy delivered to energy input required to build and operate a system during its lifetime, or

$$\text{EROI} = \frac{\text{Electricity generated during the system lifetime}}{\text{Cumulative primary energy required}} \tag{2.21}$$

In the case of electric generation, the EROI requires a comparison of the electric power generated to the amount of primary energy used in manufacturing, transporting, constructing, operating, and decommissioning the system. Comparing all of the energy requirements with the amount of electricity produced during the lifetime of the system yields the EROI.

Net energy analysis compares the costs and benefits of an energy system in energy terms. The numerator of the EROI is simply the total useful energy produced by the system over its lifetime. The denominator of the EROI ratios is the direct plus indirect energy used by the energy system. Direct energy refers to the fuel actually burned in the construction and operation of the system, such as the diesel fuel burned by an oil drilling rig or the electricity used by a parabolic trough tracking system. Indirect energy costs are the energies used elsewhere in the economy to produce the capital and labor used in building the energy system. An example of indirect energy cost is the energy used to produce the aluminum, which is used in the construction of a solar parabolic trough collector, or the extra glass in a double-glazed window to conserve energy compared to a single pane.

A key difference between the EROI for fossil or nuclear generating systems and renewable technologies, such as solar or wind, is the fact that the renewable systems depend on a primary energy source that not only varies with time of day and time of year, but also with the location at which the system is deployed. Therefore, the evaluation of the energy produced has to be carried out in considerably more detail for a renewable system than for a fossil or nuclear system. In the following chapters, we will present detailed analytical and experimental methods for estimating the energy output of renewable systems. Here, we will simply specify the capacity factor, L, as well as system lifetime, and multiply the design output by that factor to obtain a time-average value of electricity or heat generation over the lifetime of the system.

Energy analyses are useful supplements to standard econometric procedures based on monetary assessments of costs and benefits for several reasons. For example, all economic analyses require assumptions about the future, such as interest rates, future price of conventional fuels, and inflation rates. Also, assumption about future government regulation, taxation rates, and political decisions can greatly influence the economic evaluation process. On the other hand, the fossil payback time of the system can be more precisely and conveniently evaluated.

2.8.1 Process Chain Analysis

There exist essentially two different methods for an energy analysis. The first is known as the *process chain analysis*. As the name implies, a complex production chain is broken into its various elements and for each of them the relevant energy inputs and outputs are identified and evaluated. At the end of this chain is the physical system in which all the energy "expenses" have been accumulated.

For a process analysis, the entire energy system is divided into processes and process steps. Then the system can be described as an energy trajectory. This trajectory begins with an energy resource and passes through several steps to some kind of energy service, such as electric power output. As shown in Figure 2.3, all inputs and outputs have to be identified for each process in the trajectory. These do not only include the energy input, but also utilities, U, services, S, materials, M, and equipment, E, as well as possibly joint products, JP, and other outputs, OO.

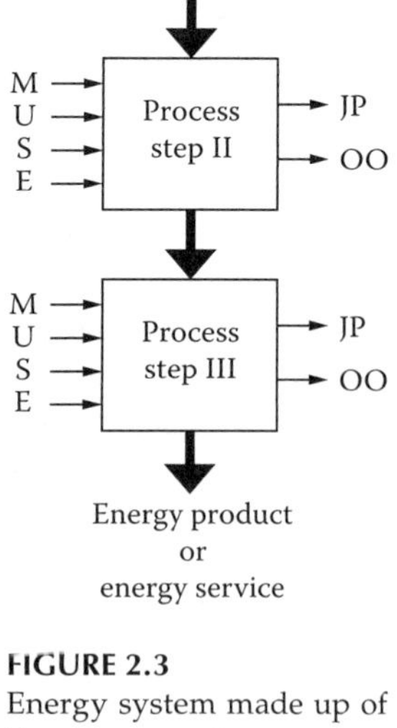

FIGURE 2.3
Energy system made up of three process steps.

Evaluating the relevant inputs and outputs for each step of the process is complicated, especially if

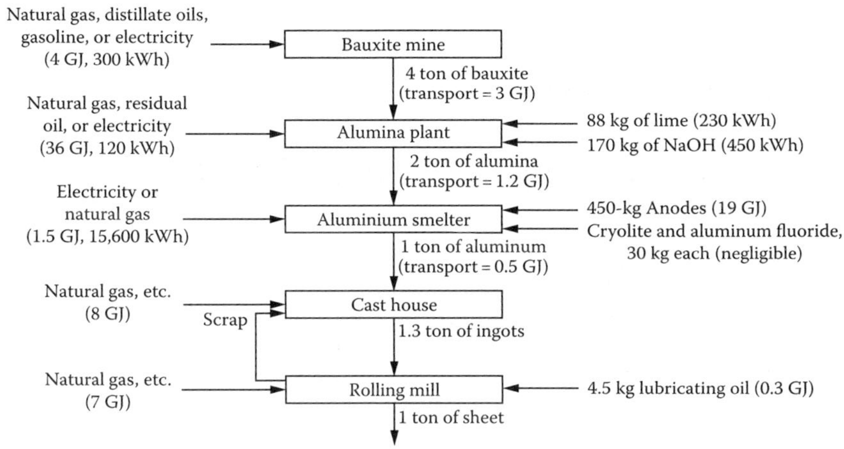

FIGURE 2.4
Energy use in the production of primary aluminum. (From Boercker, S.W., *Energy Use in the Production of Primary Aluminum*, ORAU/IEA-78-14, Oak Ridge Associated Universities, Oak Ridge, TN, 1978.)

a step is not based on existing technology. It is for this reason that, depending upon the assumptions made, the results of a process analysis can vary widely. Figure 2.4 illustrates the energy use in the production of primary aluminum. For details of the assumptions and possible pitfalls, the reader is referred to [8].

Figure 2.5 shows a schematic representation of a process analysis that would be typical of what is required for a simplified study of a renewable energy system. In practice, first the direct energy subsidies are determined, that is, the energy inputs other than the principal energy resource. Examples of such energy subsidies are the fuel used for the trucks or the energy in cleaning the heliostats of a solar power plant. Second, materials, other utilities, services, and equipment (MOUSE) have to be identified. To calculate the energy value of these inputs, one can construct a process chain such as shown in Figure 2.5, and then sum the energy subsidies and the MOUSE inputs. In practice, only certain items that have to be chosen judiciously would be used and second order approximations are sufficient. It is apparent that the process analysis is quite tedious and may require inputs that are not readily available.

2.8.2 Input–Output (I/O) Analysis

Another method to perform an energy analysis uses an input–output table, which shows in monetary units the interdependence of *all* sectors of the national economy. To convert from dollars to energy units in this approach, one attaches energy flows, which can be obtained from a national energy

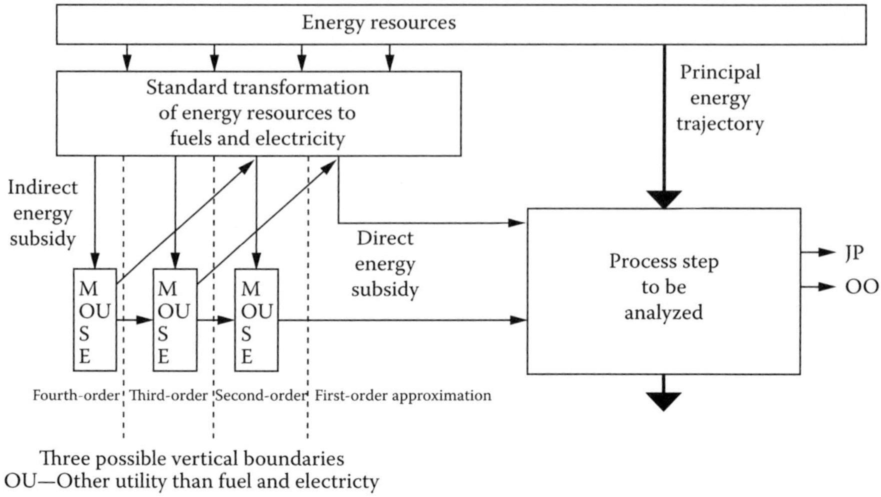

FIGURE 2.5
Schematic of process chain analysis. (Progress Energy Florida signs contract for new, advanced-design nuclear plant, *Nuclear Power Daily*, January 16, 2009, http://www.nuclear-powerdaily.com/reports/Progress_Energy_Florida_Signs_Contract_For_New_Advanced_Design_Nuclear_Plant_999.html (retrieved May 18, 2009); Statistical Abstract of the USA, 2007.)

balance, to the monetary value flows within such an input–output table. Then, an energy matrix based on the physical flows and their respective monetary values can be formulated [9]. This energy matrix finally reveals the amount of primary energy (Btu or kWh) necessary for producing one unit of monetary value in a given economic sector of the input–output matrix, including all previous energy expenditures. For example, the construction of a power plant, be it solar, nuclear, or fossil, consumes material and services from different economic sectors. If all contributions from the individual sectors of the national economy are known, the total energy input for the construction of the system can be calculated from the energy-consumption matrix. Similarly, the energy input required to maintain and operate the system can be evaluated.

It should be noted that for each intermediate and final product step, energies of different forms and quality are needed, for example, coal, oil, or electricity. For a meaningful comparison of the final energy consumptions it is therefore necessary to convert the respective energy contents of the energy flow streams to primary energy units by appropriate conversion factors and efficiencies. For example, energy input in the form of electric power produced from coal with an efficiency of 40% must be multiplied by 2.5 to obtain the primary energy input.

If this procedure is used to convert the energy output of a plant to primary energy units, one can determine an energy PP, i.e., the time that the plant must operate to repay the primary fossil fuel subsidy invested in the construction and operation of the plant. When dealing with a renewable energy

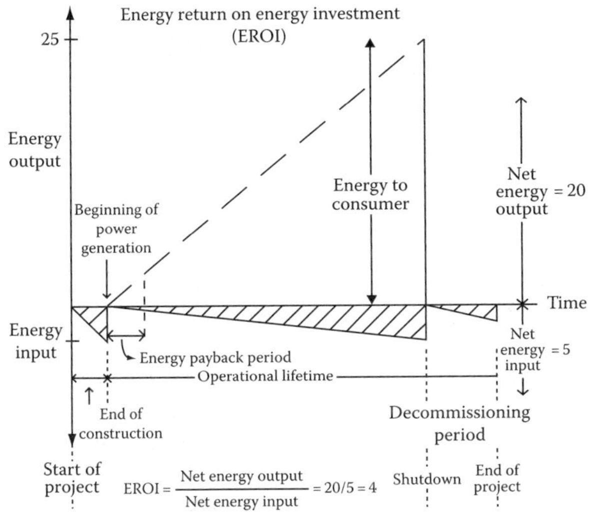

FIGURE 2.6
Energy input and output for a renewable energy generation system during its lifetime.

conversion system, for example, a solar parabolic trough thermal system or a wind power system, the operation energy is small compared to the energy used in construction, and for an energy conservation system it is virtually nil. But, a more useful concept for sustainable energy analysis is the energy return on energy invested ratio (EROI) defined as net energy output during the lifetime of the system (E_d) divided by the energy required to build, operate, and decommission the system (E_c). For a renewable energy system:

$$\text{EROI} = \frac{E_d}{E_c} \tag{2.22}$$

Figure 2.6 illustrates schematically the meaning of this for a hypothetical system. During the construction phase, 2 units of energy are used to build the system; during its operational life, and 2 units or energy are used; finally, 1 unit of energy is used to decommission the system. The net energy output for this system is 20. Thus, the EROI for this example is 4. In Figure 2.6, the shaded area of energy input during the operational lifetime includes the energy of the fuel input for a fossil or nuclear plant. If this were a renewable that does not require energy to operate and decommission, the EROI would be 12.5.

When a solar system supplies only part of the total energy requirement of a process, the energy supplied by the renewable systems can also be interpreted as savings in primary energy for the operation of the process. This interpretation permits direct comparison of the EROI of a solar system with that of an energy conservation measure, such as, for example, a waste heat recovery system.

The proposed energy analysis method can also be used to compare renewable and fossil energy systems by defining the EROI ratio as

$$\text{EROI} = \frac{(E_d - E_o - E_l)}{(E_{mp} + E_c + E_t)} \tag{2.23}$$

where

E_d is the delivered energy over the system lifetime (kWh or Btu)
E_o is the energy, including fuel, used to operate and maintain the system (O&M)
E_l is the energy lost in transmission to point of use, if necessary
E_{mp} is the energy required to mine, process, and deliver the fuel
E_c is the energy invested to construct plant initially and decommission it
E_t is the energy used to construct transmission systems, if necessary

***Note*:** All energy terms are to be converted into primary energy contents.

In an I/O analysis, indirect energy costs are calculated by first making a list of all the materials used in the construction, installation, and operation of the energy system, and their associated dollar costs. The dollar value of a particular material input is multiplied by an energy intensity factor (Btu/$ or kWh/$) to give the energy required to produce that input. That number is a measure of the energy cost of that particular material input to the energy system. The energy intensity factors were calculated for an energy-based input–output model of the U.S. economy in [9,10]. They are shown for key elements of an energy system in Table 2.8, adapted for 2008 dollars from [10], which is based on 1977 dollars. The energy input–output model traces the flow of both energy and dollars between sectors of the economy. By quantifying all the energy and dollars flowing into particular sectors, say the steel industry, it is possible to estimate the amount of energy the steel industry uses in the form of energy embodied in all the inputs it purchases from other sectors. The ratio of all such energies to the dollar value of the steel produced is the energy intensity in Btu/dollar or kWh/dollar of steel production purchases from other sectors.

In Table 2.8, five energy intensities are given for each sector, indicating the sale of (1) coal, (2) crude oil and natural gas, (3) oil refinery products, (4) electricity, and (5) utility gas. More specifically, these energy intensities indicate the outputs of each of the five energy sectors induced throughout the economy (directly or indirectly) by the production of one unit of output in the sector indicated. For the first five energy sectors, the output unit is the Btu, and the energy of intensities are therefore Btu/Btu, or, dimensionless. For the remaining sectors in Table 2.8, the energy intensities are given in Btu/$.

For example, using the table to find the energy intensity of motors, we locate motors in the table and find that, on the average, producing one dollar's worth of motor products requires 33,556 Btu (or 9.83 kWh) per 1977 dollar.

TABLE 2.8
Embedded Energy in Various Materials and Commodities

Commodity	1977 Values Energy Intensity (BTU/$)	1977 Values Energy Intensity (kWh/$)	Adjusted 2008 Values Energy Intensity (kWh/$)
Alum castings	70,000	20.5098	6.6821
Air transport	80,846	23.6877	7.7174
Aircraft	21,288	6.2373	2.0321
Aircraft engines	29,238	8.5667	2.7910
Alum rolling	106,723	31.2696	10.1876
Arch metal work	45,552	13.3466	4.3483
Asbestos product	72,140	21.1368	6.8864
Asphalt	184,121	53.9470	17.5759
Auto repair	22,055	6.4621	2.1053
Banking	9,820	2.8772	0.9374
Bearings	42,638	12.4928	4.0702
Blowers	35,077	10.2775	3.3484
Boatbuilding	23,933	7.0123	2.2846
Brass othr cast	50,508	14.7987	4.8214
Bricks	127,994	37.5019	12.2181
Canned fruit, vegetable	40,981	12.0073	3.9120
Carbon products	108,949	31.9218	10.4001
Cement	216,631	63.4723	20.6792
Ceramic tile	63,362	18.5649	6.0484
Chem mineral min	128,720	37.7146	12.2874
Clay products	119,323	34.9613	11.3904
Communications	6,887	2.0179	0.6574
Computing mach	21,306	6.2426	2.0338
Concrete blocks	70,374	20.6194	6.7178
Concrete product	50,789	14.8810	4.8482
Const machinery	34,534	10.1184	3.2966
Conveyers	31,042	9.0952	2.9632
Copper mining	86,570	25.3648	8.2638
Copper rolling	70,649	20.7000	6.7440
Doctors, dentists	8,488	2.4870	0.8103
Eat and drink places	23,620	6.9206	2.2547
Elec h'wares	36,919	10.8172	3.5242
Elec ind apparat	31,648	9.2728	3.0211
Elec meas instr	21,561	6.3173	2.0582
Electric lamps	28,390	8.3182	2.7101
Electrical equip	38,118	11.1685	3.6387
Electronic comp	31,502	9.2300	3.0071
Elevators	33,497	9.8145	3.1976

(continued)

TABLE 2.8 (continued)
Embedded Energy in Various Materials and Commodities

Commodity	1977 Values Energy Intensity (BTU/$)	1977 Values Energy Intensity (kWh/$)	Adjusted 2008 Values Energy Intensity (kWh/$)
Engine elec eq	29,349	8.5992	2.8016
Fab metal prod	47,163	13.8186	4.5021
Fab struc steel	49,469	14.4943	4.7222
Fab wire product	71,355	20.9068	6.8114
Farm machinery	32,417	9.4981	3.0945
Fed govt enterp	16,946	4.9651	1.6176
Feed grains	66,423	19.4618	6.3406
Fertilizers	17,9710	52.6546	17.1548
Food prod mach	27,346	8.0123	2.6104
For, grhouse, nurs	47,697	13.9751	4.5531
Forest fish prod	43,390	12.7132	4.1419
General ind mach	28,793	8.4363	2.7485
Glass containers	80,487	23.5825	7.6832
Glass products	58,274	17.0741	5.5627
Guided missles	14,272	4.1817	1.3624
Gypsum products	99,292	29.0923	9.4783
Hardware	38,859	11.3856	3.7094
Heating equip	33,553	9.8309	3.2029
Hoists, cranes	30,233	8.8582	2.8860
Hospitals	23,123	6.7750	2.2073
Hotels	33,939	9.9440	3.2398
Ind controls	23,412	6.8597	2.2349
Indus furnaces	27,255	7.9856	2.6017
Industrial truck	34,245	10.0337	3.2690
Inorg-org chem	171,572	50.2701	16.3780
Int combust eng	31,345	9.1840	2.9921
Ir. stl forging	75,110	22.0070	7.1699
Iron ore mining	92,650	27.1462	8.8442
Light fixtures	37,594	11.0149	3.5887
Lime	235,675	69.0522	22.4972
Local transport	32,804	9.6115	3.1314
Mach shop prod	27,116	7.9449	2.5884
Man-made fibers	123,386	36.1518	11.7782
Manufactured ice	61,718	18.0832	5.8915
Measuring pumps	29,195	8.5541	2.7869
Meat products	42,456	12.4395	4.0528
Meat, animal prod	46,311	13.5690	4.4208
Medical instr	29,339	8.5962	2.8007
Met cutting tool	23,163	6.7867	2.2111

TABLE 2.8 (continued)

Embedded Energy in Various Materials and Commodities

Commodity	1977 Values Energy Intensity (BTU/$)	1977 Values Energy Intensity (kWh/$)	Adjusted 2008 Values Energy Intensity (kWh/$)
Met forming tool	28,021	8.2101	2.6748
Met working mach	30,373	8.8992	2.8994
Metal stampings	52,502	15.3829	5.0118
Mineral wool	79,474	23.2857	7.5865
Mining machinery	31,541	9.2414	3.0109
Misc bus service	10,000	2.9300	0.9546
Misc chem prod	95,178	27.8869	9.0855
Misc leather	32,165	9.4243	3.0704
Misc metal work	65,247	19.1172	6.2284
Misc plastics	63,281	18.5412	6.0407
Misc rubber prod	53,271	15.6083	5.0852
Motor transport	29,196	8.5544	2.7870
Motor veh and parts	35,846	10.5028	3.4218
Motor, bicycles	36,373	10.6572	3.4721
Motors, generator	33,556	9.8318	3.2032
NCNST highways	59,333	17.3844	5.6638
NCST dams, resv	44,509	13.0410	4.2488
NCST elect utility	30,648	8.9798	2.9256
NCST gar, srv, sta	38,198	11.1919	3.6463
NCST gas utility	28,039	8.2154	2.6766
NCST indust. Bldg	32,103	9.4061	3.0645
NCST railroads	35,264	10.3323	3.3662
Newspapers	30,902	9.0542	2.9499
Nonclay refract	69,607	20.3947	6.6446
Nonfer casting	61,395	17.9886	5.8607
Nonfer forging	64,560	18.9159	6.1628
Nonfer rolling	81,118	23.7674	7.7434
Nonfer wire	66,253	19.4120	6.3244
Nonferr mining	53,225	15.5948	5.0808
Nonmet min prod	56,700	16.6130	5.4125
Oil bearing crop	23,352	6.8421	2.2291
Oil field mach	30,547	8.9502	2.9160
Optical instr	28,458	8.3381	2.7166
Organic fibers	99,470	29.1444	9.4952
Paint products	75,217	22.0384	7.1801
Paperboard cont	64,427	18.8769	6.1501
Paving	181,838	53.2781	17.3580
Personal service	24,405	7.1506	2.3297

(continued)

TABLE 2.8 (continued)

Embedded Energy in Various Materials and Commodities

Commodity	1977 Values Energy Intensity (BTU/$)	1977 Values Energy Intensity (kWh/$)	Adjusted 2008 Values Energy Intensity (kWh/$)
Photographic eq	35,148	10.2983	3.3552
Pipe	37,637	11.0275	3.5928
Pipe line transp	67,173	19.6815	6.4122
Plastics	126,087	36.9432	12.0361
Plumb fittings	36,093	10.5752	3.4454
Plumbing fixture	60,760	17.8025	5.8001
Power trans eq	31,596	9.2575	3.0161
Prim nonfer met	109,365	32.0437	10.4398
Primary aluminum	158,201	46.3525	15.1016
Primary battery	41,578	12.1822	3.9690
Primary copper	112,850	33.0648	10.7725
Primary lead	130,108	38.1213	12.4199
Primary zinc	137,242	40.2115	13.1009
Pumps, compressors	27,731	8.1251	2.6472
Radio, TV sets	24,898	7.2950	2.3767
Railroad	46,041	13.4899	4.3950
Railroad equip	38,825	11.3756	3.7062
Ready-mix concr	80,909	23.7061	7.7234
Real estate	6,525	1.9118	0.6229
Refrig mach	36,589	10.7205	3.4927
Retail trade	19,433	5.6938	1.8550
R-TV commun eq	18,243	5.3452	1.7414
Scien instr	24,659	7.2250	2.3539
Screw mach prod	44,247	12.9643	4.2237
Semiconductors	29,749	8.7164	2.8398
Sheet metal work	56,289	16.4925	5.3733
Shipbuilding	29,048	8.5110	2.7729
St. loc govt entr	31,395	9.1987	2.9969
Steel prod	11,5724	33.9068	11.0468
Steel springs	59,220	17.3513	5.6530
Stone product	39,667	11.6223	3.7865
Storage battery	69,024	20.2238	6.5889
Switchgear	25,726	7.5377	2.4558
Syn rubber	14,3451	42.0308	13.6936
Tanks	39,964	11.7093	3.8149
Temp controls	32,331	9.4729	3.0863
Textile goods	65,964	19.3273	6.2968
Textile mach	31,319	9.1764	2.9897
Tires	55,977	16.4011	5.3435
Trans services	17,678	5.1796	1.6875

TABLE 2.8 (continued)

Embedded Energy in Various Materials and Commodities

	1977 Values	1977 Values	Adjusted 2008 Values
Commodity	**Energy Intensity (BTU/$)**	**Energy Intensity (kWh/$)**	**Energy Intensity (kWh/$)**
Transformers	42,591	12.4791	4.0657
Transport equip	44,317	12.9848	4.2304
Truck, bus bodies	35,634	10.4407	3.4016
Vegt, misc crops	29,798	8.7307	2.8445
Water transport	11,369	3.3311	1.0853
Water, sanit ser	89,974	26.3621	8.5888
Wholesale trade	18,326	5.3695	1.7494
Wiring devices	40,586	11.8916	3.8743
Wood products	52,030	15.2447	4.9667
X-ray equipment	26,598	7.7931	2.5390

But if the motor was installed in the year 2008, inflation has diminished the value of the dollar.

The effect of inflation is evaluated by means of the CPI, which is a measure of the average price of consumer goods and services purchased. The CPI is published regularly by the Bureau of Labor Statistics [12] and yearly average for the years 1977 to 2008 are shown in Table 2.8. The CPI for 1977 is 61 and for 2008 it is 215. The ratio of 2008s CPI to 1977s CPI is 3.52 and therefore, the energy intensity adjusted for inflation, as shown in the third column of Table 2.8 is about 2.8 kWh per 2008 dollar.

In summary, for an NEA, first the dollar cost for each item of the energy production system is listed, then each material input to the entire system is assigned an energy intensity factor from Table 2.8, calculated for the sector, which produced that material. The dollar value of the input times its energy intensity is the indirect energy cost of that item. Summing the indirect energy costs for all the items of the system gives the indirect energy cost for the entire energy system. The following example illustrates the procedure.

Example 2.5: Double-Paned Windows

An architect designed in 2008 a building in Boulder, CO that is to have 500 3′ × 5′ windows. He wanted to know how much energy and money, if any, would be saved if double-paned windows are used instead of single-paned windows. He asks you to determine the following:

- The annual reduction in heat loss if double-pane (double-glazed) windows are used instead of single-pane windows.
- An estimated payback time for the extra energy used to make double-paned instead of single-paned windows.

- The total reduction in CO_2 emissions if double-paned windows are used (assume electric heating).
- The financial PPs with an estimate that includes the time value of money at an interest rate of 7% and an inflation rate of 3%.

Assume the lifetime of the windows to be 30 years, an average indoor temperature of 20°C. For complete analysis, average outdoor temperatures can be obtained from http://www.wrcc.dri.edu/cgi-bin/cliMONtavt.pl?coboul. For the purposes of this simplified analysis, use 9.13°C for the average outdoor temperature.

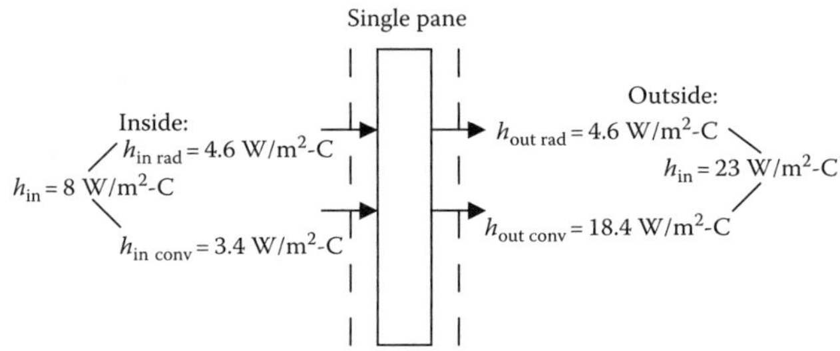

Pane thickness = 4 mm
k_{glass} = 0.9 W/m-C

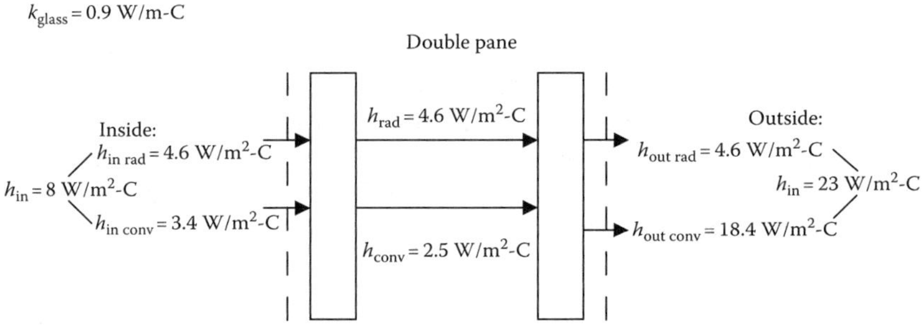

Price Distribution for Windows:

	Single Pane (%)	Double Pane (%)
Glass	16	24
Metal	12	18
Rubber	5	8
Production/installation	67	50

First, find the annual reduction in heat loss.

For single-paned windows:

$$U_1 = \frac{1}{\frac{1}{h_{in}} + \frac{d}{k} + \frac{1}{h_{out}}} = 5.8\ \text{W/m}^2\text{C}$$

$$Q_1 = \frac{U_1 A_{\text{total}} t(T_{\text{in}} - T_{\text{out}})}{1{,}000} = 315{,}000\ \text{kWh}$$

$$A_{\text{total}} = 696.8\ \text{m}^2$$

$$(T_{\text{in}} - T_{\text{out}}) = 20 - 9.13 = 10.87°\text{C}$$

$$t = 24\frac{\text{h}}{\text{day}} \cdot 30\frac{\text{days}}{\text{month}} \cdot 10\ \text{months} = 7200\ \text{h}$$

Similarly, for double-paned windows:

$$U_2 = \frac{1}{\dfrac{1}{h_{\text{in}}} + \dfrac{1}{h_{\text{mid}}} + \dfrac{2d}{k} + \dfrac{1}{h_{\text{out}}}} = 3.1\ \text{W/m}^2\text{C}$$

$$Q_2 = \frac{U_2 A_{\text{total}} t(T_{\text{in}} - T_{\text{out}})}{1{,}000} = 171{,}000\ \text{kWh}$$

Annual difference in heat loss = 315,000 kWh/year − 171,000 kWh/year = 144,000 kWh/year

Next, find the EROI PP.

Single pane price: $149 per window

Energy intensity calculations for single-paned windows:

Glass

$$\left(58{,}274\frac{\text{Btu}}{1977\$}\right)\left(\frac{60.6}{215.303}\frac{1977\$}{2008\$}\right)\left(149\frac{2008\$}{\text{window}}\right)$$
$$\times(500\ \text{windows})(0.16)\left(0.00029\frac{\text{kWh}}{\text{Btu}}\right) = 56{,}700\ \text{kWh}$$

Metal

$$\left(115{,}724\frac{\text{Btu}}{1977\$}\right)\left(\frac{60.6}{215.303}\frac{1977\$}{2008\$}\right)\left(149\frac{2008\$}{\text{window}}\right)$$
$$\times(500\ \text{windows})(0.12)\left(0.00029\frac{\text{kWh}}{\text{Btu}}\right) = 84{,}500\ \text{kWh}$$

Rubber

$$\left(53{,}271\frac{\text{Btu}}{1977\$}\right)\left(\frac{60.6}{215.303}\frac{1977\$}{2008\$}\right)\left(149\frac{2008\$}{\text{window}}\right)$$

$$\times(500\text{ windows})(0.05)\left(0.00029\frac{\text{kWh}}{\text{Btu}}\right)=16{,}200\text{ kWh}$$

Production and installation labor

$$\left(10{,}000\frac{\text{Btu}}{1977\$}\right)\left(\frac{60.6}{215.303}\frac{1977\$}{2008\$}\right)\left(149\frac{2008\$}{\text{window}}\right)$$

$$\times(500\text{ windows})(0.67)\left(0.00029\frac{\text{kWh}}{\text{Btu}}\right)=40{,}700\text{ kWh}$$

Total energy intensity for single-paned windows = 198,100 kWh
  Double pane price: $194 per window
Energy intensity calculations for double-paned windows:

Glass

$$\left(58{,}274\frac{\text{Btu}}{1977\$}\right)\left(\frac{60.6}{215.303}\frac{1977\$}{2008\$}\right)\left(194\frac{2008\$}{\text{window}}\right)$$

$$\times(500\text{ windows})(0.24)\left(0.00029\frac{\text{kWh}}{\text{Btu}}\right)=110{,}700\text{ kWh}$$

Metal

$$\left(115{,}724\frac{\text{Btu}}{1977\$}\right)\left(\frac{60.6}{215.303}\frac{1977\$}{2008\$}\right)\left(194\frac{2008\$}{\text{window}}\right)$$

$$\times(500\text{ windows})(0.18)\left(0.00029\frac{\text{kWh}}{\text{Btu}}\right)=164{,}900\text{ kWh}$$

Rubber

$$\left(53{,}271\frac{\text{Btu}}{1977\$}\right)\left(\frac{60.6}{215.303}\frac{1977\$}{2008\$}\right)\left(194\frac{2008\$}{\text{window}}\right)$$

$$\times(500\text{ windows})(0.08)\left(0.00029\frac{\text{kWh}}{\text{Btu}}\right) = 33{,}700\text{ kWh}$$

Production and installation labor

$$\left(10{,}000\frac{\text{Btu}}{1977\$}\right)\left(\frac{60.6}{215.303}\frac{1977\$}{2008\$}\right)\left(194\frac{2008\$}{\text{window}}\right)$$

$$\times(500\text{ windows})(0.50)\left(0.00029\frac{\text{kWh}}{\text{Btu}}\right) = 39{,}600\text{ kWh}$$

Total energy intensity for double-paned windows = 348,900 kWh

Extra energy to produce double-paned windows = 348,900 kWh − 198,100 kWh = 150,800 kWh

Energy payback time = $\dfrac{198{,}100}{144{,}000} = 1.38$ years

Thus, the EROI = $\dfrac{30\text{ years}}{1.38\text{ years}} = 22$

Note how energy efficient conservation measures are.
To calculate CO_2 reduction, make the following assumptions:
Coal plant efficiency − 35%

Thus, the annual extra energy = $\dfrac{144{,}000}{0.35} = 411{,}428$ kWh

Heating value of coal = 7.75 kWh/kg
%Carbon = 40%
$MW_C = 12$ $MW_{CO_2} = 44$
Extra coal for single pane = 411,428/7.75 = 53,087 kg
$C + O_2 \rightarrow CO_2$
Mass carbon = 53065 (.4) = 21226 kg
Moles carbon = 21,266(1,000)/12 = 1,768,827 mol
Moles CO_2 = 1,768,827 mol
Mass CO_2 = 1,768,827*44/1,000 = 7,7828 kg
Lifetime of windows = 30 years

Total CO_2 reduction = 30(77,828) = 2,334,851 kg
Financial payback of double-paned windows

Extra cost for double pane = 500(194 − 149) = $22,500
$i = 7\%$ $j = 3\%$

Future value of P = \$22,500 after x years:

$x = 1$, $P((1 + i)/(1 + j))x$ = \$23,374
$x = 2$, $P((1 + i)/(1 + j))x$ = \$24,282
$x = 3$, $P((1 + i)/(1 + j))x$ = \$25,224
Annual electric savings = 144,000 kWh (\$0.08/kWh) = \$11,722

Energy cost savings P = \$11,722 for given year x:

$x = 1$, $P(1 + j)(x - 1)$ = \$11,722 = S1
$x = 2$, $P(1 + j)(x - 1)$ = \$12,093 = S2
$x = 3$, $P(1 + j)(x - 1)$ = \$12,435 = S3

Future value of invested energy cost savings after x years:

$x = 1$, S1 = \$11,722
$x = 2$, S1$((1 + i)/(1 + j))$ + S2 = \$24,250 = F1
$x = 3$, F1$((1 + i)/(1 + j))$ + S3 = \$37,627
After 2 years the value of \$22,500 invested is \$24,282
After 2 years the value of annual energy savings is \$24,250
After 3 years the value of annual energy savings is \$37,627

PP for double pane = 2 + (24,282 − 24,250)/(37,627 − 24,250) = 2.0 years

2.9 EROI for a Wind Energy System

A detailed NEA for three wind power systems has been made by White and Kulsiwski [11], who calculated the EROI, as well as the air pollution emitted. The project for which the most detailed information was available was built in 1997 and is located near the city of DePere in Wisconsin and will be referred to as the DePere Wind Project (DPWP). It consists of two Taike 600e turbines that have a combined electric power capacity of 1.2 MWe with a capacity factor of 31% (predicted), and a predicted life for each of the nacelles of 20 years.

The other two wind projects were designed for a maximum power of 25 MWe (BR-I) with 73 turbines and 107 MWe (BR-II) with 143 turbines. Their respective predicted capacity factors were 33% and 35%, and the predicted nacelle life was 25 years for both.

Table 2.9 shows the energy requirements to construct the two DePere wind turbines. These requirements are obtained from the cost data shown in column 1, for example, craning cost was \$75,000. Then the energy per dollar in

TABLE 2.9

Energy Requirements to Construct Two Turbine Wind Farms at DePere, Wisconsin

	1997$	I/O Sector[a]	Btu/1977$[b]	BTU	GJ_{th}/2 Turbines
Craning	$75,000	Hoists, cranes	30,233	8.56E+08	903
Labor	$25,000	Misc. business services	10,000	9.44E+07	100
Local equipment rental	$3,000	Construction machinery	34,534	3.91E+07	41
Lodging and food for employees	$8,000	AVI[c]	28,780	8.69E+07	92
Electrical grounding	$12,000	NC, electrical utilities	30,648	1.39E+08	147
				Total: 1283 GJ_{th}/2 turbines	

Source: White, S.W. and Kulcinski, G.L. *Net Energy Payback and CO_2 Emissions from Wind-Generated Electricity in the Midwest*, Fusion Technology Institute, University of Wisconsin, Madison, WI. December, 1998.

[a] I/O sector data is from [10].

[b] 1997$ were calculated from the CPI by the scale 1977/1997 = 60.6/160.5 = 0.3776 [12].

[c] Sector AVI is an average of the I/O sectors eat and drink places and hotels.

1977 dollars is obtained from the I/O sector data in Table 2.8, for example, 30,233 Btu/1977$ for hoists and cranes. But the time at which the craning cost was $75,000 was 1997, while the I/O tables are for 1977$.

Hence we must adjust the cost figure for inflation, which for these periods can be obtained from Table 2.2. Conversion from 1997 dollars (the year of construction), to 1977 dollars (the year the I/O tables were developed) was done using the CPI ratio 1977:1997 = 60.6:160.5, obtained from the CPI of the Bureau of Labor Statistics [12]. Multiplying the cost times the price index ration and the Btu/$ gives the embodied energy, i.e., 8.56×10^8 Btu. A similar analysis for the other items in column 1 yields the total energy input required to make the turbines. A similar analysis for the energy embodied in on-site construction is shown in Table 2.10.

The lifetime and annual energy investments in the DPWP are shown in Table 2.11. Using these data, the *total energy input* per installed 1.2 MWe is 13,933 GJ_{th} or 3.89×10^6 kWh_{th}. The power output is 1.2×10^3 kWe $\times$ 0.31 = 0.372×10^3 kWe, and for the assumed lifetime of the system of 20 years, the *total power output* is 0.372×10^3 kWe $\times$ 20 year $\times$ 365 day/year $\times$ 24 h/day = 65.2 kWh_e. Hence, EROI = 65.2/3.89 = 16.75 years.

It must be noted that to compare wind with baseload technologies, one would need to include the energy requirements and CO_2 emissions from

TABLE 2.10

Costs and Energy Requirements for On-Site Construction of the DePere Wind Power Plant

	1997$	I/O Sector	Btu	GJ
Foundation preparation	$73,000	Ready mix concrete	2.23E+09	2353
Transformer (2 at $10K per)	$20,000	Transformers	3.22E+08	339
Craning	$75,000	Hoists, cranes	8.56E+08	903
Labor	$25,000	Misc. business services	9.44E+07	100
Local equipment rental	$3,000	Construction machinery	3.91E+07	41
Lodging and food for employees	$8,000	AVI	8.69E+07	92
Electrical grounding	$12,000	NC, electric utilities	1.39E+08	147
				1282

Source: White, S.W. and Kulcinski, G.L. *Net Energy Payback and CO_2 Emissions from Wind-Generated Electricity in the Midwest,* Fusion Technology Institute, University of Wisconsin, Madison, WI. December, 1998.

TABLE 2.11

Lifetime and Annual Energy Investments for the DePere Wind Power Plant

		Total Energy per Installed 1.2 MW_e	Annual Energy per GW_ey
Process	**Source**	**GJ_{th}/Power Plant**	**GJ_{th}/GW_ey**
Wind turbine (embodied)			
Blades	PCA	154	
Nacelles	PCA	2,273	
Inverter	I/O	339	
Wiring	PCA	209	
Tower	PCA	4,890	
Foundations	PCA	1,925	
Materials total		9,790	326
Transportation			
Blades		21	1
Nacelles		257	9
Towers		178	6
Concrete		23	
Control cabinets		0.29	0.01
Transportation totals		480	15
Construction	I/O	1,282.28	43
Maintenance	I/O	1,636	55
Decommissioning		641	21
Total required energy		13,933	464

energy storage units for the wind power plant. Stand-alone wind-generated electric power cannot fully compete with baseload technologies until it can supply electricity at all times. A comparison of baseload technologies to wind without energy storage favors wind because the inclusion of energy storage units would decrease the EROI of wind due to both increased energy requirements and decreased overall efficiency. Currently, energy storage is not used with commercial wind farms. Similarly, if the wind farm were located at a place where new transmission lines would have to be built to deliver the power to a user, additional energy input would be required.

The EROI of wind energy has increased substantially as the technology has improved. Table 2.12 shows the EROI for various wind systems. The average for systems built in the year 1983 was about 2.5, while for systems built in 1998 and 1999 the average was about 23, a ninefold increase. It is therefore not surprising that wind energy has become economically competitive with conventional fossil power plants.

2.10 EROI for Nuclear Power

The EROI for nuclear power today is difficult to evaluate because no nuclear plant has been built in the United States since 1971. An estimate for the average EROI of nuclear power can be obtained by a simplified analysis [7] based on the existing 104 nuclear power plants in the United States, which have a rated capacity of about 100 GW electric [13]. The installed new cost of nuclear plants is estimated at $7 billion per 1 GW electric unit [6,7]. The cost of installing 100 GW of nuclear power would therefore be of the order of $700 billion. A schematic diagram of the basic nuclear fuel cycle is shown in Figure 2.7.

To estimate the embedded energy in 100 GW of nuclear power, consider that the American industrial sector consumed 33% of the total energy of about 100 quads in 2004, while the GDP in that year equaled $11.7 nominal dollars [13]. Since the American industrial sector constituted 20% of the GDP [14], the average American energy intensity was approximately 8500 Btu/$. However, industrial energy intensity, such as building power plants, is larger than other sectors, estimated by a factor of 33/20, resulting in an energy intensity for nuclear power plants of about 14,000 Btu/$ [7]. Thus, the embedded energy in an expenditure of $700 for new nuclear power plants would be 98×10^{14} Btu (9.8 quads).

The energy embedded in the nuclear fuel is a relatively small component of the cost of energy delivered. According to [15], the market price is 0.5 cents per kWh. With a capacity factor of 90%, the electric output of all the nuclear

TABLE 2.12
EROI for Wind Systems

Year of Study	Location	EROI	CO_2 Intensity (g CO_2/kWh)	Power Rating (kW)	Lifetime (Year)	Load Factor (%)	Analysis Type	Turbine Type	On/Off Shore	Rotor Diameter (m)	Hub Height (m)	Wind Speed (m/s)	Remarks
1999	United States	23	14.4	342.5	30	24	I/O	Kenetech KVS-33	On	32.9	36.6		25 MW farm
1999	United States	17	20.2	600	20	31	I/O	Tacke 600e	On	46.0	60.0	6.1	1.2 MW farm
1999	United States	39	8.9	750	25	35	I/O	Zond Z-46	On	46.0	48.5		107.25 MW farm
1998	Germany	21.74		1500	20	31	PA	3 blades		66	67		Enercon E-66
1998	Germany	23.81		500	20	29.6	PA	3 blades		40.3	44		Enercon E-40
1998	Germany	15.38		500	20	29.6	I/O	3 blades		40.3	44		Enercon E-40
1996	Germany	8.33	17	100	20	31.4	PA	3 blades		20	30		
1998	Germany	14.08		1500	20	31	I/O	3 blades		66	67		Enercon E-66
1994	Germany	14.71	8.1	500	20	36.5	PA	2/3 blades		39	41		
1991	Japan	3.97	71.7e	100	20	31.5	I/O						
1992	Japan	2.90	95.6e	100	20	31.5	I/O						10% auxiliary power

1996	Japan	2.29	123.6e	100	30	20	I/O					Downwind propeller
1983	Germany	3.45		6	15	45.7	I/O					Average values
1983	Germany	5.00		12.5	15	45.7	I/O					Average values
1983	Germany	8.33		32.5	15	45.7	I/O					Average values
1983	Germany	1.27		3000	20	45.7	I/O	2 blades	100	100		GROWIAN prototype
1997	Denmark	8.33		15	20	20.5	I/O	1980	10	18		Vintage model
1997	Denmark	8.13		22	20	19.9	I/O	1980	10.5	18		Vintage model
1997	Denmark	10.00		30	20	19	I/O	1980	11	19		Vintage model
1997	Denmark	15.15		55	20	20.6	I/O	1980	16	20		Vintage model
1997	Denmark	27.03		600	20	26.5	I/O	3 blades	47	50	15	
1991	Germany	11.76		30	20	14.4	PA	2 blades	12.5	14.8	13	Hsw-30
1991	Germany	20.41		33	20	29.4	PA	2 blades	14.8	22	11	MAN-Aeromann

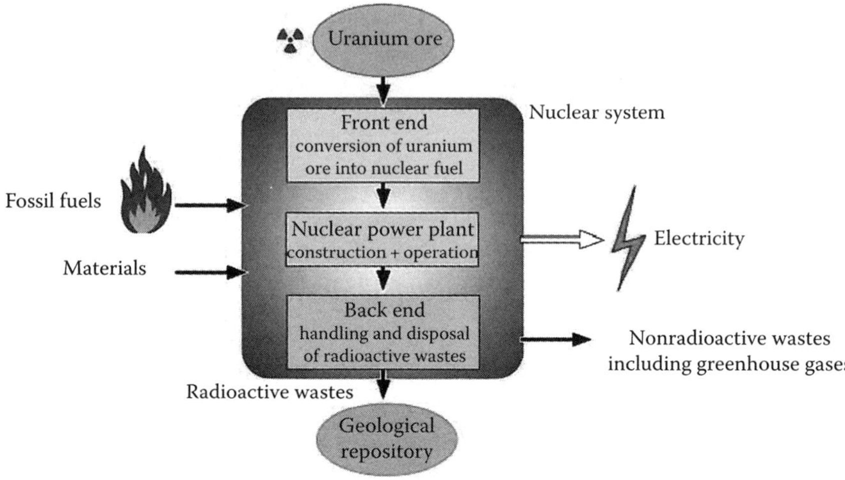

FIGURE 2.7
Basic nuclear fuel cycle.

plants working for 8760 h per year would equal about 800 trillion Wh per year. At 0.5 cents per kWh, the cost of the nuclear fuel would be $4 billion/year. At an embedded energy of 14,000 Btu/$, this equates to 56 trillion Btu/year or 0.056 quads per year. Hence, the embedded energy in all the fuel for the nuclear plants during their 40 year life is about 0.056 quads/year × 40 years, for an embedded energy in all nuclear fuels of about 2.25 quads. Therefore, the total embedded energy of the nuclear power production of 100 plants including fuel and construction, which is approximately 10 quads, during 40 years of operation amounts to 12.25 quads.

U.S. primary energy consumption for producing electricity in 2004 was about 41 quads, according to the EIA Annual Energy Review [16]. With an overall efficiency for all power production at about 33% [17], 13.5 quads of electric energy was produced in 2004. Since nuclear plants produced 20% of that, that is, 2.7 quads of electricity per year, the total electricity produced over 40 years was 108 quads, yielding an EROI (electric power out to primary fossil in) of 108/12 or 9. This estimate does not include O&M, storage of spent fuel, the social cost of mining uranium, and dismantling of the plant, which may reduce the EROI substantially. A previous analysis performed by the Oak Ridge National Laboratory in the 1970s [17] gave an EROI for nuclear power, including disposal and dismantling of the plant, of about 6. An extensive analysis of the energy balance of nuclear power was made in 2001 and was updated in February of 2008 by the CEEDATA consultancy in Shaam, the Netherlands [18] and is available at http://www.stormsmith.nl/. This lengthy and detailed report provides a background for the total cost of nuclear power, which includes all the social and environmental costs. It concluded that the EROI on that basis was only about 2.3. A rebuttal to this document was prepared by the nuclear industry, the World Nuclear Association (WNA, world-nuclear.org) and a response to the rebuttal was added to the final Web site. Another important publication regarding the nuclear power prospect was prepared in June 2009 by the Institute for Energy and the Environment of the Vermont Law School [19]. The latter of these two reports approaches the question of nuclear power mostly from an economic perspective. In both cases, the nuclear industry responded with strongly worded rebuttals, and in the absence of hard data, it is difficult to reach a conclusion about the economic and environmental consequences of nuclear power. However, investment in nuclear power requires a large commitment in money and energy and before such an undertaking on a large scale is initiated, additional careful analysis for a sustainable future is important.

Problems

2.1 The doubling time of any quantity n is related to the continuous growth rate R_{cont} by the relation $2 = \exp n\ R_{cont}$. Derive an expression for n in terms of the annual growth rate and then estimate what the doubling time for a population growth rate is averaging about 2% per year.

2.2 What is the present value of a $10,000 payment in 2015 if the discount rate is 7%?

2.3 Repeat the previous problem in constant dollars for the $10,000 payment if the inflation rate is 4%/year.

2.4 Calculate the annual payments for a solar system whose initial extra cost is $10,000. Assume that the interest rate is 8% and the mortgage term is 15 years.

2.5 Assuming that the price of fuel is $5.00/GJ and is increasing at a rate of 4%/year while the discount rate is 6%, calculate the equivalent levelized price over 20 years.

2.6 A home buyer obtains a mortgage of $100,000 at an interest rate of 8% over 20 years. Estimate the annual payments to repay this loan.

2.7 A rooftop solar hot water heater with collector areas of 8 m^2 and storage of 640 L is installed in Phoenix, Arizona, on a house that uses 160 L/day of hot water at 60°C. Seventy percent of the total energy required is supplied by solar energy (the solar fraction is 70%) and 30% is supplemented with an electric heater. The solar system costs $3000 for installation, whereas an electric water heater and tank for the same supply costs $800 for installation. Assuming that the electricity cost is $0.10/kWh and does not change over time, estimate the simple payback time for the solar hot water heater. The temperature of the water in the main from which it is to be heated is 10°C.

2.8 Repeat the above problem, but assume that the cost of installation is to be repaid at 7% interest over 15 years and that the electricity costs escalate at 10% per year. Compare the NPV of the solar system in Problem 2.7 to that of an electric heating system over a 15 year period, then discuss whether or not the solar system is a good investment relative to the payback time. Also, prepare a spreadsheet showing the NPV for each year.

2.9 Repeat the previous problem, but assume that the cost of the equipment is reduced from a government grant by 20%.

2.10 Compare the capital costs and the levelized costs of operation for a 1000 MW nuclear power plant and a 1000 MW PV system constructed in the Mojave desert in California. The PV system is to use thin film technology currently available. Neglect the cost of nuclear fuel. Comment on environmental impact and availability of water.

2.11 A solar system is to be installed in Phoenix, Arizona. The estimated amounts of energy from the system as a function of collector area are shown in the tabulation below. If the solar system costs $195/m^2 and is to be paid off over 25 years at 10% interest, prepare the total cost curve and specify the cost optimal system. Assume that the back-up fuel is fuel oil at $7/GJ and that the fuel price is inflating at 8%/year. The total energy demand for which the solar system is constructed is estimated to be 1000 GJ/year.

2.12 Calculate and tabulate the annual cash flows associated with a solar system for a 10 year period of analysis if the extra initial solar cost is $6000 and if

Interest rate = 9%

Power cost = $30/year escalating at 10%/year

Property tax = 0

Income tax bracket = 32%

Maintenance = ½%/year

Scrap value = 50% of initial cost

If the solar system saves $550/year in conventional fuel (escalating at 10%/year), is it cost effective? Work the problem in current dollars.

2.13 Compute the annual payment of a self-amortizing loan for the solar system with the characteristics tabulated below:

Factor	Specification
Expected system lifetime t (year)	20
Effective interest rate (%)	8
Collector area A_c (m^2)	20
Collector cost (\$/m^2)	100
Storage cost (\$/m^2)	6.25
Cost of control system (\$)	100
Miscellaneous costs (e.g., pipes, pumps, motors, heat exchangers) (\$)	200 + (5A_c)

If the initial extra cost of a solar system is \$5,000, what are the annual payments of a self-amortization loan if the interest rate is 12%, the inflation rate is 4%, and the mortgage term is 15 years.

2.14 A factory uses 700 40 W standard fluorescent lamps for its lighting. The lamps are to be replaced by 34/W high-efficiency lamps that can operate on the existing standard ballasts. The standard and high-efficiency fluorescent lamps cost \$1.80 and \$2.50 each, respectively. The factory operates 3000 h/year and all of the lamps operate during that period. Assuming the unit cost of electricity is 6 cents/kWh and the ballasts consume 10% of the rated power of the lamps, calculate how much energy and money will be saved as a result of switching to the high-efficiency fluorescent lamps. Then, estimate the PP assuming an interest rate of 6% and an inflation rate of 3%. Finally, calculate the levelized cost of the saved energy assuming the lamps last for 2 years.

2.15 The hallways of a university are lit by 40 fluorescent lights, each containing 2 lamps rated at 60 W each. The lights are on all day and night during the year, but the building is only used from 7 a.m. to 7 p.m. 5 days per week during the year. If the price of electricity is 7 cents/kWh, estimate the amount of energy and money that could be saved by installing two motion sensors that turn off the lights when the building is not being used. Then, estimate the simple payback and the payback with a 6% effective interest rate if the price of a sensor is \$50 and the cost of installation is \$40.

Appendix 2.A Capital Recovery Factor

In many economic analyses a *series* of annual or monthly sums is invested or used to pay off a loan. The method can easily be extended from a single payment to a series of payments. If a sum P_{sum} is invested each year at interest rate i, the present worth of the *sum* of these payments S is

$$S = \frac{P_{ann}}{1+i} + \frac{P_{ann}}{(1+i)^2} + \frac{P_{ann}}{(1+i)^3} + \cdots,$$

or

$$S = P_{ann}\left[(1+i)^{-1} + (1+i)^{-2} + (1+i)^{-3} + \cdots\right]$$

The expression in brackets is a geometric series with first term $P_{ann}/(1 + i)$ and ratio $(1 + i)^{-1}$. It follows from the expression for the sum of such a series that

$$P_{ann} = S\left[\frac{i}{1-(1+i)^{-1}}\right]$$

Alternatively, the annual payment P_{ann} required to form a total amount S, for example, to purchase a solar system, is

$$\mathrm{CRF}(i,t) \equiv \frac{P_{ann}}{S} = \frac{i}{1-(1+i)^{-t}}$$

The term in brackets in the above equation is called the *capital-recovery factor* (CRF(i, t)):

$$\mathrm{CRF}(i,t) \equiv \frac{P_{ann}}{S} = \frac{i}{1-(1+i)^{-t}}$$

Table 2.A.1 is a summary tabulation of CRFs. The term P_{ann} is the annual payment on a self-amortizing loan of value S. Each payment is a mix of interest and principal repayment. Early payments are mostly interest because of the large outstanding balance; later payments are primarily principal repayment. Most solar systems owned by private firms or individuals are purchased with self-amortizing loans.

TABLE 2.A.1

Capital-Recovery Factors

	I									
n	0%	2%	4%	6%	8%	10%	12%	15%	20%	25%
1	1.00000	1.02000	1.04000	1.06000	1.08000	1.10000	1.12000	1.15000	1.20000	1.25000
2	0.50000	0.51505	0.53020	0.54544	0.56077	0.57619	0.59170	0.16152	0.65455	0.69444
3	0.33333	0.34675	0.36035	0.37411	0.38803	0.40211	0.41635	0.43798	0.47473	0.51230
4	0.25000	0.26262	0.27549	0.28859	0.30192	0.31547	0.32923	0.35027	0.38629	0.42344
5	0.20000	0.21216	0.22463	0.23740	0.25046	0.26380	0.27741	0.29832	0.33438	0.37184
6	0.16667	0.17853	0.19076	0.20336	0.21632	0.22961	0.24323	0.26424	0.30071	0.33882
7	0.14286	0.15451	0.16661	0.17914	0.19207	0.20541	0.21912	0.24036	0.27742	0.31634
8	0.12500	0.13651	0.14853	0.16101	0.17401	0.18744	0.20130	0.22285	0.26061	0.30040
9	0.11111	0.12252	0.13449	0.14702	0.16008	0.17364	0.18768	0.20957	0.24808	0.28876
10	0.10000	0.11133	0.12329	0.13587	0.14903	0.16275	0.17698	0.19925	0.23852	0.28007
11	0.09091	0.10218	0.11415	0.12679	0.14008	0.15396	0.16842	0.19107	0.23110	0.27349
12	0.08333	0.09156	0.10655	0.11928	0.13270	0.14676	0.16144	0.18148	0.22526	0.26845
13	0.07692	0.08812	0.10014	0.11296	0.12652	0.14078	0.15568	0.17911	0.22062	0.26454
14	0.07143	0.08260	0.09467	0.10758	0.12130	0.13575	0.15087	0.17469	0.21689	0.26150
15	0.06667	0.07783	0.08994	0.10296	0.11683	0.13147	0.14682	0.17102	0.21388	0.25912

(continued)

TABLE 2.A.1 (continued)

Capital-Recovery Factors

	I									
n	0%	2%	4%	6%	8%	10%	12%	15%	20%	25%
16	0.06250	0.07365	0.08582	0.09895	0.11298	0.12782	0.14339	0.16795	0.21144	0.25724
17	0.05882	0.06997	0.08220	0.09544	0.10963	0.12466	0.14046	0.16537	0.20944	0.25576
18	0.05556	0.06670	0.07899	0.09236	0.10670	0.12193	0.13794	0.16319	0.20781	0.25459
19	0.05263	0.06378	0.07614	0.08962	0.10413	0.11955	0.13576	0.16134	0.20646	0.25366
20	0.05000	0.06116	0.07358	0.08718	0.10185	0.11746	0.13388	0.15976	0.20536	0.25292
25	0.04000	0.05122	0.06401	0.07823	0.09368	0.11017	0.12750	0.15470	0.20212	0.25095
30	0.03333	0.04465	0.05783	0.07265	0.08883	0.10608	0.12414	0.15230	0.20085	0.25031
40	0.02500	0.03656	0.05052	0.06646	0.08386	0.10226	0.12130	0.15056	0.20014	0.25003
50	0.02000	0.03182	0.04655	0.06344	0.08174	0.10086	0.12042	0.15014	0.20002	0.25000
100	0.01000	0.02320	0.04081	0.06018	0.08004	0.10001	0.12000	0.15000	0.20000	0.25000
∞		0.02000	0.04000	0.06000	0.08000	0.10000	0.12000	0.15000	0.20000	0.25000

[a] For interest rates i from 0% to 25% and for periods of analysis n from 1 to 100 years.

References

1. Rueg, R.T. (1994) Economic methods, in: Kreith, F. and West, R.E. (eds.), *CRC Handbook of Energy Efficiency*, Chap. 3, CRC Press, Boca Raton, FL.
2. Tester, J.W., Drake, E.M., Golay, M.W., Driscoll, M.J., and Peters, W.A. (2005) *Sustainable Energy: Choosing among Options*, MIT Press, Cambridge, MA.
3. Larson, R.W., Vignola, F., and West, R.E. (December 1992) *Economics of Solar Energy Technologies*, ASES, Boulder, CO.
4. Rabl, A. and Spadaro, J.V. (2007) Environmental impacts and cost of energy, in: Kreith, F. and Goswami, D.Y. (eds.), *CRC Handbook of Energy Efficiency and Renewable Energy*, Chap. 4, CRC Press, Boca Raton, FL.
5. Short, W., Packey, D.J., and Holt, T. (1995) *A Manual for the Economic Evaluation of Energy Efficiency and Renewable Energy Technologies*, U.S. Department of Energy, NREL, Golden, CO.
6. Nuclear Power Daily (2009) *Progress Energy Florida Signs Contract for New, Advanced-Design Nuclear Plant*, January 16, 2009. http://www.nuclearpowerdaily.com/reports/Progress_Energy_Florida_Signs_Contract_For_New_Advanced_Design_Nuclear_Plant_999.html (retrieved May 18, 2009).
7. *Nuclear Power Joint Fact Finding* (June 2007) The Keystone Center, Keystone, CO.
8. Spreng, D.T. (1988) *Net Energy Analysis and the Energy Requirements of Energy Systems*, Praeger Publishing Company, New York.
9. Ballard, C.W., Penner, P.S., and Pilati, D.A. (November 1978) Net energy analysis—Handbook for combining process and input-output analysis, *Resources and Energy* 1, 267–313.
10. Hannon, B., Casler, D.S., and Blasek, T. (1985) *Energy Intensity for the U.S. Economy-1977*, Energy Research Group, University of Illinois Press, Urbana, IL.
11. White, S.W. and Kulsiwski, G.L. (December 1998) Net energy payback and CO_2 emissions from wind-generated electricity in the Midwest, Fusion Tech. Inst., University of Wisconsin, Madison, WI. http://fti.neep.wisc.edu/pdf/fdm1092.pdf
12. Bureau of Labor Statistics (1998) Consumer price index—All urban consumers, March 13, 1998. http://146.142.4.24/cgi-bin/surveymost?cu
13. Kennedy III, R.G. (2008) Member of ASME Energy Committee, private communication, April 12, 2008.
14. Statistical Abstract of the United States, 2007, accessed at http://www.census.gov/compendia/statab.
15. *The Economist/Pocket World in Figures*, 2004 edition, Economists Book, 2004.
16. *EIA Annual Energy Review*, 2004, accessed at http://www.eiq.doe.gov/aer/
17. Rotty, R.M., Perry, A.M., and Reister, D.B. (May 1976). *Net Energy from Nuclear Power*, Oak Ridge Associated Universities, Inc., Oak Ridge, TN.
18. Storm van Leeuwen, J.W., and Smith, P. (2008, update), Nuclear power and the energy balance. http://www.stormsmith.nl/ (accessed August 22, 2008).
19. Cooper, M., 2009. The economics of nuclear reactors: Renaissance or relapse? Institute for Energy and the Environment, Vermont Law School, South Royalton, VT.

3

Wind Energy

> First, there is the power of the Wind, constantly exerted over the globe.... Here is an almost incalculable power at our disposal, yet how trifling the use we make of it! It only serves to turn a few mills, blow a few vessels across the ocean, and a few trivial ends besides. What a poor compliment do we pay to our indefatigable and energetic servant!
>
> **Henry Thoreau, "Paradise, Paradise (To Be) Regained" (1843)**

> ... As yet, the wind is an untamed, and unharnessed force; and quite possibly one of the greatest discoveries hereafter to be made, will be the taming, and harnessing of the wind.
>
> **Abraham Lincoln, "Discoveries and Inventions" (1858)**

3.1 Wind Power in a Nutshell

Put simply, wind turbines convert the kinetic energy of the wind into electrical power. This process begins when incoming wind strikes the rotor. The rotor consists of blades, typically three in a modern large-scale turbine, attached to a hub. The rotor assembly can be seen in Figure 3.1, which shows a large wind turbine manufactured by Clipper Windpower.

The shape of the turbine blades creates a lifting force, which is the same as that of an airplane wing. The lifting force is directed at an angle causing the rotor to rotate. The rotor is typically connected to a low-speed shaft. The spinning low-speed shaft is in turn connected to a gearbox which increases the low-speed shaft rotation speed (50–100 rpm in a "large" wind turbine) to a high-speed rotation (typically 2000–3000 rpm). A high-speed shaft on one end of the gearbox is then connected to an electrical generator which produces the desired electrical energy. The details of the hub and gearbox, as well as the general layout of a wind energy system, can be seen schematically in Figure 3.2.

As one can imagine, it takes quite a bit of engineering work to go from this simple description of a wind turbine to an actual turbine with a hub height of 100 m, producing 5 MW of power. In this chapter, we will explore some of the

FIGURE 3.1
(See color insert.) 2.5 MW Liberty wind turbine (Clipper Windpower) near Medicine Bow, Wyoming. (Courtesy of DOE/NREL, Golden, CO.)

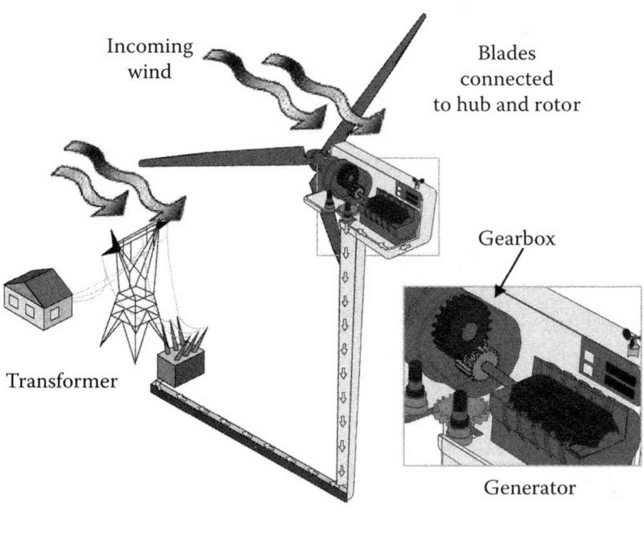

FIGURE 3.2
Schematic of typical wind turbine layout. (From Energy Efficiency & Renewable Energy (EERE) (DOE), How does a wind turbine work?, 2010, http://www1.eere.energy.gov/windandhydro/wind_animation.html)

key elements involved in turning this simple concept into reality by focusing on commercial size (1–5 MW) wind turbines. These elements include

- Fundamentals of power and energy
- Frequently asked questions (FAQs)
- Turbine types and their difference
- Aerodynamics (blade design)
- Where wind comes from, how it is modeled, and wind turbine performance
- How turbine design affects the cost of electricity produced
- Wind turbine farms and offshore wind turbines

3.2 Power and Energy

Power and energy are common terms that are often misused. Power is the rate at which energy is used or produced. Light bulbs use power; wind turbines generate power. Units of power are expressed in work (energy) over an amount of time:

$$\text{Power} = \frac{\text{Work}}{\text{Time}} = \frac{\text{Energy}}{\text{Time}} = \frac{\text{N}\cdot\text{m}}{\text{s}} = \frac{\text{J}}{\text{s}} = \text{W} \tag{3.1}$$

Wind turbines are rated in terms of power. Large commercial wind turbines have power ratings in the 1–5 MW range while a wind turbine used for a single home is likely rated in the 3–15 kW range. This means commercial-sized wind turbines produce power at a rate 100 times greater than a residential-sized wind turbine (WT):

$$\frac{\text{Commercial}}{\text{Residential}} = \frac{1\text{ MW}}{10\text{ kW}} = \frac{1\times10^{6}\text{ W}}{10\times10^{3}\text{ W}} = 100 \tag{3.2}$$

Energy, on the other hand, is a measure of how much work can be done by a force. Units on energy are the product of force and distance:

$$\text{Energy} = \text{N}\cdot\text{m} = \text{J} \tag{3.3}$$

The link between power and energy comes from the rate of power usage. In other words, energy equals the product of power and time or power is energy divided by time. If a 100 W light bulb is turned on for 24 h, the energy usage is

$$\text{Energy usage} = 100\ \text{W} \cdot 24\ \text{h} = 2400\ \text{Wh} = 2.4\ \text{kWh} = 2.4\ \text{kWh} \tag{3.4}$$

$$\text{Checking units:}\quad 2400\ \text{Wh} = \left(\frac{2400\ \text{J}}{\text{s}}\right)\cdot\left(24\ \text{h}\cdot\frac{3600\text{s}}{\text{h}}\right) = 2.074\times 10^{8}\ \text{J} \tag{3.5}$$

Turbines are denoted by their rated power production (e.g., 100 kW, 2.5 MW, etc.), while we pay for electricity based on how much energy we use (e.g., $0.06 per kWh).

In 2006, the per capita U.S. annual energy consumption was approximately 98 MWh [1]. A 10 kW wind turbine running 24 h/day for 365 days (this is not realistic, but rather illustrative) would produce approximately 88 MWh of energy. In comparison, a gallon of gasoline has an energy content of approximately 36 kWh (130 MJ). If our only energy source was gasoline, then the average American would have consumed the equivalent of 2722 gal of gasoline during the year or 7.5 gal of gasoline/day. Creating more sources of renewable energy will expand the options available for the world's energy needs.

We will now use these fundamental concepts of power and energy to determine the power contained in the wind moving toward a wind turbine. The kinetic energy, KE, of a mass of air, m (in kg), moving at speed, U (in m/s), is

$$\text{KE} = \frac{1}{2} m U^2 \tag{3.6}$$

$$\text{Unit check:}\quad \text{kg}\left(\frac{\text{m}}{\text{s}}\right)^2 = \frac{\text{kg}\cdot\text{m}^2}{\text{s}^2} = \left(\frac{\text{kg}\cdot\text{m}}{\text{s}^2}\right)\cdot\text{m} = \text{N}\cdot\text{m} = \text{J} \tag{3.7}$$

Now, imagine a horizontal cylinder of air of area A (in m^2) and velocity U, moving toward a wind turbine, as seen in Figure 3.3. Note that the area, A, corresponds to the area swept out by the rotating wind turbine blades. The mass m of the air column is volume V times the density, where ρ represents the air density in kg/m^3. Mass can then also be expressed as

$$m = \rho V = \rho A l = \rho A U t \tag{3.8}$$

where
- ρ is the air density, kg/m^3
- V is the volume, m^3
- A is the area, m^2
- l is the length, m
- U is the wind speed, m/s
- t is the time, s

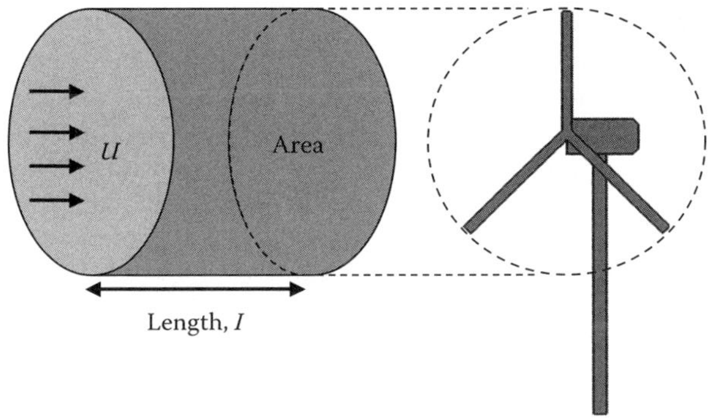

FIGURE 3.3
Incoming air column and wind turbine.

where the length of the column of air, l, is equal to the distance the wind travels in a given time interval, t, and is found by multiplying the wind speed, U, by the time. To check the units of the mass expression, Equation 3.8, above

$$\text{mass} = \frac{\text{kg}}{\text{m}^3} \cdot \text{m}^2 \cdot \frac{\text{m}}{\text{s}} \cdot \text{s} = \text{kg} \tag{3.9}$$

Combining Equation 3.6 with Equation 3.8, the kinetic energy becomes

$$\text{KE} = \frac{1}{2} mU^2 = \frac{1}{2}(\rho AUt)U^2 = \frac{1}{2}\rho AtU^3 \tag{3.10}$$

As discussed above, power is equal to energy divided by time, so simply by dividing the KE by time yields the power of a moving column of air:

$$\text{Power} = \frac{\text{KE}}{t} = \frac{1}{2}\rho AU^3 \tag{3.11}$$

In other words, the power in a column of wind is linearly proportional to the air density and area of the air column. Of even more significance, the power is also proportional to the cube of the wind speed. We will see later that a key reason wind turbines have grown in height over time is to place the wind turbine rotor blades higher in the atmosphere where the wind speed is greater and thus harnessing a much higher kinetic energy output.

Since the air density cannot readily be changed to increase power output, this leaves the area of the wind turbine blades as the only variable (besides the wind turbine height) that can be modified on a macroscale to affect the power output. Since the wind turbine power output is linearly proportional to the area, a large-scale commercial wind turbine designer would want to increase the blade length (i.e., radius) to increase the power output. Since the circular area swept out by the turbine blades is proportional to the blade radius squared (R^2), turbine power is also proportional to the blade radius squared (R^2). Turbine blades sweep out a circular area.

$$\text{Power} = \frac{1}{2}\rho A U^3 = \frac{1}{2}\rho\pi R^2 U^3 \tag{3.12}$$

In summary, longer blade lengths and taller turbines (providing access to higher wind speeds) are options to increasing wind turbine power output. However, this performance increase comes at a price. Longer blades need more sophisticated manufacturing techniques to build blades strong enough to withstand wind gusts and extreme wind speeds (in excess of 100 mph). Taller towers must be strong enough to withstand the forces transmitted from the blades, weight of the nacelle, and other parts while still being cost effective.

In contrast, why make small and/or shorter wind turbines? Some applications, for example, remote homes off the electrical grid require more modest amounts of power. The design challenges here are just as significant: structures (blades, tower, etc.) must be strong enough to withstand wind forces, but also must be light, cost effective, and reliable.

3.3 Fact or Fiction: Common Questions about Wind Turbines

Other challenges facing wind turbines beyond common design tasks are strength, reliability, and performance. As in any technology, there may be other unanticipated challenges. Some common questions that can cause controversy about wind turbines are

- How noisy are wind turbines?
- What are the lifetime environmental impacts of wind turbines?
- What about bird and bat kills?

3.3.1 Noise Issues

Noise issues related to wind energy have been reported since the 1980s. Problems were initially found with the mechanical design as well as blade design and operation. For example, the 2 MW MOD-1 turbine developed by NASA in the late 1970s was a two-bladed turbine (61 m rotor diameter) [2]. It operated with the turbine blades downstream of a lattice tower structure. Low-frequency noise was produced as the turbine blades moved in and out of the "tower shadow" caused by wind flowing around and through the

TABLE 3.1

Loudness Levels of Common Noises

Noise Level (dB)	Equivalent Activity
120	Jet plane takeoff, amplified rock music at 4–6 ft, car stereo, band practice
110	Rock music, model airplane
100	Snowmobile, chain saw, pneumatic drill
90	Lawnmower, shop tools, truck traffic, subway
80	Alarm clock, busy street
70	Vacuum cleaner
60	Conversation, dishwasher
50	Average home
40	Quiet library
30	Quiet bedroom

lattice tower. Noise of significant magnitude was produced, causing complaints from local residents.

Today, most large commercial wind turbines have the blades operating upstream of the tower and use a monopole tower. Modern commercial wind turbines have noise levels of approximately 35–45 dB at 750–1000 ft. In contrast, typical noise or loudness levels in everyday life are given in Table 3.1 [3]. As seen in Table 3.1, the modern turbine (at distances of 750–1000 ft) has a noise level comparable to a library or an average home (40–50 dB). When locating turbines, the distance between residences and wind turbines needs to be considered. Small wind turbines can actually have higher noise levels due to higher rates of rotation.

The sound typically heard from a wind turbine comes from the blades. Careful aerodynamic design of the blades, as well as understanding the maximum allowable rotational speed of the blades has reduced the aerodynamic noise level from previous levels. Similarly, mechanical components have been redesigned to virtually eliminate mechanical noise, leaving blade noise as the predominant noise source, albeit at a low noise level [4].

3.3.2 Lifetime Environmental Impact

An important consideration for any technology is the greenhouse gas emissions for the total fuel cycle. Each phase of the fuel cycle, including resource extraction, facility construction, and facility operation, must be evaluated in order to quantify and compare emissions from different technologies [5]. When the operations needed to manufacture wind turbines and build wind farms are included in a CO_2 emissions analysis, the use of wind power compared to coal power reduces CO_2 emissions by 99%. Similarly, using wind power instead of natural gas reduces CO_2 emissions by 98% [6].

Other environmental impacts from wind turbines do exist. For example, construction of a wind farm could cause erosion due to the construction and use of roads needed to build the farm (truck and crane traffic, electrical line/ equipment installation, etc.) but this impact can be mitigated using standard environmental reclamation techniques.

In Europe, "shadow flicker" created by the shadow of rotating wind turbine blades on residences has occurred due to placing wind turbines too close to the residences. The resulting flicker can be very annoying. Proper planning and accounting for latitude allows the necessary distance between turbine and residence to be calculated so no turbine shadow falls on the residence [7]. This distance should also take into account the noise level of the operating turbine (see above). The turbine should be located a sufficient distance away (based on sun angle) to eliminate any shadow flicker or noise concerns.

3.3.3 Bird and Bat Kills

As can be seen in Figure 3.4, the largest single cause of bird deaths are buildings and/or windows. A recent study [8] indicated that about 7000 bird

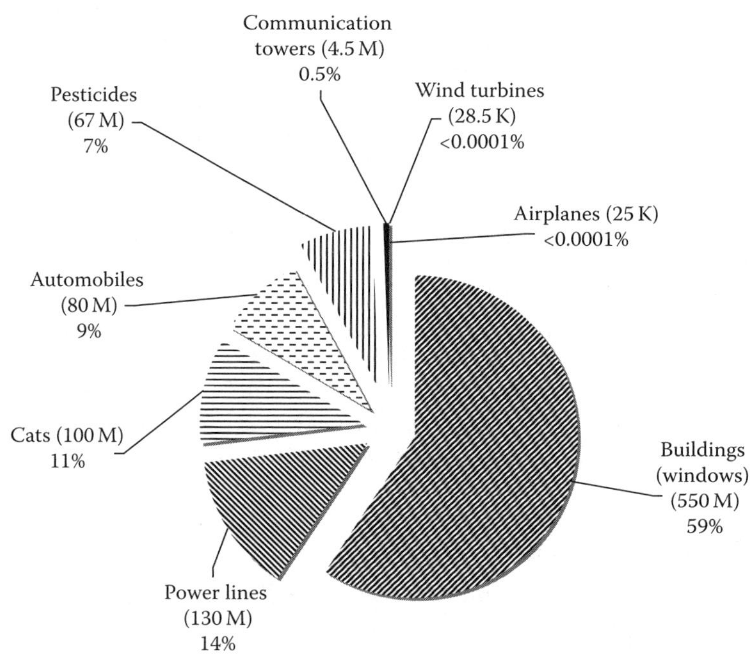

FIGURE 3.4
Most common causes of bird fatalities. Bird deaths are shown in millions (M) or thousands (K) with an annual estimated total near 1 billion birds. (From Erickson, W. et al., A summary and comparison of bird mortality from anthropogenic causes with an emphasis on collisions, USDA Forest Service General Technical Report PSW-GTR-191, 2002, http://www.fs.fed.us/psw/publications/documents/psw_gtr191/Asilomar/pdfs/1029–1042)

fatalities occurred in 2006 due to wind farms, while nuclear plants and fossil-fueled power stations are estimated to kill 46 and 2070 times more birds, respectively. A study by Erickson [9] provides an annual estimate of the sources of bird fatalities (see Figure 3.4) including deaths from wind turbines. Bird deaths from wind turbines are estimated to be much less than 0.0001% of the total number of projected deaths, on the order of 30,000/year. In contrast, bird deaths from cats or buildings are projected to be 3.5 and 19 million times, respectively, more likely than from wind turbines. If wind turbines are sited properly (e.g., away from migratory paths) or even shut down briefly with the help of radar to determine when migratory birds are approaching, the fatalities can be minimized [55]. Increasing the number of wind farms would likely increase the number of deaths and this topic remains an area of research in order to minimize the effect on avian mortality.

While bird fatalities occur due to birds being struck by blades or running into towers, bat fatalities have been found to be due to a different cause. Researchers proposed that decompression killed bats by barotraumas, which is a rapid reduction in air pressure near rotating turbine blades. Study results showed that, of the bats affected by wind turbines, "90% of bat fatalities involved internal hemorrhaging consistent with barotrauma, and that direct contact with turbine blades only accounted for about half of the fatalities" [54]. Studies have shown a higher rate of fatalities of bats with wind turbines than birds.

Bird fatalities per turbine ranged from 0.6 to 7.7 while bat fatalities ranged from 3.4 to 47.5 per turbine [10]. Research is underway to understand the higher fatality rate as well as ways to stop/minimize bat fatalities as well as predict and avoid high-risk sites [11].

One wind power project conducted a study to determine the effect of turning off the wind turbines at night during an 11 week bird migratory period corresponding to low winds. From this project, fatalities declined by approximately 50%. However, if used, this method would result in an annual loss in generating capacity of 1% if the turbines were turned off when they could have been generating electricity [12]. Further study and work is needed to ensure wind turbine technology is developed to coexist with environmental issues.

3.4 History of Wind Turbine Development: HAWTs and VAWTs

Much work has occurred over the last three decades leading to optimization of performance and cost in wind turbine design. Currently, the exclusive commercial wind turbine design (e.g., 1 MW and larger) is a horizontal

FIGURE 3.5
An example of a horizontal axis wind turbine.

axis wind turbine (HAWT) in which the axis of the rotor is horizontal to the ground, as shown in Figure 3.5, with the turbine blades spinning in the same orientation as if they are on a plane. Conversely, vertical axis wind turbines (VAWTs) have blades that spin about a vertical axis, as shown in Figure 3.6. The biggest advantage of this design is that it does not have to be pointed in the direction of the wind.

Earliest indication of wind devices (anything that converts wind energy to other forms of energy; e.g., rotational and mechanical) was in Persia around 900–1000 AD or earlier. The first practical vertical axis windmills arose in the seventh century in Sistan, Afghanistan to process grain and draw up water. Horizontal axis wind machines like that shown in Figure 3.5 appeared in the eleventh and twelfth centuries in Northwest Europe to grind grain. The Dutch built thousands of windmills over the twelfth and nineteenth centuries to move water. Clearly, early development of wind machines occurred all over the world and this trend continued as wind turbine technology continued to mature.

In Scotland, Professor James Blyth built the first wind turbine for electricity production and successfully generated electricity in July 1887 [13,14]. Shortly afterward, the first automatically operating wind turbine in the world was made by Charles F. Brush in Cleveland, Ohio in the

FIGURE 3.6
An example of a Darrieus VAWT, Quebec Canada.

winter of 1887–1888. The HAWT turbine produced 12 kW and had a 17 m (50 ft) rotor diameter [15].

The first MW-sized turbine to supply electricity to the local electrical grid occurred in 1941 on a hill Grandpa's Knob in Vermont. The 1.25 MW turbine was designed by Palmer Cosslett Putnam and was 53 m in diameter and had two blades. A failure of a metal blade in 1945 after only 1100 h of operation brought the project to an end [16,17].

The modern era in wind turbines began in the late 1970s as a result of the oil crises earlier in the decade. An increase in turbine development occurred in multiple countries around the world. For example, Danish manufacturers Kuriant, Vestas, Nordtank, and Bonus began producing 20–30 kW machines while NASA, at the Lewis (now Glenn) Research Center in Cleveland, Ohio, developed a government research and development program producing the first of the "Mod" series of wind turbines ranging from 0.2 to 2 MW [18].

As we have seen, VAWTs have a long history, while HAWTs have a shorter history. This raises the question of why today's commercial wind turbines are almost exclusively horizontal axis, two- or three-bladed, upwind turbines. To help answer this question, we first look at a summary of the qualitative differences between the two technologies and then have a discussion of quantitative differences in power-generating ability between the two designs.

3.4.1 VAWT

Strengths

- Turbine always aligned with the wind direction
- May start producing electricity at lower speeds
- Simpler design without large structures in the air (house mechanical and electrical components at ground level)
- May be lower noise than HAWT

Disadvantages

- Typically produce 50% less energy than HAWT (due to drag of blades through full rotation)
- Wind speeds are lower at lower heights meaning the energy potential of a VAWT at a given location is lower than a HAWT having a taller tower height at the same location
- Typically not self-starting
- Difficult to maintain parts which are under weight of tower
- Cannot control power output or rotational speed through blade settings
- Can require high tip-speed ratios (TSRs) causing increased noise (Darrieus)

3.4.2 HAWT

Advantages

- Highest operating efficiency (see Figure 3.7)
- Turbine blades can be placed in a higher wind speed regime (more power) on top of a tower
- Power output and blade speed can be controlled and maximized by varying the angle of the blades

Disadvantages

- Transportation costs of long (50–90 m) blades and tall towers can be significant portion (20%) of installation cost
- Installing tall HAWTs (tower, nacelle, blades, etc.) require skilled operators, expensive cranes
- Tall height means proximity to residences must be considered for visual impact

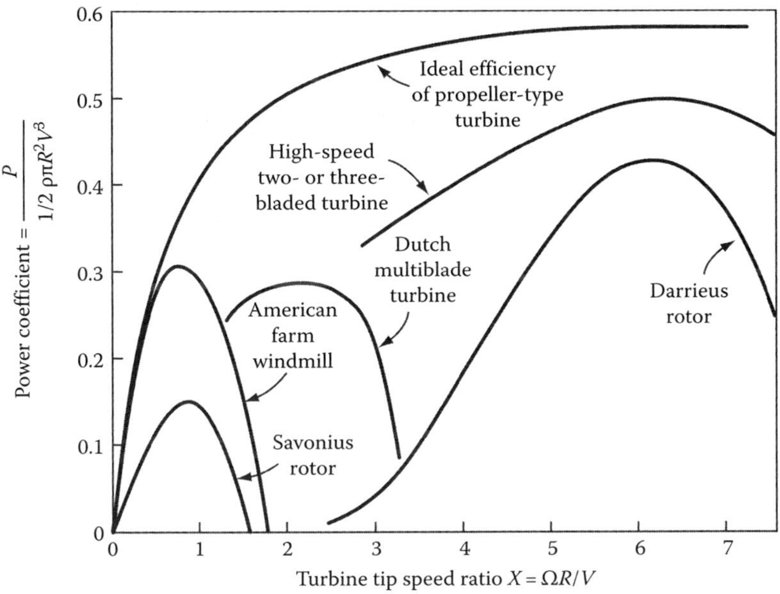

FIGURE 3.7
Various turbine power coefficients versus TSR. (From Wilson, R.E. and Lissaman, P.B.S., *Applied Aerodynamics of Wind Power Machines*, Oregon State University, Corvallis, OR, 1974, http://ir.library.oregonstate.edu/jspui/handle/1957/8140)

- Sophisticated control systems required to point turbine blades into wind, prevent overspeed conditions, etc.
- Readers interested in more information on a variety of wind turbine concepts are directed to [19,20].

3.5 Introduction to Wind Turbine Performance

A key performance measure in wind turbine design is the coefficient of performance, C_p, which is a ratio of the aerodynamic power extracted by the wind turbine, P_{aero}, divided by the total power in the wind, P_{wind}, and is calculated as

$$C_p = \frac{P_{aero}}{P_{wind}} = \frac{P_{aero}}{\frac{1}{2}\rho A U^3} = \frac{P_{aero}}{\frac{1}{2}\rho \pi R^2 U^3} \tag{3.13}$$

The theoretical maximum turbine efficiency was derived by Albert Betz in 1920 [21], as well as others, see [22] and is

$$C_p = \frac{16}{27} \approx 0.59 = \text{Betz limit} \tag{3.14}$$

In addition, C_p is a function of how fast the tips of the wind turbine blades are rotating compared to the incoming wind. The ratio of the blade's rotational speed divided by the incoming wind speed is known as the TSR:

$$\text{Turbine tip speed ratio} = \frac{\Omega R}{V} \tag{3.15}$$

where
 Ω is the turbine blade rotational speed, rad/min
 R is the rotor radius, m
 V is the wind speed, m/s

In Figure 3.7 [23], the power coefficient is plotted as a function of tip-speed ratio for different wind turbine configurations. Several key points can be seen in Figure 3.7, starting from the top. The highest line shows how the theoretical maximum power coefficient increases with increasing TSR. The Betz limit value of 0.59 is approached asymptotically as the TSR increases toward a value of seven (TSR = 7). The next highest curve denotes the performance of today's modern high-speed horizontal axis, two- or three-bladed turbines. The maximum theoretical performance coefficient for these turbines is approximately $C_p = 0.5$ while the Darrieus vertical axis rotor has a maximum C_p of approximately 0.40–0.45. Thus the modern two- and three-bladed HAWTs are approximately 10% more efficient than a Darrieus (VAWT) machine and thus have the potential to make 10% more energy and money.

Not only are the large commercial wind turbines configured horizontally, they are also predominantly three-bladed upwind (blades rotate in front of the tower). Why are HAWT mainly upwind? The blades on a downwind turbine experience cyclic loading (alternating higher forces and lower forces) due to the blades passing behind the tower which may cause fatigue failure of the blades as well as noise. Three-bladed turbines have also been found to be more aesthetically pleasing than two-bladed turbines due to a blade "disappearing" when one of the two blades is aligned with the tower.

3.6 Aerodynamics

Examining the aerodynamics of wind turbine blades helps explain why different wind turbine configurations have such different performance coefficients, C_p [24,25]. Figure 3.8 shows a wind turbine blade rotating (in the plane

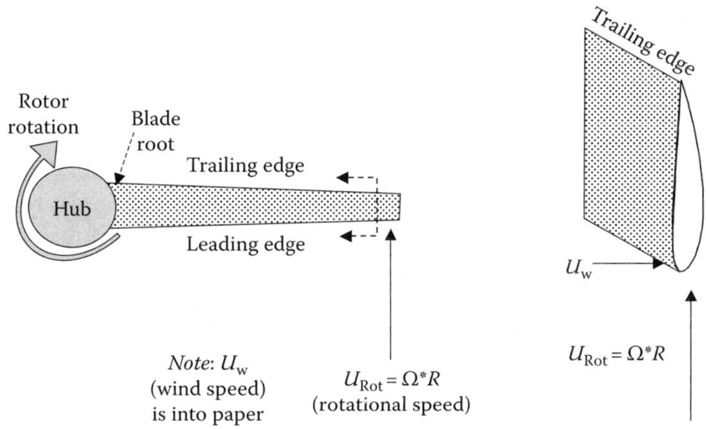

FIGURE 3.8
Rotor blade diagram.

of the paper) in a clockwise direction. The atmospheric wind, U_w, is blowing toward the blade (i.e., into the plane of the paper). A section view of the blade is then taken near the tip of the blade looking in toward the root of the blade (Figure 3.8). The cross-sectional view of the blade airfoil is shown in the right hand side of the figure. The two airflows acting on the blade are the incoming wind speed, U_w, which is perpendicular to the plane of the rotating blades, and the rotational wind speed, $U_{Rot} = \Omega R$, created by the turbine blade rotating at an angular speed, Ω, at the radius of the cross section, R. Note that the side of the airfoil facing the wind (designated the front side) is relatively flat compared to the curved rear side of the airfoil. The leading edge of the airfoil hits the airflow arising from the blade rotation first causing the airflow to separate into two streams over the sides of the airfoil. Finally, the trailing edge is where the airflow over the two sides of the airfoil is rejoined.

In Figure 3.9, a blade cross section near the tip of the blade is shown with the resultant forces created by the passing wind, the lift force, F_{Lift}, and the drag

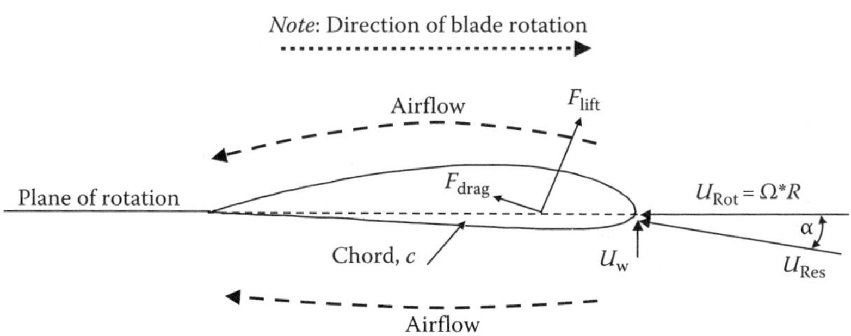

FIGURE 3.9
Lift and drag forces on a blade cross section near the tip.

force, F_{Drag}. The straight line connecting the front (leading) and rear (trailing) edges of the airfoil is known as the chord line and the distance between the two points is designated as the chord, *c*. Also shown are the resultant velocity component, U_{Res}, between the wind speed, U_w, and the wind arising from the blade's rotational speed, U_{Rot}. Finally, the angle between the resultant wind speed, U_{Res}, and the chord line is α, the angle of attack.

Via Bernoulli's law, Equation 3.16, the velocity is highest across the curved (upper) surface of any airfoil (e.g., airplane wing). In addition, Bernoulli's law tells us the pressure is lowest where the velocity is highest. In the case of a wind turbine blade, the pressure is lower over the rear (i.e., upper) surface of the airfoil compared to the front (i.e., lower) airfoil surface.

$$p + \frac{1}{2}\rho U^2 = \text{const.} \tag{3.16}$$

In summary, the lowest pressure is near the leading edge ("nose") of the airfoil, where the wind velocity is highest, and over the rear (or upper) curved surface. This results in a net force, F_{lift}, on the airfoil, causing it to be drawn forward, in the direction of the "nose" of the airfoil and hence rotating. By convention, the resultant forces on the airfoil are drawn perpendicular, F_{Lift}, and parallel, F_{Drag}, to the resultant velocity, U_{Res}. The higher the lift force (and conversely the lower the drag force) along a wind turbine blade, the greater the rotational force, which means the wind turbine is capturing and converting more of the wind's energy and converting it to electricity. In other words, the higher the lift-to-drag ratio (*L/D*) is, the more efficient the blade will be at capturing the wind's energy.

Note: It is now possible to better understand the performance of the different wind turbine configurations shown in Figure 3.7. It turns out that a Savonius VAWT has a poor *L/D* ratio and thus a very low maximum C_p, as shown in Figure 3.7. The American farm windmill and Dutch multiblade turbines have higher *L/D* ratios and hence higher maximum C_p values. Lastly, machines with high *L/D* ratios are the Darrieus VAWT rotor and the modern high-speed two- or three-bladed turbines, with the modern two- and three-bladed turbines having the highest C_p, as noted earlier.

Wind turbine blades have very complicated shapes that vary along the blade. The shape variation is needed to not only help start the wind turbine rotating but also to operate very efficiently as the wind speed increases. The airfoil is twisted near the blade root to provide the required torque to start the turbine rotating at low wind speeds. Once rotating, the optimized blade profiles in the outer portions of the blade provide the necessary aerodynamic forces to keep the blade rotating at design speeds.

Recall that Figure 3.9 illustrates a cross section of a blade in the outer portion of the blade. Figure 3.10 illustrates the lift and drag forces on a blade cross

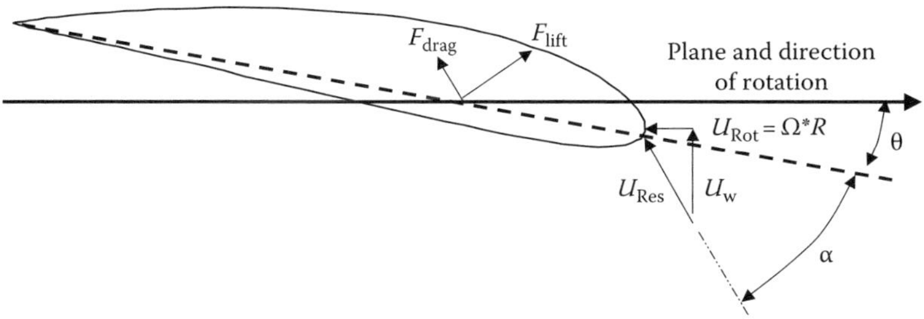

FIGURE 3.10
Lift and drag forces on a blade cross section near the blade root.

section near the root of the blade. Note the blade is twisted, the chord line now forms an angle θ with the plane of rotation of the blade. This twist angle, θ, is a function of turbine blade radius and has a larger value near the blade root and a decreasing value with increasing blade radius. Since the resultant velocity, U_{Res}, has decreased, this increases the angle between the chord line and U_{Res}, and results in an increased angle of attack, α, approaching the blade root.

While the root portion of the blade encounters the same incoming wind speed, U_w, the rotational speed, $U_{Rot} = \Omega R_{Root}$, is lower. This occurs because, while the angular rate of rotation, Ω, is the same along the blade, the radius at the blade root, R_{Root}, is of course smaller. The resulting lift force, F_{Lift}, per unit area is lower at the blade root compared to the blade tip. However, the direction of the force is better aligned to apply a torque to begin rotating the blades. In addition, the surface area of the blade is considerably larger at the root. This provides sufficient area for the lower magnitude aerodynamic forces to act over, resulting in sufficient torque to start rotation at low wind speeds. Finally, the larger surface area is needed to accommodate the necessary structural strength to handle the cantilevered loads arising from wind forces on the blade, weight of the blade, etc.

Figure 3.11 shows the vector diagrams from Figures 3.9 and 3.10 side by side. Assuming a constant wind speed at both the blade root and tip cross sections, we will consider the case where the root cross section is located at a radial location 1/5 of the blade tip cross section. This last assumption means the rotation speed at the blade root, $U_{Rot\text{-}Root}$, is 1/5 the speed at the blade tip, $U_{Rot\text{-}Tip}$. Note that $U_{Rot\text{-}Root}$ and $U_{Rot\text{-}Tip}$ are drawn to scale. The resultant velocity vector at the tip is much larger due to the contribution of the rotational wind speed. This results in a larger lift force per unit area at the tip. At the blade root, the lift force is better aligned to provide a start-up torque. While the blade root lift force is lower per unit area, it is applied over a large surface area. Interestingly, at the tip, the contribution of the wind speed to the resultant velocity is small compared to the contribution from the rotational wind speed when the turbine is at operational conditions.

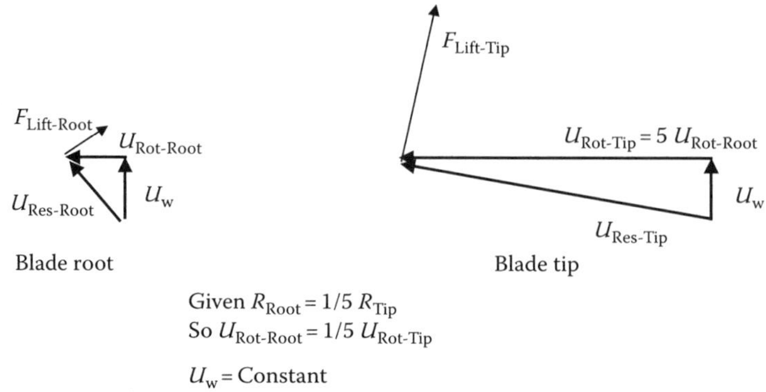

FIGURE 3.11
Comparison of velocity vector diagrams for blade root and tip cross sections.

3.7 Wind Characteristics

3.7.1 Wind Generation

Now we will turn our attention to the atmospheric wind: generation, distribution, characteristics, etc. The motion of air, otherwise known as wind, is caused by uneven heating of the Earth by solar radiation and the rotation of the Earth. The predominant wind direction varies with latitude and altitude above the Earth's surface, as shown in Figure 3.12 [25–27]. Energy is transferred from the equator to the poles via wind. As the wind moves, the Earth's rotation influences the wind direction in two ways. First, Coriolis forces accelerate a particle of air to the right in the Northern Hemisphere (and to the left in the Southern Hemisphere). Secondly, each air particle has angular momentum from west to the east due to the Earth's rotational direction. As the air particle moves toward the pole, conservation of angular momentum requires its west to east velocity component to increase. This effect is pronounced in the midlatitudes, in the region called the westerlies, near the ground [25]. In general, the westerlies are the best-suited region for wind energy applications. In addition, the polar jet stream (7–12 km or 20,000–40,000 ft above sea level) can have a profound impact on the westerlies, moving them north or south, resulting in lower or higher energy production from year to year.

Of course, the most critical part of the atmosphere for a wind turbine is relatively close to the ground. The atmospheric boundary layer (ABL) is the lowest part of the atmosphere and varies from the ground up to 500–2000 m in height depending on time of day, temperature, and pressure. In the ABL, the mean wind speed varies from zero at or near the ground up to the free stream velocity of the atmosphere at the top of the ABL. In addition, the velocity field exhibits turbulent fluctuations. In other words, each of the

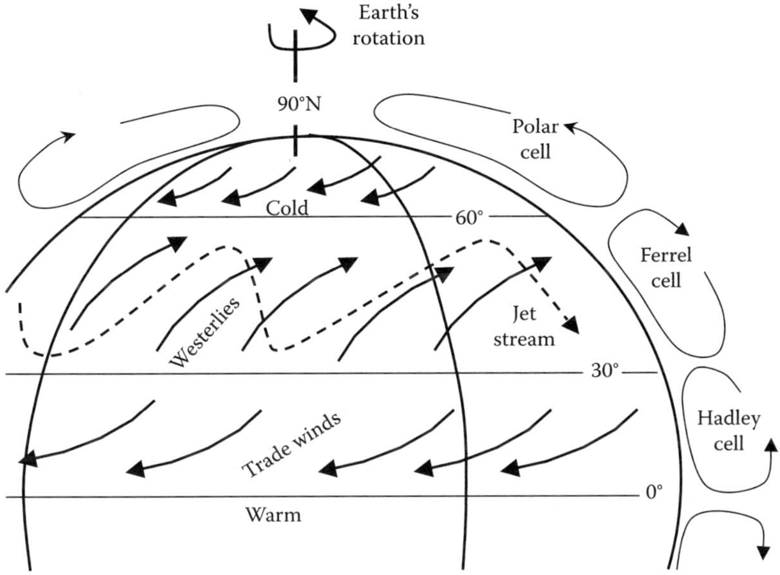

FIGURE 3.12
Wind flow in the Northern Hemisphere.

three velocity components for an air particle has a mean as well as fluctuating component. The turbulent velocity component can be quite large and by nature is time varying. This results in unanticipated fluctuating forces (high velocity causes high blade loading), which can lead to fatigue failures of wind turbine components such as blades, gearboxes, and tower. The turbulent portion of the flowfield is typically described statistically. Methods of accurately simulating turbulent flowfields for a wind turbine design are an active field of research [28].

3.7.2 Distribution of Wind

Now we transition from a global view of wind to a local area view of the wind resource. The wind resource is the mean wind speed as well as how many hours per year the wind blows at a given speed. A wind farm developer needs high confidence that the wind speeds at a given location are "high" and consistent from year to year. How much of a concern is this? If the average wind speed is only 5% lower than predicted, the maximum possible power will be 14% lower than anticipated! High confidence in the wind resource allows the developer to estimate the probable energy output and compare that to the investment needed to layout a wind site and purchase or finance, install, and operate the wind farm (consisting of tens of turbines).

Wind resource maps have been developed for the entire world. For example, the United States wind resource map is shown in Figure 3.13 [29]. Smaller

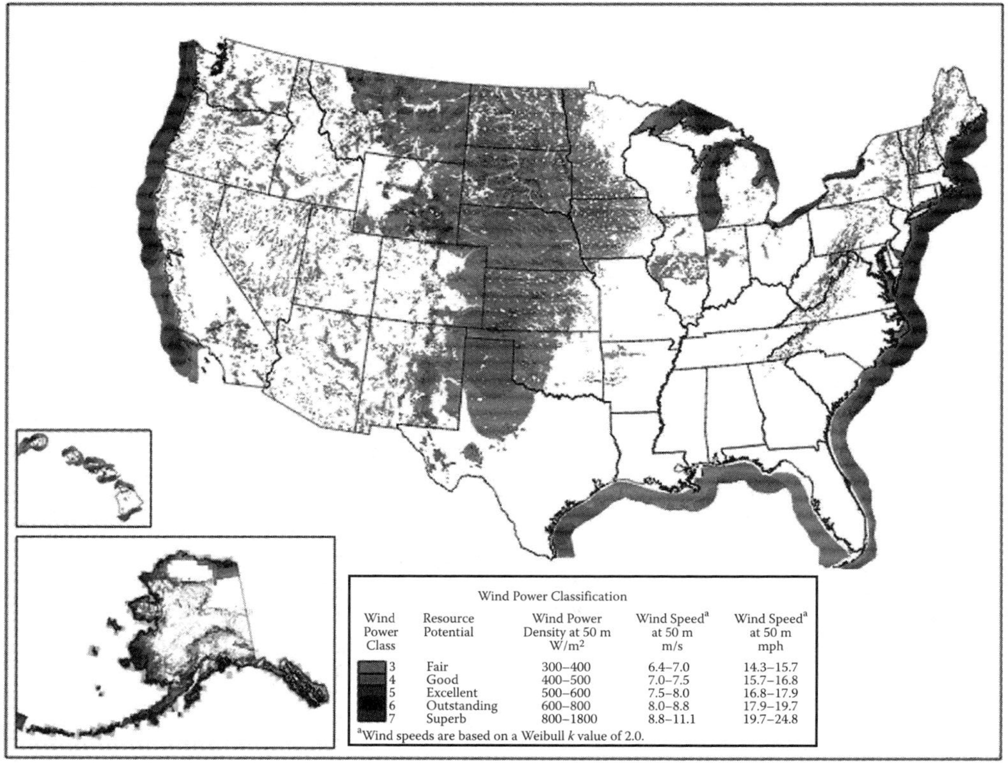

FIGURE 3.13
Wind resource map of the United States. This map shows the annual average wind power estimates at a height of 50 ms, It is a combination of high resolution and low resolution datasets produced by NREL and other organization. The data was screemed to eliminate areas unlikely to be developed onshore due to land use or environmental issues. In many states, the wind resource on this map is visually enhanced to better show the distribution on ridge crests and other features. (Courtesy of U.S. Department of Energy, National Renewable Energy Laboratory, Washington, DC.)

TABLE 3.2

Rating Wind Energy Resource Potential

Wind Power Class	Resource Potential	Wind Power Density at 50 m (W/m²)	Mean Wind Speed[a] at 50 m (m/s)	Mean Wind Speed[a] at 50 m (mph)
1	Poor	0–200	0.0–5.9	0.0–3.2
2	Marginal	200–300	5.9–6.7	13.2–15.0
3	Fair	300–400	6.7–7.4	15.0–16.6
4	Good	400–500	7.4–7.9	16.6–17.7
5	Excellent	500–600	7.9–8.4	17.7–18.8
6	Outstanding	600–800	8.4–9.3	18.8–20.8
7	Superb	>800	>9.3	>20.8

Note: Enlarged from Figure 3.13.

[a] Wind speeds are based on a Weibull k value of 2.0 (Rayleigh).

scale maps are also available, including each state in the United States [29]. Similar maps are available for all parts of the world.

Figure 3.13 color codes the estimated annual average wind power 50 m above the ground, as shown in Table 3.2. For an enlarged view of the relationship between wind power class (denoted by color code in Table 3.2) and mean wind speed, (in m/s), consult Table 3.2. Several items stand out in the U.S. map. First, the highest class wind power (or highest mean wind speeds) regions are offshore along the East and West coasts, and are Classes 6 and 7 (which represents Outstanding and Superb wind resource, respectively). Second, the middle of the country has a wind power classification of Fair to Good running from the Canadian border into Northern Texas.

A small but significant region of Outstanding/Superb wind power can be seen in southeastern Wyoming. Another valuable "offshore" resource exists in the Great Lakes, which have a power classification of Good to Outstanding. Similar regions of high annual wind power can be found in many areas of the country but are not visible at this resolution. This highlights a key point, not surprisingly, that the wind resource is highly dependent on local conditions such as vegetation (forests), terrain (e.g., hills, mountains), and water (ponds, rivers, lakes, oceans). Given sufficient resolution, one can search for and decide on a site to locate a wind turbine. In order to predict the annual power output of a turbine we must understand the variation of wind speed (and thus power) with height.

3.7.3 Wind Speed Increasing with Height

The wind resource maps in Figure 3.13 show, on a large scale, where winds suitable for commercial wind turbines exist. The next step is to develop a wind speed profile as a function of height. As mentioned earlier, wind speed increases with height and thus power increases with height.

Using local data from airports and weather stations, developers gather the data needed to get site specific wind speed data. The data is then processed to arrive at an equation expressing the wind speed as a function of height. Two common approaches are the log law and the power law [30].

3.7.4 Log Law Wind Speed Profile

The log law wind speed profile has a physical basis. It is assumed that the shear stress in the wind is constant with height, the pressure gradient near the Earth's surface is small, and the Earth's rotation is ignored. The height of this constant shear stress layer is 100–200 m and is known as the "surface layer" of the ABL (see, for example, [31]). For a smooth surface, the mixing length, l, of the turbulent wind is given as

$$l = \kappa z \tag{3.17}$$

where

κ is von Karman's constant ($\kappa = 0.4$)
z is the height above the ground

The velocity profile as a function of height can be found to be

$$U(z) = \frac{U^*}{\kappa} \ln\left(\frac{z}{z_0}\right) \tag{3.18}$$

where

$U(z)$ is the wind speed at a height, z
U^* is the friction velocity arising from shear stress
z_0 is the surface roughness length

Values for z_0 are shown in Table 3.3.

TABLE 3.3

Surface Roughness for Various Outdoor Surfaces

Terrain Type	Roughness Length, z_0(m)
Cities, forests	0.7
Suburbs, wooded countryside	0.3
Countryside with trees and hedges	0.1
Open farmland	0.03
Flat grassy plains	0.01
Flat desert, rough sea	0.001

Sources: Burton, T. et al., *Wind Energy Handbook*, Wiley, Chichester, U.K., 2001; Manwell, J.F. et al., *Wind Energy Explained, Theory, Design and Application*, Wiley Chichester, U.K., 2002.

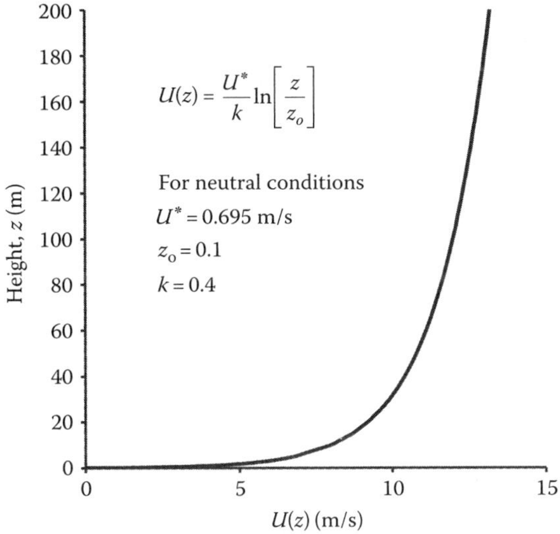

FIGURE 3.14
Log law velocity distribution.

There are two unknowns in this equation, U^* and z_0. Substituting typical values for U^* and z_0, the shape of the resulting atmospheric velocity profile is illustrated in Figure 3.14 for a neutrally stable atmosphere. One sees a predicted gradual change in wind speed through the first 20 m or so of the ABL and then an exponential growth approaching a steady-state value above 200 m.

There are three stability states of the atmosphere: neutral, stable, and unstable [32]. Neutral stability occurs when there is no change in temperature with height through the ABL due to adiabatic cooling. This state is typically accompanied by high winds. A stable atmosphere occurs when the air temperature on the ground is cold and increases with increasing altitude. This results in a steep increase in the wind speed with height (i.e., large wind shear). An unstable atmosphere occurs when the air temperature near the ground is warmer than the temperature at higher altitudes. The air on the ground rises while the colder air above sinks creating turbulent wind conditions.

If one wants to determine an unknown wind speed at a given height, it can be estimated from a known wind speed at another height using this equation. This approach is especially valuable because wind speed data is typically known at heights of 50 m or less (reference height); however, turbines operate at much higher heights where wind speed data is unknown. Using this method, wind speeds can be estimated at hub height from data available at lower heights.

So, taking Equation 3.18 at a height z, for the wind speed $U(z)$ and then dividing it by the same equation for a reference height z_r at the reference wind speed $U(z_r)$, yields

$$\frac{U(z)}{U(z_r)} = \frac{\ln(z/z_0)}{\ln(z_r/z_0)} \tag{3.19}$$

This equation has only one unknown, z_0. Consider a meteorological tower taking measurements at 40 m (z_r = 40 m), measuring a wind speed of $U(z_r)$ = 7.2 m/s at z_r, located in a wooded countryside, z_0 = 0.3, the wind speed, $U(z)$, at a hub height of z = 95 m is

$$U(z) = 7.2 \cdot \frac{\ln(95/0.3)}{\ln(40/0.3)} = 9.7 \text{ m/s} \tag{3.20}$$

In other words, the wind speed is predicted to increase by approximately 35% from 72 to 95 m if the atmospheric assumptions in the model hold. In this case, the wind power would increase by $(9.7/7.2)^3$ = 244%. Height does indeed matter when it comes to wind turbines.

3.7.5 Power Law Wind Speed Profile

Another model commonly used in engineering wind studies is the power law profile. It is used to estimate wind speeds given a reference height, z_r, to some desired height, z. The power law profile is used when one does not know the surface roughness, z_0, and/or stability information (i.e., earlier defined friction velocity, U^*) is unknown. The power law profile is given in Equation 3.21 as

$$U(z) = U(z_r) \cdot \left[\frac{z}{z_r}\right]^{\alpha} \tag{3.21}$$

where α is the power law or shear exponent.

Typical values for α are

$$\text{For neutral flows:} \quad \alpha \cong \frac{1}{7} = 0.143$$

$$\text{For highly sheared flows:} \quad \alpha > 1$$

A comparison between the log law and power law wind speed profiles is shown in Figure 3.15. This figure shows that the wind speed profile of the power law profile for a value of α = 0.17 is similar in shape to the log wind profile plotted earlier for neutral atmospheric conditions. In any case, it should be noted that the value of α is highly dependent on temperature, weather, geographic location (elevation and terrain), time (of day and season). The power law profile is typically used in engineering studies.

Consider Table 3.4 that shows the calculated wind speeds for two different values of α at a height of 100 m, which is typical hub height for modern commercial wind turbines. The change in predicted power based on these

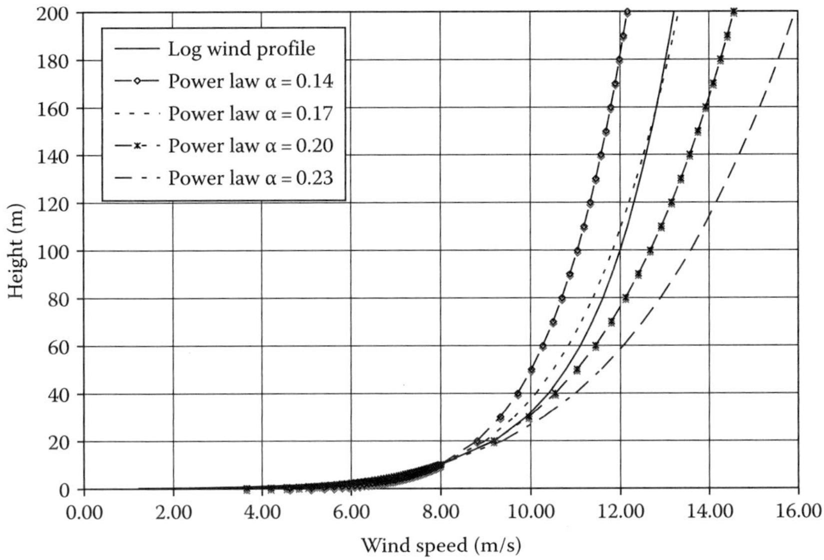

FIGURE 3.15
Log law and power law velocity profile comparison.

TABLE 3.4

Variation in Wind Speed with α at a Height of 100 m

α	Wind Speed $U(z)$ (m/s)
0.14	11.0
0.23	13.6

wind speeds is the ratio $(13.6/11)^3 = 1.89$. This indicates the predicted power can nearly double at 100 m if the value of α increases from 0.14 to 0.23. This has large economic implications. In other words, determining the correct value of α for the power law profile (or using an entirely different method to determine wind speed as a function of height) for a given location and time period is critical when making choices about where to build wind farms. The best option is to gather long-term meteorological data at hub height in a sufficient number of locations at a potential wind turbine site (if the cost can be afforded).

Consequently, methods have been developed, and continue to be developed, to choose the best value of α given a set of wind data and environmental wind factors [30].

3.7.6 Probability of Observing a Given Wind Speed

As has been discussed, the wind speed (and hence power available) varies. The next step is to determine the probability of a given wind speed occurring over a desired time period using a probability distribution function (PDF). A common PDF is the normal or Gaussian PDF, which can be seen in Figure 3.16. Many natural events can be described by a normal PDF, for example, the height distribution of 20 year old students in the United States. When

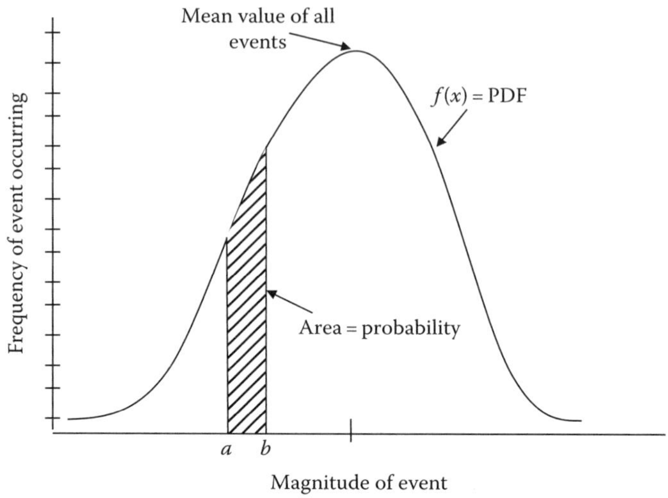

FIGURE 3.16
Illustration of a Gaussian or normal probability density function.

graphed, this Gaussian PDF yields the commonly known "bell-shaped" curve. For all PDF's, the vertical axis is the observed frequency of a given event within a certain range or magnitude, for example, number of students between the heights of 5′0″ and 5′1″. The horizontal axis contains the range of all events, e.g., 4′0″ to 6′10″. The peak of any PDF curve (where the curve is denoted by $f(x)$) is the mean magnitude of all occurring events.

The area under any section of a PDF curve corresponds to the probability, P, of a range of events occurring. As shown in Figure 3.16, the area represents the probability of having an event occur with a magnitude between a and b. The probability, P, of a range events occurring is obtained by integrating the PDF, $f(x)$, between the range of the events, a and b.

$$P = \int_a^b f(x)\,dx \tag{3.22}$$

It should also be noted that integrating the PDF between the limits of 0 to ∞ yields a value of 1. In other words, the probability of having an event occur of any magnitude between the limits of all events is 100%.

$$P = \int_0^\infty f(x)\,dx = 1 \tag{3.23a}$$

Two commonly used PDFs, Weibull and Rayleigh, are used to describe the probability of observing a given wind speed. These PDFs are typically determined at a desired height, for example, hub height. These distributions are, of course, only defined for values greater than zero. Integrating the PDF will provide the probability, $P(U)$ of wind speed, U, occurring between two wind speeds U_a and U_b.

$$P = \int_{U_a}^{U_b} f(U)dU \tag{3.23b}$$

Several other important statistical relationships can be found given a wind speed PDF, $f(U)$, including

$$\text{Mean wind speed:} \quad \bar{U} = \int_0^{\infty} Uf(U)dU \tag{3.24}$$

$$\text{Standard deviation:} \quad \sigma = \sqrt{\int_0^{\infty} (U - \bar{U})^2 f(U)dU} \tag{3.25}$$

Lastly, one is likely to want to know the probability (or the percent of time) that the wind will be equal to or less than a given wind speed. For example, what percent of a year will the wind speed be too low for the turbine to operate? This is known as the cumulative distribution function, $C(U)$, and is found by integrating from 0 m/s to a desired wind speed, U m/s:

$$C(U) = \int_0^{U} f(U)dU \tag{3.26}$$

The Weibull PDF and cumulative distribution function are given by

$$f(U) = \left(\frac{k}{c}\right)\left(\frac{U}{c}\right)^{k-1} \exp\left[-\left(\frac{U}{c}\right)^k\right] \tag{3.27}$$

$$C(U) = 1 - \exp\left[-\left(\frac{U}{c}\right)^k\right] \tag{3.28}$$

where

k is the shape parameter

c is the scale parameter

Both k and c are functions of the mean wind speed, $\bar{U}$, and the standard deviation of the wind speed, σ. The Rayleigh PDF and cumulative distribution function are given by

$$f(U) = \frac{\pi}{2}\left(\frac{U}{\bar{U}^2}\right)\exp\left[-\frac{\pi}{4}\left(\frac{U}{\bar{U}^2}\right)^2\right] \tag{3.29}$$

$$C(U) = 1 - \exp\left[-\frac{\pi}{4}\left(\frac{U}{\bar{U}}\right)^2\right] \tag{3.30}$$

The Rayleigh PDF only depends on one parameter, the mean wind speed, $\bar{U}$, making it much easier to use (but not necessarily more accurate!). Note that the Rayleigh distribution is a special case of the Weibull distribution when $k = 2$ and $c = \bar{U}$. An example of the Rayleigh distribution for different annual mean wind speeds is shown in Figure 3.17. Figure 3.17 shows that as the mean speed increases there is a larger probability of having higher wind speeds at a given site and hence the annual power production has the potential to be higher.

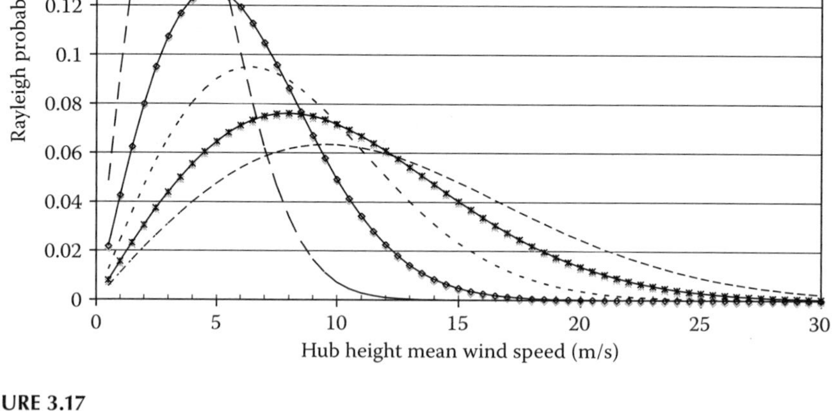

FIGURE 3.17
Rayleigh wind distributions.

Example 3.1: Rayleigh Wind Speed Distribution Calculations

An analysis of wind speed data (10 min interval average, taken over a 1 year period) has yielded an average speed of $\bar{U} = 6\,\text{m/s}$ for a potential wind turbine site. It has been determined that a Rayleigh wind speed distribution gives a good fit to the wind data. Recall the Rayleigh PDF is given by Equation 3.29.

A. Estimate the number of hours per year that the wind speed will be between $u_B = 10.5$ and $u_A = 9.5\,\text{m/s}$ during the year.

   To do this, we will find the probability the wind speed will be in a given range during the year, $p(u_B \le u \le u_A)$, and multiply the probability by the number of hours in a year.

$$P(U_B \le U \le U_A) = \int_{U_A}^{U_B} f(U)dU = \int_0^{U_B} f(U)du - \int_0^{U_A} f(U)dU = C(U_B) - C(U_A)$$

For a Rayleigh distribution: $C(U) = 1 - \exp\left[-\frac{\pi}{4}\left(\frac{U}{\bar{U}}\right)^2\right]$

So $C(U_B) - C(U_A) = \exp\left[-\frac{\pi}{4}\left(\frac{U_A}{\bar{U}}\right)^2\right] - \exp\left[-\frac{\pi}{4}\left(\frac{U_B}{\bar{U}}\right)^2\right] = p(U_B \le U \le U_A)$

Substituting $\bar{U} = 6\,\text{m/s}$, $U_B = 10.5\,\text{m/s}$, $U_A = 9.5\,\text{m/s}$ yields

$$P(U_B \le U \le U_A) = 0.049$$

In other words, the wind speed will be between U_B and U_A m/s 4.9% of the year.

Thus the total number of hours the wind speed is between u_B and u_A in a year is

$$\frac{\#\text{h}}{\text{year}} = 0.0494 \cdot \frac{24\,\text{h}}{\text{day}} \cdot \frac{365\,\text{day}}{\text{year}} = 432\,\text{h/year}$$

B. Estimate the number of hours per year that the wind speed is above 16 m/s. This problem is the same as part A with the wind speed ranges being

$$U_B = \infty \quad \text{and} \quad U_A = 16\,\text{m/s}.$$

$$p(\infty \le u \le 16) = \int_{16}^{\infty} p(u)du = C(\infty) - C(16)$$

$$C(\infty) - C(16) = 1 - \exp[-\infty] - \left[1 - \exp\left[-\frac{\pi}{4}\left(\frac{u_A}{\bar{U}}\right)^2\right]\right] = \exp\left[-\frac{\pi}{4}\left(\frac{u_A}{\bar{U}}\right)^2\right] = 0.0038$$

Thus, the total number of hours the wind speed is greater than 16 m/s (i.e., between $U_B = \infty$ and $U_A = 16$ m/s) in a year is

$$\frac{\#\text{h}}{\text{year}} = 0.0038 \cdot \frac{24\ \text{h}}{\text{day}} \cdot \frac{365\ \text{day}}{\text{year}} = 33\ \text{h/year}$$

3.8 Turbine Performance

Given a wind turbine site, selected by examining long-term wind speed data, a turbine can be picked and/or designed for the site. In order to do this, detailed engineering work has to be conducted to design the blades, tower, electrical systems, etc. In this section, we will focus on understanding the basics of the blade aerodynamics to design an efficient wind turbine. Aerodynamics is the study of bodies moving through gases. Its use allows the blade and tower shape to be optimized as well as optimizing blade rotational speed. The goal is to maximize the extraction of power from the incoming wind stream and minimize the effect of wind loads (both steady and unsteady).

The efficiency of a wind turbine is the power extracted by a wind turbine from the power contained in the wind flowing through the area of the blades. Figure 3.18 contains a schematic comparing this relationship. Note that the power contained in the wind is always greater than the power extracted by the turbine.

The turbine power curve contains several interesting points. The turbine does not start producing power until the wind speed reaches the cut-in wind speed. This is the lowest wind speed where the rotor's aerodynamic power is

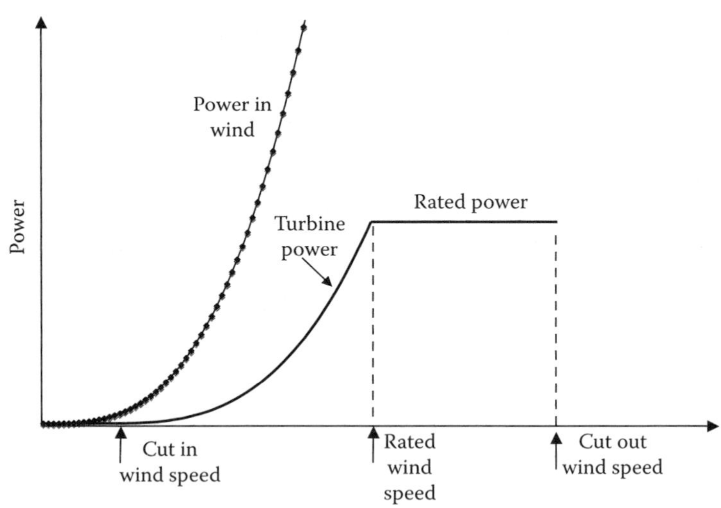

FIGURE 3.18
Wind power and turbine power curves versus wind speed.

greater than the sum of equipment loads and mechanical and electrical inefficiencies. The cut-out wind speed is the speed at which additional energy production are less than the cost of structural strength to withstand incremental fatigue damage. Lastly, the rated wind speed is the speed at which the turbine is operating near its peak efficiency. The goal is to operate the wind turbine near or at its maximum efficiency.

Earlier, we saw that the standard measure of a wind turbine's efficiency is known as the power coefficient, C_p, and is defined as

$$C_p = \frac{P_{aero}}{P_{wind}} = \frac{Q\Omega}{\frac{1}{2}\rho A U^3} \tag{3.31}$$

where
Q is the aerodynamic torque
Ω is the rotor rotational speed

A homework problem will show that the maximum theoretical limit of C_p is given by the Betz limit, $C_{p,max} = C_{p,Betz} = 16/27 \approx 0.5926$. In other words, the maximum power a wind turbine should be able to extract from the wind is approximately 59% of the power crossing the area swept by the wind turbine blades.

Another heavily used design metric is the TSR which is defined as

$$\text{TSR} = \lambda = \chi = \frac{\text{Tip speed}}{U_{wind}} = \frac{\Omega R}{U_{wind}} \tag{3.32}$$

where
R is the rotor radius
Ω is the rotor rotational speed
U_{wind} is the incoming wind speed

Design guidelines currently suggest limiting the tip speed to 90 m/s to minimize noise issues.

Example 3.2: Tip Speed Ratio Calculation

What is the TSR for a 90 m diameter turbine rotating at 15 rpm at a wind speed of 10 m/s?

$$R = \frac{D}{2} = 45\text{ m} \quad \text{and} \quad U_{wind} = 10\text{ m/s}$$

$$\Omega = 15\text{ rpm} = 15\frac{\text{rev}}{\text{min}} \cdot 2\pi\frac{\text{rad}}{\text{rev}} \cdot \frac{\text{min}}{60\text{ s}} = 1.57\text{ rad/s}$$

$$\text{TSR} = \frac{\Omega R}{U_{wind}} = \frac{1.57\text{ rad/s} \cdot 45\text{ m}}{10\text{ m/s}} = 7.07$$

Actuator disk theory allows one to find simple relationships for power and thrust. Another more comprehensive theory beyond the scope of this chapter allows one to design the blade shape to optimize power and thrust. This theory, called blade element momentum theory (BEM), is covered in detail in many of the comprehensive wind energy textbooks in the references [25,32].

The first assumption in actuator disk theory is that the area swept by the rotor blades is a permeable disk, shown in Figure 3.19, with an area swept out by the turbine blades. As the wind flows through the disk, momentum (i.e., power and thrust) is extracted.

Figure 3.20 illustrates a top view of an actuator disk (the outline of a tower and nacelle are omitted). The wind flows from the left to the right. As the wind passes through the actuator disk (area swept by the turbine blades), momentum is extracted and the wind speed is reduced, hence

$$U_{\text{wind}} > U_{\text{disk}} > U_{\text{wake}}$$

From conservation of mass, one can see that as the velocity decreases the affected area of the column of wind increases (i.e., the mass flowrate remains constant).

Note the area of the wind column increases after passing through the actuator disk.

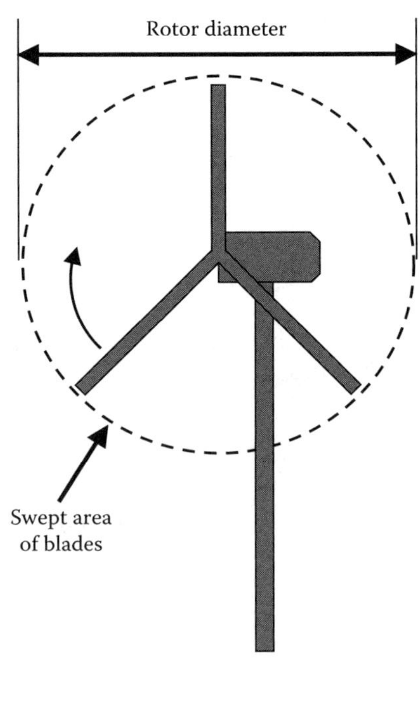

FIGURE 3.19
Actuator disk.

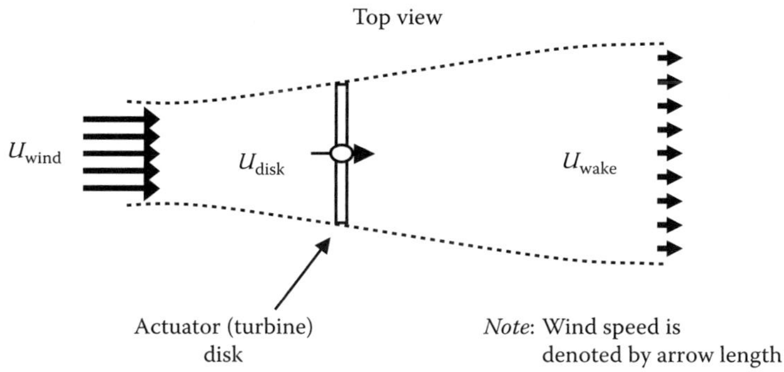

FIGURE 3.20
Actuator disk (top view), showing incoming wind passing through actuator disk.

Additional assumptions are used to develop this theory. Key ones include

- Homogeneous, incompressible, inviscid, and steady-state flow
- Static pressure far upstream and downstream are equivalent
- Velocity on either side of the actuator disk is the same
- Infinite number of blades that are infinitesimally thin
- Uniform thrust over actuator disk
- Nonrotating wake
- Velocity across the disk is constant

$$\text{Bernoulli's theory is valid: } P + (1/2)\rho U^2 = \text{const.} \tag{3.33}$$

More complicated theories such as BEM ease some of these restrictions.

Next, a momentum balance will be made on the actuator disk (see Figure 3.21) to find the power coefficient, C_p.

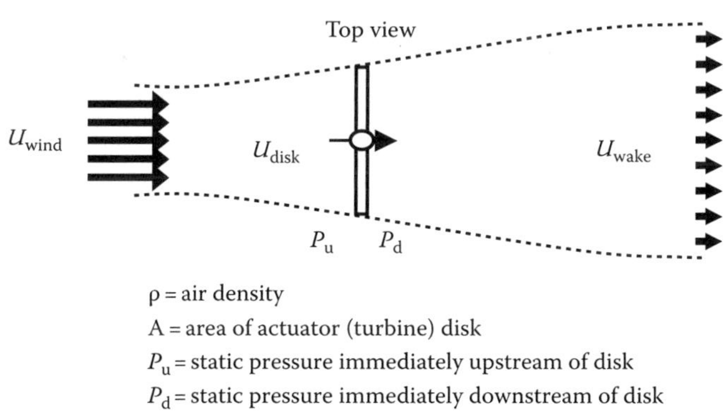

FIGURE 3.21
Actuator disk (top view) with quantities needed for momentum balance.

Defining x as the wind flow direction and using momentum theory:

$$\sum F_x = (\rho A U_{disk})(U_{wind} - U_{wake}) \tag{3.34}$$

$$A(P_{up} - P_{down}) = (\rho A U_{disk})(U_{wind} - U_{wake}) \tag{3.35}$$

Defining the axial induction factor, a, to be

$$a = \frac{U_{wind} - U_{disk}}{U_{wind}} \tag{3.36}$$

or

$$U_{disk} = (1-a)U_{wind} \tag{3.37}$$

In other words, the actuator disk induces a reduction in the wind speed at the disk, U_{disk}, compared to the free stream wind speed, U_{wind}.

From the assumptions, one can find

$$U_{wake} = (1-2a)U_{wind} \tag{3.38}$$

To find the power coefficient, C_p, we first find the aerodynamic power P_{aero}:

$$\begin{aligned} P_{aero} &= \sum F_{disk} U_{disk} \\ &= \left(2\rho A U_{wind}^2 a(1-a)\right) U_{disk} \\ &= \left(2\rho A U_{wind}^2 a(1-a)\right) U_{wind}(1-a) \\ P_{aero} &= 2\rho A U_{wind}^3 a(1-a)^2 \end{aligned} \tag{3.39}$$

and substituting P_{aero} into the relationship for C_p (see Equation 3.13)

$$C_p = \frac{P_{aero}}{P_{wind}} = \frac{2\rho A V_{wind}^3 a(1-a)^2}{\frac{1}{2}\rho A U_{wind}^3} \tag{3.40}$$

Simplifying yields

$$C_p = \frac{P_{aero}}{P_{wind}} = 4a(1-a)^2 \tag{3.41}$$

It's left as problem to find the Betz limit of $C_{pmax} = 16/27 = 0.59$ from this relationship.

In a similar fashion, a thrust coefficient, C_T, can be found which characterizes the axial thrust on the actuator disk. Summing forces on the disk yields the thrust, *T*:

$$T = \sum F_{disk} \tag{3.42}$$

$$T = 2\rho A U_{wind}^2 a(1-a) \tag{3.43}$$

Nondimensionalizing the thrust by P_{wind} yields the thrust coefficient

$$C_T = \frac{2\rho A U_{wind}^2 a(1-a)}{\frac{1}{2}\rho A U_{wind}^2} \tag{3.44}$$

and simplifying yields

$$C_T = 4a(1-a) \tag{3.45}$$

The power and thrust coefficients are plotted in Figure 3.22 as function of the axial induction factor, *a*. Note that the maximum power and thrust occur at different axial induction factors. The maximum power output of $C_{pmax} = C_{pBetz} = 0.59$ occurs at $a = 1/3$, while the maximum nondimensional thrust of 1.0 occurs at a value of $a = 0.5$. When power is maximum at $a = 1/3$, $C_T = 8/9$.

Also note that the axial induction factor theory is not valid for $a > 0.5$. Equation 3.38 shows that the speed of the wake, U_{wake}, behind the turbine decreases as the axial induction factor increases from 0.0 to 0.5. At $a = 0.5$, the

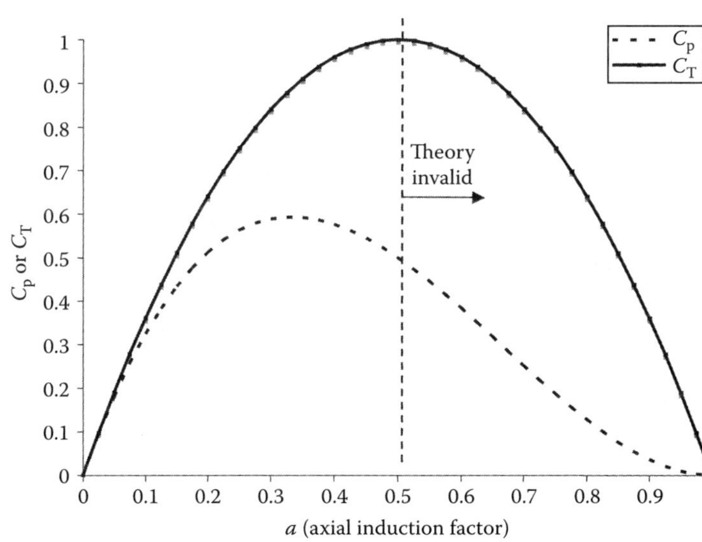

FIGURE 3.22
Power and thrust coefficients versus induction factor showing maximum power coefficient, $C_{p,max}$.

FIGURE 3.23
(See color insert.) Flow visualization tests conducted in the NASA Ames wind tunnel using smoke emitted from the tips of the turbine helped researches determine the extent of the wake under a controlled set of conditions. (Courtesy of DOE/NREL, Golden, CO.)

wake speed equals zero. For values of $a > 0.5$, the wake speed would be negative (Figure 3.23), violating the assumptions of this simple theory. Figure 3.23 illustrates the wake tip vortex produced by a rotating wind turbine located in a wind tunnel.

Actuator disk theory provides a theoretical limit for the maximum power coefficient of a wind turbine, the Betz limit. In practice, C_p is actually a function of many variables including wind speed, pitch angle, and TSR. Values of C_p are obtained through experiment or numerical modeling.

The aerodynamic power of the wind is determined by rearranging

$$C_p = \frac{P_{aero}}{P_{wind}} \tag{3.46}$$

to yield

$$P_{aero} = C_p P_{wind} = \frac{1}{2} C_p \rho A V_{wind}^3 \tag{3.47}$$

Given C_p and the expected probability of wind distribution, the annual energy production can be estimated. This in turn will allow the cost of energy (COE) to be calculated, as shown in the Section 3.9.

Example 3.3: Determine Annual Energy Production

Given a wind turbine with the following parameters, calculate the annual energy production: $C_p = C_{pmax} = 0.48$ (C_p value assumed maximum and constant for all wind speeds)

Rated power = 5000 kW = 5.0 MW
Annual average wind speed = 8.0 m/s
Rated wind speed = 13.0 m/s
Cut-out wind speed = 25 m/s
Rotor radius = 48.0 m
Area = 7238.2 m^2
Air density = 1.20 kg/m^3

The annual energy production for all wind speeds will be illustrated by first calculating the annual energy production at a single wind speed. Then, the methodology will be applied to all wind speeds to arrive at the annual energy production for all wind speeds.

First, determine the power production at a single wind speed, say 10 m/s. Then a Rayleigh distribution will be calculated to determine the number of hours the wind speed is at 10 m/s. The power and number of hours are then multiplied to determine the yearly energy production for a 10 m/s wind speed.

Power production: For a wind speed of 10 m/s, the power produced for the values shown above is

$$P_{\text{aero}} = \frac{1}{2}C_p A V_{\text{wind}}^3 = \frac{1}{2}\cdot 0.48 \cdot \left(1.2\,\frac{\text{kg}}{\text{m}^3}\right)\cdot(7238.2\ \text{m}^2)\cdot\left(10.0\,\frac{\text{m}}{\text{s}}\right)^3$$

$$= 2{,}084{,}600\ \text{W} = 2{,}084.6\ \text{kW}$$

A plot of the power produced for all wind speeds is shown in Figure 3.24.

Number of hours at 10 m/s: Assuming a Rayleigh PDF with an average wind speed of 8 m/s, the annual probability of a 10 m/s wind is

$$f(10) = \frac{\pi}{2}\left(\frac{U}{\bar{U}^2}\right)\exp\left[-\frac{\pi}{4}\left(\frac{U}{\bar{U}}\right)^2\right] = \frac{\pi}{2}\left(\frac{10}{8^2}\right)\exp\left[-\frac{\pi}{4}\left(\frac{10}{8}\right)^2\right] = 0.0719 = 7.19\%$$

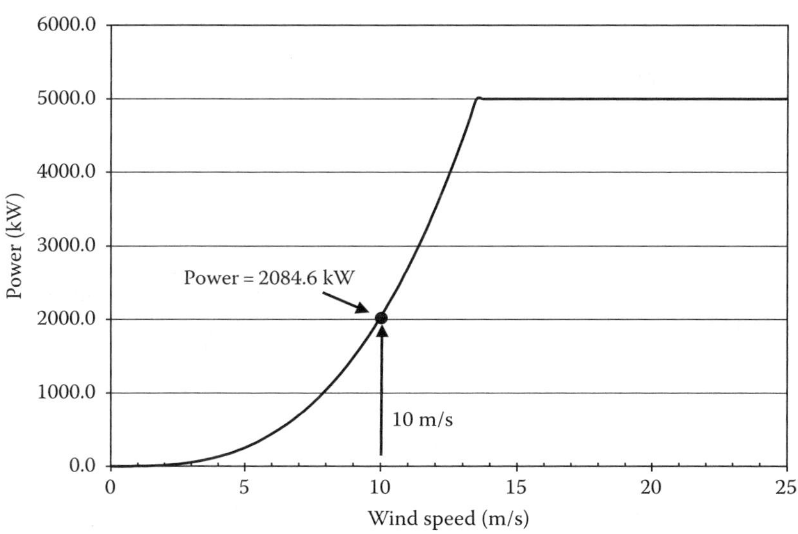

FIGURE 3.24
Power versus wind speed for a 5000 kW turbine with $C_p = C_{pmax} = 0.48$.

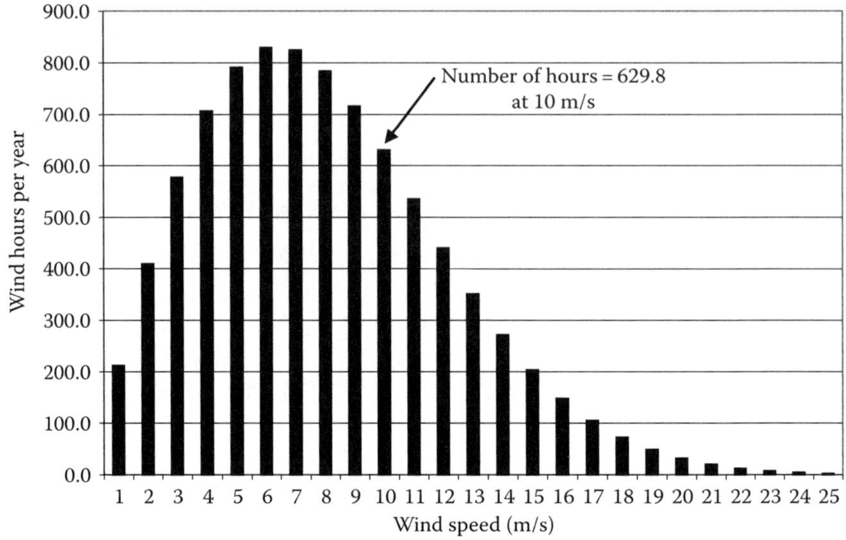

FIGURE 3.25
Number of hours at a given wind speed for a Rayleigh distribution with an annual average wind speed of 8.0 m/s.

Thus, in a given year the number of hours the wind will blow at 10 m/s at this site is the probability times the number of hours in a year:

$$0.0719 \cdot \frac{24\text{ h}}{\text{day}} \cdot \frac{365\text{ day}}{\text{year}} = 629.8\text{ h}$$

A plot of the number of hours the wind blows at a given wind speed for all wind speeds is shown in Figure 3.25.

Finally, the annual energy production for this wind turbine, at this site, at a wind speed of 10 m/s, is found by multiplying the power production times the number of hours the power is produced, in this case:

$$\text{Annual energy production at 10 m/s} = 2084.6\text{ kW} \cdot 629.8\text{ h} = 1312.9\text{ MWh}$$

In a similar fashion, this is done for all wind speeds up to cut-out. The energy production from all wind speeds is then summed to determine the annual energy production estimate. A summary table of the calculations is shown in Table 3.5.

Examination of Table 3.5 shows that the annual energy produced at this site is calculated to be 13,873 MWh. If electricity costs 10 cents/kWh ($0.10/kWh), the annual value of the energy produced in this example would be

$$\text{Annual value} = 13{,}873{,}070\text{ kWh} \cdot \frac{\$0.10}{\text{kWh}} = \$1.4\text{ million}$$

Now the rationale to build wind farms comprised of dozens of wind turbines with a similar energy (cash) output can be seen. The wind is literally producing money!

TABLE 3.5

Power and Annual Energy Calculation

Wind Speed (m/s)	Power (kW)	Probability (%)	# of Hours/Year	Energy (kWh)
1	2.1	2.42	212.4	442.7
2	16.7	4.67	409.4	6,827.6
3	56.3	6.59	577.6	32,507.8
4	133.4	8.07	706.7	94,283.4
5	260.6	9.03	791.0	206,114.7
6	450.3	9.47	829.3	373,429.3
7	715.0	9.42	824.9	589,808.3
8	1067.3	8.95	784.2	837,017.8
9	1519.7	8.17	716.1	1,088,282.6
10	2084.6	7.19	630.2	1,313,739.7
11	2774.6	6.12	535.7	1,486,478.3
12	3602.2	5.03	440.7	1,587,567.2
13	4579.9	4.01	351.3	1,608,943.1
14	5000.0	3.10	271.6	1,358,130.2
15	5000.0	2.33	203.9	1,019,404.1
16	5000.0	1.70	148.7	743,288.9
17	5000.0	1.20	105.4	526,756.5
18	5000.0	0.83	72.6	362,992.5
19	5000.0	0.56	48.7	243,323.3
20	5000.0	0.36	31.7	158,710.6
21	5000.0	0.23	20.2	100,758.5
22	5000.0	0.14	12.5	62,274.8
23	5000.0	0.09	7.5	37,478.8
24	5000.0	0.05	4.4	21,967.4
25	5000.0	0.03	2.5	12,541.7
		Total: 99.76		Total: 1,387,3070.1 kWh

Note that in Table 3.5 the power produced above 13 m/s (rated wind speed) is a constant value of 5000 kW. Control schemes are typically implemented to limit the maximum amount of power produced by a turbine to the rated turbine power (5 MW in this case) at the rated wind speed (13 m/s). Limiting the power production is important from both a structural standpoint as well as an electrical generation standpoint. In both cases, wind turbine components must be sized to operate under maximum defined loads and thus cannot handle larger loads.

Another measure of a wind turbine's performance is the capacity factor. Since the wind never blows continuously, a turbine will never be able to operate continuously at full power. The capacity factor is the ratio of the energy produced to the theoretical maximum the turbine could produce (i.e., rated power production, 24 h/day, 7 day/week for a year):

$$\text{Capacity factor} = \frac{\text{Energy generated}}{\text{Rated power} \times (\#\text{h/year})} = \frac{\text{Energy generated}}{\text{Rated power} \times 8760\ \text{h}} \quad (3.48)$$

For this example, the capacity factor is found to be

$$\text{Capacity factor} = \frac{13{,}880\ \text{MWh}}{5\ \text{MW} \times 8760\ \text{h}} = 31.7\%$$

For comparison, typical wind turbine capacity factors are 20%–40% [33].

Finally, the calculations shown in Table 3.5 would typically be done at wind speed increments of 0.10 or 0.25 m/s. Note that slight differences in values between the spreadsheet table and the example calculations are due to round-off errors in the example calculations.

3.8.1 Control Schemes

As a comparison, we will briefly examine the major control schemes currently used in large commercial wind turbines. In the first case, the rotor speed is held fixed or constant while the blade pitch is variable. In the second case, both the rotor speed and the blade pitch are variable. The fixed rotor speed turbine has a simpler mechanical design, the control scheme is easier to implement, and it also offers advantages by simplifying the electrical power generation. The variable speed and variable pitch control scheme is more complicated, both mechanically and electrically. Is the more complicated variable speed, variable blade pitch turbine worth it from an energy production standpoint?

Figure 3.26 illustrates the power output versus wind speed for both wind turbine control methods. The operating parameters cited in Example 3 were also used for Figure 3.26. In addition, the results in Example 3 are the same shown here for the variable speed/variable pitch turbine.

The key point is that the variable speed/variable pitch (VSVP) turbine has a higher power output. Two regions can be seen where the variable speed/variable pitch turbine power output is higher than the constant speed/variable pitch (CSVP) turbine. This increased power output will lead to a higher annual energy production and capacity factor. This is due to being able to operate the VSVP turbine at $C_p = C_{pmax}$ at all wind conditions. The CSVP runs at a variable C_p. Since the turbine cannot vary rotor speed, it cannot run at $C_p = C_{pmax}$ at all conditions.

Note that both turbines eventually reach a rated power of 5 MW. However, the CSVP reaches the 5 MW turbine rating at a higher wind speed, 15 m/s,

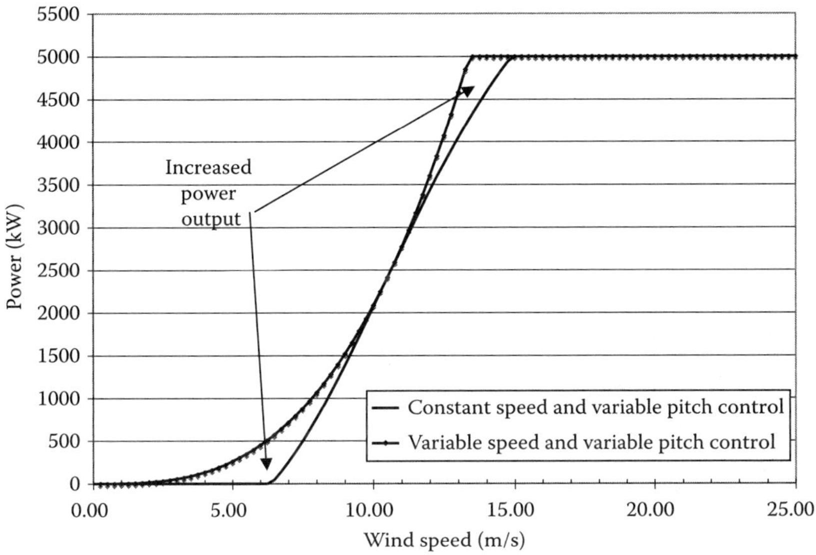

FIGURE 3.26
Power output for constant speed/variable pitch and variable-speed and pitch control schemes.

while the VSVP turbine reaches 5 MW rated power at 13.5 m/s. This is because the CSVP turbine is operating at a lower C_p and thus needs a higher wind speed to reach the higher 5 MW power rating. For the purposes of this illustration, C_p for the CSVP was assumed to be a function of TSR (ΩR/(wind speed)) and a fixed pitch angle of 0°:

$$C_p(\lambda, \beta) = 0.5176\left(\frac{116}{\lambda_i} - 0.4\beta - 5\right)e^{-\frac{21}{\lambda_i}} + 0.0068\lambda \tag{3.49}$$

where

$$\lambda = \text{Tip speed ratio} = \frac{\Omega R}{\text{Wind speed}}$$

$$\frac{1}{\lambda_i} = \frac{1}{\lambda + 0.08\beta} - \frac{0.035}{\beta^3 + 1}$$

β = Blade pitch angle

Examining Table 3.6, we can see the VSVP-controlled turbine produces over 15% more energy (i.e., 15% more yearly revenue), and thus also has a capacity factor 15% higher, than the CSVP-controlled turbine. The goal of wind turbine manufacturers is to design and build turbines, such as the VSVP turbine, that produce more power while minimizing the cost to design and build these turbines, to achieve the highest level of performance possible.

TABLE 3.6

Energy Production and Capacity Factor for VSVT and CSVP Wind Turbines

	CS/VP	VS/VP	Percent Increase from CS/VP to VS/VP (%)
Energy (MWh)	12,000	13,880	15.3
Capacity factor	27.5%	31.7%	15.3

3.9 Cost of Energy

A successful wind turbine design must be strong, reliable, and produce energy at a cost competitive with other technologies. Among others, the United States National Renewable Energy Laboratory's (NREL) National Wind Technology Center (NWTC) has developed a wind energy electricity cost estimating tool. The tool focuses on wind turbine configurations that are currently the most common: three-bladed, upwind, pitch-controlled, variable-speed, and for both land-based and offshore wind turbines [34]. Results are in 2002 dollars and cost data is based on a mature technical design and mature component production for a 50 MW wind farm installation.

The COE of a single wind turbine is found using

$$\mathrm{COE} = \frac{(\mathrm{FCR} \cdot \mathrm{ICC})}{\mathrm{AEP_{net}}} + \mathrm{AOE}$$

where

COE is the levelized cost of energy (in constant \$) (\$/kWh)

FCR is the fixed charge rate (in constant \$. Interest rate in %/year expressed as decimal value/year)

ICC = initial capital cost (\$)
= Turbine cost + balance of station

$\mathrm{AEP_{net}}$ = Net annual energy production (kWh/year)
= Capacity factor · rated turbine power (kW) · 8766 (h/year)

where

$$\text{Capacity factor} = \frac{\text{Energy capture (kWh)}}{\text{Rated turbine power (kW)} \times 8766\ \text{h/year}}$$

Note: The number of hours in a year is typically defined in one of two ways:

1. 24 h/day × 365 day/year = 8760 h/year

or

2. To take in account a leap year every fourth year, 6 h/year is added to account for 1/4 of the leap day: 24 h/day × 365 day/year + 6 h/year = 8766 h/year

The NREL COE tool uses the second definition, 8766 h/year, for no. of hours in a year.

AOE = annual operating expenses ($/kWh)

$$AOE = LLC + \frac{(O\&M + LRC)}{AEP_{net}}$$

where
LLC is the land lease cost ($/kWh)
O&M is the levelized operation and maintenance cost ($/year)
LRC is the levelized replacement/overhaul cost ($/year)

3.9.1 Definitions

Fixed charge rate (FCR) is the interest rate needed to cover the capital cost, a return on debt and equity, and various other fixed charges.

The initial capital cost (ICC) is the sum of the turbine system cost and the balance of station cost: ICC = turbine system cost + balance of station

The components in these systems are
Turbine system (rotor, drive train, nacelle) cost:

Rotor: blades, hub, pitch mechanisms and bearings, spinner, nose cone

Drive train and nacelle: Low-speed shaft, bearings, gearbox, mechanical brake, high-speed coupling, and associated components, generator, variable-speed electronics, yaw drive and bearing, main frame, electrical connections, hydraulic and cooling systems, nacelle cover

Control, safety system, and condition monitoring

Tower

Note: Total wind turbine cost = turbine system + control + tower

The total turbine cost is often expressed as a turbine cost per kW (\$/kW) multiplied by the turbine rating (kW)

For example: Given a typical \$1100/kW turbine cost per kW for a 5 MW turbine

$$\text{total turbine cost} = \$1100 \times 5000 = \$5.5 \text{ million}$$

Balance of station

Foundation/support structure
Transportation
Roads, civil work
Assembly and installation
Electrical interface/connections, cables, transformers
Engineering permits

Offshore turbines have additional costs:

Marinization (added cost to handle marine environments)
Port and staging equipment
Personal access equipment
Scour protection
Surety bond (to cover decommissioning)
Offshore warranty premium

***Note*:** Currently, a typical balance of station cost for a land-based turbine is approximately \$700,000 per turbine.

LLCs, one component of the annual operating expense, are the rental or lease fees charged (e.g., land or ocean bottom lease cost) for the turbine installation.

Operation and maintenance (O&M) costs are a component of the annual operating expense that is larger than the LLC. The O&M cost normally includes

- Labor, parts, and supplies for scheduled and unscheduled turbine maintenance
- Parts and supplies for equipment and facilities maintenance
- Labor for administration and support

Levelized replacement/overhaul cost (LRC) distributes the cost of major replacements and overhauls over the life of the wind turbine.

Finally, the AOE includes land or ocean bottom lease cost (LLC), levelized O&M cost, and LRC.

COE can then be expanded to show the annual cost is due to three major components:

$$\text{COE} = \frac{(\text{FCR} \times \text{ICC})}{\text{AEP}_{\text{net}}} + \text{AOE} = \text{LLC} + \frac{(\text{FCR} \times \text{ICC})}{\text{AEP}_{\text{net}}} + \frac{(\text{O\&M} + \text{LRC})}{\text{AEP}_{\text{net}}}$$

In other words, the COE is due to

1. Costs of leasing land (if land is purchased this cost is $0)
2. ICCs to buy/construct all wind turbine parts
3. Ongoing costs due to yearly operations, maintenance, overhauls, and replacements

Typical values used to calculate COE are

Fixed charge rate (FCR) = 10%–15%
Turbine cost/kW = $800–1100/kW (for MW size turbines)
Capacity factor = 25%–40%
O&M costs (scheduled maintenance) = $0.005–$0.010/year

Finally, typical component costs for a MW-scale wind turbine are shown in Figure 3.27. Figure 3.27 shows BOS (21%), O&M (19%), and LRC (11%) sum to 51% of the total COE cost, while the drive train (27%) has the largest single cost of the turbine components. The rotor (12%), tower (9%), and control

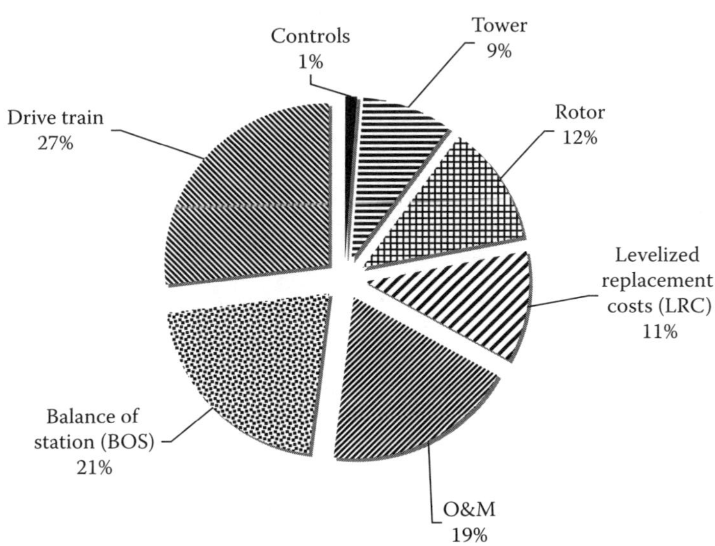

FIGURE 3.27
Individual component costs as % of COE for a land-based, MW-scale wind turbine.

system (1%) sum to only 22% of the total COE. When the drive train is added to these components, the sum of all wind turbine components is 49% of the total COE cost. In summary, installing, maintaining, and overhauling/replacing components is approximately half (51%) of the cost of electricity for a wind turbine which illustrates the need for an efficient installation method, and a well-designed, reliable wind turbine.

Example 3.4: Cost of Energy (COE) Calculation

What is the COE for a wind turbine with the following characteristics?

Turbine rating = 5 MW= 5,000 kW
CF = capacity factor = 40.00%
TC = turbine cost = $1100/kW
O&M cost/year = $87,600/year
FCR = fixed charge rate = 10.00%
BOS = balance of station = $2,360,000
LRC = levelized replacement costs = $123,000/year
LLC = land lease costs= $0/year

Answer:

$$\begin{aligned}\mathrm{AEP_{net}} &= \text{Annual energy production, Net (kWh/year)}\\ &= \mathrm{CF}\cdot\text{turbine rating}\cdot 8766 = 17{,}532{,}000\ \text{kWh/year}\end{aligned}$$

$$\begin{aligned}\mathrm{ICC} &= \text{Initial capital cost} = \text{Turbine cost} + \text{Balance of station}\\ &= \text{Rating}\cdot \mathrm{TC} + \mathrm{BOS} = \$7{,}860{,}000\end{aligned}$$

$$\mathrm{AOE} = \mathrm{LLC} + \frac{(\mathrm{O\&M} + \mathrm{LRC})}{\mathrm{AEP_{net}}} = 0 + \frac{(87{,}600 + 123{,}000)}{17{,}532{,}000} = \$0.012$$

$$\mathrm{COE} = \frac{(\mathrm{FCR}\cdot !\mathrm{CC})}{\mathrm{AEP_{net}}} + \mathrm{AOE} = \frac{(0.1\cdot 7{,}860{,}000)}{17{,}532{,}000} + 0.012 = 0.045 + 0.012 = \$0.057/\text{kWh}$$

% of total COE due to ICC = 79%
% of total COE due to AOE = 21%

In summary, for this example, 79% of the yearly cost is due to the actual wind turbine and installation of the turbine. The remaining 21% of the yearly cost is due to expenses needed to keep the turbine running.

This analysis assumes costs are known for the turbine of interest. If one is planning to design a turbine, the NREL report contains cost relationships for turbine components, for example, estimated cost of a wind turbine blade as a function of length. This allows one to estimate the turbine, control system, tower, balance of station, and offshore turbine costs. Knowledge of maintenance costs, replacement costs, and actual wind turbine site information (for capacity factor) allows the COE for a wind turbine at a site to be determined and evaluated for possible commercial development.

3.10 Wind Farms

As noted earlier, a wind farm is a group of wind turbines producing electricity in a single location (see Figure 3.28). In order to have a successful wind farm, selecting a proper location is key [35,36]. The most important item is selecting a site with the desired wind resource. The winds must be consistent and have a sufficient speed to generate the desired energy. Wind resource maps (see Section 3.7.2) are a typical starting point. Meteorological data from "met" towers erected by companies interested in a site give a detailed view of the wind resource at a specific site.

Potential impacts of wind turbine farms also need to be considered. As noted earlier, bird/bat activity and avian migration need to be considered to prevent negative effects on the environment. Noise effects can also have an impact in the area local to a proposed wind farm site. In addition, larger questions are also studied, for example: Will the turbines create interference for local air traffic? These additional factors must be weighed in addition to the wind resource when selecting a wind turbine site.

Besides wind resource and environmental effects, other factors that must be considered include land access (lease or purchase, zoning and permits

FIGURE 3.28
Wind farm located in Tehachapi, California, operational since early 1980s.

must be obtained) and sufficient capital to fund the project (approximately $1 million per MW of installed generating capacity) (AWEA b). The wind farm would ideally be located near existing transmission lines to limit the cost (up to $1 million/mile) of installing large lengths of new transmission lines which drive the balance of station (BOS) costs up. In addition, there has to be a market available and an agreement in place to buy the electric power produced. Assuming other factors are favorable, based on wind resource and turbine costs, the wind farm developer selects a wind turbine model. Additional steps include checking with the manufacturer for availability and considering a service provider for annual maintenance.

Another consideration is the issue of intermittency. The power output from a single wind turbine can be intermittent since the wind speed can fluctuate below and above rated wind speed. On the other hand, a wind farm, with wind turbines distributed over a large, geographically diverse area, will have less chance of the wind speed being below rated wind speed for all wind turbines in the wind farm. A study by Archer and Jacobson [37] indicates that when multiple wind farms are interconnected together in an array through the electric transmission grid, the correlation between the wind speeds at the sites decreases. Consequently, the array of wind farms behaves more like a single wind farm experiencing a steady wind speed and thus producing a constant power output. In addition, the benefits of interconnecting wind farms continued to increase as the number of connected wind farms increased.

On the other hand, distributing wind turbines in a wind farm over a large geographic area drives up BOS costs due to the increased cost of electrical interconnects and the potential need for new transmission lines. Wind farms are typically arranged with a rectangular grid pattern of turbines with each successive row staggered, as shown in Figure 3.29. Turbines are staggered by a specified number of rotor diameters apart, both side to side (crosswind spacing) and between rows (downwind spacing) of turbines. This distance varies with the wind farm owner, site conditions, and wind resource. If only BOS costs are considered, the turbines should be placed closer together. From a power output standpoint, if 100 turbines rated at 2 MW each are to be placed at this site, the total power output could be anticipated to be 200 MW. Unfortunately, this is not the case.

Consider a typical wind farm development scenario. Once transmission, land use issues, and a power purchase agreement are in place, the wind farm development can proceed. The wind resource estimates from the met tower for a potential site are anticipated to be sufficient for power production; however, wind farms do not always produce the expected power. Several factors could be responsible for this unexpected reduction in power output.

For example, the met tower used to collect data to make the wind resource predictions can only measure wind at one location. The wind resource could change throughout the site depending on terrain such as ridges and hills or on surface features like a forest. As turbines are built farther away from the met tower, the resource prediction becomes less accurate.

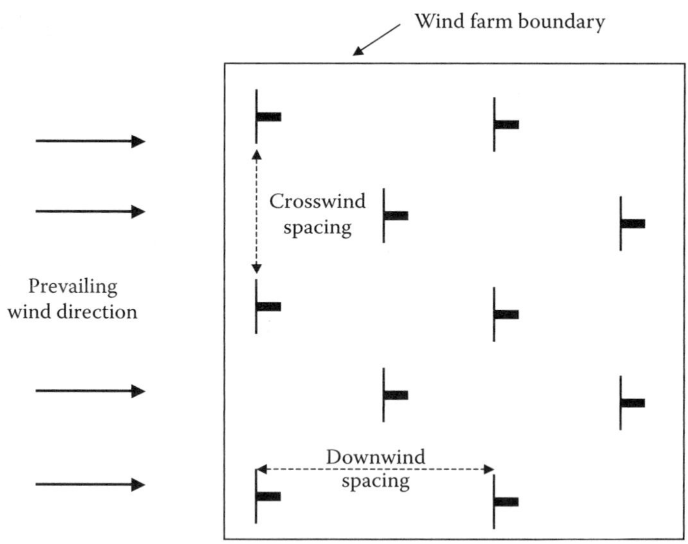

FIGURE 3.29
Schematic (top view) of typical wind farm layout.

Another suspect for producing less than expected power is wake effects or array effects. Depending on the locations of the turbines positioned on the site, and the direction of the wind, some turbines can be located downwind of other turbines. As a turbine extracts energy from the wind, the flow downstream becomes slower and more turbulent. This slower, more turbulent flow persisting downwind of a turbine is known as a wake. Turbines built in the wakes of other turbines have a diminished wind resource and do not produce the power levels of the upwind turbines. This will result in the production of less energy which means a reduced monetary return for investors. Downstream of a wind turbine (perhaps 15 diameters or much more depending on conditions) the wind speed in the wake will begin to approach the free stream velocity allowing higher energy production by downstream turbines. Unfortunately, increasing the spacing between turbines increases BOS costs (e.g., more construction and materials) resulting in a lowered return on investment. Several projects into studying wake effects strive to quantify wake losses to better design wind farms.

Extensive turbine wake modeling has been conducted by efforts such as the Efficient Development of Offshore Wind farms (ENDOW) project and UpWind projects. As part of the ENDOW project, Rados [38] reviews six wake models and compares the results to experimental data for the decease in wind speed behind the turbine (wake velocity deficit) and turbulence intensity (a measure of the variability of the wind speed). The results show the models consistently overpredict the wake velocities and turbulence intensity. In a review of current wind farm models, Barthelmie [39] recognizes that "wind farm models are lacking one or more components which account for the modification of the overlying boundary layer by the reduced wind speed, high turbulence atmosphere generated by large wind farms."

A study by Stovall [40] using computational fluid dynamics (CFD) and large eddy simulations (LES) looked at the overlying boundary layer effects on wake propagation. Two turbines, with 100 m rotor diameters, are modeled as being inline. The incoming wind speed to the upstream turbine is approximately 8–10 m/s. The objective is to determine the effect the wake has on the power output of the downstream turbine compared to power output of the upstream turbine. In this study, the turbines have a 9 diameter downwind spacing. The power coefficient for the turbines in this simulation is approximately $C_p = 0.50$.

Prior to modeling the power output of the wind turbines, a simulation of the ABL is created. Once the ABL simulation is validated, turbines are imposed into the domain with the turbine rotors modeled via actuator disks. The actuator disks impose a pressure drop on the flowfield proportional to the power extracted by a turbine.

Results from the numerical simulation show the diminished wind resource caused by the wake-forming downstream of a wind turbine. Figure 3.30 illustrates the power in the wind calculated as the area integral of velocity cubed (Equation 3.11). Distances between the turbines are shown in rotor diameters (D). The inlet of the computational domain used in this numerical simulation is located at $D = +2$. The first turbine is located at $D = 0$ while the second turbine is located 9 diameters downstream of the first turbine at

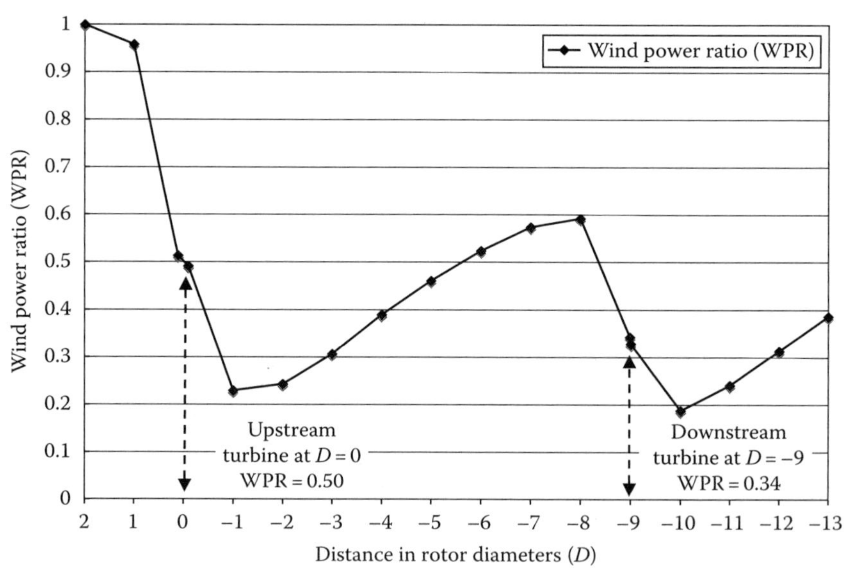

FIGURE 3.30
Wind power wake reduction: Local wind power normalized by wind power level at inlet. (From Stovall, T. D., Simulations of wind turbine wake interactions in OpenFOAM, MS thesis, University of Colorado, Boulder, CO, 2009.)

$D = -9$. All wind powers are expressed as a wind power ratio (WPR) of the wind power at a given location divided by the inlet wind power (located at $D = +2$). This results in a WPR = 1.0 at the inlet ($D = +2$).

Notice the nondimensional wind power extracted by the first turbine is half of the wind power at the inlet, yielding a WPR = 0.50, as would be expected since the power coefficient is $C_p = 0.50$.

Downstream of the first turbine, the wake is represented by a further reduction in wind power due to turbulent losses. As distance increases downstream of the first turbine, the wake begins to recover (D increases from $-1 \rightarrow -8$) as surrounding higher speed (i.e., higher energy) wind mixes with the wake. The power in the wake, however, does not achieve free stream power levels prior to reaching the second turbine.

Since the incoming power of the wind is lower immediately upstream of the second turbine, the power output of the second turbine should be lower than the power output of the first turbine. Indeed, the WPR of the second turbine in this simulation is WPR = 0.34, indicating the second turbine was only able to extract 34% of the energy available in the incoming stream at $D = +2$. The wake from the first turbine adversely impacts the power output of the second turbine. Another way to view this is the ratio of the power extracted by a second turbine (located 9*D* downstream of the first turbine in this study) to the first turbine is approximately 67% (i.e., 0.34/0.50 = 0.67%). In economic terms, the second turbine will only produce 67% of the income of the first turbine in this scenario.

Increasing the spacing between the turbines in the downwind direction reduces the wake effect but increases the BOS costs. A trade-off between these two competing economic factors remains an active area of research. Although the results from the Stovall study match up with experimental data from wind farms, the simulation is not yet commercially viable for developing wind farms. The computational resources required for modeling an entire wind farm (dozens of turbines) are currently too high for industry to optimize wind farm layouts. Existing models such as Riso DTU [41] and Garrad Hassan [42] are designed for wind farm optimization, but lack the accuracy of full CFD simulations. Significant work still remains to develop a robust, accurate, and fast optimization tool for wind farm wake effects. With such a tool, BOS costs can be optimized simultaneously with turbine location resulting in maximum energy (income) production.

3.11 Offshore Wind Energy

Offshore wind turbine farms (see Figure 3.31) address limits of land-based wind farms such as transmission line access and capacity [43], space for wind turbines (more of an issue in higher density population areas such as

FIGURE 3.31
Offshore wind farm. (Courtesy of Hans Hillewaert.)

Europe), while providing access to a vast wind resource. Significant development still remains for wind farms in water deeper than 30 m while balance of station costs can double in an offshore installation compared to land-based installation. Still, the complexity of offshore technology development is more than offset by the need and desire to harness the higher wind resource available offshore (ocean as well as large inland lakes).

Offshore wind energy has a number of benefits including

- Better wind resources
    - Reduced turbulence resulting in steadier winds
    - Higher wind speed resulting in higher energy production
    - Higher capacity factors resulting in better electric load matching
- Closer proximity to electric load centers resulting in
    - Lower transmission costs and constraints
    - Ability to serve high cost regions
- Avoids land-based size limits
    - Shipping—no land-based roadway limits
    - Assembly—sea-based cranes are taller
    - Larger machines are more economical

These benefits are significant throughout the world. For example, the United States has no offshore wind farms as of 2009 while 9 farms are proposed for the future [36]. Twenty-eight of the 48 contiguous United States have a coastal boundary and these 28 states use 78% of the electricity in the United States [44].

Since these coastal load centers are typically located hundreds if not thousands of miles from land-based wind energy sources, offshore wind power provides a nearby source of energy. Of these 28 states, only 6 have a sufficient land-based wind energy resource to meet more than 20% of their electric requirements through wind power. If shallow water offshore potential (less than 30 m in depth) is included in the wind resource mix, 26 of the 28 states would have the wind resources to meet at least 20% of their electric needs. Several of these states would have sufficient offshore wind resources to meet 100% of their electric needs [45]. The goal of the United States achieving 20% of its energy from wind energy by 2030 [36] assumes significant contributions from offshore turbines to meet this goal.

Studies have been conducted [46], which illustrate the potential available offshore wind power as a function of depth and coastal location for the United States. Figure 3.32 shows the location of the offshore U.S. wind power (Atlantic and Pacific Coasts, Gulf of Mexico, Great Lakes, and Hawaii). Table 3.7 quantifies the amount of power available by water depth. For example, approximately 430 GW of power is available in shallow waters (depth of 30 m or less). Similarly, the mid-Atlantic region has a potential of 347.2 GW of power available in waters 60 m or less in depth. Considering all seven regions (except Hawaii) 970 GW or nearly 1 TW (where 1 TW = 1,000 GW = 1,000,000 MW) of power is available in water depths of 60 m or less.

In contrast, countries in the European Union, as of September 2009, lead the world in developing offshore wind farms with over 25 operational offshore wind farms (most installed in water less than 22 m deep) and more than 15 additional offshore wind farms under construction. Dozens of offshore wind farms are proposed in countries throughout the world including Canada, China, Germany, Netherlands, United Kingdom, and the United States, among others.

The European Wind Energy Association's (EWEA) statistics [47] for January 2009 cite an operational wind farm capacity of 1.4 GW for Europe, with the United Kingdom and Denmark combining for over 67% of the total capacity, as shown in Figure 3.33.

Proposed European offshore wind farms for 2015, cited in the same EWEA report, show an enormous growth in proposed capacity (see Figure 3.34) from 1471 MW in 2009 to a proposed capacity of 37,444 MW, a 25 times increase! If plans come to fruition, Germany would jump to number one in offshore capacity with a 900 times increase from 24 to 10,928 MW, resulting in a 30% share of the total proposed offshore capacity. The United Kingdom would have the next largest share with 23%, resulting from a 14.8 times increase in offshore capacity (590.8–8,755.8 MW). The number of European countries with offshore capacity would likewise increase from 8 countries in 2009 to 13 countries in 2015.

As noted earlier, benefits of offshore turbines include using larger blades since the wind speeds are higher yet steadier offshore, allowing for better

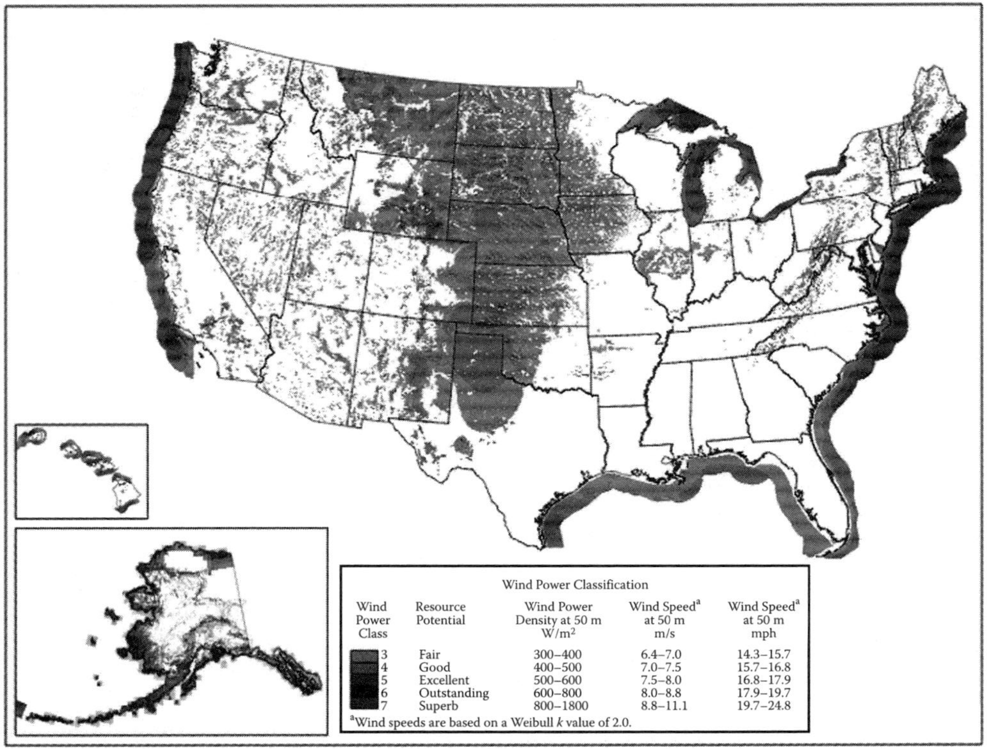

FIGURE 3.32
United States wind power resource map. This map shows the annual average wind power estimates at a height of 50 ms. It is a combination of high resolution and low resolution datasets produced by NREL and other organizations. The data was screemed to eliminate areas unlikely to be developed onshore due to land use or environmental issues. In many states, the wind resource on this map is visually enhanced to better show the distribution on ridge crests and other features. (Courtesy of U.S. Department of Energy, National Renewable Energy Laboratory, Washington, DC.)

TABLE 3.7

Offshore U.S. Wind Power (GW) for Wind Class 5 or Greater by Water Depth

Coastline	0–30 m	30–60 m	60–900 m	>900 m
New England	59.2	127.7	273.4	0
Mid-Atlantic	165.6	181.6	59.7	56.6
South Atlantic Bight	28.4	58.2	13.7	0
Gulf of Mexico	0	12.3	54.7	0
California	2.3	4.8	130.5	277.9
Pacific Northwest	7.5	19.2	188.1	121
Great Lakes	166.6	137	813.2	0
Total	429.6	540.8	1,533.30	455.5
Total power all depths = 2,959.2 GW (=2,959,200 MW = 2.96 TW)				
% Power of total power	14.50	18.30	51.80	15.40
Hawaii (not included in totals)	0.8	1.4	24.9	123.6

Source: Musial, W., *Mar. Technol. Soc. J.*, 41(3), 24, 2007.
Note: Power is in GW, not MW!

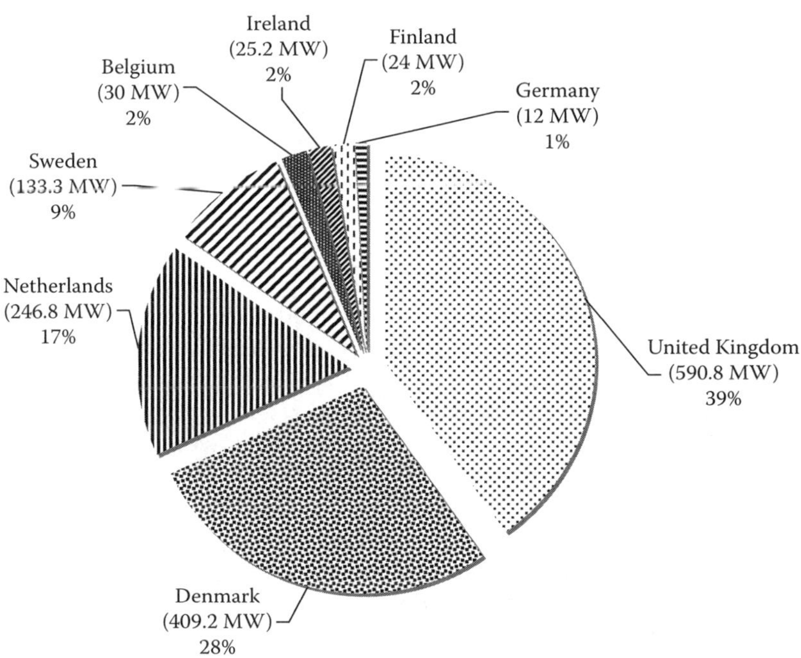

FIGURE 3.33
Operational offshore wind power in Europe as of January 2009. Total capacity of 1471 MW. (From European Wind Energy Association (EWEA), Offshore statistics for January, 2009, http://www.ewea.org/fileadmin/ewea_documents/documents/statistics/Offshore_Wind_Farms _2008.pdf)

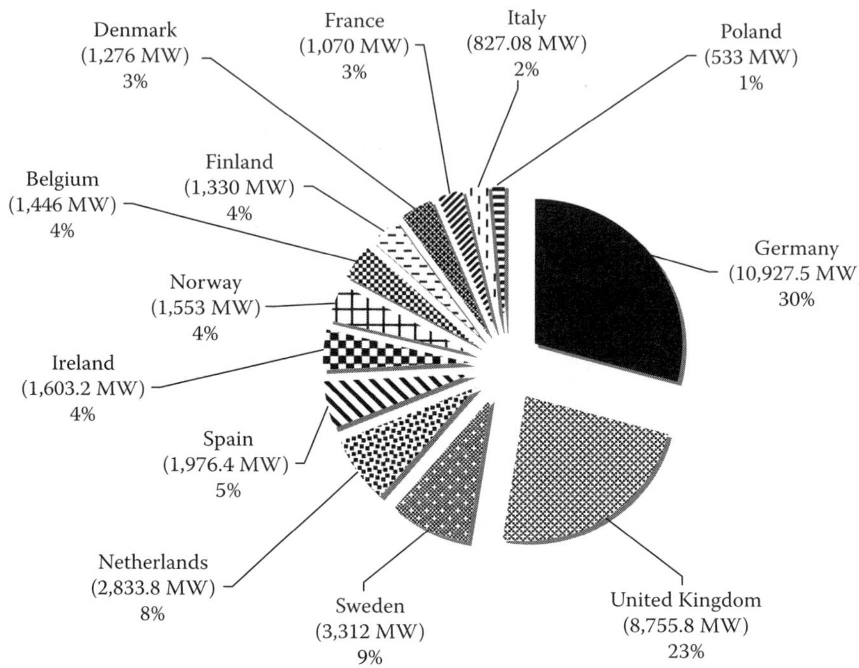

FIGURE 3.34
Proposed offshore wind power in Europe in 2015. Total capacity of 37,444 MW. (From European Wind Energy Association (EWEA), Offshore statistics for January, 2009, http://www.ewea.org/fileadmin/ewea_documents/documents/statistics/Offshore_Wind_Farms _2008.pdf)

economics. Significant technical challenges remain, especially in water deeper than 30 m. From Table 3.7, we can see that in waters 0–30 m and 30–60 m deep, the wind power potential off the coast of California is 2.3 and 4.8 GW, respectively, while in water 60–900 m deep, the potential wind power is 130.5 GW. Significant technology developments remain to make offshore wind power viable to the state of California, which contained 12% of the U.S. population in 2008 [48] and used 7% of the electricity in the United States in 2007 [49], and other parts of the world with deep water close to shore.

Figure 3.35 illustrates the depths for which wind turbine technology is needed. In waters of 30 m or less, proven technology is available. In waters between 30 and 60 m, investigations are currently beginning to develop the generation of technology. In waters deeper than 60 m, significant technology challenges exist to develop floating wind turbines.

In Figure 3.36, three concepts of different technology approaches for developing floating wind turbines are shown. Modeling a floating wind turbine is a challenging task because it must include effects from wind and wave interaction with the turbine and floating structure, a nonstationary and rotating turbine,

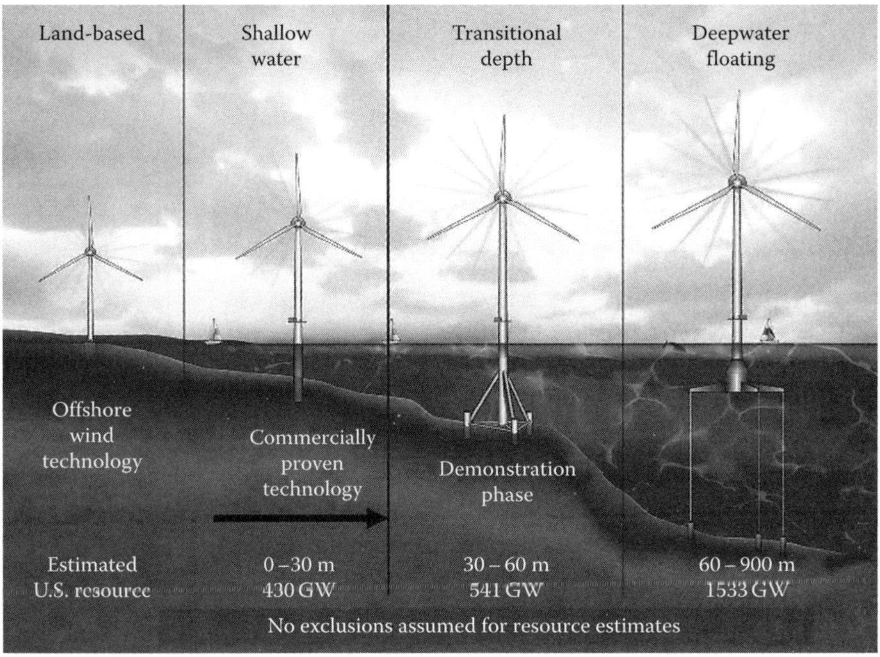

FIGURE 3.35
U.S. offshore wind power resource estimate as a function of depth and status of the needed wind turbine technology. (From Robinson, M. and Musial, W., Offshore wind technology overview, NREL/PR-500-40462, 2006, http://www.nrel.gov/docs/gen/fy07/40462.pdf)

and a control system, which monitors and controls the turbine for these effects. State-of-the-art research is underway using codes such as FAST [50].

The winning technology design will likely include elements from these three concepts as well as other designs.

3.12 Additional Topics for Study

A single chapter cannot, of course, completely cover a topic as rich, diverse, and rapidly changing as wind turbines. Topics omitted for brevity include (but are not limited to!) statistical representation of wind, vibration and fatigue, strength of materials, manufacturing (blades, towers, and nacelles), detailed aerodynamic design, gearbox, generators, power converters, control systems, connection to electrical grid, and transmission. Discussion of all these topics and more can be found in wind energy texts such as Manwell et al. [30], Burton et al. [32], or Hau [25].

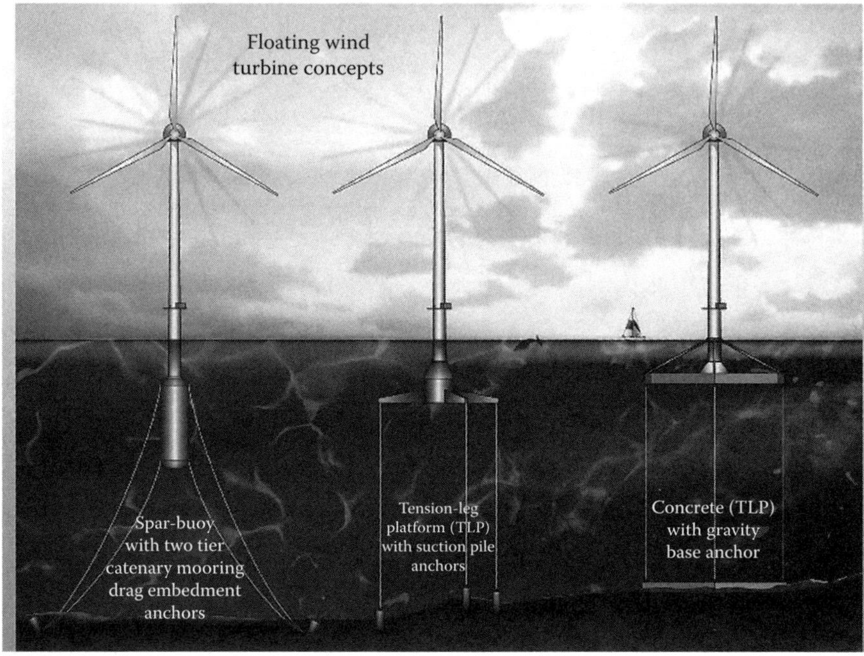

FIGURE 3.36
Floating wind turbine concepts for deepwater (>60 m) wind resource locations. (From Butterfield, S. et al., Engineering challenges for floating offshore wind turbines, NREL/CP-500-38776, 2007, http://www.nrel.gov/wind/pdfs/38776.pdf)

Acknowledgment

My thanks to Tim Stovall for his contributions to Section 3.10.

Problems

3.1 While completing this homework, imagine the study space you are using employs a space heater with a rating of 30 kW, a refrigerator (for snacks) that is rated for 5 kW and a light fixture with 2100 W bulbs, all operating at a constant rate for the 1.5 h it takes you to finish.

a. What is the total power consumed at any given moment?
b. How much energy do you consume to complete the task?

c. Assume your local power company, XEnergy, charges 6.5 cents per unit of energy in part b, how much did it cost you to complete the assignment?

d. If you wanted to pay XEnergy \$5, what would be the power rating of an additional appliance you'd need to turn on?

3.2 You have assembled a modern three-bladed upwind wind turbine with a rotor radius of 5 m at sea level where the density is 1.225 kg/m^3.

a. If the wind is blowing at an average speed of 10 m/s, what is the power of the wind traveling through the swept area of the rotor?

b. What is the average actual power produced by a turbine of this type at this wind speed? (*Hint:* Find C_p)

Hint: Find the maximum power coefficient, C_p, based on the equation below for axial induction.

$$C_p = 4a(1-a)^2$$

3.3 An investor comes to you and tells you that a company wants to invest in a new turbine design with the following performance characteristics:

- 112 m rotor diameter
- 13 m/s wind speed
- 2.1 MW
- 1.223 kg/m^3 air density

a. What is the claimed rotor power coefficient, C_p, for this turbine design?

b1. Do you encourage the investor to invest the company's \$1 million? Answer in complete sentence(s)

b2. Why? Answer in complete sentence(s)

3.4 Weibull distribution: From an analysis of wind speed data (hourly interval average, taken over a 1 year period), the Weibull parameters are determined to be $c = 6$ m/s and $k = 1.8$ at a potential wind turbine site.

The Weibull PDF is given by

$$p(U) = \left(\frac{k}{c}\right)\left(\frac{U}{c}\right)^{k-1} \exp\left[-\left(\frac{U}{c}\right)^k\right]$$

a. Estimate the number of hours per year that the wind speed will be between $u_B = 6.5$ and $u_A = 7.5$ m/s during the year.

b. Estimate the number of hours per year that the wind speed is above 16 m/s.

3.5 Given a wind turbine with the following parameters, calculate the annual energy production assuming a Rayleigh probability distribution function:

$C_p = C_{pmax} = 0.40$ (C_p value assumed maximum and constant for all wind speeds)

Rated power = 3000 kW = 3.0 MW

Annual average wind speed = 7.0 m/s

Rated wind speed = 12.5 m/s

Cut-out wind speed = 25 m/s

Rotor radius = 45.0 m

Area = 6361.73 m^2

Air density = 1.20 kg/m^3

Wind Speed (m/s)	Power (kW)	Probability (%)	# of Hours/Year	Energy (kWh)
1	1.5			
2	12.2			
3	41.2			
4	97.7			
5	190.9			
6	329.8			
7	523.7			
8	781.7			
9	1113.0			
10	1526.8			
11	2032.2			
12	2638.3			
13	3000.0			
14	3000.0			
15	3000.0			
16	3000.0			
17	3000.0			
18	3000.0			
19	3000.0			
20	3000.0			
21	3000.0			
22	3000.0			
23	3000.0			
24	3000.0			
25	3000.0			
	Total:		Total:	kWh
				MWh

3.6 (a) What is the cost of energy for a turbine with the following assumptions:

- 5 MW rating
- 31% capacity factor
- $1,250/kW turbine cost
- $91,000 operations and maintenance cost/year
- Fixed charge rate of 9%
- Balance of station costs of $1,999,999
- Levelized replacement costs of $135,000/year

(b) If a gearbox replacement was required every 5 years at a cost of $600,000 what would the COE be?

How much (in percent) does the COE increase due to the gearbox replacement?

References

1. Energy Information Administration (EIA). (2008). Per capita (per person) total primary energy consumption (million Btu per person), United States 1949–2007. http://www.eia.doe.gov/emeu/international/energyconsumption.html
2. Winds of Change. (2006). American Federal Projects 1975–1985. http://www.windsofchange.dk/WOC-usastat.php
3. American Speech-Language-Hearing Association (ASHA). (1997). Noise and hearing loss. http://www.asha.org/public/hearing/disorders/noise.htm http://www.sengpielaudio.com/TableOfSoundPressureLevels.htm
4. American Wind Energy Association (AWEA). (2010). Wind power myths vs. facts. http://www.awea.org/pubs/factsheets/MythsvsFacts-FactSheet.pdf
5. San Martin, R.L. (1989). Environmental emissions from energy technology systems: The total fuel cycle. Energy Citations Database. http://www.osti.gov/energycitations/product.biblio.jsp?osti_id=860715
6. American Wind Energy Association (AWEA). (1996). Wind web tutorial. http://www.awea.org/faq/wwt_environment.html#How%20does%20wind%20stack%20up%20on%20greenhouse%20gas%20emissions
7. Danish Wind Industry Association (DWIA). (1997). Shadow casting from wind turbines. http://www.talentfactory.dk/en/tour/env/shadow/index.htm
8. Sovacool, B.K. (2009). Contextualizing avian mortality: A preliminary appraisal of bird and bat fatalities from wind, fossil-fuel, and nuclear electricity, *Energy Policy* 37(6), 2241–2248.
9. Erickson, W., Johnson, G., and Young, D. Jr. (2002). A summary and comparison of bird mortality from anthropogenic causes with an emphasis on collisions. USDA Forest Service General Technical Report PSW-GTR-191. http://www.fs.fed.us/psw/publications/documents/psw_gtr191/Asilomar/pdfs/1029–1042.

10. National Wind Coordinating Committee (NWCC). (2004). Wind turbine interactions with birds and bats: A summary of research results and remaining questions. http://www.nationalwind.org/publications/wildlife/wildlife_factsheet.pdf
11. Cryan, P.M. (2008). Overview of issues related to bats and wind energy: Web version of presentation to the *Wind Turbine Guidelines Advisory Committee Technical Workshop & Federal Advisory Committee Meeting*, Washington, DC, February 26, 2008.
12. Vestel, L.B. (2009). Study finds reduction in turbine bat kills. http://greeninc.blogs.nytimes.com/2009/05/18/study-finds-reduction-in-turbine-bat-kills/
13. Nebraska Wind and Solar. (2008). History of wind power. http://nebraskawindandsolar.com/history.aspx
14. Price, T.J. (2005). James Blyth—Britain's first modern wind power pioneer. *Wind Engineering* 29(3), 191–200.
15. Danish Wind Industry Association (DWIA). (1997). A wind energy pioneer: Charles F. Brush. http://www.talentfactory.dk/en/pictures/brush.htm
16. Gipe, P. (2003). Smith-Putnam industrial photos. http://www.wind-works.org/photos/Smith-PutnamPhotos.html
17. Renewable Energy Vermont. The story of Grandpa's Knob: How Vermont made wind energy history. http://www.revermont.org/grandpa_knob.pdf
18. Gipe, P. (1995). *Wind Energy Comes of Age*, John Wiley & Sons, Inc., New York.
19. Eldridge, F.R. (1980). *Wind Machines*, 2nd edn., Van Nostrand Reinhold, New York.
20. Nelson, V. (1996). *Wind Energy and Wind Turbines*, Alternative Energy Institute, Canyon, TX.
21. Betz, A. (1920). Das maximum der theoretisch möglichen ausnützung des windes durch windmotoren, *Zeitschrift für das gesamte Turbinenwesen* 26, 307–309.
22. Gijs, A.M. van Kuik. (2007). The Lanchester–Betz–Joukowsky limit, *Wind Energy* 10, 289–291.
23. Wilson, R.E. and Lissaman, P.B.S. (1974). *Applied Aerodynamics of Wind Power Machines*, Oregon State University, Corvallis, OR. http: ir.library.oregonstate.edu/jspui/handle/1957/8140.
24. Stiesdal, H. (1999). The wind turbine: Components and operation. Bonus Energy A/S. *Bonus Info Newsletter* (autumn).
25. Hau, E. (2006). *Wind Turbines—Fundamentals, Technologies, Application, Economics*, Springer, Berlin, Germany.
26. Frost, W.D. and Asphiden, D. (1994). Characteristics of the wind, in: Spera, D.A. (ed.), *Wind Turbine Technology*, ASME Press, New York.
27. Pidwirny, M. (2006). *Fundamentals of Physical Geography*, 2nd edn, Chap. 7(p). http://www.physicalgeography.net/fundamentals/7p.html
28. Kelley, N.D. and Jonkman, B.J. (2007). Overview of the TurbSim stochastic inflow turbulence simulator version 1.21 (revised February 1, 2007). National Renewable Energy Laboratory Technical Report, NREL/TP-500-41137. http://wind.nrel.gov/designcodes/preprocessors/turbsim/TurbSimOverview.pdf
29. Wind Powering America. (2009). Wind resource maps. http://www.windpoweringamerica/wind_maps.asp
30. Manwell, J.F., McGowan, J.G., and Rogers, A.L. (2002). *Wind Energy Explained, Theory, Design and Application*, Wiley, Chichester, U.K.

31. Holton, J. (2004). The planetary boundary layer, in: *Dynamic Meteorology*, International Geophysics Series, vol. 88, Chap. 5, Elsevier Academic Press, Burlington, MA.
32. Burton, T., Sharpe, D., Jenkins, N., and Bossanyi, E. (2001). *Wind Energy Handbook*, Wiley, Chichester, U.K.
33. Renewable Energy Research Laboratory: U. Massachusetts at Amherst. (2006). Capacity factor, intermittency, and what happens when the wind doesn't blow. http://www.ceere.org/rerl/about_wind/RERL_Fact_Sheet_2a_Capacity_Factor.pdf
34. Fingersh, L., Hand, M., and Laxson, A. (2006). Wind turbine design cost and scaling model. Technical Report NREL/TP-500-40566. www.nrel.gov/wind/pdfs/40566.pdf
35. American Wind Energy Association (AWEA). (2010). 10 steps in building a wind farm. http://www.awea.org/pubs/factsheets/10stwf_fs.pdf
36. Department of Energy (DOE). (2008). 20% Wind energy by 2030 increasing wind energy's contribution to U.S. electricity supply. http://www.nrel.gov/docs/fy08osti/41869.pdf
37. Archer, C.L. and Jacobson, M.Z. (2007). Supplying baseload power and reducing transmission requirements by interconnecting wind farms. *Journal of Applied Meteorology and Climatology* 46(11), 1701–1717.
38. Rados, K., Larsen, G., Barthelmie, R., Schlez, W., Lange, B., Schepers, G., Hegberg, T., and Magnisson, M. (2001). Comparison of wake models with data for offshore windfarms, *Wind Engineering* 25, 271–280.
39. Barthelmie, R. et al. (2007). Modeling and measurements of wakes in large wind farms. *Journal of Physics: Conference Series* 75, 012049.
40. Stovall, T.D. (2009). Simulations of wind turbine wake interactions in OpenFOAM. MS thesis, University of Colorado, Boulder, CO.
41. Riso DTU. (2003). Wind Atlas Analysis and Application Program (WAsP). www.wasp.dk.
42. Garrad Hassan & Partners Ltd. (2009). GH WindFarmer 4.0 www.garradhassan.com/products/ghwindfarmer/
43. Piwko, R., Osborn, D., Gramlich, R., Jordan, G., Hawkins, D., and Porter, K. (2005). Wind energy delivery issues: Transmission planning and competitive electricity market operation, *IEEE Power and Energy Magazine* 3(6), 47–56.
44. Energy Information Administration (EIA). (2006). Spread sheet state sales. www.eia.doe.gov/cneaf/electricity/epa/sales_state.xls
45. Musial, W. and Ram, B. (2007). Large scale offshore wind deployments: Barriers and opportunities. NREL, Golden, CO.
46. Musial, W. (2007). Offshore wind electricity: A viable energy option for the coastal United States, *Marine Technology Society Journal* 41(3), 24–35.
47. European Wind Energy Association (EWEA). (2009). Offshore statistics for January 2009. www.ewea.org/fileadmin/ewea_documents/documents/statistics/Offshore_Wind_Farms _2008.pdf
48. U.S. Census Bureau. (2009). National and state population estimates: Annual population estimates 2000 to 2008. http://www.census.gov/popest/states/NST-ann-est.html
49. Energy Information Administration (EIA). (2008). Electricity consumption estimates by sector, 2007. http://www.eia.doe.gov/states/sep_fuel/html/pdf/fuel_use_es.pdf

50. Jonkman, J.M. and Buhl, M.L. Jr. (2007). Development and verification of a fully coupled simulator for offshore wind turbines. NREL/CP-500-40979. www.nrel.gov/wind/pdfs/40979.pdf
51. Energy Efficiency & Renewable Energy (EERE) (DOE). (2010). How does a wind turbine work? http://www1.eere.energy.gov/windandhydro/wind_animation.html
52. Robinson, M. and Musial, W. (2006). Offshore wind technology overview. NREL/PR-500-40462. http://www.nrel.gov/docs/gen/fy07/40462.pdf
53. Butterfield, S., Musial, W., Jonkman, J., and Sclavounos, P. (2007). Engineering challenges for floating offshore wind turbines. NREL/CP-500-38776. www.nrel.gov/wind/pdfs/38776.pdf
54. Baerwald, E., D'Amours, G., Klug, B., and Barclay, R. (2008). Barotrauma is a significant cause of bat fatalities at wind turbines. *Current Biology* 18(16), R695–R696 (August).
55. Goldenberg, S. (2009). Texas wind farm pioneers radar technology to protect migrating birds. guardian.co.uk. http://www.guardian.co.uk/environment/2009/may/01/wind-farm-bird-radar

4

Capturing Solar Energy through Biomass

> The use of vegetable oils for engine fuels may seem insignificant today, but such oils may become in the course of time as important as the petroleum and coal tar products of the present time.
>
> **—Rudolph Diesel (1912)**

4.1 Introduction

Biorenewable resources, sometimes referred to as *biomass*, are organic materials of recent biological origin [1]. This definition is deliberately broad with the intent of only excluding fossil fuel resources from the wide variety of organic materials that arise from the biotic environment.

Biorenewable resources can be converted into a variety of gaseous and liquid fuels using both thermal and biological processes. The choice of process and product depends upon the nature of the biomass feedstock and the market into which it is being sold. Gaseous bioenergy products include biogas from anaerobic digestion and producer gas or syngas from thermal gasification; and hydrogen, methane, ammonia, and dimethyl ether from upgrading of the primary products of anaerobic digestion or gasification. Liquid bioenergy products are conveniently classified according to the intermediate substrate derived from physically, chemically, or thermally processing biomass: carbohydrates, triglycerides, syngas, and bio-oil/biocrude. Liquid fuel production from carbohydrates has traditionally focused on ethanol although butanol, isoprenes, furans, and even alkanes can also be produced. Liquid fuel production from triglycerides has traditionally focused on methyl esters (biodiesel) although conversion to alkanes and propane via hydrotreating/hydrocracking is receiving increasing attention. Liquid fuels from syngas include Fischer–Tropsch liquids, methanol, ethanol, and methanol-to-gasoline (MTG). Liquid fuels from upgrading bio-oil and biocrude are a range of hydrocarbons suitable as gasoline, diesel fuel, or aviation fuel. Finally, biomass can be converted to electricity, which offers an alternative approach to providing energy for transportation.

4.2 Biomass Feedstocks

4.2.1 Introduction

Biorenewable resources are generally classified as either *wastes* or *dedicated energy crops*. A waste is a material that has been traditionally discarded because it has no apparent value or represents a nuisance or even a pollutant to the local environment. Dedicated energy crops are plants grown specifically for production of biobased products; that is, for purposes other than food or feed.

4.2.2 Waste Materials

Categories of waste materials that qualify as biorenewable resources include municipal solid wastes (MSWs), agricultural residues, agricultural processing by-products, and manure. MSWs are whatever is thrown out in the garbage, and clearly includes materials that do not qualify as biorenewable resources, such as glass, metal, and plastics. Agricultural residues are simply that part of a crop discarded after harvest such as corn stover (husks and stalks), rice hulls, wheat straw, and bagasse (fibrous material remaining after the milling of sugarcane). Food processing waste is the effluent from a wide variety of industries ranging from breakfast cereal manufacturers to alcohol breweries. These wastes may be dry solids or watery liquids. The recent concentration of animals into giant livestock facilities has led to calls to treat animal wastes in a manner similar to that for human wastes.

Waste materials share few common traits other than the difficulty of characterizing them because of their variable and complex composition. Thus, waste biomass presents special problems to engineers who are tasked with converting this sometimes unpredictable feedstock into reliable power or high-quality fuels and chemicals. The major virtue of waste materials is their low cost. By definition, waste materials have little apparent economic value and often can be acquired for little more than the cost of transporting the material from its point of origin to a processing plant. Increasing costs for solid waste disposal and sewer discharges and restrictions on landfilling certain kinds of wastes allow some wastes to be acquired at negative cost; that is, a biorenewable resource processing plant is paid by a company seeking to dispose of a waste stream. For this reason, many of the most economically attractive opportunities in biorenewable resources involve waste feedstocks.

Clearly, a waste material that can be used as feedstock for an energy-conversion process is no longer a waste material. As demand for these new-found feedstocks increases, those that generate it come to view themselves as suppliers and may demand payment for the one-time waste: a negative feedstock cost becomes a positive cost. Such a situation developed in the California biomass power industry during the 1980s [2]. Concerns about air pollution in California led to restrictions on open-field burning of agricultural residues, a practice designed to control infestations of pests. With no means for getting rid of these residues, an enormous reserve of biomass feedstocks materialized. These feedstocks were so inexpensive that independent power producers recognized that even small, inefficient power plants using these materials as fuel would be profitable. A number of plants were constructed and operated on agricultural residues. Eventually, the feedstock producers had plant operators bidding up the cost of their once nuisance waste material. In the end, many of these plants were closed because of the escalating cost of fuel.

4.2.3 Energy Crops

Energy crops are defined as plants grown specifically as an energy resource. It is important to note that firewood obtained from cutting down an old-growth forest does not constitute an energy crop. An energy crop is planted and harvested periodically. Harvesting may occur on an annual basis, as with sugar beets or switchgrass, or on a 5–7 year cycle, as with certain strains of fast-growing trees such as hybrid poplar or willow. The cycle of planting and harvesting over a relatively short time period assures that the resource is used in a sustainable fashion, that is, the resource will be available for future generations.

Energy crops contain significant quantities of one or more of four important energy-rich components: oils, sugars, starches, and lignocellulose (fiber). Crops rich in the first three have historically been grown for food and feed: oils from soybeans and nuts; sugars from sugar beets, sorghum, and sugarcane; and starches from corn and cereal crops. Oil, sugars, and starches are easily metabolized. On the other hand, lignocellulose is indigestible by humans although certain domesticated animals with specialized digestive tracts are able to break down the polymeric structure of lignocellulose and use it as an energy source. From this discussion, it might appear that the best strategy for developing biomass resources is to grow crops rich in oils, sugars, and starches. However, even for "oil crops" or "starch crops" the largest single constituent is invariably lignocellulose, which is the structural (fibrous) material of the plant: stems, leaves, and roots. If oils, sugars, and starches are harvested and the lignocellulose is left behind as an agricultural residue rather than used as fuel, the greatest portion of the biomass crop is wasted.

Research has shown that energy yields (kilojoules per hectare per year) are usually greatest for plants that are mostly "roots and stems"; in other words, plant resources are directed toward the manufacture of lignocellulose rather than oils, sugars, and starches. As a result, there has been a bias toward the development of energy crops that focus on lignocellulosic biomass, which is reflected in the discussion that follows.

Dedicated energy crops are typically high-fiber crops grown specifically for their high productivity of holocellulose (cellulose and hemicellulose).

Harvesting may occur on an annual basis, as with switchgrass, or on a 5–7 year cycle, as with certain strains of fast-growing trees such as hybrid poplar. Lignocellulosic crops are conveniently divided into herbaceous energy crops (HEC) and short rotation woody crops (SRWC) [3].

Herbaceous crops are plants that have little or no woody tissue. The above ground growth of these plants usually lives for only a single growing season. However, herbaceous crops include both annuals and perennials. Annuals die at the end of a growing season and must be replanted in the spring. Perennials die back each year in temperate climates but reestablish themselves each spring from rootstock. Both annual and perennial HEC are harvested on at least an annual basis, if not more frequently, with yields averaging 5.5–11 Mg/ha/year, with maximum yields between 20 and 25 Mg/ha/year in temperate regions [3]. As with trees, yields can be much higher in tropical and subtropical regions.

Herbaceous crops more closely resemble hardwoods in their chemical properties than they do softwoods. Their low lignin content makes them relatively easy to delignify, which improves accessibility of the carbohydrate in the lignocellulose. The hemicellulose contains mostly xylan, which is highly susceptible to acid hydrolysis, compared to the cellulose. As a result, agricultural residues are susceptible to microbial degradation, destroying their processing potential in a matter of days if exposed to the elements. Herbaceous crops have relatively high silica content compared to woody crops, which can present problems during processing.

SRWC is used to describe woody biomass that is fast growing and suitable for use in dedicated feedstock supply systems. Desirable SRWC candidates display rapid juvenile growth, wide site adaptability, and pest and disease resistance. Woody crops grown on a sustainable basis are harvested on a rotation of 3–10 years.

Woody crops include hardwoods and softwoods. Hardwoods are trees classified as angiosperms, which are also known as flowering plants. Examples include willow, oak, and poplar. Hardwoods can be regrown from stumps, a process known as coppicing, which reduces their production costs compared to softwoods. Advantages of hardwoods in processing include: high density for many species; relative ease of delignification and accessibility of wood carbohydrates; the presence of hemicellulose high in xylan, which can be removed relatively easily; low content of ash, particularly silica, compared to softwoods and herbaceous crops; and high acetyl content compared to most softwoods and herbaceous crops, which is an advantage in the recovery of acetic acid.

Softwoods are trees classified as gymnosperms, which encompass most trees known as evergreens. Examples include pine, spruce, and cedar. Softwoods are generally fast growing but their carbohydrate is not as accessible for chemical processing as the carbohydrates in hardwood. Since softwoods have considerable value as construction lumber and pulpwood, they are more readily available as waste material in the form of logging and manufacturing residues than are hardwoods. Logging residues, consisting of a high proportion of branches and tops, contain considerable high-density compression wood, which is not easily delignified. Logging residues are more suitable as boiler fuel or other thermochemical treatments than as feedstock for chemical or enzymatic processing.

4.2.4 Size of Resource Base

The amount of biomass that could be produced annually in the United States is very difficult to estimate. In addition to the uncertainties in estimating yields on the wide diversity of landscapes that could be planted to biomass crops, a variety of social, political, economic, and environmental factors influence decisions on putting land into biomass production. Since these factors are not static, assessments of land available can change with time. It is not surprising that there is a wide opinion on the size of the biomass supply.

Two studies in the 1990s illustrate the divergence of these views. Graham [4] published an analysis of potential biomass supply in 1994 that foresaw 159 million hectares of land suitable for dedicated energy crops. Since "suitable" was defined as lands capable of producing in excess of 11.2 Mg/ha/year of dry biomass, the total supply of biomass was projected to be 1.8 billion Mg of dry biomass. Assuming heating value of 18 MJ/kg, this analysis, which ignores the contribution of agricultural residues to biomass supply, yields 32 billion GJ of energy.* In contrast, a study by Wyman and Goodman [5] a year earlier estimated dedicated energy crops could supply 26–52 billion GJ of energy. They also included agricultural residues, forestry residues, and MSWs in their analysis, which contributed another 8 billion GJ of biomass energy, yielding a total annual energy supply from biomass of up to 60 billion GJ. Since energy consumption in the United States is about 98 billion GJ per year [6], the Graham study suggests that biomass could supply up to 33% of U.S. energy demand while the Wyman and Goodman study places this percentage as high as 60%.

More recently, the United States Department of Agriculture sponsored a study of biomass supply, including both dedicated crops and agricultural residues [7]. The conclusion of this study is that in excess of 1.2 billion Mg of dry biomass, representing 25 billion GJ of energy could be produced in a sustainable manner. Since this report is the most recent and takes into consideration information from earlier studies, it will be examined here in more detail.

In 2003, the harvested biomass supply in the United States was 172 million dry Mg in the form of wood residues and pulping liquors from the forest

* In the United States, the unit of national energy consumption is the quad, which is 1 quadrillion British Thermal Units (BTU), equal to 1.054 billion GJ. Thus, in the above discussions, a billion GJ can be thought of as one quad.

TABLE 4.1
Present Biomass Supply in the United States (2003)

	Annual Biomass Supply (Million Dry Mg/Year)
Forestry products industry	
Wood residues	40
Pulping liquors	47
Urban woods and processing residues	32
Fuel wood	32
Grains for biofuels	16
Bioproducts	5
Biomass total	172

products industry, urban wood and processing residues, fuel wood, grains for biofuels, and bioproduct [6]. Table 4.1 details U.S. biomass sources, with pulping liquors representing the single largest source (27% of the total). With an average heating value of 18 MJ/kg, biomass contributed nearly 3 billion GJ to the nation's energy supply, which is only 3% of the total U.S. consumption. This is only a small fraction of what could be produced and harvested in a sustainable fashion.

The land base of the United States encompasses nearly 2263 million acres. About 33% is classified as forest land, 26% as grassland pasture and range, 20% as cropland, 8% special uses (e.g., public facilities), and 13% miscellaneous such as urban areas, swamps, and deserts [8,9]. The USDA study considered how the first three categories of land could be employed in biomass production, carefully excluding inaccessible and environmentally sensitive lands from consideration. It also accounted for use of these lands in the production of conventional forest products and agricultural commodities.

The study included some forward-looking projections on agricultural technology: small grain yields increases by 50%; residue ratio for soybeans increases to 2:1; harvest technology recovers 75% of crop residues (when sustainable removal is possible); all cropland is managed by no-till methods; 55 million acres of cropland, idle cropland, and pasture are dedicated to perennial bioenergy crops; and all manure in excess of allowable on-farm soil application is used for biofuel.

Table 4.2 summarizes the findings of the USDA study, which projects at least 1.2 billion Mg of dry biomass could be harvested annually in the United States. Forestry resources, in the form of fuelwood, milling residues, urban wood residue, logging residues, and wood recovered from forest thinning (for forest fire control) could yield 333 million Mg of dry wood per year or 27% of the total. Agricultural resources, in the form of crop residues, perennial crops, grains for biofuels, and processing residues and manure, could yield 904 million Mg of dry biomass per year. Crop residues and perennial crops represent 33% and 28%, respectively, of the total biomass supply.

TABLE 4.2

Potential Biomass Supply for the United States

	Annual Biomass Supply (Million Dry Mg/Year)
Logging and other residue	58
Fuel treatments	54
Urban wood residues	43
Wood processing residues	64
Pulping liquor	67
Fuelwood	47
Perennial crops	342
Crop residues	405
Process residues	79
Grain-to-ethanol	79
Total	1237

Assuming an average heating value of 18 MJ/kg, this 1.2 billion Mg supply of biomass represents 22 billion GJ of energy, or 21% of the total U.S. energy consumption. As the report points out, this could fulfill one-third of U.S. demand for transportation fuel. Thus, biomass could be a significant contribution to the future U.S. energy supply.

4.2.5 Important Properties of Biomass

Evaluation of biomass resources as potential feedstocks generally requires information about plant composition, heating value, bulk density, and production yields. Compositional information can be reported in terms of organic components, proximate analysis, or ultimate analysis. This information is widely scattered and often shows wide variations. A table of properties for a few representative kinds of fibrous biomass is given in Table 4.3. Compilations of relevant biomass properties are found in Brown [10], Klass [11], Jenkins et al. [12], and Hofbauer [13].

Analysis in terms of organic components reports the kinds and amounts of plant compounds including proteins, oils, sugars, starches, and lignocellulose (fiber). The components vary widely among plant parts. For example, corn grain is mostly starch (72 wt.%) with a relatively small amount of fiber (13 wt.%) while cornstover, that part of the crop left on the field, is mostly fiber (84%) with very little starch content. Often, the fiber, which is a polymeric composite of cellulose, hemicellulose, and lignin, is reported in terms of these three constituents. Knowledge of organic components is important in designing both biochemical and thermochemical processes since it will influence the kinds of products that can be produced.

Proximate analysis is important in developing thermochemical conversion processes for biomass. Proximate analysis reports the yields (% mass basis)

TABLE 4.3

Properties of Selected Biomass

	Organic Composition (wt.%)				Elemental Analysis (Dry wt.%)					Proximate Analysis (Dry wt.%)					
Feedstock	**Cellulose**	**Hemi-cellulose**	**Lignin**	**Other**	**C**	**H**	**O**	**N**	**Ash**	**Volatile Matter**	**Fixed C**	**Ash**	**HHV (MJ/kg)**	**Bulk Density (kg/m^3)**	**Yield (Mg/ha)**
Corn stover	53	15	16	16	44	5.6	43	0.6	6.8	75	19	6	17.7	160–300	8,400
Herbaceous crop	45	30	15	10	47	5.8	42	0.7	4.5	81	15	4	18.7	160–300	14,000
Woody crop	50	23	22	5	48	5.9	44	0.5	1.6	82	16	1.3	19.4	280–480	14,000

of various products obtained upon heating the material under controlled conditions; these products include moisture, volatile matter, fixed carbon, and ash. Since moisture content of biomass is so variable and can be easily determined by gravimetric methods (weighing, heating at 100°C, and reweighing), the proximate analysis of biomass is commonly reported on a dry basis. Volatile matter is that fraction of biomass that decomposes and escapes as gas upon heating a sample at moderate temperatures (about 400°C) in an inert (nonoxidizing) environment. Knowledge of volatile matter is important in designing burners and gasifiers for biomass. The remaining fraction is a mixture of solid carbon (fixed carbon) and mineral matter (ash), which can be distinguished by further heating the sample in the presence of oxygen: the carbon is converted to carbon dioxide leaving only the ash.

Ultimate analysis is simply the (major) elemental composition of the biomass on a gravimetric basis: carbon, hydrogen, oxygen, nitrogen, sulfur, and chlorine along with moisture and ash. Sometimes this information is presented on a dry, ash-free (daf) basis. This information is very important in performing mass balances on biomass conversion processes. Compared to fossil fuels, biomass is characterized by relatively high oxygen content of biomass (typically 40–45 wt.%), which detracts from its heating value and represents new challenges in converting these compounds into substitutes for the hydrocarbons that currently dominate our economy. In many instances, generic molecular formula based on one mole of carbon is convenient for performing mass balances on a process. Cellulose and starch have the generic molecular formula $CH_{1.7}O_{0.83}$, hemicellulose can be represented by $CH_{1.6}O_{0.8}$, and wood is $CH_{1.4}O_{0.66}$ [11].

Heating value is the net enthalpy released upon reacting a particular fuel with oxygen under isothermal conditions (the starting and ending temperatures are the same). If water vapor formed during reaction condenses at the end of the process, the latent enthalpy of condensation contributes to what is known as the higher heating value (HHV). Otherwise, the latent enthalpy does not contribute and the lower heating value (LHV) prevails. These measurements are typically performed in a bomb calorimeter and yield the HHV for the fuel. Heating values of biomass can be conveniently estimated from the percent of carbon in the biomass on a dry basis using the empirical relationship [11]:

$$\text{HHV in MJ/dry kg} = 0.4571 \times (\%\text{C on dry basis}) - 2.70 \tag{4.1}$$

For estimating purposes, assuming an HHV of 18 MJ/kg is a good approximation for many kinds of (dry) biomass.

Enthalpies of formation are very useful in thermodynamic calculations but this data is rarely tabulated for biomass because of the wide variability in its composition. However, if HHV for a biomass fuel has been determined in a bomb calorimeter, its enthalpy of formation can be determined by summing the enthalpies of formation of the products of combustion and subtracting from this sum the HHV. For example, suppose a sample of switchgrass has an elemental analysis that gives a generic molecular formula of $CH_{1.4}O_{0.8}$ (molecular weight of 26.2 kg/kmol) and its HHV is measured to be 18.1 MJ/kg. The combustion of 1 kmol of switchgrass can be represented by

$$CH_{1.4}O_{0.8} + 0.95O_2 \rightarrow CO_2 + 0.7H_2O\,(\text{liq}) \tag{4.2}$$

The enthalpy of reaction, ΔH_R, is calculated from the various enthalpies of formation, $\mathring{h}_f$:

$$\begin{aligned}\Delta H_R &= \mathring{h}_{f\,CO_2} + 0.7\mathring{h}_{f\,H_2O(liq)} - \left(\mathring{h}_{f\,CH_{1.4}O_{0.8}} + 0.95\mathring{h}_{f\,O_2}\right) - 18.1\frac{\text{MJ}}{\text{kg}}\left(26.2\frac{\text{kg}}{\text{kmol}}\right) \\ &= -393.5\frac{\text{MJ}}{\text{kmol}} + 0.7\left(-285.8\frac{\text{MJ}}{\text{kmol}}\right) - \left(\mathring{h}_{f\,CH_{1.4}O_{0.8}} + 0\right)\end{aligned} \tag{4.3}$$

Solving for the enthalpy of formation for $CH_{1.4}O_{0.8}$ yields

$$\mathring{h}_{f\,CH_{1.4}O_{0.8}} = -119.3\frac{\text{MJ}}{\text{kmol}} \tag{4.4}$$

This value can be used in various thermodynamic calculations for this switchgrass sample, whether gasification to hydrogen or hydrolysis to fermentable sugars.

Bulk density is determined by weighing a known volume of biomass that is packed or baled in the form anticipated for its transportation or use. Clearly, solid logs will have higher bulk density than the same wood chipped. Bulk density will be an important determinant of transportation costs and the size of fuel storage and handling equipment. Volumetric energy content is also important in transportation and storage issues. Volumetric energy content, which is simply the enthalpy content of fuel per unit volume, is calculated by multiplying the HHV of a fuel by its bulk density.

Knowledge of the annual yields of biomass crops (kilograms per hectare) is important in planning a biomass energy project. Yield depends on many factors including plant variety, crop management (fertilization and pest control), soil type, landscape, climate, weather, and water drainage. Thus, each project will require site-specific information obtained through discussions with state extension agents and local agronomists in combination with field trials in advance of detailed manufacturing plant design.

4.2.6 Summary

Biomass feedstocks are raw material for industrial processes and are classified as either waste materials or energy crops. Biomass waste is material of recent biological origin that has been discarded or has no value for a given process. Examples include agricultural residues, yard waste, MSW, and even sewage. Waste material is very heterogeneous and can contain a large variety of organics that can cause challenges for conversion processes. The major advantage of waste material is their typical low cost, even negative cost in some cases, although waste material costs can increase with usage. Energy crops are plants grown periodically for energy purposes. These plants can be harvested in cycles ranging from annually to 5–7 years and serve additional purposes such as food, feed, and fiber. Energy crops contain varying amounts of oil, sugars, starches, and fiber. Fiber is the dominant component for most kinds of biomass including "starch crops." A convenient classification of energy crops consists of woody and herbaceous crops. Woody energy crops are typically harvested every 5–7 years and are therefore known as short-rotation woody crops. Herbaceous crops are either annual or perennial, which are distinguished by whether they need to be replanted annually in the case of the former, or grow-back from their rootstock as the later.

Feedstock composition is an important consideration for an energy resource, and there are various analyses available to characterize energy feedstock. Biochemical analysis reports plant chemicals such as proteins, oils, sugars, starches, and fiber. Proximate analysis consists of the yield of various products after thermal heating of the material. These products consist of moisture, volatile matter, fixed carbon, and ash. Ultimate analysis is simply the (major) elemental composition of the biomass on a gravimetric basis: carbon, hydrogen, oxygen, nitrogen, sulfur, and chlorine along with moisture and ash. Other properties that are typically considered in energy applications are feedstock heating value, and inorganic composition.

Problems

4.1 Define waste biomass and list some examples.

4.2 Explain a scenario where a waste product is no longer considered "waste."

4.3 What are the main components and their typical proportions in MSW (not including yard waste)?

4.4 What is a common trait of waste materials?

4.5 What is an energy crop?

4.6 What measures can be taken to ensure that energy crops are employed in a sustainable fashion?

4.7 What are the major plant compounds in energy crops, and which one is the largest fraction?

4.8 Describe the two main categories of energy crops.

4.9 What is the approximate yield of SRWC (ton/ha)?

4.10 What are the three main analyses used to describe biomass feedstock.

4.11 Describe the process by which proximate analysis of a given material can be determined.

4.12 What is volatile matter and why is it important?

4.13 How are HHV and LHV differentiated?

4.14 Given the enthalpy of formation for food waste to be −394.7 MJ/kmol, use the combustion reaction from Equation 4.2 to calculate its HHV (assume the molecular formula is the same for switchgrass). Repeat this calculation for redwood (assume enthalpy of formation of −50.7 MJ/kmol).

4.15 What are some important variables that affect biomass yields?

4.3 Conversion of Biomass to Gaseous Fuels

4.3.1 Introduction

Some biomass conversion processes generate significant quantities of gaseous products. These are not usually of sufficient quality to be used directly as pipeline gas or transportation fuel but they can be employed in the production of heat and power. Anaerobic digestion and thermal gasification are the most important processes that yield gases as a primary product. In the case of anaerobic digestion, bacteria decompose organic waste to various gases while heat drives thermal gasification. Anaerobic digestion is usually applied to high moisture biomass while thermal gasification requires moisture content of biomass to be less than about 20 wt.%. The upgrading of these gaseous products to liquids or fuels suitable as transportation fuels is discussed in Section 4.4.

4.3.2 Anaerobic Digestion

Anaerobic digestion is the decomposition of organic waste to gaseous products by bacteria in an oxygen-free environment [14]. The process occurs in stages to successively break down the organic matter into simpler organic compounds. The desired product, known as biogas, is a mixture of methane (CH_4), carbon dioxide (CO_2), and some trace gases. Most digestion systems produce biogas that is between 55% and 75% methane by volume. Biogas, once treated to remove sulfur compounds, can substitute for natural gas in many applications, including stationary power generation. Biogas is a suitable fuel for engine generator sets, small gas turbines, and some kinds of fuel cells.

The biological processes within an anaerobic digester that lead to biogas, as summarized in Figure 4.1, are relatively complicated [15]. The process consists of three basic steps: hydrolysis and fermentation, transitional acetogenic dehydrogenation, and methanogenesis. Hydrolysis and fermentation involve hydrolytic and fermentative bacteria that break down proteins, carbohydrates, and fats into simpler acids, alcohols, and neutral compounds. Hydrogen and carbon dioxide are also produced. This step is followed by transitional digestion through the action of acid-forming bacteria. Products of fermentation that are too complex for methane-forming bacteria to consume are further degraded in this step to acetate, hydrogen, and carbon dioxide. Traces of oxygen in the feedstock are consumed in this step, which benefits oxygen-sensitive, methane-forming bacteria. The final step, methanogenesis, converts acetate to methane by the action of methane-forming bacteria.

Basic anaerobic digestion systems are relatively simple, as illustrated in Figure 4.1. Whether batch or steady flow processing, two effluent streams result: biogas and sludge. The biogas may have to be scrubbed in advance of combustion to remove hydrogen sulfide that would otherwise appear as the pollutant sulfur dioxide in the exhaust gas.

Examples of anaerobic reactor designs include the simple batch reactor, plug-flow reactors, continuously stirred tank reactors (CSTR), upflow reactors, and two-tank reactor systems. The batch reactor is a single vessel design in which waste is added and removed in batches and all steps of the digestion process take place in the same environment. More advanced reactor designs aim to improve waste contact with active bacteria and/or to separate and control the environments for acid-forming and methane-forming bacteria. Relatively new two-tank reactor technology physically separates the acid formation and methane formation phases of digestion so that each takes place under optimal conditions [16]. Benefits of this technology are numerous: hydrolysis and acidification occur quicker than in conventional systems, the common problems of foaming in single-tank systems are reduced by the destruction of biochemical foaming agents before they reach the methane forming reactor, and the biogas produced is typically rich in methane.

The biological conversion process of anaerobic digestion can be used for a wide range of feedstocks of medium-to-high moisture content. Sewage sludge is most commonly used as feedstock for anaerobic digesters, but MSW, food-processing wastes, agricultural wastes, Napier grass, kelp, bagasse, and water hyacinth can also be digested. Digesters are designed to maintain optimal conditions for a specific type of waste. Waste pretreatment, heating, mixing, nutrient addition, specialized bacteria addition, and pH adjustment can be manipulated to control digester performance.

Typical thermodynamic efficiency in converting the chemical energy of dry matter to methane is about 60% with methane yields ranging between

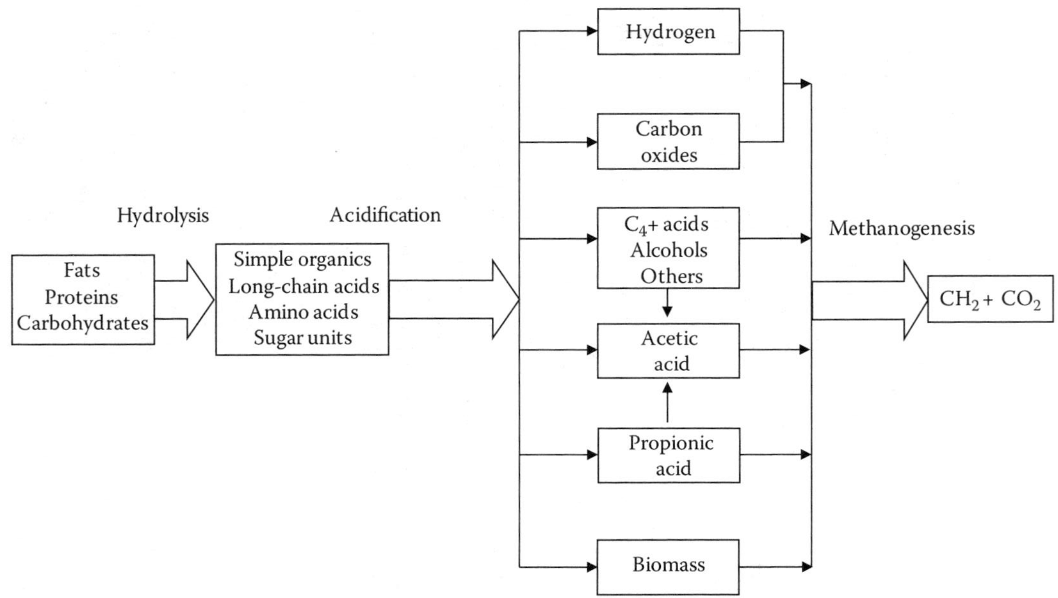

FIGURE 4.1
Microbial phases in anaerobic digestion. (Adapted from Klass, D., *Biomass for Renewable Energy, Fuels, and Chemicals*, Academic Press, San Diego, CA, p. 462, 1998.)

8000–9000 ft^3/ton of volatile solids added to the digester. However, thermodynamic efficiencies approaching 90% with methane yields of 11,000–13,000 ft^3/ton volatile solids have been achieved in research programs.

4.3.3 Thermal Gasification

Gasification is the high-temperature (750°C–850°C) conversion of solid, carbonaceous fuels into flammable gas mixtures, sometimes known as producer gas, consisting of carbon monoxide (CO), hydrogen (H_2), methane (CH_4), nitrogen (N_2), carbon dioxide (CO_2), and smaller quantities of higher hydrocarbons. The overall process is endothermic and requires either the simultaneous burning of part of the fuel or the delivery of an external source of heat to drive the process [17].

Gasification was placed into commercial practice as early as 1812 when coal was converted to gas for illumination (known as manufactured gas or town gas) in England. This technology was widely adopted in industrialized nations and was employed in the United States as late as the 1950s when interstate pipelines made inexpensive supplies of natural gas available [18]. Some places, such as China, still manufacture gas from coal.

The high volatile content of biomass (70–90 wt.%) compared to coal (typically 30–40 wt.%) and the high reactivity of its char make biomass an ideal gasification fuel. However, issues of cost and convenience of biomass gasification has limited its applications to special situations and niche markets. For example, in response to petroleum shortages during World War II, over 1 million small-scale wood gasifiers were built to supply low enthalpy gas for automobiles and steam boilers [17]. These were abandoned soon after the war.

Not only can producer gas be used for generation of heat and power, it can serve as feedstock for production of liquid fuels and chemicals. Because of this flexibility of application, gasification has been proposed as the basis for "energy refineries" that would provide a variety of energy and chemical products, including electricity and transportation fuels.

Figure 4.2 illustrates the several steps of gasification: heating and drying, pyrolysis, gas-solid reactions that consume char and gas-phase reactions that adjust the final chemical composition of the producer gas (see Figures 4.3 and 4.4) [19]. Drying and pyrolysis are similar to the corresponding processes for direct combustion described in the previous section. Pyrolysis begins between 300°C and 400°C and produces the intermediate products of char, gases (mainly CO, CO_2, H_2, and light hydrocarbons) and condensable vapor (including water, methanol, acetic acid, acetone, and heavy hydrocarbons). The distribution of these products depends on the chemical composition of the fuel and the heating rate and temperature achieved in the reactor. However, the total pyrolysis yield of pyrolysis products and the amount of char residue can be roughly estimated from the proximate analysis of the fuel. The fuel's volatile matter roughly corresponds to the pyrolysis yield while the combination of fixed carbon and ash content can be used to estimate the char yield.

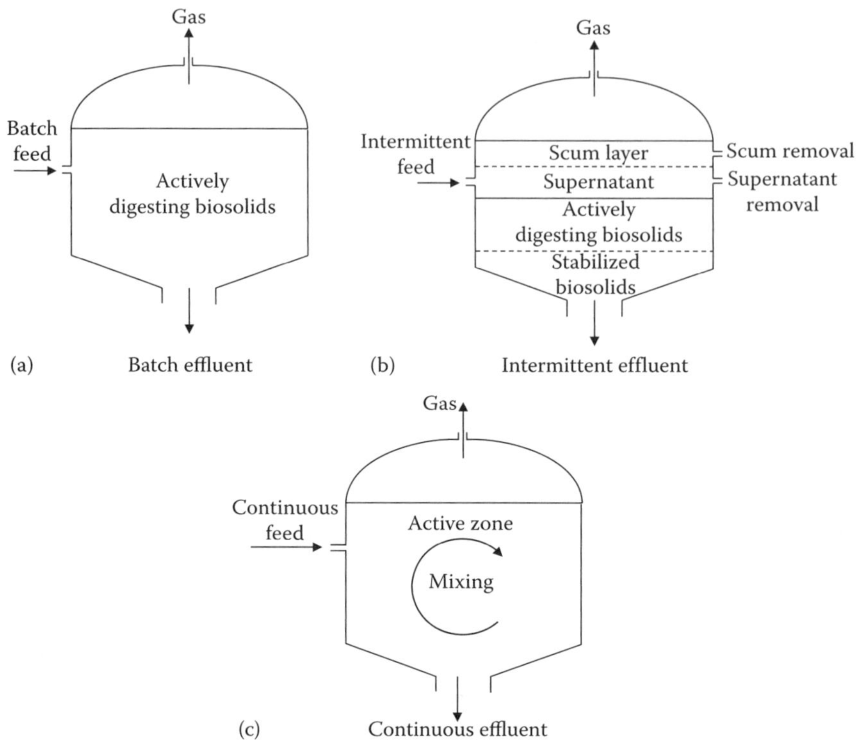

FIGURE 4.2
Types of anaerobic digesters: (a) batch fed, (b) intermittently fed, and (c) continuously stirred and fed.

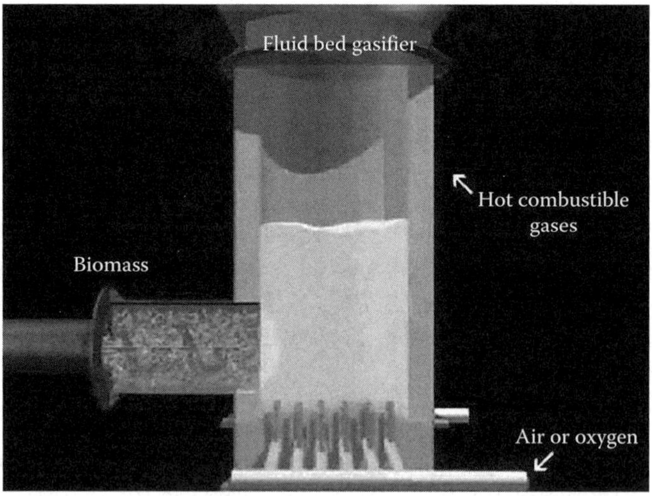

FIGURE 4.3
(See color insert.) Cross section of fluid plate gasifier.

FIGURE 4.4
(See color insert.) Frontline low-temperature gasifier.

Heating and drying are endothermic processes, requiring a source of heat to drive them. This heat can be supplied by an external source in a process called indirectly heated gasification. More typically, a small amount of air or oxygen (typically not more than 25% of the stoichiometric requirement for complete combustion of the fuel) is admitted for the purpose of partial oxidation, which releases sufficient heat for drying and pyrolysis as well as for the subsequent endothermic chemical reactions described below.

The third step of gasification is gas-solid reactions. These reactions convert solid carbon into gaseous CO, H_2, and CH_4 [20]:

$$\text{Carbon–oxygen reaction:} \quad C + \tfrac{1}{2}O_2 \leftrightarrow CO \qquad \Delta H_R = -110.5 \text{ MJ/kmol}$$

$$\text{Boudouard reaction:} \quad C + CO_2 \leftrightarrow 2CO \qquad \Delta H_R = 172.4 \text{ MJ/kmol}$$

$$\text{Carbon–water reaction:} \quad C + H_2O \leftrightarrow H_2 + CO \qquad \Delta H_R = 131.3 \text{ MJ/kmol}$$

$$\text{Hydrogenation reaction:} \quad C + 2H_2 \leftrightarrow CH_4 \qquad \Delta H_R = -74.8 \text{ MJ/kmol}$$

The first of these, known as the carbon–oxygen reaction, is strongly exothermic and is important in supplying the energy requirements for drying, pyrolysis, and endothermic solid–gas reactions. The hydrogenation reaction also contributes to the energy requirements of the gasifier, although significantly more char reacts with oxygen than hydrogen in the typical air-blown gasifier.

The fourth step of gasification is gas-phase reactions, which determine the final mix of gaseous products [20]:

Water–gas shift reaction: $CO + H_2O \leftrightarrow H_2 + CO_2 \quad \Delta H_R = -41.1\ \mathrm{MJ/kmol}$

Methanation: $CO + 3H_2 \leftrightarrow CH_4 + H_2O \quad \Delta H_R = -206.1\ \mathrm{MJ/kmol}$

The final gas composition is strongly dependent on the amount of oxygen and steam admitted to the reactor as well as the time and temperature of reaction. For sufficiently long reaction times, chemical equilibrium is attained and the products are essentially limited to the light gases CO, CO_2, H_2, and CH_4 (and nitrogen if air was used as a source of oxygen). Analysis of the chemical thermodynamics of these six gasification reactions reveals that low temperatures and high pressures favor the formation of CH_4 whereas high temperatures and low pressures favor the formation of H_2 and CO.

Often, gasifier temperatures and reaction times are not sufficient to attain chemical equilibrium and the producer gas contains various amounts of light hydrocarbons such as C_2H_2 and C_2H_4, as well as up to 10 wt.% heavy hydrocarbons that condense to a black, viscous liquid known as "tar" [21]. This latter product is undesirable as it can block valves and filters and interferes with downstream conversion processes. Steam injection and addition of catalysts to the reactor are sometimes used to shift products toward lower-molecular-weight compounds. Gasifiers are generally classified according to the method of contacting fuel and gas.

The four classes of gasifiers are updraft (countercurrent), downdraft (concurrent), fluidized bed, and entrained flow [22]. These are illustrated in Figure 4.5 while their performance characteristics are summarized in Table 4.4.

Updraft gasifiers, illustrated in Figure 4.6a, are the simplest as well as the first type of gasifiers developed. They were a natural evolution from charcoal kilns, which yield smoky, yet flammable, gas as a waste product, and blast furnaces, which generate product gas that reduces ore to metallic iron. Updraft gasifiers are little more than grate furnaces with chipped fuel admitted from above and insufficient air for complete combustion entering from below. Above the grate, where air first contacts the fuel, combustion occurs and very high temperatures are produced. Although the gas flow is depleted of oxygen higher in the fuel bed, hot H_2O and CO_2 from combustion near the grate reduce char to H_2 and CO. These reactions cool the gas, but temperatures are still high enough to heat, dry, and pyrolyze the fuel moving down toward the grate. Of course, pyrolysis releases both condensable and noncondensable gases, and the producer gas leaving an updraft gasifier contains large quantities of tars, on the order of $50\ \mathrm{g/m^3}$. As a result, updraft gasifiers are generally not strong candidates for biomass energy applications.

In downdraft gasifiers, fuel and gas move in the same direction. Downdraft gasifiers appear to have been developed near the end of the nineteenth century after the introduction of induced draft fans allowed air to be drawn

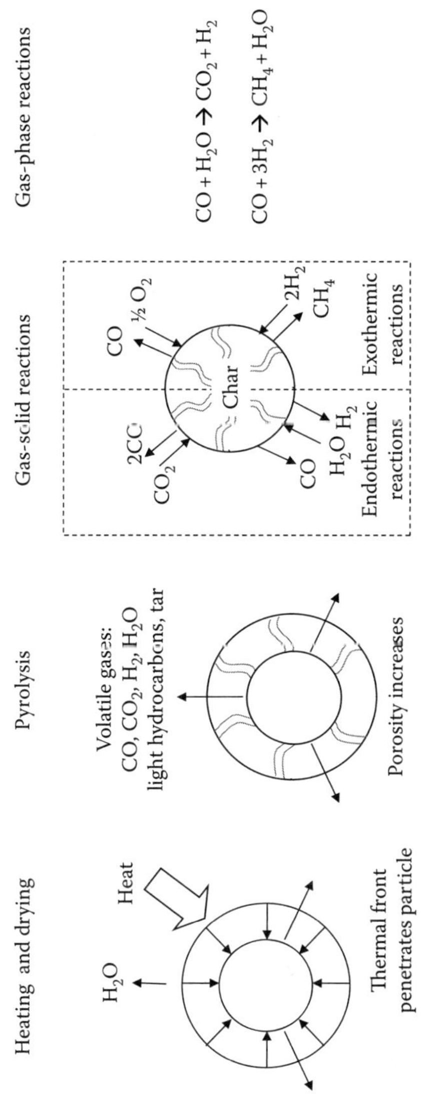

FIGURE 4.5
Processes of thermal gasification.

TABLE 4.4

Producer Gas Composition from Various Kinds of Gasifiers

Gasifier Type	Gaseous Constituents (Vol.% Dry)					HHV	Gas Quality	
	H_2	CO	CO_2	CH_4	N_2	(MJ/m^3)	Tars	Dust
Air-blown updraft	11	24	9	3	53	5.5	High (~10 g/m^3)	Low
Air-blown downdraft	17	21	13	1	48	5.7	Low (~1 g/m^3)	Medium
Air-blown fluidized bed	9	14	20	7	50	5.4	Medium (~10 g/m^3)	High
Oxygen-blown downdraft	32	48	15	2	3	10.4	Low (~1 g/m^3)	Low
Indirectly heated fluidized bed	31	48	0	21	0	17.4	Medium (~10 g/m^3)	High

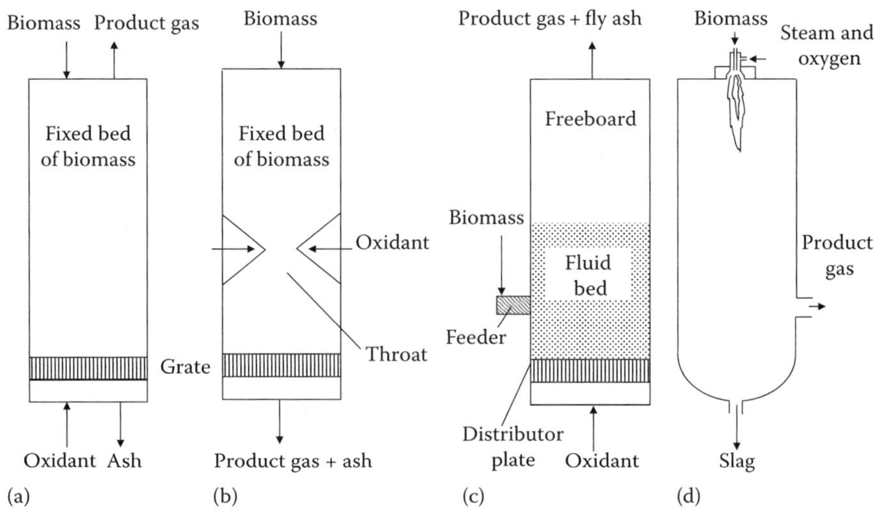

FIGURE 4.6
Common types of biomass gasifiers: (a) updraft, (b) downdraft, (c) fluidized bed, and (d) entrained flow.

downward through a gasifier in the same direction as the gravity-fed fuel. As shown in Figure 4.6b, contemporary designs usually add an arrangement of tuyeres that admit air or oxygen directly into a region known as the throat where combustion forms a bed of hot char. This design assures that condensable gases released during pyrolysis are forced to flow through the hot char bed, where tars are cracked. The producer gas is relatively free of tar (<1 g/m^3), making it a satisfactory fuel for engines. A disadvantage is the need for tightly controlled fuel properties (particles sized to between 1 and 30 cm, low ash content, and moisture less than 30%). Another disadvantage is a tendency for slagging or sintering of ash in the concentrated oxidation zone. Rotating ash grates or similar mechanisms can solve this problem. Furthermore, the incorporation of a throat region limits the maximum size of downdraft gasifiers to about 400 kg/h.

In a fluidized bed gasifier, illustrated in Figure 4.6c, a gas stream passes vertically upward through a bed of inert particulate material to form a turbulent mixture of gas and solid. Fuel is added at such a rate that it is only a few percent by weight of the bed inventory. Unlike the updraft and downdraft gasifiers, no segregated regions of combustion, pyrolysis, and tar cracking exist. The violent stirring action makes the bed uniform in temperature and composition with the result that gasification occurs simultaneously at all locations in the bed. Typically, a fluidized gasifier operates in the range of 700°C–850°C. By injecting fuel in the base of the bed, much of the tar can be cracked within the fluidized bed. However, a large insulated space above the bed, known as the freeboard, is usually included to promote additional tar cracking as well as more complete conversion of char. Nevertheless, tar production is intermediate between updraft and downdraft gasifiers (about 10 g/m^3). Fluidized beds are attractive for biomass gasification. They are able to process a wide variety of fuels including those of high moisture and small size. They are easily scaled to large sizes suitable for electric power production. Disadvantages include relatively high power consumption to move gas through the fluidized bed; high exit gas temperatures, which complicates efficient energy recovery; and relatively high particulate burdens in the gas due to the abrasive forces acting within the fluidized bed.

A fourth kind of gasifier, illustrated in Figure 4.6d, employs finely pulverized fuel in an entrained flow reactor. This reactor was developed for steam-oxygen gasification of coal at temperatures of 1200°C–1500°C. These high temperatures assure excellent char conversion (approaching 100%) and low tar production, and convert the ash to molten slag, which drains from the bottom of the reactor. The technology is attractive for advanced coal power plants but has not been widely explored for biomass because of the expense of finely dividing biomass, the difficulty of reaching high temperatures with biomass, and the presence of alkali in biomass, which leads to severe ash sintering. Despite these difficulties, entrained flow reactors are being considered for gasification of pretreated biomass (such as torrefied biomass and pyrolysis liquids) [23].

The fuel/air ratio is the single most important parameter for determining gasifier performance. Downdraft gasifiers can have better conversion efficiency and producer gas quality than fluidized bed gasifiers because they utilize a higher fuel/air ratio. Fluidized bed gasifiers, on the other hand, can generally handle a wider range of biomass feedstocks with higher moisture content.

Recent research in biomass gasification has focused on improving the heating value of product gas. Conventional gasification admits sufficient air or oxygen to the reactor to burn part of the fuel thus releasing heat to support pyrolysis of the rest of the fuel. Gas produced in air-blown biomass gasifiers typically has heating value that is only 15%–20% that of natural gas. This low heating value is largely the result of nitrogen diluting the fuel gas.

Oxygen can be used as the fluidization agent, but high capital costs preclude this from consideration except for relatively large-scale gasification systems.

Indirectly heated gasification can improve gas heating value by physically separating combustion and pyrolysis [22]. As a result, the products of combustion do not appear in the fuel gas. HHVs of 14,200 kJ/m^3 or higher are expected. Several schemes have been suggested for transporting heat from the combustion reactor to the pyrolysis reactor. These include transferring hot solids from the combustor to the pyrolyzer, transferring a chemically regenerative heat carrier between two reactors, transferring heat through a wall common to the reactors, and storing heat in high-temperature phase-change material.

Depending upon the kind of gasifier, product gas can be contaminated by tar and particulate matter. Tar cannot be tolerated in many downstream applications, including internal combustion engines, fuel cells, and chemical synthesis reactors. Tar can deposit on surfaces in filters, heat exchangers, and engines where they reduce component performance and increase maintenance requirements. A well-designed gasifier has a high-temperature zone through which pyrolysis products pass, allowing tars to thermally decompose ("crack") into the low molecular gases predicted by equilibrium theory. The production of particulate matter, a combination of ash and unreacted char from the biomass, varies considerably from one type of gasifier to another. Particulate matter can also deposit on critical power system components, degrading performance even at relatively low concentrations in some machinery.

Gas cleaning usually involves separate processes for removing solid particles and tar vapor or aerosols. A gas cyclone operated above the condensation temperature for the least volatile tar constituent can remove most ash and char particles larger than about 5 μm in diameter. If removal of finer particles is required, then additional filtration is required, such as ceramic candle filters or moving bed granular filters.

Two approaches are available for removing tar from product gas [21]. Scrubbing the gas stream with a fine mist of water or oil removes both tar and particles from the gas stream. Gas scrubbers are widely used in the chemical industry and are relatively inexpensive. However, cooling the gas and removing organic compounds reduce both sensible energy and chemical energy, which decreases overall energy conversion efficiency. Gas scrubbing also produces a waste-water stream contaminated with sometimes toxic tar, which complicates waste disposal.

The other approach to removing tar converts it into low-molecular-weight compounds, which has the advantage of increasing the heating value of the producer gas while eliminating a waste-disposal problem. This "cracking" of tar into CO and H_2 can be accomplished by raising the gas to temperatures exceeding 1000°C. Such high temperatures are not easily achieved in most biomass gasifiers. The use of catalysts allows tar cracking to occur at much lower temperatures, in the range of 600°C–800°C. The gas is passed through a packed bed reactor containing catalyst with steam or oxygen sometimes used to promote reaction.

The thermodynamic efficiency of gasifiers is strongly dependent on the kind of gasifier and how the product gas is employed [22]. Some high-temperature, high-pressure gasifiers are able to convert 90% of the chemical energy of solid fuels into chemical and sensible heat of the product gas. However, these high efficiencies come at high capital and operating costs. Most biomass gasifiers have conversion efficiencies ranging between 70% and 90%. In some applications, such as process heaters or driers, both the chemical and sensible heat of the product gas can be utilized. In many power applications, though, the hot product gas must be cooled before it is utilized; thus, the sensible heat of the gas is lost. In this case, "cold gas" efficiency can be as low as 50%–70%. Whether the heat removed from the product gas can be recovered for other applications, like steam raising or fuel drying, ultimately determines which of these conversion efficiencies is most meaningful. Gas cleaning to remove tar and particulate matter also has a small negative impact on gasifier efficiency since it removes flammable constituents from the gas (tar and char particles) and generally requires a small amount of energy to run pumps.

4.3.4 Summary

Biomass conversion can generate significant quantities of gaseous products that can be employed for heat or power generations. Anaerobic digestion and thermal gasification are broad classifications for the conversion of biomass into gas products. Anaerobic digestion produces a mixture of gases called biogas composed of methane, carbon dioxide, and trace gases. Biogas can substitute for natural gas if proper care is taken to remove sulfur. Thermal gasification on the other hand, converts solid fuels to a gaseous mixture known as producer gas. Applications for producer gas include heat and power production, as well as conversion to liquid fuels.

Anaerobic digestion is a complicated biological process that employs three major steps: hydrolysis and fermentation, transitional acetogenic dehydrogenation, and methanogenesis. Various reactor designs are available to help facilitate contact between bacteria and feedstock, and advanced designs can control the environment to optimize bacterial activity.

Thermal gasification consists of two processes in parallel: combustion and pyrolysis. Combustion is an exothermic process that provides heat to drive endothermic pyrolysis reactions. Combustion rapidly oxidizes fuel while pyrolysis degrades solid fuel into intermediate gas, and liquid organic products, which later form gasification products. Thermal gasification is a temperature-dependent process with different reactions dominating at temperatures of 100°C, 400°C, and 700°C. At high temperatures, char reacts with oxygen, carbon dioxide, and water vapor, providing heat for a series of equilibrium reactions. Gasifiers are generally classified according to the method of contacting fuel and gas with the four main types being counterflow, coflow, fluidized bed, and entrained flow. Gasification streams must be cleaned of various contaminants including sulfur, tars, and other components depending on the end-use application.

Problems

4.16 What are the two common names for the mixture of gases produced by anaerobic digestion and thermal gasification?

4.17 List common compounds found in gases produced by anaerobic digestion and thermal gasification.

4.18 What are the products of anaerobic digestion?

4.19 Advanced reactor designs employ more than one vessel for anaerobic digestion; how is the digestion process different in two tank designs and what are the advantages?

4.20 What is the typical thermodynamic efficiency conversion of dry matter to methane?

4.21 Describe the exothermic and endothermic processes that occur during gasification.

4.22 Describe the formation and decomposition of char during biomass gasification.

4.23 What are the major constituents of gasification and what reactions produce them?

4.24 What gasification conditions favor the formation of methane? List, and describe, the four main gasifier designs.

4.25 Entrained flow gasifiers are very attractive to process coal, but not biomass—what are two main reasons that make entrained flow gasifiers unattractive for biomass gasification?

4.26 How does a fluid bed gasifier operate, and what are its advantages and disadvantages?

4.27 What is tar, and how is it typically removed from the gas stream?

4.4 Transportation Fuels from Biomass

4.4.1 Introduction

Almost 25% of energy consumption in the United States goes toward transportation. Approximately half of this amount comes from imported petroleum. Thus, development of transportation fuels from biorenewable resources is a priority if decreased dependence on foreign sources of energy is to be achieved.

Table 4.5 lists properties of both traditional and biobased transportation fuels [24–27]. Traditional transportation fuels are classified as gasoline, diesel fuel, or jet fuel. Gasoline is intended for spark-ignition (Otto cycle) engines;

TABLE 4.5

Comparison of Transportation Fuels

	Fossil Fuel-Derived		Biomass-Derived							
Fuel Type	Gasoline[24]	No. 2 Diesel Fuel[24]	Methanol[24]	Ethanol[24]	Methyl Ester[24] (from Soybean Oil)	Fischer–Tropsch A[24]	Hydrogen[25]	Methane[25]	Ammonia[26]	Dimethyl Ether[27]
Specific gravity[a]	0.72–0.78	0.85	0.796	0.794	0.886	0.770	0.071 (liq)	0.422 (liq)	0.682 (liq)	0.660
Kinematic viscosity at 20°C–25°C (mm^2/s)	0.8	2.5	0.75	1.51	3.9	2.08	105 [106]	16.5 [106]	14.7 [107]	0.227
Boiling point range (°C)	30–225	210–235	65	78	339	164–352	−253	−162	−33.4 [107]	−24.9
Flash point (°C)	−43	52	11	13	188	58.5		−184	—	—
Autoignition temperature (°C)	370	254	464	423	—	—	566–582	540	630	235
Octane no. (research)	91–100	—	109	109	—	—	>130	>120	110	—
Octane no. (motor)	82–92	—	89	90	—	—	—	—	—	—
Cetane no.	<15	37–56	<15	<15	55	74.6	—	—	—	>55
Heat of vaporization (kJ/kg)	380	375	1185	920	—	—	447	509	1371 [107]	402 [108]
LHV (MJ/kg)	43.5	45	20.1	27	37	43.9	120	49.5	18.8	28.88

Note: Superscripted numbers are references to the literature.

[a] Measured at 16°C except for liquefied gases, which are saturated liquids at their respective boiling points.

thus, it is relatively volatile but resistant to autoignition during compression. Diesel fuel is intended for use in compression-ignition (Diesel cycle) engines; thus, it is less volatile compared to gasoline and more susceptible to autoignition during compression. Jet fuel is designed for use in gas turbine (Brayton cycle) engines, which are not limited by autoignition characteristics but otherwise have very strict fuel specifications for reasons of safety and engine durability.

Gasoline is a mixture of hundreds of different hydrocarbons obtained from a large number of refinery process streams that contain between 4 and 12 carbon atoms with boiling points in the range of 25°C–225°C [25]. Most of the mixture consists of alkanes with butanes and pentanes added to meet vapor pressure specifications. A few percent of aromatic compounds are added to increase octane number, the figure of merit used to indicate the tendency of a fuel to undergo premature detonation within the combustion cylinder of an internal combustion engine. The higher the octane number, the less likely a fuel will detonate until exposed to an ignition source (electrical spark). Premature denotation is responsible for the phenomenon known as engine knock, which reduces fuel economy and can damage an engine. Various systems of octane rating have been developed, including research octane and motor octane numbers. Federal regulation in the United States requires gasoline sold commercially to be rated using an average of the research and motor octane numbers. Gasoline rated as "regular" has a commercial octane number of about 87 while premium grade is 93.

Diesel fuel, like gasoline, is also a mixture of light distillate hydrocarbons but with lower volatility and higher viscosity. Because diesel fuel is intended to be ignited by compression rather than by a spark, its autoignition temperature is lower than for gasoline [28]. The combustion behavior of diesel fuels are conveniently rated according to cetane number, an indication of how long it takes a fuel to ignite (ignition delay) after it has been injected under pressure into a diesel engine. A high cetane number indicates short ignition delay; for example, No. 2 diesel fuel has cetane number of 37–56 while gasoline has a cetane number less than 15.

Jet fuel is designated as either Jet A fuel or Jet B fuel. Jet A fuel is a kerosene type of fuel with relatively high flash point whereas Jet B fuel is a wide boiling range fuel, which more readily evaporates.

Biobased transportation fuels, also known as biofuels, are currently dominated by ethanol and biodiesel. Ethanol, by virtue of its high octane number, is suitable for use in spark-ignition engines, while the high cetane numbers of biodiesel, which are methyl or ethyl esters formulated from vegetable or animal fats, make them suitable for use in compression-ignition engines. However, there are other candidate liquid biofuels including methanol, mixed alcohols, Fischer–Tropsch liquids, and alkanes from hydrocracked bio-oil and lipids, as well as gaseous biofuels including hydrogen, methane, ammonia, and dimethyl ether, which will also be discussed.

4.4.2 Grain Ethanol

The biological production of ethanol and its distillation have been practiced for thousands of years. This fact combined with the wide availability of suitable fermentation feedstocks (starch and sugar crops) explains the early entry of ethanol into the biofuels market. In fact, Henry Ford had intended to fuel the Model T with ethanol, but inexpensive gasoline soon replaced it.

Ethanol has some shortcomings compared to gasoline. On a volumetric basis, it only has 66% of the heating value of gasoline. However, fuel economy depends on many complex interactions between a fuel and the combustion environment within an engine, which some argue improves the relative performance of ethanol [29]. For example, the higher octane number for ethanol compared to gasoline (109 vs. 91–101) allows engines to be designed to run at higher compression ratios, which improves both power and fuel economy. Estimates for efficiency improvements in engines optimized for ethanol instead of gasoline range are 15%–20%, resulting in a driving range approaching 80% of that of gasoline [30].

A significant problem with ethanol–gasoline blends is water-induced phase separation. Water contaminating a storage tank or pipeline is readily absorbed by ethanol, resulting in a lower water-rich layer and an upper hydrocarbon-rich layer, which interferes with proper engine operation. Water contamination is a problem that has not been fully addressed by the refining, blending, and distribution industries; thus, the transportation of ethanol–gasoline blends in pipelines is not permitted in the United States, and long-term storage is to be avoided [31].

A variety of carbohydrates (sugars, starches, hemicellulose, and cellulose) can serve as feedstock in ethanol fermentation as long as they can be broken down to sugars that can be fermented by yeasts or bacteria. Plant materials high in sugar, such as sugarcane, sugar beets, and sweet sorghum, are the simplest to convert to ethanol since these sugars can be directly fermented. Sugarcane, for example, is the basis of Brazil's fuel ethanol industry.

The process becomes more difficult as the carbohydrate becomes more complex. Starch, a storage carbohydrate for plants, requires an additional processing step before it can be used in fermentation. Starch is essentially long chains of glucose connected by hydrogen bonds, which can be decomposed into simple sugars by breaking these bonds. The process, known as hydrolysis, is accomplished by cooking milled starch granules in a dilute acid solution or by reacting the starch with thermophyllic (heat tolerant) enzymes. The cost of hydrolyzing starch is a very small proportion of the cost of producing ethanol. The cost of starch crops dominates process economics. Starch crops suitable as feedstock include grains and potatoes. Corn starch has been the basis for the United States fuel ethanol industry for over two decades.

Inulin is another storage carbohydrate suitable for fermentation to ethanol. Like starch, inulin is essentially chains of sugars that can be hydrolyzed

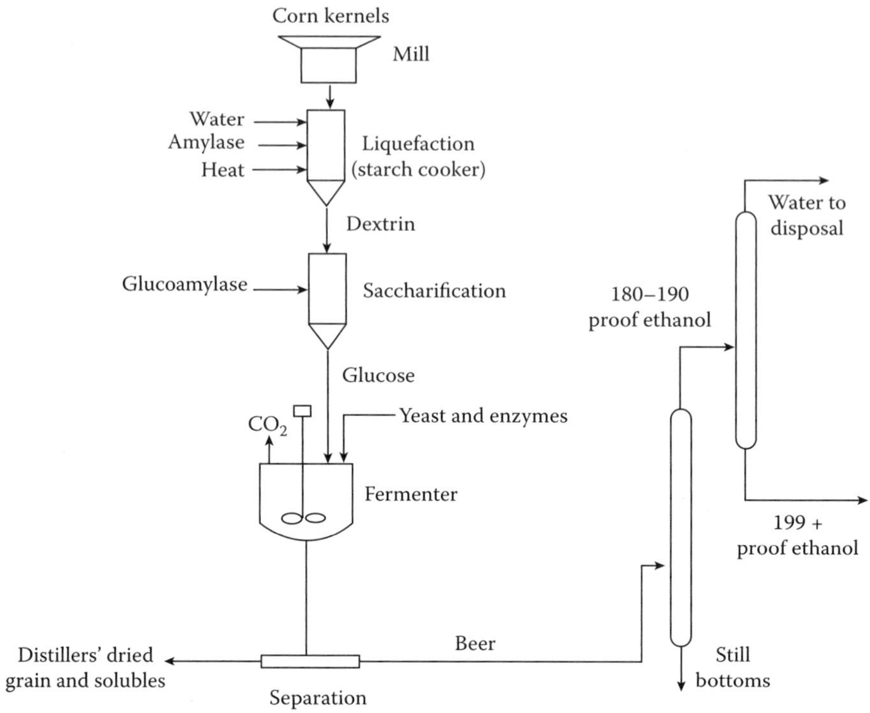

FIGURE 4.7
Dry milling of corn.

to simple sugars by acid or enzymes. Jerusalem artichokes and dahlias are notable for their high inulin content.

Grain ethanol plants are generally categorized as either dry grind or wet-milling plants depending on the pretreatment method. Dry grind plants, shown in Figure 4.7, grind the whole kernel while wet-milling plants soak the grain with water and acid to separate the corn germ, fiber, gluten, and starch components before mechanical grinding [32]. The capital investment for dry grind is less than that for a comparably sized wet-milling plant. However, the higher value of its by-products, greater product flexibility, and simpler ethanol production can make a wet-milling plant a more profitable investment.

Dry grind ethanol takes place in four major steps: pretreatment, cooking, fermentation, and distillation. Pretreatment consists of grinding the corn kernel into flour "meal," which is mixed with water, enzymes, and ammonia. This mixture ("mash") is then "cooked" to reduce bacteria levels. After cooling, the mash is sent to the fermenter where it remains for 40 h or more. The beer resulting from fermentation consists of a mixture of ethanol and stillage. Energy-intensive distillation of the beer is necessary to separate the stillage and water content from the ethanol and achieve maximum concentrations of 95%. This is followed by further purification to 100% ethanol using molecular sieves.

A modern dry grind plant will produce over 2.7 gal of ethanol per bushel of corn processed. Yields of coproducts per bushel of corn are 7.7–8.2 kg (17–18 lb) of dry distillers grain with solubles (DDGS) and 7.3–7.7 kg (16–17 lb) of carbon dioxide evolved from fermentation, the latter of which can be sold to the carbonated beverage industry. As a rule of thumb, the three products are produced in approximately equal weight per bushel.

Costs for ethanol from corn grain have been developed by the USDA [33] for the dry mill process. Capital costs for a 40 million gallon per year ethanol plant were estimated at $46.7 million with fuel production costs of $1.03 per gallon (2006).

4.4.3 Cellulosic Ethanol

Much of the carbohydrate in plant materials is structural polysaccharides, providing shape and strength to the plant. This structural material, known as lignocellulose, is a composite of cellulose fibers embedded in a cross-linked lignin–hemicellulose matrix [34]. Depolymerization to basic plant components is difficult because lignocellulose is resistant to both chemical and biological attack [35].

Cellulose to ethanol consists of four steps: pretreatment, enzymatic hydrolysis, fermentation, and distillation [32]. Of these, pretreatment is the most costly step, accounting for about 33% of the total processing costs [36]. An important goal of all pretreatments is to increase the surface area of lignocellulosic material, making the polysaccharides more susceptible to hydrolysis. Thus, comminution, or size reduction, is an integral part of all pretreatments. Some pretreatments are thought to reduce crystallinity of cellulose, which improves reactivity, but this does not appear to be the key for many successful pretreatments.

Three basic methods for hydrolyzing structural polysaccharides in plant cell walls to fermentable sugars are available: concentrated acid hydrolysis, dilute acid hydrolysis, and enzymatic hydrolysis [35,37]. The two acid processes hydrolyze both hemicellulose and cellulose with very little pretreatment beyond comminution of the lignocellulosic material to particles of about 1 mm size. The enzymatic process must be preceded by extensive pretreatment to separate the cellulose, hemicellulose, and lignin fractions.

Enzymatic hydrolysis was developed to better utilize both cellulose and hemicellulose from lignocellulosic materials. Pretreatment solubilizes hemicellulose under milder conditions than those required for acid hydrolysis of cellulose. Subsequent enzymatic hydrolysis of the cellulose does not degrade pentoses released during prehydrolysis. Cellulose is a homopolysaccharide of glucose linked by β-1,4′-glycosidic bonds. Thus, enzymatic hydrolysis of cellulose proceeds in several steps to break glycosidic bonds by the action of a system of enzymes known as cellulase. The system of enzymes also usually contains hemicellulase to hydrolyze any hemicellulose not solubilized by prehydrolysis.

Simultaneous saccharification and fermentation (SSF) has been developed for fermenting sugars released from lignocelluloses [35,37]. The SSF process

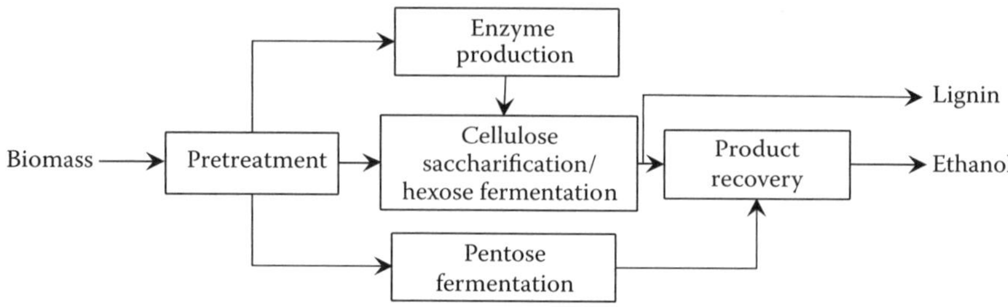

FIGURE 4.8
Cellulose-to-ethanol processing (via enzymatic hydrolysis).

combines hydrolysis (saccharification) and fermentation to overcome end product inhibition that occurs during hydrolysis of cellobiose. By combining hydrolysis and fermentation in the same reactor, glucose is rapidly removed before it can inhibit further hydrolysis. The SSF process, illustrated in Figure 4.8, begins with the grinding of the biomass followed by prehydrolysis to yield a mixture of pentoses, primarily xylose and arabinose, and fiber. The pentose is separated from the undigested fibers and fermented with appropriate yeast or bacteria to yield ethanol. The pretreated fibers are mixed with cellulases, which are either purchased commercially or produced on site, along with yeast and nutrients. The cellulose is solubilized to hexose (glucose), which is simultaneously fermented to ethanol. The rate-limiting step is the hydrolysis of cellulose to glucose. Lignin is separated from the mixture and used as boiler fuel. The beer is distilled to ethanol in a process identical to that employed after sugar or starch fermentations.

Although thermodynamic efficiencies for conversion of carbohydrates to ethanol can be calculated, it is more typical to report the volumetric yield of ethanol per unit mass of feedstock. The yield of ethanol from energy crops varies considerably. Among sugar crops, sweet sorghum yields 80 L/ton, sugar beets yield 90–100 L/ton, and sugarcane yields 75 L/ton. Among starch and inulin crops, the ethanol yield is 350–400 L/ton of corn, 400 L/ton of wheat, 90 L/ton of Jerusalem artichoke. Among lignocellulosic crops, the potential ethanol yield is 400 L/ton of hybrid poplar, 450 L/ton for corn stover, 510 L/ton for corn cobs, and 490 L/ton for wheat straw.

Researchers at the National Renewable Energy Laboratory [38] developed a 2012 market target design report showing capital costs of $114 million and operating costs of $1.07 per gallon (2000) of ethanol for a 69 million gallon per year ethanol plant.

4.4.4 Other Fuels from Sugar Fermentation

Ethanol has several limitations as a transportation fuel, including its affinity for water, which prevents it from being fully compatible with the existing fuel infrastructure, and its low volumetric heating value, which is only two-thirds that of gasoline. For this reason, fermentations that produce

TABLE 4.6

Butanol Fuel Properties in Comparison to Gasoline and Other Alcohols

Fuel	Formula	Energy Content (MJ/L)	Motor Octane Number	Air-to-Fuel Ratio	Vapor Pressure (psi at 100°F)
Methanol	CH_3OH	16	104	6.6	4.6
Ethanol	C_2H_5OH	19.6	102	9	2
Butanol	C_4H_9OH	29.2	78	11–12	0.33
Gasoline	Hydrocarbons	32	81–99	12–15	4.5

metabolites other than ethanol have been proposed. Alternative fermentations could produce hydrophobic molecules that are less oxidized than ethanol including higher alcohols (most prominently butanol), fatty acids, fatty alcohols, esters, alkanes, alkenes, and isoprenes.

Butanol is a three-carbon sugar with superior fuel properties compared to either methanol or ethanol. As shown in Table 4.6, butanol has volumetric energy content that is 91% of gasoline. Furthermore, it can be burned with an air-to-fuel ratio similar to that for gasoline and fuel tank emissions will be very low compared to the alternatives due to its low vapor pressure. Butanol's one disadvantage is an octane rating below gasoline. However, because it is fully compatible with gasoline, it could be blended to desired octane numbers.

At one time, butanol was fermented from sugar using the bacteria *Clostridium acetobutylicum* in a process known as acetone, butanol, and ethanol (ABE) fermentation. It was subsequently displaced by petroleum-derived butanol, but growing interest in biobased products has renewed interest in this bacterial fermentation [39]. Among the major challenges is increasing butanol yields, which rarely exceeds 20 g/L compared to 150 g/L for traditional ethanol fermentations. Butanol is highly toxic to microorganisms, which presents a major challenge in improving yields even through metabolical engineering.

Researchers have developed a cost estimate for a high-yield (21 g/L) butanol process [40]. The capital cost of a 136 million liters butanol plant is estimated as $110 million with operating costs of $68.4 million.

Hydrocarbons, the basis of our current transportation system, could potentially be synthesized in microbial fermentations. Hydrocarbon synthesis is widespread among both prokaryotes and eukaryotes [41]. Even the common yeast *Saccharomyces* can synthesize *n*-alkanes and unsaturated and branched hydrocarbons from C10 to C34 yielding as much as 10.2 wt.% of dry biomass under anaerobic conditions [42]. The highest reported yield of hydrocarbons in wild-type bacteria is the halotolerant bacterium *Vibrio furnissii*, which produced C10 to C34 hydrocarbons at 60 wt.% of dry biomass [43]. More typically, hydrocarbon yields are just a few percent weight of dry biomass.

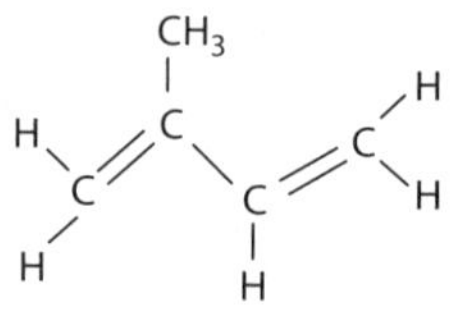

FIGURE 4.9
Structural formula for isoprene

Metabolic engineering could improve yields of hydrocarbons by mapping metabolic fluxes associated with their production and identifying enzymes responsible for hydrocarbon synthesis. For example, isoprene synthesis has been promoted in both eukaryotes and prokaryotes through metabolic engineering [44]. The isoprene molecule, illustrated in Figure 4.9, is characterized by pairs of conjugated double bonds [1], making it ideal for refining into fuels and commodity chemicals. Although a number of start-up companies are pursuing this approach to advanced biofuels, very little cost data is available in the public domain to evaluate its economic prospects.

4.4.5 Ethanol from Syngas

Efforts in Germany during World War II to develop alternative motor fuels discovered that iron-based catalysts could yield appreciable quantities of water-soluble alcohols from syngas, especially ethanol [18]:

$$CO + 3H_2 \rightarrow CH_3CH_2OH \quad (4.5)$$

These early efforts yielded liquids containing as much as 45%–60% alcohols of which 60%–70% was ethanol. Working at pressures of around 50 bar and temperatures in the range of 220°C–370°C, researchers have developed catalysts with selectivity to alcohols of over 95%, but production of pure ethanol has been elusive.

Some researchers have advocated the use of "mixed alcohols" as transportation fuels because the product typically contains a mixture of methanol, ethanol, 1-propanol, and 2-propanol. One advantage is the ability to use lower H_2:CO ratios than is required for methanol or Fischer–Tropsch synthesis [45]:

$$nCO + 2nH_2 \rightarrow C_nH_{2n+1}OH + (n-1)H_2O \quad (4.6)$$

with n typically ranging from 1 to 8. The process was commercialized in Germany between 1935 and 1945 but eventually abandoned because of the increased availability of inexpensive petroleum. Working at pressures of around 50 bar and temperatures in the range of 220°C–370°C, researchers have developed catalysts with selectivity to alcohols of over 95%, but production of pure ethanol has been elusive.

An alternative approach to obtaining neat ethanol from syngas is to first synthesize methanol and subsequently reacting this product with additional syngas [46,47]:

$$CH_3OH + 2CO + H_2 \rightarrow CH_3CH_2OH + CO_2 \quad (4.7)$$

Direct carbonylation of methanol has the advantage of yielding ethanol without coproduct water, which would eliminate energy-intensive distillations. The cost-effectiveness of this approach to ethanol synthesis has not been proven.

In addition to this catalytic route for converting syngas to ethanol, syngas can be biocatalytically converted to ethanol. In a process known as syngas fermentation, microorganisms consume the CO, CO_2, and H_2 in syngas to form a variety of products including carboxylic acids, alcohols, and esters [48]. Syngas fermentation has several advantages compared to catalytic upgrading of syngas [49]. Most catalysts used in the petrochemical industry are readily poisoned by sulfur-bearing gases whereas syngas-consuming anaerobes are sulfur tolerant. Unlike metal catalysts, biocatalysts do not require high pressure or specific CO/H_2 stoichiometries. Gas-phase catalysts typically employ temperatures of several hundreds of degree centigrade and at least 10 atm of pressure, whereas syngas fermentation proceeds at near-ambient conditions.

Several barriers must be overcome before syngas fermentation is commercially viable including growth rate and production of microorganisms, product inhibition by acids and alcohols, and limitations of syngas mass transfer [49]. Detailed economic analyses of syngas fermentation to ethanol have not appeared in the published literature, although Coskata, Inc., which is commercializing the process, reports that production cost (minus cost of capital recovery) will be less than $1 per gallon [50].

4.4.6 Ethanol from Bio-Oil

Bio-oil is a mixture of oxygenated organic compounds produced from fast pyrolysis, which is the rapid thermal decomposition of plant materials in the absence of oxygen at temperatures approaching 500°C. Although bio-oil is not usually considered a substrate for fermentation, it represents an alternative to enzymatic hydrolysis for the production of fermentable sugars from cellulose. Bio-oil fermentation is based on the observation that under rapid pyrolytic conditions, pure cellulose yields an anhydrosugar known as levoglucosan, which can be hydrolyzed to glucose. However, most biomass contains alkali, which serves as a catalyst for reactions that yield degradation products unsuitable for fermentation. If alkali is removed from biomass by washing in dilute acid before pyrolysis, cellulose can be thermally converted to fermentable sugar.

So and Brown [51] have performed one of the only techno-economic analyses of bio-oil fermentation. Their total capital investment estimate is $69 million with operating costs of $1.57 per gallon for an ethanol plant based on bio-oil fermentation producing 25 million gallons.

4.4.7 Methanol

Methanol is one of the ten most produced chemicals in the world. Methanol is a clear, odorless, flammable liquid that can be generated from any hydrocarbon. Methanol production is a mature process with process efficiencies of 99% or higher. Almost all methanol synthesis is derived from natural gas because of lower processing costs [52]. It is an intermediate chemical in the production of formaldehyde, dimethyl ether, acetic acid, and others. Methanol can also be blended with gasoline for use as a clean transportation fuel.

As a transportation fuel, it has many of the same advantages and disadvantages as ethanol [53]. The fuel properties of methanol are similar to those of ethanol: narrow boiling point range, high heat of vaporization, and high octane number. It has only 49% of the volumetric heating value of gasoline. However, methanol is considerably more toxic than ethanol. Recent rulings by the U.S. Environmental Protection Agency (EPA) are likely to ban the closely related and similarly toxic methyl tert-butyl ether (MTBE) as a fuel additive because of concerns about ground-water contamination.

A promising application for methanol is in fuel cell technology. Methanol is a leading candidate to provide the hydrogen necessary to power fuel cell vehicles, the next generation of automotive technology [47].

Methanol synthesis development correlates closely with Fischer–Tropsch research. Methanol is currently a commercial product produced mostly from natural gas [52]. Methanol is formed by the exothermic reaction of one mole of carbon monoxide with two moles of hydrogen [54]:

$$CO + 2H_2 \rightarrow CH_3OH \tag{4.8}$$

Low temperatures and high pressures thermodynamically favor the production of methanol. Current commercial operations use a fixed catalytic bed operated at 250°C and 60–100 atm with gas recycle to remove the large amount of heat released by this exothermic reaction. More recently, liquid-phase slurry reactors have been introduced to improve contact between syngas and catalyst as well as enhance the removal of heat from the reactor [55].

Methanol and by-products water and higher alcohols leave the reactor in the gas-phase along with noncondensable gases. Methanol is recovered by distillation. The remaining syngas is recycled to the reactor increasing the conversion efficiency significantly. Liquid-phase methanol synthesis could reduce investment costs of methanol plants [56].

4.4.8 Mixed Alcohols

Biomass gasification followed by catalytic synthesis can yield a mixture of alcohols from CO and H_2. Compared to methanol, mixed alcohols have some attractive fuel properties such as HHV, improved water tolerance, and lower vapor pressure. The composition of mixed alcohols can vary based on process conditions, which would affect its fuel properties although an ethanol fraction of 71 wt.% has been assumed in some design studies [57]. Because the higher molecular weight alcohols have significant value in markets outside of transportation, they are likely to be separated from the alcohol mixture before it is sold as fuel.

Mixed alcohols synthesis consists of an initial production of methanol and its subsequent homologation to ethanol, propanol, butanol, and other higher alcohols [58]. The overall reaction has been described as follows:

$$nCO + 2nH_2 \rightarrow C_nH_{2n+1}OH + (n-1)H_2O \tag{4.9}$$

where n typically ranges between 1 and 8 [45]. Mixed alcohol synthesis combines reactions from other fuel synthesis pathways such as methanol synthesis, Fischer–Tropsch, and water-gas shift reactions. Methanol synthesis is favored at low temperatures and high pressure, and higher alcohols are produced as the temperature is increased. Process conditions for high-temperature, high-pressure synthesis catalysts range between 300°C and 425°C and 12.5–30 MPa using modified methanol catalysts; modified Fischer–Tropsch catalysts may be employed with different process conditions.

Mixed alcohol synthesis has not been widely commercialized due to poor product selectivity and low syngas conversion. On a single pass, about 10% of syngas gets converted with most of the product composed of methanol. Recycling methanol aids the production of higher alcohols. Generating revenue from higher alcohols will be important for the successful commercialization of this process.

NREL researchers suggest that ethanol and mixed alcohol synthesis could be cost competitive with corn ethanol by 2012 [57]. Their estimates show capital costs of $137 million and operating costs of $1.01 per gallon for a 72.6 million gallon per year mixed alcohol biomass plant.

4.4.9 Fischer–Tropsch Liquids

Fischer–Tropsch liquids from biomass have antecedents in the coal-to-liquids industry. Germany extensively developed the Fischer–Tropsch process during World War II when it was denied access to petroleum-rich regions of the world. Likewise, when South Africa faced a world oil embargo during their era of apartheid, they employed Fischer–Tropsch technology to sustain its national economy. A comprehensive bibliography of Fischer–Tropsch literature can be found on the Web [59].

Fischer–Tropsch catalysis produces a large variety of hydrocarbons including light hydrocarbon gases, paraffinic waxes, and alcohols according to the generalized reaction:

$$CO + 2H_2 \rightarrow -CH_2- + H_2O \tag{4.10}$$

Fischer–Tropsch liquids composition depends on the process selectivity. Process selectivity is affected by various factors including catalyst and feed gas properties. The Anderson–Schulz–Flory (ASF) distribution describes the probability of hydrocarbon chain growth where the molar yield for a carbon chain can be calculated using the following equation [60]:

$$C_n = \alpha^{n-1}(1-\alpha) \tag{4.11}$$

where α is the chain growth probability of a hydrocarbon of length n. Light hydrocarbons (mostly methane) can be fed into a gas turbine to provide power. Fischer–Tropsch liquids can be separated into various products in a process similar to petroleum distillation.

Product distribution is a function of temperature, pressure, feed gas composition (H_2/CO), and catalyst type and composition [61]. Depending on the types and quantities of Fischer–Tropsch products desired, either low (200°C–240°C)- or high-temperature (300°C–350°C) synthesis at pressures ranging between 10 and 40 bar are used. For example, high gasoline yield can be achieved using high process temperatures and an iron catalyst. Fischer–Tropsch synthesis requires careful control of the H_2/CO ratio to satisfy the stoichiometry of the synthesis reactions as well as avoid deposition of carbon on the catalysts (coking). An optimal H_2/CO ratio of 2:1 is maintained through the water-gas shift reaction.

Senden et al. [62] calculated a theoretical limit for gasoline yield selectivity to be 48%. Fischer–Tropsch diesel's selectivity limit is estimated to be even lower at 40%. Remaining fractions can be used to substitute for petroleum in the production of commodity chemicals. Because selectivity toward alkanes suitable for gasoline and diesel is low, it will be important to find markets for the by-product hydrocarbons.

Cobalt and iron are the main active catalyst ingredients used for Fischer–Tropsch synthesis. Cobalt catalysts last longer than iron catalysts and have improved carbon conversion to products due to reduced CO_2 generation. Iron, on the other hand, suffers from increased carbon deposition. Since cobalt catalysts can operate at lower pressures, they are commonly used as an alternative to iron catalysts. The costs for cobalt catalysts are about 230 times more expensive than iron, but this cost is offset by lower operating costs. Ruthenium is the most active Fischer–Tropsch catalyst but has a prohibitive cost. Much research has been done on improving catalyst activity, lifetime, and selectivity. Adding promoters, improving catalyst preparation, and using zeolites among other methods, can achieve this.

Fischer–Tropsch catalyst can have limited lifetime due to loss of catalyst activity. Poisoning can cause significant catalyst deactivation although the problem can be controlled through gas cleaning. Sulfur, the main Fischer–Tropsch catalyst poison, occurs at relatively low concentrations in biomass but it must be removed to levels as low as 60 ppb. Other poisons include nitrogen compounds and halides. Moisture in the syngas can also have an inhibiting effect due to its water gas shift activity.

Tijmensen et al. developed one of the few analyses on Fischer–Tropsch liquid fuels from biomass [61]. Their analysis found capital costs of

$341 million and operating costs of $1.45 per gallon for a 35 million gallon per year biomass plant.

4.4.10 Fuels from Pyrolysis Oils

Pyrolysis is the thermal decomposition of organic compounds in the absence of oxygen [63]. The resulting product streams depend on the rate and duration of heating. Liquid yields exceeding 70% are possible under conditions of fast pyrolysis, which is characterized by rapid heating rates (up to 1000°C/s), moderate reactor temperatures (450°C–600°C), short vapor residence times (<0.5 s), and rapid cooling at the end of the process. Rapid cooling is essential if high-molecular-weight liquids are to be condensed rather than further decomposed to low molecular weight gases.

Pyrolysis liquid, also known as bio-oil, is a low viscosity, dark-brown fluid with up to 15%–20% water, which contrasts with the black, tarry liquid resulting from slow pyrolysis [64]. Fast pyrolysis liquid is a complicated mixture of organic compounds arising from thermal degradation of carbohydrate and lignin polymers in biomass [65]. The liquid is highly oxygenated, approximating the elemental composition of the feedstock, which makes it highly unstable. Among the constituents are alcohols, aldehydes, carbohydrates, carboxylic acids, phenols, and lignin oligomers. The liquid, despite its high water content, shows no appreciable phase separation. The low pH of pyrolysis liquids, which arises from organic acids, makes the liquid highly corrosive. The liquid often contains particulate ash and char. The HHV of pyrolysis liquids ranges between 17 and 20 MJ/kg with liquid densities of about 1280 kg/m^3. Assuming conversion of 72% of the biomass feedstock to liquid on a weight basis, yields of pyrolysis oil are about 135 gal/ton.

Pyrolysis liquids can be used directly as a substitute for heating oil. In some circumstances, they are also suitable as fuel for combustion turbines or modified diesel engines. Recovery of high-value chemicals is another possibility, suggesting an integrated approach to the production of both chemicals and fuel.

The production of pyrolysis oils and its coproducts involves several steps [66], which is illustrated in Figure 4.10. Lignocellulosic feedstock, such as wood or agricultural residues, is milled to a fine powder to promote rapid reaction. The particles are augured into the pyrolysis reactor where they are rapidly heated and converted into condensable vapors, liquid aerosols, noncondensable gases, and charcoal. These products are transported out of the reactor into a cyclone operating above the condensation point of pyrolysis vapors where the charcoal is removed. Vapors and gases are transported to a quench vessel where a spray of pyrolysis liquid cools vapors sufficiently for them to condense. The noncondensable gases, which include flammable carbon monoxide, hydrogen, and methane, are burned in air to provide heat for the pyrolysis reactor. A number of schemes have been developed for indirectly heating the reactor, including transport of solids into fluidized beds or cyclonic configurations to bring the particles in contact with hot surfaces.

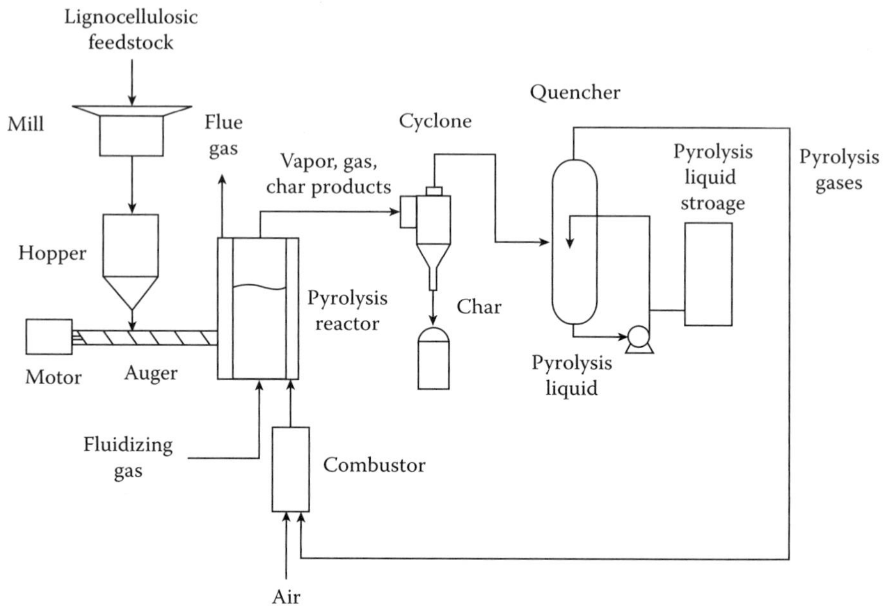

FIGURE 4.10
Bio-oil production via fast pyrolysis.

There are several problems with pyrolysis liquids [67]. Phase-separation and polymerization of the liquids and corrosion of containers make storage of these liquids difficult. The high oxygen and water content of pyrolysis liquids makes them incompatible with conventional hydrocarbon fuels. Furthermore, pyrolysis liquids have much lower quality than even Bunker C heavy fuel oil. Thus, upgrading pyrolysis liquids to more conventional hydrocarbon fuels such as gasoline is highly desirable.

Either traditional hydrotreatment/hydrocracking or emerging zeolite technology can accomplish upgrading. The process of hydrotreating/hydrocracking bio-oil into hydrocarbons suitable as transportation fuel is similar to the process for refining petroleum [67]. Hydrotreating is the reaction of organic compounds with hydrogen over a catalyst at high pressure to remove sulfur, nitrogen, oxygen, and other contaminants. Whereas oxygen is a relatively minor contaminant in petroleum, it represents a major constituent in bio-oil, which may require catalyst formulations optimized for deoxygenation of bio-oil. Hydrocracking is the reaction of hydrogen with organic compounds to break long-chain molecules into lower molecular weight compounds. Although fast pyrolysis is a depolymerization process, a number of carbohydrate and lignin oligomers remain in the oil, which hydrocracking can convert into more desirable paraffinic or naphthenic compounds. Hydrotreating has the advantage of yielding high-quality products at maximum liquid yield. Disadvantages of hydrotreating include high hydrogen consumption and the need to operate at elevated pressures (13–17 MPa), which adversely affect the economics of the process. Upgrading can also be accomplished at atmospheric pressure without reducing gases by dehydration and decarboxylation over acidic zeolites. A yield of 17% of C_5–C_{10} hydrocarbons has been reported in a study of upgrading of pyrolytic liquids from poplar wood [68].

Recent cost estimates for producing gasoline from bio-oil are $1.80 and $2.01 per gallon from corn stover and wood, respectively [69]. Bio-oil production cost estimates vary between $0.41 and $2.46 per gallon [70,71].

4.4.11 Fuels from Biocrude

Hydrothermal processing (HTP) describes the thermal treatment of wet biomass to produce carbohydrate, liquid hydrocarbons, or gaseous products depending upon the reaction conditions [72,73]. HTP liquids are commonly referred to as biocrude. As the desired reaction temperature increases, higher pressures are required to prevent boiling of the water in the wet biomass. Thus, processing conditions range from hot, compressed water at 200°C to supercritical water above 374°C.

At 200°C and 20 bar pressure, fibrous biomass undergoes a fractionation process to yield cellulose, lignin, and pentose (from hemicellulose). Low-temperature HTP has potential application as a pretreatment for enzymatic hydrolysis of cellulose.

At 300°C–350°C and 120–180 bar pressure and reaction times of 5–20 min, biomass undergoes more extensive chemical reaction. Reaction products include a hydrocarbon-rich fraction known as biocrude representing as much as 45 wt.% of the product. Products also include 25 wt.% noncondensable gas, which is mostly CO_2, and 30% aqueous phase, which is about 30 wt.% dissolved organic compounds such as acetic acid and ethanol. The biocrude is attractive for its low oxygen content (10–18 wt.% compared to 45 wt.% for the original biomass) and high LHV (30–35 MJ/kg).

At 600°C–650°C and 300 bar pressures, supercritical conditions prevail. For reaction times of 0.5–2.0 min, gas is the primary reaction product and the process is referred to as supercritical gasification. Typical volume fractions of gas constituents are: H_2—56 vol.%; CH_4—47 vol.%; CO_2—33 vol.%; and CO—4 vol.%. The combination of relatively low temperature and high pressure makes inevitable high methane yields, which is problematic for Fischer–Tropsch synthesis and steam reforming of the gas.

4.4.12 Fuels from Lipids

Lipids are a large group of hydrophobic, fat-soluble compounds produced by plants and animals for high-density energy storage. Triglycerides of fatty acids, commonly known as fats and oils depending upon their melting points, are among the most familiar form of lipids and have been widely used in recent years for the production of diesel fuel substitutes. Compared to petroleum-based diesel fuels, fats and oils have higher viscosity and lower volatility, which result in fouling of engine valves and less favorable combustion performance, especially in direct-injection engines [74]. The solution to this problem is to convert the triglycerides into methyl esters or ethyl esters of the fatty acids, known as biodiesel, and the by-product 1,2,3-propanetriol (glycerol).

The fuel properties of biodiesel are very similar to petroleum-based diesel. The specific gravity and viscosity of biodiesel are only slightly higher than for diesel while the cetane numbers and heating values are comparable. Significantly higher flash points for biodiesel represent greater safety in storage and transportation. Biodiesel can be used in unmodified diesel engines with no excess wear or operational problems. Tests in light and heavy trucks showed few differences other than a requirement for more frequent oil changes because of the build-up of ester fuel in engine crankcases [74].

Triglycerides are esters of glycerol and fatty acids, which are long-chain carboxylic acids containing even numbers of carbon atoms [75]. The acid fractions of triglycerides can vary in chain length and degree of saturation. Fats, which are solid or semisolid at room temperature, have a high percentage of saturated acids, whereas oils, which are liquid at room temperature, have a high percentage of unsaturated acids. Plant-derived triglycerides are typically oils containing unsaturated fatty acids, including oleic, linoleic, and linolenic acids.

A wide variety of plant species produce triglycerides in commercially significant quantities, most of it occurring in seeds [76]. Average oil yields range from 150 L/ha for cottonseed to 814 L/ha for peanut oil although intensive cultivation might double these numbers. Soybeans are responsible for more than 50% of world production of oilseed, representing 48–82 million bbl/year. The average oil yield for soybeans is 383 L/ha [76]. A higher yielding crop is oil palm, already grown in plantations for vegetable oil production [77]. Although oil palm yields are 10 times higher than soybeans, some environmentalists are concerned that its cultivation for fuel production will encourage rainforest destruction. Several oil seed crops have been identified that could be grown on waste land or even saline soils, which reduces concerns about competition for food crops and rainforest destruction. These alternative oil seed crops include jatropha, Chinese tallow tree, and salicornia [78,79].

Finally, lipid-rich microalgae might be grown in brackish water or even in seawater [80]. Algae can produce as much as 60% of their body weight as lipids when deprived of key nutrients such as silicon for diatoms or nitrogen for green algae. They employ relatively low substrate concentrations, on the order of 10–40 g/L.

Algal biomass uses carbon dioxide as its carbon source and sunlight as its energy source. About 1.8 kg of carbon dioxide is fixed for every kg of algal biomass, which contains up to 50% carbon by dry weight. Controlled production of renewable fuels from algae has been proposed in either raceway ponds or photobioreactors. Raceway ponds consist of open, shallow recirculation channels with mechanical flow control and surfaces that enhance light retention. Raceway ponds are inexpensive, but relatively inefficient when compared to photobioreactors. There are various photobioreactor designs with the common goal of maintaining a monoculture of algae that is efficiently exposed to sunlight and carbon dioxide. A common design employs arrays of tubes arranged vertically to minimize land use and oriented north–south to maximize light exposure. Reported areal productivity for photobioreactors is 0.048 kg/m^2/day vs. 0.035 kg/m^2/day for raceway ponds [81]. Based on studies by the U.S. Department of Energy in the 1980s and adjusting for inflation [82], capital costs for open raceway ponds are expected to be $100,000 per hectare. Unfortunately, photobioreactors are expected to be 10 times more expensive or about $1 million per hectare, which detracts from its advantages [83,84].

Given that algae do not require fresh water or fertile soils, waste lands have been suggested as potential locations to grow algae. One suggestion is to build algae ponds in the desert Southwest United States where inexpensive flat land, abundant sunlight, water from alkaline aquifiers, and carbon dioxide from power plants could be combined to generate renewable fuels.

Recovery of lipids from oil seed crops is relatively straightforward. The seeds are crushed to release the oil. Mechanical pressing is used to extract oil from seeds with oil content exceeding 20%. Solvent extraction is required for seeds of lower oil content. The residual seed material, known as meal, is used in animal feed.

Recovery of lipids from microalgae has yet to be worked out for commercial applications. Although microalgae growth and (biomass) production rates are very high, microalgae thrive at relatively low population densities. Thus, unlike the seasonal harvests of traditional agriculture, microalgae must be harvested continuously to obtain the high biomass yields projected for it (nature achieves this by populating the oceans with predators that continuously graze the microalgae "fields"). Thus, open raceways and photobioreactors are continuously pumping algae-enriched water through biomass recovery equipment that must concentrate the algae by a factor of five hundred. Just to harvest the microalgae is energy intensive and expensive. The lipid must then be extracted from algal cells, which is more difficult than crushing oil seeds. Finally, uses must be found for the non-lipid portion of the biomass, which represents more than half of the algal biomass. Although it has been known for almost 50 years that algae contain proteins of quality comparable to conventional vegetable proteins, recovery for food or feed applications has yet to be commercialized because of technical difficulties and cost [85]. These complications of algae harvest and lipid recovery suggest its use as feedstock for the production of methane by anaerobic digestion [86]. Although not a liquid fuel, methane could be used for either compressed natural gas transportation fuel [87] or electric power production.

The higher viscosity and lower volatility of triglycerides compared to diesel fuel leads to coking of the injectors and rings of diesel engines. Chemical modification of triglycerides to methyl or ethyl esters yields excellent

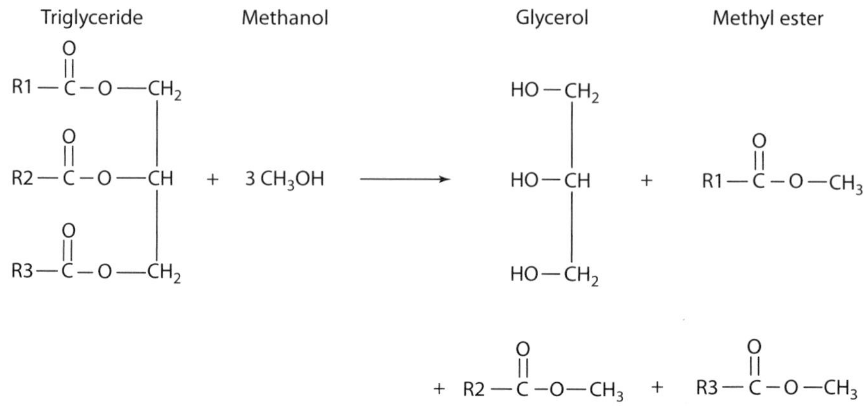

FIGURE 4.11
Transesterification reaction.

diesel-engine fuel. Biodiesel is the generic name given to these modified vegetable oils and animal fats. Transesterification describes the process by which triglycerides are reacted with methanol or ethanol to produce methyl esters and ethyl esters, respectively, along with the coproduct glycerol [88]. For example, as shown in Figure 4.11, one triglyceride molecule reacts with three methanol molecules to produce one molecule of 1,2,3-propanetriol (glycerol) and three ester molecules.

Near quantitative yields of methyl (or ethyl) esters can be produced in 1 h at room temperature using 6:1 molar ratios of alcohol and oil when catalyzed by 1% lye (NaOH or KOH). The lye also serves as a reactant in the conversion of esters into salts of fatty acids. These salts are familiarly known as soaps and the process is called saponification (soap-forming). Small amounts of soap are also produced by the reaction of lye with fatty acids. Upon completion, the glycerol and soap are removed in a phase separator. A flow sheet for a biodiesel production facility is given in Figure 4.12.

Biodiesel can be substituted for diesel fuel in many applications although it has some prominent shortcomings. Fatty acid methyl esters are subject to microbial or oxidative attack, making them unsuitable in applications requiring long-term fuel storage. Low temperature performance of biodiesel is sometimes problematic. Also, glycerol by-product of biodiesel production is contaminated with salts, which reduces its economic value.

For these reasons, hydrogenation is being evaluated as a replacement for transesterification in the production of lipid-based biofuels [89]. Hydrogenation includes a number of reactions: large molecules are broken into smaller molecules; carbon–carbon double bonds are converted into more stable single bonds; molecular structures are rearranged; and undesirable atoms such as sulfur, nitrogen, and oxygen are removed from the hydrogenated compounds. In the case of lipids, hydrogenation yields alkanes, which are highly desirable fuel molecules.

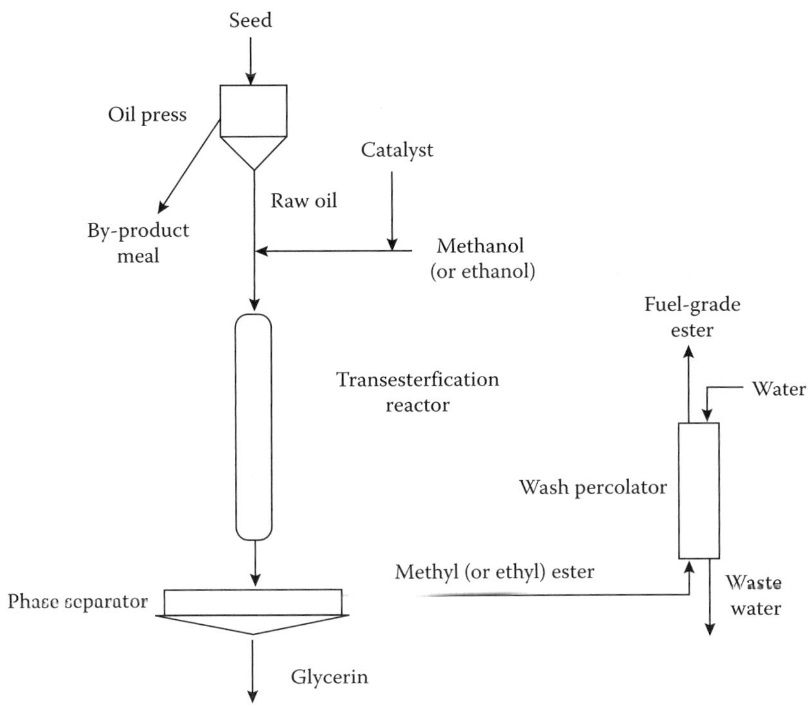

FIGURE 4.12
Conversion of triglycerides to methyl (or ethyl) esters and glycerol.

4.4.13 Methylated Furans

Furans are heterocyclic aromatic ethers consisting of a ring of four carbon atoms and one oxygen atom (see Figure 4.13 for some examples). Furans are colorless, water-insoluble, flammable liquids with volatility comparable to hydrocarbons of similar molecular weight. They were originally derived from the destructive distillation of wood although a number of synthesis routes are possible [90]. Methylated furans have heating values and octane numbers comparable to gasoline making them potential transportation fuel [91]. However, neither the fuel properties nor toxicity of these compounds

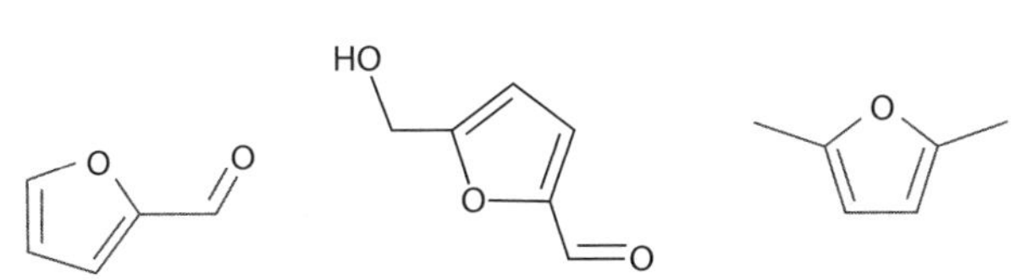

FIGURE 4.13
Furans relevant to the production of transportation fuels by thermochemical processing of sugars.

have been much studied raising questions as to their ultimate practicality as transportation fuel.

Furans can be produced from both the hexose and pentose content of lignocellulosic biomass and other carbohydrate-rich plant materials. Treatment with acid and/or heat can drive dehydration reactions that ultimately yield furans. Catalysts can improve yields by making furan-producing pathways more selective among the large number of competing reactions that can occur during pyrolysis of biomass. 2,5-Dimethyl furan in particular has received recent interest because new catalytic synthesis routes from sugars have been developed [92,93].

Sugars consisting of five-member rings like xylose (a pentose) or fructose (a ketohexose) are readily dehydrated to the five-member rings of furan compounds [94]. Dehydration of xylose yields furfural (an aldehyde) while fructose yields 5-hydroxymethylfurfural (a dimethylated furan). A diagram of proposed reaction pathways to 5-hydroxymethylfurfural is shown in Figure 4.14. All described methods of the synthesis of 5-hydroxymethylfurfural require thermal dehydration in acidic medium, which presents some challenges in isolating this unstable compound. Aldohexoses like glucose can also be dehydrated to 5-hydroxymethylfurfural but with considerable more difficulty. Xylose is obtained from the hemicellulose content of biomass while fructose is obtained from sucrose or the isomerization of glucose.

Neither furfural nor 5-hydroxymethylfurfural is suitable as motor fuel and must be further upgraded. Furfural can be methylated by reaction with methanol over zeolite catalyst to yield methyl furan, dimethyl furan, trimethyl furan, and tetramethyl furan [91,95].

Dumesic and coworkers used a biphasic reactor to convert fructose to HMF [92]. The fructose and an acid catalyst were dissolved in the water phase

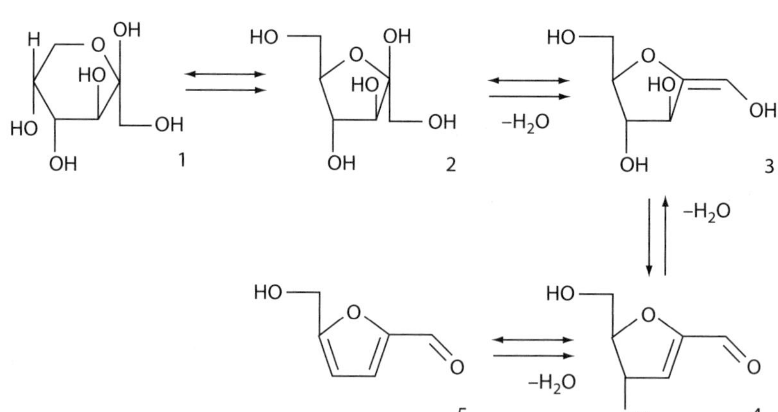

FIGURE 4.14
Proposed reaction pathway for conversion of fructose to hydroxymethylfurfural (HMF): fructopyranose (1), fructofuranose (2), two intermediate stages of dehydration (3, 4) and HMF (5).

of the reactor. HMF was extracted as it is formed using a low-boiling-point solvent (methyl isobutyl ketone). In an optimized system, fructose conversion is 77% with half the HMF ending up in the organic phase. Dumesic and coworkers have improved upon previous HMF synthesis by increasing yields and improving extraction. The HMF is converted to DMF using a copper-based catalyst. This conversion removes two oxygen atoms lowering the boiling point to a range that makes it suitable as motor fuel.

Zhao and coworkers have developed an alternative approach for converting sugars into HMF. Instead of an aqueous solution and acid catalyst, they use ionic liquids as solvent and metal chlorides as catalysts. The absence of an aqueous phase prevents secondary reactions that introduce contaminants to the desired product. The process also works well on both fructose (90% yield) and glucose (70% yield) [93].

4.4.14 Compressed Gases as Transportation Fuel

The ideal transportation fuel is a stable liquid at ambient temperature and pressure that can be readily vaporized and burned within an engine. However, some gaseous compounds are potential transportation fuels by increasing their density through compression. Among these gaseous transportation fuels are hydrogen, methane, ammonia, and dimethyl ether.

Hydrogen can be manufactured from syngas via the water-gas shift reaction. This moderately exothermic reaction is best performed at relatively low temperatures in one or more stages with the aid of catalysts. Although this might be one of the most cost-effective ways to produce clean-burning biofuel, the physical characteristics of hydrogen present challenges in its use as transportation fuel. In particular, its low density even under cryogenic or high-pressure conditions limits onboard storage of this fuel. Its wide flammability range also presents unique safety problems in its use in transportation systems [96].

Methane can be the main product of gasification under conditions known as hydrogasification [17]:

$$C + 2H_2 \rightarrow CH_4$$

$$CO + 3H_2 \rightarrow CH_4 + H_2O$$

These exothermic reactions require low temperatures, high pressures, and large quantities of hydrogen to favor complete conversion to methane. Thus, the process requires separate generation of hydrogen and catalysts to achieve reasonable reaction rates. It has been demonstrated at the commercial scale using coal as the carbonaceous fuel. Although methane is more easily pressurized or liquefied than hydrogen, its density is still too low to be an attractive transportation fuel except in some urban mass transit applications [97].

Anhydrous ammonia (NH_3) is a gas at ambient conditions but is readily liquefied at room temperature by storage at 10 bar pressure, achieving 87% of the density of gasoline [27]. In fact, it has nearly the same density, boiling point, and octane number as propane, which has been widely employed as a portable fuel source.

Ammonia is produced by the Haber process at 200 bar and 500°C [98]:

$$N_2 + 3H_2 \rightarrow 2NH_3$$

As a widely employed agricultural fertilizer, the United States already has in place production, storage, and distribution infrastructure for its use.

Ammonia has been tested as fuel in spark ignition engines, diesel engines, and gas turbines. In tests dating back to the 1960s, near-theoretical performance was achieved with ammonia if it was partially dissociated to achieve 1% hydrogen concentration at the engine intake [99]. Somewhat surprising for this nitrogen-rich fuel, nitrogen oxide emissions were lower than those obtained from octane fuel.

Despite these advantages, it is far from an ideal transportation fuel. Its low flammability requires the addition of an ignition promoter, which complicates engine design. Because of its lower volumetric heating value, an ammonia-fueled vehicle would require a fuel tank about 2.4 times larger than for a propane-fueled vehicle. Ammonia is highly volatile and extremely toxic, further detracting from its use as transportation fuel.

Dimethyl ether, like liquefied petroleum gas (LPG) is a nontoxic, flammable gas at ambient conditions that is easily stored as a liquid under modest pressures [100]. It is currently used as an aerosol propellant in the cosmetic industry, but has excellent potential as a fuel for heating, cooking, and power. It is particularly attractive as a substitute for petroleum-based diesel fuel since it has comparable cetane number but yields essentially zero particulate emissions and low NO_x emissions. It is produced either directly from syngas or indirectly through the dehydration of methanol by reactions at high pressure over catalysts. Of course, this syngas route also allows it to be produced from fossil fuels, and much of the current interest in this alternative fuel arises from the possibility of manufacturing it from inexpensive stranded natural gas.

4.4.15 Well-to-Wheel Analysis of Transportation Fuels

A well-to-wheel (WTW) analysis accounts for inefficiencies in the various energy-conversion steps required to collect, process, and consume liquid fuels. A simplified WTW analysis employs two separate factors to calculate the overall efficiency. These factors are the well-to-tank (WTT) and tank-to-wheel (TTW) efficiencies. Careful assumptions for the extent of system boundaries and energy flows are required to calculate these factors, and these are not the subject of this chapter. Published results for key transportation pathways are provided in Table 4.7 [101].

TABLE 4.7

Well-to-Wheel Analysis of Transportation Fuels

Fuel	Power Train	WTT Efficiency (%)	WTT Energy Consumption (MJ/MJ)	TTW Energy Consumption (MJ/km)	WTW Energy Consumption (MJ/km)	WTW GHG Emissions (g CO_2/MJ)	WTW GHG Emissions (g CO_2/km)
Gasoline	Gasoline ICE	88	1.14	3.25	3.7	89	289
	Gasoline HEV	88	1.14	2.47	2.8	89	219
Diesel	Diesel ICE	79	1.27	2.62	3.3	92	241
	Diesel HEV	79	1.27	2.21	2.8	92	203
Ethanol	Ethanol ICE	38	2.63	3.08	8.1	54	168
	Ethanol HEV	38	2.63	2.33	6.1	54	127
CNG	CNG ICE	92	1.09	2.95	3.2	67	196
	CNG HEV	92	1.09	2.37	2.6	67	158
GH2 (coal)	Coal H2 ICE	36	2.75	2.71	7.5	253	684
	Coal H2 HEV	36	2.75	2.03	5.6	253	513
GH2 (grid)	Grid H2 ICE	21	4.86	2.71	13.2	486	1318
	Grid H2 HEV	21	4.86	2.03	9.9	486	987
Electric	Grid Li-ion battery	34	2.97	1.14	3.4	297	339
Electric	NG CC Li-ion battery	47	2.11	1.14	2.4	114	130
Electric	Hydropower Li-ion battery	93	1.08	1.14	1.2	0	0

Notes: WTT, wells-to-tank; TTW, tank-to-wheels; WTW, wells-to-wheels; GHG, greenhouse gases; ICE, internal combustion engine; HEV, hybrid electric vehicle; CNG, compressed natural gas; Grid, electricity obtained from the existing electric grid; NG CC, natural gas combined cycle power.

WTT efficiency is the ratio of energy output from a plant or refinery to the energy input required to collect, convert, and deliver fuel. For example, WTT efficiency of ethanol production is the ratio of energy in the liquid fuel (and by-products) to energy inputs including fertilizer to grow crops, fuel to run harvesting machinery, and energy required to operate the plant. An estimated 2.63 MJ of energy input is required for every MJ of ethanol produced, which represents a WTT efficiency of 38%.

TTW efficiency recognizes thermodynamic limits in the conversion of fuel to kinetic power. Energy losses take place in the combustion engine, through the transmission, and on the pavement. TTW is conveniently expressed here as MJ/km traveled. An internal combustion engine employing ethanol (E-85) consumes 3.08 MJ/km traveled. The overall WTW efficiency for an ethanol-powered internal combustion engine is 8.1 MJ/km. Efficiencies for a gasoline-powered vehicle are 1.14 MJ/MJ of delivered gasoline (88% efficiency), 3.25 MJ/km traveled TTW efficiency, and a WTW efficiency of 3.7 MJ/km. Similar methodology can be employed to estimate WTW greenhouse gas (GHG) emissions.

4.4.16 Summary

Biochemical production of biofuels consists of fermentation of sugar to ethanol or other metabolic products suitable as fuel. The sugar feedstock can be obtained directly from sugar crops or through hydrolysis of carbohydrate polymers like starch and cellulose. Both starch and cellulosic biomass must be pretreated to improve the accessibility of enzymes to these carbohydrate polymers.

Grain ethanol plants are categorized as either dry grind or wet milling depending upon the pretreatment method. Three major products from the dry grind fermentation process consist of CO_2, DDGS, and ethanol.

Advanced biochemical technologies based on lignocellulosic biomass employ expensive pretreatment processes to overcome the recalcitrance of lignocellulosic biomass. Concentrated acid, dilute acid, and enzymatic hydrolysis are the three main pretreatment processes currently under consideration. Simultaneous saccharification and fermentation schemes combine pretreatment and conversion processes to overcome end product inhibition and increase yields.

Because ethanol has limitations as a transportation fuel, fermentations that produce metabolites other than ethanol have been proposed. Alternative fermentations could produce hydrophobic molecules that are less oxidized than ethanol including higher alcohols (most prominently butanol), fatty acids, fatty alcohols, esters, alkanes, alkenes, and isoprenes. Thermochemical conversion processes can be classified as gasification, fast pyrolysis, and hydrothermal processing. Gasification produces a gaseous product known as syngas that can be catalytically converted to liquid fuels. Fast pyrolysis yields a primarily liquid product, bio-oil, which can be upgraded to liquid fuels. Hydrothermal processing products vary based on process conditions that also require further processing to be suitable for transportation fuel.

Syngas, composed primarily of CO and H_2, can be generated from a large number of carbonaceous feedstocks. The ratio of CO to H_2 is an important factor during catalytic conversion. Metal-based catalysts will yield different products depending on the process conditions and syngas composition including methanol, ethanol, mixed alcohols, and Fischer–Tropsch liquids, as well as gaseous biofuels including hydrogen, methane, ammonia, and dimethyl ether. The flexibility of catalytic synthesis allows for the production of various liquid fuels suitable for transportation applications. An alternative to catalytic synthesis is hybrid thermochemical/biochemical technology employing microorganisms to ferment syngas into various products including ethanol.

WTW analysis is a method employed to calculate the efficiency of collecting, converting, and consuming transportation fuels. This efficiency can be calculated from the WTT and TTW efficiency. Similarly, fuel cycle GHG emissions can be estimated using this methodology.

Problems

4.28 What do starch and cellulose have in common? Why is cellulose more difficult to convert into fermentable sugars?

4.29 What are the products of a dry grind ethanol plant and what are their approximate proportions?

4.30 Which three basic methods can be employed to convert lignocellulosic material to fermentable sugars?

4.31 List the four main processes involved in converting lignocellulose into ethanol.

4.32 Describe some other biologically synthesized compounds that could potentially serve as alternatives to ethanol as transportation fuels.

4.33 What are triglycerides? Described two alternatives for upgrading them into suitable transportation fuels.

4.34 What are the favorable syngas H_2 to CO ratios for production of Fischer–Tropsch liquids and methanol?

4.35 Assuming a chain growth probability of 0.8, calculate the mole yields for hydrocarbons with 1–24 carbon atoms using the Anderson–Schulz–Flory distribution. How does this distribution compare to that of a 0.9 chain growth probability.

4.36 Calculate the heat of reaction for ethanol (heat of formation –277.7 MJ/kmol) synthesis with the mixed alcohol process and with the methanol (heat of formation –238.4 MJ/kmol) conversion process.

4.37 Assuming nominal annual yields of 7,000, 55,000, and 38,000 kg/ha for corn grain, sugarcane, and sweet sorghum, respectively, which conversion pathway is the most productive use of land? Use conversion yield values given in Section 4.4.3 and units of liters per hectare.

4.38 Describe the water-gas shift reaction and explain why it is important in many thermochemical conversion processes.

4.39 Describe various thermochemical routes to ethanol.

4.40 Define fast pyrolysis and describe its several products.

4.41 Describe the composition of bio-oil. How can bio-oil be upgraded into transportation fuel?

4.42 Assuming an 8400 kg/ha yield for corn stover, calculate the volumetric yield (liters per hectare) of corn stover pyrolysis oil.

4.43 What are furans and how are they produced from biomass?

4.44 Describe the various options for gaseous transportation fuels including their respective advantages and disadvantages.

4.45 Using an ethanol fuel content of 21.2 MJ/L and TTW efficiency of 3.08 MJ/km calculate the theoretical travel distance (units of km per hectare of land) for an E-85 vehicle powered by corn grain ethanol. How does this number compare to sugarcane ethanol and cellulosic ethanol? *Hint:* employ appropriate assumptions for crop yields and process conversion efficiency.

4.46 Assuming a 10% improvement in the production of H_2 from coal, what is the new WTW efficiency for a H_2 ICE, and what is the reduction in CO_2 emissions per kilometer?

4.5 Conversion of Biomass to Electricity

4.5.1 Introduction

The chemical energy of biomass can be converted into electric power by a number of routes. Direct combustion of biomass releases heat that can be used in Stirling engines or Rankine steam power cycles. Fast pyrolysis and thermal gasification, described in earlier sections, yield bio-oil and syngas, respectively, which are suitable for firing in gas turbines or even fuel cells.

4.5.2 Direct Combustion

Combustion is the rapid oxidation of fuel to obtain energy in the form of heat. Since biomass fuels are primarily composed of carbon, hydrogen, and oxygen, the main oxidation products are carbon dioxide and water although

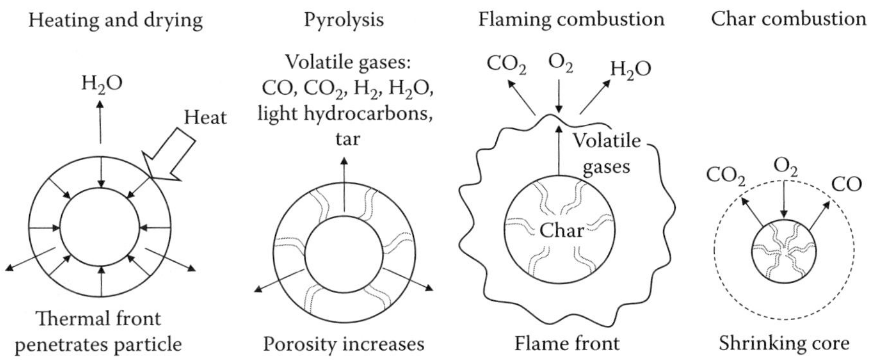

FIGURE 4.15
Processes of solid fuel combustion.

fuel-bound nitrogen and sulfur can be significant sources of sulfur oxide and nitrogen oxide emissions. Depending on the heating value and moisture content of the fuel, the amount of air used to burn the fuel, and the construction of the furnace, flame temperatures can exceed 1650°C.

Solid-fuel combustion consists of four steps, illustrated in Figure 4.15: heating and drying, pyrolysis, flaming combustion, and char combustion [102]. Heating and drying of the fuel particle is normally not accompanied by chemical reaction. Water is driven from the fuel particle as the thermal front advances into the interior of the particle. As long as water remains, the temperature of the particle does not rise high enough to initiate pyrolysis, the second step in solid fuel combustion.

Pyrolysis is a complex series of thermally driven chemical reactions that decompose organic compounds in the fuel [103]. Pyrolysis proceeds at relatively low temperatures, which depend on the type of plant material. Hemicellulose begins to pyrolyze at temperatures between 150°C and 300°C, cellulose pyrolyzes between 275°C and 350°C, and lignin pyrolysis is initiated between 250°C and 500°C.

The resulting decomposition yields a large variety of volatile organic and inorganic compounds, the types and amounts dependent on the fuel and the heating rate of the fuel. Pyrolysis products include carbon monoxide (CO), carbon dioxide (CO_2), methane (CH_4), and high molecular weight compounds that condense to a tarry liquid if cooled before they are able to burn. Fine droplets of these condensable compounds represent much of the smoke associated with smoldering fires. Pyrolysis follows the thermal front through the particle, releasing volatile compounds and leaving behind pores that penetrate to the surface of the particle.

Pyrolysis is very rapid compared to the overall burning process for solid carbonaceous fuels like biomass. Pyrolysis may be as short as a second for small particles of fuel but can extend to many minutes in wood logs. Although the net result of combustion is oxidation of fuel molecules and the release of heat, neither of these processes occurs to a significant extent during pyrolysis. Indeed, if pyrolysis is to proceed at all, heat must be added to the fuel. Oxygen is excluded from the pyrolysis zone by the large gaseous outflux of pyrolysis products from the surface of the fuel particle. Only after pyrolysis gases escape the particle and diffuse into the surrounding air are they able to burn. There remains upon completion of pyrolysis a porous carbonaceous residue known as char.

Both the volatile gases and the char resulting from pyrolysis can be oxidized if sufficient oxygen is available to them. Oxidation of the volatile gases above the solid fuel results in flaming combustion. The ultimate products of volatile combustion are CO_2 and H_2O although a variety of intermediate chemical compounds can exist in the flame, including CO, condensable organic compounds, and long chains of carbon known as soot. Indeed, hot, glowing soot is responsible for the familiar orange color of wood fires.

Combustion intermediates will be consumed in the flame if sufficient temperature, turbulence, and time are allowed. High combustion temperature assures that chemical reactions will proceed at high rates. Turbulent or vigorous mixing of air with the fuel makes certain that every fuel molecule comes into contact with oxygen molecules. Long residence times for fuel in a combustor allow the fuel to be completely consumed. In the absence of good combustion conditions, a variety of noxious organic compounds can survive the combustion process including CO, soot, polycyclic aromatic hydrocarbons (PAH), and the particularly toxic families of chlorinated hydrocarbons known as furans and dioxins. In some cases, a poorly operated combustor can produce pollutants from relatively benign fuel molecules.

The next step in combustion of solid fuels is solid-gas reactions of char, also known as glowing combustion, familiar as red-hot embers in a fire. Char is primarily carbon with a small amount of mineral matter interspersed. Char oxidation is controlled by mass transfer of oxygen to the char surface rather than by chemical kinetics, which is very fast at the elevated temperatures of combustion. Depending on the porosity and reactivity of the char and the combustion temperature, oxygen may react with char at the surface of the particle or it may penetrate into the pores before oxidizing char inside the particle. The former situation results in a steadily shrinking core of char whereas the latter situation produces a constant diameter particle of increasing porosity. Both CO and CO_2 can form at or near the surface of burning char [102]:

$$C + \tfrac{1}{2}O_2 \rightarrow CO$$

$$CO + \tfrac{1}{2}O_2 \rightarrow CO_2$$

These gases escape the immediate vicinity of the char particle where CO is oxidized to CO_2 if sufficient oxygen and temperature are available; otherwise it appears in the flue gas as a pollutant.

4.5.3 Combustion Equipment

A combustor is a device that converts the chemical energy of fuels into high-temperature exhaust gases. Heat from the high-temperature gases can be employed in a variety of applications, including space heating, drying, and power generation. However, with the exception of kilns used by the cement industry, most solid-fuel combustors today are designed to produce either low-pressure steam for process heat or high-pressure steam for power generation. Combustors integrated with steam-raising equipment are called boilers. In some boiler designs, distinct sections exist for combustion, high-temperature heat transfer, and moderate-temperature heat transfer: these are called the furnace, radiative, and convective sections of the boiler, respectively. In other designs, no clear separation between the processes of combustion and heat transfer exists.

Solid-fuel combustors, illustrated in Figure 4.16, can generally be categorized as grate-fired systems, suspension burners, or fluidized beds [104]. Grate-fired systems were the first burner systems to be developed, evolving during the late nineteenth century and early twentieth century into a variety of automated systems. The most common system is the spreader-stoker, consisting of a fuel feeder that mechanically or pneumatically flings fuel onto a moving grate where the fuel burns. Much of the ash falls off the end of the moving grate although some fly ash appears in the flue gas. Grate systems rarely achieve combustion efficiencies exceeding 90%.

Suspension burners were introduced in the 1920s as a means of efficiently burning large quantities of coal pulverized to less than 50 μm particle sizes.

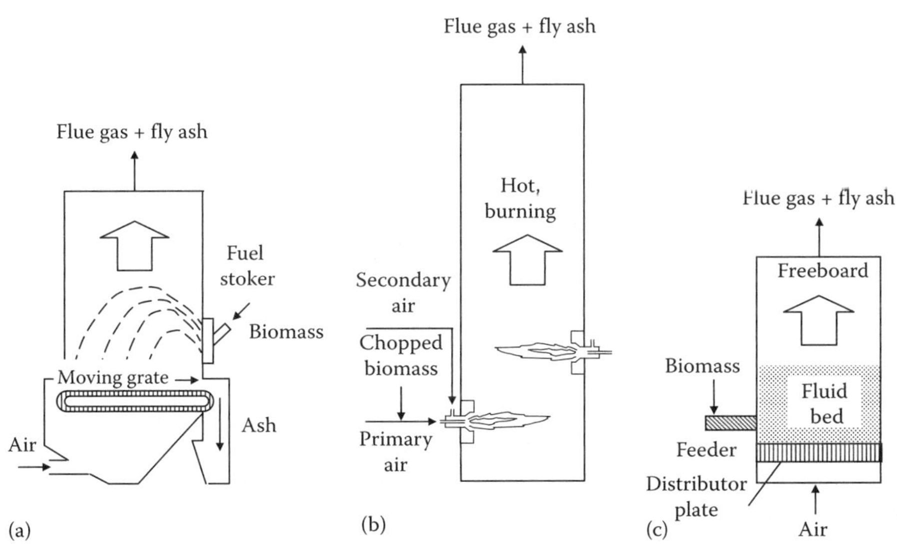

FIGURE 4.16
Common types of combustors: (a) grate-fired, (b) suspension, and (c) fluidized bed.

Suspension burners suspend the fuel as fine powder in a stream of vertically rising air. The fuel burns in a fireball and radiates heat to tubes that contain water to be converted into steam. Suspension burners, also known as pulverized coal (PC) boilers, have dominated the U.S. power industry since World War II because of their high volumetric heat release rates and their ability to achieve combustion efficiencies, often exceeding 99%. However, they are not well suited to burning coarse particles of biomass fuel and they are notorious generators of nitrogen oxides. Biomass is fed from a bunker through pulverizers designed to reduce fuel particle size enough to burn in suspension. The fuel particles are suspended in the primary airflow and fed to the furnace section of the boiler through burner ports where it burns as a rising fireball. Secondary air injected into the boiler helps complete the combustion process. Heat is absorbed by steam tubes arrayed in banks of heat exchangers (waterwall, superheaters, and economizer) before exiting through a bag house designed to capture ash released from the fuel. Steam produced in the boiler is part of a Rankine power cycle.

Fluidized bed combustors are a recent innovation in boiler design. Air injected into the bottom of the boiler suspends a bed of sand or other granular refractory material producing a turbulent mixture of air and sand. The high rates of heat and mass transfer in this environment are ideal for efficiently burning a variety of fuels. Furthermore, the large thermal mass of the sand bed allows the unit to be operated as low as 850°C, which lowers the emission of nitrogen compounds. A commercial market for fluidized bed boilers developed during the 1980s, especially for industrial applications.

Whole-tree burners have been proposed for electrical utility steam raising. The advantage of this approach is the minimization of field processing and handling and the elimination of fuel chipping, which can save about 35% of the cost of harvesting and handling woody fuels.

A variety of biomass materials have proven suitable for direct combustion, including whole trees, wood chips, forestry residue, agricultural residue, pulp and paper refuse, food processing wastes, MSW, and straws and grasses. Wood and wood wastes dominate the biomass-to-power market in the United States and much of the rest of the world. However, a variety of agricultural wastes have been utilized through direct combustion in California, and Denmark has a number of commercial plants that burn straw.

Direct combustion has the advantage that it employs commercially well-developed technology. There are a number of vendors who supply turnkey systems and considerable operating experience exists both in the United States and abroad. However, there are three prominent disadvantages with direct firing. These include penalties associated with burning high-moisture fuels, agglomeration and ash fouling due to alkali compounds in biomass, and relatively low thermodynamic efficiencies for steam power plants of the size appropriate to biomass power.

The moisture in biomass degrades boiler performance for two reasons. First, the energy required to vaporize fuel moisture can only be recovered if exhaust gases are cooled sufficiently to condense the water vapor in this gas. Although this procedure would result in very high thermal efficiencies for a boiler burning even very high moisture fuels, it is not commonly done in practice because the condensate is often corrosive to boiler tubes. Accordingly, most direct-combustion systems are penalized by moisture in the fuel. Second, high moisture fuels simply do not burn well because the process of fuel drying suppresses fuel temperatures below that required for ignition. Water contents exceeding 30% are unacceptable in most boilers for this reason. Fluidized-bed combustors, however, because of the enormous thermal mass associated with the hot bed material, can accept biomass with moisture content as high as 50% although even fluidized beds are penalized in thermal efficiency by the presence of this moisture. Depending on moisture content, as much as 15% of the heating value of biomass is required to dry the fuel. Obviously, field drying of biomass is desired to reduce both transportation costs and heating penalties within the boiler.

Alkali in biomass fuels presents a difficult problem for direct combustion systems. Compounds of alkali metals, such as potassium and sodium salts, are common in rapidly growing plant tissues. Annual biomass crops contain large quantities of alkali while the old-growth parts of perennial biomass contain relatively small quantities of alkali. These alkali compounds appear as oxides in the residue left after combustion of volatiles and char. Alkali vapors may combine with sulfur and silica to form low-melting-point compounds [12]. These sticky compounds bind ash particles to fuel grates and heat exchanger surfaces. Boiler performance degrades as airflow and heat transfer are restricted by ash deposits. Boilers that fire straw, which contain both high silica and alkali, have experienced serious problems in slagging and fouling as well as high-temperature corrosion unless steam temperatures are kept relatively low (less than 500°C).

As an alternative to completely replacing coal with biomass fuel in a boiler, mixtures of biomass and coal can be burned together in a process known as co-firing [105]. Co-firing offers several advantages for industrial boilers. Industries that generate large quantities of biomass wastes, such as lumber mills or pulp and paper companies, can use co-firing as an alternative to costly landfilling of wastes. Federal regulations also make co-firing attractive. The New Source Performance Standards, which limits particulate emissions from large coal-fired industrial boilers to 0.05 lb/MMBtu, doubles this allowance in co-fired boilers in which the capacity factor for biomass exceeds 10%. Adopting co-firing is a good option for companies that are slightly out of compliance with their coal-fired boilers. Similarly, a relatively inexpensive method for a company to reduce sulfur emissions from its boilers is to co-fire with biomass, which contains much less sulfur than coal. Co-firing capability also provides fuel flexibility and reduces ash-fouling problems associated with using only biomass as fuel.

The principal disadvantages of co-firing relate to the characteristics of biomass fuels. Because of the lower energy density and higher moisture content of biomass, the steam-generating capacity of co-fired boilers is often reduced. Also, the elemental composition of fly ash from biomass is distinct from that of coal fly ash. Utilities are concerned that comingled biomass and coal ash will not meet the American Society for Testing and Materials (ASTM) definition of fly ash that is acceptable for concrete admixtures, thus eliminating an important market for this combustion by-product.

It is generally recommended that total fuel (single or mix) alkali content be limited to less than 0.17–0.34 kg/GJ (0.4–0.8 lb/MMBtu), which translates to only 5%–15% co-firing of biomass with coal [12]. Also, furnace temperatures should be kept below 980°C to help prevent the buildup of alkali-containing mineral combinations known as eutectics. At higher temperatures, molten eutectics bind dirt and other particulates to form slag and fouling deposits.

The best wood-fired power plants, which are typically 20–100 MW in capacity, have heat rates exceeding 12,500 Btu/kWh. In contrast, large, coal-fired power plants have heat rates of only 10,250 Btu/kWh. The relatively low thermodynamic efficiency of steam power plants at the sizes of relevance to biomass power systems may ultimately limit the use of direct combustion to convert biomass fuels to useful energy.

4.5.4 Summary

Steam power involves the direct combustion of fuel to raise steam, which is fed into a turbine generator to produce electricity. Coal and natural gas account for most of the feedstock employed for power generation, but biomass may be directly fired or cofired in boilers. Co-firing, the simultaneous combustion of biomass and fossil fuel, provides various benefits to traditional coal-fired plants. Co-firing allows power plants to fall back into compliance with federal regulations, increases feedstock flexibility, and emits lower sulfur emissions.

Different combustors designs for solid fuels include grate-fired systems, suspension burners, and fluidized beds. Combustors can reach temperatures of 1650°C and achieve efficiencies of 99%. The Rankine power cycle can take advantage of the heat generated by direct fuel combustion. Biomass ash and alkali present challenges due to deposits in the burner equipment. Co-firing can mitigate this undesirable effect.

Problems

4.47 Describe the four steps in the combustion of biomass.

4.48 Describe three different solid-fuel combustor designs.

4.49 What are the several advantages of co-firing biomass with coal?

4.50 What problems arise in firing biomass in boilers?

4.51 Describe the various simple cycles for converting chemical energy into electric power.

4.52 Describe how biomass can be used as fuel in a gas turbine cycle.

4.53 Heat rate is commonly employed to report the efficiency of power generation plants. The inverse of heat rate is thermal efficiency. Compare the thermal efficiencies of wood-fired plants to coal plants using the heat rate values provided in this section.

4.54 What makes alkali and ash content important for combustors?

4.6 Conclusions

First-generation biofuels based on grain ethanol and soydiesel will continue to be important sources of alternative transportation fuels for several years because significant investment has been made in their infrastructure and continuing political support of government incentives. However, significant expansion of the biofuels industry will require alternatives to corn and soybean crops as feedstocks. These possibilities include both lignocellulose and lipids from alternative crops. Each of the biofuel and bioenergy options considered in this chapter has its advantages and disadvantages in terms of market, opportunities, environmental benefits, and process economics. Accordingly, no single biofuel is likely to dominate transportation in the same manner as currently exists for petroleum-based gasoline and diesel. Niche markets and regional considerations will heavily influence the best choice of fuel for specific applications. Finally, increased research efforts could raise new possibilities that change the outlook for biofuels.

Although biochemical processing dominates current-generation biofuels, thermochemical processing is likely to play an important role in advanced biofuels because of its technical maturity and potential to efficiently produce hydrocarbon fuels. However, it is too early to designate winners and losers among the many biofuel options. What is certain is that biofuels will evolve as a result of several forces including new technologies, market demand, environmental concerns, and government policies.

Additional Problems

Investigate four ways of using biomass to create power. You can think of each of the four as an option for a power plant and you are attempting to decide which option is the best for you. The four power plant options are listed below.

Plant Options:

Option 1

The plant uses biomass in its solid form directly in a boiler that is operating a Rankine Cycle steam plant to generate electricity.

Assumptions:

- Based on your previous thermodynamics class you should make an assumption about the efficiency of this process. Describe your thought process in the work you hand in.

Plant output. Electricity (MW).

Capital cost. Between \$1400 and \$1800 per kW capacity.

Option 2

The plant gasifies the biomass and uses the biogas in the existing natural gas system, where the existing system consists of both natural gas-fired power plants and natural gas pipeline for residential/commercial heat needs.

Assumptions:

- Use your knowledge of thermodynamics to assume an efficiency of a natural gas–fired power plant. Describe your thought process.
- Assume that the biomass can be converted to gas with 90% efficiency.
- Assume the biogas has an HHV of 17.4 MJ/m^3 and a density of 1 kg/m^3.
- Assume that 80% of the gas goes to produce electricity and the rest goes into the pipeline for residential/commercial needs (this assumes an indirectly heated fluidized bed).

Plant output. Electricity (MW) and methane (MMBtu/h).

Capital cost. Between \$2000 and \$3000 per kW capacity of electricity.

Option 3

This plant separates the biomass into two components: starch and the rest. The starch component is converted into sugar and fermented to create ethanol. The rest is burned similarly to option 1.

Assumptions:

- Assume that the rest has an HHV of 15 MJ/kg.
- Assume that you are doing about as good as theoretically possible at converting the starch to ethanol, that is, you are converting 45% of the starch mass to ethanol mass.

Plant output. Electricity (MW) and ethanol (barrels/day).

Capital cost. $30,000 per barrel/day capacity of ethanol plus the electric power cost.

Option 4

This plant separates the biomass into four components: starch, cellulose/hemicellulose, lignin, and oils. The starch is converted into sugar, which is fermented to create ethanol. The cellulose/hemicellulose is also converted into sugar and fermented to create ethanol. The lignin is combusted as described in option 1. The oil is converted into biodiesel.

Assumptions:

- Assume that you are doing about as good as theoretically possible at converting the starch to ethanol, that is, you are converting 45% of the starch mass to ethanol mass.
- Conversion of cellulose/hemicellulose to fermentable sugars is only still in its infancy, so let us assume that we are only 60% effective at this, so combined with the theoretical maximum sugar to ethanol fermentation efficiency, we will only be able to convert 25% of the cellulose/hemicellulose mass to ethanol mass.
- Assume that the conversion of oils to diesel is 80% efficient.

Plant output. Electricity (MW), ethanol (barrels/day), and biodiesel (barrels/day).

Capital cost. $35,000 per barrel/day capacity of ethanol, plus the cost of electric power—do not worry about the diesel cost as we will assume that the excess production capacity John Pellegrino mentioned in class makes this cheap.

Plant Size

Assume that the biomass needs to be grown within 50 mi of the processing/power plant to be economical. A 50 mi radius circle would contain a little more than 2 million hectare, but given spatial needs for roads and the plant, we can say that the plant has 2 million hectares of land at its disposal.

1. Determine the output (electricity, methane, ethanol, biodiesel) for each of the four plants for two of the following five crops: wheat, sugarcane, switch grass, corn, and hybrid poplar.
2. What is the annual transportation fuel use and transportation fuel cost for the plant (should be the same for all plants)? Use units of gallons of diesel and dollars.

3. How does this transportation fuel cost compare to the capital cost?
4. Write a discussion about the different bioenergy harvesting options that you investigated in questions 1–3. Comment on what you choose. (This is open-ended by design.)

References

1. Brown, R.C. (2003) *Biorenewable Resources: Engineering New Products from Agriculture*, Iowa State Press, A Blackwell Publishing Company, Ames, IA, pp. 59–75.
2. Morris, G. (2003) *The Status of Biomass Power Generation in California, July 31, 2003*, NREL/SR-510-35114, National Renewable Energy Laboratory, Golden, CO.
3. Wright, L. and Hohenstein, W. (1994) Dedicated feedstock supply systems: Their current status in the USA, *Biomass and Bioenergy* 6(3), 159–241.
4. Graham, R. (1994) An analysis of the potential land base for energy crops in the conterminous United States, *Biomass and Bioenergy* 6(3), 175–189.
5. Wyman, C. and Goodman, B. (1993) Biotechnology for production of fuels, chemicals, and materials from biomass, *Applied Biochemistry and Biotechnology* 39(1), 41–59.
6. Energy Information Administration (2004) *Annual Energy Outlook 2004: With Projections to 2025*, January 2004, Energy Information Administration, Washington, DC.
7. Perlack, R., Wright, L., Turhollow, A., Graham, R., Stokes, B., and Erbach, D. (2005) *Biomass as Feedstock for a Bioenergy and Bioproducts Industry: The Technical Feasibility of a Billion-Ton Annual Supply*, A357634, Oak Ridge National Laboratory, Oak Ridge, TN.
8. Vesterby, M. and Krupa, K. (2001) *Major Uses of Land in the United States, 1997*.
9. Alig, R., Plantinga, A., Ahn, S., and Kline, J. (2003) *Land Use Changes Involving Forestry in the United States: 1952 to 1997, with Projections to 2050*, General Technical Report PNW-GTR-587, USDA Forest Service, Pacific Northwest Research Station, Portland, OR, p. iii + 92 pp.
10. Brown, R.C. (2003) *Biorenewable Resources: Engineering New Products from Agriculture*, Iowa State Press, A Blackwell Publishing Company, Ames, IA, pp. 59–75.
11. Klass, D. (1998) *Biomass for Renewable Energy, Fuels, and Chemicals*, Academic Press, New York, pp. 72–90.
12. Jenkins, B., Baxter, L., and Miles, T. (1998) Combustion properties of biomass, *Fuel Processing Technology* 54(1–3), 17–46.
13. Hofbauer, H. (2006) *BIOBIB—A Database for Biofuels*, Institute of Chemical Engineering, Fuel an Environmental Technology, University of Technology, Vienna, Austria.
14. Chynoweth, D., Owens, J., and Legrand, R. (2001) Renewable methane from anaerobic digestion of biomass, *Renewable Energy* 22(1–3), 1–8.
15. Klass, D. (1998) *Biomass for Renewable Energy, Fuels, and Chemicals*, Academic Press, San Diego, CA, p. 356.

16. Demirel, B. and Yeniguen, O. (2002) Two-phase anaerobic digestion processes: A review, *Journal of Chemical Technology and Biotechnology* 77(7), 743–755.
17. Reed, T. (1981) *Biomass Gasification: Principles and Technology*, Noyes Data Corp., Park Ridge, NJ.
18. Klass, D. (1998) *Biomass for Renewable Energy, Fuels, and Chemicals*, Academic Press, San Diego, CA, pp. 274–276.
19. Hall, D. and Overend, R. (1987) *Biomass: Regenerable Energy*, John Wiley & Sons, New York.
20. Higman, C. and Van Der Burgt, M. (2008) *Gasification*, Gulf Professional Publishing, Burlington, MA.
21. Milne, T., Evans, R., and Abatzaglou, N. (1998) *Biomass Gasifier "Tars": Their Nature, Formation, and Conversion*, NREL/TP-570-25357, National Renewable Energy Laboratory, Golden, CO.
22. Bridgwater, A. (1995) The technical and economic feasibility of biomass gasification for power generation, *Fuel* 74(5), 631–653.
23. Henrich, E. and Weirich, F. (2004) Pressurized entrained flow gasifiers for biomass, *Environmental Engineering Science* 21(1), 53–64.
24. Borman, G.L. and Ragland, K.W. (1998) *Combustion Engineering*, McGraw-Hill Company, New York.
25. National Renewable Energy Laboratory, Alternative fuels, general table of fuel properties. Washington, DC, Available from: http://www.eere.energy.gov/afdc/altfuel/fuel_properties.html
26. MacKenzie, J. and Avery, W. (1996) Ammonia fuel: The key to hydrogen-based transportation, in: *Energy Conversion Engineering Conference, IECEC 96*, Washington, DC, August 11–16, 1996, pp. 1761–1766.
27. Teng, H., McCandless, J., and Schneyer, J. (2001) *Thermochemical Characteristics of Dimethyl Ether—An Alternative Fuel for Compression-Ignition Engines*, SAE Technical Paper 2001-01-0154.
28. Borman, G.L. and Ragland, K.W. (1998) *Combustion Engineering*, McGraw-Hill Company, New York, p. 34.
29. Bailey, B. (1996) Performance of ethanol as a transportation fuel, in: Wyman, C. (ed.), *Handbook on Bioethanol*, Taylor & Francis, Washington, DC, pp. 37–58.
30. Lynd, L., Cushman, J.H., Nichols, R.J., and Wyman, C.E. (1991) Fuel ethanol from cellulosic biomass, *Science* 251(4999), 1318–1323.
31. Klass, D. (1998) *Biomass for Renewable Energy, Fuels, and Chemicals*, Academic Press, San Diego, CA, pp. 383–443.
32. Brown, R.C. (2003) *Biorenewable Resources: Engineering New Products from Agriculture*, Iowa State Press, A Blackwell Publishing Company, Ames, IA, pp. 159–194.
33. Kwiatkowski, J., McAloon, A., Taylor, F., and Johnston, D. (2006) Modeling the process and costs of fuel ethanol production by the corn dry-grind process, *Industrial Crops & Products* 23(3), 288–296.
34. Rowell, R., Young, R., and Rowell, J. (1997) *Paper and Composites from Agro-Based Resources*, CRC Press, Boca Raton, FL.
35. Schell, D.J., McMillian, J.D., and Philippidis, G.P. (1992) Ethanol from lignocellulosic biomass, *Advances in Solar Energy* 7, 373–448.
36. Lynd, L.R. (1996) Overview and evaluation of fuel ethanol from cellulosic biomass: Technology, economics, the environment and policy, *Annual Reviews in Energy and the Environment* 21(1), 403–465.

37. Wayman, M. and Parekh, S. (1990) *Biotechnology of Biomass Conversion: Fuels and Chemicals from Renewable Resources*, Open University Press, Philadelphia, PA.
38. Aden, A., Ruth, M., Ibsen, K., Jechura, J., Neeves, K., Sheehan, J., Wallace, B., Montague, L., and Slayton, A. (2002) *Lignocellulosic Biomass to Ethanol Process Design and Economics Utilizing Co-Current Dilute Acid Prehydrolysis and Enzymatic Hydrolysis for Corn Stover*, NREL/TP-510-32438, National Renewable Energy Laboratory, Golden, CO.
39. Wang, B., Ezeji, T., Shi, Z., Feng, H., and Blaschek, H.P. (2008) *Production of Acetone-Butanol-Ethanol (ABE) with Distiller's Dried Grains with Solubles*, American Society of Agricultural and Biological Engineers, Providence, RI.
40. Qureshi, N. and Blaschek, H. (2000) Economics of butanol fermentation using hyper-butanol producing *Clostridium beijerinckii* BA101, *Food and Bioproducts Processing/Official Journal of the European Federation of Chemical Engineering: Part B* 78(C3), 139–144.
41. Ladygina, N., Dedyukhina, E., and Vainshtein, M. (2006) A review on microbial synthesis of hydrocarbons, *Process Biochemistry* 41(5), 1001–1014.
42. Baraud, J., Cassagne, C., Genevois, L., and Joneau, M. (1967) Présence d'Hydrocarbures dans l'Insaponifiable des Lipides de Levures, *Comptes Rendus de l' Academie des Sciences (Paris) Series D* 265, 83–85.
43. Jones, J. (1969) Studies on lipids of soil micro-organisms with particular reference to hydrocarbons, *Journal of General Microbiology* 59(2), 145–152.
44. Wu, S., Schalk, M., Clark, A., Miles, R., Coates, R., and Chappell, J. (2006) Redirection of cytosolic or plastidic isoprenoid precursors elevates terpene production in plants, *Nature Biotechnology* 24, 1441–1447.
45. Forzatti, P., Tronconi, E., and Pasquon, I. (1991) Higher alcohol synthesis, *Catalysis Reviews* 33(1), 109–168.
46. Zeikus, J.G., Kerby, R., and Krzycki, J.A. (1985) Technology: Catalyst converts methanol to dry ethanol, *Chemical & Engineering News* 60(38), 41.
47. Anon. (2007) Beyond the internal combustion engine. American Methanol Institute, Arlington, VA, June 2007,Available from: http://www.methanol.org/fuelcell/special/ami.pdf
48. Brown, R.C. (2007) Hybrid thermochemical/biological processing, *Applied Biochemistry and Biotechnology* 137(1), 947–956.
49. Grethlein, A. and Jain, M. (1992) Bioprocessing of coal-derived synthesis gases by anaerobic bacteria, *Trends in Biotechnology* 10(12), 418–423.
50. Coskata Inc. Lowest cost of production. Warrenville, Illinois, Available from: http://www.coskata.com/ProcessLowestCost.asp (cited December 26, 2008).
51. So, K.S. and Brown, R.C. (1999) Economic analysis of selected lignocellulose-to-ethanol conversion technologies, *Applied Biochemistry and Biotechnology* 77, 633–640.
52. Davenport, B. (2002) Methanol. Report number 674.5000, SRI International, Menlo Park, CA.
53. Gray Jr., C. and Alson, J. (1989) The case for methanol, *Scientific American* 261(5), 108–115.
54. Beenackers, A. and Van Swaaij, W. (1984) Methanol from wood I. Process principles and technologies for producing methanol from biomass, *International Journal of Sustainable Energy* 2(5), 349–367.
55. Cybulski, A. (1994) Liquid-phase methanol synthesis: Catalysts, mechanism, kinetics, chemical equilibria, vapor-liquid equilibria, and modeling—A review, *Catalysis Reviews* 36(4), 557–615.

56. Heydorn, E., Diamond, B., and Lilly, R. (2003) Commercial-scale demonstration of the liquid phase methanol (LPMEOH) process. Final Report to the National Energy Technology Laboratory, U.S. Department of Energy under Cooperative Agreement No. DE-FC22-92PC90543, Washington, DC.
57. Phillips, S., Aden, A., Jechura, J., Dayton, D., and T. E. (2007) *Thermochemical Ethanol Via Indirect Gasification and Mixed Alcohol Synthesis of Lignocellulosic Biomass*, NREL/TP-510-41168, National Renewable Energy Laboratory, Golden, CO.
58. Quarderer, G.J. (1986) *Mixed Alcohols from Synthesis Gas*, American Institute of Chemical Engineers, New York.
59. Anon. (2005) Fischer–Tropsch archive. Available from: www.fischer-tropsch.org
60. Schulz, H. (1999) Short history and present trends of Fischer–Tropsch synthesis, *Applied Catalysis A: General* 186(1–2), 3–12.
61. Tijmensen, M., Faaij, A., Hamelinck, C., and van Hardeveld, M. (2002) Exploration of the possibilities for production of Fischer–Tropsch liquids and power via biomass gasification, *Biomass and Bioenergy* 23(2), 129–152.
62. Senden, M.M.G., Sie, S.T., Post, M.F.M., and Ansorge, J. (1992) Engineering aspects of the conversion of natural gas into middle distillates, in: De Lasa, H.I., Doğu, G., and Ravella, A. (eds.), *Chemical Reactor Technology for Environmentally Safe Reactors and Products*, Kluwer Academic Publishers, Dordrecht, the Netherlands, pp. 227–247.
63. Bridgwater, A. and Peacocke, G. (2000) Fast pyrolysis processes for biomass, *Renewable and Sustainable Energy Reviews* 4(1), 1–73.
64. Czernik, S. and Bridgwater, A. (2004) Overview of applications of biomass fast pyrolysis oil, *Energy and Fuels* 18(2), 590–598.
65. Piskorz, J., Scott, D.S., and Radlein, D. (1988) Composition of oils obtained by the fast pyrolysis of different woods. *Pyrolysis Oils from Biomass: Producing, Analyzing, and Upgrading*, ACS Publications, Washington, DC, pp. 167–178.
66. Bridgwater, A. (2003) Renewable fuels and chemicals by thermal processing of biomass, *Chemical Engineering Journal* 91(2–3), 87–102.
67. Oasmaa, A. and Peacocke, C. (2001) *A Guide to Physical Property Characterisation of Biomass-derived Fast Pyrolysis Liquids*, Technical Research Centre of Finland, VTT Publications, Espoo, Finland.
68. Carlson, T., Vispute, T., and Huber, G. (2008) Green gasoline by catalytic fast pyrolysis of solid biomass derived compounds, *ChemSusChem* 1(5), 397.
69. Holmgren, J., Nair, P., Elliot, D., Bain, R., and Marinangeli, R. (2008) Converting pyrolysis oils to renewable transport fuels: Processing challenges & opportunities, in: *National Petrochemical & Refiners Association Annual Meeting*, San Diego, CA, March 9–11, 2008, pp. 9–11.
70. Arthur J. Power & Associates, Inc. (1991) Feasibility Study: One thousand tons per day feedstock wood to crude pyrolysis oils plant 542,000 pounds per year using fast pyrolysis biomass process. Prepared for Solar Energy Research Institute, Golden, CO.
71. Solantausta, Y., Beckman, D., Bridgwater, A., Diebold, J., and Elliott, D. (1992) Assessment of liquefaction and pyrolysis systems, *Biomass and Bioenergy* 2(1), 279–297.
72. Elliott, D., Beckman, D., Bridgwater, A., Diebold, J., Gevert, S., and Solantausta, Y. (1991) Developments in direct thermochemical liquefaction of biomass: 1983–1990, *Energy & Fuels* 5(3), 399–410.

73. Elliott, D., Neuenschwander, G., Hart, T., Butner, R., Zacher, A., Engelhard, M., Young, J., and McCready, D. (2004) Chemical processing in high-pressure aqueous environments. 7. Process development for catalytic gasification of wet biomass feedstocks, *Industrial and Engineering Chemistry Research* 43(9), 1999–2004.
74. Krawczyk, T. (1999) Specialty oils, *INFORM-International News on Fats, Oils, and Related Materials* 10(6), 552.
75. Ouellette, R. and Bailey, M. (1998) *Organic Chemistry, a Brief Introduction*, Prentice Hall, Upper Saddle River, NJ.
76. Lipinsky, E.S., McClure, T. A., Kresovich, S., Otis, J. L., and Wagner, C. K. (1984) Fuels and chemicals from oilseeds, in: Shultz Jr., E. B. and Morgan, R. P. (ed.), *AAAS Selected Symposium*, Westview Press, Boulder, CO.
77. Openshaw, K. (2000) A review of *Jatropha curcas*: An oil plant of unfulfilled promise, *Biomass and Bioenergy* 19(1), 1–15.
78. Glenn, E., O'Leary, J., Watson, M., Thompson, T., and Kuehl, R. (1991) *Salicornia bigelovii* Torr.: An oilseed halophyte for seawater irrigation, *Science* 251(4997), 1065–1067.
79. Kumar, A. and Sharma, S. (2008) An evaluation of multipurpose oil seed crop for industrial uses (*Jatropha curcas* L.): A review, *Industrial Crops & Products* 28(1), 1–10.
80. Klass, D. (1998) *Biomass for Renewable Energy, Fuels, and Chemicals*, Academic Press, San Diego, CA, pp. 341–344.
81. Chisti, Y. (2007) Biodiesel from microalgae. *Biotechnology Advances* 25(3), 294–306.
82. Sheehan, J., Dunahay, T., Benemann, J., and Roessler, P. (1998) *A Look Back at the US Department of Energy's Aquatic Species Program—Biodiesel from Algae*, NREL/TP-580-24190, National Renewable Energy Laboratory, Golden, CO,
83. Weissman, J., Goebel, R., and Benemann, J. (1988) Photobioreactor design: Comparison of open ponds and tubular reactors, *Biotechnology and Bioengineering* 31, 336–344.
84. Benemann, J., Goebel, R., Weissman, J., and Augenstein, D. (1982) Microalgae as a source of liquid fuels, in: *SERI Biomass Program Principal Investigators' Review Meeting*, Washington, DC, DOE/ER/30014-T1.
85. Becker, E. (2007) Micro-algae as a source of protein, *Biotechnology Advances* 25(2), 207–210.
86. Golueke, C., Oswald, W., and Gotaas, H. (1957) Anaerobic digestion of algae, *Applied and Environmental Microbiology* 5(1), 47–55.
87. Persson, M., Jönsson, O., and Wellinger, A. (2006) *Biogas Upgrading to Vehicle Fuel Standards and Grid Injection*, IEA Bioenergy Task 37.
88. Ma, F. and Hanna, M. (1999) Biodiesel production: A review, *Bioresource Technology* 70(1), 1–15.
89. Kram, J.W. (2009) Aviation alternatives, *Biodiesel Magazine*, January 2009.
90. Moreau, C., Naceur Belgacem, M., and Gandini, A. (2004) Recent catalytic advances in the chemistry of substituted furans from carbohydrates and in the ensuing polymers, *Topics in Catalysis* 27(1–4), 11–30.
91. Hanniff, M.I. and Dao, L.H. (1987) Conversion of biomass carbohydrates into hydrocarbon products, in: *Proceedings of IGT/CBETS Conference on Energy from Biomass and Wastes X*, Washington, DC, Institute of Gas Technology, Chicago, IL.
92. Román-Leshkov, Y., Barrett, C.J., Liu, Z.Y., and Dumesic, J.A. (2007) Production of dimethylfuran for liquid fuels from biomass-derived carbohydrates, *Nature* 447, 982–985.

93. Zhao, H., Holladay, J., Brown, H., and Zhang, Z. (2007) Metal chlorides in ionic liquid solvents convert sugars to 5-hydroxymethylfurfural, *Science* 316(5831), 1597.
94. Lewkowski, J. (2001) Synthesis, chemistry and applications of 5-hydroxymethylfurfural and its derivatives, *Arkivoc* 1, 17–54.
95. Diebold, J. and Evans, R. (1988) Process for producing furan from furfural aldehyde, The United States of America as represented by the United States (Washington, DC), US Patent 4,764,627.
96. Romm, J. (2005) *The Hype about Hydrogen: Fact and Fiction in the Race to Save the Climate*, Island Press, Washington, DC.
97. Vieira de Carvalho, A.J. (1982) Natural gas and other alternative fuels for transportation purposes, *Energy* 10(2), 187–215.
98. Satterfield, C. (1991) *Heterogeneous Catalysis in Industrial Practice*, McGraw-Hill Book Co., New York.
99. Starkman, E., Newhall, H., Sutton, R., and Maguire, T. (1966) Ammonia as a spark ignition engine fuel: Theory and application, SAE Paper No. 660155, pp. 765–784.
100. Sorenson, S. (2001) Dimethyl ether in diesel engines: Progress and perspectives, *Journal of Engineering for Gas Turbines and Power* 123, 652.
101. Simpson, A. (2004) Full cycle assessment of alternative fuels for light duty road vehicles in Australia, 19th World Energy Congress, Sydney, pp. 5–9.
102. Tillman, D. (1991) *The Combustion of Solid Fuels and Wastes*, Academic Press, Bellevue, WA.
103. Kumar, J.V. and Pratt, B.C. (1996) Compositional analysis of some renewable biofuels, *American Laboratory* 28(8), 15–20.
104. Bain, R., Overend, R., and Craig, K. (1998) Biomass-fired power generation, *Fuel Processing Technology* 54(1–3), 1–16.
105. Sami, M., Annamalai, K., and Wooldridge, M. (2001) Co-firing of coal and biomass fuel blends, *Progress in Energy and Combustion Science* 27(2), 171–214.
106. Munson, B. et al. (1994) *Fundamentals of Fluid Mechanics*, 2nd edn, John Wiley & Sons, New York, Table 1.6.
107. Perry, R. and Green, W. (1984) *Perry's Chemical Engineers' Handbook*, 6th edn., McGraw-Hill, New York, Chap. 3.
108. Kajitani, S. et al. (1997) Engine performance and exhaust characteristics of direct-injection diesel engine operated with DME, Society of Automotive Engineers Inc., Warrendale, PA, p. 35.

5

Fundamentals of Solar Radiation

> Having been admonished by the Holy Office [the Inquisition] entirely to abandon the false option that the Sun was the center of the universe and immovable, and that the Earth was not the center of the same and that it moved…I abjure…
>
> **—Galileo**

5.1 The Physics of the Sun and Its Energy Transport

The nature of energy generation in the sun is still an unanswered question. Spectral measurements have confirmed the presence of nearly all the known elements in the sun. However, 80% of the sun is hydrogen and 19% helium. Therefore, the remaining 100-plus observed elements make up only a tiny fraction of the composition of the sun. It is generally accepted that a hydrogen-to-helium thermonuclear reaction is the source of the sun's energy. Yet, because such a reaction has not been duplicated in the laboratory, it is unclear precisely what the reaction mechanism is, what role the turbulent flows in the sun play, and how solar prominences and sunspots are created.

The nature of the energy-creation process is of no importance to terrestrial users of the sun's radiation. Of interest is the amount of energy, its spectral and temporal distribution, and its variation with time of day and year. These matters are the main subject of this chapter.

The sun is a 13.9×10^5 km diameter sphere comprised of many layers of gases, which are progressively hotter toward its center. The outermost layer, that from which energy is radiated into the solar system, is approximately at an equivalent black-body temperature of 5760 K (10,400°R). The center of the sun, however, may be at 20×10^6 K. The rate of energy emission from the sun is 3.8×10^{23} kW, which results from the conversion of 4.3×10^9 g/s (4.7×10^6 ton/s) of mass to energy. Of this total, only a tiny fraction, approximately 1.7×10^{14} kW, is intercepted by the earth, which is located about 150 million km from the sun (Figure 5.1).

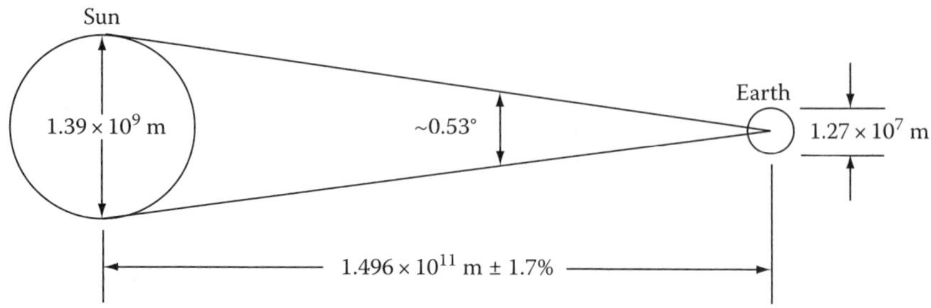

FIGURE 5.1
Relationship between the sun and the earth.

Solar energy is the world's most abundant permanent source of energy. The amount of solar energy intercepted by the planet earth is 5000 times greater than the sum of all other inputs (terrestrial, nuclear, geothermal, and gravitational energies, and lunar gravitational energy). Of this amount, 30% is reflected to space, 47% is converted to low-temperature heat and reradiated to space, and 23% powers the evaporation/precipitation cycle of the biosphere. Less than 0.5% is represented in the kinetic energy of the wind and waves and in photosynthetic storage in plants.

Total terrestrial radiation is only about one-third of the extraterrestrial total during a year, and 70% of that falls on the oceans. However, the remaining 1.5×10^{17} kWh that falls on land is a prodigious amount of energy—about 6000 times the total energy usage of the United States in 2000. However, only a small fraction of this total can be used because of physical and socioeconomic constraints, as described in Chapter 1.

5.2 Thermal Radiation Fundamentals

The material presented in this section has been selected from textbooks on heat transfer and radiation (for example, Refs. [1–4]). It provides the background needed to understand the nature of solar radiation for the engineering analysis of solar energy systems.

To begin with, all radiation travels at the speed of light which is equal to the product of the wavelength and the frequency of radiation. The speed of light in a medium equals the speed of light in a vacuum divided by the refractive index of the medium through which it travels:

$$c = \lambda \nu = \frac{c_0}{n}, \tag{5.1}$$

where

- λ is the wavelength in m (or μm, 1 μm = 10^{-6} m)
- ν is the frequency in s^{-1}
- c is the speed of light in a medium (m/s)
- c_0 is the speed of light in a vacuum (m/s)
- n is the index of refraction of the medium

Thermal radiation is one kind of electromagnetic energy and all bodies emit thermal radiation by virtue of their temperature. When a body is heated, its atoms, molecules, or electrons are raised to higher levels of activity called excited states. However, they tend to return to lower energy states, and in this process energy is emitted in the form of electromagnetic waves. Changes in energy states result from rearrangements in the electronic, rotational, and vibrational states of atoms and molecules. Since these rearrangements involve different amounts of energy changes and these energy changes are related to the frequency, the radiation emitted by a body is distributed over a range of wavelengths. A portion of the electromagnetic spectrum is shown in Figure 5.2. The wavelengths associated with the various mechanisms are not sharply defined; thermal radiation is usually considered to fall within the band from about 0.1 to 100 μm, whereas solar radiation has most of its energy between 0.1 and 3 μm.

For some problems in solar energy engineering, the classical electromagnetic wave theory is not suitable. In such cases, for example, in photovoltaic or photochemical processes, it is necessary to treat the energy transport from the point of view of quantum mechanics. In this view, energy is transported

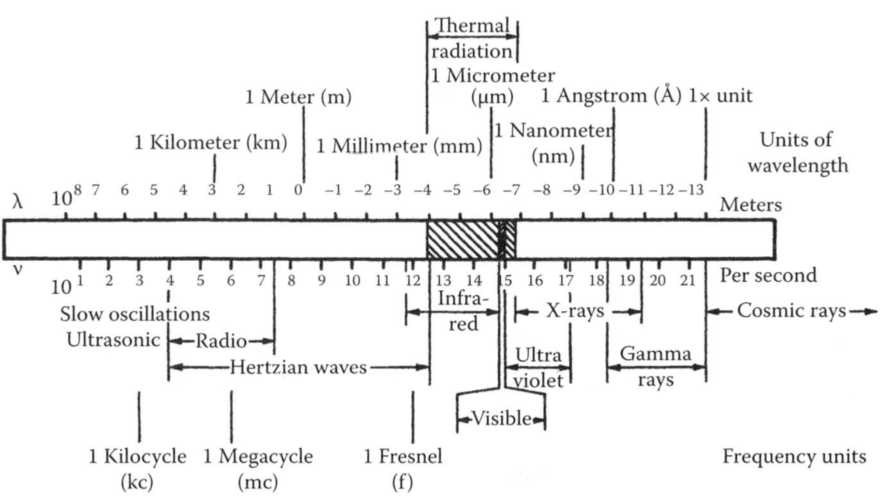

FIGURE 5.2
Electromagnetic radiation spectrum.

by particles or *photons*, which are treated as energy units or quanta rather than waves. The energy of a photon, E_p, of frequency ν_p is

$$E_p = h\nu_p, \tag{5.2}$$

where h = Planck's constant (6.625×10^{-34} J · s).

5.2.1 Black-Body Radiation

The energy density of the radiation emitted at a given wavelength (monochromatic) by a perfect radiator, usually called a black body, is given according to the relation

$$E_{b\lambda} = \frac{C_1}{(e^{C_2/\lambda T} - 1)\lambda^5 n^2}, \tag{5.3}$$

where

$C_1 = 3.74 \times 10^8$ W · μm^4/m^2 (1.19×10^8 Btu · μm^4/h · ft^2)
$C_2 = 1.44 \times 10^4$ μm · K (2.59×10^4 μm · °R)
n = Refractive index of the medium = 1.0 for vacuum; n is taken to be approximately equal to 1 for air

The quantity $E_{b\lambda}$ has the units of W/m^2 · μm (Btu/h · ft^2 · μm) and is called the monochromatic emissive power of a black body, defined as the energy emitted by a perfect radiator per unit wavelength at the specified wavelength per unit area and per unit time at the temperature T.

The total energy emitted by a black body, E_b, can be obtained by integration over all wavelengths. This yields the Stefan–Boltzmann law

$$E_b = \int_0^\infty E_{b\lambda} d\lambda = \sigma T^4, \tag{5.4}$$

where

σ is the Stefan–Boltzmann constant = 5.67×10^{-8} W/m^2 · K^4 (0.1714×10^{-8} Btu/h · ft^2 · R^4)
T is the absolute temperature in K (or R = 460 + °F)

The concept of a black body, although no such body actually exists in nature, is very convenient in engineering because its radiation properties can readily be related to those of real bodies.

5.2.2 Radiation Function Tables

Engineering calculations of radiative transfer are facilitated by the use of radiation function tables, which present the results of Planck's law in a more

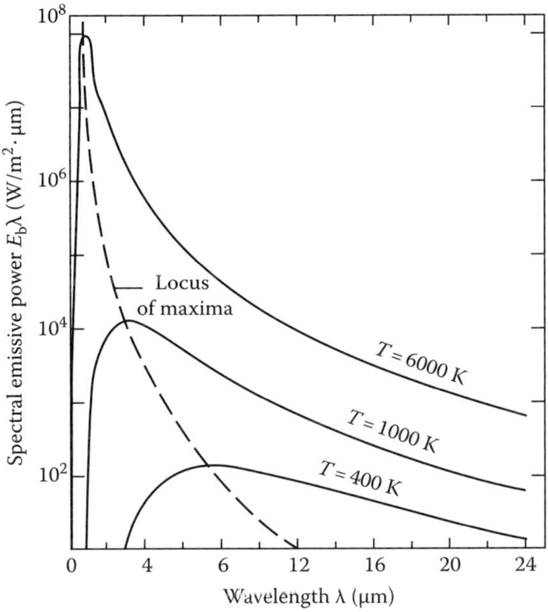

FIGURE 5.3
Spectral distribution of black-body radiation.

convenient form than Equation 5.3. A plot of the monochromatic emissive power of a black body as a function of wavelength as the temperature is increased is given in Figure 5.3. The emissive power shows a maximum at a particular wavelength. These peaks, or inflection points, are uniquely related to the body temperature. By differentiating Planck's distribution law (Equation 5.3) and equating to zero, the wavelength corresponding to the maximum value of $E_{b\lambda}$ can be shown to occur when

$$\lambda_{max} T = 2897.8\ \mu\text{m K} \,(5215.6\ \mu\text{m R}). \tag{5.5}$$

Frequently, one needs to know the amount of energy emitted by a black body within a specified range of wavelengths. This type of calculation can be performed easily with the aid of the radiation functions mentioned previously. To construct the appropriate radiation functions in dimensionless form, note that the ratio of the black-body radiation emitted between 0 and λ and between 0 and ∞ can be made a function of the single variable (λT) by using Equation 5.3 as shown below (for $n = 1$):

$$\frac{E_{b,0-\lambda}}{E_{b,0-\infty}} = \frac{\int_0^{\lambda} E_{b\lambda} d\lambda}{\sigma T^4} = \int_0^{\lambda T} \frac{C_1 d(\lambda T)}{\sigma(\lambda T)^5 (e^{C_2/\lambda T} - 1)}. \tag{5.6}$$

The above relation is plotted in Figure 5.4 and the results are also shown in tabular form in Table 5.1. In this table, the first column is the ratio of

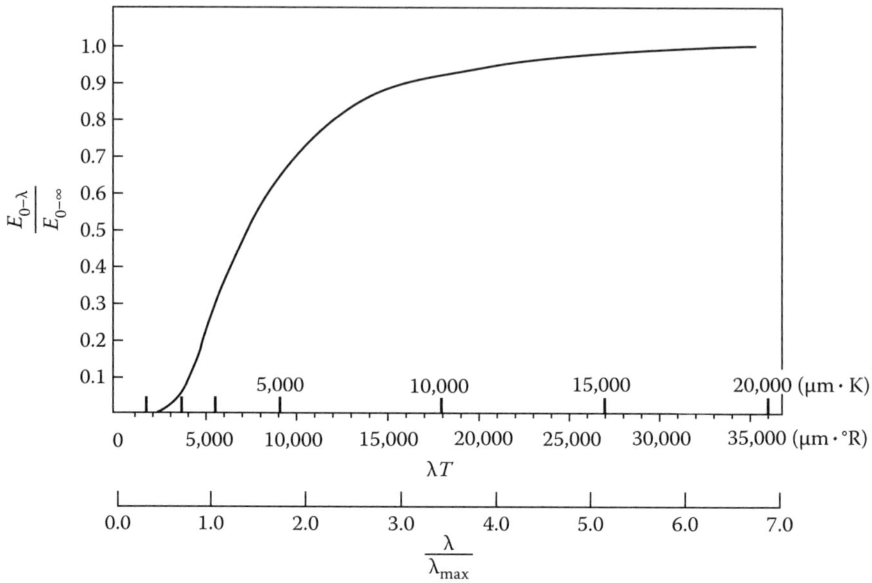

FIGURE 5.4
Fraction of total emissive power in spectral region between $\lambda = 0$ and λ as a function of λT and λ/λ_{max}.

λ to λ_{max} from Equation 5.5, and the third column the ratio of $E_{b,0-\lambda}$ to σT^4 from Equation 5.6. For use on a computer, Equation 5.6 can be approximated by the following polynomials:

$$\nu \geq 2 \frac{E_{b,0-\lambda}}{\sigma T^4} = \frac{15}{\pi^4} \sum_{m=1,2,\ldots} \frac{E^{-m\upsilon}}{m^4} \{[(m\nu + 3)m\nu + 6]m\nu + 6\}, \tag{5.7a}$$

and

$$\nu < 2 \frac{E_{b,0-\lambda}}{\sigma T^4} = \frac{15}{1-\pi^4} \nu^3 \left(\frac{1}{3} - \frac{\nu}{8} - \frac{\nu^2}{60} - \frac{\nu^4}{5,040} + \frac{\nu^6}{272,160} - \frac{\nu^8}{13,305,600} \right), \tag{5.7b}$$

where $\nu = C_2/\lambda T$.

5.2.3 Intensity of Radiation and Shape Factor

The emissive power of a surface gives the total radiation emitted in all directions. To determine the radiation emitted in a given direction we must define another quantity, the radiation intensity I. This quantity is defined as the radiant energy passing through an imaginary plane in space per unit area

TABLE 5.1

Thermal Radiation Functions[a]

$\frac{\lambda}{\lambda_{max}}$	$\frac{E_{b\lambda}}{E_{b\lambda,max}}$	$\frac{E_{b\lambda,0-\lambda}}{\sigma T^4}$
0.00	0.0000	0.0000
0.20	0.0000	0.0000
0.25	0.0003	0.0000
0.30	0.0038	0.0001
0.35	0.0187	0.0004
0.40	0.0565	0.0015
0.45	0.1246	0.0044
0.50	0.2217	0.0101
0.55	0.3396	0.0192
0.60	0.4664	0.0325
0.65	0.5909	0.0499
0.70	0.7042	0.0712
0.75	0.8007	0.0960
0.80	0.8776	0.1236
0.85	0.9345	0.1535
0.90	0.9725	0.1849
0.96	0.9936	0.2172
1.00	1.0000	0.2501
1.05	0.9944	0.2829
1.10	0.9791	0.3153
1.15	0.9562	0.3472
1.20	0.9277	0.3782
1.25	0.8952	0.4081
1.30	0.8600	0.4370
1.35	0.8231	0.4647
1.40	0.7854	0.4911
1.45	0.7477	0.5163
1.50	0.7103	0.5403
1.55	0.6737	0.5630
1.60	0.6382	0.5846
1.65	0.6039	0.6050
1.70	0.5710	0.6243
1.75	0.5397	0.6426
1.80	0.5098	0.6598
1.85	0.4815	0.6761
1.90	0.4546	0.6915
1 95	0.4293	0.7060
2.00	0.4054	0.7197

(continued)

TABLE 5.1 (continued)

Thermal Radiation Functions[a]

$\frac{\lambda}{\lambda_{max}}$	$\frac{E_{b\lambda}}{E_{b\lambda,max}}$	$\frac{E_{b\lambda,0-\lambda}}{\sigma T^4}$
2.05	0.3828	0.7327
2.10	0.3616	0.7449
2.15	0.3416	0.7565
2.20	0.3229	0.7674
2.25	0.3053	0.7777
2.30	0.2887	0.7875
2.35	0.2731	0.7967
2.40	0.2585	0.8054
2.45	0.2447	0.8137
2.50	0.2318	0.8215
2.55	0.2197	0.8290
2.60	0.2083	0.8360
2.65	0.1976	0.8427
2.70	0.1875	0.8490
2.75	0.1780	0.8550
2.80	0.1691	0.8607
2.85	0.1607	0.8661
2.90	0.1528	0.8713
2.95	0.1454	0.8762
3.00	0.1384	0.8809
3.10	0.1255	0.8895
3.20	0.1141	0.8974
3.30	0.1038	0.9045
3.40	0.0947	0.9111
3.50	0.0865	0.9170
3.60	0.0792	0.9225
3.70	0.0726	0.9275
3.80	0.0667	0.9320
3.90	0.0613	0.9362
4.00	0.0565	0.9401
4.20	0.0482	0.9470
4.40	0.0413	0.9528
4.60	0.0356	0.9579
4.80	0.0308	0.9622
5.00	0.0268	0.9660
6.00	0.0142	0.9790
7.00	0.0082	0.9861
8.00	0.0050	0.9904
9.00	0.0033	0.9930

TABLE 5.1 (continued)

Thermal Radiation Functions[a]

$\frac{\lambda}{\lambda_{max}}$	$\frac{E_{b\lambda}}{E_{b\lambda,max}}$	$\frac{E_{b\lambda,0-\lambda}}{\sigma T^4}$
10.00	0.0022	0.9948
20.00	0.0002	0.9993
40.00	0.0000	0.9999
50.00	0.0000	1.0000

[a] λ = wavelength in μm.
λ_{max} = wavelength at $E_{b\lambda,max}$ in μm = $2898/T$.
$E_{b\lambda}$ = monochromatic emissive power in W/m² · μm
= $374.15 \times 10^6/\lambda^5\,[\exp(14{,}387.9/\lambda T) - 1]$,
$E_{b\lambda,max}$ = maximum monochromatic emissive power in W/m² · μm
= $12.865 \times 10^{-12} T^5$,
$E_{b\lambda,0-\lambda} = \int_0^{\lambda} E_{b\lambda} d\lambda$,
$\sigma T^4 = E_{b\lambda/0-\infty} = 5.670 \times 10^{-8} T^4$ W/m²
T = absolute temperature in K.

per unit time and per unit solid angle perpendicular to the plane, as shown in Figure 5.5. I is defined by the relation

$$I = \lim_{\substack{dA' \to 0 \\ d\omega \to \infty}} = \frac{dE}{dA' d\omega} \tag{5.8}$$

Radiation intensity has both magnitude and direction. It can be related to the radiation flux, defined as the radiant energy passing through an imaginary

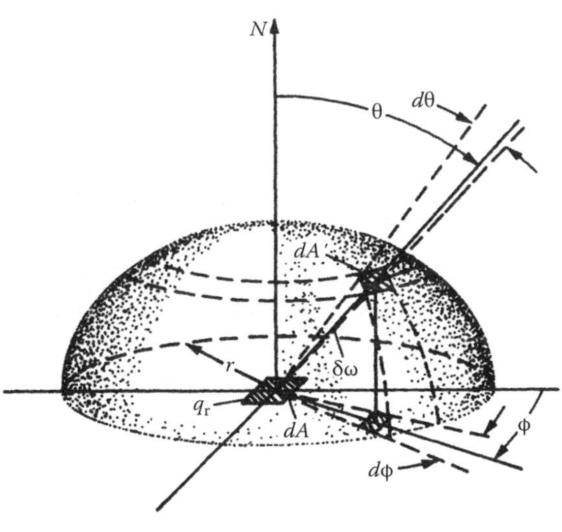

FIGURE 5.5
Schematic diagram illustrating radiation intensity and flux.

plane per unit area per unit time in all directions. Note that, whereas for the intensity, the area dA' is perpendicular to the direction of the radiation, for the flux the area dA is at the base in the center of a hemisphere through which all of the radiation passes. Recalling that the definition for the solid angle between dA' and dA is $d\omega = dA'/r^2$, the radiation flux q_r emanating from dA can be obtained by integrating the intensity over the hemisphere. As shown in Figure 5.5, the unit projected area for I is $dA \cos\theta$ and the differential area dA' on the hemisphere is $r^2 \sin\theta \, d\theta \, d\phi$; thus

$$q_r = \int_0^{2\pi} \int_0^{\pi/2} I \cos\theta \sin\theta \, d\theta \, d\phi. \tag{5.9}$$

If the area dA is located on a surface, the emissive power E can also be obtained from Equation 5.9. For the special case of a diffuse surface, for which I is the same in all directions, Equation 5.9 gives

$$q_r = \pi I. \tag{5.10}$$

Since all black surfaces are diffuse,

$$E_b = \pi I_b \tag{5.11}$$

Equation 5.11 can, of course, also be written for monochromatic radiation as

$$E_{b\lambda} = \pi I_{b\lambda} \tag{5.12}$$

In the evaluation of the rate of radiation heat transfer between two surfaces, not only their temperatures and their radiation properties but also their geometric configurations and relationships play a part. The influence of geometry in radiation heat transfer can be expressed in terms of the *radiation shape factor* between any two surfaces 1 and 2 defined as follows:

$F_{1\text{-}2}$ = fraction of radiation leaving surface 1 that reaches surface 2
$F_{2\text{-}1}$ = fraction of radiation leaving surface 2 that reaches surface 1

In general, F_{m-n} = fraction of radiation leaving surface "m" that reaches surface "n." If both surfaces are black, the energy leaving surface "m" and arriving at surface "n" is $E_{bm}A_mF_{m-n}$ and the energy leaving surface "n" and arriving at "m" is $E_{bn}A_nF_{n-m}$. If both surfaces absorb all the incident energy, the net rate of exchange $q_{m\Leftrightarrow n}$ will be

$$q_{m\Leftrightarrow n} = E_{bm}A_mF_{m-n} - E_{bn}A_nF_{n-m} \tag{5.13}$$

If both surfaces are at the same temperature, $E_{bm} = E_{bn}$ and the net exchange is zero, $q_{m\Leftrightarrow n} = 0$. This shows that the geometric radiation shape factor must obey the reciprocity relation

$$A_m F_{m-n} = A_m F_{n-m} \tag{5.14}$$

The net rate of heat transfer can therefore be written in two equivalent forms:

$$q_{m\Leftrightarrow n} = A_m F_{m-n}(E_{bm} - E_{bn}) = A_n F_{n-m}(E_{bm} - E_{bn}). \tag{5.15}$$

The evaluation of geometric shape factors is in general quite involved. For a majority of solar energy applications, however, only a few special cases are of interest. One of these is a small convex object of area A_1 surrounded by a large enclosure A_2. Since all radiation leaving A_1 is intercepted by A_2, $F_{1\text{-}2} = 1$ and $F_{2\text{-}1} = A_1/A_2$.

Another case is the exchange of radiation between two large parallel surfaces. If the two surfaces are near each other, almost all of the radiation leaving A_1 reaches A_2 and vice versa. Thus, $F_{1\text{-}2} = F_{2\text{-}1} = 1.0$, according to the definition of the shape factor. A third case of importance is the exchange between a small surface ΔA_1 and a portion of space A_2, for example, the exchange between a flat-plate solar collector tilted at an angle β from the horizontal and the sky it can see. For this situation, we refer to the definition of radiation flux (see Figure 5.5). The portion of the radiation emitted by ΔA_1 that is intercepted by the surrounding hemisphere depends on the angle of tilt. When the surface is horizontal, $F_{1\text{-}2} = 1$; when it is vertical, $F_{1\text{-}2} = 1/2(\beta = 90°)$. For intermediate values, it can be shown that [4]:

$$F_{1\text{-}2} = \frac{1}{2}(1 + \cos\beta) = \cos^2\left(\frac{\beta}{2}\right). \tag{5.16}$$

If the diffuse sky radiation is uniformly distributed and assumed to be black, then a small black area A_1 receives radiation at the rate

$$A_1 F_{1-\text{sky}} E_{\text{sky}} = \frac{A_1}{2}(1 + \cos\beta)\sigma T_{\text{sky}}^4 \tag{5.17}$$

whereas the net radiation heat transfer is given by

$$q_{\text{sky}\Leftrightarrow 1} = A_1 F_{1-\text{sky}} \sigma\left(T_{\text{sky}}^4 - T_1^4\right). \tag{5.18}$$

If the receiving area is gray with an absorptance $\bar{\alpha}$ equal to the emittance $\bar{\epsilon}$ the net exchange is given by

$$q_{\text{sky}\Leftrightarrow 1} = A_1 F_{1-\text{sky}} \bar{\alpha}\sigma\left(T_{\text{sky}}^4 - T_1^4\right). \tag{5.19}$$

5.2.4 Transmission of Radiation through a Medium

When radiation passes through a transparent medium such as glass or the atmosphere, the decrease in intensity can be described by Bouger's law, which assumes that the attenuation is proportional to the local intensity in the medium. If $I_\lambda(x)$ is monochromatic intensity after the radiation has traveled a distance x, the law is expressed by the equation

$$-dI_\lambda(x) = I_\lambda(x)K_\lambda dx, \tag{5.20}$$

where K_λ is the monochromatic extinction coefficient assumed to be a constant of the medium. If the transparent medium is a slab of thickness L and the intensity at $x = 0$ is designated by the symbol $I_{\lambda,0}$, the monochromatic transmittance τ_λ is equal to the ratio of the intensity at $x = L$ to $I_{\lambda,0}$. An expression for $I_\lambda(L)$ can be obtained by integrating Equation 5.20 between 0 and L, which gives

$$\ln\frac{I_\lambda(L)}{I_{\lambda,0}} = -K_\lambda L \quad \text{or} \quad I_\lambda(L) = I_{\lambda,0}e^{-K_\lambda L} \tag{5.21}$$

Then

$$\tau_\lambda = \frac{I_\lambda(L)}{I_{\lambda,0}} = e^{-K_\lambda L} \tag{5.22}$$

The extinction coefficient K_λ is a complex property of the medium since it combines the effects of absorption, emission, and scattering by the molecules and particles that make up the medium. Fortunately, for materials such as glass and plastics with known compositions, this coefficient can be determined accurately. Transmission of radiation through such materials is discussed further in Chapter 7. In this chapter, we are concerned about the transmission of solar radiation through the atmosphere. The atmosphere consists of the molecules of gases in it, such as N_2, O_2, CO_2, H_2O, etc., and aerosols such as dust particles, water droplets, and ice crystals. The extinction processes of the atmosphere consist of (a) absorption and emission by the molecules and aerosols, (b) scattering by the molecules, and (c) scattering by aerosols.

Since the atmosphere consists of a large number of components whose concentration changes as a function of time and location, determining the extinction coefficient of the atmosphere presents a formidable challenge. A major research effort is underway by scientists trying to predict global climate change. Some early attempts for the estimation of extinction coefficient for "average atmospheric conditions" were combined with an empirical approach [5] to use the above equation for the estimation of terrestrial solar radiation resource. This approach is described later in this chapter.

5.3 Sun–Earth Geometric Relationship

Figure 5.6 shows the annual orbit of the earth around the sun. The distance between the earth and the sun changes throughout the year, the minimum being 1.471×10^{11} m at winter solstice (December 21) and the maximum being 1.521×10^{11} m at summer solstice (June 21). The year-round average earth–sun distance is 1.496×10^{11} m. The amount of solar radiation intercepted by the earth, therefore, varies throughout the year, the maximum being on December 21 and the minimum on June 21.

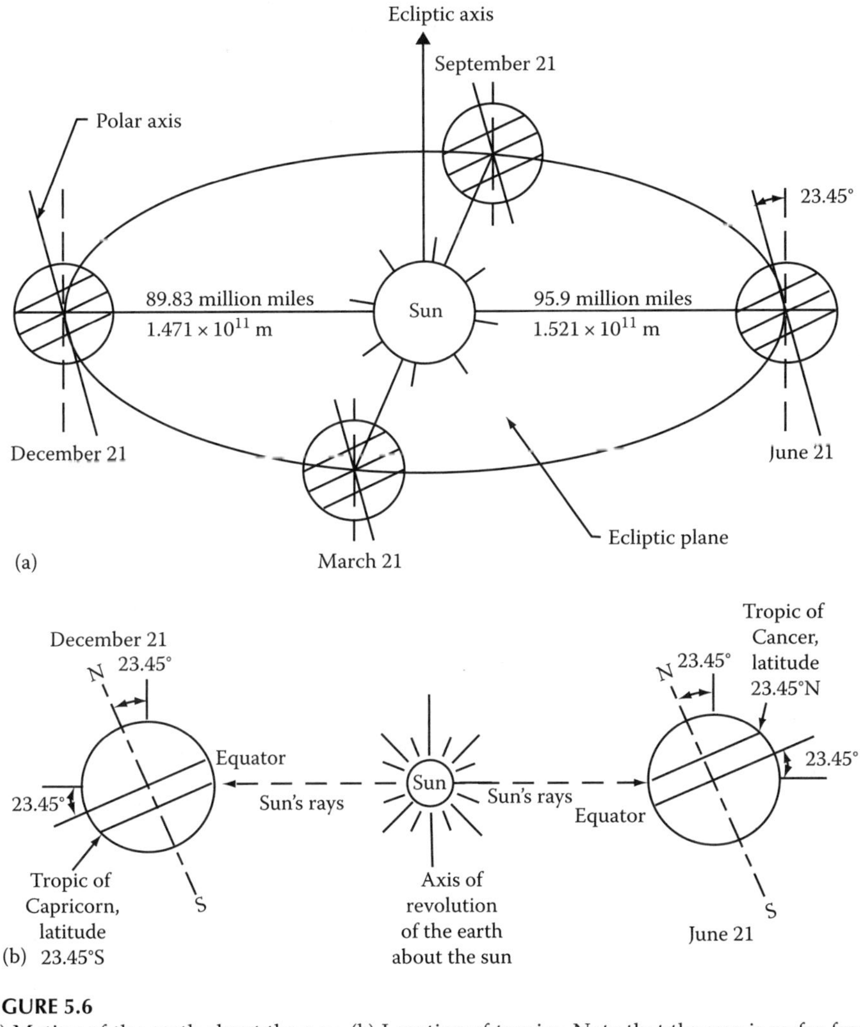

FIGURE 5.6
(a) Motion of the earth about the sun. (b) Location of tropics. Note that the sun is so far from the earth that all the rays of the sun may be considered as parallel to one another when they reach the earth.

The axis of the earth's daily rotation around itself is at an angle of 23.45° to the axis of its ecliptic orbital plane around the sun. This tilt is the major cause of the seasonal variation of the solar radiation available at any location on the earth. The angle between the earth–sun line (through their centers) and the plane through the equator is called the *solar declination*, δ_s. The declination varies between −23.45° on December 21 to +23.45° on June 21. Stated another way, the declination has the same numerical value as the latitude at which the sun is directly overhead at solar noon on a given day. The tropics of Cancer (23.45°N) and Capricorn (23.45°S) are at the extreme latitudes where the sun is overhead at least once a year, as shown in Figure 5.6. The Arctic and Antarctic circles are defined as those latitudes above which the sun does not rise above the horizon plane at least once per year. They are located, respectively, at 66½°N and 66½°S. Declinations north of the equator (summer in the northern hemisphere) are positive; those south, negative. The solar declination may be estimated by the relation:*

$$\delta_s = 23.45° \sin\left[\frac{360(284+n)}{365}\right]^{\circ}, \tag{5.23}$$

where n is the day number during a year with January 1 being $n = 1$. Approximate values of declination may also be obtained from Table 5.2 or Figure 5.7. For most calculations, the declination may be considered constant during any given day.

For the purposes of this book, the Ptolemaic view of the sun's motion provides a simplification to the analysis that follows. It is convenient to assume the earth to be fixed and to describe the sun's apparent motion in a coordinate system fixed to the earth with its origin at the site of interest. Figure 5.8 shows an apparent path of the sun to an observer. The position of the sun can be described at any time by two angles, the altitude and azimuth angles, as shown in Figure 5.8. The *solar altitude angle*, α, is the angle between a line collinear with the sun's rays and the horizontal plane. The *solar azimuth angle*, α_s, is the angle between a due south line and the projection of the site to sun line on the horizontal plane. The sign convention used for azimuth angle is positive west of south and negative east of south. The *solar zenith angle*, z, is the angle between the site to sun line and the vertical at the site:

$$z = 90° - \alpha. \tag{5.24}$$

The solar altitude and azimuth angles are not fundamental angles. Hence, they must be related to the fundamental angular quantities *hour angle*, *latitude*, and *declination*. The three angles are shown in Figure 5.9. The solar hour

* A more accurate relation is $\sin \delta_s = \sin(23.45°) \sin[360(284 + n)/365]°$. Because the error is small, Equation 5.23 is generally used.

TABLE 5.2

Summary Solar Ephemeris[a]

Date		Declination		Equation of Time		Date		Declination		Equation of Time	
		Deg	Min	Min	s			Deg	Min	Min	s
January	−1	−23	−4	−3	−14	February	−1	−17	−19	−13	−34
	−5	−22	−42	−5	−6		−5	−16	−10	−14	−2
	−9	−22	−13	−6	−50		−9	−14	−55	−14	−17
	−13	−21	−37	−8	−27		−13	−13	−37	−14	−20
	−17	−20	−54	−9	−54		−17	−12	−15	−14	−10
	−21	−20	−5	−11	−10		−21	−10	−50	−13	−50
	−25	−19	−9	−12	−14		−25	−9	−23	−13	−19
	−29	−18	−9	−13	−5						
March	−1	−7	−53	−12	−38	April	1	+4	14	−4	−12
	−5	−6	−21	−11	−48		5	5	46	−3	−1
	−9	−5	−48	−10	−51		9	7	17	−1	−52
	−13	−3	−14	−9	−49		13	8	46	−0	−47
	−17	−1	−39	−8	−42		17	10	12	+0	13
	−21	−0	−5	−7	−32		21	11	35	1	6
	25	+1	30	−6	−20		25	12	56	1	53
	29	3	4	−5	−7		29	14	13	2	33
May	1	+14	50	+2	50	June	1	+21	57	2	27
	5	16	2	3	17		5	22	28	1	49
	9	17	9	3	35		9	22	52	1	6
	13	18	11	3	44		13	23	10	+0	18
	17	19	9	3	44		17	23	22	−0	−33
	21	20	2	3	24		21	23	27	−1	−25
	25	20	49	3	16		25	23	25	−2	−17
	29	21	30	2	51		29	23	17	−3	−7
July	1	+23	10	−3	−31	August	1	+18	14	−6	−17
	5	22	52	−4	−16		5	17	12	−5	−59
	9	22	28	−4	−56		9	16	6	−5	−33
	13	21	57	−5	−30		13	14	55	−4	−57
	17	21	21	−5	−57		17	13	41	−4	−12
	21	20	38	−6	−15		21	12	23	−3	−19
	25	19	50	−6	−24		25	11	2	−2	−18
	29	18	57	−6	−23		29	9	39	−1	−10
September	1	+8	35	−0	15	October	1	−2	53	−10	−1
	5	7	7	+1	2		5	4	26	−11	−17
	9	5	37	2	22		9	5	58	−12	−27
	13	4	6	3	45		13	7	29	−13	−30
	17	2	34	5	10		17	8	58	−14	−25
	21	1	1	6	35		21	10	25	−15	−10
	25	−0	−32	8	0		25	11	50	−15	−46
	29	−2	−6	9	22		29	13	12	−16	−10

(*continued*)

TABLE 5.2 (continued)

Summary Solar Ephemeris[a]

Date		Declination		Equation of Time		Date		Declination		Equation of Time	
		Deg	Min	Min	s			Deg	Min	Min	s
November	1	−14	−11	+16	21	December	1	−21	−41	11	16
	5	−15	−27	16	23		5	−22	−16	9	43
	9	−16	−38	16	12		9	−22	−45	8	1
	13	−17	−45	15	47		13	−23	−6	6	12
	17	−18	−48	15	10		17	−23	−20	4	47
	21	−19	−45	14	18		21	−23	−26	2	19
	25	−20	−36	13	15		25	−23	−25	+0	20
	29	−21	−21	11	59		29	−23	−17	−1	39

[a] Since each year is 365.25 days long, the precise value of declination varies from year to year. *The American Ephemeris and Nautical Almanac*, published each year by the U.S. Government Printing Office, contains precise values for each day of each year.

angle h_s is based on the nominal time of 24h required for the sun to move 360° around the earth or 15° per hour. Therefore, h_s is defined as

$$h_s = 15°/\text{h} \cdot (\text{hours from local solar noon})$$

$$= \frac{\text{Minutes from local solar noon}}{4\ \text{min/degree}}. \tag{5.25}$$

Again, values east of due south, that is, morning values, are negative; and values west of due south are positive.

The latitude angle L is the angle between the line from the center of the earth to the site and the equatorial plane. The latitude may be read from an atlas and is considered positive north of the equator and negative south of the equator.

5.3.1 Solar Time and Angles

The sun angles are obtained from the local solar time, which differs from the local standard time. The relationship between the local solar time and the local standard time (LST) is

$$\text{Solar time} = \text{LST} + \text{ET} + (l_{st} - l_{local}) \cdot 4\ \text{min/degree}. \tag{5.26}$$

where

ET is the equation of time, which is a correction factor that accounts for the irregularity of the speed of earth's motion around the sun

l_{st} is the standard time meridian

l_{local} is the local longitude

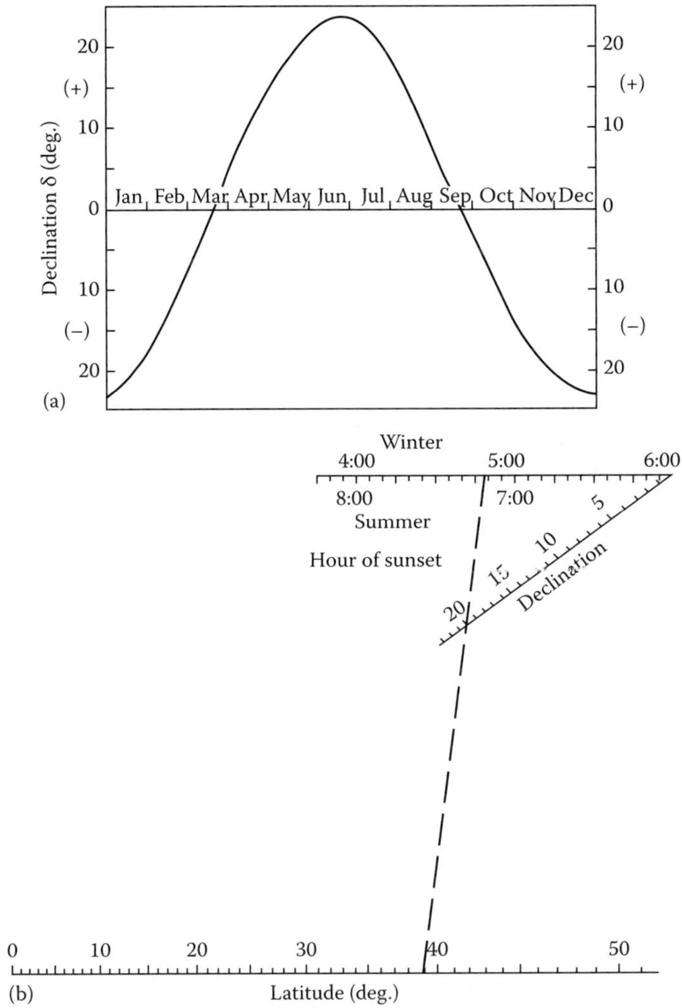

FIGURE 5.7
(a) Graph to determine the solar declination. (b) Sunset nomograph example. Example (b) shows determination of sunset time for summer (7:08 p.m.) and winter (4:52 p.m.) when the latitude is 39°N and the solar declination angle is 20°.

ET may be estimated from Table 5.2 or calculated from the following empirical equation:

$$\text{ET (in min)} = 9.87 \sin 2B - 7.53 \cos B - 1.5 \sin B \tag{5.27}$$

where $B = 360(n - 81)/364°$.

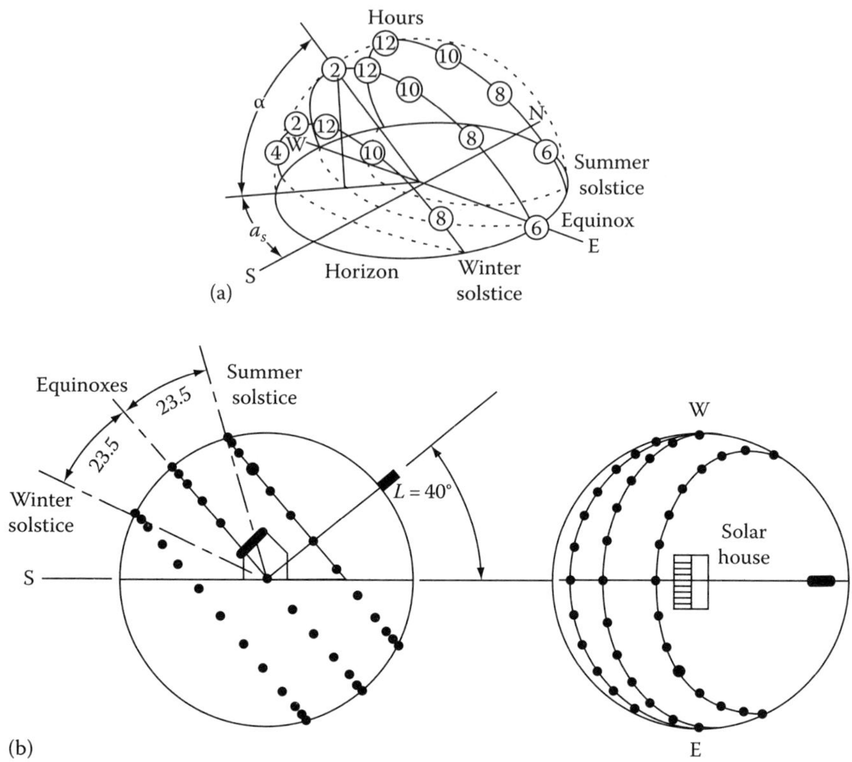

FIGURE 5.8
Sun paths for the summer solstice (6/21), the equinoxes (3/21 and 9/21), and the winter solstice (12/21) for a site at 40°N: (a) isometric view; (b) elevation and plan views.

The solar altitude angle, α, can be found from the application of the law of cosines to the geometry of Figure 5.9 and can be simplified as

$$\sin\alpha = \sin L \sin\delta_s + \cos L \cos\delta_s \cos h_s. \tag{5.28}$$

Using a similar technique, the solar azimuth angle, a_s, can be found as

$$\sin a_s = \frac{\cos\delta_s \sin h_s}{\cos\alpha} \tag{5.29}$$

At local solar noon, $h_s = 0$; therefore, $\alpha = 90 - |L - \delta_s|$, and $a_s = 0$.

In calculating the solar azimuth angle from Equation 5.29, a problem occurs whenever the absolute value of a_s is greater than 90°. A computational

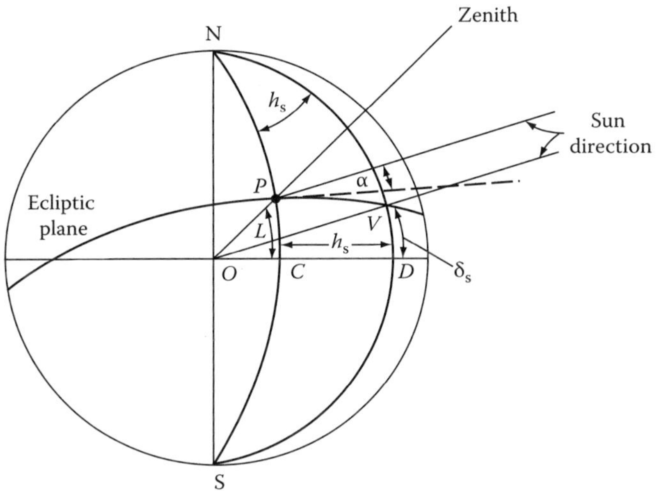

FIGURE 5.9
Definition of solar hour angle h_s (CND), solar declination δ_s (VOD), and latitude L (POC); p, site of interest. (Modified from Kreider, J.F. and Kreith, F., *Solar Heating and Cooling*, Hemisphere Publ. Corp., Washington, DC, 1982.)

device usually calculates the angle as less than 90° since $\sin a_s = \sin(180 - a_s)$. The problem can be solved in the following way:

For $L > \delta_s$, the solar times when the sun is due east (t_E) or due west (t_W) can be calculated by t_E or t_W = 12:00 Noon ± $\cos^{-1}[\tan \delta_s/\tan_L]°/(15°/h)$ (− for t_E, + for t_W).

For solar times earlier than t_E or later than t_W, the sun would be north (south in the southern hemisphere) of the east–west line and the absolute value of a_s would be greater than 90°. Then, a_s may be calculated as a_s is $a_s = \pm(180° - |a_s|)$.

For $L \le \delta_s$, the sun remains north (south in the southern hemisphere) of the east–west line and the true value of a_s is greater than 90°.

Sunrise and *sunset* times can be estimated by finding the hour angle for $\alpha = 0$. Substituting $\alpha = 0$ in Equation 5.28 gives the hour angles for sunrise (h_{sr}) and sunset (h_{ss}) as

$$h_{ss} \text{ or } h_{sr} = \pm\cos^{-1}[-\tan L \cdot \tan \delta_s]. \quad (5.30)$$

It should be emphasized that Equation 5.30 is based on the center of the sun at the horizon. In practice, sunrise and sunset are defined as the times when the upper limb of the sun is on the horizon. Because the radius of the sun is 16′, the sunrise would occur when $\alpha = -16'$. Also, at lower solar elevations, the sun will appear on the horizon when it is actually 34′ below the horizon. Therefore, for apparent sunrise or sunset, $\alpha = -50'$.

Example 5.1

Find the solar altitude and azimuth angles at solar noon in Gainesville, Florida on February 1. Also find the sunrise and sunset times in Gainesville on that day.

Solution
For Gainesville,

$$\text{Latitude } L = 29° + 41'\text{N} \quad \text{or} \quad 29.68°\text{N}.$$

$$\text{Longitude } l_{\text{local}} = 82° + 16'\text{W} \quad \text{or} \quad 82.27°\text{W}.$$

On February 1, day number
$n = 32$.
Therefore, declination

$$\delta_s = 23.45 \sin\left[\frac{360(284+32)}{365}\right]^\circ = -17.5°$$

At solar noon, $h_s = 0$. Therefore,

$$\begin{aligned}\sin\alpha &= \cos L \cos\delta_s \cos h_s + \sin L \sin\delta_s \\ &= \cos(29.68°)\cos(-17.5°)\cos(0) + \sin(29.68°)\sin(-17.5°),\end{aligned}$$

or

$$\alpha = 42.82°$$

$$\sin a_s = \frac{\cos(-17.5°)\sin(0)}{\cos(42.8°)} = 0,$$

or

$$a_s = 0.$$

At solar noon, α can also be found as

$$\begin{aligned}\alpha &= 90 - |L - \delta_s|^\circ \\ &= 90 - |29.68 + 17.5| = 42.82° \\ h_{ss} \text{ or } h_{sr} &= \pm\cos^{-1}\left[-\tan L \cdot \tan\delta_s\right] \\ &= \pm\cos^{-1}\left[-\tan(29.68°)\tan(-17.5°)\right] \\ &= \pm 79.65°.\end{aligned}$$

$$\text{Time from solar noon} = \pm(79.65°)(4\,\text{min/degree})$$
$$= \pm 319\,\text{min or} \pm (5\text{ h }19\,\text{min}).$$
$$\text{Sunrise time} = 12{:}00\text{ noon} - (5\text{ h }19\,\text{min})$$
$$= 6\text{ h }41\text{min a.m. (solar time)}.$$
$$\text{Sunset time} = 12{:}00\text{ noon} + (5\text{ h }19\,\text{min})$$
$$= 5\text{ h }19\text{min p.m. (solar time)}.$$

To convert these times to local times, we need to find ET:

$$\text{ET} = (9.87\sin 2B - 7.53\cos B - 1.5\sin B)\,\text{min}.$$

$$B = \frac{360}{364}(n - 81) = \frac{360}{364}(32 - 81) = -48.46°.$$

Therefore,

$$\text{ET} = -13.67\,\text{min}.$$

$$\text{LST} = \text{solar time} - \text{ET} - 4(l_{st} - l_{local}).$$

Gainesville, Florida is in the eastern standard time (EST) zone, where l_{st} = 75°W. Therefore,

$$\text{LST} = \text{Solar time} - (-13.67\,\text{min}) - 4(75 - 82.27)\,\text{min}$$
$$= \text{Solar time} + 42.75\,\text{min}.$$

Therefore,

$$\text{Sunrise time} = 6{:}41\text{ a.m.} + 43\text{min} = 7{:}24\text{ a.m. EST},$$

and

$$\text{Sunset time} = 5{:}19\text{ p.m.} + 43\text{ min} = 6{:}02\text{ p.m. EST}.$$

Note: Since the sunrise and sunset times are calculated when the center of the sun is at the horizon, they differ from the apparent times. If we use $\alpha = -50'$, the apparent sunrise and sunset times would be 7:20 a.m. EST and 6:06 p.m. EST, respectively.

Knowledge of the solar angles is helpful in the design of passive solar buildings, especially the placement of windows for solar access and the roof overhang for shading the walls and windows at certain times of the year. The following example illustrates this point.

Example 5.2

Find the roof overhang L of a south-facing window of height H = 1 m, such that the window is completely shaded at solar noon on April 1 and not shaded at all at noon on November 1. Assume that the roof extends far beyond the window on either side. Location: Gainesville, Florida. Also, find the overhang if S = 1.3 m.

Solution
From the geometry of the figure:

$$L = \frac{S}{\tan\alpha_2}.$$

Also,

$$L = \frac{(S+H)}{\tan\alpha_1}.$$

On April 1,

$$n = 91.$$

$$\delta_s = 23.45°\sin\left[\frac{360(284+91)}{365}\right] = 4.02°$$

Therefore, at solar noon,

$$\alpha_1 = 90 - |29.68 - 4.02| = 64.34°.$$

On November 1,

$$n = 305.$$

$$\delta_s = 23.45°\sin\left[\frac{360(284+305)}{365}\right]^{\circ} = -15.4°.$$

Therefore,

$$\alpha_2 = 90 - |29.68 + 15.4| = 44.9°.$$

$$L = \frac{H}{\tan\alpha_1 - \tan\alpha_2} = \frac{1}{\tan(64.34°) - \tan(44.9°)} = 0.92\,\text{m}.$$

$$S = L\tan\alpha_2 = 0.92 \times \tan(44.9°) = 0.92\,\text{m}.$$

If $S = 1.3$ m, then:

$$L = \frac{1.3}{\tan 44.9°} = 1.3\ \text{m}.$$

Also,

$$L = \frac{2.3}{\tan 64.34°} = 1.1\,\text{m}.$$

Therefore, $1.1\ \text{m} \le L \le 1.3\ \text{m}$.

5.3.2 Sun-Path Diagram

The projection of the sun's path on the horizontal plane is called a *sun-path diagram*. Such diagrams are very useful in determining shading phenomena associated with solar collectors, windows, and shading devices. As shown earlier, the solar angles (α, a_s) depend upon the hour angle, declination, and latitude. Since only two of these variables can be plotted on a two-dimensional graph, the usual method is to prepare a different sun-path diagram for each latitude with variations of hour angle and declination shown for a full year. A typical sun-path diagram is shown in Figure 5.10 for 30°N latitude.

Sun-path diagrams for a given latitude are used by entering them with appropriate values of declination δ_s and hour angle h_s. The point at the intersection of the corresponding δ_s and h_s lines represents the instantaneous location of the sun. The solar altitude can then be read from the concentric circles in the diagram; the azimuth, from the scale around the circumference of the diagram. A complete set of sun-path diagrams is contained in Figure A.5.1.

Example 5.3

Using Figure 5.10, determine the solar altitude and azimuth for March 8 at 10 a.m. Compare the results to those calculated from the basic equations (Equations 5.28 and 5.29).

Solution

On March 8, the solar declination is −5°; therefore the −5° sun path is used. The intersection of the 10 AM line and the −5° declination line in the diagram represents the sun's location; it is marked with a heavy dot in Figure 5.10.

Declination	Approx. dates
+23° 27′	June 22
+20°	May 21, July 24
+15°	May 1, Aug. 12
+10°	Apr. 16, Aug. 28
+5°	Apr. 3, Sep. 10
0°	Mar. 21, Sep. 23
−5°	Mar. 8, Oct. 6
−10°	Feb. 23, Oct. 20
−15°	Feb. 9, Nov. 3
−20°	Jan. 21, Nov. 22
−23° 27′	Dec. 22

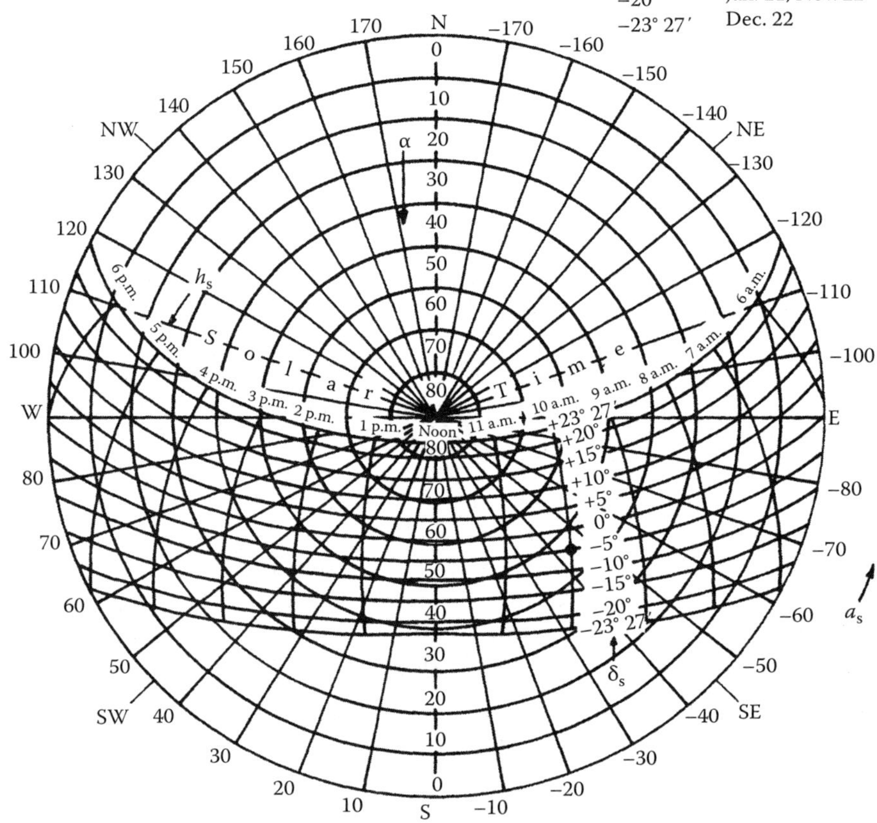

FIGURE 5.10
Sun-path diagram for 30°N latitude showing altitude and azimuth angles. (Modified from Kreider, J.F., and Kreith, F., *Solar Heating and Cooling*, revised 1st edn., Hemisphere Publ. Corp., Washington, DC, 1977.)

The sun's position lies midway between the 40° and 50° altitude circles, say at 45°, and midway between the −40° and −50° azimuth radial lines, say at −45°. So $\alpha \cong 45°$ and $a_s \cong -45°$. Equations 5.28 and 5.29 give precise values for α and a_s:

$$\sin\alpha = \sin(30°)\sin(-5°) + \cos(30°)\cos(-5°)\cos(-30°)$$

$$\alpha = 44.7°$$

$$\sin a_s = \frac{\cos(-5°)\sin(-30°)}{\cos(44.7°)}$$

$$a_s = -44.5°$$

Therefore, the calculated values are within ±0.5° (1%) of those read from the sun-path diagram.

5.3.3 The Shadow-Angle Protractor

The shadow-angle protractor used in shading calculations is a plot of solar altitude angles, projected onto a given plane, versus solar-azimuth angle. The projected altitude angle is usually called the *profile angle* γ. It is termed as the angle between the normal to a surface and the projection of the sun's rays on a vertical plane normal to the same surface. The profile angle is shown in Figure 5.11a with the corresponding solar altitude angle. The profile angle, which is always used in sizing shading devices, is given by

$$\tan\gamma = \sec a \tan\alpha, \tag{5.31}$$

where a is the solar azimuth angle with respect to the wall normal.

Figure 5.11b shows the shadow-angle protractor to the same scale as the sun-path diagrams in Figure 5.10 and Appendix A. It is used by plotting the limiting values of profile angle γ and azimuth angle a, which will start to cause shading of a particular point. The shadow-angle protractor is usually traced onto a transparent sheet so that the shadow map constructed on it can be placed over the pertinent sun-path diagram to indicate the times of day and months of the year during which shading will take place. The use of the shadow-angle protractor is best illustrated by an example.

Example 5.4

A solar building with a south-facing collector is sited to the north–northwest of an existing building. Prepare a shadow map showing what months of the year and what part of the day point C at the base of the solar collector will be shaded. Plan and elevation views are shown in Figure 5.12. Latitude = 40°N.

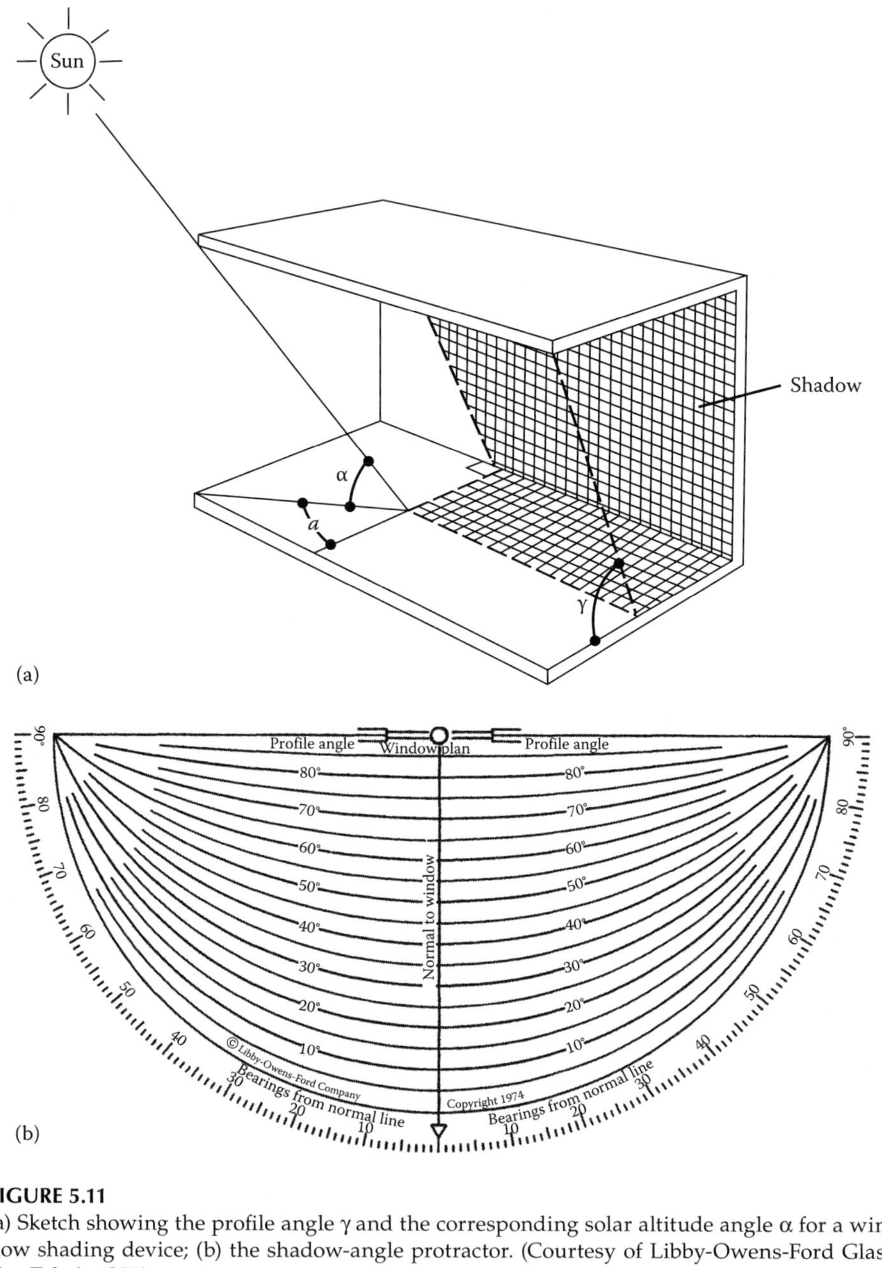

FIGURE 5.11
(a) Sketch showing the profile angle γ and the corresponding solar altitude angle α for a window shading device; (b) the shadow-angle protractor. (Courtesy of Libby-Owens-Ford Glass Co., Toledo, OH.)

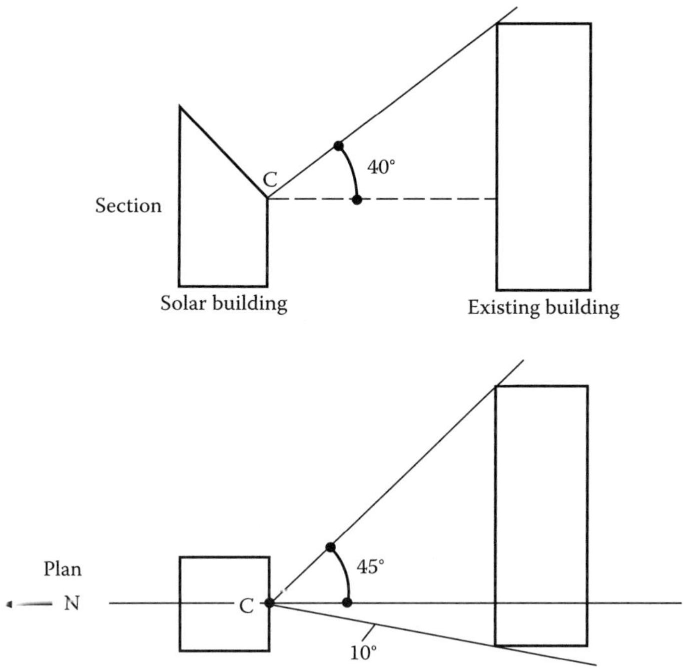

FIGURE 5.12
Plan and elevation view of proposed solar building and existing building, which may shade solar collector at point C.

Solution
The limiting profile angle for shading is 40° and the limiting azimuth angles are −45° and +10°, as shown in Figure 5.12. These values are plotted on the shadow-angle protractor (Figure 5.13a). The shadow map, when superimposed on the sun-path diagram (Figure 5.13b), shows that point C will be shaded during the following times of day for the periods shown:

Declination	Date	Time of Day
−23°27′	December 22	8:45 a.m.–12:40 p.m.
−20°	January 21, November 22	8:55 a.m.–12:35 p.m.
−15°	February 9, November 3	9:10 a.m.–12:30 p.m.

In summary, during the period from November 3 to February 9, point C will be shaded between 3 and 4 h. It will be shown later that this represents about a 50% loss in collector performance for point C, which would be unacceptable for a collector to be used for heating a building in winter.

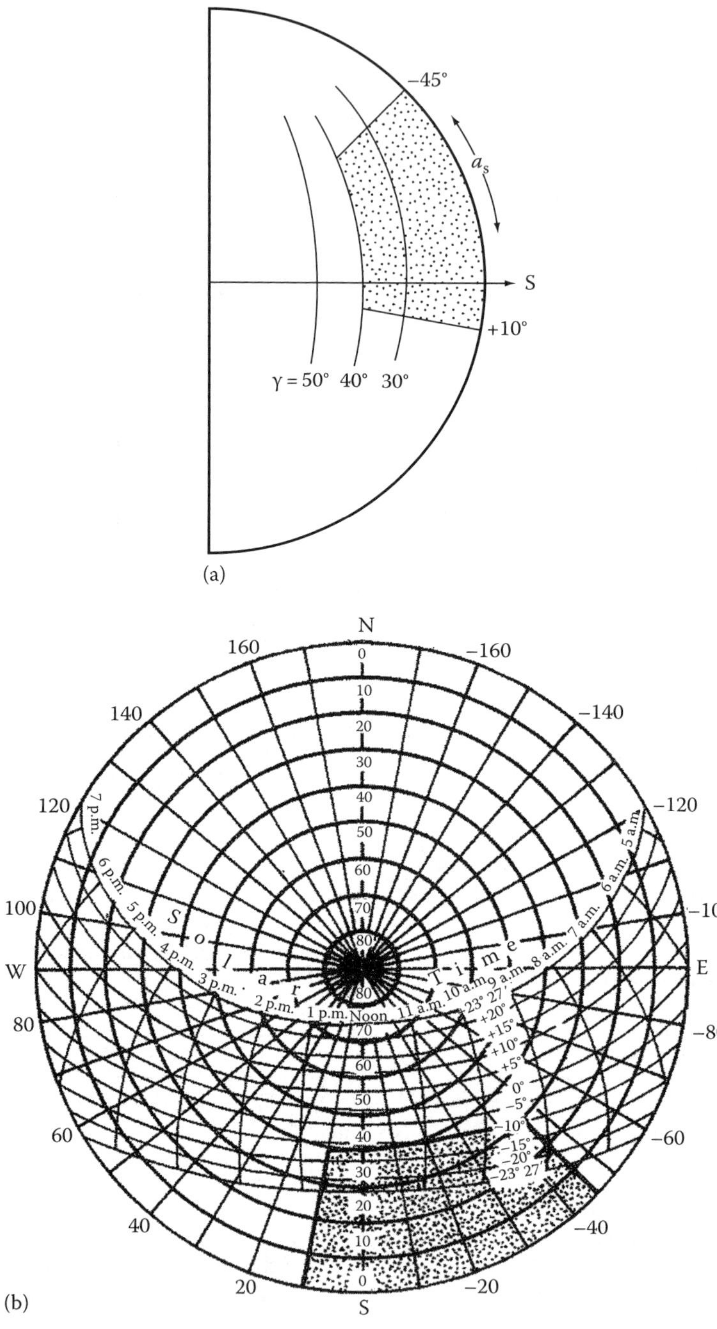

FIGURE 5.13
(a) Shadow map constructed for the example shown in Figure 2.13; (b) shadow map superimposed on sun-path diagram.

5.4 Solar Radiation

Detailed information about solar radiation availability at any location is essential for the design and economic evaluation of a solar energy system. Long-term measured data of solar radiation are available for a large number of locations in the United States and other parts of the world. Where long-term measured data are not available, various models based on available climatic data can be used to estimate the solar energy availability. Solar energy is in the form of electromagnetic radiation with the wavelengths ranging from about 0.3 μm (10^{-6} m) to over 3 μm, which correspond to ultraviolet (less than 0.4 μm), visible (0.4 and 0.7 μm), and infrared (over 0.7 μm). Most of this energy is concentrated in the visible and the near-infrared wavelength range (see Figure 5.14). The incident solar radiation, sometimes called *insolation*, is measured as irradiance, or the energy per unit time per unit area (or power per unit area). The units most often used are Watts per square meter (W/m²), British thermal units per hour per square foot (Btu/h · ft²), and Langleys per minute (calories per square centimeter per minute, cal/cm² · min).

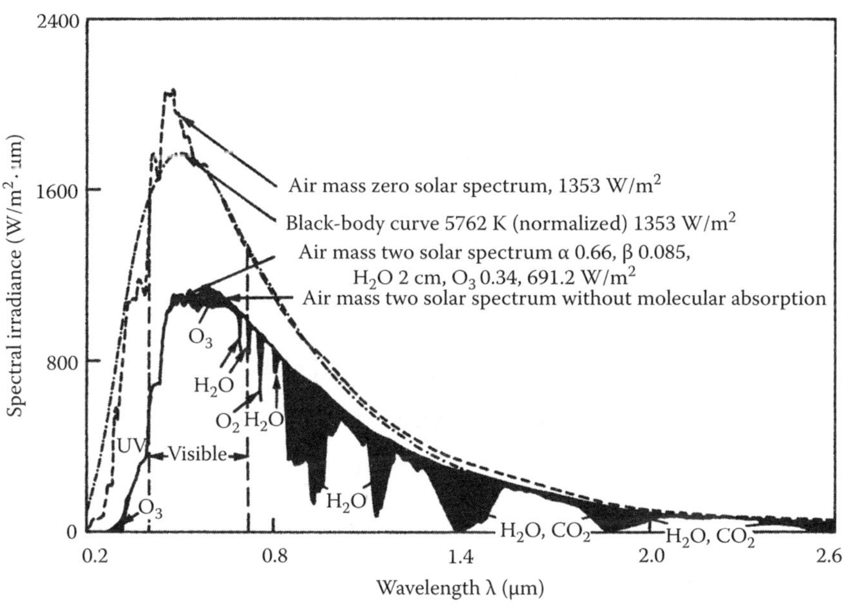

FIGURE 5.14
Extraterrestrial solar radiation spectral distribution. Also shown are equivalent black-body and atmosphere-attenuated spectra.

5.4.1 The Extraterrestrial Solar Radiation

The average amount of solar radiation falling on a surface normal to the rays of the sun outside the atmosphere of the earth (extraterrestrial) at mean earth–sun distance (D_0) is called the *solar constant*, I_0. Measurements by NASA indicated the value of the solar constant to be 1353 W/m² (±1.6%), 429 Btu/h · ft² or 1.94 cal/cm² · min (Langleys/min). This value was revised upward by Fröhlich et al. [6] to 1377 W/m² or 437.1 Btu/h · ft² or 1.974 Langleys/min, which was the value used in compiling SOLMET data in the United States [7,8]. At present, there is no consensus on the value of the solar constant. However, considering that the difference between the two values is about 1.7% and the uncertainties in estimation of terrestrial solar radiation are 10% or higher, either value may be used. A value of 1367 W/m² is also used by many references.

The variation in seasonal solar radiation availability at the surface of the earth can be understood from the geometry of the relative movement of the earth around the sun. Since the earth's orbit is elliptical, the earth–sun distance varies during a year, the variation being ±1.7% from the average. Therefore, the extraterrestrial radiation, I, also varies by the inverse square law as below:

$$I = I_0\left(\frac{D_0}{D}\right)^2, \tag{5.32}$$

where

D is the distance between the sun and the earth

D_0 is the yearly mean earth–sun distance (1.496×10^{11} m)

The $(D_0/D)^2$ factor may be approximated as [9]

$$\left(\frac{D_0}{D}\right)^2 = 1.00011 + 0.034221\cos(x) + 0.00128\sin(x) + 0.000719\cos(2x) + 0.000077\sin(2x), \tag{5.33}$$

where

$$x = \frac{360(n-1)}{365^\circ}, \tag{5.34}$$

and n = Day number (starting from January 1 as 1). The following approximate relationship may also be used without much loss of accuracy:

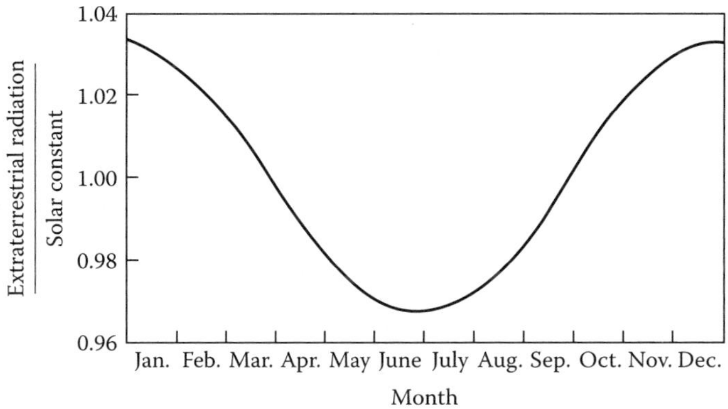

FIGURE 5.15
Effect of the time of year on the ratio of extraterrestrial radiation to the nominal solar constant.

$$I = I_0\left[1 + 0.034\cos\left(\frac{360n}{365.25}\right)^{\circ}\right]. \tag{5.35}$$

Figure 5.15 also shows the relationship of the extraterrestrial solar radiation to the solar constant. For many solar energy applications, such as photovoltaics and photocatalysis, it is necessary to examine the distribution of energy within the solar spectrum. Figure 5.15 shows the spectral irradiance at the mean earth–sun distance for a solar constant of 1353 W/m^2 as a function of wavelength according to the standard spectrum data published by NASA in 1971. The data are also presented in Table 5.3 and their use is illustrated in the following example.

Example 5.5

Calculate the fraction of solar radiation within the visible part of the spectrum, that is, between 0.40 and 0.70 μm.

Solution

The first column in Table 5.3 gives the wavelength. The second column gives the averaged solar spectral irradiance in a band centered at the wavelength in the first column. The fourth column, $D\lambda$, gives the percentage of solar total radiation at wavelengths shorter than the value of λ in the first column. At a value of 0.40 μm, 8.7% of the total radiation occurs at shorter wavelengths. At a wavelength of 0.70, 46.88% of the radiation occurs at shorter wavelengths. Consequently, 38% of the total radiation lies within the band between 0.40 and 0.70 μm, and the total energy received outside the earth's atmosphere within that spectral range is 517 W/m^2 (163 Btu/h · ft^2).

TABLE 5.3

Extraterrestrial Solar Irradiance[a]

λ (μm)	$E\lambda$[b] (W/m²·μm)	$E\lambda$[b] (Btu/h·ft²·μm)	$D\lambda$[c] (%)	λ (μm)	$E\lambda$ (W/m²·μm)	$E\lambda$ (Btu/h·ft²·μm)	$D\lambda$ (%)	λ (μm)	$E\lambda$ (W/m²·μm)	$E\lambda$ (Btu/h·ft²·μm)	$D\lambda$ (%)
0.115	0.007	0.002	1×10^{-4}	0.43	1639	520	12.47	0.90	891	283	63.37
0.14	0.03	0.010	5×10^{-4}	0.44	1810	574	13.73	1.00	748	237	69.49
0.16	0.23	0.073	6×10^{-4}	0.45	2006	636	15.14	1.2	485	154	78.40
0.18	1.25	0.397	1.6×10^{-3}	0.46	2066	655	16.65	1.4	337	107	84.33
0.20	10.7	3.39	8.1×10^{-3}	0.47	2033	645	18.17	1.6	245	77.7	88.61
0.22	57.5	18.2	0.05	0.48	2074	658	19.68	1.8	159	50.4	91.59
0.23	66.7	21.2	0.10	0.49	1950	619	21.15	2.0	103	32.7	93.49
0.24	63.0	20.0	0.14	0.50	1942	616	22.60	2.2	79	25.1	94.83
0.25	70.9	22.5	0.19	0.51	1882	597	24.01	2.4	62	19.7	95.86
0.26	130	41.2	0.27	0.52	1833	581	25.38	2.6	48	15.2	96.67
0.27	232	73.6	0.41	0.53	1842	584	26.74	2.8	39	12.4	97.31
0.28	222	70.4	0.56	0.54	1783	566	28.08	3.0	31	9.83	97.83
0.29	482	153	0.81	0.55	1725	547	29.38	3.2	22.6	7.17	98.22
0.30	514	163	1.21	0.56	1695	538	30.65	3.4	16.6	5.27	98.50

0.31	689	219	1.66	0.57	1712	543	31.91	3.6	13.5	4.28	98.72
0.32	830	263	2.22	0.58	1715	544	33.18	3.8	11.1	3.52	98.91
0.33	1059	336	2.93	0.59	1700	539	34.44	4.0	9.5	3.01	99.06
0.34	1074	341	3.72	0.60	1666	528	35.68	4.5	5.9	1.87	99.34
0.35	1093	347	4.52	0.62	1602	508	38.10	5.0	3.8	1.21	99.51
0.36	1068	339	5.32	0.64	1544	490	40.42	6.0	1.8	0.57	99.72
0.37	1181	375	6.15	0.66	1486	471	42.66	7.0	1.0	0.32	99.82
0.38	1120	355	7.00	0.68	1427	453	44.81	8.0	0.59	0.19	99.88
0.39	1098	348	7.82	0.70	1369	434	46.88	10.0	0.24	0.076	99.94
0.40	1429	453	8.73	0.72	1314	417	48.86	15.0	0.0048	0.015	99.98
0.41	1751	555	9.92	0.75	1235	392	51.69	20.0	0.0015	0.005	99.99
0.42	1747	554	11.22	0.80	1109	352	56.02	50.0	0.0004	0.0001	100.00

Source: Adapted from Iqbal, M., *Sol. Energy*, 23, 169, 1979; Thekaekara, M.P., *Sol. Energy*, 14, 109, 1973.

[a] Solar constant = 429 Btu/h·ft^2 = 1353 W/m^2.

[b] $E\lambda$ is the solar spectral irradiance averaged over a small bandwidth centered at λ.

[c] $D\lambda$ is the percentage of the solar constant associated with wavelengths shorter than λ.

5.5 Estimation of Terrestrial Solar Radiation

As extraterrestrial solar radiation, *I*, passes through the atmosphere, a part of it is reflected back into space, a part is absorbed by air and water vapor, and some gets scattered by molecules of air, water vapor, aerosols, and dust particles (Figure 5.16). The part of solar radiation that reaches the surface of the earth with essentially no change in direction is called *direct* or *beam radiation*. The scattered diffuse radiation reaching the surface from the sky is called the *sky diffuse radiation*.

Although extraterrestrial radiation can be predicted with certainty,* radiation levels on the earth are subject to considerable uncertainty resulting from local climatic interactions. The most useful solar radiation data is based on long-term (30 years or more) measured average values at a location, which unfortunately are not available for most locations in the world. For such locations, an estimation method (theoretical model) based on some measured climatic parameters may be used. This chapter describes several ways of estimating terrestrial solar radiation; all have large uncertainties (as much as ±30%) associated with them.

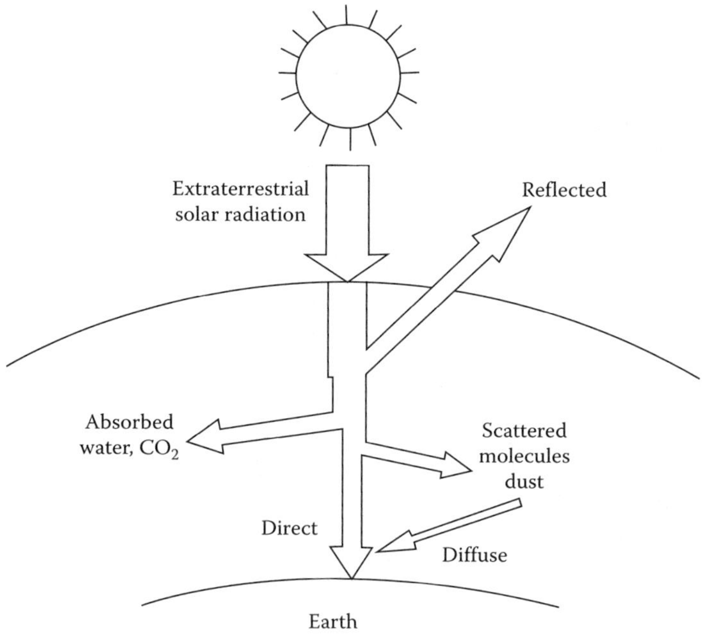

FIGURE 5.16
Attenuation of solar radiation as it passes through the atmosphere.

* The effect of sunspots, which may cause up to 0.5% variation, is neglected.

5.5.1 Atmospheric Extinction of Solar Radiation

As solar radiation I travels through the atmosphere, it is attenuated due to absorption and scattering. If K is the local extinction coefficient of the atmosphere, the beam solar radiation at the surface of the earth can be written according to Bouger's law Equation 5.21 as

$$I_{b,N} = Ie^{-\int Kdx}, \tag{5.36}$$

where

$I_{b,N}$ is the instantaneous beam solar radiation per unit area normal to the sun's rays

x is the length of travel through the atmosphere

If L_o is the vertical thickness of the atmosphere and

$$\int_0^{L_o} Kdx = k, \tag{5.37}$$

the beam normal solar radiation for a solar zenith angle of z will be

$$I_{b,N} = Ie^{-k\sec z} = Ie^{-k/\sin\alpha} = Ie^{-km}, \tag{5.38}$$

where m is a dimensionless path length of sunlight through the atmosphere, sometimes called the *air mass ratio* (Figure 5.17). When solar altitude angle is 90° (sun is overhead), $m = 1$.

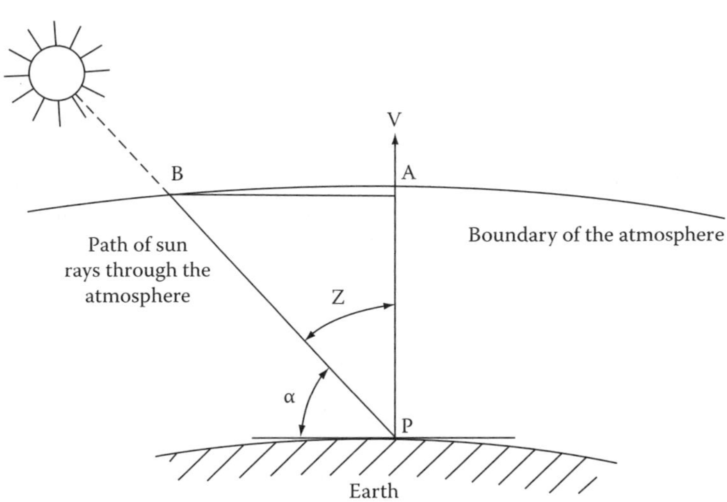

FIGURE 5.17
Air mass definition; air mass m = BP/AP = cosec α, where α is the altitude angle. The atmosphere is idealized as a constant thickness layer.

TABLE 5.4

Average Values of Atmospheric Optical Depth (k) and Sky Diffuse Factor (C) for 21st Day of Each Month, for Average Atmospheric Conditions at Sea Level for the United States

Month	1	2	3	4	5	6	7	8	9	10	11	12
k	0.142	0.144	0.156	0.180	0.196	0.205	0.207	0.201	0.177	0.160	0.149	0.142
C	0.058	0.060	0.071	0.097	0.121	0.134	0.136	0.122	0.092	0.073	0.063	0.057

Source: Threlkeld, J.L. and Jordan, R.C., *ASRAE Trans.*, 64, 45, 1958; Islam, M.R. *RERIC Int. Energy J.*, 16(2), 103113, 1994.

Threlkeld and Jordan [5] estimated values of k (also known as optical depth) for average atmospheric conditions at sea level with a moderately dusty atmosphere and the amount of precipitable water vapor equal to the average value for the United States for each month. These values are given in Table 5.4. To account for the differences in local conditions from the average sea-level conditions, Equation 5.38 is modified by a parameter called clearness number C_n, introduced by Threlkeld and Jordan [5]:

$$I_{\mathrm{b,N}} = C_n I e^{-k/\sin\alpha}. \tag{5.39}$$

5.5.2 Solar Radiation on Clear Days

Total instantaneous solar radiation on a horizontal surface (see Figure 5.18), I_{h}, is a sum of the beam or direct radiation, $I_{\mathrm{b,h}}$ and the sky diffuse radiation $I_{\mathrm{d,h}}$:

$$I_{\mathrm{h}} = I_{\mathrm{b,h}} + I_{\mathrm{d,h}}. \tag{5.40}$$

According to the Threlkeld and Jordan model [5], the sky diffuse radiation on a clear day is proportional to the beam normal solar radiation and can be estimated by using an empirical sky diffuse factor C. Therefore, I_{h} can be estimated as

$$\begin{aligned} I_{\mathrm{h}} &= I_{\mathrm{b,N}} \cos z + C I_{\mathrm{b,N}} \\ &= I_{\mathrm{b,N}} \sin\alpha + C I_{\mathrm{b,N}} \\ &= C_n I e^{-k/\sin\alpha}(C + \sin\alpha). \end{aligned} \tag{5.41}$$

Values of C are given in Table 5.4.

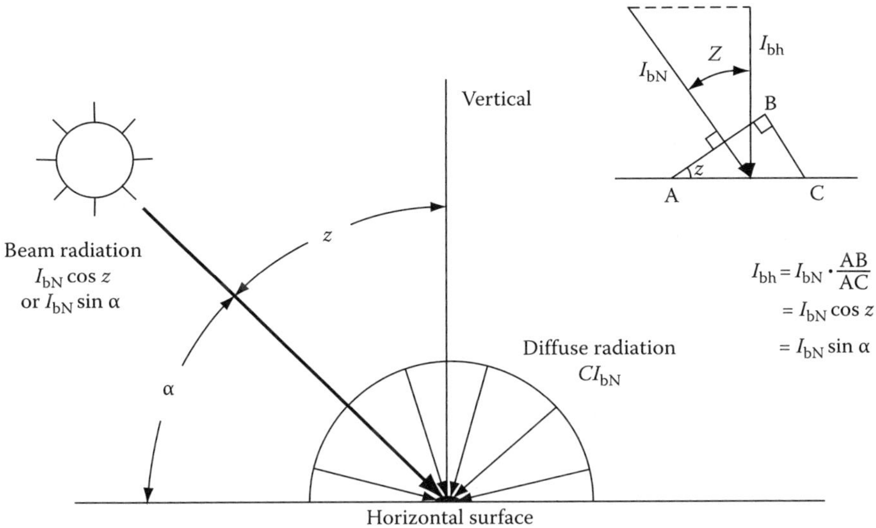

FIGURE 5.18
Solar radiation on a horizontal surface.

5.5.3 Solar Radiation on a Tilted Surface

Solar radiation on an arbitrary tilted surface having a tilt angle of β from the horizontal and an azimuth angle of a_w (assumed + west of south), as shown in Figure 5.19, is the sum of components consisting of beam ($I_{b,c}$), sky diffuse ($I_{d,c}$), and ground-reflected solar radiation ($I_{r,c}$):

$$I_c = I_{b,c} + I_{d,c} + I_{r,c}. \tag{5.42}$$

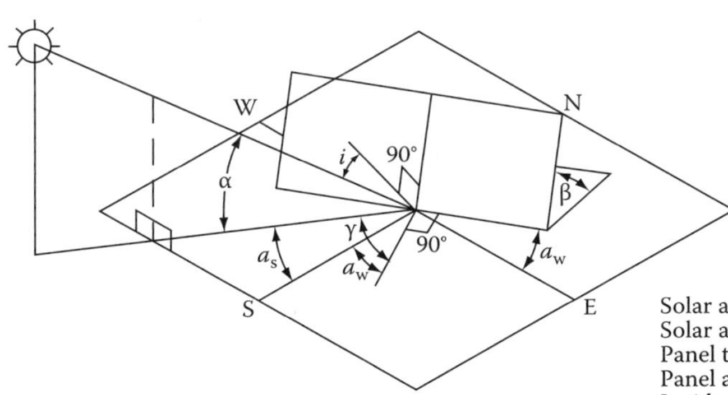

FIGURE 5.19
Definitions of solar angles for a tilted surface.

If i is the *angle of incidence* of the beam radiation on the tilted surface, it is simple to show that the instantaneous beam radiation on the surface per unit area is

$$I_{b,c} = I_{b,N} \cos i. \tag{5.43}$$

From the geometry in Figure 5.19, it can be shown that the angle of incidence i for the surface (angle between the normal to the surface and a line collinear with the sun's rays) is related to the solar angles as

$$\cos i = \cos\alpha \cos(a_s - a_w)\sin\beta + \sin\alpha\cos\beta. \tag{5.44}$$

The diffuse radiation on the surface ($I_{d,c}$) can be obtained by multiplying the sky diffuse radiation on a horizontal surface by the view factor between the sky and the surface*:

$$\begin{aligned} I_{d,c} &= I_{d,h} \frac{(1+\cos\beta)}{2} \\ &= CI_{b,N} \frac{(1+\cos\beta)}{2} \\ &= CI_{b,N} \cos^2\left(\frac{\beta}{2}\right) \end{aligned} \tag{5.45}$$

The ground-reflected solar radiation can be found from the total solar radiation incident on a horizontal surface and the ground reflectance ρ as

$$I_{r,c} = I_h \rho. \tag{5.46}$$

The part of $I_{r,c}$. intercepted by the tilted surface can be found by multiplying the ground-reflected radiation by the view factor between the surface† and the ground:

$$\begin{aligned} I_{r,c} &= \rho I_h \frac{(1-\cos\beta)}{2} = \rho I_h \sin^2\left(\frac{\beta}{2}\right) \\ &= \rho I_{b,N}(\sin\alpha + C)\sin^2\left(\frac{\beta}{2}\right). \end{aligned} \tag{5.47}$$

For ordinary ground or grass, ρ is approximately 0.2 and for snow covered ground it can be taken as approximately 0.8.

---

* The surface has been assumed infinitely large for this view factor. See Section 5.2.3.

† The tilted surface and the ground in front of it have been assumed to be infinitely large for this view factor.

Example 5.6

Find the instantaneous solar radiation at 12:00 noon EST on a solar collector surface ($\beta = 30°$, $a_w = +10°$) on February 1 in Gainesville, Florida.

Solution

From Example 5.1, for February 1:

$$n = 32, \quad \delta_s = -17.5°, \quad \text{and} \quad \text{ET} = -13.7 \text{ min}$$

For finding the values of solar radiation on the collector, we will need to calculate angles α, a_s, h_s, and i.

$$\begin{aligned}\text{Solar time} &= \text{LST} + \text{ET} + 4(l_{st} - l_{local}) \\ &= 12{:}00 - 13.7 \text{ min} + 4(75° - 82.27°) \\ &= 11{:}17.2 \text{ a.m.}, \\ h_s &= \frac{\text{Minutes from solar noon}}{4 \text{ min/deg}}, \\ &= \frac{-42.8}{4} = -10.7°(-\text{before noon}).\end{aligned}$$

From Equation 5.28,

$$\begin{aligned}\alpha &= \sin^{-1}\left(\sin(29.68°)\sin(-17.5°) + \cos(29.68°)\cos(-17.5°)\cos(-10.7°)\right) \\ &= 41.7°.\end{aligned}$$

From Equation 5.29,

$$\begin{aligned}a_s &= \frac{\sin^{-1}\left(\cos(-17.5°)\sin(-10.7°)\right)}{\cos(41.7°)} \\ &= -13.7°.\end{aligned}$$

Angle of incidence i for the solar collector is given by Equation 5.44:

$$\begin{aligned}\cos i &= \cos(41.7°)\cos(-13.7° - 10°)\sin(30°) + \sin(41.7°)\cos(30°) \\ i &= 23.4°\end{aligned}$$

To calculate the solar radiation using Equations 5.42, 5.43, 5.45, and 5.47, we need to find $I_{b,N}$ and I.

Extraterrestrial solar radiation

$$I = I_0\left[1+0.034\cos\left(\frac{360n}{365.25}\right)^{\circ}\right]$$

$$= 1353\ \text{W/m}^2\left[1+0.034\cos\left(\frac{360\times 32}{365.25)}\right)\right]$$

$$= 1392\ \text{W/m}^2,$$

$$I_{b,N} = C_n I e^{-k/\sin\alpha}$$

(Find k from Table 5.4. Assume $C_n = 1$)

$$= 1392e^{-0.144/\sin(41.7^{\circ})}$$

$$= 1121\,\text{W/m}^2.$$

Beam radiation on the collector (Equation 5.43):

$$I_{b,c} = I_{b,N}\cos i$$

$$= (1121)\cos(23.4^{\circ})$$

$$= 1029\ \text{W/m}^2.$$

Sky diffuse radiation on the collector (Equation 5.45):

$$I_{d,c} = CI_{b,N}\cos^2\left(\frac{\beta}{2}\right)$$

$$= (0.060)(1121)\cos^2\left(\frac{30}{2}\right)$$

(Find C from Table 5.4.)

$$63\ \text{W/m}^2.$$

Ground-reflected radiation on the collector (Equation 5.47):

$$I_{r,c} = \rho I_{b,N}(\sin\alpha + C)\sin^2\left(\frac{\beta}{2}\right)$$

Assume

$$\rho = 0.2$$

$$I_{r,c} = (0.2)(1121)(\sin 41.7^{\circ} + 0.060)\sin^2(15^{\circ})$$

$$= 11\,\text{W/m}^2.$$

Total insolation on the collector:

$$I_c = 1029 + 63 + 11 = 1103\ \text{W/m}^2.$$

A convenient Web site that will provide the same information obtained in the preceding example can be found at http://www.builditsolar.com/Tools/RadOnCol/radoncol.htm. The model requires merely inputting the latitude, collector tilt, collector azimuth, altitude of location, the month, and the day. With this information, the program provides the direct, diffuse, and total radiation perpendicular to the sun in the third, fourth, and fifth columns. The sixth column provides the total radiation on a horizontal surface. The seventh column gives the incident angle of the sun on the collector. The 9th, 10th, and 11th columns give the direct, diffuse, and total radiation falling on the collector surface. From this program, which was updated in 2009, the radiation falling on the collector at noon is 1037 W/m², which is about 4% less than the value obtained by the empirical procedure outlined in the book. The printout from the model for the day is shown in Table 5.5.

5.5.4 Monthly Solar Radiation Estimation Models

One of the earliest methods of estimating solar radiation on a horizontal surface was proposed by the pioneer spectroscopist Angström. It was a simple linear model relating average horizontal radiation to clear-day radiation and to the sunshine level, that is, percent of possible hours of sunshine. Since the definition of a clear day is somewhat nebulous, Page [11] refined the method and based it on extraterrestrial radiation instead of the ill-defined clear day:

$$\begin{aligned}\bar{H}_h &= \bar{H}_{o,h}\left(a + b\frac{\bar{n}}{\bar{N}}\right)\\ &= \bar{H}_{o,h}\left(a + b\frac{\overline{PS}}{100}\right)\end{aligned} \tag{5.48}$$

where

$\bar{H}_h$ and $\bar{H}_{o,h}$ are the horizontal terrestrial and horizontal extraterrestrial radiation levels averaged for a month

$\overline{PS}$ is the monthly averaged percent of possible sunshine (that is, hours of sunshine/maximum possible duration of sunshine × 100)

a and b are constants for a given site

$\bar{n}$ and $\bar{N}$ are the monthly average numbers of hours of bright sunshine and day length, respectively

The ratio $\bar{n}/\bar{N}$ is also equivalent to the monthly average percent sunshine ($\overline{PS}$). $\bar{H}_{o,h}$ can be calculated by finding $H_{o,h}$ from Equation 5.49, using

TABLE 5.5

Solar Radiation Output for Example 5.6 Using the Online Calculator

Solar radiation on collector for day of: 2/1
Collector area: 1 (ft^2)
Collector azimuth: 0 (degree) measured from South + is to East
Collector tilt: 40 (degree) measured from horizontal
Latitude: 30 (degree)
Altitude: 0 (ft) above sea level
Sun rise: 6.7 (h) sunrise in solar time

			Sun					Collector		
Time (h)	Az:	Elev:	Dir Normal	Diffuse	Total	Horz. Tot:	Incid Ang:	Direct	Diffuse	Total
4	88.9	−34.1	0	0	0	0	90	0	0	0
5	82	−21.2	0	0	0	0	90	0	0	0
6	75.1	−8.4	0	0	0	0	87	0	0	0
7	67.7	3.9	47	3	50	6	73	14	2	16
8	59.2	15.5	228	13	241	74	59	119	12	131
9	48.8	26	280	16	297	139	44	201	15	216
10	35.6	34.8	302	18	320	190	30	262	16	278
11	19.1	40.8	312	18	330	222	16	300	16	316
12	0	42.9	315	18	333	233	7	312	16	329
13	−19.1	40.8	312	18	330	222	16	300	16	316
14	−35.6	34.8	302	18	320	190	30	262	16	278
15	−48.8	26	280	16	297	139	44	201	15	216
16	−59.2	15.5	228	13	241	74	59	119	12	131
17	−67.7	3.9	47	3	50	6	73	14	2	16
18	−75.1	−8.4	0	0	0	0	87	0	0	0
19	−82	−21.2	0	0	0	0	90	0	0	0
20	−88.9	−34.1	0	0	0	0	90	0	0	0
Day total			2653	156	2809	1496		2102	138	2240

Angles in degrees
Radiations in Btu/h
Day total in Btu/day
1 Btu/h · ft^2 = 3.152 W/m^2

Equations 5.35 and 5.44 and averaging $I_{o,h}$ for the number of days in each month; or data in Table A.5.2 can be used.

$$H_{o,h} = \int_{t_{sr}}^{t_{ss}} I \sin\alpha \, dt \tag{5.49}$$

Some typical values of *a* and *b* are given in Table 5.5 [12]. Additional values for worldwide locations are given in Table A.5.4.

Example 5.7

Using the predictive method of Angstrom–Page, estimate the monthly horizontal solar radiation for Miami, Florida, located at a latitude of 25° N. Percentages of possible sunshine and extraterrestrial radiation can be taken from tables in Appendix A.

Solution
Using the climate data given, the expected monthly average horizontal radiation for Miami is calculated in the following table using a = 0.42 and b = 0.22 from Table 5.6.

TABLE 5.6

Coefficients a and b in the Angström–Page Regression Equation

		Sunshine Hours in Percentage of Possible			
Location	**Climate**[a]	**Range**	**Average**	a	b
Albuquerque, New Mexico	BS-BW	68–85	78	0.41	0.37
Atlanta, Georgia	Cf	45–71	59	0.38	0.26
Blue Hill, Massachusetts	Df	42–60	52	0.22	0.50
Brownsville, Texas	BS	47–80	62	0.35	0.31
Buenos Aires, Argentina	Cf	47–68	59	0.26	0.50
Charleston, South Carolina	Cf	60–75	67	0.48	0.09
Dairen, Manchuria	Dw	55–81	67	0.36	0.23
El Paso, Texas	BW	78–88	84	0.54	0.20
Ely, Nevada	BW	61–89	77	0.54	0.18
Hamburg, Germany	Cf	11–49	36	0.22	0.57
Honolulu, Hawaii	Af	57–77	65	0.14	0.73
Madison, Wisconsin	Df	40–72	58	0.30	0.34
Malange, Angola	Aw-BS	41–84	58	0.34	0.34
Miami, Florida	Aw	56–71	65	0.42	0.22
Nice, France	Cs	49–76	61	0.17	0.63
Poona, India (monsoon)	Am	25–49	37	0.30	0.51
(Dry)		65–89	81	0.41	0.34
Stanleyville, Congo	Af	34–56	48	0.28	0.39
Tamanrasset, Algeria	BW	76–88	83	0.30	0.43

Source: Löf, G.O.G. et al., World distribution of solar energy, University of Wisconsin, Madison, WI, Engineering Experiment Station Report 21, 1996.

[a] Af, tropical forest climate, constantly moist, rainfall all through the year; Am, tropical forest climate, monsoon rain, short dry season, but total rainfall sufficient to support rain forest; Aw, tropical forest climate, dry season in winter; BS, steppe or semiarid climate; BW, desert or arid climate; Cf, mesothermal forest climate, constantly moist, rainfall all through the year; Cs, mesothermal forest climate, dry season in winter; Df, microthermal snow forest climate, constantly moist, rainfall all through the year; Dw, microthermal snow forest climate, dry season in winter.

		$\bar{H}_{o,h}$ [a]		$\bar{H}_h$	
Month	$\overline{PS}/100$	Wh/m² · Day	kJ/m² · Day	kJ/m² · Day	Btu/ft² · Day
January	0.66	6,656	23,962	13,543	1193
February	0.68	7,769	27,968	15,931	1403
March	0.74	9,153	32,951	19,204	1691
April	0.76	10,312	37,123	21,799	1919
May	0.72	10,936	39,370	22,771	2005
June	0.68	11,119	40,028	22,800	2008
July	0.72	10,988	39,557	22,880	2015
August	0.71	10,484	37,742	21,747	1915
September	0.7	9,494	34,178	19,618	1728
October	0.7	8,129	29,264	16,798	1479
November	0.67	6,871	24,736	14,035	1236
December	0.63	6,284	22,622	12,637	1113

[a] Monthly averaged, daily extraterrestrial radiation.

5.6 Models Based on Long-Term Measured Horizontal Solar Radiation

Long-term measured solar radiation data is usually available as monthly averaged total solar radiation per day on horizontal surfaces. In order to use this data for tilted surfaces, the total solar radiation on a horizontal surface must first be broken down into beam and diffuse components. A number of researchers have proposed models to do that, prominent among them being Liu and Jordan, Collares-Pereira and Rabl, and Erbs, Duffie, and Klein.

5.6.1 Monthly Solar Radiation on Tilted Surfaces

In a series of papers, Liu and Jordan [13–17] have developed an essential simplification in the basically complex computational method required to calculate long-term radiation on tilted surfaces. This is called the LJ method. The fundamental problem in such calculations is the decomposition of long-term measured total horizontal radiation into its beam and diffuse components.

If the decomposition can be computed, the trigonometric analysis presented earlier can be used to calculate incident radiation on any surface in a straightforward manner. Liu and Jordan (LJ) correlated the diffuse-to-total radiation ratio ($\bar{D}_h/\bar{H}_h$) with the *monthly clearness index* $\bar{K}_T$, which is defined as

$$\bar{K}_T = \frac{\bar{H}_h}{\bar{H}_{o,h}}, \tag{5.50}$$

where

- $\bar{H}_h$ is the monthly averaged terrestrial radiation per day on a horizontal surface
- $\bar{H}_{o,h}$ is the corresponding extraterrestrial radiation, which can be calculated from Equation 5.49 by averaging each daily total for a month

The original LJ method was based upon the extraterrestrial radiation at mid-month, which is not truly an average.

The LJ correlation predicts the monthly diffuse ($\bar{D}_h$) to monthly total $\bar{H}_h$ ratio. It can be expressed by the empirical equation

$$\frac{\bar{D}_h}{\bar{H}_h} = 1.390 - 4.027\bar{K}_T + 5.531\bar{K}_T^2 - 3.108\bar{K}_T^3. \quad (5.51)$$

Note that the LJ correlation is based upon a solar constant value of 1394 W/m^2 (442 Btu/h · ft^2), which was obtained from terrestrial observations, whereas the newer value, based on satellite data, is 1377 W/m^2 (437 Btu/h · ft^2). The values of $\bar{K}_T$ must be based on this earlier value of the solar constant to use the LJ method. Collares-Pereira and Rabl [8] conducted a study and concluded that although Liu and Jordan's approach is valid, their correlations would predict significantly smaller diffuse radiation components. They also concluded that Liu and Jordan were able to correlate their model with the measured data because they used the measured data which was not corrected for the shade ring (see solar radiation measurements). Collares-Pereira and Rabl (C-P&R) also introduced the sunset hour angle h_{ss} in their correlation to account for the seasonal variation in the diffuse component. The C-P&R correlation is:

$$\frac{\bar{D}_h}{\bar{H}_h} = 0.775 + 0.347\left(h_{ss} - \frac{\pi}{2}\right) - \left[0.505 + 0.02619\left(h_{ss} - \frac{\pi}{2}\right)\right]\cos\left(2\bar{K}_T - 1.8\right) \quad (5.52)$$

where h_{ss} is the sunset hour angle in radians. The C-P&R correlation agrees well with the correlations for India [19], Israel [20] and Canada [21] and is, therefore, preferred to Equation 5.51.

The monthly average beam component $\bar{B}_h$ on a horizontal surface can be readily calculated by simple subtraction since $\bar{D}_h$ is known:

$$\bar{B}_h = \bar{H}_h - \bar{D}_h. \quad (5.53)$$

It will be recalled on an instantaneous basis from Equations 5.41 and 5.43 and Figure 5.19 that

$$I_{b,N} = \frac{I_{b,h}}{\sin\alpha}, \quad (5.54)$$

$$I_{b,c} = I_{b,N} \cos i, \tag{5.43}$$

where $I_{b,h}$ is the instantaneous horizontal beam radiation. Solving for $I_{b,c}$, the beam radiation on a surface,

$$I_{b,c} = I_{b,h}\left(\frac{\cos i}{\sin \alpha}\right). \tag{5.55}$$

The ratio in parentheses is usually called the beam radiation *tilt factor* R_b. It is a purely geometric quantity that converts instantaneous horizontal beam radiation to beam radiation intercepted by a tilted surface.

Equation 5.55 cannot be used directly for the long-term beam radiation $\bar{B}_h$. To be strictly correct, the instantaneous tilt factor R_b should be integrated over a month with the beam component $I_{b,h}$ used as a weighting factor to calculate the beam tilt factor. However, the LJ method is used precisely when such short-term data as $I_{b,h}$ are not available. The LJ recommendation for the monthly mean tilt factor $\bar{R}_b$ is simply to calculate the monthly average of cos i and divide it by the same average of sin α. In equation form for south-facing surfaces, this operation yields:

$$\bar{R}_b = \frac{\cos(L-\beta)\cos\delta_s \sin h_{sr} + h_{sr}\sin(L-\beta)\sin\delta_s}{\cos L \cos\delta_s \sin h_{sr}(\alpha=0) + h_{sr}(\alpha=0)\sin L \sin\delta_s}, \tag{5.56}$$

where the sunrise hour angle $h_{sr}(\alpha = 0)$ in radians is given by Equation 5.30 and h_{sr} is the min $\left[|h_s(\alpha=0)|, |h_s(i=90°)|\right]$, respectively, and are evaluated at midmonth. Non-south-facing surfaces require numerical integration or iterative methods to determine $\bar{R}_b$. The long-term beam radiation on a tilted surface $\bar{B}_c$ is then,

$$\bar{B}_c = \bar{R}_b \bar{B}_h, \tag{5.57}$$

which is the long-term analog of Equation 5.43. Values of $\bar{R}_b$ are tabulated in Table A.5.5.

Diffuse radiation intercepted by a tilted surface differs from that on a horizontal surface, because a tilted surface does not view the entire sky dome, which is the source of diffuse radiation. If the sky is assumed to be an isotropic source of diffuse radiation, the instantaneous and long-term tilt factors for diffuse radiation, R_d and $\bar{R}_d$ respectively, are equal and are simply the radiation view factor from the plane to the visible portion of a hemisphere. In equation form:

$$R_d = \bar{R}_d = \cos^2\frac{\beta}{2} = \frac{(1+\cos\beta)}{2}. \tag{5.58}$$

In some cases where solar collectors are mounted near the ground, some beam and diffuse radiation reflected from the ground can be intercepted by the collector surface. The tilt factor $\bar{R}_r$ for reflected total radiation ($\bar{D}_h + \bar{B}_h$) is then calculated to be

$$\bar{R}_r = \frac{\bar{R}}{\bar{D}_h + \bar{B}_h} = \rho \sin^2 \frac{\beta}{2} = \frac{\rho(1-\cos\beta)}{2}, \tag{5.59}$$

in which ρ is the diffuse reflectance of the surface south of the collector assumed uniform and of infinite extent.

For snow, $\rho \cong 0.75$; for grass and concrete, $\rho \cong 0.2$. A more complete list of reflectances is provided in Table A.5.7. The total long-term radiation intercepted by a surface $\bar{H}_c$ is then the total of beam, diffuse, and diffusely reflected components:

$$\bar{H}_c = \bar{R}_b \bar{B}_h + \bar{R}_d \bar{D}_h + \bar{R}_r(\bar{D}_h + \bar{B}_h). \tag{5.60}$$

Using Equations 5.58 and 5.59, we have

$$\bar{H}_c = \bar{R}_b \bar{B}_h + \bar{D}_h \cos^2 \frac{\beta}{2} + (\bar{D}_h + \bar{B}_h)\rho \sin^2 \frac{\beta}{2}, \tag{5.61}$$

in which $\bar{R}_b$ is calculated from Equation 5.56.

Example 5.8

Using the average horizontal monthly radiation, find the monthly average insolation per day on a south-facing solar collector tilted at an angle of 25° in Miami, Florida, for January and compare it with the value from NREL (http://rredc.nrel.gov/solar/old_data/nsrdb/redbook/atlas/). Assume a ground reflectivity of 0.2.

Solution

The solution below is for the month of January. Values for the other months can be found by following the same method.

$$\bar{H}_h = 13{,}543 \frac{\text{kJ}}{\text{m}^2 \cdot \text{day}}$$

From Table A.5.2:

$$\bar{H}_{o,h} = 23{,}962 \frac{\text{kJ}}{\text{m}^2 \cdot \text{day}}$$

Therefore,

$$\bar{K}_T = \frac{\bar{H}_h}{\bar{H}_{o,h}} = 0.565$$

δ_s and h_{sr} can be found for the middle of the month (January 16):

$$\delta_s = 23.45° * \sin\left[\frac{360\times(284+16)}{365}\right]$$
$$= -21.1°$$

$$h_{sr}(\alpha = 0) = -\cos^{-1}(\tan L \tan\delta)$$
$$= -79.6° = -1.389 \text{ rad}$$
$$h_{ss} = 1.389 \text{ rad}$$

Using equation,

$$\frac{\bar{D}_h}{\bar{H}_h} = 0.775 + 0.347\times(1.389-1.5708) - [0.505 + 0.0261\times(1.389-1.5708)]$$
$$\times\cos(2\times 0.678 - 1.8)$$
$$= 0.212$$

Therefore,

$$\bar{D}_h = 0.212\times 13543 = 2867\frac{\text{kJ}}{\text{m}^2\cdot\text{day}}$$

and

$$\bar{B}_h = \bar{H}_h - \bar{D}_h = 10{,}676\frac{\text{kJ}}{\text{m}^2\cdot\text{day}}$$

Insolation on a tilted surface can be found from Equation 5.60. We need to find $\bar{R}_b$ from Equation 5.56.

Therefore,

$$\bar{R}_b = \frac{\cos(0°)\times\cos(-21.1°)\times\sin(-79.6°) - 1.389\times\sin(0°)\times\sin(-21.1°)}{\cos(25°)\times\cos(-21.1°)\times\sin(-79.6°) - 1.389\times\sin(25°)\times\sin(-21.1°)}$$
$$= 1.47$$

$$\bar{R}_d = \cos^2\left(\frac{25}{2}\right) = 0.953$$

$$\bar{R}_r = \rho \times \sin^2\left(\frac{\beta}{2}\right) \quad \text{(Assume } \rho = 0.2\text{)}$$

$$= 0.2 \times \sin^2(12.5°)$$

$$= 0.009$$

Therefore,

$$\bar{H}_c = (\bar{R}_b \times \bar{B}_h) + (\bar{R}_d \times \bar{D}_h) + (\bar{R}_r \times \bar{H}_h)$$

$$= (1.47 \times 10,676) + (0.953 \times 2,867) + (0.009 \times 13,543)$$

$$= 18,548 \frac{\text{kJ}}{\text{m}^2 \cdot \text{day}}$$

Using the NREL program, yields a value of approximately 18,000 kJ/m$^2 \cdot$day from the map, which is in kWh/m^2/day.

5.6.2 Circumsolar or Anisotropic Diffuse Solar Radiation

The models described in the above sections assume that the sky diffuse radiation is isotropic. However, this assumption is not true because of circumsolar radiation (brightening around the solar disk). Although the assumption of isotropic diffuse solar radiation does not introduce errors in the diffuse values on horizontal surfaces, it can result in errors of 10%–40% in the diffuse values on tilted surfaces. A number of researchers have studied the anisotropy of the diffuse solar radiation because of circum solar radiation. Temps and Coulson [22] introduced an anisotropic diffuse radiation algorithm for tilted surfaces for clear-sky conditions. Klucher [23] refined the Temps and Coulson algorithm by adding a cloudiness function to it:

$$R_d = \frac{1}{2}(1 + \cos\beta)M_1 M_2, \tag{5.62}$$

where

$$M_1 = 1 + F\sin^3\left(\frac{\beta}{2}\right), \tag{5.63}$$

$$M_2 = 1 + F\cos^2 i \sin^3(z), \tag{5.64}$$

and

$$F = 1 - \left(\frac{D_h}{H_h}\right)^2. \tag{5.65}$$

Examining F, we find that under overcast skies ($D_h = H_h$), R_d in Equation 5.62 reduces to the isotropic term of Liu and Jordan. The Klutcher algorithm reduces the error in diffuse radiation to about 5%.

In summary, monthly averaged, daily solar radiation on a surface is calculated by first decomposing total horizontal radiation into its beam and diffuse components using Equation 5.51 or 5.52. Various tilt factors are then used to convert these horizontal components to components on the surface of interest.

5.6.3 Daily Solar Radiation on Tilted Surfaces

Prediction of daily horizontal total solar radiation for sites where solar data are not measured can be done using the Angström–Page model. Instead of monthly values, however, daily values are used for percent sunshine *PS* and extraterrestrial radiation I_{day}. The results of using this simple model would be expected to show more scatter than monthly values, however.

All U.S. National Weather Service (NWS) stations with solar capability report daily horizontal total (beam and diffuse) radiation. Liu and Jordan have extended their monthly method described above to apply to daily data. The equation, analogous to Equation 5.51, used to calculate the daily diffuse component $\overline{I}_{d,h}$ is [24]

$$\frac{\overline{I}_{d,h}}{\overline{I}_h} = 1.0045 + 0.04349K_T - 3.5227K_T^2 + 2.6313K_T^3, \quad \text{and} \qquad K_T \le 0.75, \tag{5.66}$$

where K_T (no overbar) is the daily clearness index analogous to the monthly $\overline{K}_T$. (In this section, overbars indicate daily radiation totals.) For values of $K_T > 0.75$, the diffuse-to-total ratio is constant at a value of 0.166. K_T is given by

$$K_T = \frac{\overline{I}_h}{\overline{I}_{o,h}}, \tag{5.67}$$

The daily extraterrestrial total radiation $\overline{I}_{o,h}$ is calculated from Equation 5.49. Note that Equation 5.66 is based on the early solar constant value of 1394 W/m^2 (442 Btu/h·ft^2).

The daily horizontal beam component $\overline{I}_{b,h}$ is given by simple subtraction:

$$\overline{I}_{b,h} = \overline{I}_h - \overline{I}_{d,h}. \tag{5.68}$$

The beam, diffuse, and reflected components of radiation can each be multiplied by their tilt factors R_b, R_d, and R_r to calculate the total radiation on a tilted surface

$$\overline{I}_c = R_b\overline{I}_{b,h} + R_d\overline{I}_{d,h} + R_r(\overline{I}_{b,h} + \overline{I}_{d,h}), \tag{5.69}$$

in which

$$R_b = \frac{\cos(L-\beta)\cos\delta_s \sin h_{sr} + h_{sr}\sin(L-\beta)\sin\delta_s}{\cos L\cos\delta_s \sin h_{sr}(\alpha=0) + h_{sr}(\alpha=0)\sin L\sin\delta_s}, \tag{5.70}$$

$$R_d = \cos^2\frac{\beta}{2}, \tag{5.71}$$

and

$$R_r = \rho\sin^2\frac{\beta}{2}, \tag{5.72}$$

by analogy with the previous, monthly analysis.

If daily solar data are available, they can be used for design, the same as monthly data. Daily calculations are necessary when finer timescale performance is required. In addition, daily data can be decomposed into hourly data, which are useful for calculations made with large, computerized solar system simulation models.

5.6.4 Hourly Solar Radiation on Tilted Surfaces

Hourly solar radiation can be predicted in several ways. Correlations between hourly total and hourly diffuse (or beam) radiation or meteorological parameters such as cloud cover or air mass may be used. Alternatively, a method proposed by Liu and Jordan based on the disaggregation of daily data into hourly data could be used [13–17]. Even if hourly NWS data are available, it is necessary to decompose these total values into beam and diffuse components depending upon the response of the solar conversion device to be used by these two fundamentally different radiation types.

Randall and Leonard [25] have correlated historical data from the NWS stations at Blue Hill, Massachusetts, and Albuquerque, New Mexico to predict *hourly beam radiation* I_b and *hourly total horizontal radiation* I_h. This method can be used to decompose NWS data into its beam and diffuse components.

The hourly beam radiation was found to be fairly well correlated by hourly *percent of possible insolation* k_t, defined as

$$k_t = \frac{I_h}{I_{o,h}}, \tag{5.73}$$

in which $I_{o,h}$ is the hourly horizontal extraterrestrial radiation, which can be evaluated from Equation 5.49 using 1 h integration periods. Carrying out the integration over a 1 h period yields

$$I_{o,h} = I_o\left(1 + 0.034\cos\frac{360n}{365}\right)(0.9972\cos L\cos\delta_s\cos h_s + \sin L\sin\delta_s), \tag{5.74}$$

where the solar hour angle h_s is evaluated at the center of the hour of interest.

The direct normal-beam correlation based on k_t is [17]

$$I_{b,N} = -520 + 1800k_t \ (\text{W/m}^2) \quad 0.85 > k_t \geq 0.30 \tag{5.75}$$

$$I_{b,N} = 0 \quad k_t < 0.30 \tag{5.76}$$

This fairly simple correlation gives more accurate $I_{b,N}$ values than the more cumbersome Liu and Jordan procedure, at least for Blue Hill and Albuquerque. Vant-Hull and Easton [26] have also devised an accurate predictive method for beam radiation.

Randall and Leonard [25] have made a correlation of *total* horizontal hourly radiation I_h(W/m²) on the basis of *opaque cloud cover* CC and *air mass m* using data for Riverside, Los Angeles, and Santa Monica, California. Cloud cover is defined as CC = 1.0 for fully overcast and CC = 0.0 for clear skies. A polynomial fit was used:

$$I_h = \frac{I_{o,h}}{100}(83.02 - 3.847m - 4.407\text{CC} + 1.1013\text{CC}^2 - 0.1109\text{CC}^3) \tag{5.77}$$

The average predictive error for this correlation was ±2.3% of the NWS data; the correlation coefficient of I_h with CC is 0.76 for the data used. Equation 5.80 was used to predict I_h for Inyokern, California, a site not used in the original correlation. Predictions of solar radiation for Inyokern were within 3.2% of NWS. The diffuse radiation can be calculated from Equations 5.75 through 5.77:

$$I_{d,h} = I_h - I_b\sin\alpha \tag{5.78}$$

The American Society of Heating, Refrigerating, and Air Conditioning Engineers (ASHRAE) has calculated hourly clear-sky values on vertical, horizontal, and tilted surfaces. One table for a full year has been prepared for each of six values of latitude spanning the continental United States. In addition, tables of solar azimuth and altitude angle have been prepared for the same six latitudes. The solar radiation data represent the maximum that could be expected on a clear day and are therefore of limited usefulness in the design of solar systems. They can be used, however, to calculate cooling loads on buildings and the like. The ASHRAE tables are contained in Appendix A.

5.7 Measurement of Solar Radiation

Solar radiation measurements of importance to most engineering applications, especially thermal applications, include total (integrated over all wavelengths) direct or beam and sky diffuse values of solar radiation on instantaneous, hourly, daily, and monthly bases. Some applications such as photovoltaics, photochemical, and daylighting require knowledge of spectral (wavelength specific) or band (over a wavelength range, e.g., ultraviolet, visible, infrared) values of solar radiation. This section describes some of the instrumentation used to measure solar radiation and sunshine, and some sources of long-term measured data for different parts of the world. Also described briefly in this section is the method of satellite-based measurements.

5.7.1 Instruments for Measuring Solar Radiation and Sunshine

There are two basic types of instruments used to measure solar radiation, *pyranometer* and *pyrheliometer.* A pyranometer has a hemispherical view of the surroundings and therefore is used to measure total, direct and diffuse, solar radiation on a surface. A pyrheliometer, on the other hand, has a restricted view (about 5°) and is, therefore, often used to measure the direct or beam solar radiation by pointing it toward the sun. Pyranometers are also used to measure the sky diffuse radiation by using a shadow band to block the direct sun view. A detailed discussion of the instrumentation and calibration standards is given by Iqbal [27] and Zerlaut [28].

A pyranometer consists of a flat sensor/detector (described later) with an unobstructed hemispherical view, which allows it to convert and correlate the total radiation incident on the sensor to a measurable signal. The pyranometers using thermal detectors for measurements can exhibit serious errors at tilt angles from the horizontal due to free convection. These errors are minimized by enclosing the detector in double hemispherical high-transmission glass domes. The second dome minimizes the error due to infrared radiative exchange between the sensor and the sky. A desiccator is usually provided

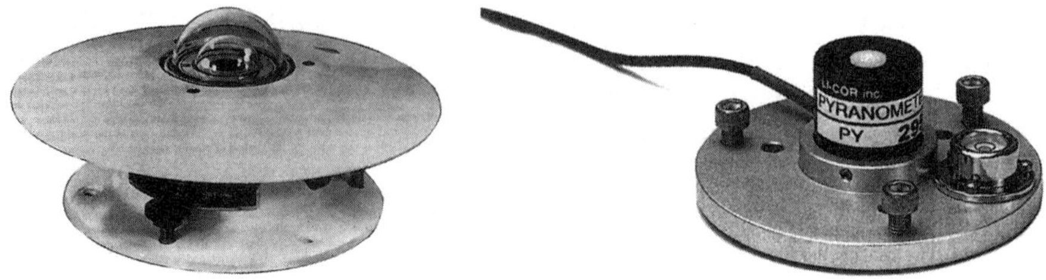

(a) (b)

FIGURE 5.20
Typical commercially available pyranometers with (a) thermal detector and (b) photovoltaic detector.

to eliminate the effect due to condensation on the sensor or the dome. Figure 5.20 shows pictures of typical commercially available precision pyranometers.

A pyranometer can be used to measure the sky diffuse radiation by fitting a shade ring to it, as shown in Figure 5.21, in order to block the beam radiation throughout the day. The position of the shade ring is adjusted periodically as the declination changes. Since the shade ring obstructs some diffuse radiation from the pyranometer, correction factors must be applied.

Geometric correction factors (GCF) that account for the part of the sky obstructed by the shade ring can be easily calculated. However a GCF assumes isotropic sky, which results in errors because of the circumsolar

FIGURE 5.21
A pyranometer with a shade ring to measure sky diffuse radiation.

TABLE 5.7

Shading Band Correction Factors due to Anisotropy

Solar Altitude Angle	k_T									
	0.0	**0.1**	**0.2**	**0.3**	**0.4**	**0.5**	**0.6**	**0.7**	**0.8**	**0.9**
<20°	0.0	0.0	0.0	0.0	0.015	0.06	0.14	0.23	0.24	0.24
20°–40°	0.0	0.0	0.0	0.0	0.006	0.05	0.125	0.205	0.225	0.225
40°–60°	0.0	0.0	0.0	0.0	0.003	0.045	0.115	0.175	0.205	0.205
60°+	0.0	0.0	0.0	0.0	0.0	0.035	0.09	0.135	0.17	0.17

Source: Gautier, C. et al., *J. Appl. Meteorol.*, 19(8), 1005, 1980.

anisotropy. Eppley Corp. recommends additional correction factors to account for anisotropy as: +7% for clear sky, +4% for partly cloudy condition, and +3% for cloudy sky. Mujahid and Turner [30] determined that these correction factors gave less than 3% errors on partly cloudy days but gave errors of −11% for clear-sky conditions and +6% on overcast days. They suggested correction factors due to anisotropy as tabulated in Table 5.7 which reduce the errors to less than ±3%. It must be remembered that these correction factors are in addition to the geometric correction factors. Recently, a sun occulting disk has been employed for shading the direct sun.

Beam or direct solar radiation is usually measured with an instrument called a pyrheliometer. Basically a pyrheliometer places the detector at the base of a long tube. This geometry restricts the sky view of the detector to a small angle of about 5°. When the tube points toward the sun the detector measures the beam solar radiation and a small part of the diffuse solar radiation within the view angle. Figure 5.22 shows the geometry of a pyrheliometer sky occulting tube.

In this figure,

The opening half angle

$$\theta_o = \tan^{-1}\frac{R}{L}. \tag{5.79}$$

The slope angle

$$\theta_p = \tan^{-1}\left[\frac{(R-r)}{L}\right]. \tag{5.80}$$

The limit half angle

$$\theta = \tan^{-1}\left[\frac{(R+r)}{L}\right]. \tag{5.81}$$

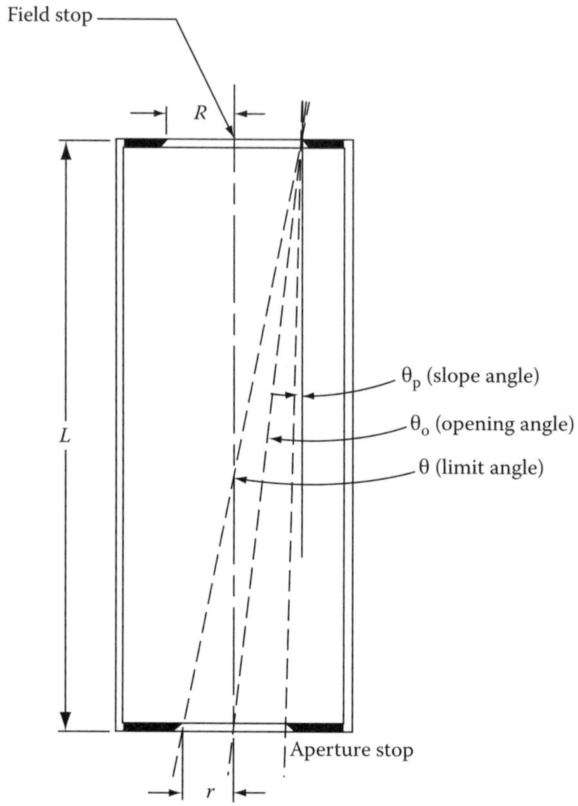

FIGURE 5.22
Geometry of a pyroheliometer sky occulting tube.

The field of view is $2\theta_o$. The World Meteorological Organization (WMO) recommends the opening half angle θ_o to be 2.5° [28] and the slope angle θ_p to be 1°.

Continuous tracking of the sun is required for the accuracy of the measurements. This is obtained by employing a tracking mechanism with two motors, one for altitude and the other for azimuthal tracking. Another problem is that the view angle of a pyrheliometer is significantly greater than the angle subtended by the solar disk (about 0.5°). Therefore the measurements using a pyrheliometer include the beam and the circumsolar radiation. These measurements may present a problem in using the data for central receiver systems which use only direct beam radiation. However, this is not a significant problem for parabolic trough concentrators which in most cases have field of view of the order of 5°.

5.7.2 Detectors for Solar Radiation Instrumentation

Solar radiation detectors are of four basic types [27,28]: thermomechanical, calorimetric, thermoelectric, and photoelectric. Of these, thermoelectric and photoelectric are the most common detectors in use today.

A *thermoelectric detector* uses a thermopile which consists of a series of thermocouple junctions. The thermopile generates a voltage proportional to the temperature difference between the hot and cold junctions which, in turn, is proportional to the incident solar radiation.

Photovoltaic detectors normally use silicon solar cells measuring the short circuit current. Such detectors have the advantage of being simple in construction. Because heat transfer is not a consideration, they do not require clear domes or other convection suppressing devices. They are also insensitive to tilt as the output is not affected by natural convection. One of the principal problems with photovoltaic detectors is their spectral selectivity. Radiation with wavelengths greater than the band gap of the photovoltaic detector cannot be measured. Silicon has a bandgap of 1.07 eV corresponding to a wavelength of 1.1 μm. A significant portion of the infrared part of solar radiation has wavelengths greater than 1.1 μm. Therefore, photovoltaic detectors are insensitive to changes in the infrared part of solar radiation.

5.7.3 Measurement of Sunshine Duration

The time duration of bright sunshine data is available at many more locations in the world than the solar radiation. That is why a number of researchers have used this data to estimate the available solar radiation. Two instruments are widely used to measure the sunshine duration. The device used by the U.S. National Weather Service is called a *sunshine switch*. It is composed of two photovoltaic cells—one shaded, the other not. During daylight a potential difference is created between the two cells, which in turn operates the recorder. The intensity level required to activate the device is that just sufficient to cast a shadow. The other device commonly used to measure the sunshine duration is called the *Campbell–Stokes sunshine recorder*. It uses a solid, clear glass sphere as a lens to concentrate the solar beam on the opposite side of the sphere. A strip of standard treated paper marked with time graduations is mounted on the opposite side of the sphere where the solar beam is concentrated. Whenever the solar radiation is above a threshold, the concentrated beam burns the paper. The length of the burned part of the strip gives the duration of bright sunshine. The problems associated with the Campbell–Stokes sunshine recorder include the uncertainties of the interpretation of burned portions of the paper, especially on partly cloudy days, and the dependence on the ambient humidity (Figure 5.23).

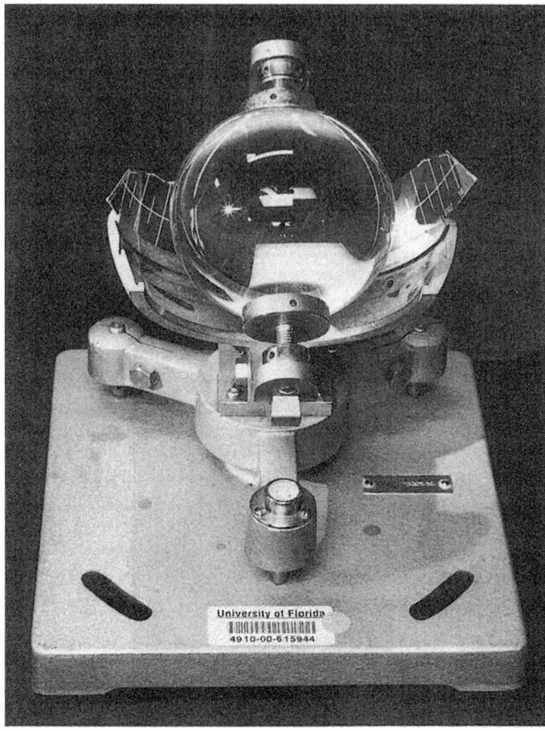

FIGURE 5.23
Campbell–Stokes sunshine recorder.

5.7.4 Measurement of Spectral Solar Radiation

Spectral solar radiation measurements are made with spectroradiometers. Full spectrum scanning is difficult, requires constant attention during operation and is therefore expensive. Zerlaut [28] has described a number of solar spectroradiometers. These instruments consist basically of a monochromator, a detector-chopper assembly, an integrating sphere, and a signal conditioning/computer package. They have the capability of measuring solar radiation in the wavelength spectrum of 280–2500 nm.

5.7.5 Solar Radiation Data and Web Sites

Measured solar radiation data is available at a number of locations throughout the world. Data for many other locations have been estimated based on measurements at similar climatic locations. Some of the available data from various locations in the world is presented in Tables A.5.3a and A.5.3b. This appendix also provides tables of modeled clear-sky data for various latitudes (Table A.5.6).

Solar radiation data for United States. are available from the National Climatic Data Center (NCDC, http://www.ncdc.noaa.gov/oa/climate/climateinventories.html) of the National Oceanic and Atmospheric

Administration (NOAA), and the National Renewable Energy Laboratory (NREL, http://www.nrel.gov/rredc/solar_resource.html). NOAA also provides data on the percent of possible sunshine for locations throughout the United States (http://www.ncdc.noaa.gov/oa/climate/online/ccd/pctpos.txt). In the mid-1970s, NOAA compiled a data base of measured hourly global horizontal solar radiation for 28 locations for the period 1952–1975 (called SOLMET) and of data for 222 additional sites (called ERSATZ) estimated from SOLMET data and some climatic parameters such as sunshine duration and cloudiness. NOAA also has two data sets of particular interest to engineers and designers: the typical meteorological year (TMY) and the Weather Year for Energy Calculations (WYEC) data sets. TMY data set represents typical values from 1952 to 1975 for hourly distribution of direct beam and global horizontal solar radiation. WYEC data set contains monthly values of temperature, direct beam and diffuse solar radiation and estimates of *illuminance* (for daylighting applications). Illuminance is solar radiation in the visible range to which the human eye responds. Recently, NREL compiled a National Solar Radiation Data Base (NSRDB, http://rredc.nrel.gov/solar/old_data/nsrdb/) for 239 stations in the United States [31,32]. NSRDB is a collection of hourly values of global horizontal, direct normal, and diffuse solar radiation based on measured and estimated values for a period of 1961–1990. Since long-term measurements were available for only about 50 stations measured data makes up only about 7% of the total data in the NSRDB. A typical meteorological year data set from NSRDB is available as TMY2. NREL also provides maps with averaged daily solar radiation per month (http://rredc.nrel.gov/solar/old_data/nsrdb/redbook/atlas/).

The data for other locations in the world is available from national government agencies of most countries of the world. Worldwide solar radiation data is also available from the World Radiation Data Center (WRDC) in St. Petersburg, Russia, based on worldwide measurements made through local weather service operations [33]. WRDC, operating under the auspices of the World Meteorological Organization (WMO), has been archiving data from over 500 stations and operates a worldwide web site in collaboration with NREL with an address of http://wrdc.mgo.nrel.gov. Data for some cities of the world from WRDC is given in Table A.5.3. An International Solar Radiation Data Base was also developed by the University of Lowell [34].

Problems

5.1 Calculate the declination, the zenith angle, and the azimuth angle of the sun for New York City (latitude 40.77°N) on October 1 at 2:00 p.m. solar time.

5.2 A solar energy system in Gainesville, Florida requires two rows of collectors facing south and tilted at a fixed 30° angle. Find the minimum normalized distance at which the second row should be placed behind the first row for no shading at noon at winter solstice. What percentage of the second row is shaded on the same day at 9:00 a.m. solar time?

5.3 Find the sunrise and sunset times for a location of your choice on September 1.

5.4 Construct a table of hourly sun angles for the 15th day of each month for a location of your choice. Also show the sunrise and sunset times for those days.

5.5 Referring to Figure 5.19, prove Equation 5.44 for the angle of incidence. (*Hint*: Use direction cosines of the sun-ray vector and a vector normal to the tilted surface to find the angle between them. Dot product of two unit vectors gives the cosine of the angle between them.)

5.6 Determine the following for a south-facing surface at 30° slope in Gainesville, Florida (latitude = 29.68°N, longitude = 82.27°W) on September 21 at noon solar time:

A. Zenith angle
B. Angle of incidence
C. Beam radiation
D. Diffuse radiation
E. Reflected radiation
F. Total radiation
G. Local time

5.7 Show that the hourly averaged, extraterrestrial radiation for a given hour is the same, to within 1%, as instantaneous radiation at the hour's midpoint. This is equivalent to deriving Equation 5.74.

5.8 Prepare shadow maps for point P on the sun-path diagrams for 35°N and 40°N for the three geometries shown in a, b, and c below. Determine the hours of shading that occur each month.

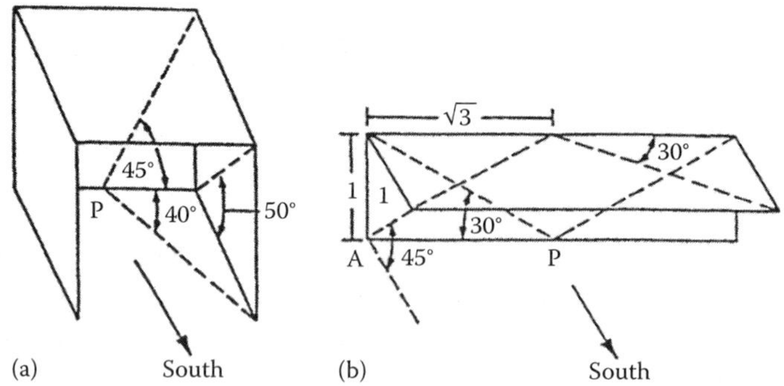

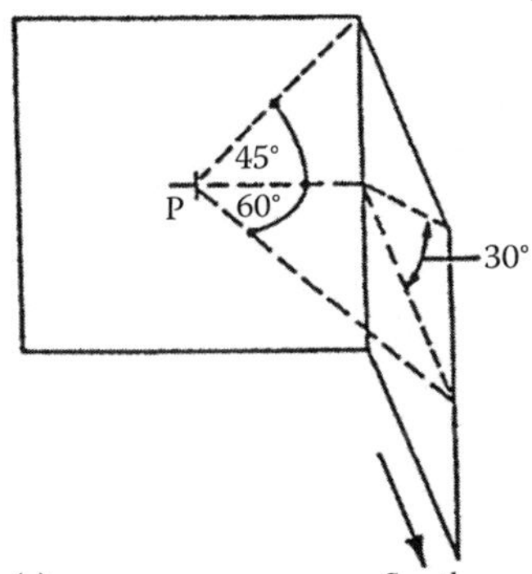

(c)

5.9 Repeat Problem 5.8c if the surface containing point P faces due west instead of due south for a 40° N location.

5.10 Calculate the incidence angle at noon and 9 a.m. on a fixed, flat-plate collector located at 40° N latitude and tilted 70° up from the horizontal. Find *i* for June 21 and December 21.

5.11 A. If the surface in Problem 5.10 faces S 45° E, what are the incidence angles?
B. If the collector in Problem 5.10 has a cylindrical surface, what are the incidence angles on June 21 and December 21?

5.12 Using a one-term Fourier cosine series, develop an empirical equation for solar declination as a function of day number counted from January 1 (see Table 5.2 and Figure 5.7).

5.13 Derive an equation for the lines of constant declination in a sun-path diagram, for example, Figure 5.10. Check your equation by plotting a few declination lines on a piece of polar coordinate graph paper.

5.14 Derive Equation 5.31 relating profile angle γ to azimuth angle α and altitude angle α.

5.15 Based on Equation 5.44, what value of β would result in the annual minimum value of the incidence angle *i*? Note that this tilt angle would result in maximum collection of beam radiation on a fixed, flat, south-facing surface. *Hint*: Use a double integration procedure.

5.16 At what time does the sun set in Calcutta (23°N) on May 1 and December 1?

5.17 What is the true solar time in Sheridan, Wyoming (107°W) at 10:00 a.m. Mountain Daylight Time on June 10? What is the true solar time at 10:00 a.m. Mountain Standard Time on January 10?

5.18 Using the Angström–Page method, calculate the average horizontal insolation in Hamburg in May and in October with $\overline{PS}$ 40% and 60%, respectively.

5.19 Equation 5.51 is based on an early solar constant value of 1394 W/m^2. Derive a modified form of Equation 5.51 based on the presently accepted value for the solar constant of 1353 W/m^2.

5.20 Predict the hourly beam and diffuse radiation on a horizontal surface for Denver (40°N) on September 9 at 9:30 a.m. on a clear day.

5.21 Derive an expression for the minimum allowable distance between east–west rows of solar collectors that will assure no shading of one row by the row immediately to the south. Use the law of sines and express the result in terms of the collector tilt and face length and the controlling value of the solar profile angle.

5.22 Using the predictive method of Angström or Page, estimate the monthly solar radiation for the North Central Sahara Desert (Tamanrasset, Algeria area) at latitude = 25°N. Percentages of possible sunshine and extraterrestrial radiation for this site are given in the table below.

5.23 Using the $\bar{H}_h$ data calculated in Example 5.7 in place of the long-term measured data for the North Central Sahara Desert at latitude 25°N, find the monthly averaged insolation per day on a south-facing solar collector tilted at an angle of 25° from the horizontal.

Appendix A.5: Calculation of True Solar Time

The altitude and azimuth of the sun are given by

$$\sin\alpha = \sin L \sin \delta_S + \cos\phi \cos\delta_S \cos h_S \quad \text{(A.5.1)}$$

and

$$\sin a_S = \frac{-\cos\delta_S \sin h_S}{\cos\alpha} \quad \text{(A.5.2)}$$

where

α is the altitude of the sun (angular elevation above the horizon)
L is the latitude of the observer
δ_S is the declination of the sun
h_S is the hour angle of sun (angular distance from the meridian of the observer)
a_S is the azimuth of the sun (measured eastward from north)

From Equations A.5.1 and A.5.2, it can be seen that the altitude and azimuth of the sun are functions of the latitude of the observer, the time of day (hour angle), and the date (declination).

Figure A.5.1a through f provides a series of charts, one for each 5° of latitude (except 5°, 15°, 75°, and 85°) giving the altitude and azimuth of the sun as a function of the true solar time and the declination of the sun in a form originally suggested by Hand. Linear interpolation for intermediate latitudes will give results within the accuracy to which the charts can be read.

On these charts, a point corresponding to the projected position of the sun is determined from the heavy lines corresponding to declination and solar time.

To find the solar altitude and azimuth:

1. Select the chart or charts appropriate to the latitude.
2. Find the solar declination δ corresponding to the date.
3. Determine the *true solar time* as follows:
   a. To the *local standard time* (zone time) add 4′ for each degree of longitude the station is east of the standard meridian or subtract 4′ for each degree west of the standard meridian to get the *local mean solar time.*
   b. To the *local mean solar time* add algebraically the equation of time; the sum is the required *true solar time.*
4. Read the required altitude and azimuth at the point determined by the declination and the true solar time. Interpolate linearly between two charts for intermediate latitudes.

It should be emphasized that the solar altitude determined from these charts is the true geometric position of the center of the sun. At low solar elevations terrestrial refraction may considerably alter the apparent position of sun. Under average atmospheric refraction the sun will appear on the horizon when it actually is about 34′ below the horizon; the effect of refraction decreases rapidly with increasing solar elevation. Since sunset or sunrise is defined as the time when the upper limb of the sun appears on the horizon, and the semidiameter of the sun is 16′, sunset or sunrise occurs under average atmospheric refraction when the sun is 50′ below the horizon. In polar regions especially, unusual atmospheric refraction can make considerable variation in the time of sunset or sunrise.

The 90°N chart is included for interpolation purposes; the azimuths lose their directional significance at the pole.

Altitude and azimuth in southern latitudes: To compute solar altitude and azimuth for southern latitudes, change the sign of the solar declination and proceed as above. The resulting azimuths will indicate angular distance from *south* (measured eastward) rather than from north.

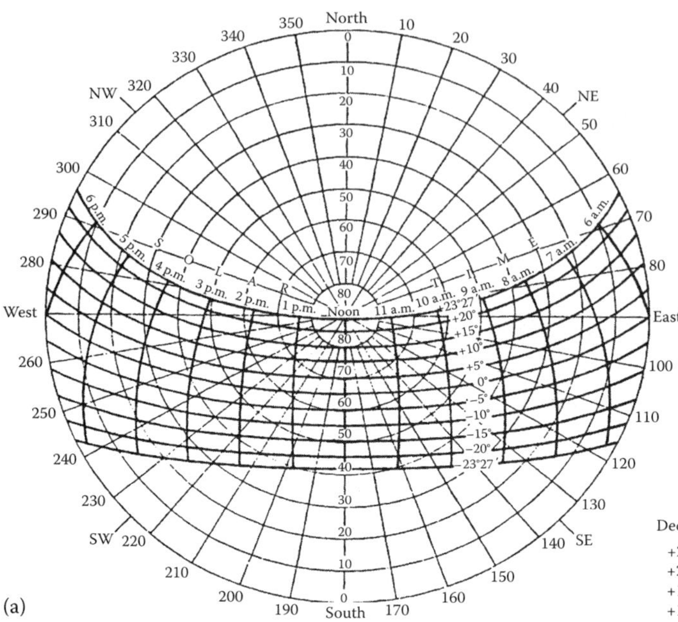

(a)

Declination	Approx. dates
+23° 27′	June 22
+20°	May 21, July 24
+15°	May 1, Aug. 12
+10°	Apr. 16, Aug. 28
+5°	Apr. 3, Sept. 10
0°	Mar. 21, Sept. 23
−5°	Mar. 8, Oct. 6
−10°	Feb. 23, Oct. 20
−15°	Feb. 9, Nov. 3
−20°	Jan. 21, Nov. 22
−23° 27′	Dec. 22

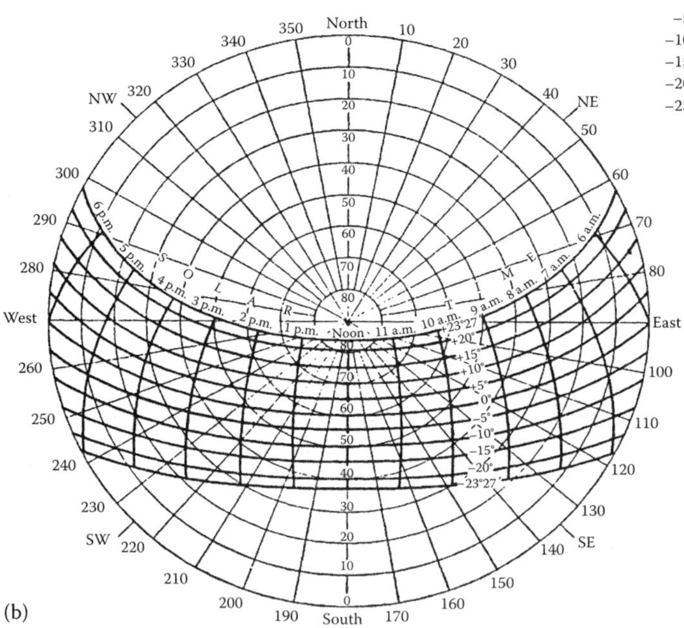

(b)

FIGURE A.5.1
Meteorological charts for computing solar altitude and azimuth angles. (a) chart, 25°N latitude; (b) chart, 30°N latitude;

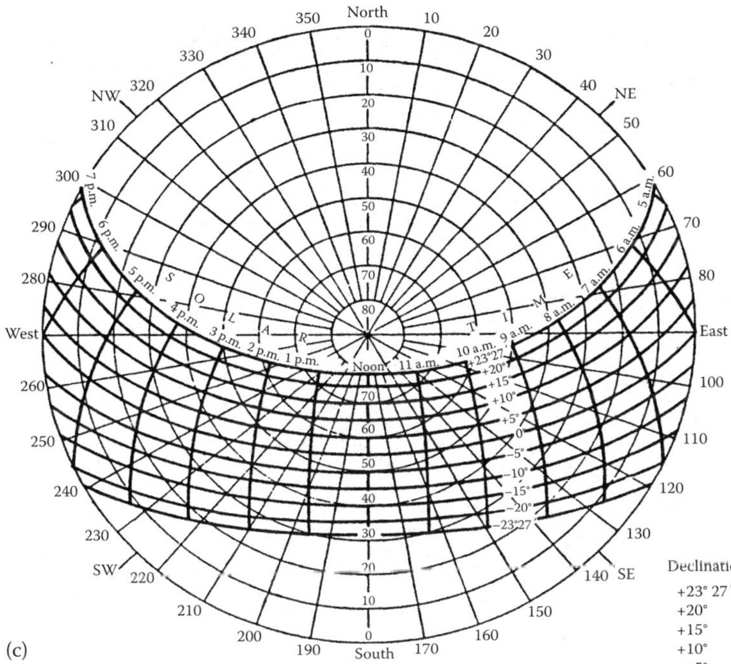

(c)

Declination	Approx. dates
+23° 27′	June 22
+20°	May 21, July 24
+15°	May 1, Aug. 12
+10°	Apr. 16, Aug. 28
+5°	Apr. 3, Sept. 10
0°	Mar. 21, Sept. 23
−5°	Mar. 8, Oct. 6
−10°	Feb. 23, Oct. 20
−15°	Feb. 9, Nov. 3
−20°	Jan. 21, Nov. 22
−23° 27′	Dec. 22

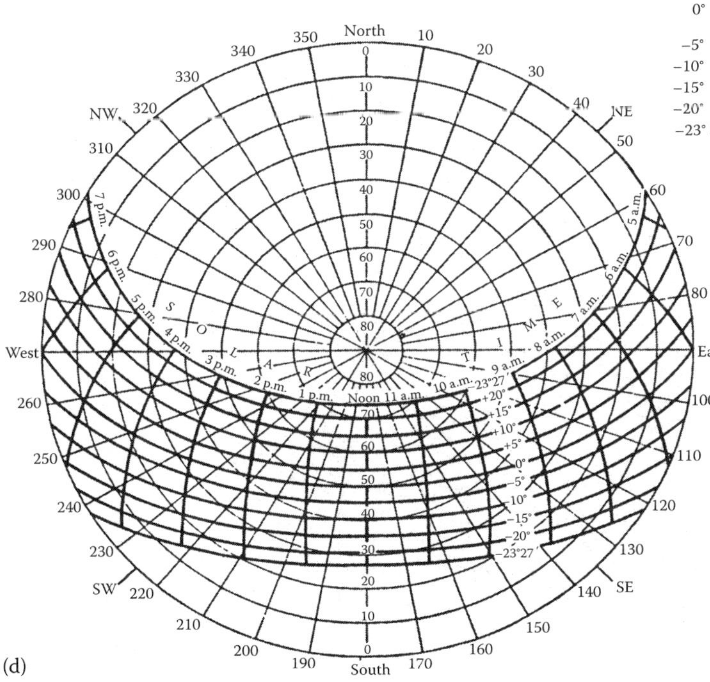

(d)

FIGURE A.5.1 (continued)
(c) chart, 35°N latitude; (d) chart, 40°N latitude;

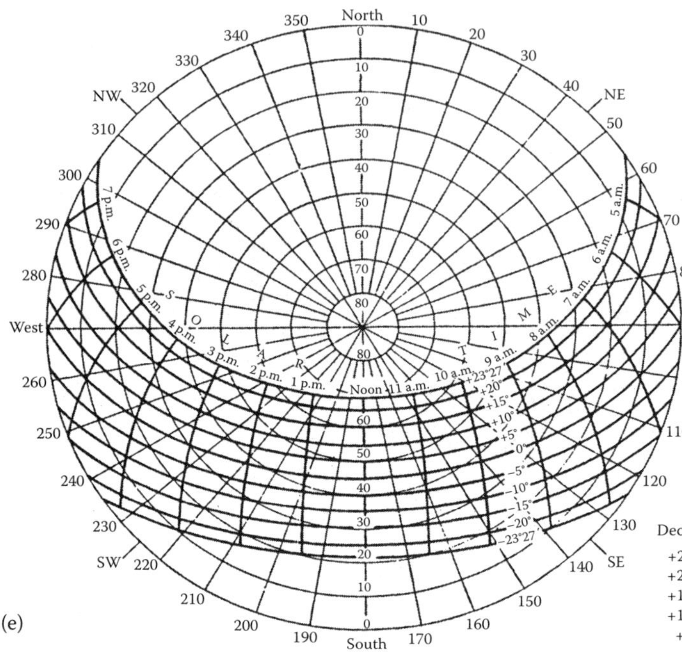

(e)

Declination	Approx. dates
+23° 27′	June 22
+20°	May 21, July 24
+15°	May 1, Aug. 12
+10°	Apr. 16, Aug. 28
+5°	Apr. 3, Sept. 10
0°	Mar. 21, Sept. 23
−5°	Mar. 8, Oct. 6
−10°	Feb. 23, Oct. 20
−15°	Feb. 9, Nov. 3
−20°	Jan. 21, Nov. 22
−23° 27′	Dec. 22

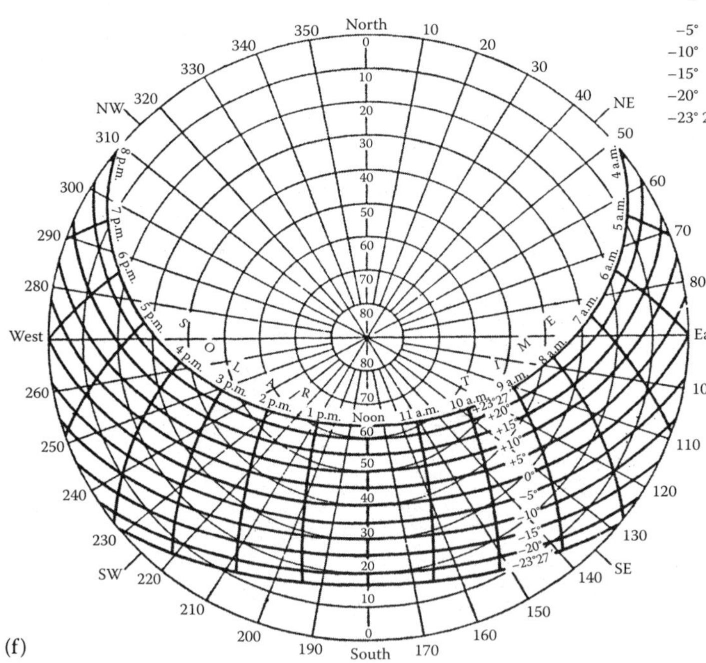

(f)

FIGURE A.5.1 (continued)
(e) chart, 45°N latitude; (f) chart, 50°N latitude. (Reproduced from *Smithsonian Meteorological Tables*, Smithsonian Institute, Washington, DC. With permission.)

TABLE A.5.2

Monthly Averaged, Daily Extraterrestrial Insolation on a Horizontal Surface (Units: Wh/m^2)

Latitude (°)	January	February	March	April	May	June	July	August	September	October	November	December
20	7415	8397	9552	10,422	10,801	10,868	10,794	10,499	9791	8686	7598	7076
25	6656	7769	9153	10,312	10,936	11,119	10,988	10,484	9494	8129	6871	6284
30	5861	7087	8686	10,127	11,001	11,303	11,114	10,395	9125	7513	6103	5463
35	5039	6359	8153	9,869	10,995	11,422	11,172	10,233	8687	6845	5304	4621
40	4200	5591	7559	9,540	10,922	11,478	11,165	10,002	8184	6129	4483	3771
45	3355	4791	6909	9,145	10,786	11,477	11,099	9,705	7620	5373	3648	2925
50	2519	3967	6207	8,686	10,594	11,430	10,981	9,347	6998	4583	2815	2100
55	1711	3132	5460	8,171	10,358	11,352	10,825	8,935	6325	3770	1999	1320
60	963	2299	4673	7,608	10,097	11,276	10,657	8,480	5605	2942	1227	623
65	334	1491	3855	7,008	9,852	11,279	10,531	8,001	4846	2116	544	97

TABLE A.5.3A

Global Horizontal Average Solar Radiation (Units: MJ/m^2-Day)

Position	Lat.	Long.	January	February	March	April	May	June	July	August	September	October	November	December
Argentina														
Buenos Aires	34.58 S	58.48 W	24.86	21.75	18.56	11.75	8.71	7.15	7.82	8.75	14.49	16.66	24.90	21.93
Australia														
Adelaide	34.93 S	138.52 E	20.99	17.50	20.15	18.27	17.98	—	18.81	19.64	20.11	20.88	20.57	20.72
Brisbane	27.43 S	153.08 E	25.36	22.22	13.25	16.61	12.23	11.52	9.70	15.10	17.61	19.89	—	—
Canberra	35.30 S	148.18 E	28.20	24.68	20.56	14.89	10.29	6.62	—	12.33	16.88	24.06	26.00	25.77
Darwin	12.47 S	130.83 E	26.92	23.40	18.13	13.62	9.30	7.89	9.41	11.15	14.85	18.87	23.43	22.34
Hobart	42.88 S	147.32 E	—	—	—	10.09	7.26	6.04	5.72	9.21	13.54	18.12	—	—
Laverton	37.85 S	114.08 E	22.96	20.42	15.59	13.40	7.48	6.10	6.54	10.43	13.24	18.76	—	—
Sydney	33.87 S	151.20 E	21.09	21.75	17.63	13.63	9.78	8.79	7.62	12.84	16.93	22.10	—	—
Austria														
Wien	48.20 N	16.57 E	3.54	7.10	8.05	14.72	16.79	20.87	19.89	17.27	12.55	8.45	3.51	2.82
Innsbruck	47.27 N	11.38 E	5.57	9.28	10.15	15.96	14.57	17.65	18.35	17.26	12.98	9.08	4.28	3.50
Barbados														
Husbands	13.15 N	59.62 W	19.11	20.23	—	21.80	19.84	20.86	21.55	22.14	—	—	18.30	16.56
Belgium														
Ostende	51.23 N	2.92 E	2.82	5.75	9.93	15.18	16.74	16.93	18.21	18.29	11.71	6.15	2.69	1.97
Melle	50.98 N	3.83 E	2.40	4.66	8.41	13.55	14.23	13.28	15.71	15.61	10.63	5.82	2.40	1.59
Brunei														
Brunei	4.98 N	114.93 E	19.46	20.12	22.71	20.54	19.74	18.31	19.38	20.08	20.83	17.51	17.39	18.12

Bulgaria														
Chirpan	42.20 N	25.33 E	6.72	6.79	8.54	13.27	17.25	17.39	19.85	14.61	12.53	8.52	5.08	5.09
Sofia	42.65 N	23.38E	4.05	6.23	7.93	9.36	12.98	19.73	19.40	17.70	14.71	6.44	—	3.14
Canada														
Montreal	45.47 N	73.75 E	4.74	8.33	11.84	10.55	15.05	22.44	21.08	18.67	14.83	9.18	4.04	4.01
Ottawa	45.32 N	75.67 E	5.34	9.59	13.33	13.98	20.18	20.34	19.46	17.88	13.84	7.38	4.64	5.04
Toronto	43.67 N	79.38 E	4.79	8.15	11.96	14.00	18.16	24.35	23.38	—	15.89	9.40	4.72	3.79
Vancouver	49.18 N	123.17 E	3.73	4.81	12.14	16.41	20.65	24.04	22.87	19.08	12.77	7.39	4.29	1.53
Chile														
Pascua	27.17 S	109.43 W	19.64	16.65	—	11.12	9.52	8.81	10.90	12.29	17.19	20.51	21.20	22.44
Santiago	33.45 S	70.70 W	18.61	16.33	13.44	8.32	5.07	3.66	3.35	5.65	8.15	13.62	20.14	23.88
China														
Beijing	39.93 N	116.28 W	7.73	10.59	13.87	17.93	20.18	18.65	15.64	16.61	15.52	11.29	7.25	6.89
Guangzhou	23.13 N	113.32 E	11.01	6.32	4.04	7.89	10.53	12.48	16.14	16.02	15.03	15.79	11.55	9.10
Harbin	45.75 N	126.77 E	5.15	9.54	17.55	20.51	20.33	17.85	19.18	16.09	13.38	14.50	10.50	6.98
Kunming	25.02 N	102.68 E	9.92	11.26	14.38	18.00	18.53	17.37	11.95	18.47	15.94	12.45	11.96	13.62
Lanzhou	36.05 N	103.88 E	7.30	12.47	10.62	18.91	17.40	20.40	20.23	17.37	13.23	10.21	8.22	6.43
Shanghai	31.17 N	121.43 E	7.44	10.31	11.78	14.36	14.23	16.79	14.63	11.85	15.96	12.03	7.73	8.70
Columbia														
Bogota	4.70 N	74.13 W	17.89	—	19.37	16.58	14.86	—	15.42	18.20	17.05	14.58	14.20	16.66
Cuba														
Havana	23.17 N	82.35 W	—	14.70	18.94	20.95	22.63	18.83	21.40	20.19	16.84	16.98	13.19	13.81

(continued)

TABLE A.5.3A (continued)

Global Horizontal Average Solar Radiation (Units: MJ/m^2-Day)

Position	Lat.	Long.	January	February	March	April	May	June	July	August	September	October	November	December
Czech														
Kucharovice	48.88 N	16.08 E	3.03	5.85	9.88	14.06	20.84	19.24	21.18	19.41	13.61	6.11	3.47	2.12
Churanov	49.07 N	13.62 E	2.89	5.82	9.24	13.18	21.32	15.68	20.51	19.49	12.84	5.68	3.36	2.99
Hradec Kralov	50.25 N	15.85 E	3.51	5.94	10.58	15.95	20.42	18.43	17.17	17.92	11.86	6.27	2.45	1.89
Denmark														
Copenhagen	55.67 N	12.30 E	1.83	3.32	7.09	11.12	21.39	24.93	—	13.92	10.10	5.20	2.81	1.23
Egypt														
Cairo	30.08 N	31.28 E	10.06	12.96	18.49	23.04	21.91	26.07	25.16	23.09	21.01	—	11.74	9.85
Mersa Matruh	31.33 N	27.22 E	8.38	11.92	18.47	24.27	24.17	—	26.67	26.27	21.92	18.28	11.71	8.76
Ethiopia														
Addis Ababa	8.98 N	38.80 E	—	11.39	—	12.01	—	—	—	6.33	9.35	11.71	11.69	11.50
Fiji														
Nandi	17.75 S	177.45 E	20.82	20.6.5	20.25	18.81	15.68	14.18	15.08	16.71	19.37	20.11	21.78	25.09
Suva	48.05 S	178.57 E	20.37	17.74	16.22	13.82	10.81	12.48	11.40	—	—	18.49	19.96	20.99
Finland														
Helsinki	60.32 N	24.97 E	1.13	2.94	5.59	11.52	17.60	16.81	20.66	15.44	8.44	3.31	0.97	0.63
France														
Agen	44.18 N	0.60 E	4.83	7.40	10.69	17.12	19.25	20.42	21.63	20.64	15.56	8.41	5.09	5.01
Nice	43.65 N	7.20 E	6.83	—	11.37	17.79	20.74	24.10	24.85	24.86	15.04	10.99	7.08	6.73
Paris	48.97 N	2.45 E	2.62	5.08	7.21	12.90	14.84	13.04	15.54	16.30	10.17	5.61	3.14	2.20

Germany														
Bonn	50.70 N	7.15 E	2.94	5.82	8.01	14.27	15.67	14.41	18.57	17.80	11.70	6.15	3.42	1.90
Nuremberg	53.33 N	13.20 E	3.23	6.92	9.08	15.69	15.71	18.21	21.14	17.98	12.43	8.15	2.79	2.51
Bremen	53.05 N	8.80 E	2.36	4.93	8.53	14.52	14.94	14.52	19.40	15.02	10.48	6.27	2.80	1.66
Hamburg	53.63 N	10.00 E	1.97	3.96	7.59	12.32	14.11	12.69	19.00	14.11	10.29	6.45	2.33	1.43
Stuttgart	48.83 N	9.20 E	3.59	7.18	9.22	15.81	17.72	17.44	22.21	19.87	12.36	7.81	3.19	2.54
Ghana														
Bole	9.03 N	2.48 W	18.29	19.76	19.71	19.15	16.61	—	—	13.68	16.29	17.27	17.33	15.93
Accra	5.60 N	0.17 W	14.82	16.26	18.27	16.73	18.15	13.96	15.86	13.49	15.32	19.14	18.16	14.23
Great Britain														
Belfast	54.65 N	6.22 W	2.00	3.60	6.85	12.00	15.41	15.09	15.46	13.56	11.49	4.63	2.34	1.24
Jersey	49.22 N	2.20 W	2.76	5.65	9.51	14.98	18.51	17.83	18 14	18.62	12.98	6.16	3.26	2.83
London	51.52 N	0.12 W	2.24	3.87	7.40	12.01	12.38	13.24	16.59	16.23	12.59	5.67	2.87	1.97
Greece														
Athens	37.97 N	23.72 E	9.11	10.94	15.70	20.91	23.85	25.48	24.21	23.08	19.03	13.29	5.98	6.64
Sikiwnia	37.98 N	22.73 E	7.60	8.16	11.99	21.06	22.62	24.32	23.56	21.73	17.30	11.75	9.45	6.35
Guadeloupe														
Le Raizet	16.27 N	61.52 W	14.88	18.10	20.55	19.69	20.26	20.65	20.65	20.24	18.47	17.79	13.49	14.38
Guyana														
Cayenne	4.83 N	52.37 W	14.46	14.67	16.28	17.57	—	14.92	17.42	18.24	20.52	—	22.69	17.04
Hong Kong														
King's Park	22.32 N	114.17 W	12.34	7.39	6.94	9.50	11.38	13.60	16.70	17.06	15.91	16.52	14.19	10.00

(continued)

TABLE A.5.3A (continued)

Global Horizontal Average Solar Radiation (Units: MJ/m^2-Day)

Position	Lat.	Long.	January	February	March	April	May	June	July	August	September	October	November	December
Hungary														
Budapest	47.43 N	19.18 E	2.61	7.46	11.14	14.46	20.69	19.47	21.46	19.72	12.88	7.96	2.95	2.47
Iceland														
Reykjavik	64.13 N	21.90 W	0.52	2.02	6.25	11.77	13.07	14.58	16.83	11.35	9.70	3.18	1.00	0.65
India														
Bombay	19.12 N	72.85 E	18.44	21.00	22.72	24.52	24.86	19.75	15.84	16.00	18.19	20.38	19.18	17.81
Calcutta	22.53 N	88.33 E	15.69	18.34	20.09	22.34	22.37	17.55	17.07	16.55	16.52	16.90	16.35	15.00
Madras	13.00 N	80.18 E	19.09	22.71	25.14	24.88	23.89	—	18.22	19.68	19.81	16.41	14.76	15.79
Nagpur	21.10 N	79.05 E	18.08	21.01	22.25	24.08	24.79	19.84	15.58	15.47	17.66	20.10	18.98	17.33
New Delhi	28.58 N	77.20 E	14.62	18.25	20.15	23.40	23.80	19.16	20.20	19.89	20.08	19.74	16.95	14.22
Ireland														
Dublin	53.43 N	6.25 W	2.51	4.75	7.48	11.06	17.46	19.11	15.64	13.89	9.65	5.77	2.93	—
Israel														
Jerusalem	31.78 N	35.22 E	10.79	13.01	18.08	23.79	29.10	31.54	31.83	28.79	25.19	20.26	12.61	10.71
Italy														
Milan	45.43 N	9.28 E	—	6.48	10.09	13.17	17.55	16.32	18.60	16.86	11.64	5.40	3.52	2.41
Rome	41.80 N	12.55 E	—	9.75	13.38	15.82	15.82	18.89	22.27	21.53	16.08	8.27	6.41	4.49

Japan														
Fukuoka	33.58 N	130.38 E	8.11	8.72	10.95	13.97	14.36	12.81	13.84	16.75	13.92	11.86	10.05	7.30
Tateno	36.05 N	140.13 E	9.06	12.17	11.00	15.78	16.52	15.26	—	—	—	9.60	8.55	8.26
Yonago	35.43 N	133.35 E	6.25	7.16	10.87	17.30	16.72	15.44	17.06	19.93	12.41	10.82	7.50	5.51
Kenya														
Mombasa	4.03 S	39.62 E	22.30	22.17	22.74	18.49	18.31	17.41	—	18.12	21.03	22.97	21.87	21.25
Nairobi	1.32 S	36.92 E	—	24.10	21.20	18.65	14.83	15.00	13.44	14.12	19.14	19.38	16.90	18.27
Lithuania														
Kaunas	54.88 N	23.88 E	1.89	4.43	7.40	12.97	18.88	18.74	21.41	15.79	10.40	5.64	1.80	1.10
Madagascar														
Antanarivo	18.80 S	47.48 E	15.94	13.18	13.07	11.53	9.25	8.21	9.32	—	—	16.43	15.19	15.62
Malaysia														
Kualalumpur	3.12 N	101.55 E	15.36	17.67	18.48	16.87	15.67	16.24	15.32	15.89	14.62	14.13	13.54	11.53
Piang	5.30 N	100.27 E	19.47	21.35	23.24	20.52	18.63	19.32	17.17	16.96	15.93	16.01	18.35	17.37
Martinique														
Le Lamentin	14.60 N	61.00 W	17.76	20.07	22.53	21.95	22.42	21.23	20.36	21.84	20.23	19.87	14.08	16.25
Mexico														
Chihuahua	28.63 N	106.08 W	14.80	—	—	—	26.94	26.28	24.01	24.22	20.25	19.55	10.57	15.79
Orizabita	20.58 N	99.20 E	19.49	23.07	27.44	27.35	26.04	25.05	—	27.53	21.06	17.85	15.48	12.93
Mongolia														
Ulan Bator	47.93 N	106.98 E	6.28	9.22	14.34	18.18	20.50	19.34	16.34	16.65	14.08	11.36	7.19	5.35
Uliasutai	47.75 N	96.85 E	6.43	10.71	14.83	20.32	23.86	20.46	21.66	17.81	15.97	10.92	7.32	5.08

(continued)

TABLE A.5.3A (continued)

Global Horizontal Average Solar Radiation (Units: MJ/m^2-Day)

Position	Lat.	Long.	January	February	March	April	May	June	July	August	September	October	November	December
Morocco														
Casablanca	33.57 N	7.67 E	11.46	12.70	15.93	21.25	24.45	25.27	25.53	23.60	19.97	14.68	11.61	9.03
Mozambique														
Maputo	25.97 S	32.60 E	26.35	23.16	19.33	20.54	16.33	14.17	—	—	—	22.55	25.48	26.19
Netherlands														
Maastricht	50.92 N	5.78 E	3.20	5.43	8.48	14.82	14.97	14.32	18.40	17.51	11.65	6.51	3.01	1.72
New Caledonia														
Koumac	20.57 S	164.28 E	24.89	21.15	16.96	18.98	15.67	14.55	15.75	17.62	22.48	15.83	27.53	26.91
New Zealand														
Wilmington	41.28 S	174.77 E	22.59	19.67	14.91	9.52	6.97	4.37	5.74	7.14	12.50	16.34	19.07	24.07
Christchurch	43.48 S	172.55 E	23.46	19.68	13.98	8.96	6.47	4.74	5.38	6.94	13.18	17.45	18.91	24.35
Nigeria														
Benin City	6.32 N	5.60 E	14.89	17.29	19.15	17.21	16.97	15.04	10.24	12.54	14.37	15.99	17.43	15.75
Norway														
Bergen	60.40 N	5.32 E	0.46	1.33	3.18	8.36	19.24	16.70	16.28	10.19	6.53	3.19	1.36	0.35
Oman														
Seeb	23.58 N	58.28 E	12.90	14.86	21.22	22.22	25.30	24.02	23.46	21.66	20.07	18.45	15.49	13.12
Salalah	17.03 N	54.08 E	16.52	16.92	18.49	20.65	21.46	16.92	8.52	11.41	17.14	18.62	16.42	—

Pakistan														
Karachi	24.90 N	67.13 E	13.84	—	—	19.69	20.31	16.62	—	—	—	—	12.94	11.07
Multan	30.20 N	71.43 E	12.29	15.86	18.33	22.35	22.57	21.65	20.31	20.44	20.57	15.91	12.68	10.00
Islamabad	33.62 N	73.10 E	10.38	12.42	16.98	22.65	—	25.49	20.64	18.91	14.20	15.30	10.64	8.30
Peru														
Puno	15.83 S	70.02 W	14.98	12.92	16.08	20.03	17.45	17.42	15.74	15.32	16.11	16.18	14.24	13.90
Poland														
Warszawa	52.28 N	20.97 E	1.73	3.83	7.81	10.53	19.22	17.11	20.18	15.00	10.65	4.95	2.39	1.68
Kolobrzeg	54.18 N	15.58 E	2.50	3.25	8.86	15.21	20.79	20.50	17.19	16.46	7.95	5.75	1.78	1.18
Portugal														
Evora	38.57 N	7.90 W	9.92	12.43	17.81	18.69	23.57	29.23	28.75	23.77	20.17	—	6.81	4.57
Lisbon	38.72 N	9.15 W	9.24	11.60	17.52	18.49	24.64	29.02	28.14	22.20	19.76	13.56	7.18	4.83
Romania														
Bucuresti	44.50 N	26.13 E	7.05	10.22	12.04	16.53	18.97	22.16	23.19	—	17.17	9.55	4.82	—
Constania	44.22 N	28.63 E	5.62	9.28	14.31	20.59	23.23	25.80	27.98	24.22	16.91	11.89	6.19	5.10
Galati	45.50 N	28.02 E	6.09	9.33	14.31	17.75	21.77	22.74	25 55	19.70	14.05	11.26	6.32	5.38
Russia														
Alexandovsko	60.38 N	77.87 E	1.34	4.17	9.16	17.05	21.83	21.34	20.26	13.05	10.16	4.68	1.71	0.68
Moscow	55.75 N	37.57 E	1.45	3.96	8.09	11.69	18.86	18.12	17.51	14.17	10.92	4.03	2.28	1.29
St. Petersburg	59.97 N	30.30 E	1.03	3.11	4.88	12.24	20.59	21.55	20.43	13.27	7.83	2.93	1.16	0.59
Verkhoyansk	67.55 N	133.38 E	0.21	2.25	7.61	15.96	19.64	—	—	14.12	7.59	3.51	0.54	—
St. Pierre and Miquelon														
St. Pierre	46.77 N	56.17 W	4.43	6.61	12.50	17.57	18.55	17.84	19.95	16.46	12.76	8.15	3.69	3.33

(continued)

TABLE A.5.3A (continued)

Global Horizontal Average Solar Radiation (Units: MJ/m^2-Day)

Position	Lat.	Long.	January	February	March	April	May	June	July	August	September	October	November	December
Singapore														
Singapore	1.37 N	103.98 E	19.08	20.94	20.75	18.20	14.89	15.22	13.92	16.66	16.51	15.82	13.81	12.67
South Korea														
Seoul	37.57 N	126.97 E	6.24	9.40	10.34	13.98	16.35	17.49	10.65	12.94	11.87	10.35	6.47	5.14
South Africa														
Cape Town	33.98 S	18.60 E	27.47	25.57	—	15.81	11.44	9.08	8.35	13.76	17.30	22.16	26.37	27.68
Port Elizabeth	33.98 S	25.60 E	27.22	22.06	19.01	15.29	11.79	11.13	10.73	13.97	18.52	23.09	23.15	27.26
Pretoria	25.73 S	28.18 E	26.06	22.43	20.52	16.09	15.67	13.67	15.19	18.65	21.62	21.75	24.82	23.43
Spain														
Madrid	40.45 N	3.72 W	7.73	10.53	15.35	21.74	22.81	22.05	26.27	22.90	18.89	10.21	8.69	5.56
Sudan														
Wad Madani	14.40 N	33.48 E	21.92	24.01	23.43	25.17	23.92	23.51	22.40	22.85	21.75	20.47	20.19	19.21
Elfasher	13.62 N	25.33 E	21.56	21.84	24.54	25.29	24.31	24.15	22.87	21.19	22.58	23.85	—	—
Shambat	15.67 N	32.53 E	23.90	27.38	—	27.45	23.21	26.15	23.55	25.46	24.05	23.51	23.82	22.53
Sweden														
Karlstad	59.37 N	13.47 E	1.26	3.13	5.02	14.01	19.90	16.70	20.92	14.14	10.52	3.98	1.47	0.94
Lund	55.72 N	13.22 E	1.97	3.47	6.66	12.48	17.83	13.38	18.74	14.99	10.39	5.45	1.82	1.21
Stockholm	59.35 N	18.07 E	1.32	2.69	4.75	13.21	15.58	14.79	20.52	14.48	10.50	4.04	1.19	0.83

Switzerland														
Geneva	46.25 N	6.13 E	2.56	7.21	9.46	17.07	20.98	19.78	22.38	20.50	13.62	8.44	3.31	2.87
Zurich	47.48 N	8.53 E	2.31	7.02	7.54	15.04	16.33	16.73	20.28	18.32	12.52	7.18	2.64	2.29
Thailand														
Bangkok	13.73 N	100.57 E	16.67	19.34	23.00	22.48	20.59	17.71	18.02	16.04	16.23	16.81	18.60	16.43
Trinidad and Tobago														
Crown Point	11.15 N	60.83 W	13.05	15.61	15.17	16.96	17.61	15.37	13.16	13.08	12.24	8.76	—	—
Tunisia														
Sidi Botizid	36.87 N	10.35 E	7.88	10.38	13.20	17.98	25.12	26.68	27.43	24.33	18.87	12.11	9.37	6.72
Tunis	36.83 N	10.23 E	7.64	9.88	14.79	31.61	25.31	26.03	26.60	20.37	19.58	12.91	9.35	7.16
Ukraine														
Kiev	50.40 N	30.45 E	2.17	4.87	11.15	12.30	20.49	—	18.99	18.55	9.72	9.84	3.72	2.52
Uzbekistan														
Tashkent	41.27 N	69.27 E	7.27	10.81	15.93	23.60	25.21	29.53	28.50	26.68	20.76	13.25	8.61	4.59
Venezuela														
Caracas	10.50 N	66.88 W	14.25	13.56	16.30	15.56	15.69	15.56	16.28	17.11	17.04	15.14	14.74	13.50
St. Antonio	7.85 N	72.45 W	11.78	10.54	10.65	12.07	12.65	21.20	14.58	15.86	16.62	15.32	12.28	11.28
St. Fernando	7.90 N	67.42 W	14.92	16.82	16.89	—	—	14.09	13.78	14.42	14.86	15.27	14.25	13.11
Vietnam														
Hanoi	21.03 N	105.85 E	5.99	7.48	8.73	13.58	19.10	21.26	19.85	19.78	20.67	14.78	12.44	13.21

(continued)

TABLE A.5.3A (continued)

Global Horizontal Average Solar Radiation (Units: MJ/m^2-Day)

Position	Lat.	Long.	January	February	March	April	May	June	July	August	September	October	November	December
Yugoslavia														
Beograd	44.78 N	20.53 E	4.92	6.27	10.64	14.74	20.95	22.80	22.09	20.27	15.57	11.24	6.77	4.99
Kopaonik	43.28 N	20.80 E	7.03	10.93	14.75	12.78	13.54	20.43	22.48	—	20.14	11.61	6.26	4.64
Portoroz	45.52 N	13.57 E	5.11	7.84	13.75	17.30	23.66	22.31	25.14	21.34	13.40	8.98	6.04	3.92
Zambia														
Lusaka	15.42 S	28.32 W	16.10	18.02	20.24	19.84	17.11	16.37	19.45	20.72	21.68	23.83	23.85	20.52
Zimbabwe														
Bulawayo	20.15 S	28.62 N	20.03	22.11	21.03	18.09	17.15	15.36	16.46	19.49	21.55	23.44	25.08	23.46
Harare	17.83 S	31.02 N	19.38	19.00	19.22	17.67	18.35	16.10	14.55	17.87	21.47	23.98	19.92	21.88

Source: Courtesy of Voeikov Main Geophysical Observatory, Russia, http://wrdc-mgo.nrel.gov/html/get_data-ap.htrnl; For Canadian data, Environment Canada, http://www.ec.gc.ca./envhome.html

Note: Data for 872 locations is available from these sources in 68 countries.

TABLE A.5.3B

Average Daily Solar Radiation on a Horizontal Surface in the United States (Units: MJ/m^2-Day)

Position	January	February	March	April	May	June	July	August	September	October	November	December	Average
Alabama													
Birmingham	9.20	11.92	13.67	19.65	21.58	22.37	21.24	20.21	17.15	14.42	10.22	8.40	16.01
Montgomery	9.54	12.49	16.24	20.33	22.37	23.17	21.80	20.56	17.72	14.99	10.90	8.97	16.58
Alaska													
Fairbanks	0.62	2.77	8.31	14.66	17.98	19.65	16.92	12.36	7.02	3.20	1.01	0.23	8.74
Anchorage	1.02	3.41	8.18	13.06	15.90	17.72	16.69	12.72	8.06	3.97	1.48	0.56	8.63
Nome	0.51	2.95	8.29	15.22	18.97	19.65	16.69	11.81	7.72	3.63	0.99	0.09	8.86
St. Paul Island	1.82	4.32	8.52	12.72	14.08	14.42	12.83	10.33	7.84	4.54	2.16	1.25	7.95
Yakutat	1.36	3.63	7.72	12.61	14.76	15.79	14.99	12.15	7.95	3.97	1.82	0.86	8.18
Arizona													
Phoenix	11.58	15.33	19.87	25.44	28.85	30.09	27.37	25.44	21.92	17.60	12.95	10.56	20.56
Tucson	12.38	15.90	20.21	25.44	28.39	29.30	25.44	24.08	21.58	17.94	13.63	11.24	20.44
Arkansas													
Little Rock	9.09	11.81	15.56	19.19	21.80	23.51	23.17	21.35	17.26	14.08	9.77	8.06	16.24
Fort Smith	9.31	12.15	15.67	19.31	21.69	23.39	23.85	24.46	17.26	13.97	9.88	8.29	16.35
California													
Bakersfield	8.29	11.92	16.69	22.15	26.57	28.96	28.73	26.01	21.35	15.90	10.33	7.61	18.74
Fresno	7.61	11.58	16.81	22.49	27.14	29.07	28.96	25.89	21.12	15.56	9.65	6.70	18.62
Long Beach	9.99	12.95	17.03	21.60	23.17	24.19	26.12	24.08	19.31	14.99	11.24	9.31	17.83
Sacramento	6.93	10.68	15.56	21.24	25.89	28.28	28.62	25.32	20.56	14.54	8.63	6.25	17.72
San Diego	11.02	13.97	17.72	21.92	22.49	23.28	24.98	23.51	19.53	15.79	12.26	10.22	18.06

(continued)

TABLE A.5.3B (continued)

Average Daily Solar Radiation on a Horizontal Surface in the United States (Units: MJ/m^2-Day)

Position	January	February	March	April	May	June	July	August	September	October	November	December	Average
San Francisco	7.72	10.68	15.22	20.44	24.08	25.78	26.46	23.39	19.31	13.97	8.97	7.04	16.92
Los Angeles	10.11	13.06	17.26	21.80	23.05	23.74	25.67	23.51	18.97	14.99	11.36	9.31	17.72
Santa Maria	10.22	13.29	17.49	22.26	25.10	26.57	26.91	24.42	20.10	15.67	11.47	9.54	18.62
Colorado													
Boulder	7.84	10.45	15.64	17.94	17.94	20.47	20.28	17.12	16.07	12.09	8.66	7.10	14.31
Colorado Springs	9.09	12.15	16.13	20.33	22.26	24.98	23.96	21.69	18.5a	14.42	9.99	8.18	16.81
Connecticut													
Hartford	6.70	9.65	13.17	16.69	19.53	21.24	21.12	18.51	14.76	10.68	6.59	5.45	13.74
Delaware													
Wilmington	7.27	10.22	13.97	17.60	20.33	22.49	21.80	19.65	15–79	11.81	7.84	6.25	14.65
Florida													
Daytona Beach	11.24	13.85	17.94	22.15	23.17	22.03	21.69	20.44	17.72	14.99	12.15	10.33	17.38
Jacksonville	10.45	13.17	17.03	21.12	22.03	21.58	21.01	19.42	16.69	14.20	11.47	9.65	16.47
Tallahassee	10.33	13.29	16.92	21.24	22.49	22.03	20.90	19.65	17.72	15.56	11.92	9.77	16.81
Miami	12.72	15.22	18.51	21.58	21.46	20.10	21.10	20.10	17.60	15.67	13.17	11.81	17.38
Key West	13.17	16.01	19.65	22.71	22.83	22.03	22.03	21.01	18.74	16.47	13.85	15.79	18.40
Tampa	11.58	14.42	18.17	22.26	23.05	21.92	20.90	19.65	17.60	16.01	12.83	11.02	17.49
Georgia													
Athens	9.43	12.38	16.01	20.21	22.03	22.83	21.80	20.21	17.26	14.42	10.45	8.40	16.29
Atlanta	9.31	12.26	16.13	20.33	22.37	23.17	22.15	20.56	17.49	14.54	10.56	8.52	16.43

Columbus	9.77	12.72	16.47	20.67	22.37	22.83	21.58	20.33	17.60	14.99	11.02	9.09	16.62
Macon	9.54	12.61	16.35	20.56	22.37	22.83	21.58	20.21	17.26	14.88	10.90	8.86	16.50
Savanna	9.99	12.72	16.81	21.01	22.37	22.60	21.80	19.76	16.92	14.65	11.13	9.20	16.58
Hawaii													
Honolulu	14.08	16.92	19.42	21.24	22.83	23.51	23.74	23.28	21.35	18.06	14.88	13.40	19.42
Idaho													
Boise	5.79	8.97	13.63	18.97	23.51	26.01	27.37	23.62	18.40	12.26	6.70	5.11	15.90
Illinois													
Chicago	6.47	9.31	12.49	16.47	20.44	22.60	22.03	19.31	15.10	10.79	6.47	5.22	13.85
Rockford	6.70	9.77	12.72	16.58	20.33	22.49	22.15	19.42	15.22	10.79	6.59	5.34	14.08
Springfield	7.50	10.33	13.40	17.83	21.46	23.51	23.05	20.56	16.58	12.26	7.72	6.13	15.10
Indiana													
Indianapolis	7.04	9.99	13.17	17.49	21.24	23.28	22.60	20.33	16.35	11.92	7.38	5.79	14.76
Iowa													
Mason City	6.70	9.77	13.29	16.92	20.78	22.83	22.71	19.76	15.33	10.90	6.59	5.45	14.31
Waterloo	6.81	9.77	13.06	16.92	20.56	22.83	22.60	19.76	15.33	10.90	6.70	5.45	14.20
Kansas													
Dodge City	9.65	12.83	16.69	21.01	23.28	25.78	25.67	22.60	18.40	14.42	10.11	8.40	17.49
Goodland	8.97	11.92	16.13	20.44	22.71	25.78	25.55	22.60	18.28	14.08	9.65	7.84	17.03
Kentucky													
Lexington	7.27	9.88	13.51	17.60	20.56	22.26	21.46	19.65	16.01	12.38	7.95	6.25	14.54
Louisville	7.27	10.22	13.63	17.83	20.90	22.71	22.03	20.10	16.35	12.38	7.95	6.25	14.76

(continued)

TABLE A.5.3B (continued)

Average Daily Solar Radiation on a Horizontal Surface in the United States (Units: MJ/m^2-Day)

Position	January	February	March	April	May	June	July	August	September	October	November	December	Average
Louisiana													
New Orleans	9.77	12.83	16.01	19.87	21.80	22.03	20.67	19.65	17.60	15.56	11.24	9.31	16.35
Lake Charles	9.77	12.83	16.13	19.31	21.58	22.71	21.58	20.33	18.06	15.56	11.47	9.31	16.58
Maine													
Portland	6.70	9.99	13.78	16.92	19.99	21.92	21.69	19.31	15.22	10.56	6.47	5.45	13.97
Maryland													
Baltimore	7.38	10.33	13.97	17.60	20.21	22.15	21.69	19.19	15.79	11.92	8.06	6.36	14.54
Massachusetts													
Boston	6.70	9.65	13.40	16.92	20.21	22.03	21.80	19.31	15.33	10.79	6.81	5.45	14.08
Michigan													
Detroit	5.91	8.86	12.38	16.47	20.33	22.37	21.92	18.97	14.76	10.11	6.13	4.66	13.63
Lansing	5.91	8.86	12.49	16.58	20.21	22.26	21.92	18.85	14.54	9.77	5.91	4.66	13.51
Minnesota													
Duluth	5.68	9.31	13.74	17.38	20.10	21.46	21.80	18.28	13.29	8.86	5.34	4.43	13.29
Minneapolis	6.36	9.77	13.51	16.92	20.56	22.49	22.83	19.42	34.65	9.99	6.13	4.88	13.97
Rochester	6.36	9.65	13.17	16.58	20.10	22.15	22.15	19.08	14.54	10.11	6.25	5.11	13.74
Mississippi													
Jackson	9.43	12.38	16.13	19.87	22.15	23.05	22.15	19.08	14.54	10.11	6.25	5.11	13.74
Missouri													
Columbia	8.06	10.90	14.31	18.62	21.58	23.62	23.85	21.12	16.69	12.72	8.29	6.70	15.56
Kansas City	7.95	10.68	14.08	18.28	21.24	23.28	23.62	20.78	16.58	12.72	8.40	6.70	15.44

Springfield	8.52	11.02	14.65	18.62	21.24	23.05	23.62	21.24	16.81	13.17	8.86	7.27	15.67
St. Louis	7.84	10.56	13.97	18.06	21.12	23.05	22.94	20.44	16.58	12.49	8.18	6.59	15.22
Montana													
Helena	5.22	8.29	12.61	17.15	20.67	23.28	25.21	21.24	15.79	10.45	6.02	4.43	14.20
Lewistown	5.22	8.40	12.72	17.15	20.33	23.05	24.53	20.78	15.10	10.22	5.91	4.32	13.97
Nebraska													
Omaha	7.50	10.33	13.97	18.06	21.24	2.40	23.51	20.56	16.01	11.81	7.61	6.13	15.10
Lincoln	7.33	10.10	13.65	16.22	19.26	21.21	22.15	18.87	15.44	11.54	7.76	6.20	14.16
Nevada													
Elko	7.61	10.56	14.42	18.85	22.71	25.67	26.69	23.62	19.31	13.63	8.29	6.70	16.58
Las Vegas	10.79	14.42	19.42	24.87	28.16	30.09	28.28	25.89	22.15	17.03	12.15	9.88	20.33
Reno	8.29	11.58	16.24	21.24	25.10	27.48	28.16	24.98	20.56	14.88	9.31	7.38	17.94
New Hampshire													
Concord	6.81	10.11	13.97	16.92	20.21	21.80	21.80	19.08	14.99	10.45	6.47	5.45	14.08
New Jersey													
Atlantic City	7.38	10.22	13.97	17.49	20.21	21.92	21.24	19.19	15.79	11.92	8.06	6.36	14.54
Newark	6.93	9.77	13.51	17.26	19.76	21.35	21.01	18.85	15.33	11.36	7.27	5.68	13.97
New Mexico													
Albuquerque	11.47	14.99	19.31	24.53	27.60	29.07	27.03	24.76	21.12	17.03	12.49	10.33	19.99
New York													
Albany	6.36	9.43	12.95	16.69	19.53	21.46	21.58	18.51	14.65	10.11	6.13	5.00	13.51
Buffalo	5.68	8.40	12.15	16.35	19.76	22.03	21.69	18.62	14.08	9.54	5.68	4.54	13.29

(continued)

TABLE A.5.3B (continued)

Average Daily Solar Radiation on a Horizontal Surface in the United States (Units: MJ/m²-Day)

Position	January	February	March	April	May	June	July	August	September	October	November	December	Average
New York City	6.93	9.88	13.85	17.72	20.44	22.03	21.69	19.42	15.56	11.47	7.27	5.79	14.31
Rochester	5.68	8.52	12.26	16.58	19.87	21.92	21.69	18.51	14.20	9.54	5.68	4.54	13.29
North Carolina													
Charlotte	8.97	11.81	15.67	19.76	21.58	22.60	21.92	19.99	16.92	13.97	9.99	8.06	16.01
Wilmington	9.31	12.15	16.24	20.44	21.92	22.60	21.58	19.53	16.69	14.08	10.56	8.52	16.13
North Dakota													
Fargo	5.79	9.09	13.17	16.92	20.56	22.37	23.17	19.87	14.31	9.54	5.68	4.54	13.74
Bismarck	6.12	9.75	13.88	17.43	21.45	23.01	24.06	20.12	15.21	10.61	6.28	4.84	14.39
Ohio													
Cleveland	5.79	8.63	12.04	16.58	20.10	22.15	21.92	18.97	14.76	10.22	6.02	4.66	13.51
Columbus	6.47	9.09	12.49	16.58	19.76	21.58	21.12	18.97	15.44	11.24	6.81	5.34	13.74
Dayton	6.81	9.43	12.83	17.03	20.33	22.37	22.37	19.65	15.90	11.47	7.04	5.45	14.20
Youngstown	5.79	8.40	11.92	15.90	19.19	21.24	20.78	18.06	14.31	10.11	6.02	4.77	13.06
Oklahoma													
Oklahoma City	9.88	1.25	16.47	20.33	22.26	24.42	24.98	22.49	18.17	14.54	10.45	8.74	17.15
Oregon													
Eugene	4.54	7.04	11.24	15.79	19.99	22.37	24.19	21.01	15.90	9.65	5.11	3.75	13.40
Medford	5.34	8.52	13.17	18.62	23.39	26.23	27.82	23.96	18.62	11.92	6.02	4.43	15.67
Portland	4.20	6.70	10.68	15.10	18.97	21.24	22.60	19.53	14.88	9.20	4.88	3.52	12.61
Pacific Islands													
Guam	16.35	17.38	19.65	20.78	20.56	19.76	18.28	17.49	17.49	16.58	15.79	15.10	17.94

Pennsylvania													
Philadelphia	7.04	9.88	13.63	17.26	19.99	22.03	21.46	19.42	15.67	11.58	7.72	6.02	14.31
Pittsburgh	6.25	8.97	12.61	16.47	19.65	21.80	21.35	18.85	15.10	10.90	6.59	5.00	13.63
Rhode Island													
Providence	6.70	9.65	13.40	16.92	19.99	21.58	21.24	18.85	15.22	11.02	6.93	5.56	13.97
South Carolina													
Charleston	9.77	12.72	16.81	21.12	22.37	22.37	21.92	19.65	16.92	14.54	11.02	9.09	16.58
Greenville	9.20	12.04	15.90	19.99	21.58	22.60	21.58	19.87	16.81	14.08	10.22	8.18	16.01
South Dakota													
Pierre	6.47	9.54	13.85	17.94	21.46	24.08	24.42	21.46	16.35	11.24	7.04	5.45	14.99
Rapid City	6.70	9.88	14.20	18.28	21.46	24.19	24.42	21.80	16.92	11.81	7.50	5.79	15.33
Tennessee													
Memphis	8.86	11.58	15.22	19.42	22.03	23.85	23.39	21.46	17.38	14.20	9.65	7.84	16.24
Nashville	8.29	11.13	14.65	19.31	21.69	23.51	22.49	20.56	16.81	13.51	8.97	7.15	15.67
Texas													
Austin	10.68	13.63	17.03	19.53	21.24	23.74	24.42	22.83	18.85	15.67	11.92	9.99	17.49
Brownsville	10.33	13.17	16.47	19–08	20.78	22.83	23.28	21.58	18.62	16.13	12.38	9.88	17.03
El Paso	12.38	16.24	20.90	25.44	28.05	28.85	26.46	24.30	21.12	17.72	13.63	11.47	20.56
Houston	9.54	12.26	15.22	18.06	20.21	21.69	21.35	20.21	17.49	15.10	11.02	8.97	15.90
San Antonio	10.88	13.53	16.26	17.35	21.10	23.87	24.92	22.81	19.22	15.52	11.50	9.98	17.24
Utah													
Salt Lake City	6.93	10.45	14.76	19.42	23.39	26.46	26.35	23.39	18.85	13.29	8.06	6.02	16.47

(continued)

TABLE A.5.3B (continued)

Average Daily Solar Radiation on a Horizontal Surface in the United States (Units: MJ/m²-Day)

Position	January	February	March	April	May	June	July	August	September	October	November	December	Average
Vermont													
Burlington	5.79	9.20	13.06	16.47	19.87	21.69	21.80	18.74	14.42	9.43	5.56	4.43	13.40
Virginia													
Norfolk	8.06	10.90	14.65	18.51	20.78	22.15	21.12	19.42	16.13	12.49	9.09	7.27	15.10
Richmond	8.06	10.90	14.76	18.62	20.90	22.49	21.58	19.53,	16.24	12.61	8.97	7.15	15.22
Washington													
Olympia	3.63	6.02	9.99	14.20	18.06	20.10	21.12	18.17	13.63	7.95	4.32	3.07	11.70
Seattle	3.52	5.91	10.11	14.65	19.08	20.78	21.80	18.51	13.51	7.95	4.20	2.84	11.92
Yakirna	4.88	7.95	12.83	17.83	22.49	24.87	25.89	22.26	16.92	10.68	5.56	4.09	17.76
West Virginia													
Charleston	7.04	9.65	13.40	17.15	20.21	21.69	20.90	18.97	15.56	11.81	7.72	6.02	14.20
Elkins	6.93	9.43	12.83	16.35	19.08	20.56	19.99	18.06	14.88	11.13	7.27	5.79	13.51
Wisconsin													
Green Bay	6.25	9.31	13.17	16.81	20.56	22.49	22.03	18.85	14.20	9.65	5.79	4.88	13.74
Madison	6.59	9.88	13.29	16.92	20.67	22.83	22.37	19.42	14.76	3.41	6.25	5.22	14.08
Milwaukee	6.47	9.31	12.72	16.69	20.78	22.94	22.60	19.42	14.88	10.22	6.25	5.11	13.97
Wyoming													
Rock Springs	7.61	10.90	15.30	19.42	23.17	26.01	25.78	22.94	18.62	13.40	8.40	6.70	16.58
Seridan	6.47	9.77	13.97	17.94	20.90	23.85	24.64	21.69	16.47	11.24	7.15	5.56	14.99

Source: Courtesy of National Renewable Energy Laboratory, Golden, CO, http://rredc.nrel.gov/solar

TABLE A.5.4

Regression Coefficients *a* and *b* for Page Model for Solar Radiation (Equation 5.52) for Worldwide Locations[a]

Country	Station	Lat.	Alt. (rad)	Climate [34]	*a*	*b*	Ref.
Egypt	Bahtim	30.13°N	200	BWh	0.220	0.550	[38]
	Cairo	30.08°N	112	BWh	0.140	0.610	[38]
	Giza	30.05°N	19	BWh	0.230	0.540	[38]
	Kharogaosts	25.45°N	78	BWh	0.520	0.230	[38]
	Mersa M.	31.33°N	20	BWh	0.170	0.590	[38]
	Tahrir	30.65°N	16	BWh	0.290	0.460	[38]
Ghana	Accra	5.33°N	~20	BS	0.290	0.470	[18]
	Bole	9.02°N	~350	Aw	0.280	0.440	[18]
	Ho	6.35°N	~300	Aw	0.210	0.460	[18]
	Kumasi	6.41°N	~400	Aw	0.250	0.440	[18]
	Saltpond	5.12°N	~50	BS	0.260	0.450	[18]
	Takoradi	4.59°N	~20	BS	0.250	0.470	[18]
	Tamale	9.25°N	~250	Aw	0.270	0.470	[18]
	Wenchi	7.42°N	~300	Aw	0.280	0.360	[18]
	Yendi	9.26°N	~300	Aw	0.320	0.410	[18]
Greece	Agrino	38.63°N	60	Csa	0.240	0.520	[39]
	Athens	37.97°N	~300	Csa	0.230	0.460	[39]
	Chania	35.50°N	50	Csa	0.220	0.580	[39]
	Kavala	40.93°N	150	Csa	0.250	0.460	[39]
	Larissa	39.63°N	65	Csa	0.230	0.560	[39]
	Mytlline	39.15°N	65	Csa	0.240	0.510	[39]
	Rhodes	36.37°N	5	Csa	0.260	0.520	[39]
Hong Kong	Hong Kong	22.37°N	~5	Caf	0.214	0.514	[14]
India	Ahmedabad	22.37°N	55	Aw	0.302	0.464	[40]
	Bombay (Mumbai)	19.12°N	14	Aw	0.292	0.464	[40]
	Calcutta	22.65°N	6	Aw	0.327	0.399	[40]
	Goa	15.48°N	55	Am	0.279	0.514	[40]
	Jodhpur	26.30°N	224	BSh	0.309	0.439	[40]
	Madras	13.00°N	16	Aw	0.340	0.399	[40]
	Nagpur	21.10°N	310	Aw	0.293	0.460	[40]
	New Delhi	28.63°N	216	Caw	0.341	0.446	[40]
	Poona	18.53°N	559	Bs	0.330	0.453	[40]
	Trivandrum	8.48°N	64	Am	0.393	0.357	[40]
	Vizagapatnam	17.72°N	3	Aw	0.286	0.467	[40]

(continued)

TABLE A.5.4 (continued)

Regression Coefficients *a* and *b* for Page Model for Solar Radiation (Equation 5.52) for Worldwide Locations[a]

Country	Station	Lat.	Alt. (rad)	Climate [34]	*a*	*b*	Ref.
Italy	Alghero	40.63°N	40	Csa	0.118	0.765	[41]
	Amendola	41.53°N	56	Csa	0.212	0.635	[41]
	Ancona	43.63°N	105	Csa	0.197	0.679	[41]
	Balzano	46.47°N	241	Csa	0.192	0.719	[41]
	Bologna	44.53°N	49	Csa	0.216	0.639	[41]
	Brindisi	40.65°N	10	Csa	0.183	0.706	[41]
	Cagliari	39.25°N	18	Csa	0.175	0.679	[41]
	Cape Mele	43.95°N	221	Csa	0.129	0.779	[41]
	Cape Palinus	40.02°N	185	Csa	0.213	0.604	[41]
	Crotone	39.07°N	158	Csa	0.266	0.546	[41]
	Gela	37.09°N	33	Csa	0.105	0.851	[41]
	Genova	44.42°N	3	Csa	0.089	0.821	[41]
	M. Cimone	44.20°N	2137	H	0.104	0.755	[41]
	M. Terminillo	42.47°N	1875	H	0.246	0.482	[41]
Italy	Messina	38.20°N	59	Csa	0.199	0.689	[41]
	Milano	45.43°N	103	Csa	0.148	0.775	[41]
	Napoli	40.85°N	72	Csa	0.181	0.709	[41]
	Olbia	40.93°N	2	Csa	0.183	0.633	[41]
	Pantelleria	36.82°N	170	Csa	0.191	0.679	[41]
	Pescara	42.43°N	18	Csa	0.188	0.669	[41]
	Pianosa	42.58°N	27	Csa	0.143	0.771	[41]
	Pisa	43.67°N	1	Csa	0.205	0.614	[41]
	Roma	41.80°N	131	Csa	0.148	0.722	[41]
	Torino	45.18°N	282	Daf	0.161	0.780	[41]
	Trapani	37.92°N	14	Csa	0.204	0.662	[41]
	Trieste	45.65°N	20	Csa	0.157	0.701	[41]
	Udine	46.03°N	92	Csa	0.167	0.666	[41]
	Ustica	38.70°N	251	Csa	0.254	0.586	[41]
	Venezia	45.50°N	6	Csa	0.144	0.782	[41]
	Vigna Divali	42.08°N	270	Csa	0.154	0.689	[41]
Malaysia	Kota Bahru	6.17°N	5	Af	0.340	0.490	[42]
	Kuala Lumpur	3.12°N	19	Af	0.340	0.490	[42]
	Penang	5.33°N	35	Af	0.350	0.570	[42]
Nigeria	Enugu	6.47°N	137	Af	0.228	0.492	[43]
	Makurdi	7.70°N	76	Aw	0.288	0.472	[43]
	Nsukka	6.80°N	147	Af	0.217	0.490	[43]
	Port Harcourt	4.85°N	6	Af	0.246	0.488	[43]

TABLE A.5.4 (continued)

Regression Coefficients *a* and *b* for Page Model for Solar Radiation (Equation 5.52) for Worldwide Locations[a]

Country	Station	Lat.	Alt. (rad)	Climate [34]	*a*	*b*	Ref.
Sierre L.	Lungi Free.	8.61°N	25	Am	0.260	0.440	[44]
Sudan	Abu Naama	12.73°N	445	BShw	0.433	0.271	[45]
	Aroma	15.83°N	430	BWh	0.460	0.208	[45]
	Dongola	19.17°N	225	BWh	0.211	0.572	[45]
	El Fesher	13.63°N	773	BShw	0.361	0.366	[45]
	El Showak	14.22°N	380	BWh	0.325	0.423	[45]
	G. Ghawazat	11.47°N	480	Aw	0.350	0.353	[45]
	Hudeiba	17.57°N	350	BWh	0.208	0.544	[45]
	Jubo	4.87°N	460	Aw	0.402	0.234	[45]
	Kadugli	11.00°N	501	Aw	0.237	0.463	[45]
	Malakai	9.55°N	387	Aw	0.339	0.359	[45]
	Port Sudan	19.58°N	5	BWh	0.315	0.402	[45]
	Shambat	15.67°N	376	BWh	0.278	0.467	[45]
	Wadi Medani	14.38°N	405	BWh	0.357	0.374	[45]
	Zalingei	12.90°N	900	BShw	0.325	0.456	[45]
Yemen	El Boun	15.73°N	2100	H	0.331	0.385	[46]
	El Khabar	14.38°N	2100	H	0.342	0.372	[46]
	El Macca	13.25°N	10	BWh	0.358	0.346	[46]
	Hodeidah	14.75°N	33	BWh	0.374	0.321	[46]
	Sana	15.52°N	2210	BSh	0.347	0.364	[46]
	Taiz	13.58°N	1400	H	0.364	0.335	[46]
Zambia	Kasama	10.22°S	1384	Aw	0.268	0.454	[47]
	Livingstone	17.82°S	986	Aw	0.187	0.613	[47]
	Luangwa	13.27°S	570	Aw	0.253	0.588	[47]
	Lusaka	15.32°S	1154	Aw	0.198	0.551	[47]
	Mansa	11.10°S	1259	Aw	0.265	0.476	[47]
	Mbala	8.37°S	1673	Aw	0.245	0.505	[47]
	Mongu	15.25°S	1053	Aw	0.188	0.556	[47]
	Ndolo	13.00°S	1270	Aw	0.288	0.386	[47]

[a] Data compiled from Ref. [35,37].

TABLE A.5.5

Solar Collector Tilt Factor ($\bar{R}_b$)

	L = 20°		L = 30°		L = 40°		L = 50°	
Month	**β = 20°**	**β = 40°**	**β = 30°**	**β = 50°**	**β = 40°**	**β = 60°**	**β = 50°**	**β = 70°**
January	1.36	1.52	1.68	1.88	2.28	2.56	3.56	3.94
February	1.22	1.28	1.44	1.52	1.80	1.90	2.49	2.62
March	1.08	1.02	1.20	1.15	1.36	1.32	1.65	1.62
April	1.00	0.83	1.00	0.84	1.05	0.90	1.16	1.00
May	0.92	0.70	0.87	0.66	0.88	0.66	0.90	0.64
June	0.87	0.63	0.81	0.58	0.79	0.60	0.80	0.56
July	0.89	0.66	0.83	0.62	0.82	0.64	0.84	0.62
August	0.95	0.78	0.93	0.76	0.96	0.78	1.02	0.83
September	1.04	0.95	1.11	1.00	1.24,	1.12	1.44	1.32
October	1.17	1.20	1.36	1.36	1.62	1.64	2.10	2.14
November	1.30	1.44	1.60	1.76	2.08	2.24	3.16	3.32
December	1.39	1.60	1.76	1.99	2.48	2.80	4.04	4.52

Source: Kreider, J.F. and Kreith, F., *Solar Heating and Cooling*, revised 1st edn., Hemisphere Publ. Corp., Washington, DC, 1977.

Notes: The solar collector tilt factor is the ratio of monthly beam insolation on a tilted surface to monthly beam insolation on a horizontal surface. Here β = collector tilt angle and L = collector latitude.

TABLE A.5.6A

Solar Position and Insolation Values for 24° North Latitude

Date	Solar Time a.m.	Solar Time p.m.	Solar Position Alt.	Solar Position Azm.	BTUH/ft² Total Insolation on Surface[a] Normal[b]	Horiz.	South Facing Surface Angle with Horiz. 14	24	34	44	90
January 21	7	5	4.8	65.6			17	21	25	28	31
	8	4	16.9	58.3	239	83	110	126	137	145	127
	9	3	27.9	48.8	288	151	188	207	221	228	176
	10	2	37.2	36.1	308	204	246	268	282	287	207
	11	1	43.6	19.6	317	237	283	306	319	324	226
	12		46.0	0.0	320	249	296	319	332	336	232
	Surface daily totals				2766	1622	1984	2174	2300	2360	1766
February 21	7	5	9.3	74.6	158	35	44	49	53	56	46
	8	4	22.3	67.2	263	116	135	145	150	151	102
	9	3	34.4	57.6	298	187	213	225	230	228	141
	10	2	45.1	44.2	314	241	273	286	291	287	168
	11	1	53.0	25.0	321	276	310	324	328	323	185
	12		56.0	0.0	324	288	323	337	341	335	191
	Surface daily totals				3036	1998	2276	2396	2436	2424	1476
March 21	7	5	13.7	83.3	194	60	63	64	62	59	27
	8	4	27.2	76.8	267	141	150	152	149	142	64
	9	3	40.2	67.9	295	212	226	229	225	214	95
	10	2	52.3	54.8	309	266	285	288	283	270	120
	11	1	61.9	33.4	315	300	322	326	320	305	135
	12		66.0	0.0	317	312	334	339	333	317	140
	Surface daily totals				3078	2270	2428	2456	2412	2298	1022

(continued)

TABLE A.5.6A (continued)

Solar Position and Insolation Values for 24° North Latitude

	Solar Time		Solar Position		BTUH/ft² Total Insolation on Surface[a]						
							South Facing Surface Angle with Horiz.				
Date	**a.m.**	**p.m.**	**Alt.**	**Azm.**	**Normal[b]**	**Horiz.**	**14**	**24**	**34**	**44**	**90**
April 21	6	6	4.7	100.6	40	7	5	4	4	3	2
	7	5	18.3	94.9	203	83	77	70	62	51	10
	8	4	32.0	89.0	256	160	157	149	137	122	16
	9	3	45.6	81.9	280	227	227	220	206	186	46
	10	2	59.0	71.8	292	278	282	275	259	237	61
	11	1	71.1	51.6	298	310	316	309	293	269	74
	12		77.6	0.0	299	321	328	321	305	280	79
	Surface daily totals				3036	2454	2458	2374	2228	2016	488
May 21	6	6	8.0	108.4	86	22	15	10	9	9	5
	7	5	21.2	103.2	203	98	85	73	59	44	12
	8	4	34.6	98.5	248	171	159	145	127	106	15
	9	3	48.3	93.6	269	233	224	210	190	165	16
	10	2	62.0	87.7	280	281	275	261	239	211	22
	11	1	75.5	76.9	286	311	307	293	270	240	34
	12		86.0	0.0	288	322	317	304	281	250	37
	Surface daily totals				3032	2556	2447	2286	2072	1800	246

June 21	6	6	9.3	111.6	97	29	20	12	12	11	7
	7	5	22.3	106.8	201	103	87	73	58	41	13
	8	4	35.5	102.6	242	173	158	142	122	99	16
	9	3	49.0	98.7	263	234	221	204	182	155	18
	10	2	62.6	95.0	274	280	269	253	229	199	18
	11	1	76.3	90.8	279	309	300	283	259	227	19
	12		89.4	0.0	281	319	310	294	269	236	22
	Surface daily totals				2994	2574	2422	2230	1992	1700	204
July 21	6	6	8.2	109.0	81	23	16	11	10	9	6
	7	5	21.4	103.8	195	98	85	73	59	44	13
	8	4	34.8	99.2	239	169	157	143	125	104	16
	9	3	48.4	94.5	261	231	221	207	187	161	18
	10	2	62.1	89.0	272	278	270	256	235	206	21
	11	1	75.7	79.2	278	307	302	287	265	235	32
	12		86.6	0.0	280	317	312	298	275	245	36
	Surface daily totals				2932	2526	2412	2250	2036	1766	246
August 21	6	6	5.0	101.3	35	7	5	4	4	4	2
	7	5	18.5	95.6	186	82	76	69	60	50	11
	8	4	32.2	89.7	241	158	154	146	134	118	16
	9	3	45.9	82.9	265	223	222	214	200	181	39
	10	2	59.3	73.0	278	273	275	268	252	230	58
	11	1	71.6	53.2	284	304	309	301	285	261	71
	12		78.3	0.0	286	315	320	313	296	272	75
	Surface daily totals				2864	2408	2402	2316	2168	1958	470

(continued)

TABLE A.5.6A (continued)

Solar Position and Insolation Values for 24° North Latitude

Date	Solar Time		Solar Position		BTUH/ft² Total Insolation on Surface[a]						
							South Facing Surface Angle with Horiz.				
	a.m.	p.m.	Alt.	Azm.	Normal[b]	Horiz.	14	24	34	44	90
September 21	7	5	13.7	83.8	173	57	60	60	59	56	26
	8	4	27.2	76.8	248	136	144	146	143	136	62
	9	3	40.2	67.9	278	205	218	221	217	206	93
	10	2	52.3	54.8	292	258	275	278	273	261	116
	11	1	61.9	33.4	299	291	311	315	309	295	131
	12		66.0	0.0	301	302	323	327	321	306	136
	Surface daily totals				2878	2194	2342	2366	2322	2212	992
October 21	7	5	9.1	74.1	138	32	40	45	48	50	42
	8	4	22.0	66.7	247	111	129	139	144	145	99
	9	3	34.1	57.1	284	180	206	217	223	221	138
	10	2	44.7	43.8	301	234	265	277	282	279	165
	11	1	52.5	24.7	309	268	301	315	319	314	182
	12		55.5	0.0	311	279	314	328	332	327	188
	Surface daily totals				2868	1928	2198	2314	2364	2346	1442

November 21	7	5	4.9	65.8	67	10	16	20	24	27	29
	8	4	17.0	58.4	232	82	108	123	135	142	124
	9	3	28.0	43.9	282	150	186	205	217	224	172
	10	2	37.3	36.3	303	203	244	265	278	283	204
	11	1	43.8	19.7	312	236	280	302	316	320	222
	12		46.2	0.0	315	247	293	315	328	332	228
	Surface daily totals				2706	1610	1962	2146	2268	2324	1730
December 21	7	5	3.2	62.6	30	3	7	9	11	12	14
	8	4	14.9	55.3	225	71	99	116	129	139	130
	9	3	25.5	46.0	281	137	176	198	214	223	184
	10	2	34.3	33.7	304	189	234	258	275	283	217
	11	1	40.4	18.2	314	221	270	295	312	320	236
	12		42.6	0.0	317	232	282	308	325	332	243
	Surface daily totals				2624	1474	1852	2058	2204	2286	1808

Source: Kreider, J.F. and Kreith, F., *Solar Heating and Cooling*, revised 1st edn., Hemisphere Publ. Corp., Washington, DC, 1977.

[a] 1 Btu/h · ft^2 = 3.152 W/m^2. Ground reflection not included on normal or horizontal surfaces.

[b] Normal insolation does not include diffuse component.

TABLE A.5.6B

Solar Position and Insolation Values for 32° North Latitude

Date	Solar Time		Solar Position		BTUH/ft^2 Total Insolation on Surfaces[a]						
							South Facing Surface Angle with Horiz.				
	a.m.	p.m.	Alt.	Azm.	Normal[b]	Horiz.	22	32	42	52	90
January 21	7	5	1.4	65.2	1	0	0	0	0	1	1
	8	4	12.5	56.5	203	56	93	106	116	123	115
	9	3	22.5	46.0	269	118	175	193	206	212	181
	10	2	30.6	33.1	295	167	235	256	269	274	221
	11	1	36.1	17.5	306	198	273	295	308	312	245
	12		38.0	0.0	310	209	285	308	321	324	253
	Surface daily totals				2458	1288	1839	2008	2118	2166	1779
February 21	7	5	7.1	73.5	121	22	34	37	40	42	38
	8	4	19.0	64.4	247	95	127	136	140	141	108
	9	3	29.9	53.4	288	161	206	217	222	220	158
	10	2	39.1	39.4	306	212	266	278	283	279	193
	11	1	45.6	21.4	315	244	304	317	321	315	214
	12		48.0	0.0	317	255	316	330	334	328	222
	Surface daily totals				2872	1724	2188	2300	2345	2322	1644
March 21	7	5	12.7	81.9	185	54	60	60	59	56	32
	8	4	25.1	73.0	260	129	146	147	144	137	78
	9	3	36.8	62.1	290	194	222	224	220	209	119
	10	2	47.3	47.5	304	245	280	283	278	265	150
	11	1	55.0	26.8	311	277	317	321	315	300	170
	12		58.0	0.0	313	287	329	333	327	312	177
	Surface daily totals				3012	2084	2378	2403	2358	2246	1276

April 21	6	6	6.1	99.9	66	14	9	6	6	5	3
	7	5	18.8	92.2	206	86	78	71	62	51	10
	8	4	31.5	84.0	255	158	156	148	136	120	35
	9	3	43.9	74.2	278	220	225	217	203	183	68
	10	2	55.7	60.3	290	267	279	272	256	234	95
	11	1	65.4	37.5	295	297	313	306	290	265	112
	12		69.6	0.0	297	307	325	318	301	276	118
	Surface daily totals				3076	2390	2444	2356	2206	1994	764
May 21	6	6	10.4	107.2	119	36	21	13	13	12	7
	7	5	22.8	100.1	211	107	88	75	60	44	13
	8	4	35.4	92.9	250	175	159	145	127	105	15
	9	3	48.1	84.7	269	233	223	209	188	163	33
	10	2	60.6	73.3	280	277	273	259	237	208	56
	11	1	72.0	51.9	285	305	305	290	268	237	72
	12		78.0	0.0	286	315	315	301	278	247	77
	Surface daily totals				3112	2582	2454	2284	2064	1788	469
June 21	6	6	12.2	110.2	131	45	26	16	15	14	9
	7	5	24.3	103.4	210	115	91	76	59	41	14
	8	4	36.9	96.8	245	180	159	143	122	99	16
	9	3	49.6	89.4	264	236	221	204	181	153	19
	10	2	62.2	79.7	274	279	268	251	227	197	41
	11	1	74.2	60.9	279	306	299	282	257	224	56
	12		81.5	0.0	280	315	309	292	267	234	60
	Surface daily totals				3084	2634	2436	2234	1990	1690	370

(continued)

TABLE A.5.6B (continued)

Solar Position and Insolation Values for 32° North Latitude

	Solar Time		Solar Position		BTUH/ft^2 Total Insolation on Surfaces[a]						
							South Facing Surface Angle with Horiz.				
Date	a.m.	p.m.	Alt.	Azm.	Normal[b]	Horiz.	22	32	42	52	90
July 21	6	6	10.7	107.7	113	37	22	14	13	12	8
	7	5	23.1	100.6	203	107	87	75	60	44	14
	8	4	35.7	93.6	241	174	158	143	125	104	16
	9	3	48.4	85.5	261	231	220	205	185	159	31
	10	2	60.9	74.3	271	274	269	254	232	204	54
	11	1	72.4	53.3	277	302	300	285	262	232	69
	12		78.6	0.0	279	311	310	296	273	242	74
	Surface daily totals				3012	2558	2422	2250	2030	1754	458
August 21	6	6	6.5	100.5	59	14	9	7	6	6	4
	7	5	19.1	92.8	190	85	77	69	60	50	12
	8	4	31.8	84.7	240	156	152	144	132	116	33
	9	3	44.3	75.0	263	216	220	212	197	178	65
	10	2	56.1	61.3	276	262	272	264	249	226	91
	11	1	66.0	38.4	282	292	305	298	281	257	107
	12		70.3	0.0	284	302	317	309	292	268	113
	Surface daily totals				2902	2352	2388	2296	2144	1934	736
September 21	7	5	12.7	81.9	163	51	56	56	55	52	30
	8	4	25.1	73.0	240	124	140	141	138	131	75
	9	3	36.8	62.1	272	188	213	215	211	201	114

	10	2	47.3	47.5	287	237	270	273	268	255	145
	11	1	55.0	26.8	294	268	306	309	303	289	164
			58.0	0.0	296	278	318	321	315	300	171
	Surface daily totals				2808	2014	228	2308	2264	2154	1226
October 21	7	5	6.8	73.1	99	19	29	32	34	36	32
	8	4	18.7	64.0	229	90	120	128	133	134	104
	9	3	29.5	53.0	273	155	198	208	213	212	153
	10	2	38.7	39.1	293	204	257	269	273	270	188
	11	1	45.1	21.1	302	236	294	307	311	306	209
			47.5	0.0	304	247	306	320	324	318	217
	Surface daily totals				3696	1654	2100	2208	2252	2232	1588
November 21	7	5	1.5	65.4	2	0	0	0	1	1	1
	8	4	21.7	56.6	196	55	91	104	113	119	111
	9	3	22.6	46.1	263	118	173	190	202	208	176
	10	2	30.8	33.2	289	166	233	252	265	270	217
	11	1	36.2	17.6	301	197	270	291	303	307	241
				0.0	304	207	282	304	316	320	249
	Surface daily totals				2406	1280	1816	1980	2084	2130	1742
December 21	8	4	10.3	53.8	176	41	77	90	101	108	107
	9	3	19.8	43.6	257	102	161	180	195	204	183
	10	2	27.6	31.2	288	150	221	244	259	267	226
	11	1	32.7	16.4	301	180	258	282	298	305	251
	12		34.6	0.0	304	190	271	295	311	318	259
	Surface daily totals				2348	1136	1704	1888	2016	2086	1794

Source: Kreider, J.F. and Kreith, F., *Solar Heating and Cooling*, revised 1st edn., Hemisphere Publ. Corp., Washington, DC, 1977.

[a] 1 Btu/h · ft^2 = 3.152 W/m^2. Ground reflection not included on normal or horizontal surfaces.

[b] Normal insolation does not include diffuse component.

TABLE A.5.6C
Solar Position and Insolation Values for 40° North Latitude

	Solar Time		Solar Position		BTUH/ft² Total Insolation on Surface[ba]						
							South Facing Surface Angle with Horiz.				
Date	a.m.	p.m.	Alt.	Azm.	Normal[b]	Horiz.	30	40	50	60	90
January 21	8	4	8.1	55.3	142	28	65	74	81	85	84
	9	3	16.8	44.0	239	83	155	171	182	187	171
	10	2	23.8	30.9	274	127	218	237	249	254	223
	11	1	28.4	16.0	289	154	257	277	290	293	253
	12		30.0	0.0	294	164	270	291	303	306	263
	Surface daily totals				2182	948	1660	1810	1906	1944	1726
February 21	7	5	4.8	72.7	69	10	19	21	23	24	22
	8	4	15.4	62.2	224	73	114	122	126	127	107
	9	3	25.0	50.2	274	132	195	205	209	208	167
	10	2	32.8	35.9	295	178	256	267	271	267	210
	11	1	38.1	18.9	305	206	293	306	310	304	236
	12		40.0	0.0	308	216	306	319	323	317	245
	Surface daily totals				2640	1414	2060	2162	2202	2176	1730
March 21	7	5	11.4	80.2	171	46	55	55	54	51	35
	8	4	22.5	69.6	250	114	140	141	138	131	89
	9	3	32.8	57.3	282	173	215	217	213	202	138
	10	2	41.6	41.9	297	218	273	276	271	258	176
	11	1	47.7	22.6	305	247	310	313	307	293	200
	12		50.0	0.0	307	257	322	326	320	305	208
	Surface daily totals				2916	1852	2308	2330	2284	2174	1484

April 21	6	6	7.4	93.9	89	20	11	8	7	7	4
	7	5	18.9	89.5	206	87	77	70	61	50	12
	8	4	30.3	79.3	252	152	153	145	133	117	53
	9	3	41.3	67.2	274	207	221	213	199	179	93
	10	2	51.2	51.4	286	250	275	267	252	229	126
	11	1	58.7	29.2	292	277	308	303	285	260	147
	12		61.6	0.0	293	287	320	313	296	271	154
	Surface daily totals				3092	2274	2412	2320	2168	1956	1022
May 21	5	7	1.9	114.7	1	0	0	0	0	0	0
	6	6	12.7	105.6	144	49	25	15	14	13	9
	7	5	24.0	96.6	216	214	89	76	60	44	13
	8	4	35.4	87.2	250	175	158	144	125	104	25
	9	3	46.8	76.0	267	227	221	206	186	160	60
	10	2	57.5	60.9	277	267	270	255	233	205	89
	11	1	66.2	37.1	283	293	301	287	264	234	108
	12		70.0	0.0	284	301	312	297	274	243	114
	Surface daily totals				3160	2552	2442	2264	2040	1760	724
June 21	5	7	4.2	117.3	22	4	3	3	2	2	1
	6	6	14.8	108.4	155	60	30	18	17	16	10
	7	5	26.0	99.7	216	123	92	77	59	41	14
	8	4	37.4	90.7	246	182	159	142	121	97	16
	9	3	48.8	80.2	263	233	219	202	179	151	47
	10	2	59.8	65.8	272	272	266	248	224	194	74
	11	1	69.2	41.9	277	296	296	278	253	221	92
	12		73.5	0.0	279	304	306	289	263	230	98
	Surface daily totals				3180	2648	2434	2224	1974	1670	610

(continued)

TABLE A.5.6C (continued)
Solar Position and Insolation Values for 40° North Latitude

Date	Solar Time		Solar Position		BTUH/ft^2 Total Insolation on Surface[b,a]						
							South Facing Surface Angle with Horiz.				
	a.m.	p.m.	Alt.	Azm.	Normal[b]	Horiz.	30	40	50	60	90
July 21	5	7	2.3	115.2	2	0	0	0	0	0	0
	6	6	13.1	106.1	138	50	26	17	15	14	9
	7	5	24.3	97.2	208	114	89	75	60	44	14
	8	4	35.8	87.8	241	174	157	142	124	102	24
	9	3	47.2	76.7	259	225	218	203	182	157	58
	10	2	57.9	61.7	269	265	266	251	229	200	86
	11	1	66.7	37.9	275	290	296	281	258	228	104
	12		70.6	0.0	276	298	307	292	269	238	111
	Surface daily totals				3062	2534	2409	2230	2006	1728	702
August 21	6	6	7.9	99.5	81	21	12	9	8	7	5
	7	5	19.3	90.9	191	87	76	69	60	49	12
	8	4	30.7	79.9	237	150	150	141	129	113	50
	9	3	41.8	67.9	260	205	216	207	193	173	89
	10	2	51.7	52.1	272	246	267	259	244	221	120
	11	1	59.3	29.7	278	273	300	292	276	252	140
	12		62.3	0.0	280	282	311	303	287	262	147
	Surface daily totals				2916	2244	2354	2258	2104	1894	978
September 21	7	5	11.4	80.2	149	43	51	51	49	47	32
	8	4	22.5	69.6	230	109	133	134	131	124	84
	9	3	32.8	57.3	263	167	206	208	203	193	132

	10	2	41.6	41.9	280	21[illegible]	262	265	260	247	168
	11	1	47.7	22.6	287	239	298	301	295	281	192
	12		50.0	0.0	290	249	310	313	307	292	200
	Surface daily totals				2708	1788	2210	2228	2182	2074	1416
October 21	7	5	4.5	72.3	48	7	14	15	17	17	16
	8	4	15.0	61.9	204	68	106	113	117	118	100
	9	3	24.5	49.8	257	126	185	195	200	198	160
	10	2	32.4	35.6	280	170	245	257	261	257	203
	11	1	37.6	18.7	291	199	283	295	299	294	229
	12		39.5	0.0	294	208	295	308	312	306	238
	Surface daily totals				2454	1348	1962	2060	2098	2074	1654
November 21	8	4	8.2	55.4	136	28	63	72	78	82	81
	9	3	17.0	44.[illegible]	232	82	152	167	178	183	167
	10	2	24.0	31.0	268	126	215	233	245	249	219
	11	1	28.6	16.1	283	153	254	273	285	288	248
	12		30.2	0.0	288	163	267	287	298	301	258
	Surface daily totals				2128	942	1636	1778	1870	1908	1686
December 21	8	4	5.5	53.0	89	14	39	45	50	54	56
	9	3	14.0	41.9	217	65	135	152	164	171	163
	10	2	20.0	29.4	261	107	200	221	235	242	221
	11	1	25.0	15.2	280	134	239	262	276	283	252
	12		26.6	0.0	285	143	253	275	290	296	263
	Surface daily totals				1978	782	1480	1634	1740	1796	1646

Source: Kreider, J.F. and Kreith, F., *Solar Heating and Cooling*, revised 1st edn., Hemisphere Publ. Corp., Washington, DC, 1977.

[a] 1 Btu/h · ft² = 3.152 W/m². Ground reflection not included on normal or horizontal surfaces.

[b] Normal insolation does not include diffuse component.

TABLE A.5.6D
Solar Position and Insolation Values for 48° North Latitude

	Solar Time		Solar Position		BTUH/ft² Total Insolation on Surfaces[a]						
							South Facing Surface Angle with Horiz.				
Date	**a.m.**	**p.m.**	**Alt.**	**Azm.**	**Normal[b]**	**Horiz.**	**38**	**48**	**58**	**68**	**90**
January 21	8	4	3.5	54.6	37	4	17	19	21	22	22
	9	3	11.0	42.6	185	46	120	132	140	145	139
	10	2	16.9	29.4	239	83	190	206	216	220	206
	11	1	20.7	15.1	261	107	231	249	260	263	243
	12		22.0	0.0	267	115	245	264	275	278	255
	Surface daily totals				1710	596	1360	1478	1550	1578	1478
February 21	7	5	2.4	72.2	12	1	3	4	4	4	4
	8	4	11.6	60.5	188	49	95	102	105	106	96
	9	3	19.7	47.7	251	100	178	187	191	190	167
	10	2	26.2	33.3	278	139	240	251	255	251	217
	11	1	30.5	17.2	290	165	278	290	294	288	247
	12		32.0	0.0	293	173	291	304	307	301	258
	Surface daily totals				2330	1080	1880	1972	2024	1978	1720
March 21	7	5	10.0	78.7	153	37	49	49	47	45	35
	8	4	19.5	66.8	236	96	131	132	129	122	96
	9	3	28.2	53.4	270	147	205	207	203	193	152
	10	2	35.4	37.8	287	187	263	266	261	248	195
	11	1	40.3	19.8	295	212	300	303	297	283	223
	12		42.0	0.0	298	220	312	315	309	294	232
	Surface daily totals				2780	1578	2208	2228	2182	2074	1632

April 21	6	6	8.6	97.8	108	27	13	9	8	7	5
	7	5	18.6	86.7	205	85	76	69	59	48	21
	8	4	28.5	74.9	247	142	149	141	129	113	69
	9	3	37.8	61.2	268	191	216	208	194	174	115
	10	2	45.8	44.6	280	228	268	260	245	223	152
	11	1	51.5	24.0	286	252	301	294	278	254	177
	12		53.6	0.0	288	260	313	305	289	264	185
	Surface daily totals				3076	2106	2358	2266	2114	1902	1262
May 21	5	7	5.2	114.3	41	9	4	4	4	3	2
	6	6	14.7	103.7	162	61	27	16	15	13	10
	7	5	24.6	93.0	219	118	89	75	60	43	13
	8	4	34.7	81.6	248	171	156	142	123	101	45
	9	3	44.3	68.3	264	217	217	202	182	156	86
	10	2	53.0	51.3	274	252	265	251	229	200	120
	11	1	59.5	28.6	279	274	296	281	258	228	141
	12		62.0	0.0	280	281	306	292	269	238	149
	Surface daily totals				3254	2482	2418	2234	2010	1728	982
June 21	5	7	7.9	116.5	77	21	9	9	8	7	5
	6	6	17.2	106.2	172	74	33	19	18	16	12
	7	5	27.0	95.8	220	129	93	77	59	39	15
	8	4	37.1	84.6	246	181	157	140	119	95	35
	9	3	46.9	71.6	261	225	216	198	175	147	74
	10	2	55.8	54.8	269	259	262	244	220	189	105
	11	1	62.7	31.2	274	280	291	273	248	216	126
	12		65.5	0.0	275	287	301	283	258	225	133
	Surface daily totals				3312	2626	2420	2204	1950	1644	874

(continued)

TABLE A.5.6D (continued)
Solar Position and Insolation Values for 48° North Latitude

Date	Solar Time		Solar Position		BTUH/ft² Total Insolation on Surfaces[a]						
							South Facing Surface Angle with Horiz.				
	a.m.	p.m.	Alt.	Azm.	Normal[b]	Horiz.	38	48	58	68	90
July 21	5	7	5.7	114.7	43	10	5	5	4	4	3
	6	6	15.2	104.1	156	62	28	18	16	15	11
	7	5	25.1	93.5	211	118	89	75	59	42	14
	8	4	35.1	82.1	240	171	154	140	121	99	43
	9	3	44.8	68.8	256	215	214	199	178	153	83
	10	2	53.5	51.9	266	250	261	246	224	195	116
	11	1	60.1	29.0	271	272	291	276	253	223	137
	12		62.6	0.0	272	279	301	286	263	232	144
	Surface daily totals				3158	2474	2386	2200	1974	1694	956
August 21	6	6	9.1	98.3	99	28	14	10	9	8	6
	7	5	19.1	87.2	190	85	75	67	58	47	20
	8	4	29.0	75.4	232	141	145	137	125	109	65
	9	3	38.4	61.8	254	189	210	201	187	168	110
	10	2	46.4	45.1	266	225	260	252	237	214	146
	11	1	52.2	24.3	272	248	293	285	268	244	169
	12		54.3	0.0	274	256	304	296	279	255	177
	Surface daily tools				2898	2086	2300	2200	2046	1836	1208
September 21	7	5	10.0	78.7	131	35	44	44	43	40	31
	8	4	19.5	66.8	215	92	124	124	121	115	90
	9	3	28.2	53.4	251	142	196	197	193	183	143

	10	2	35.4	37.8	269	181	251	254	248	236	185
	11	1	40.3	19.8	278	205	287	289	284	269	212
	12		42.0	0.0	280	213	299	302	296	283	221
	Surface daily totals				2568	1522	2102	2118	2070	1966	1546
October 21	7	5	2.0	71.9	4	0	1	1	1	1	1
	8	4	11.2	60.2	165	44	86	91	95	95	87
	9	3	19.3	47.4	233	94	167	176	180	178	157
	10	2	25.7	33.1	262	133	228	239	242	239	207
	11	1	30.0	17.1	274	157	266	277	281	276	237
	12		31.5	0.0	278	166	279	291	294	288	247
	Surface daily totals				2154	1022	1774	1860	1890	1866	1626
November 21	8	4	3.6	54.7	36	5	17	19	21	22	22
	9	3	11.2	42.7	179	46	117	129	137	141	135
	10	2	17.1	29.5	233	83	186	202	212	215	201
	11	1	20.9	15.1	255	107	227	245	255	258	238
	12		22.2	0.0	261	115	241	259	270	272	250
	Surface daily totals				1668	596	1336	1448	1518	1544	1442
December 21	9	3	8.0	40.9	140	27	87	98	105	110	109
	10	2	13.6	28.2	214	63	164	180	192	197	190
	11	1	17.3	14.4	242	86	207	226	239	244	231
	12		18.6	0.0	250	94	222	241	254	260	244
	Surface daily totals				1444	446	1136	1250	1326	1364	1304

Source: Kreider, J.F. and Kreith, F., *Solar Heating and Cooling*, revised 1st edn., Hemisphere Publ. Corp., Washington, DC, 1977.

[a] 1 Btu/h · ft^2 = 3.152 W/m^2. Ground reflection not included on normal or horizontal surfaces.

[b] Normal insolation does not include diffuse component.

TABLE A.5.6E

Solar Position and Insolation Values for 56° North Latitude

	Solar Time		Solar Position		BTUH/ft² Total Insolation on Surfaces[a]						
							South Facing Surface Angle with Horiz.				
Date	a.m.	p.m.	Alt.	Azm.	Normal[b]	Horiz.	46	56	66	76	90
January 21	9	3	5.0	41.8	78	11	50	55	59	60	60
	10	2	9.9	28.5	170	39	35	146	154	156	53
	11	1	12.9	14.5	207	58	183	197	206	208	201
	12		14.0	0.0	217	65	198	214	222	225	217
	Surface daily totals				1126	282	934	1010	1058	1074	1044
February 21	8	4	7.6	59.4	129	25	65	69	72	72	69
	9	3	14.2	45.9	214	65	151	159	162	161	151
	10	2	19.4	31.5	250	98	215	225	228	224	208
	11	1	22.8	16.1	266	119	254	265	268	263	243
	12		24.0	0.0	270	126	268	279	282	276	255
	Surface daily totals				1986	740	1640	1716	1742	1716	1598
March 21	7	5	8.3	77.5	128	28	40	40	39	37	32
	8	4	16.2	64.4	215	75	119	120	117	111	97
	9	3	23.3	50.3	253	118	192	193	189	180	154
	10	2	29.0	34.9	272	151	249	251	246	234	205
	11	1	32.7	17.9	282	172	285	288	282	268	236
	12		34.0	0.0	284	179	297	300	294	280	246
	Surface daily totals				2586	1268	2066	2084	2040	1938	1700
April 21	5	7	1.4	108.8	0	0	0	0	0	0	0
	6	6	9.6	96.5	122	32	14	9	8	7	6

	7	5	18.0	84.1	201	81	74	66	57	46	29
	8	4	26.1	70.9	239	129	143	135	123	108	82
	9	3	33.6	56.3	260	169	208	200	186	167	133
	10	2	39.9	39.7	272	201	259	251	236	214	174
	11	1	44.1	20.7	278	220	292	284	268	245	200
	12		45.6	0.0	280	227	303	295	279	255	209
	Surface daily totals				3024	1892	2282	2186	2038	1830	1458
May 21	4	8	1.2	125.5	0	0	0	0	0	0	0
	5	7	8.5	113.4	93	25	10	9	8	7	6
	6	6	16.5	101.5	175	71	28	17	15	13	11
	7	5	24.8	89.3	219	119	88	74	58	41	16
	8	4	33.1	76.3	244	163	153	138	119	98	63
	9	3	40.9	61.6	259	201	212	197	176	151	109
	10	2	47.6	44.2	268	231	259	244	222	194	146
	11	1	52.3	23.4	273	249	288	274	251	222	170
	12		54.0	0.0	275	255	299	284	261	231	178
	Surface daily totals				3340	2374	2374	2188	1962	1682	1218
June 21	4	8	4.2	127.2	21	4	2	2	2	2	1
	5	7	11.4	115.3	122	40	14	13	11	10	8
	6	6	19.3	103.6	185	86	34	19	17	15	12
	7	5	27.6	91.7	222	132	92	76	57	38	15
	8	4	35.9	78.8	243	175	154	137	116	92	55
	9	3	43.8	64.1	257	212	211	193	170	143	98
	10	2	50.7	46.4	265	240	255	238	214	184	133
	11	1	55.6	24.9	269	258	284	267	242	210	156
	12		57.5	0.0	271	264	294	276	251	219	164
	Surface daily totals				3438	2526	2388	2166	1910	1606	1120

(continued)

TABLE A.5.6E (continued)

Solar Position and Insolation Values for 56° North Latitude

Date	Solar Time		Solar Position		BTUH/ft² Total Insolation on Surfaces[a]						
							South Facing Surface Angle with Horiz.				
	a.m.	p.m.	Alt.	Azm.	Normal[b]	Horiz.	46	56	66	76	90
July 21	4	8	1.7	125.8	0	0	0	0	0	0	0
	5	7	9.0	113.7	91	27	11	10	9	8	6
	6	6	17.0	101.9	169	72	30	18	16	14	12
	7	5	25.3	89.7	212	119	88	74	58	41	15
	8	4	33.6	76.7	237	163	151	136	117	96	61
	9	3	41.4	62.0	252	201	208	193	173	147	106
	10	2	48.2	44.6	261	230	254	239	217	189	142
	11	1	52.9	23.7	265	248	283	268	245	216	165
	12		52.9	0.0	267	254	293	278	255	225	173
	Surface daily totals				3240	2372	2342	2152	1926	1646	1186
August 21	5	7	2.0	109.2	1	0	0	0	0	0	0
	6	6	10.2	97.0	112	34	16	11	10	9	7
	7	5	18.5	84.5	187	82	73	65	56	45	28
	8	4	26.7	71.3	225	128	140	131	119	104	78
	9	3	34.3	56.7	246	168	202	193	179	160	126
	10	2	40.5	40.0	258	199	251	242	227	206	166
	11	1	44.8	20.9	264	218	282	274	258	235	191
	12		46.3	0.0	266	225	293	285	269	245	200
	Surface daily totals				2850	1884	2218	2118	1966	1760	1392

September 21	7	5	8.3	77.5	107	25	36	36	34	32	28
	8	4	16.2	64.4	194	72	111	111	108	102	89
	9	3	23.3	50.3	233	114	181	182	178	168	147
	10	2	29.0	34.9	253	146	236	237	232	221	193
	11	1	32.7	17.9	263	166	271	273	267	254	223
	12		34.0	0.0	266	173	283	285	279	265	233
	Surface daily totals				2368	1220	1950	1962	1918	1820	1594
October 21	8	4	7.1	59.1	104	20	53	57	59	59	57
	9	3	13.8	45.7	193	60	138	145	148	147	138
	10	2	19.0	31.3	231	92	201	210	213	210	195
	11	1	22.3	16.0	248	112	240	250	253	248	230
	12		23.5	0.0	253	119	253	263	266	261	241
	Surface daily totals				1804	688	1516	1586	1612	1588	1480
November 21	9	3	5.2	41.9	76	12	49	54	57	59	58
	10	2	10.0	28.5	165	39	132	143	149	152	148
	11	1	13.1	14.5	201	58	179	193	201	203	196
	12		14.2	0.0	211	65	194	209	217	219	211
	Surface daily totals				1094	284	914	986	1032	1046	1016
December 21	9	3	1.9	40.5	5	0	3	4	4	4	4
	10	2	6.6	27.5	113	19	86	95	101	104	103
	11	1	9.5	13.9	166	37	141	154	163	167	164
	12		10.6	0.0	180	43	159	173	182	186	182
	Surface daily totals				748	156	620	678	716	734	722

Source: Kreider, J.F. and Kreith, F., *Solar Heating and Cooling*, revised 1st edn., Hemisphere Publ. Corp., Washington, DC, 1977.

[a] $1\,\text{Btu/h} \cdot \text{ft}^2 = 3.152\,\text{W/m}^2$. Ground reflection not included on normal or horizontal surfaces.

[b] Normal insolation does not include diffuse component.

TABLE A.5.6F

Solar Position and Insolation Values for 56° North Latitude

Date	Solar Time a.m.	Solar Time p.m.	Solar Position Alt.	Solar Position Azm.	BTUH/ft² Total Insolation on Surfaces[a] Normal[b]	Horiz.	South Facing Surface Angle with Horiz. 54	64	74	84	90
January 21	10	2	2.8	28.1	22	2	17	19	20	20	20
	11	1	5.2	14.1	81	12	72	77	80	81	81
	12		6.0	0.0	100	16	91	98	102	103	103
	Surface daily totals				306	45	268	290	302	306	304
February 21	8	4	3.4	58.7	35	4	17	19	19	19	19
	9	3	8.6	44.8	147	31	103	108	111	110	107
	10	2	12.6	30.3	199	55	170	178	181	178	173
	11	1	15.1	15.3	222	71	212	220	223	219	213
	12		16.0	0.0	228	77	225	235	237	232	226
	Surface daily totals				1432	400	1230	1286	1302	1282	1252
March 21	7	5	6.5	76.5	95	18	30	29	29	27	25
	8	4	20.7	62.6	185	54	101	102	99	94	89
	9	3	18.1	48.1	227	87	171	172	169	160	153
	10	2	22.3	32.7	249	112	227	229	224	213	203
	11	1	25.1	16.6	260	129	262	265	259	246	235
	12		26.0	0.0	263	134	274	277	271	258	246
	Surface daily totals				2296	932	1856	1870	1830	1736	1656
April 21	5	7	4.0	108.5	27	5	2	2	2	1	1
	6	6	10.4	95.1	133	37	15	9	8	7	6
	7	5	17.0	81.6	194	76	70	63	54	43	37

	8	4	23.3	67.5	228	112	136	128	116	102	91
	9	3	29.0	52.3	248	144	197	189	176	158	145
	10	2	33.5	36.0	260	169	246	239	224	203	188
	11	1	36.5	18.4	266	184	278	270	255	233	216
	12		97.6	0.0	268	190	289	281	266	243	225
	Surface daily totals				2982	1644	2176	2082	1936	1736	1594
May 21	4	8	5.8	125.1	51	11	5	4	4	3	3
	5	7	11.6	112.1	132	42	13	11	10	9	8
	6	6	17.9	99.1	185	79	29	16	14	12	11
	7	5	24.5	85.7	218	117	86	72	56	39	28
	8	4	30.9	71.5	239	152	148	133	115	94	80
	9	3	36.8	56.1	252	182	204	190	170	145	128
	10	2	41.6	38.9	261	205	249	235	213	186	167
	11	1	44.9	20.1	265	219	278	264	242	213	193
	12		46.0	0.0	267	224	228	274	251	222	201
	Surface daily totals				3470	2236	2312	2124	1898	1624	1436
June 21	3	9	4.2	139.4	21	4	2	2	2	2	1
	4	8	9.0	126.4	93	27	10	9	8	7	6
	5	7	14.7	113.6	154	60	16	15	13	11	10
	6	6	21.0	87.5	194	96	34	19	17	14	13
	7	5	27.5	73.3	221	132	91	74	55	36	23
	8	4	34.0	57.8	239	166	150	133	112	88	73
	9	3	39.9	40.4	251	195	204	187	164	137	119
	10	2	44.9	20.9	258	217	247	230	206	177	157
	11	1	48.3	0.0	262	231	275	258	233	202	181
	12		49.5		236	235	284	267	242	211	189
	Surface daily totals				3650	2488	2342	2118	1862	1558	1356

(continued)

TABLE A.5.6F (continued)

Solar Position and Insolation Values for 56° North Latitude

Date	Solar Time		Solar Position		BTUH/ft² Total Insolation on Surfaces[a]						
							South Facing Surface Angle with Horiz.				
	a.m.	p.m.	Alt.	Azm.	Normal[b]	Horiz.	54	64	74	84	90
July 21	4	8	6.4	125.3	53	13	6	5	5	4	4
	5	7	12.1	112.4	128	14	14	13	11	10	9
	6	6	18.4	99.4	179	81	30	17	16	13	12
	7	5	25.0	86.0	211	118	86	72	56	38	28
	8	4	31.4	71.8	231	152	146	131	113	91	77
	9	3	37.3	56.3	245	182	201	186	166	141	124
	10	2	42.2	39.2	253	204	245	230	208	181	162
	11	1	45.4	20.2	257	218	273	258	236	207	187
	12		46.6	0.0	259	223	282	267	245	216	195
	Surface daily totals				3372	2248	2280	2090	1864	1588	1400
August 21	5	7	4.6	108.8	29	6	3	3	2	2	2
	6	6	11.0	95.5	123	39	16	11	10	8	7
	7	5	17.6	81.9	181	77	69	61	52	42	35
	8	4	23.9	67.8	214	113	132	123	112	97	87
	9	3	29.6	52.6	234	144	190	182	169	150	138
	10	2	34.2	36.2	246	168	237	229	215	194	179
	11	1	37.2	18.5	252	183	268	260	244	222	205
	12		38.3	0.0	254	188	278	270	255	232	215
	Surface daily totals				2808	1646	2108	1008	1860	1662	1522

September 21	7	5	6.5	76.5	77	16	25	25	24	23	21
	8	4	12.7	72.6	163	51	92	92	90	85	81
	9	3	18.1	48.1	206	83	159	159	156	147	141
	10	2	22.3	32.7	229	108	212	213	209	198	189
	11	1	25.1	16.6	240	124	246	248	243	230	220
	12		26.0	0.0	244	129	258	260	254	241	230
	Surface daily totals				2074	892	1726	1736	1696	1608	1532
October 21	8	4	3.0	58.5	17	2	9	9	10	10	10
	9	3	8.1	44.6	122	26	86	91	93	92	90
	10	2	12.1	30.2	176	50	152	159	161	159	155
	11	1	14.6	15.2	201	65	193	201	203	200	195
	12		15.5	0.0	208	71	207	215	217	213	208
	Surface daily totals				1238	358	1088	1136	1152	1134	1106
November 21	10	2	3.0	28.1	23	3	18	20	21	21	21
	11	1	5.4	14.2	79	12	70	76	79	80	79
	12		6.2	0.0	97	17	89	96	100	101	100
	Surface daily totals				302	46	266	286	298	302	300
December 21	11	1	1.8	13.7	4	0	3	4	4	4	4
	12		2.6	0.0	16	2	14	15	16	17	17
	Surface daily totals				24	2	20	22	24	24	24

Source: Kreider, J.F. and Kreith, F., *Solar Heating and Cooling*, revised 1st edn., Hemisphere Publ. Corp., Washington, DC, 1977.

[a] 1 Btu/h · ft^2 = 3.152 W/m^2. Ground reflection not included on normal or horizontal surfaces.

[b] Normal insolation does not include diffuse component.

TABLE A.5.7

Reflectivity Values for Characteristic Surfaces (Integrated over Solar Spectrum and Angle of Incidence)

Surface	Average Reflectivity
Snow (freshly fallen or with ice film)	0.75
Water surfaces (relatively large incidence angles)	0.07
Soils (clay, loam, etc.)	0.14
Earth roads	0.04
Coniferous forest (winter)	0.07
Forests in autumn, ripe field crops, plants	0.26
Weathered blacktop	0.10
Weathered concrete	0.22
Dead laves	0.30
Dry grass	0.20
Green grass	0.26
Bituminous and gravel roof	0.13
Crushed rock surface	0.20
Building surfaces, dark (red brick, dark paints, etc.)	0.27
Building surfaces, light (light brick, light paints, etc.)	0.60

Sources: Hunn, B.D. and Calafell, D.O., *Solar Energy*, 19, 87, 1977; see also List, R.J., *Smithsonian Meterological Tables*, 6th edn., Smithsonian Institution Press, Washington, DC, pp. 442–443.

References

1. Howell, J.R. and R. Siegel. (1992) *Thermal Radiation Heat Transfer*, McGraw-Hill Book Co., New York.
2. Kreith, F. (1962) *Radiation Heat Transfer for Spacecraft and Solar Power Plant Design*, International Textbook Co., Scranton, PA.
3. Kreith, F. and Bohn, M. (1993) *Principles of Heat Transfer*, 5th edn., West Publishing Co., St. Paul, MN.
4. Sparrow, E.M. and Cess, R.D. (1978) *Radiation Heat Transfer*, Wadsworth Publ. Co., Belmont, CA.
5. Threlkeld, J.L. and Jordan, R.C. (1958) Direct radiation available on clear days, *Trans. ASHRAE* 64, 45.
6. Frohlich, C. et al. (1973) The third international comparison of pyroheliometers and a comparison of radiometric scales. *Solar Energy* 14, 157–166.
7. Quinlan, F.T. (ed.) (1977) *SOLMET Vol. 1: Hourly Solar Radiation Surface Meteorological Observations*, NOAA, Asheville, NC.
8. Quinlan, F.T. (ed.) (1979) *SOLMET Vol. 2: Hourly Solar Radiation Surface Meteorological Observations*, NOAA, Asheville, NC.
9. Kimura, K. and Stephenson, D.G. (1969) Solar radiation on cloudy days, *Trans. ASHRAE* 75, 227–234.
10. Thekaekara, M.P. (1973) Solar energy outside the Earth's atmosphere, *Sol. Energy* 14, 109–127.

11. Page, J.K. (1966) The estimation of monthly mean values of daily total short-wave radiation on vertical and inclined surfaces from sunshine records or latitudes 40°N–40°S, *Proc. U.N. Conf. New Sources Energy* 4, 378.
12. Löf, G.O.G. et al. (1966) World distribution of solar energy, University of Wisconsin, Madison, WI, Eng. Expt. Station Rept. 21.
13. Liu, B.Y.H. and Jordan, R.C. (1961) Daily insolation on surfaces titled toward the equator, *Trans. ASHRAE* 67, 526–541.
14. Liu, B.Y.H. and Jordan, R.C. (1961) Daily insolation on surface titled toward the equator, *Trans. ASHRAE* 3(10), 53–59.
15. Liu, B.Y.H. and Jordan, R.C. (1967) Availability of solar energy for flat-plate solar heat collectors, in: *Low Temperature Engineering of Solar Energy*, Chap. 1, ASHRAE, New York; see also 1977 revision.
16. Liu, B.Y.H. and Jordan, R.C. (1963) A rational procedure for predicting the long-term average performance of flat-plate solar energy collectors, *Sol. Energy* 7, 53–74.
17. Liu, B.Y.H. and Jordan, R.C. (1960) The interrelationship and characteristic distribution of direct, diffuse and total solar radiation, *Sol. Energy* 4, 1–19. See also Liu, B.Y.H. (1960) Characteristics of solar radiation and the performance of flat plate solar energy collectors, PhD dissertation, University of Minnesota, Minneapolis, MN.
18. Collares-Pereira, M. and Rabl, A. (1979) The average distribution of solar radiation-correlation between diffuse and hemispherical, *Sol. Energy* 22, 155–166.
19. Choudhury, N.K.O. (1963) Solar radiation at New Delhi, *Solar Energy* 7, 44.
20. Stanhill, G. (1966) Diffuse sky and cloud radiation in Israel, *Sol. Energy* 10, 66.
21. Ruth, D.W. and Chant, R.E. (1976) The relationship of diffuse radiation to total radiation in Canada, *Sol. Energy* 18, 153.
22. Temps, R.C. and Coulson, K.L. (1977) Solar radiation incident upon slopes of different orientations, *Sol. Energy* 19(2), 179–184.
23. Klucher, T.M. (1979) Evaluation of models to predict insolation on tilted surfaces, *Sol. Energy*, 23(2), 111–114.
24. Budde, W. (1983) *Physical Detectors of Optical Radiation*, Vol. 4 of *Optical Radiation Measurements*, Academic Press, New York.
25. Randall, C.M. and Leonard, S.L. (1974) Reference insolation data base: A case history, with recommendations, in: *Rept. Recommendations Solar Energy Data Workshop*, pp. 93–103.
26. Vant-Hull, L. and Easton, C.R. (1975) Solar thermal power systems based on optical transmission, NTIS Rept. PB253167.
27. Iqbal, M. (1983) *An Introduction to Solar Radiation*, Academic Press, New York.
28. Zerlaut, G. (1989) Solar radiation instrumentation, In: *Solar Resources*, Roland L. Hulstrom (ed.), MIT Press, Cambridge, MA.
29. Gautier, C., Diak, G., and Masse, S. (1980) A simple physical model to estimate incident solar radiation at the surface from GOES satellite data, *J. Appl. Meteorol.* 19(8), 1005.
30. Mujahid, A. and Turner, W.D. (1979) Diffuse sky measurement and model, ASME Pap. no. 79-WA/Sol-5.
31. Maxwell, E.L. (1998) METSTAT—The solar radiation model used in the production of the national solar radiation data base (NSRDB), *Sol. Energy* 62, 263–279.
32. NSRDB. (1992) *Volume 1: Users Manual: National Solar Radiation Data Base (1961–1990), Version 1.0*, National Renewable Energy Laboratory, Golden, CO.

33. Voeikov Main Geophysical Observatory. (1999) Worldwide daily solar radiation, Available from Worldwide Web: http://www.mgo.rssi.ru.
34. University of Lowell Photovoltaic Program. (1990) *International Solar Irradiation Database*, Version 1.0, University of Lowell Research Foundation, Lowell, MA.
35. Whillier, A. (1965) Solar radiation graphs, *Sol. Energy* 9, 165–166.
36. Kreider, J.F. and Kreith, F. (1982) *Solar Heating and Cooling*, Hemisphere Publishing Corp., Washington, DC.
37. Akinoglu, B.G. (1991) A review of sunshine-based models used to estimate monthly average global solar radiation, *Renewable Energy* 1(3), 479–499.
38. Jain, S. and Jain, P.G. (1985) A comparison of the angstrom type correlations and the estimation of monthly average daily global irradiation, International Centre for Theoretical Physics, Internal Report, No. IC85-269.
39. Zabara, K. (1986) Estimation of the global solar radiation in Greece, *Solar Wind Technol.* 3, 267.
40. Fritz, S. (1949) Solar radiation in the United States, *Heat. Vent.* 16, 61–64, July.
41. Garg, H.P. and Garg, S.N. (1985) Correlation of the monthly average daily global, diffuse and beam radiation with bright sunshine hours, *Energy Convers. Manage.* 25(4), 409–417.
42. Cano, D. et al. (1986) Method for the determination of the global solar radiation from meteorological satellite data, *Solar Energy* 37(1), 31–39.
43. Dedieu, G., Deschamps, P.Y., and Kerr, Y.H. (1987) Satellite estimation of solar irradiance at the surface of the earth and of surface albedo using a physical model applied to METEOSAT data, *J. Clim. Appl. Meteorol.* 26, 79–87.
44. Massaquoi, J.G.M. (1988) Global solar radiation in Sierre Leone (West Africa), *Solar Wind Technol.* 5, 281.
45. Jain, P.C. (1986) Irradiation estimation for Italian localities, *Solar Wind Technol.* 3, 323.
46. Khogali, A. (1983) Star radiation over Sudan: Comparison of measured and predicted data, *Solar Energy* 31, 45.
47. Gopinathan, K. (1988) A general formula for computing the coefficients of the correlation connecting global solar radiation to sunshine duration, *Solar Energy* 41, 499.
48. Hunn, B.D. and Calafell, D.O. (1977) Determination of average ground reflectivity for solar collectors, *Solar Energy*, 19, 87.
49. List, R.J. (1949) *Smithsonian Meteorological Tables*, 6th edn., Smithsonian Institution Press, Washington, DC, pp. 442–443.

6

Photovoltaics

> We do not inherit the earth from our ancestors, we borrow it from our children.
>
> **—Chief Seattle**

Photovoltaic conversion is the direct conversion of sunlight into electricity with no intervening heat engine. Photovoltaic devices are solid-state devices; therefore, they are rugged and simple in design and require very little maintenance. Perhaps, the biggest advantage of solar photovoltaic devices is that they can be constructed as stand-alone systems to give outputs from microwatts to megawatts. That is why they have been used as the power sources for calculators, watches, water pumping, remote buildings, communications, satellites and space vehicles, and even megawatt-scale power plants. Photovoltaic panels can be made to form components of building skin, such as roof shingles and wall panels. With such a vast array of applications, the demand for photovoltaics is increasing every year. The market for photovoltaic solar cells is growing exponentially, increasing from 227 MW in 2000 to 3800 MW in 2008 and is expected to continue increasing exponentially. An example of a photovoltaic residential rooftop system is shown in Figure 6.1.

In the early days of solar cells in the 1960s and 1970s, more energy was required to produce a cell than it could ever deliver during its lifetime. Since then, dramatic improvements have taken place in the efficiencies and manufacturing methods. The energy payback periods have been reduced to about 2.5–5 years, depending on the location of use [1], while panel lifetimes were increased to over 25 years. The costs of photovoltaic panels have come down to less than $5 per peak watt over the last two decades and are targeted to reduce to around $1.00 per peak watt soon.

Historically, the photoelectric effect was first noted by Becquerel in 1839 when light was incident on an electrode in an electrolyte solution [2]. Adams and Day first observed the effect in solids in 1877 while working with selenium. Early work was done with selenium and copper oxide by pioneers such as Schottky, Lange, and Grandahl. In 1954, researchers at RCA and Bell Laboratories reported achieving efficiencies of about 6% by using devices made of p and n types of semiconductors. The space race between the United States and the Soviet Union resulted in dramatic improvements in photovoltaic devices. Ref. [3] gives a review of the early developments in photovoltaic conversion.

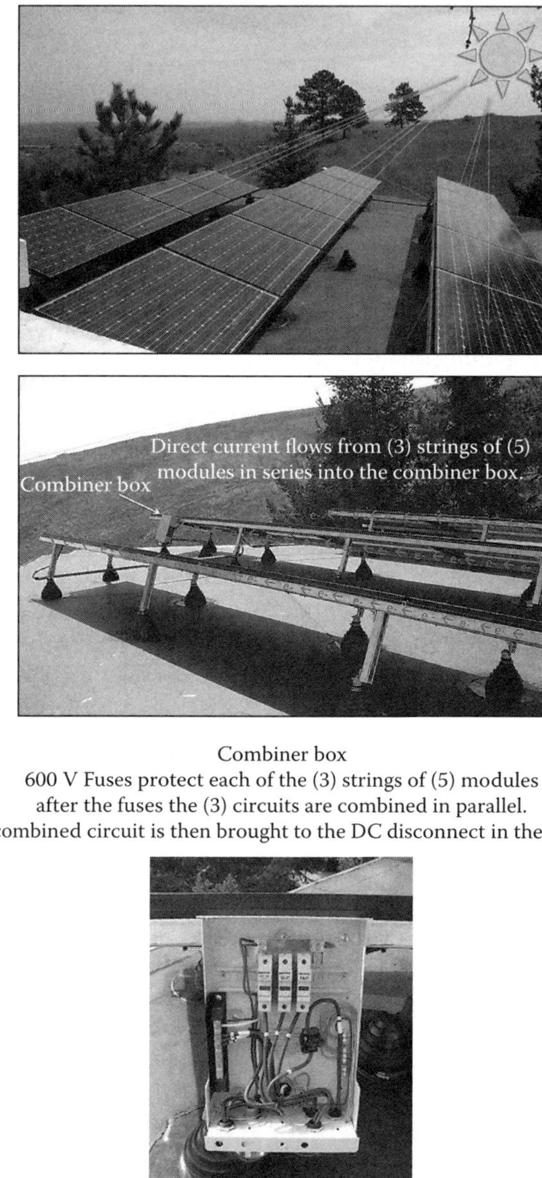

FIGURE 6.1
(See color insert.) Kreith residence, 3 kW Sanyo high efficiency photovoltaic system. (Courtesy of Chris Klinga, Lighthouse Solar.)

6.1 Semiconductors

A basic understanding of the atomic structure is quite helpful in understanding the behavior of semiconductors and their use as the photovoltaic energy conversion devices. Any fundamental book on physics or chemistry generally gives adequate background for basic understanding. Reference [3] presents an in-depth treatment of a number of topics in semiconductor physics.

For any atom, the electrons arrange themselves in orbitals around the nucleus so as to result in the minimum amount of energy. Table 6.1 shows the distribution of the electrons in various shells and subshells in light elements. In elements that have electrons in multiple shells, the innermost electrons have the minimum energy and, therefore, require the maximum amount of externally imparted energy to overcome the attraction of the nucleus and become free. Electrons in the outermost band of subshells are the only ones that participate in the interaction of an atom with its neighboring atoms. If these electrons are very loosely attached to the atom, they may attach themselves with a neighboring atom to give that atom a negative charge, leaving the original atom as a positive-charged ion. The positive and negatively charged ions become attached by the force of attraction of the charges thus forming *ionic bonds.* If the electrons in the outermost band do not fill the band completely but are not loosely attached either, they arrange themselves so that neighboring atoms can share them to make the outermost bands full. The bonds thus formed between the neighboring atoms are called *covalent bonds.*

Since electrons in the outermost band of an atom determine how an atom will react or join with a neighboring atom, the outermost band is called the *valence band.* Some electrons in the valence band may be so energetic that they jump into a still higher band and are so far removed from the nucleus that a small amount of impressed force would cause them to move away from the atom. Such electrons are responsible for the conduction of heat and electricity, and this remote band is called a *conduction band.* The difference in the energy of an electron in the valence band and the innermost subshell of the conduction band is called the *band gap* or the forbidden gap.

Materials whose valence bands are full have very high band gaps (>3 eV). Such materials are called *insulators.* Materials, on the other hand, that have relatively empty valence bands and may have some electrons in the conduction band are good *conductors.* Metals fall in this category. Materials with valence bands partly filled have intermediate band gaps (≤3 eV). Such materials are called *semiconductors* (Figure 6.2). Pure semiconductors are called *intrinsic semiconductors,* while semiconductors doped with very small amounts of impurities are called *extrinsic semiconductors.* If the dopant material has more electrons in the valence band than the

TABLE 6.1

Electronic Structure of Atoms

Principal Quantum Number n				**1**	**2**		**3**		**4**		**5**
Azimuthal Quantum Number l				**0**	**0**	**1**	**0**	**1**	**2**	**0**	**1**
Letter Designation of State				**1s**	**2s**	**2p**	**3s**	**3p**	**3d**	**4s**	**4p**
Z	**Symbol**	**Element**	**V_i (V)**								
1	H	Hydrogen	13.60	1							
2	He	Helium	24.58	2							
3	Li	Lithium	5.39		1						
4	Be	Beryllium	9.32		2						
5	B	Boron	8.30		2	1					
6	C	Carbon	11.26	Helium core	2	2					
7	N	Nitrogen	14.54		2	3					
8	O	Oxygen	13.61		2	4					
9	F	Fluorine	17.42		2	5					
10	Ne	Neon	21.56		2	6					
11	Na	Sodium	5.14				1				
12	Mg	Magnesium	7.64				2				
13	Al	Aluminum	5.98				2	1			
14	Si	Silicon	8.15	Neon core			2	2			
15	P	Phosphorus	10.55				2	3			
16	S	Sulfur	10.36				2	4			
17	Cl	Chlorine	13.01				2	5			
18	A	Argon	15.76				2	6			
19	K	Potassium	4.34							1	
20	Ca	Calcium	6.11							2	
21	Sc	Scandium	6.56						1	2	
22	Ti	Titanium	6.83						2	2	
23	V	Vanadium	6.74						3	2	
24	Cr	Chromium	6.76						5	1	
25	Mn	Manganese	7.43						5	2	
26	Fe	Iron	7.90						6	2	
27	Co	Cobalt	7.86						7	2	
28	Ni	Nickel	7.63	Argon core					8	2	
29	Cu	Copper	7.72						10	1	
30	Zn	Zinc	9.39						10	2	
31	Ga	Gallium	6.00						10	2	1
32	Ge	Germanium	7.88						10	2	2
33	As	Arsenic	9.81						10	2	3
34	Se	Selenium	9.75						10	2	4
35	Br	Bromine	11.84						10	2	5
36	Kr	Krypton	14.00						10	2	6

Source: Moore, C.E., *Atomic Energy Levels*, vol. 2, National Bureau of Standards Circular 467, U.S. Government Printing Office, Washington, DC, 1952.

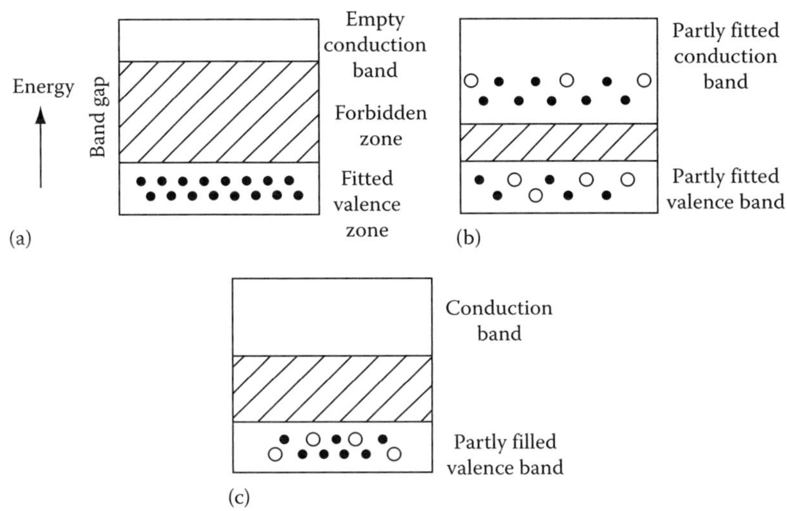

FIGURE 6.2
Electrical conduction is described in terms of allowed and forbidden energy bands. Band gap for insulators (a) is the highest, followed by semiconductors (c) and metals (b), respectively.

semiconductor, the doped material is called an *n*-type of semiconductor. Such a material seems to have excess electrons available for conduction even though the material is electronically neutral. For example, silicon has four electrons in the valence band. Atoms of pure silicon arrange themselves in such a way that, to form a stable structure, each atom shares two electrons with each neighboring atom with covalent bands. If phosphorous, which has five valence electrons (one more than Si), is introduced as an impurity in silicon, the doped material seems to have excess electrons even though it is electrically neutral. Such a doped material is called *n*-type silicon. If on the other hand, silicon is doped with boron which has three valence electrons (one less than Si), there seems to be a positive hole (missing electrons) in the structure, even though the doped material is electrically neutral. Such material is called *p*-type silicon. Thus, *n*- and *p*-type semiconductors make it easier for the electrons and holes, respectively, to move in the semiconductors.

6.1.1 *p–n* Junction

As explained earlier, an *n*-type material has some impurity atoms with more electrons than the rest of the semiconductor atoms. If those excess electrons are removed, the impurity atoms will fit more uniformly in the structure formed by the main semiconductor atoms; however, the atoms

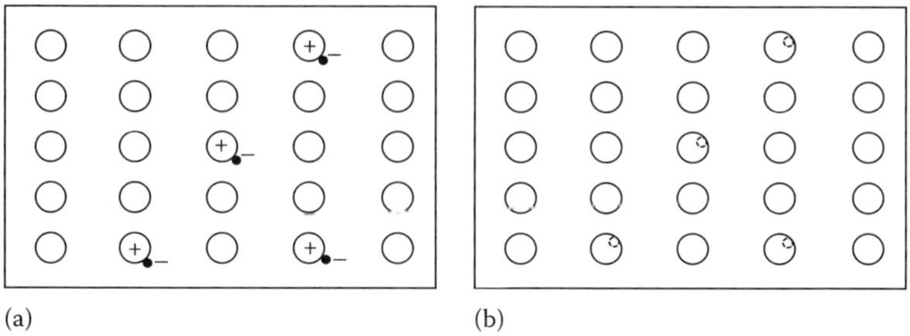

FIGURE 6.3
Representation of *n*- and *p*-type semiconductors: (a) *n*-type showing "excess" electrons as dots; (b) *p*-type showing "excess" positive holes as ○.

will be left with positive charges. On the other hand, a *p*-type material has some impurity atoms with fewer electrons than the rest of the semiconductor atoms. Therefore, these atoms seem to have holes that could accommodate excess electrons even though the atoms are electrically neutral (Figure 6.3). If additional electrons could be brought to fill the holes, the impurity atoms would fit more uniformly in the structure formed by the main semiconductor atoms, however, the atoms will be negatively charged.

The above scenario occurs at the junction when a *p*- and an *n*-type of material are joined together, as shown in Figure 6.4. As soon as the two materials are joined, "excess" electrons from the *n* layer jump to fill the "holes" in the *p* layer. Therefore, close to the junction, the material has positive charges on the *n*-side and negative charges on the *p*-side. The negative charges on the *p*-side restrict the movement of additional electrons from the *n*-side to the *p*-side, while the movement of additional electrons from the *p*-side to the *n*-side is made easier because of the positive charges at the junction on the *n*-side. This restriction makes the *p*–*n* junction behave like a diode. This diode character of a *p*–*n* junction is made use of in solar photovoltaic cells, as explained below.

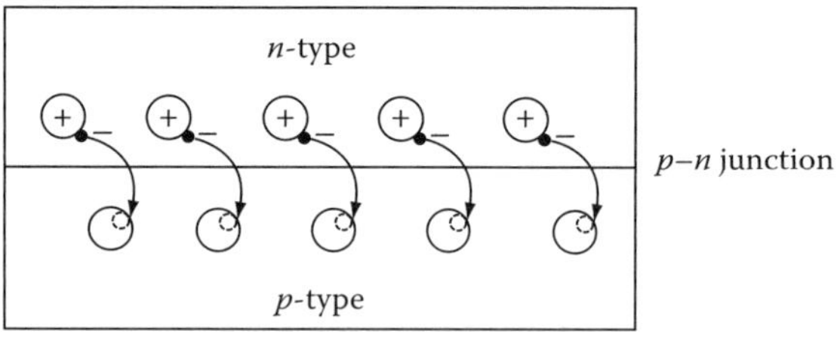

FIGURE 6.4
"Excess" electrons from *n* material jump to fill "excess" holes on the *p*-side of a *p*–*n* junction, leaving the *n*-side of the junction positively charged and the *p*-side negatively charged.

6.1.2 The Photovoltaic Effect

When a photon of light is absorbed by a valence electron of an atom, the energy of the electron is increased by the amount of energy of the photon. If the energy of the photon is equal to or more than the band gap of the semiconductor, the electron with excess energy will jump into the conduction band where it can move freely. If, however, the photon energy is less than the band gap, the electron will not have sufficient energy to jump into the conduction band. In this case, the excess energy of the electrons is converted to excess kinetic energy of the electrons, which manifests in increased temperature. If the absorbed photon had more energy than the band gap, the excess energy over the band gap simply increases the kinetic energy of the electron. It must be noted that a photon can free up only one electron even if the photon energy is a lot higher than the band gap. This fact is a big reason for the low conversion efficiency of photovoltaic devices. The key to using the photovoltaic effect for generating useful power is to channel the free electrons through an external resistance before they recombine with the holes. This is achieved with the help of the *p*–*n* junction.

Figure 6.5 shows a schematic of a photovoltaic device. As free electrons are generated in the *n* layer by the action of photons, they can either pass through an external circuit, recombine with positive holes in the lateral direction, or move toward the *p* layer. The negative charges in the *p* layer at the *p*–*n* junction restrict their movement in that direction. If the *n* layer is made extremely thin, the movement of the electrons and, therefore, the probability of recombination within the *n* layer is greatly reduced unless the external circuit is open. If the external circuit is open, the electrons generated by the action of photons eventually recombine with the holes, resulting in an increase in the temperature of the device.

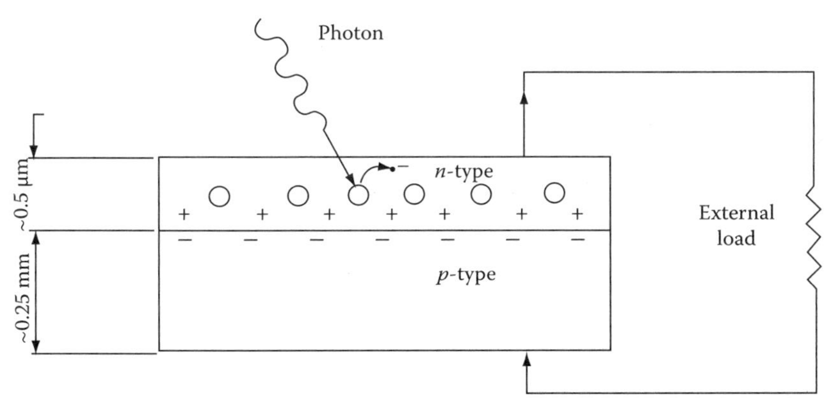

FIGURE 6.5
Schematic of a photovoltaic device.

In a typical crystalline silicon cell, the *n* layer is about 0.5 μm thick and the *p* layer is about 0.25 mm thick. As explained in Chapter 5, energy contained in a photon E_p is given by

$$E_p = h\nu, \tag{6.1}$$

where

h is Planck's constant (6.625×10^{-34} J-s)

ν is the frequency which is related to the wavelength λ and the speed of light *c* by

$$\nu = \frac{c}{\lambda}.$$

Therefore,

$$E_p = \frac{hc}{\lambda}. \tag{6.2}$$

For silicon, which has a band gap of 1.11 eV, the following example shows that photons of solar radiation of wavelength 1.12 μm or less are useful in creating electron–hole pairs. This spectrum represents a major part of the solar radiation. Table 6.2 lists some candidate semiconductor materials for photovoltaic cells along with their band gaps.

Example 6.1

Calculate the wavelength of light capable of forming an electron–hole pair in silicon.

Solution

The band gap energy of silicon is 1.11 eV. From Equation 6.2, we can write

$$\lambda = \frac{hc}{E_p}.$$

For $c = 3 \times 10^8$ m/s, $h = 6.625 \times 10^{-34}$ J-s, and 1 eV = 1.6×10^{-19} J, the above equation gives the required wavelength as

$$\lambda = \frac{(6.625\times10^{-34}\ \text{J-s})(3\times10^{8}\ \text{m/s})}{(1.11)(1.6\times10^{-19}\ \text{J})} = 1.12\ \mu\text{m}.$$

Example 6.2

A monochromatic red laser beam emitting 1 mW at a wavelength of 638 nm is incident on a silicon solar cell. Find

TABLE 6.2

Energy Gap for Some Candidate Materials for Photovoltaic Cells

Material	Band Gap (eV)
Si	1.11
SiC	2.60
$CdAs_2$	1.00
CdTe	1.44
CdSe	1.74
CdS	2.42
$CdSnO_4$	2.90
GaAs	1.40
GaP	2.24
Cu_2S	1.80
CuO	2.00
Cu_2Se	1.40
$CuInS_2$	1.50
$CuInSe_2$	1.01
$CuInTe_2$	0.90
InP	1.27
In_2Te_3	1.20
In_2O_3	2.80
Zn_3P_2	1.60
ZnTe	2.20
ZnSe	2.60
AlP	2.43
AlSb	1.63
As_2Se_3	1.60
Sb_2Se_3	1.20
Ge	0.67
Se	1.60

Source: Garg, H.P., *Advances in Solar Energy Technology*, vol. 3, D. Reidel Publishing Company, Dordrecht, the Netherlands, 1987.

a. The number of photons per second incident on the cell
b. The maximum possible efficiency of conversion of this laser beam to electricity

Solution

a. The intensity of light in the laser beam (I_p) is equal to the energy of all the photons in it. If the number of photons is N_{ph}, then

$$I_p = N_{ph} \cdot E_p$$
$$1 \times 10^{-3}\ \mathrm{W} = N_{ph} \cdot E_p (1\ \mathrm{W} = 1\ \mathrm{J/s}) \tag{6.3}$$

$$E_p = \frac{hc}{\lambda}$$

$$= \frac{(6.625\times10^{-34}\ \text{J-s})\cdot 3\times10^{8}\ \text{m/s}}{638\times10^{-9}\ \text{m}}$$

$$= 3.12\times10^{19}\ \text{J}$$

$$\therefore N_{ph} = \frac{1\times10^{-3}\ \text{J/s}}{3.12\times10^{-19}\ \text{J}} = 3.21\times10^{15}\ \text{photons/s}$$

b. Assuming that each photon is able to generate an electron, a total number of N_{ph} electrons will be generated. Therefore, the electrical output will be equal to N_{ph} (n). Therefore, the maximum possible efficiency is

$$\eta_{max} = \frac{(N_{ph})\cdot(\text{B.G.})}{(N_{ph})\cdot E_p} = \frac{\text{B.G.}}{E_p}$$

$$= \frac{1.11\times1.6\times10^{-19}\ \text{J}}{3.12\times10^{-19}\ \text{J}} = 0.569\% \quad \text{or} \quad 56.9\%. \tag{6.4}$$

From the above examples, it is clear that for a silicon solar cell, none of the photons of the sunlight over 1.12 μm wavelength will produce any electricity. However, photons of sunlight at a wavelength of 1.12 μm may be converted to electricity at a maximum efficiency of 100%, while photons at lower wavelengths will be converted at lower efficiencies. The overall maximum efficiency

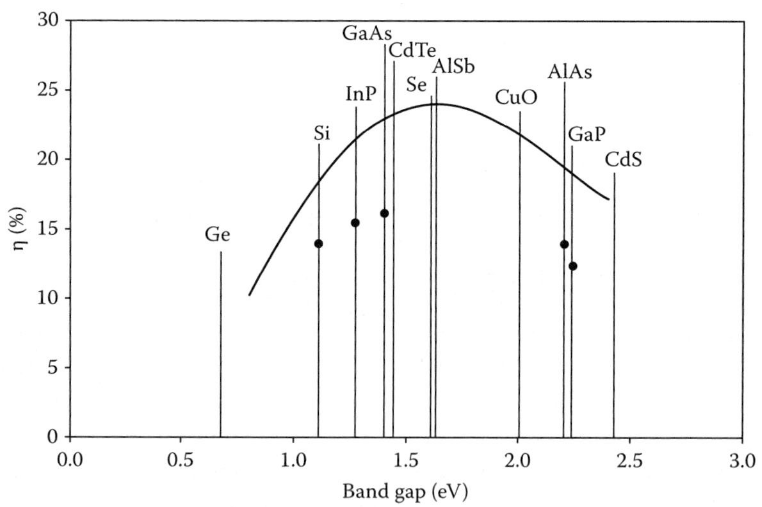

FIGURE 6.6
The maximum solar energy conversion efficiency as a function of the energy gap of the semiconductor. The measured efficiency for various materials is denoted by a solid square energy gap of the material. The curve has been calculated for an ideal junction outside the atmosphere.

of a cell can be found by integrating the efficiency at each wavelength over the entire solar spectrum:

$$\eta = \frac{\int \eta_\lambda I_\lambda d\lambda}{\int I_\lambda d\lambda} \tag{6.5}$$

In addition, other factors such as probability of electron–hole recombination reduce the theoretical maximum achievable efficiency of a solar cell. Figure 6.6 shows a comparison of the maximum energy conversion of cells using different materials.

Figure 6.6 shows that the optimum band gap for terrestrial solar cells is of the order of 1.5 eV.

6.2 Analysis of Photovoltaic Cells

This section presents an electrical analysis of photovoltaic cells, which will be useful in the design of photovoltaic devices for various applications. The physics leading to the expressions for the number density of electrons and holes in n and p materials at a temperature T will not be presented here. For such details, the reader is referred to books such as Refs. [2,3]. It would suffice to point out here that at the p–n junction a current is generated called the *junction current*. The junction current J_j is the net current due to the J_o from the p-side to the n-side (called the *dark current* or the *reverse saturation current*) and *a light-induced recombination current* J_r from the n-side to the p-side. Based on the temperature T, a certain number of electrons in the p material exist in the conduction band. These electrons can easily move to the n-side to fill the holes created at the p–n junction, generating a current J_o. Normally, the electrons occupying the conduction band due to the temperature in the n material do not have enough potential energy to cross the p–n junction to the p-side. However, if a forward-bias voltage V is applied, which in a photovoltaic cell is due to the action of the photons of light, some of the electrons thus generated have enough energy to cross over and recombine with the holes in the p region. This gives rise to a light-induced recombination current J_r, which is proportional to J_o and is given by

$$J_r = J_o \exp\left(\frac{e_o V}{kT}\right). \tag{6.6}$$

where

e_o = charge of an electron
  = 1.602×10^{-19} C or J/V, place on line above
k = Boltzmann's constant = 1.381×10^{-23} J/K

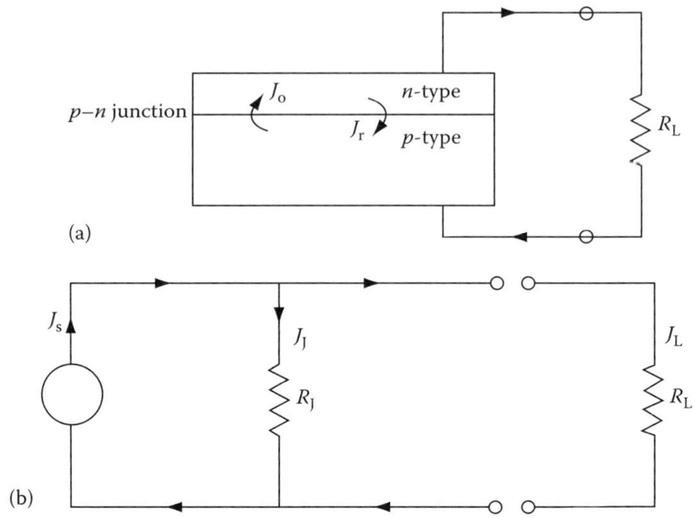

FIGURE 6.7
Equivalent circuit of a photovoltaic cell. (a) *p–n* junction. (b) Electrical circuit.

The junction current J_j is the net current due to J_r and J_o.

$$J_j = J_r - J_o = J_o\left[\exp\left(\frac{e_o V}{kT}\right) - 1\right]. \tag{6.7}$$

Referring to Figure 6.7, it is clear that the current generated in the cell has two parallel paths, one through the junction and the other through the external resistance, R_L. Figure 6.7 shows an equivalent circuit of a photovoltaic cell. It must be pointed out here that the current generated in a photovoltaic cell, including the junction current, is proportional to the area of the cell. Therefore, it is appropriate to analyze in terms of the current density J (current per unit area) instead of the current I. The relationship between the two is

$$I = J \cdot A. \tag{6.8}$$

Referring to Figure 6.7, we can write

$$\begin{aligned} J_L &= J_s - J_j \\ &= J_s - J_o\left[\exp\left(\frac{e_o V}{kT}\right) - 1\right], \end{aligned} \tag{6.9}$$

where J_s is the short circuit current.

For short circuit, $V = 0$ and $J_L = J_s$.

For open circuit, $J_L = 0$ and $V = V_{oc}$,

which gives

$$0 = J_s - J_o\left[\exp\left(\frac{e_o V_{oc}}{kT}\right) - 1\right], \tag{6.10}$$

or

$$V_{oc} = \frac{kT}{e_o}\ln\left(\frac{J_s}{J_o} + 1\right).$$

Figure 6.8 shows a typical performance curve ($I - V$) of a solar cell. The power output is the product of the load current and voltage, and is a function of the load resistance

$$\begin{aligned} P_L &= AJ_L V = I_L V \\ &= I_L^2 R_L, \end{aligned} \tag{6.11}$$

where A is the area of the cell.

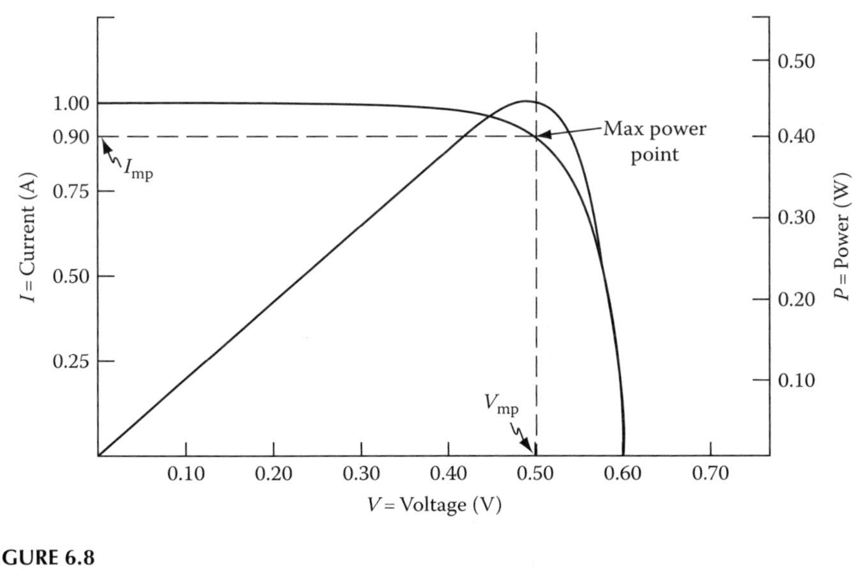

FIGURE 6.8
Typical current, voltage, and power characteristics of a solar cell.

The power output exhibits a maximum. To find the condition for the maximum power output (P_{max}), differentiate P with respect to V and equate it to zero:

$$\exp\left(\frac{e_o V_m}{kT}\right)\left(1+\frac{e_o V_m}{kT}\right)=1+\frac{J_s}{J_o}, \tag{6.12}$$

where V_m stands for voltage at maximum power. The current at max power condition $J_{L,m}$ and the maximum power P_{max} can be found from Equations 6.9 and 6.11, respectively.

$$J_{L,m} = J_s - J_o\left[\exp\left(\frac{e_o V_m}{kT}\right)-1\right]. \tag{6.13}$$

Combining Equations 6.12 and 6.13, $J_{L,m}$ is found to be

$$J_{L,m} = \frac{e_o V_m/kT}{1+(e_o V_m/kT)}(J_s + J_o), \tag{6.14}$$

$$P_{max} = \frac{e_o V_m^2/kT}{1+(e_o V_m/kT)}(J_s + J_o)\cdot A. \tag{6.15}$$

Figure 6.9 shows the effect of illumination intensity and the load resistance on the performance of a silicon cell. Temperature also affects the performance in such a way that the voltage and thus the power output decreases with increasing temperature.

Example 6.3

The dark current density for a silicon solar cell at 40°C is 1.8×10^{-8} A/m² and the short circuit current density is 200 A/m². Calculate

a. Open circuit voltage
b. Voltage at maximum power
c. Current density at maximum power
d. Maximum power
e. Maximum efficiency
f. The cell area required for an output of 25 W when exposed to solar radiation of 900 W/m²

Solution
Given:

$$J_o = 1.8\times10^{-8}\ \text{A/m}^2$$

$$J_s = 200\ \text{A/m}^2,$$

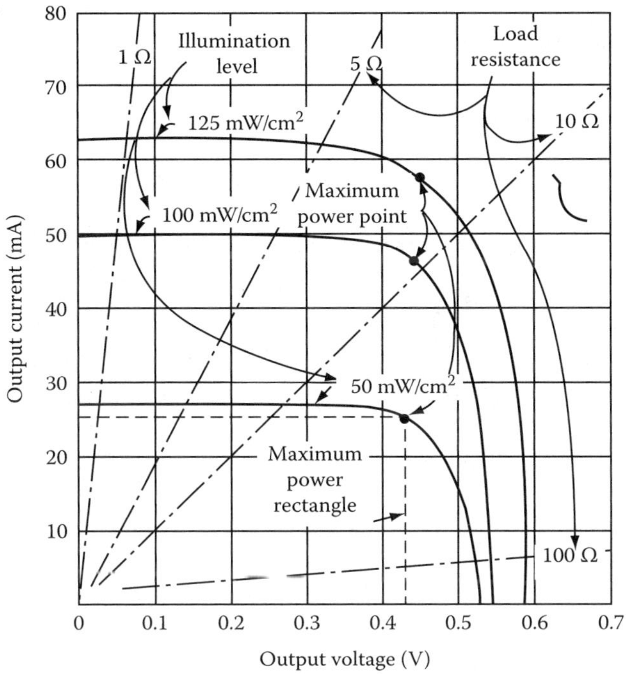

FIGURE 6.9
Typical current–voltage characteristics of a silicon cell showing the effects of illumination level and load resistance.

and

$$T = 40°\text{C} = 313\ \text{K}.$$

a. Using Equation 6.10,

$$V_{oc} = \frac{kT}{e_o}\ln\left(\frac{J_s}{J_o}+1\right).$$

Since (e_o/kT) will be needed for other parts of the problem solution, it will be evaluated separately:

$$\frac{e_o}{kT} = \frac{1.602\times10^{-19}\ \text{J/V}}{(1.381\times10^{-23}\ \text{J/K})(313\ \text{K})} = 37.06\ \text{V}^{-1}.$$

Therefore,

$$V_{oc} = \frac{1}{37.06}\ln\left(\frac{200}{1.8\times10^{-8}}+1\right) = 0.624\ \text{V}.$$

b. Voltage at maximum power condition can be found from Equation 6.12 by an iterative or trial-and-error solution:

$$\exp(37.06V_m)(1+37.06V_m) = 1+\frac{200}{1.8\times10^{-8}}$$

$$\text{or} \quad V_m - 0.542\ \text{V}.$$

c. Current density at maximum power can be found from Equation 6.14:

$$J_{L,m} = \frac{e_o V_m/kT}{1+(e_o V_m/kT)}(J_s + J_o)$$

$$= \frac{(37.06)\cdot(0.542)}{1+(37.06)\cdot(0.542)}(200+1.8\times10^{-8})\ \text{A/m}^2$$

$$= 190.5\ \text{A/m}^2.$$

d.

$$P_{max} = V_m \cdot J_m \cdot A$$

$$\frac{P_{max}}{A} = (0.542V)\cdot(190.5\ \text{A/m}^2)$$

$$= 103.25\ \text{W/m}^2.$$

e.

$$\eta_{max} = \frac{103.25\ \text{W/m}^2}{900\ \text{W/m}^2} = 11.5\%.$$

f. Cell area required:

$$A = \frac{P_{out}}{P_{max}/A} = \frac{25\ \text{W}}{103.25\ \text{W/m}^2}$$

$$= 24.2\ \text{cm}^2.$$

6.2.1 Efficiency of Solar Cells

Theoretical limitation on the efficiency of a single-layer solar cell can be calculated using Equation 6.5. These efficiency limitations and the practical efficiencies of some of the cells are shown in Figure 6.6. Some of the reasons for the actual efficiency being lower than the theoretical limitation are

1. Reflection of light from the surface of the cell. This can be minimized by antireflection (AR) coating. For example, AR coatings can reduce the reflection from a Si cell to 3% from more than 30% from an untreated cell.

2. Shading of the cell due to current collecting electrical contacts. This can be minimized by reducing the area of the contacts and/or making them transparent; however, both of these methods will increase the resistance of the cell to current flow.
3. Internal electrical resistance of the cell.
4. Recombination of electrons and holes before they can contribute to the current. This effect can be reduced in polycrystalline and amorphous cells by using hydrogen alloys.

6.2.2 Multijunction Solar Cells

The limits imposed on single-layer solar cells due to band gap can be partially overcome by using multiple layers of solar cells stacked on top of each other, with each layer having a band gap higher than the layer below it. For example (Figure 6.10), if the top layer is made from a cell of material A (band gap corresponding to λ_A), solar radiation with wavelengths less than λ_A would be absorbed to give an output equal to the hatched area A. The solar radiation with wavelength greater than λ_A would pass through A and be converted by the bottom layer cell B (band gap corresponding to λ_B) to give an output equal to the hatched area B. The total output and therefore the efficiency of this tandem cell would be higher than the output and the efficiency of each single cell individually. The efficiency would increase with the number of layers. For this concept to work, each layer must be as thin as possible, which puts a very difficult if not an insurmountable constraint on

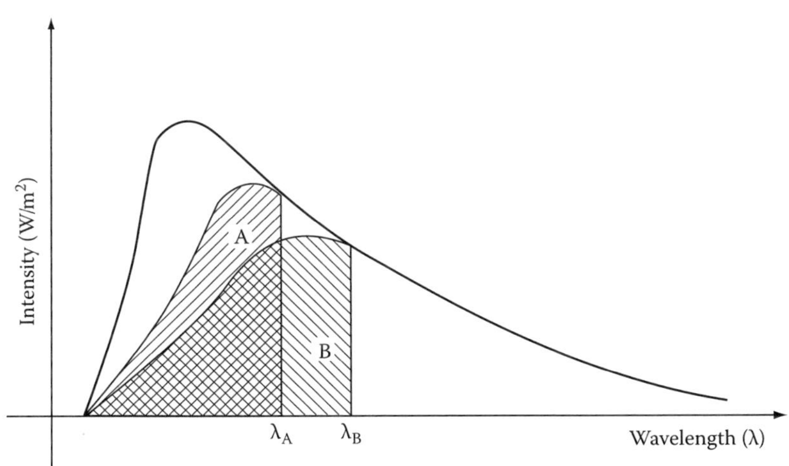

FIGURE 6.10
Energy conversion from a two-layered stacked cell. (From Goswami, Y. et al., *Principles of Solar Engineering*, 2nd edn., Taylor & Francis, Philadelphia, 2000.)

crystalline and polycrystalline cells to be made multijunction. As a result, this concept is being developed mainly for other materials, including thin-film amorphous solar cells.

6.2.3 Design of a Photovoltaic System

Solar cells may be connected in series, parallel, or both to obtain the required voltage and current. When similar cells or devices are connected in series, the output voltages and current are as shown in Figure 6.11. A parallel connection results in the addition of currents, as shown in Figure 6.12. If the cells or devices 1 and 2 have dissimilar characteristics, the output characteristics will be as shown in Figure 6.13. Cells are connected to form modules, modules are connected to form panels, and panels are connected to form arrays. Principles shown in Figures 6.11 and 6.12 apply to all of these connections.

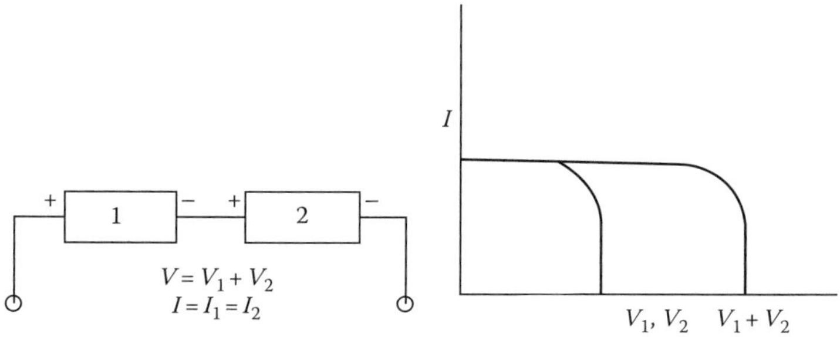

FIGURE 6.11
Characteristics of two similar cells connected in series.

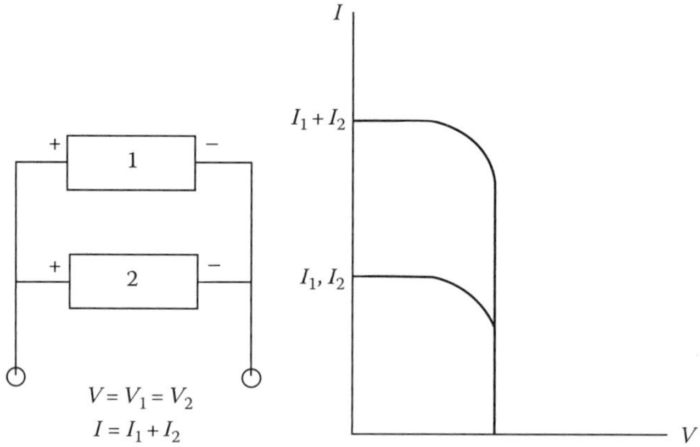

FIGURE 6.12
Characteristics of two similar cells connected in parallel.

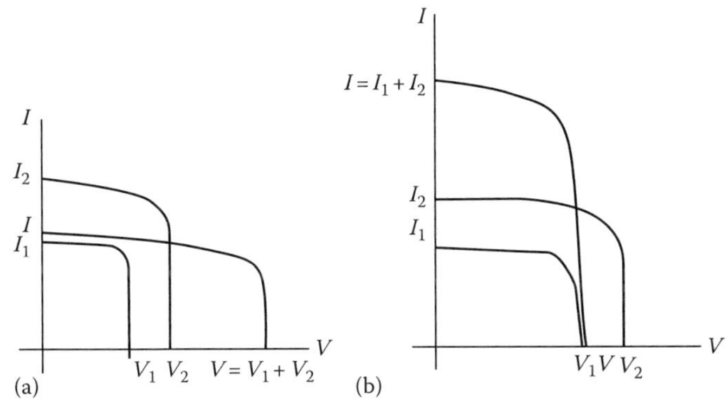

FIGURE 6.13
Characteristics of two dissimilar cells connected in (a) series and (b) parallel.

Example 6.4

An application requires 300 W at 28 V. Design a PV panel using solar cells from Example 6.3 each with an area of 6 cm².

Solution
Assuming that the cell will be operated at maximum power conditions, the voltage and current from each cell are

$$V_m = 0.542\ \text{V},\quad I_m = \left(190.5\ \frac{\text{A}}{\text{m}^2}\right)(6\times 10^{-4}\ \text{m}^2)$$
$$= 0.1143\ \text{A},$$

$$\text{Power/cell} = 0.542 \times 0.1143 = 0.062\ \text{W},$$

$$\text{Number of cells required} = \frac{300\ \text{W}}{0.062\ \text{W/cell}} = 4840,$$

$$\text{Number of cells in series} = \frac{\text{System voltage}}{\text{Voltage/cell}} = \frac{28\ \text{V}}{0.542\ \text{V}} = 52,$$

$$\text{Number of rows of 52 cells connected in parallel} = \frac{4840}{52} \approx 93.1.$$

Since the number of rows must be a whole number, we may increase the number to 94 rows, which will give 303 W output (Figure 6.14).

A blocking diode is used in series with a module or an array to prevent the current from flowing backward, for example, from the battery to the cells under dark conditions. A bypass diode is used in parallel with a module in an array to bypass the module if it is shaded. A photovoltaic system may be connected to a DC or an AC load, as shown in Figure 6.15.

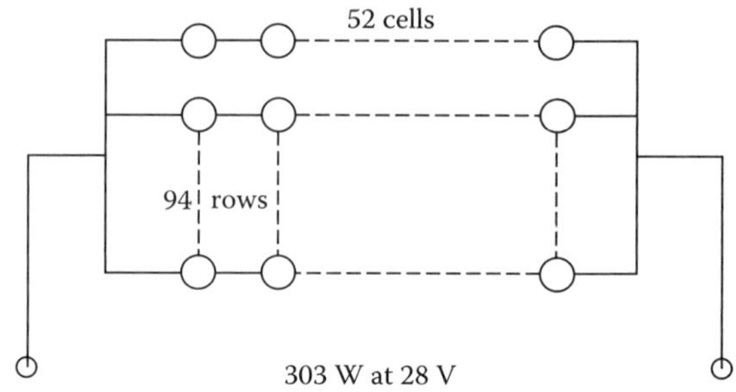

FIGURE 6.14
Connection of cells in rows and columns for Problem 6.4.

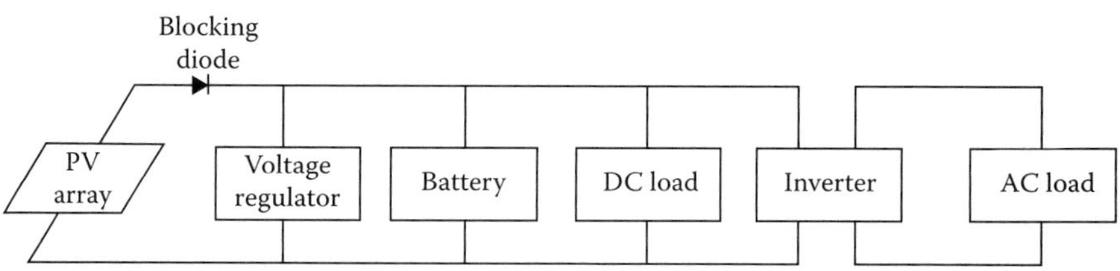

FIGURE 6.15
Schematic of a photovoltaic system.

6.3 Manufacture of Solar Cells and Panels

Manufacture of crystalline silicon solar cells is an outgrowth of the manufacturing methods used for microprocessors. A major difference is that silicon used in microprocessors is ultra pure, which is not needed for photovoltaic cells. Therefore, a major source of feedstock for silicon solar cells has been the waste material from the microelectronics industry. Solar cells are also manufactured as polycrystalline and thin films. Below are some of the common methods of manufacture of silicon solar cells.

6.3.1 Single Crystal and Polycrystalline Cells

Single crystal silicon cells are produced by a series of processes: (1) growing crystalline ingots of *p*-silicon, (2) slicing wafers from the ingots, (3) polishing and cleaning the surface, (4) doping with *n* material to form the *p*–*n* junction, (5) deposition of electrical contacts, (6) application of AR coating, and (7) encapsulation. Figure 6.16 illustrates the process.

Czochralski method (Figure 6.17a) is the most common method of growing single-crystal ingots. A seed crystal is dipped in molten silicon doped with a *p*-material (boron) and drawn upward under tightly controlled conditions of

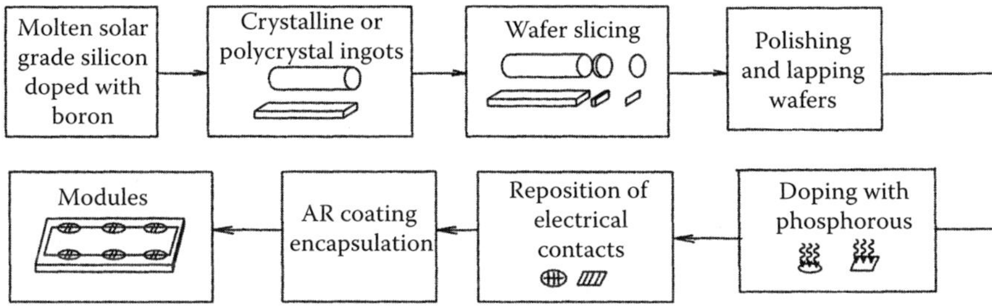

FIGURE 6.16
Series of processes for the manufacture of crystalline L polycrystalline cells.

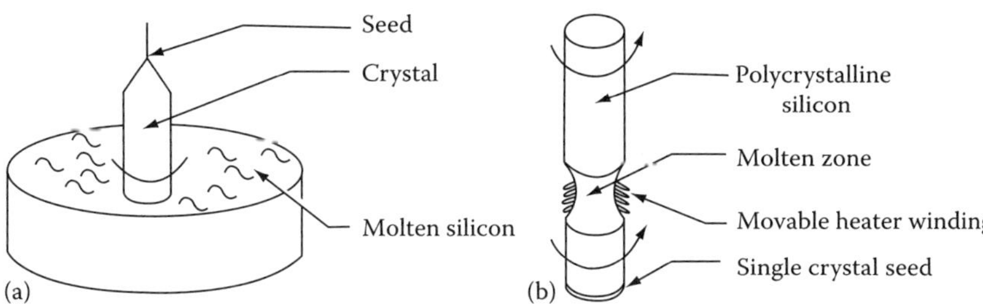

FIGURE 6.17
Crystalline silicon ingot production methods. (a) Czochralski method. (b) Float zone method.

linear and rotational speed, and temperature. This process produces cylindrical ingots of typically 10 cm diameter, although ingots of 20 cm diameter and more than 1 m long can be produced for other applications. An alternative method is called the *float zone* method (Figure 6.17b). In this method, a polycrystalline ingot is placed on top of a seed crystal and the interface is melted by a heating coil around it. The ingot is moved linearly and rotationally, under controlled conditions. This process has the potential to reduce the cell cost.

Polycrystalline ingots are produced by casting silicon in a mold of preferred shape (rectangular), as shown in Figure 6.18. Molten silicon is cooled slowly in a mold along one direction in order to orient the crystal structures and grain boundaries in a preferred direction. In order to achieve efficiencies of greater than 10%, grain sizes greater than 0.5 mm are needed and the grain boundaries must be oriented perpendicular to the wafer. Ingots as large as 400 × 40 × 40 can be produced by this method.

Ingots are sliced into wafers by internal diameter (ID) saws or multiwire saws impregnated with diamond abrasive particles. Both of these methods result in high wastage of valuable crystalline silicon.

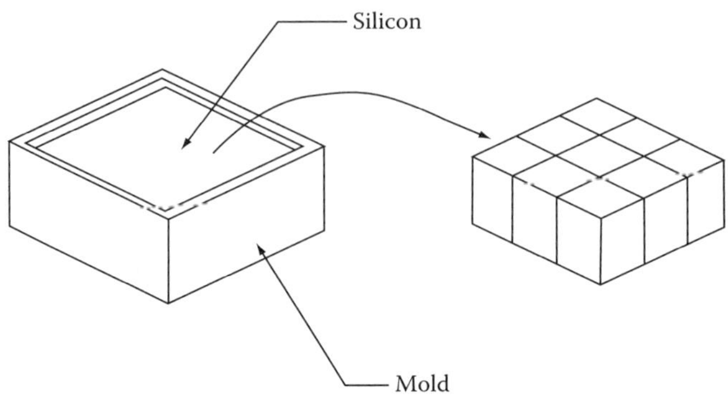

FIGURE 6.18
Polycrystalline ingot production.

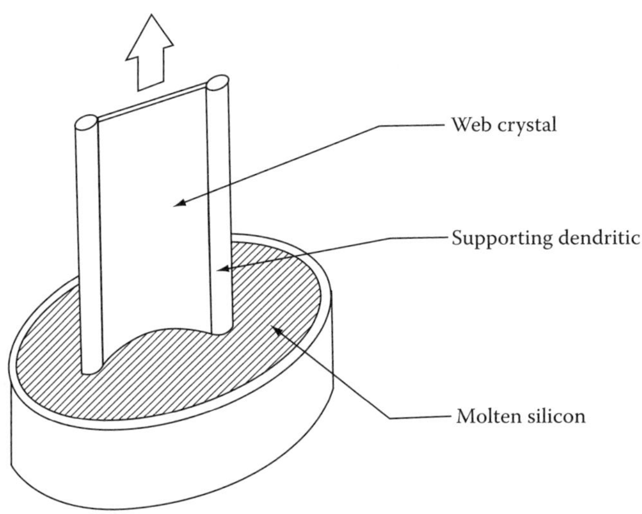

FIGURE 6.19
Thin-film production by dendritic web growth.

Alternative methods that reduce wastage are those that grow polycrystalline *thin films*. Some of the thin-film production methods include dendritic web growth (Figure 6.19), edge-defined film-fed growth (EFG) (Figure 6.20), ribbon against drop (RAD) method, supported web method, and ramp-assisted foil casting technique (RAFT) (Figure 6.21).

A *p–n* junction is formed in the cell by diffusing a small amount of *n* material (phosphorous) in the top layer of a *p*-silicon wafer. The most common

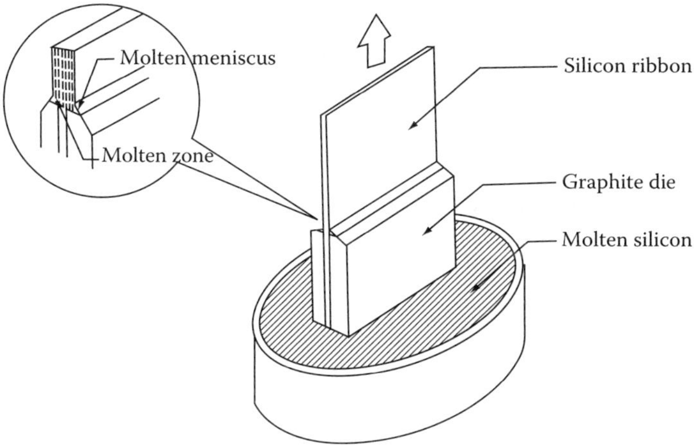

FIGURE 6.20
Thin-film production by edge-defined film-fed growth (EFG).

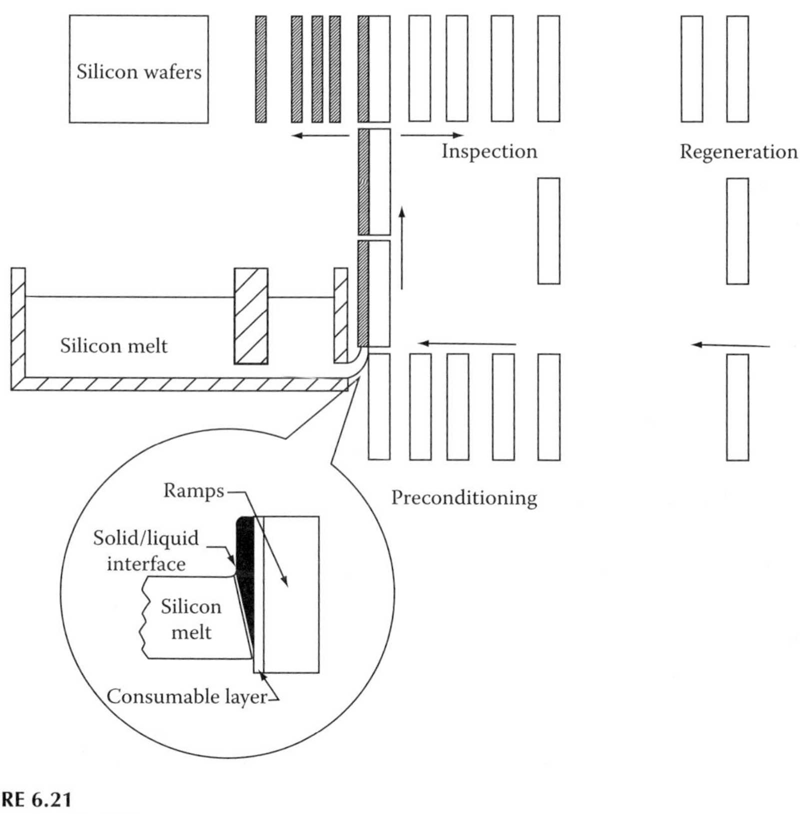

FIGURE 6.21
Schematic of RAFT processing.

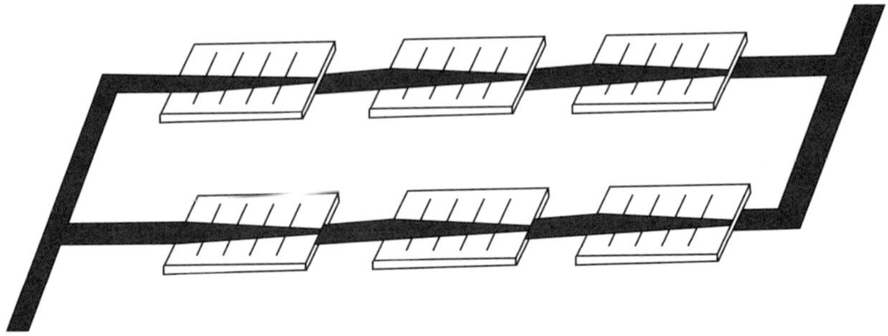

FIGURE 6.22
Assembly of solar cells to form a module.

method is diffusion of phosphorous in the vapor phase. In this case, the backside of the wafer must be covered to prevent the diffusion of vapors from that side. An alternate method is to deposit a solid layer of the dopant material on the top surfaces followed by high temperature (800°–900°) diffusion.

Electrical contacts are attached to the top surface of the cell in a grid pattern to cover no more than 10% of the cell surface and a solid metallic sheet is attached to the back surface. The front grid pattern is made by either vacuum metal vapor deposition through a mask or by screen printing. Figure 6.22 shows how cells are connected to form modules.

AR coatings of materials such as silicon dioxide (SiO_2), titanium dioxide (TiO_2), and tantalum pentaoxide (Ta_2O_5) are deposited on the cell surface to reduce reflection from more than 30% for untreated Si to less than 3%. AR coatings are deposited by vacuum vapor deposition, sputtering, or chemical spraying. Finally, the cells are encapsulated in a transparent material to protect them from the environment. Encapsulants usually consist of a layer of either polyvinyl butyryl (PVB) or ethylene vinyl acetate (EVA) and a top layer of low iron glass.

6.3.2 Amorphous Silicon and Multijunction Thin-Film Fabrication

Amorphous silicon (*a*-Si) cells are made as thin films of *a* Si:H alloy doped with phosphorous and boron to make *n* and *p* layers, respectively. The atomic structure of an *a*-Si cell does not have any preferred orientation. The cells are manufactured by depositing a thin layer of *a*-Si on a substrate (glass, metal, or plastic) from glow discharge, sputtering, or chemical vapor deposition (CVD) methods. The most common method is by an RF glow discharge decomposition of silane (SiH_4) on a substrate heated to a temperature of 200°C–300°C. To produce *p*-silicon, diborane (B_2H_6) vapor is introduced with the silane vapor. Similarly, phosphene (PH_3) is used to produce *n*-Silicon. The cell consists of an *n*-layer, an intermediate undoped *a*-Si layer, and a *p*-layer on a substrate. The cell thickness is about 1 μm. The manufacturing

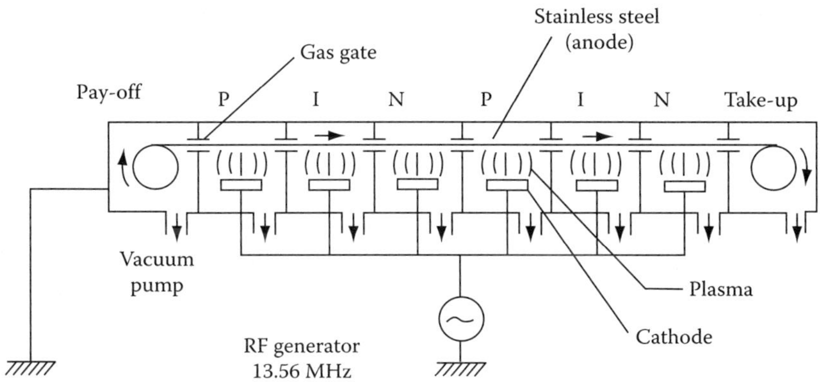

FIGURE 6.23
A schematic diagram of a roll-to-roll plasma CVD machine.

process can be automated to produce rolls of solar cells from rolls of substrate. Figure 6.23 shows an example of roll-to-roll *a*-Si cell manufacturing equipment using a plasma CVD method. This machine can be used to make multifunction or tandem cells by introducing the appropriate materials at different points in the machine.

6.4 Design for Remote Photovoltaic Applications

Photovoltaic power may be ideal for a remote application requiring a few watts to hundreds of kilowatts of electrical power. Even where a conventional electrical grid is available, for some applications, where uninterruptible or emergency standby power is necessary, photovoltaic power would be appropriate. Some examples of remote PV applications include water pumping for potable water supply and irrigation, power for remote houses, street lighting, battery charging, telephone and radio communication relay stations, and weather stations. Examples of some other applications include, electrical utility switching stations, peak electrical utility power where environmental quality is a concern, data acquisition systems, and speciality applications such as ventilation fans and vaccine refrigeration.

The design of a PV system is based on some basic considerations for the application:

1. Which is more important, the daily energy output or the power (average or peak)?
2. Is a backup energy source needed and/or available?

3. Is energy storage important? What type—battery, pumped water, etc.?
4. Is the power needed as AC or DC? What voltage?

There are three basic steps in the design of a PV system:

1. Estimation of load and load profile
2. Estimation of available solar radiation
3. Design of PV system, including area of PV panels, selection of other components, and electrical system schematic

Each of these steps will be explained in the following examples. These examples are based on Refs. [1,5].

6.4.1 Estimation of Loads and Load Profiles

Precise estimation of loads and their timings (load profile) are important for PV systems since the system is sized as the minimum required to satisfy the demand over a day. For example, if power is needed for five different appliances requiring 200, 300, 500, 1000, and 1500 W, respectively, so that only one appliance is on at any one time and each appliance is on for an average of 1 h a day, the PV system would be sized based on 1500 W peak power and 3500 Wh of daily energy requirement. The multiple loads on a PV system are intentionally staggered to use the smallest possible system, since the capital costs of a PV system are the most important as opposed to the energy costs in a conventional fuel-based system.

Example 6.5: Daily Load Calculations

How much energy per day is used by a remote weather station given the following load characteristics?

Load	Load Power (W)	Run Time (h/Day)
Charge controller	2.0	8
Data gathering	4.0	3
Modem (standby)	1.5	22.5
Modem (send/receive)	30.0	1.5

Solution

$$\begin{aligned}\text{Daily energy} &= (2.0\ \text{W})(8\ \text{h}) + (4.0\ \text{W})(3\ \text{h}) + (1.5\ \text{W})(22.5\ \text{h}) + (30\ \text{W})(2.5\ \text{h})\\ &= 106.75\ \text{Wh}\end{aligned}$$

Daily energy use is about 107 Wh per day.

Example 6.6: Load Calculations

An owner of a remote cabin wants to install a PV power system. The loads in the home are described below. Assume that all lights and electronics are powered by AC. Find the daily and weekly peak and average energy use estimates. The system used is a 24 V DC system with an inverter.

Lights	4, 23 W compact fluorescent bulbs	On at night for 5 h
Lights	6, 13 W compact fluorescent bulbs	2 h each (daytime)
Stereo	110 W (amplifier), 15 W (other)	On for 8 h per week
Water pump	55 W (3.75 A start current)	Runs for 2 h per day
Computer	250 W (monitor included)	On for 1½ h daily (weekend nights only)
Bathroom fan	40 W (3.5 A start current)	On for 1 h per day
Microwave	550 W (AC)–1000 W surge	On for 30 min per day

Solution

Loads need to be broken down according to (1) run time, (2) peak power, (3) night or day use, and (4) AC or DC loads. The load profile is as follows:

Load Name Description	Power (W)		Run Time (h)		Energy (Wh)	
	Average	Peak	Day	Week	Day	Week
Lights (AC)	(4) (23)	(4) (23)	5.0	35	460	3220
Lights (AC)	(6) (13)	(6) (13)	2.0	14	156	1092
Stereo (AC)	(1) (110)	(1) (110)	—	8	—	880
Pump (DC)	(1) (55)	(3.75 A) (24 V)	2.0	14	110	770
Computer (AC)	(1) (250)	(1) (250)	1.5	3	—	750
Fan (DC)	(1) (40)	(3.5 A) (24 V)	1.0	7	40	280
Microwave (AC)	(1) (550)	(1) (1000)	0.5	3.5	275	1925

Average DC load: [770 + 280]/7 = 150 Wh/day.
Average AC load: [3220 + 1092 + 880 + 750 + 1925]/7 = 1124 Wh/day.
Peak DC load: max [{(3.5)(24) + 40}:: {(3.75)(24) + 55}] = 145 W.*
Peak AC load: (1000) + max [(4)(23):: (6)(13)] + 250 + 100 = 1442 W.†

6.4.2 Estimation of Available Solar Radiation

Methods of estimation of available solar radiation are described in Chapter 5. If long-term measured solar radiation values are available at a location, Equations 5.50 through 5.61 can be used to estimate the monthly

* It can be assumed that the pump and fan will not start precisely at the same instant.
† It is assumed that the night and day lighting loads will not be on simultaneously.

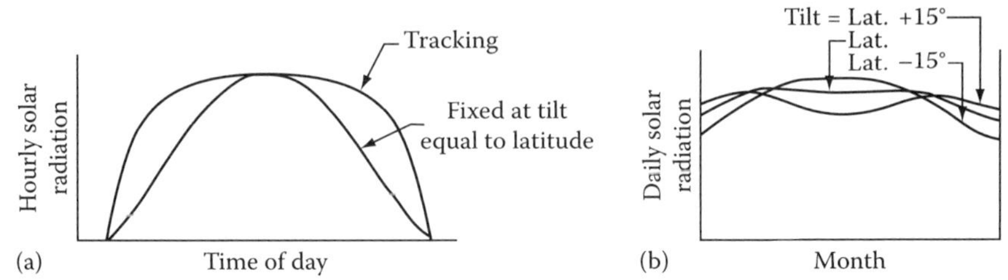

FIGURE 6.24
Solar radiation on panels at different tilt angles. (a) Hourly. (b) Monthly.

average solar radiation per day. Otherwise, clear day data can be used along with the percent sunshine data to calculate the radiation. In addition, appropriate Web sites such as http://www.builditsolar.com/Tools/RadOnCol/radoncol.htm or http://rredc.nrel.gov/solar/old_data/nsrdb/redbook/atlas/ are useful resources for very accurate approximations. For designing a PV system, a decision is made whether the PV panel will be operated as tracking the sun or will be fixed at a certain tilt and azimuth angle. For fixed panels, a tilt angle of latitude +15° works best for winter and latitude −15° for summer. To keep the panel fixed year round, an angle equal to the latitude provides the maximum yearly energy (see Figure 6.24).

6.4.3 PV System Sizing

If meeting the load at all times is not critical, PV systems are usually sized based on the average values of energy and power needed, available solar radiation, and component efficiencies. This is known as the *heuristic approach*. It is important to note that a system designed by this approach will not give the best design but may provide a good start for a detailed design. A detailed design accounts for the changes in the efficiencies of the components depending on the load and the solar radiation availability and whether the system is operating in a PV-to-load, PV-to-storage, or storage-to-load mode.

Example 6.7: Heuristic Approach to PV-System Sizing

A PV system using 50 W, 12 V panels with 6 V, 125 Ah batteries is needed to power a home in Farmington, New Mexico, with a daily load of 1700 Wh. System voltage is 24 V.

There are an average of 5 daylight hours in the winter. Specify the collector and storage values for the system using the heuristic approach.

Solution

Load = 1700 Wh/day

Daylight hours = 5 h/day

Average panel output = 50 W

$$\text{Number of panels} = \frac{1700\,\text{Wh/Day}}{(5\,\text{h/day})(50\,\text{W/panel})} = 6.8\text{, round off to 7 panels}$$

Since the system voltage is 24 V, but each panel produces only 12 V, an even number of panels will be needed. Therefore number of panels = 8.

Farmington, New Mexico, is a very sunny location, so 3 days of storage are sufficient. Assuming a battery efficiency of 75% and a maximum depth of discharge of 70%,

$$\text{Storage} = (1700)(3)/(0.75 \times 0.7) = 9714\,\text{Wh},$$

$$\text{Number of batteries} = (9714/\text{Wh})/(125\,\text{Ah} \times 6\,\text{V}) = 13 \text{ (rounded off to next whole number).}$$

Since the system voltage is 24 V, and each battery provides 6 V, the number of batteries is increased to 16. In a detailed design, the efficiencies of battery storage, inverter, and the balance of system (BOS) must be accounted for. The following example shows how these efficiencies increase the energy requirements of the PV panel.

Example 6.8: System Operating Efficiency

Using the cabin electrical system from Example 6.6, calculate the overall system efficiency for each operating mode possible for the system. Estimate the amount of energy required per day for the system. When load timing (day or night), assume half of the load runs during the day and half runs at night. The inverter used has a component efficiency of 91%, the battery efficiency is 76%, and the distribution system efficiency is 96%.

Solution

From the example, the loads are

Average DC load: 150 Wh/day,
Average AC load: 1124 Wh/day.

The various system efficiencies are

PV to load (DC):	0.96 (day, DC),
Battery to load (DC):	(0.76) (0.96) = 0.73 (night, DC),
PV to load (AC):	(0.96) (0.91) = 0.874 (day, AC),
Battery to load (AC):	(0.76) (0.91) (0.96) = 0.664 (night, AC).

Expected day and night loads are

Day (DC):	(0.5) (110) + (0.5) (40) = 75 Wh/day,
Night (DC):	(0.5) (110) + (0.5) (40) = 75 Wh/day,
Day (AC):	(156) + (0.5) (880 + 750)/7 + (0.5) (275) = 409.9 Wh/day,
Night (AC):	(460) + (0.5) (880 + 750)/7 + (0.5) (275) = 713.9 Wh/day.

Without considering system efficiency, the daily energy requirement is

$$E_{\text{day}} = (150) + (1124) = 1274 \text{ Wh/day}.$$

The expected daily energy requirement is

$$E_{\text{day}} = \frac{75}{0.96} + \frac{75}{0.73} + \frac{409.9}{0.874} + \frac{713.9}{0.664}$$

$$E_{\text{day}} = 1725 \text{ Wh/day}$$

The actual energy requirement is 35% higher than that obtained in the approximate calculation in the previous example.

6.4.4 Water Pumping Applications

Water pumping for drinking water or irrigation at remote locations is an important application of PV. For a simple schematic shown in Figure 6.25, the power needed to pump water at a volumetric rate $\dot{V}$; is given by $P = \rho \dot{V} g H/\eta_p$ where ρ is the density of water, g is the acceleration due to gravity, and η_p is the pump efficiency. The static head H_s is (A + B). In case the water level is drawn down, the static head would be (A + B + C). The pump must work against the total head H, which includes the dynamic head:

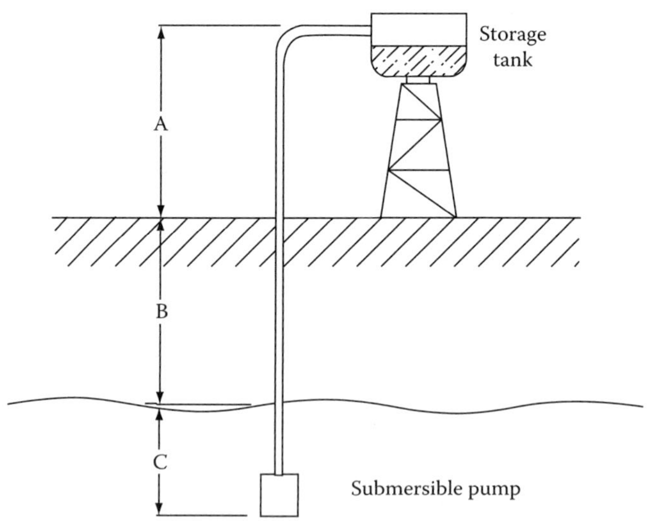

FIGURE 6.25
Water pumping using a submersible pump.

$$H = H_f + \frac{v^2}{2g}$$

where
H_f is the frictional head loss in the pipe and the bends
v is the velocity of the water at the pipe outlet

The pump efficiency η_p is a function of the load (head and flowrate) and is available as a characteristic curve from the manufacturer. There are two basic types of pumps: centrifugal and positive displacement. These pumps can be driven by AC or DC motors. DC motors are preferable for the PV applications, because they can be directly coupled to the PV array output. Centrifugal pumps with submersible motors are the optimum for PV applications because of their efficiency, reliability, and economy. However, for deep wells, Jack pumps may be necessary. Jack pumps are the piston type of positive displacement pumps that move chunks of water with each stroke. They require very large currents; therefore, they are connected through batteries.

6.5 Thin-Film PV Technology

During the past few years, enormous advances have been made in the development of thin-film PV technologies. Although the market share worldwide for thin-film PV in 2006 was less than 6%, the market share of thin-film PV in the United States has grown rapidly and was reported at more than 44% of all photovoltaics in 2006 and expected to rise significantly. As thin-film PV is developing, a number of applications are being pursued commercially, including building integrated photovoltaics (BIPV), rooftop applications, and utility scale applications.

Thin-film PV cells require multiple layers of precisely deposited thin films. An important criteria for the production of thin-film PV in a continuous process, such as a roller coater, is the percentage of the output that has an acceptable efficiency. For all of the thin-film technologies, it is possible to produce a so-called champion cell under laboratory conditions designed to achieve the highest cell efficiency possible (see Figure 6.26). These champion cells are indicative of what may be possible with a given technology, but as produced by laboratories such as NREL, they are not cost effective. However, they represent a goal which mass production facilities aim to achieve. The three primary thin-film PV technologies currently on the market are based on amorphous silicon (*a*-Si), cadmium telluride (CdTe), and CIGS (copper indium gallium diselenide). Each of these technologies has advantages and disadvantages, and the differences between these technologies are described briefly below.

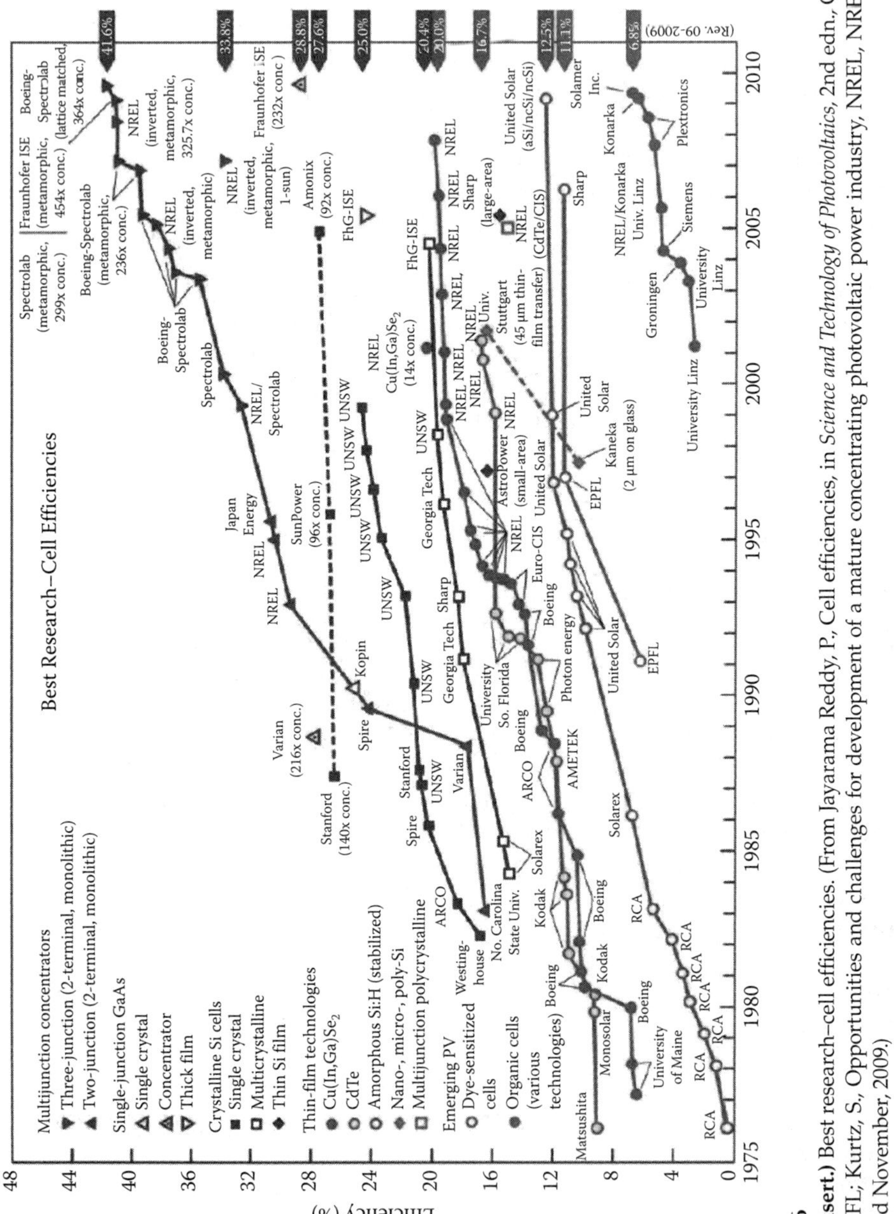

FIGURE 6.26
(See color insert.) Best research–cell efficiencies. (From Jayarama Reddy, P., Cell efficiencies, in *Science and Technology of Photovoltaics*, 2nd edn., CRC Press, Boca Raton, FL; Kurtz, S., Opportunities and challenges for development of a mature concentrating photovoltaic power industry, NREL/TP-520-43208, revised November, 2009.)

Amorphous silicon PV technology utilizes a widely available material, but the cells have a relatively low efficiency. An amorphous silicon cell prepared within a laboratory environment at the National Renewable Energy Laboratory (NREL) achieved an efficiency of 12.3%. However, commercial production efficiencies for amorphous silicon typically only range from 6% to 8%. Furthermore, the capital equipment costs for manufacturing thin-film amorphous silicon are generally higher than for other thin-film technologies.

Cadmium-telluride (CdTe) thin-film PV modules have made impressive advances. As commercial-scale production has been approached, the manufacturing costs have dropped dramatically from \$2.94/W in 2004 to about \$1.25/W in 2009 with manufacturing target costs expected to be less than \$1/W as a result of improvements in productivity, module efficiency, and yield. If this goal is met, CdTe thin-film PV modules would be competitive price-wise with average grid electricity. Colorado State University recently announced a spin-off plan for a manufacturing plant entitled, "Abound Solar," which can manufacture 3 million thin-film photovoltaic solar panels a year. The technology is based on CdTe, which is commonly found in copper/zinc mines. The cheap material coupled with an efficient process allows it to convert sheets of glass into solar panels at a cost of about \$1/W at peak.

CdTe champion cells produced by NREL have achieved efficiencies as high as 16.5%. Commercially the production efficiencies for CdTe PV modules ranges from 9% to 11%. CdTe thin films have the advantages of low material costs and handling requirements associated with the manufacturing process. However, current production processes are limited to glass substrates, which are more limited in their applications than flexible films that can be used in buildings/integrated photovoltaic products, and can often be used to replace conventional building materials in some parts of a building.

An analysis of energy return on energy investment of thin-film CdTe photovoltaic modules has recently been conducted under the order of the European research project, PVACCEPT. In addition to the EROI, the study also investigated the environmental impact of CdTe thin-film modules. The following assumptions were made for the analysis:

- The expected lifetime of modules were assumed to be 20 years.
- The average insolation was assumed to be at 1700 kWh/(m^{2*}/annum).
- A 20% efficiency loss was assumed with respect to nominal values for the modules.
- All waste materials generated in the production were assumed to be recycled and/or safely disposed of.
- Module decommissioning was not included because of a lack of data.

Using an efficiency of 8%, which is guaranteed at current by manufacturers, 10% for CIGS and 14% for poly-Si, which are typical values for the current

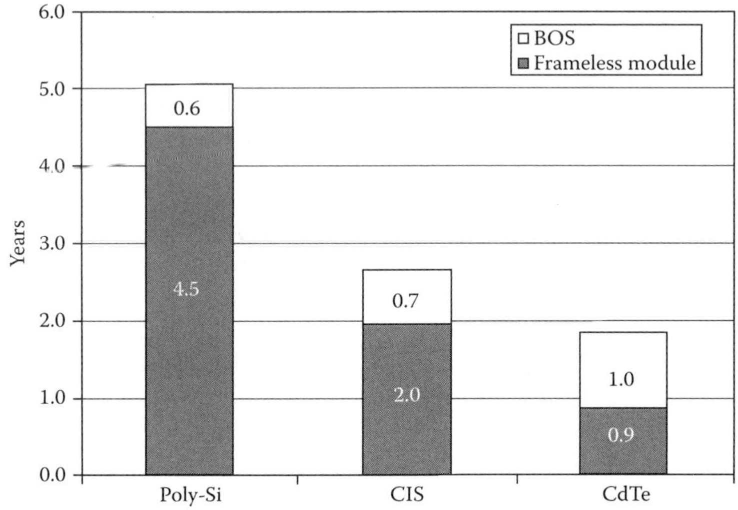

FIGURE 6.27
Payback times for three types of PV with frameless module and balance of systems. (From Raugei, M. et al., Energy and life cycle assessment of thin flim CdTe photovoltaic moduler, accessed at http://www.abound.com/pdf/NREL_PV_Embodied_Energy.pdf)

state of actual production, the payback period for all three types of thin-film modules in years is shown in Figure 6.27.

The shaded area is the payback period for the module and the white area accounts for the balance of systems (BOS). The chart shows that the payback for the CdTe module itself is about 0.9 years, whereas the payback for the entire system is about 1.9 years. This compares to a payback period for the CIS system of 2.5 years and 4 years for the poly-Si system. Assuming a lifetime of 20 years, these payback periods result in EROIs of 10.5 for CdTe, 8 for CIGS, and 4 for poly-Si systems. The CdTe modules also have a very favorable environmental impact, generating only about 32 g of CO_2 equivalent per kWh of electricity produced.

CIGS thin-film photovoltaic products are available worldwide. The technology, however, is still in the development stage and as many as 10 different deposition methods for growing the thin CIGS absorber layers are being pursued. *CIGS PV modules* have the best efficiency characteristics of the three technologies. An NREL champion cell achieved an efficiency of 20%. CIGS technology utilizes less raw material than the other two technologies, and, therefore, achieves lower film thicknesses. Moreover, CIGS films have high light absorption capabilities. This technology appears to combine the highest solar conversion radiation efficiency with low manufacturing costs and the potential of being mass-produced with flexible substrates (see Figure 6.28).

The production of CIGS solar cells requires deposition of molybdenum, copper, indium, gallium, selenide, cadmium sulfide, and a transparent conductive

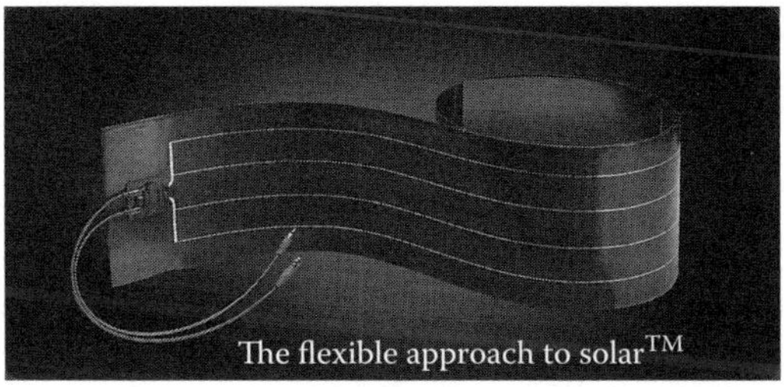

FIGURE 6.28
(See color insert.) Thin film flexible photovoltaic CIGS module. (Courtesy of SoloPower, Inc., San Jose, CA.)

oxide (TCO) onto a substrate. These manufacturing techniques are very demanding in accuracy, and a variety of different processes are under development. All current CIGS processes use sputtering during the molybdenum and TCO deposition steps. But, the other CIGS compounds can be deposited with different technologies, including thermal evaporation, electrochemical plating, nanoparticle printing, or sputtering. The details of these technologies are under active development and their eventual success will determine the cell efficiencies and cost achievable. A brief description of each of the four major deposition technologies is given below.

Sputtering: Sputtering is used extensively in the semiconductor industry to deposit thin films of various materials. The technology is a physical vapor deposition (PVD) method which deposits thin films by ejecting (called "sputtering") material from a "target" that is a source which then deposits onto a substrate such as a silicon wafer. Sputtering is used for high-volume and high-production manufacturing and has been proven effective in industrial-scale processes of coating glass and plasma display production. In this process, ions or reactive gases dislodge atoms from a solid plate of elements, which are deposited on an adjacent substrate. Targets in the PV industry are made up of materials such as CIGS and are attached to a reusable black back-plate installed within the PVD system.

Thermal evaporation: This is the most commonly used CIGS deposition technique. For this process, elements are heated in a vacuum and the resulting condensing vapor transfers these elements onto a substrate. Materials for this process are easy to obtain and use, but the process is often inefficient as far as material utilization is concerned and it is difficult to control for the uniformity of deposition on large areas. Hence, the percent of spoilage can be large, although recent developments may improve this. The thermal evaporation process is also costly.

Electrochemical plating: In this process, a base metal is covered with a thin film of CIGS by submersing the substrate in a chemical solution through which an electric current flows. Although electrochemical plating has a high material utilization, the process may be slow and often results in nonuniform coverage. Hence, costly disposal of waste by-products from the process may be necessary.

Nanoparticle printing: This process is not yet in commercial production and poses considerable challenges for high-production operation. Ink-based technologies that can be deposited by low-cost inkjet or screen printing approaches have the potential for low cost with adequate resolution for PV applications. Using nanoparticulate precursors as the base for the ink may permit a degree of control in the composition process needed to meet electrical and stability requirements. Ink-depositing processes "print" CIGS semiconductor ink on a conductive foil substrate. The inks currently used for the CIGS absorber layer consist of oxides or selenide nanoparticles of the metals copper, indium, and gallium dispersed as a colloidal suspension in a solvent. The nanoparticles are intended to form a solid layer of CIGS. The majority of flexible substrates in use today are metal foils, but using polymer fibers such as polyimide that allow for high-speed roll-to-roll processing are gaining in acceptance.

6.6 Multilayer PV Technology

Advances in photovoltaic energy conversion have recently been summarized by Dan E. Arvizu, the director of the NREL [11]. As noted previously, the efficiency of a single-gap solar cell is limited as a result of its inability to efficiently convert the full range of energy of the photons emanating from the sun. Photons with energy below the band gap of the cell material either pass through the cell or are converted to heat within the material. Photons with energy above the band gap are also lost since only the energy necessary to generate the hole–electron pair is utilized while the remainder merely generates heat. The limitations of a single band gap solar cell can be overcome, however, by utilizing multiple junctions with several band gaps that can utilize a much broader range of photon energies from the sun. In the construction of multilayer cells, the layers are optically in series with the highest band gap material at the top. The first junction receives all of the spectrum where photons above the band gap of the first junction are absorbed. Photons below the band gap of the first layer pass through it to the second layer and are absorbed there. This is repeated for the two layers and the third layer then absorbs as much of the remaining photons as possible. As many as five layers seem feasible.

Multijunction photovoltaic cells use several layers of epitaxy deposited films. Epitaxy is the method of depositing a monocrystalline film on a monocrystalline substrate with the deposited film called the epitaxial layer. Triple junction high-efficiency solar cells for terrestrial photovoltaic applications have been commercialized by a number of companies, including EMCORE. The cells are currently much more expensive than single band gap photovoltaic cells, but they can be combined with various types of concentrating devices that can capture solar radiation over a large area and then redirect it onto a small, high-efficiency solar cell. Figure 6.29 shows such a concentrating scheme with solar energy passing through a Fresnel lens that directs the

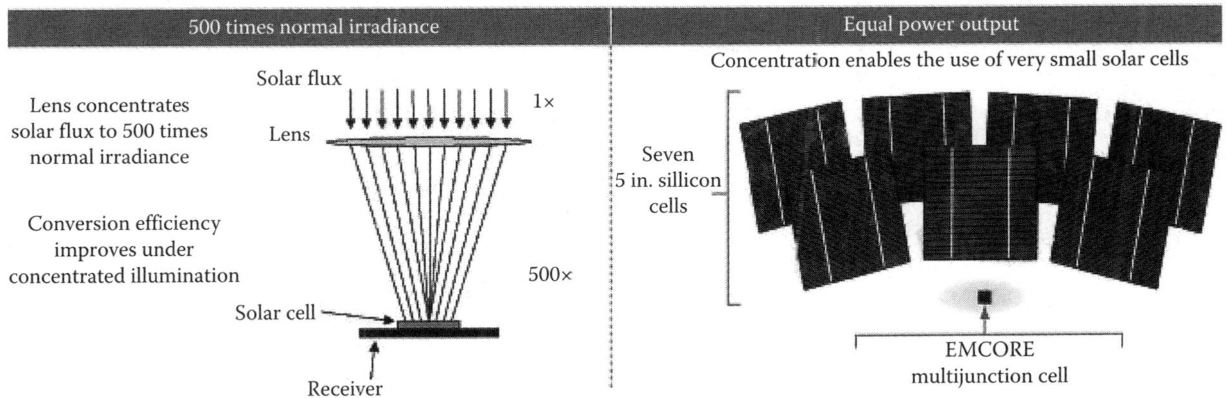

FIGURE 6.29
Schematic of multijunction solar cell with Fresnel concentrating lens arrangement. (Courtesy of EMCORE, http://www.emcore.com/solar_photovoltaics/terrestrial_concentrator_photovoltaic_arrays. With permission.)

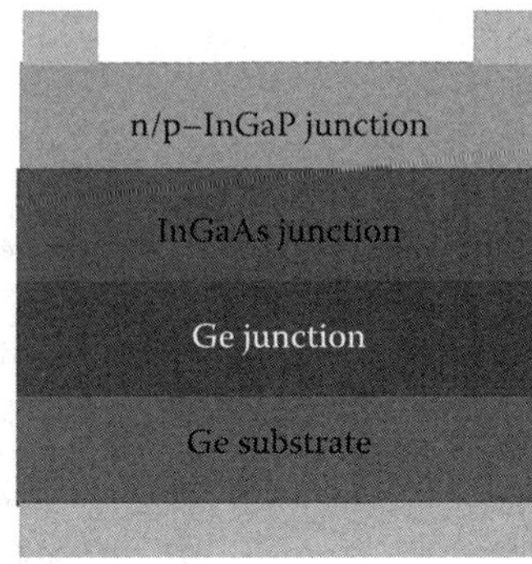

FIGURE 6.30
Physical cell structure. (Courtesy of EMCORE, http://www.emcore.com/solar_photovoltaics/terrestrial_concentrator_photovoltaic_arrays. With permission.)

radiation to a small, three-layer cell. A typical example of a three-layer cell is the EMCORE T1000 cell that has achieved an efficiency better than 39% of a solar radiation concentration ratio of 503.

The physical cell structure is shown in Figure 6.30. It consists of an outer layer of antireflective coating that provides a low reflectance over wavelength in the range from 0.3 to 1.8 μm. The next three layers are an indium gallium phosphorous junction, followed by an indium gallium arsenide layer, and a Ge junction on a Ge substrate.

The quantum efficiency of this type of three-layer cell is shown in Figure 6.31 as a function of wavelength and the overall efficiency is shown as a function of concentration in Figure 6.32.

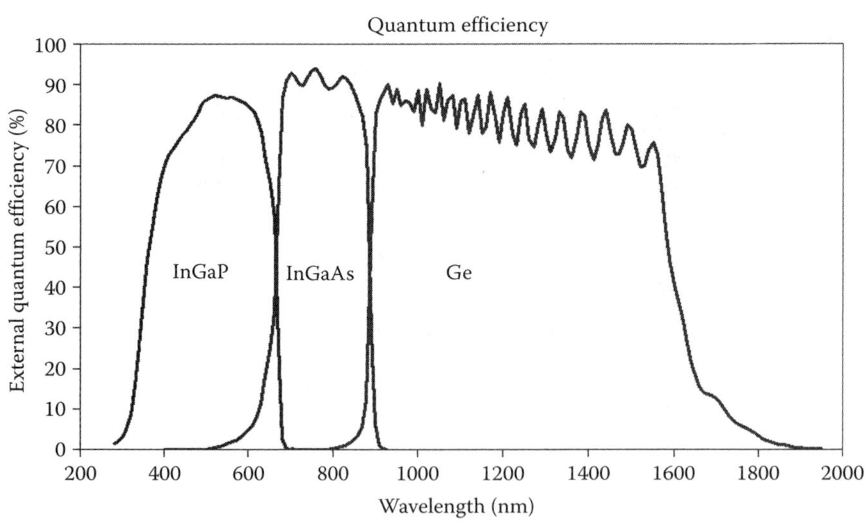

FIGURE 6.31
Quantum efficiency as a function of wavelength. (Courtesy of EMCORE, http://www.emcore.com/solar_photovoltaics/terrestrial_concentrator_photovoltaic_arrays. With permission.)

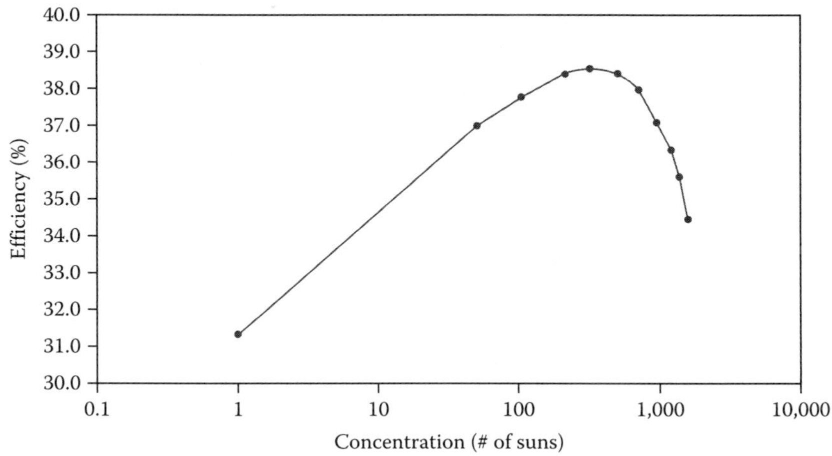

FIGURE 6.32
Overall cell efficiency as a function of concentration. (Courtesy of EMCORE, http://www.emcore.com/solar_photovoltaics/terrestrial_concentrator_photovoltaic_arrays. With permission.)

It can be seen from the data in Figure 6.32 that the efficiency increases from 31.4% at 1× concentration to 39% at 503× concentration, where the peak occurs, and then decreases to about 33% at 1150× concentration. Typical current vs. voltage curves (*I*–*V* curves) at varying concentration levels and 25°C are shown in Figure 6.33.

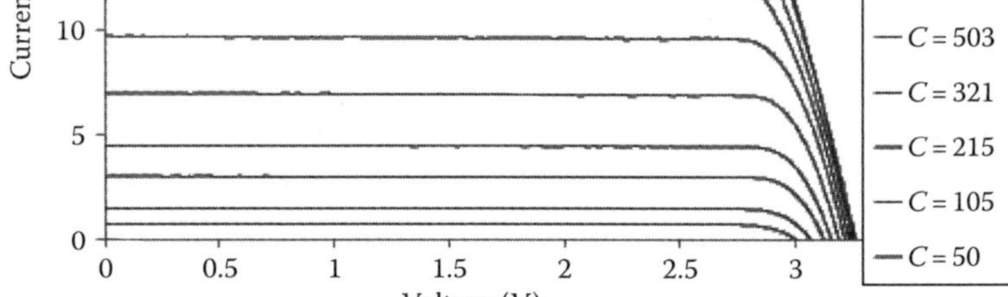

FIGURE 6.33
Typical *I*–*V* curves at varying concentration levels, 25°C. (Courtesy of EMCORE, http://www.emcore.com/solar_photovoltaics/terrestrial_concentrator_photovoltaic_arrays. With permission.)

(a)

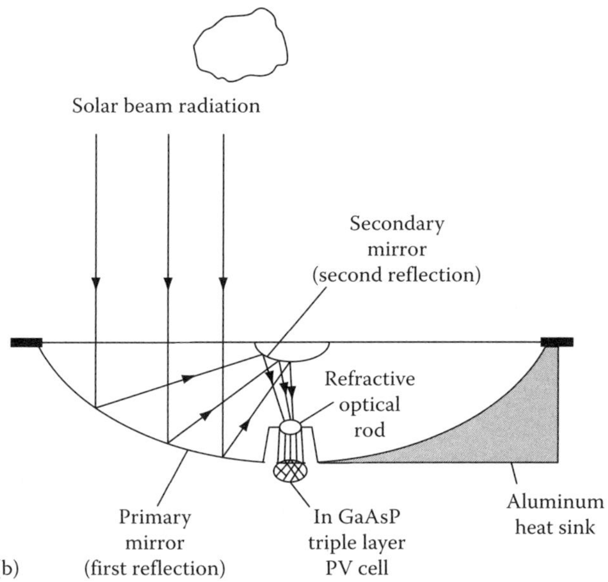

(b)

FIGURE 6.34
Dual-reflection-single-refraction PV concentrator system. (a) Photograph of SolFocus module; (b) Schematic of optical arrangement. (Courtesy of SolFocus, Inc., Mountain View, CA. With permission.)

There are a number of methods available for concentrating sunlight. Figure 6.34 shows a photo and a schematic of a PV concentration system using two reflections that direct the solar beam into a refractive rod that guides the radiation toward a triple layer high-efficiency PV cell. This arrangement can achieve a concentration of 650 with a dual-axis tracker according to Ref. [16]. Another arrangement (Figure 6.35a) with only one reflection and a secondary refractive (Figure 6.35b) lens to guide the radiation toward the PV cell is shown in the color photo with a novel dual-axis tracker developed at the University of Colorado (Figure 6.35c). This arrangement is may be able to achieve a concentration as high as 1000.

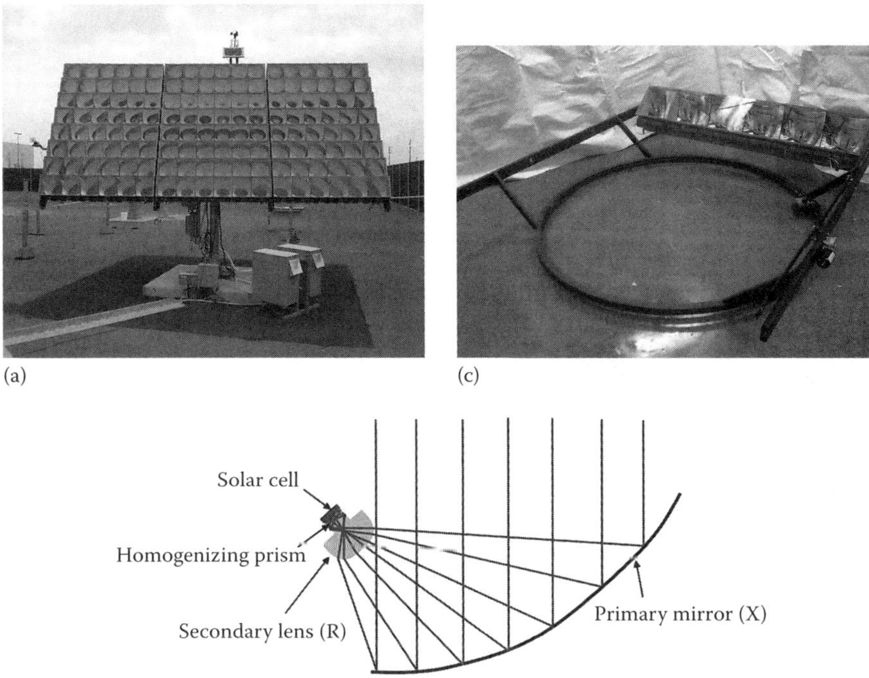

FIGURE 6.35
(See color insert.) Dual-axis rooftop tracker for Boeing high-concentration PV system developed at the University of Colorado, Boulder, Colorado. (a) Solar array mounted on conventional pedestal. (Courtesy of Adam Plesniak, Guy Martins, John Hall, Andy Messina and other members of the Boeing CPV Engineering Team.) (b) Ray trace pattern of the off-axis SMS3D non-imaging optical system in the Boeing CPV design. (Courtesy of Pablo Benitez, Professor, Universidad Polytechnic de Madrid and Light Prescription Innovators, LLC.) (c) Advanced dual axis tracker with four of the six string Boeing high concentration modules installed. Tracker was developed by Senior Design Teams at the University of Colorado, Boulder, Colorado, Dr. Frank Kreith, faculty advisor. (Photo courtesy of Kane Chinnel and Nate Bailey).

The optimum concentration ratio is determined by economic, rather than optical, considerations. For a cell such as EMCORE T1000, the external dimensions are 12.58 mm × 12.58 mm while the total active area is only 108 mm^2 with a total cell thickness is 0.16 mm. The conversion efficiency of the design developed at the University of Colorado can be improved by a finned heat pipe dissipation technique with the heat pipe attached to the bottom of the Ge substrate to avoid shading from the sun. Cooling of such a device is important because the efficiency decreases at about 0.06% per °C increase in operating temperature.

Improvements in multilayer photovoltaic cells are continuing and in April 2008, the National Renewable Energy Laboratory announced that a world record in solar cell efficiency of 40.8% had been achieved at a concentration

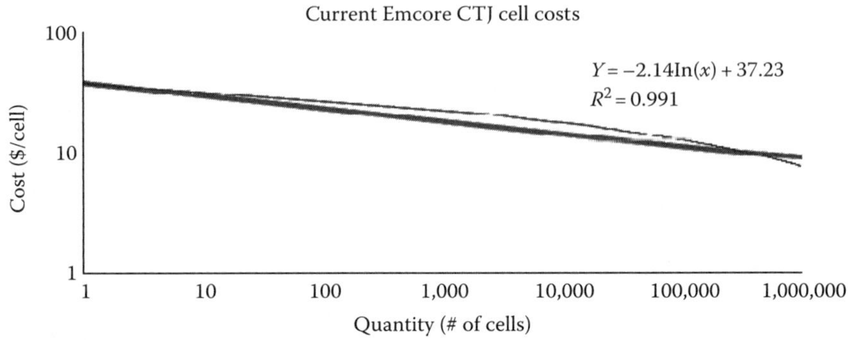

FIGURE 6.36
Multilayer cell costs as a function of production quantity. (Courtesy of EMCORE representative, Personal communication, Albuquerque, NM.)

ratio of 326 suns with an inverted, metaphoric, triple junction solar cell. This new design used a composition of a gallium indium phosphide and gallium indium arsenide to split the solar spectrum into three equal parts that are absorbed by each of the cells' three junctions for a higher efficiency. The construction was accomplished by growing the solar cell on a gallium arsenide wafer, flipping it over, and then removing the wafer [News release NR-2708 "NREL solar cell sets world efficiency record at 40.8%"]. The NREL record, however, was soon surpassed by another type of triple cell developed at the Frauenhof Institute for Solar Energy Systems, which announced in January 2009 that an efficiency of 41.1% had been obtained at a concentration ratio of 545. Details of all of these triple cells, their manufacturing methodology and costs are still privileged information.

The cost of three layer cells is still high and justifies use of highly accurate CPC design with concentration ratios as high as 1000. But, as shown in Figure 6.36, preliminary estimates of cell cost as a function of production by EMCORE indicate that cell cost is likely to decrease and concentrations beyond 500 suns may become counterproductive. Furthermore, since cell efficiency decreases with increase in cell temperature by 0.06% per °C, cooling may be important. Since higher concentration will also result in higher heat flux on the cell cooling becomes more difficult for CPC high concentration designs.

Problems

6.1 Find the wavelength of radiation whose photons have energy equal to the band gap of GaAs.

6.2 What is the theoretical maximum efficiency of conversion if blue light of wavelength 0.45 μm is incident on a GaAs solar cell?

6.3 Find the theoretical maximum overall efficiency of GaAs solar cells in space.

6.4 The reverse saturation current I_o of a silicon cell at 40°C is 1.8×10^{-7} A. The short circuit current when exposed to sunlight is 5 A. From this information, compute

a. Open circuit voltage.
b. Max. power output of the cell.
c. The number of 4 × 4 cm cells needed to supply 100 W at 12 V. How must the cells be arranged?

6.5 At what efficiency is a photovoltaic array running if insolation on the collector is 650 W/m^2, the total collector area is 10 m^2, the voltage across the array is 50 V, and the current being delivered is 15 A?

6.6 If a PV array has a maximum power output of 10 W under an insolation level of 600 W/m^2, what must the insolation be to achieve a power output of 17 W? Would you expect the open circuit voltage to increase or decrease? Would you expect the short-circuit current to increase or decrease?

6.7 A PV battery system has an end-to-end efficiency of 77%. The system is used to run an all-AC load that is run only at night. The charge controller efficiency is 96% and the inverter efficiency is 85%. How much energy will need to be gathered by the PV array if the load is 120 W running for 4 h per night?

6.8 If the average output of the PV system in Problem 6.7 is 200 W, the load is changed to run during the day, how much PV output energy is needed for the same load conditions? Assume that the battery bank is at 100% charge and that input efficiency is equal to output efficiency.

6.9 For the system in Problem 6.7, how many hours of sunlight are needed to ensure that the battery bank is at 100% charge at the end of the day assuming the same load?

Problems 6.10–6.14 The owner of a small cabin would like to convert her home to PV power. She has the following equipment and associated run times:

Household Equipment	Power (W)	Run Time (Day) (h)	Run Time (Night) (h)
Lighting (DC)	25	2	4
Stereo (AC)	40	3	2
Refrigerator (DC)	125	3[a]	3[a]
Water pump (DC)	400	1.5	0.5
Alarm clock (DC)	8	12	12
Computer (AC)	250	3	0
Printer	175	0.25	0
Outdoor safety lights	48	0	8
Answering machine (AC)	7	12	12
Coffee pot (AC)	1200	0.25	0

[a] The refrigerator is assumed to run 25% of the time.

6.10 a. What is the homeowner's daily energy requirement as measured from the load?
b. If she replaces her alarm clock with a wind-up clock, how much energy per day will she avoid using?
c. What would you suggest she do to cut back her daily load?

6.11 How many 50 W panels will the owner require assuming battery storage is 75% efficient and all loads are DC (no inverter)?
a. For a stationary system "seeing" 5 h of sunlight per day?
b. For a tracking system "seeing" 8 h of sunlight per day?

6.12 For the loads listed:
a. What size inverter (peak watts) should she purchase?
b. If the inverter is 88% efficient, how much more daily energy is required from the PV array as compared to an all-DC system?

6.13 The homeowner decides to hire you to design a system for her. She has arranged with a local solar supplier for the following equipment. Specify the system and provide a line diagram showing system connection.

PV panels: 42 W, nominal 12 V,
Batteries: 125 Ah, 6 V, end-to-end efficiency = 72%,
Charge controller: 95% efficient, 12 V, and
Inverter: 90% efficient, sizes of 500, 1000, 2000, and 4000 W available, 12 V input.

6.14 Redesign the system in problem 6.14 with the following equipment:

PV panels: 51 W, nominal 12 V,
Batteries: 200 Ah, 12 V, end-to-end efficiency = 78%,
Charge controller: 97% efficient, 24 V, and
Inverter: 91% efficient, sizes of 500, 1200, 2500, and 5000 W available, 24 V input.

6.15 A flashing beacon is mounted on a navigation buoy in the shipping channel at a port at 30° N latitude. The load consists of a single lamp operating 1.0 s on and 3.6 s off during the hours of darkness. Hours of darkness vary from 9.8 h in July to 13.0 h in December. The lamp draws 2 A at 12 V when lighted. A flasher controls the lamp and draws 0.22 A when the lamp is on. There is a surge current of 0.39 A each time the flasher turns on. This current flows approximately 1/10 of the time the flasher is on. The design has 14 days of battery capacity. Provision has to be made to disconnect the load if the battery voltage drops below 11 V.

The available module has a rated voltage of 17.2 V at 25°C (15 V at 55°C) and 2.3 A at 1 kW/m^2. The available battery has a rated capacity of 105 Ah at 12 V. Assume maximum depth of discharge is 0%. Design the PV system.

6.16 Design a PV system for the following application: A refrigerator/freezer unit for vaccine storage in a remote island of Roatan, Honduras (16°N lat., 86°W long., Temp, range 15–30°C).
Two compressors—one each for refrigerator and freezer.
Each compressor draws 5 A at 12 V.

Compressors Remain on for	Summer (h/day)	Winter (h/day)
Refrigerator	9	5
Freezer	7	4

Design the PV system using the panels and the batteries described in Problem 6.15.

6.17 A homeowner in Santa Fe, New Mexico, is interested in having a solar PV system installed at his house. He would like the system to be able to cover all of his annual energy needs. Below is a table from his utility provider showing the monthly energy use in the household. His roof with solar access is facing due East at a 20° pitch with no shading. In addition, the local electric company is offering a rebate on residential solar installations of $2.00 per installed Watt, as well as a renewable energy credit rebate based on predicted system performance. The REC rebate is $2.50/W, which is then multiplied by the system performance. Since the installation is East facing, the predicted performance efficiency is 83% (with a system facing due south having a performance efficiency of 100%).

Consumption:

	kWh
January	400
February	400
March	600
April	600
May	800
June	1000
July	1000
August	900
September	800
October	600
November	600
December	400

a. Design a system that will offset 100% of the home's consumption.
b. Assuming $0.10/W electricity expense with a 5% annual inflationary rate increase, and a $7.50/W installed cost what will the payback be for this system after the rebate from the utility provider?

c. How do you wire the solar panel? Specifically, how many cells would you need and how would you wire them (series versus parallel) if the solar panel is to have a working voltage of 24.46 V and an amperage of 7.6 A? Assume the individual cells produce.98 V and 3.8 A.

Resources

The SMA string sizing program—found on their Web site—http://www.smaamerica.com/home

PV Watts—found on the NREL Web site.

6.18 A village in Antigua (17°N, 61°W, 15–30°C temperature, range), West Indies, requires 5000 gal of water per day for community water supply. Assuming a year round average insolation of 8 kWh/day, design the system using the following components:
PV Panels: Solarex panels 17.5 V and 3.6 A at 1000 W/m^2 and 25°C
Pump: Grundfos multistage pump input 105 V DC, 9 A, 30% efficiency

6.19 Prepare a preliminary design for a solar PV system to provide 1 kW of stand-alone 24 h/day power to a travel trailer. Make a schematic diagram of the overall system, a cost estimate, a cell array design, and a battery storage. Then discuss the assumptions you made for your design.

References

1. Nijs, J. et al. (1997) Energy payback time of crystalline silicon solar modules, in: K.W. Böer (ed.), *Advances in Solar Energy*, vol. 11, Chap. 6, American Solar Energy Society, Boulder, CO, pp. 291–327.
2. Angrist, S.W. (1976) *Direct Energy Conversion*, 3rd edn., Allyn and Bacon, Boston, MA.
3. Bube, R.H. (1960) *Photoconductivity of Solids*, John Wiley & Sons, New York.
4. Garg, H.P. (1987) *Advances in Solar Energy Technology*, vol. 3, D. Reidel Publishing Company, Dordrecht, the Netherlands.
5. Post, H.N. and Risser, V.V. (1995) *Stand-Alone Photovoltaic Systems—A Handbook of Recommended Design Practices*, Sandia National Laboratory Report SAND-87-7023, Sandia National Laboratories, Albuquerque, NM.
6. Crossley, P.A., Noel, G.T., and Wolf, M. (1968) Review and evaluation of past solar cell development efforts, RCA Astro-Electronics Division Report, AED R-3346, Contract NASW-1427, NASA, Washington, DC.
7. Florida Solar Energy Center. (1991) Photovoltaic system design, FSEC-GP-31-86, Florida Solar Energy Center, Cocoa Beach, FL.
8. Lasnier, F. and Ang, T.G. (1990) *Photovoltaic Engineering Handbook*, A. Hilger Publishing, New York.
9. Masia, S. (2010) Concentrating photovoltaics, *Solar Today*, January/February, p. 42.

10. Raugei, M., Bargigli, S., and Ulgaiati, S. (2007) Energy and life cycle assessment of the thin flim CdTe photovoltaic modules, accessed at http://www.abound.com/pdf/NREL_PV_Embodied_Energy.pdf
11. Arvizu, D.E. (March 2009) Let the solar revolution begin, *Solar Today* 23(2), 24–27.
12. Jayarama Reddy, P. (2009) Cell efficiences, in: *Science and Technology of Photovoltaics*, 2nd edn., CRC Press, Boca Raton, FL.
13. Kurtz, S. (2009) Opportunities and challenges for development of a mature concentrating photovoltaic power industry, NREL, NREL/TP-520-43208, revised November 2009.
14. Goswami, Y. et al. (2000) *Principles of Solar Engineering*, 2nd edn., Taylor & Francis, Philadelphia.

Suggested Readings

Böer, K.W. (1990) *Survey of Semiconductor Physics*, Van Nostrand Reinhold, New York.

Crandall, R. and Luft, W. (1995) The future of amorphous silicon photovoltaic technology. NREL/TP-441-8019, National Renewable Energy Laboratory, Golden, CO.

Loferski, J.J. (1963) Recent research on photovoltaic solar energy converters. *Proceedings of the IEEE* 51, 667–674.

Moore, C.E. (1952) *Atomic Energy Levels*, vol. 2, National Bureau of Standards Circular 467, U.S. Government Printing Office, Washington, DC.

Taylor, M. and Kreider, J. (1998) *Solar Energy Applications*, 3rd edn., University of Colorado Bookstore, Boulder, CO.

7

Solar Heating and Cooling of Buildings

I dreamed my genesis in sweat and death, fallen twice in the feeding sea, grown stale of Adam's brine until, vision of new man strength, I seek the sun.

—Dylan Thomas

Converting the sun's radiant energy to heat is a common and well-developed solar conversion technology. The temperature level and amount of this converted energy are the key parameters that must be known to match a conversion scheme to a specified task effectively.

Heating and cooling of buildings offers great opportunities for energy conservation as well as utilization of solar energy. A first step in designing an energy-efficient building should be an assessment of the various options for conservation. These should include energy-efficient lighting, double or triple glazed windows, shading, increased insulation, and elimination of unnecessary air leakage to maintain interior comfort. These measures are usually the domain of the architect, but recently specialized energy conservation engineering for buildings has become a part of the architectural design.

The basic principle of solar thermal collection is that when solar radiation strikes a surface, part of it is absorbed, thereby increasing the temperature of the surface. The efficiency of that surface as a solar collector depends not only on the absorption efficiency, but also on how the thermal and reradiation losses to the surroundings are minimized and how the energy from the collector is removed for useful purposes. Various solar thermal collectors range from unglazed flat-plate-type solar collectors operating at about 5°C–10°C above the ambient, to central receiver concentrating collectors operating at above 1000°C. Table 7.1 lists various types of solar thermal collectors and their typical temperature and concentration ranges. This chapter and the next chapter analyze the thermal and optical performance of several solar-thermal collectors. They range from non-concentrating, flat-plate types to compound-curvature, continuously tracking types with concentration ratios up to 1000 or more. Applications of the energy converted by solar thermal collectors are described in this chapter for buildings and in Chapter 8 for power and process heat. The next section describes some fundamental radiative properties of materials, knowledge of which helps in the design of solar thermal collectors.

See Figures 7.1 and 7.2 for examples of some flat plate solor collectors.

TABLE 7.1

Types of Solar Thermal Collectors and Their Typical Temperature Range

Type of Collector	Concentration Ratio	Typical Working Temperature Range (°C)
Flat-plate collector	1	≤70
High efficiency flat-plate collector	1	60–120
Fixed concentrator	3–5	100–150
Parabolic trough collector	10–50	150–350
Parabolic dish collector	200–500	250–700
Central receiver	500–>3000	500–>1000

7.1 Radiative Properties and Characteristics of Materials

When radiation strikes a body, a part of it is reflected, a part is absorbed and, if the material is transparent, a part is transmitted, as shown in Figure 7.3. The fraction of the incident radiation reflected is defined as the reflectance ρ, the fraction absorbed as the absorptance α, and the fraction transmitted as the transmittance τ. According to the first law of thermodynamics these three components must add up to unity, or

$$\alpha + \tau + \rho = 1 \tag{7.1}$$

Opaque bodies do not transmit any radiation and $\tau = 0$.

The reflection of radiation can be *specular* or *diffuse*. When the angle of incidence is equal to the angle of reflection, the reflection is called specular; when the reflected radiation is uniformly distributed into all directions it is called diffuse (see Figure 7.4). No real surface is either specular or diffuse, but a highly polished surface approaches specular reflection, whereas a rough surface reflects diffusely.

Another important radiative property is called emittance, ε, which is the ratio of the radiative emissive power of a real surface to that of an ideal "black" surface.

All of the radiative properties of materials, α, τ, ρ, and ε can be functions of the wavelength and direction. In fact, such dependence is used in the design of solar energy devices and systems. For example, selective absorbers are used for solar collectors and passive heating systems, and glazing materials for daylighting and solar collectors.

The monochromatic directional emittance of a surface, $\varepsilon_\lambda(\theta, \phi)$, in a direction signified by an azimuth angle ϕ and a polar angle θ, is

$$\varepsilon_\lambda(\theta, \phi) = \frac{I_\lambda(\theta, \phi)}{I_{b\lambda}} \tag{7.2}$$

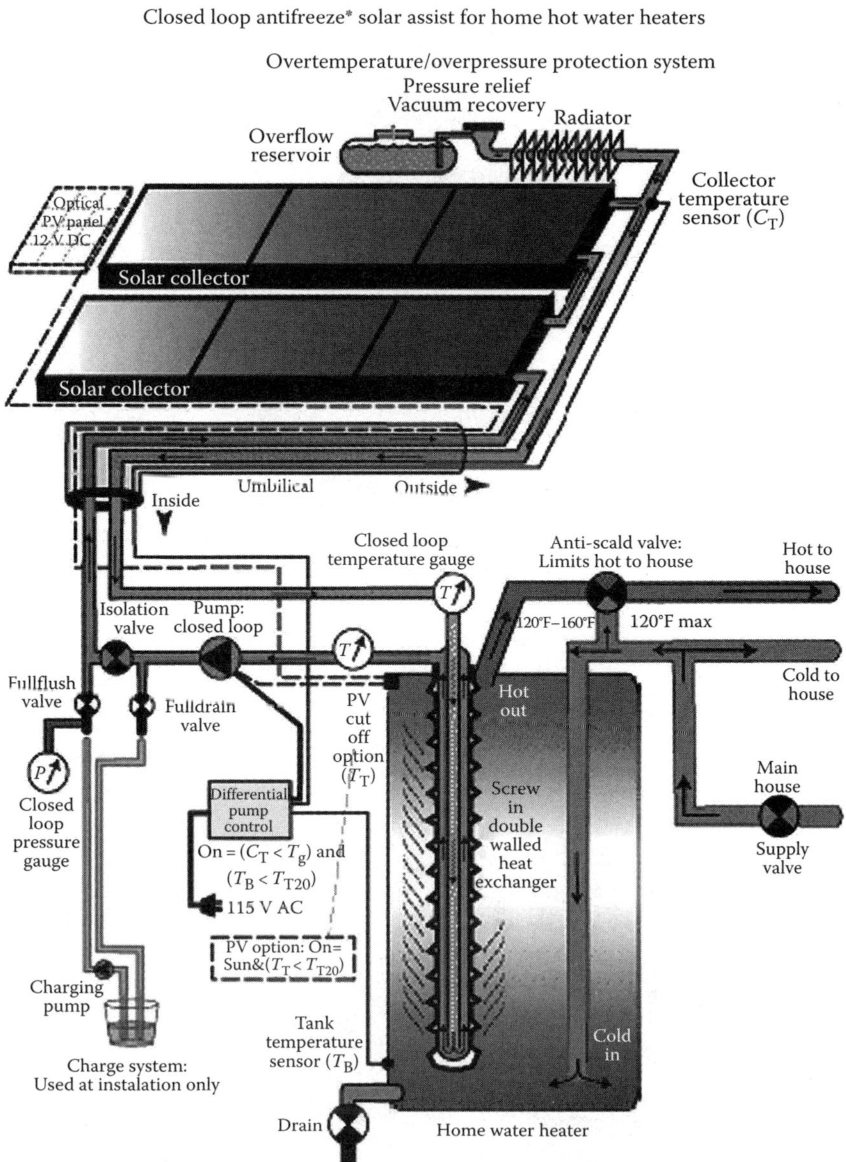

(a)

FIGURE 7.1
(See color insert.) (a) Schematic of an active flat-plate solar collector with freeze protection.

(continued)

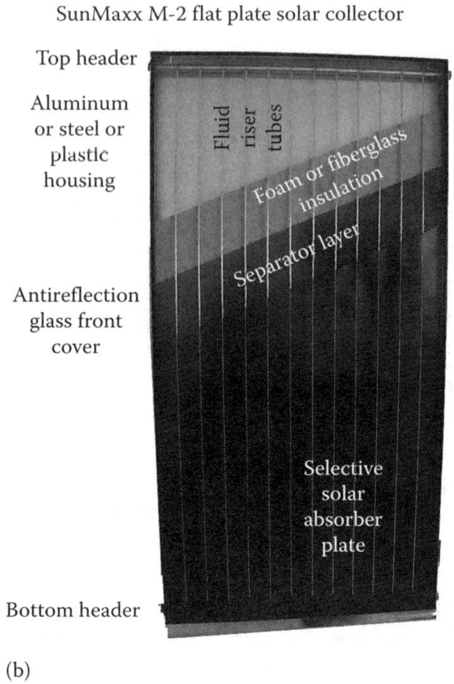

FIGURE 7.1 (continued)
(See color insert.) (b) Photograph of an active flat-plate solar collector. (Courtesy of Barry Butler, PhD, Butler Sun Solutions, Inc., Solana Beach, CA.)

FIGURE 7.2
(See color insert.) Photo of active flat-plate solar collector on Kreith residence.

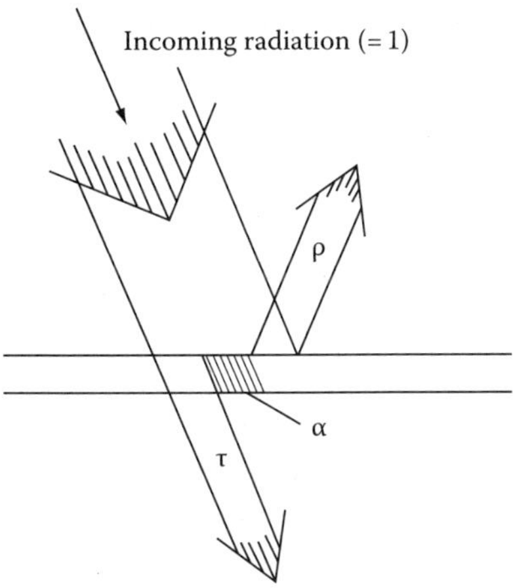

FIGURE 7.3
Schematic representation of transmittance τ, absorptance α, and reflectance ρ.

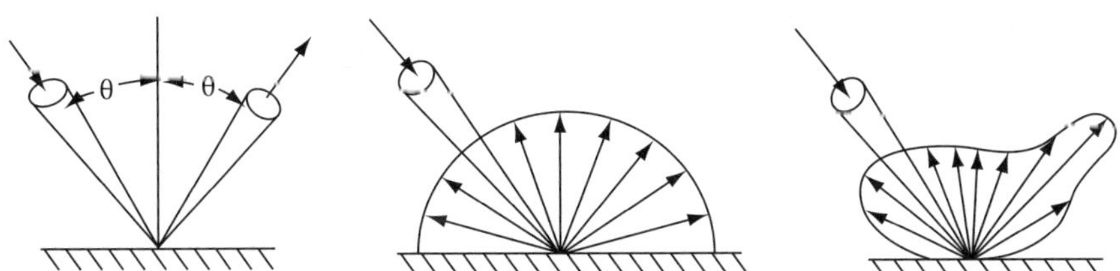

FIGURE 7.4
Reflections from (a) ideal specular, (b) ideal diffuse, and (c) real surfaces.

From the above, the total directional emittance $\varepsilon(\theta,\phi)$ over all the wavelengths or the monochromatic hemispherical emittance ε_λ can be obtained by the integration of $\varepsilon_\lambda(\theta, \phi)$ over all the wavelengths or the entire hemispherical space, respectively. The overall emittance ε, is found by integrating the hemispherical emittance over all the wavelengths.

$$\varepsilon = \frac{1}{\sigma T^4}\int_0^\infty \varepsilon_\lambda E_{b\lambda} d\lambda. \tag{7.3}$$

Observe that both ε_λ and ε are properties of the surface.

The next most important surface characteristic is the absorptance. We begin by defining the monochromatic directional absorptance as the fraction of the incident radiation at wavelength λ from the direction θ, ϕ that is absorbed, or

$$\alpha_\lambda(\theta,\phi) = \frac{I_{\lambda,\mathrm{a}}(\theta,\phi)}{I_{\lambda,\mathrm{i}}(\theta,\phi)}, \tag{7.4}$$

where the subscripts "a" and "i" denote absorbed and incident radiation, respectively. The monochromatic directional absorptance is also a property of the surface.

More important than $\alpha_\lambda(\theta, \phi)$ is the overall directional absorptance $\alpha(\theta, \phi)$, defined as the fraction of the total radiation from the direction θ, ϕ that is absorbed, or

$$\alpha(\theta,\phi) = \frac{\int_0^\infty \alpha_\lambda(\theta,\phi) I_{\lambda,i}(\theta,\phi) d\lambda}{\int_0^\infty I_{\lambda,i}(\theta,\phi) d\lambda} = \frac{1}{I_i(\theta,\phi)} \int_0^\infty \alpha_\lambda(\theta,\phi) I_{\lambda,i}(\theta,\phi) d\lambda. \tag{7.5}$$

The overall absorptance is a function of the characteristics of the incident radiation and is, therefore, unlike the monochromatic absorptance, not a property of a surface alone. It is this characteristic that makes it possible to have selective surfaces that absorb the radiation from one source at a higher rate than from another. In other words, even though according to Kirchhoff's law the monochromatic emittance at λ must equal the monochromatic absorptance

$$\alpha_\lambda(\theta,\phi) = \epsilon_\lambda(\theta,\phi), \tag{7.6}$$

the overall emittance is not necessarily equal to the overall absorptance unless thermal equilibrium exists, and the incoming and outgoing radiations have the same spectral characteristics.

The effect of incidence angle on the absorptance is illustrated in Table 7.2 where the angular variation of the absorptance for a nonselective black surface, typical of those used on flat-plate collectors, is shown. The absorptance of this surface for diffuse radiation is approximately 0.90.

The third characteristic to be considered is the reflectance. Reflectance is particularly important for the design of focusing collectors. As mentioned

TABLE 7.2

Angular Variation of Absorptance of Lampblack Paint

Incidence Angle i(°)	Absorptance $\alpha(i)$
0–30	0.96
30–40	0.95
40–50	0.93
50–60	0.91
60–70	0.88
70–80	0.81
80–90	0.66

Source: Adapted from Löf, G.O.G. and Tybout, R.A., *Sol. Energy*, 16, 9, 1974; Löf, G.O.G. and Tybout, R.A., Model for optimizing solar heating design, ASME Paper 72-WA/SOL-8.

previously, there are two limiting types of reflection: specular and diffuse. As illustrated in Figure 7.4a, when a ray of incident radiation at an angle θ is reflected at the same polar angle and the azimuthal angles differ by 180°, as for a perfect mirror, the reflection is said to be specular. The reflection is said to be diffuse if the incident radiation is scattered equally in all directions, as shown in Figure 7.4b.

7.1.1 Selective Surfaces

Two types of special surfaces of great importance in solar collector systems are selective and reflecting surfaces. Selective surfaces combine a high absorptance for solar radiation with a low emittance for the temperature range in which the surface emits radiation. This combination of surface characteristics is possible because 98% of the energy in incoming solar radiation is contained within wavelengths below 3 μm, whereas 99% of the radiation emitted by black or gray surfaces at 400 K is at wavelengths longer than 3 μm. The dotted line in Figure 7.5 illustrates the spectral reflectance of an ideal, selective semi-gray surface having a uniform reflectance of 0.05 below 3 μm, but 0.95 above 3 μm. Real surfaces do not approach this performance. Table 7.3 lists properties of some selective coatings.

7.1.2 Reflecting Surfaces

Concentrating solar collectors require the use of reflecting surfaces with high specular reflectance in the solar spectrum or refracting devices with high transmittance in the solar spectrum. Reflecting surfaces are usually highly polished metals or metal coatings on suitable substrates. With opaque substrates, the reflective coatings must always be front-surfaced,

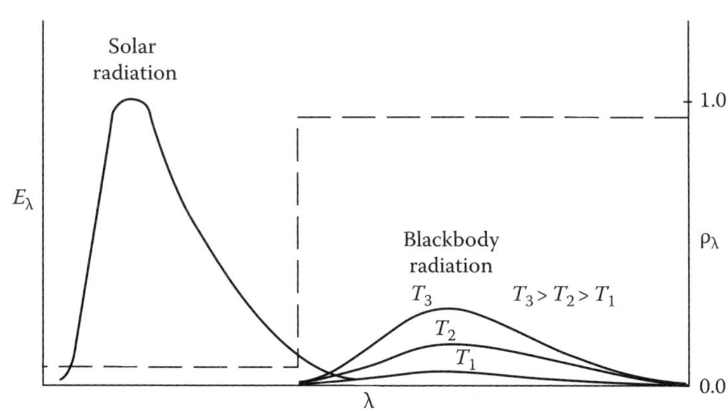

FIGURE 7.5
Illustrations of reflectance characteristics of an ideal selective surface. Shows radiation from ideal surfaces at different temperatures and solar radiation.

TABLE 7.3

Properties of Some Selective Plated Coating Systems

Coating[a]	Substrate	$\bar{\alpha}_s$	$\bar{\varepsilon}_i$	Durability: Breakdown Temperature (°F)	Durability: Humidity-Degradation MIL STD 810B
Black nickel on nickel	Steel	0.95	0.07	>550	Variable
Black chrome on nickel	Steel	0.95	0.09	>800	No effect
Black chrome	Steel	0.91	0.07	>800	Completely rusted
	Copper	0.95	0.14	600	Little effect
	Galvanized steel	0.95	0.16	>800	Complete removal
Black copper	Copper	0.88	0.15	600	Complete removal
Iron oxide	Steel	0.85	0.08	800	Little effect
Manganese oxide	Aluminum	0.70	0.08		
Organic overcoat on iron oxide	Steel	0.90	0.16		Little effect
Organic overcoat on black chrome	Steel	0.94	0.20		Little effect

Source: U.S. Dept. of Commerce, Optical coatings for flat plate solar collectors, NTIS No. PN-252–383, Honeywell, Inc., Morristown, NJ, 1975.

[a] Black nickel coating plated over a nickel-steel substrate has the best selective properties ($\bar{\alpha}_s = 0.95$, $\bar{\varepsilon}_i = 0.07$), but these degraded significantly during humidity tests. Black chrome plated on a nickel-steel substrate also had very good selective properties ($\bar{\alpha}_s = 0.95$, $\bar{\varepsilon}_i = 0.09$) and also showed high resistance to humidity.

for example, chrome plating on copper or polished aluminum. If a transparent substrate is used, however, the coating may be front- or back-surfaced. In any back-surfaced reflector, the radiation must pass through the substrate twice and the transmittance of the material becomes very important.

Table 7.4 presents typical values for the normal specular reflectance of new surfaces for beam solar radiation.

TABLE 7.4

Specular Reflectance Values for Solar Reflector Materials

Material	ρ
Silver (unstable as front surface mirror)	0.94 ± 0.02
Gold	0.76 ± 0.03
Aluminized acrylic, second surface	0.86
Anodized aluminum	0.82 ± 0.05
Various aluminum surfaces range	0.82–0.92
Copper	0.75
Back-silvered water-white plate glass	0.88
Aluminized type-C Mylar (from Mylar side)	0.76

7.1.3 Transparent Materials

The optical transmission behavior can be characterized by two wavelength-dependent physical properties—the index of refraction n and the extinction coefficient k. The index of refraction, which determines the speed of light in the material, also determines the amount of light reflected from a single surface, while the extinction coefficient determines the amount of light absorbed in a substance in a single pass of radiation.

Figure 7.6 defines the angles used in analyzing reflection and transmission of light. The angle i is called the *angle of incidence*. It is also equal to the angle at which a beam is specularly reflected from the surface. Angle θ_r is the *angle of refraction*, which is defined, as shown in the figure. The incidence and refraction angles are related by Snell's law:

$$\frac{\sin(i)}{\sin(\theta_r)} = \frac{n'_r}{n'_i} = n_r, \tag{7.7}$$

where

n'_i and n'_r are the two refractive indices

n_r is the index ratio for the two substances forming the interface

Typical values of refractive indices for various materials are shown in Table 7.5. For most materials of interest in solar applications, the values range from 1.3 to 1.6, a fairly narrow range.

By having a gradual change in the index of refraction, reflectance losses are reduced significantly. The reflectance of a glass-air interface common in

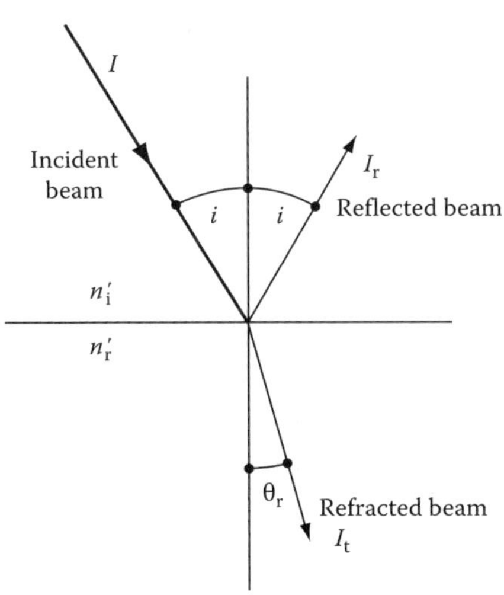

FIGURE 7.6
Diagram showing incident, reflected, and refracted beams of light and incidence and refraction angles for a transparent medium.

TABLE 7.5

Refractive Index for Various Substances in the Visible Range Based on Air

Material	Index of Refraction
Air	1.000
Clean polycarbonate	1.59
Diamond	2.42
Glass (solar collector type)	1.50–1.52
Plexiglass[a] (polymethyl methacrylate, PMMA)	1.49
Mylar[a] (polyethylene terephthalate, PET)	1.64
Quartz	1.55
Tedlar[a] (polyvinyl fluoride, PVF)	1.45
Teflon[a] (polyfluoroethylenepropylene, FEP)	1.34
Water–liquid	1.33
Water–solid	1.31

[a] Trademark of the DuPont Company, Wilmington, Delaware.

solar collectors may be reduced by a factor of four by an etching process. If glass is immersed in a silica supersaturated fluosilicic acid solution, the acid attacks the glass and leaves a porous silica surface layer. This layer has an index of refraction intermediate between glass and air. Figure 7.7 shows the spectral reflectance of a pane of glass before and after etching.

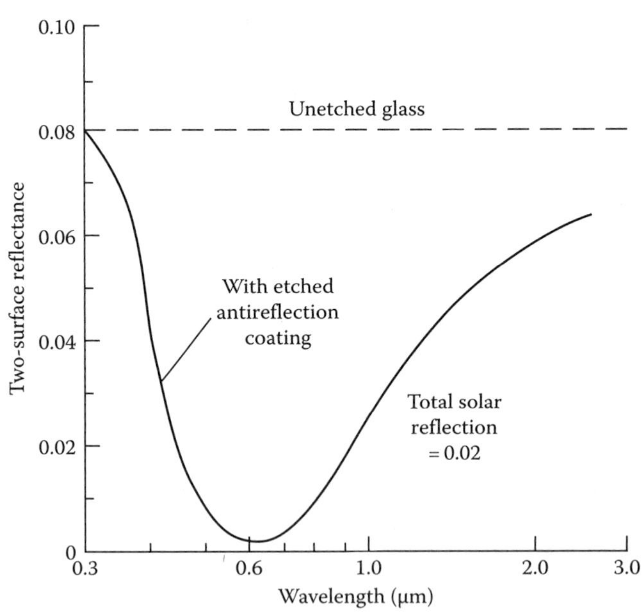

FIGURE 7.7
Reflection spectra for a sample of glass before and after etching.

7.2 Flat-Plate Collectors

A simple flat-plate collector consists of an absorber surface (usually a dark thermally conducting surface); a trap for reradiation losses from the absorber surface (such as glass, which transmits shorter wavelength solar radiation but blocks the longer wavelength radiation from the absorber); a heat transfer medium such as air, water etc.; and some thermal insulation behind the absorber surface. Flat-plate collectors are used typically for temperature requirements up to 75°C although higher temperatures can be obtained from high efficiency collectors. These collectors are of two basic types based on the heat transfer fluid:

Liquid type: where heat transfer fluid may be water, mixture of water and antifreeze oil, etc.;

Air type: where heat transfer medium is air, used mainly for drying and some space heating.

7.2.1 Liquid-Type Collectors

Figure 7.8 shows a typical liquid-type flat-plate collector. In general it consists of

1. *Glazing*: One or more covers of transparent material like glass, plastics, etc. Glazing may be left out for some low-temperature applications.
2. *Absorber*: A plate with tubes or passages attached to it for the passage of a working fluid. The absorber plate is usually painted flat black or electroplated with a selective absorber.

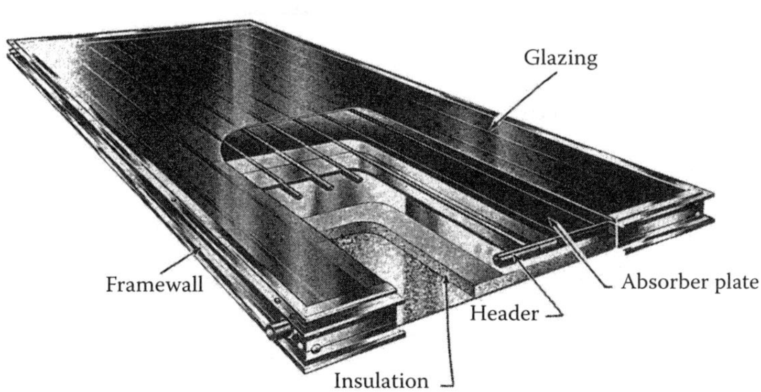

FIGURE 7.8
Typical liquid-type flat-plate collector. (Courtesy of Morning Star Corporation, Orange Park, FL.)

3. *Headers or manifolds*: To facilitate the flow of heat transfer fluid.
4. *Insulation*: To minimize heat loss from the back and the sides.
5. *Container*: Box or casing.

7.2.2 Air-Type Collectors

Air types of collectors are more commonly used for agricultural drying and space heating applications. Their basic advantages are low sensitivity to leakage and no need for an additional heat exchanger for drying and space heating applications. However, because of the low heat capacity of the air and the low convection heat transfer coefficient between the absorber and the air, a large heat transfer area and high flow rates are needed. Figure 7.9 shows some common configurations of air heating collectors. Common absorber materials include corrugated aluminum or galvanized steel sheets, black metallic screens, or simply any black painted surface.

Unglazed, transpired solar air collectors offer a low-cost opportunity for some applications such as preheating of ventilation air and agricultural drying and curing [2]. Such collectors consist of perforated absorber sheets that are exposed to the sun and through which air is drawn. The perforated absorber sheets are attached to the vertical walls, which are exposed to

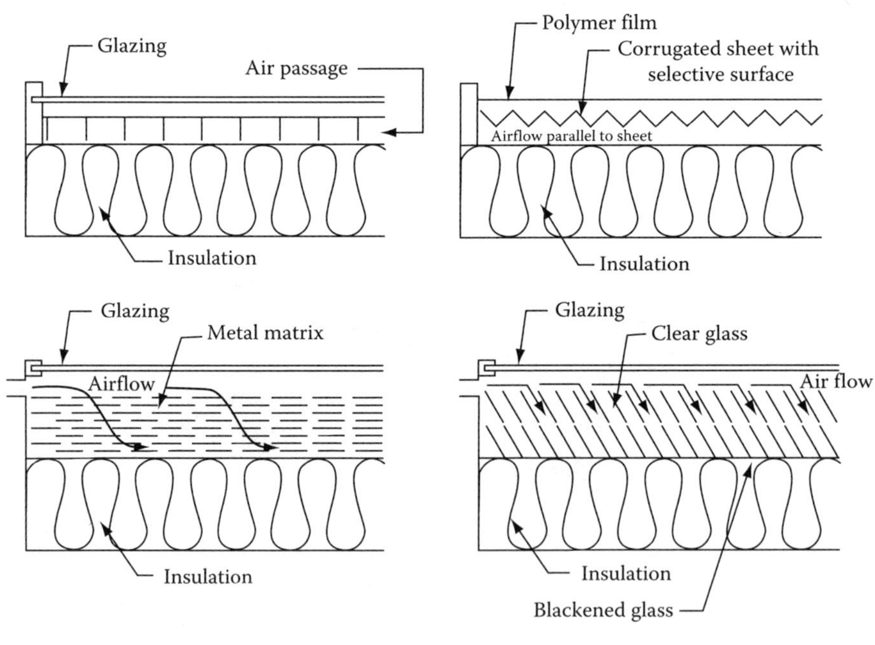

FIGURE 7.9
Some common configurations of air heating collectors.

FIGURE 3.1
2.5 MW Liberty wind turbine (Clipper Windpower) near Medicine Bow, Wyoming. (Courtesy of DOE/NREL, Golden, CO.)

FIGURE 3.23
Flow visualization tests conducted in the NASA Ames wind tunnel using smoke emitted from the tips of the turbine helped researches determine the extent of the wake under a controlled set of conditions. (Courtesy of DOE/NREL, Golden, CO.)

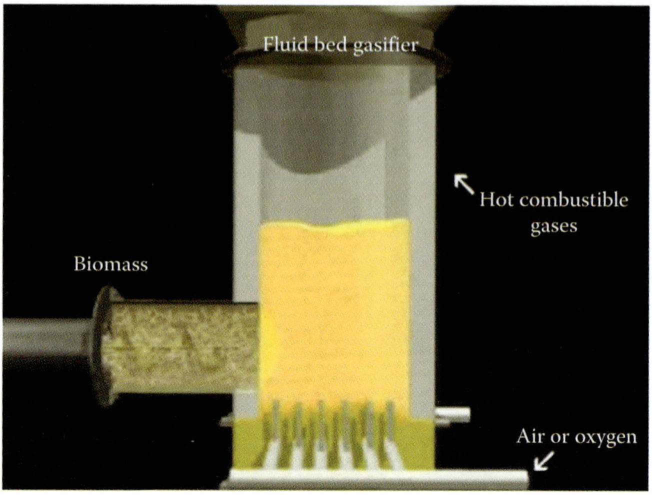

FIGURE 4.3
Cross section of fluid plate gasifier.

FIGURE 4.4
Frontline low-temperature gasifier.

Combiner box
600 V Fuses protect each of the (3) strings of (5) modules
after the fuses the (3) circuits are combined in parallel.
The combined circuit is then brought to the DC disconnect in the garage.

FIGURE 6.1
Kreith residence, 3 kW Sanyo high efficiency photovoltaic system. (Courtesy of Chris Klinga, Lighthouse Solar.)

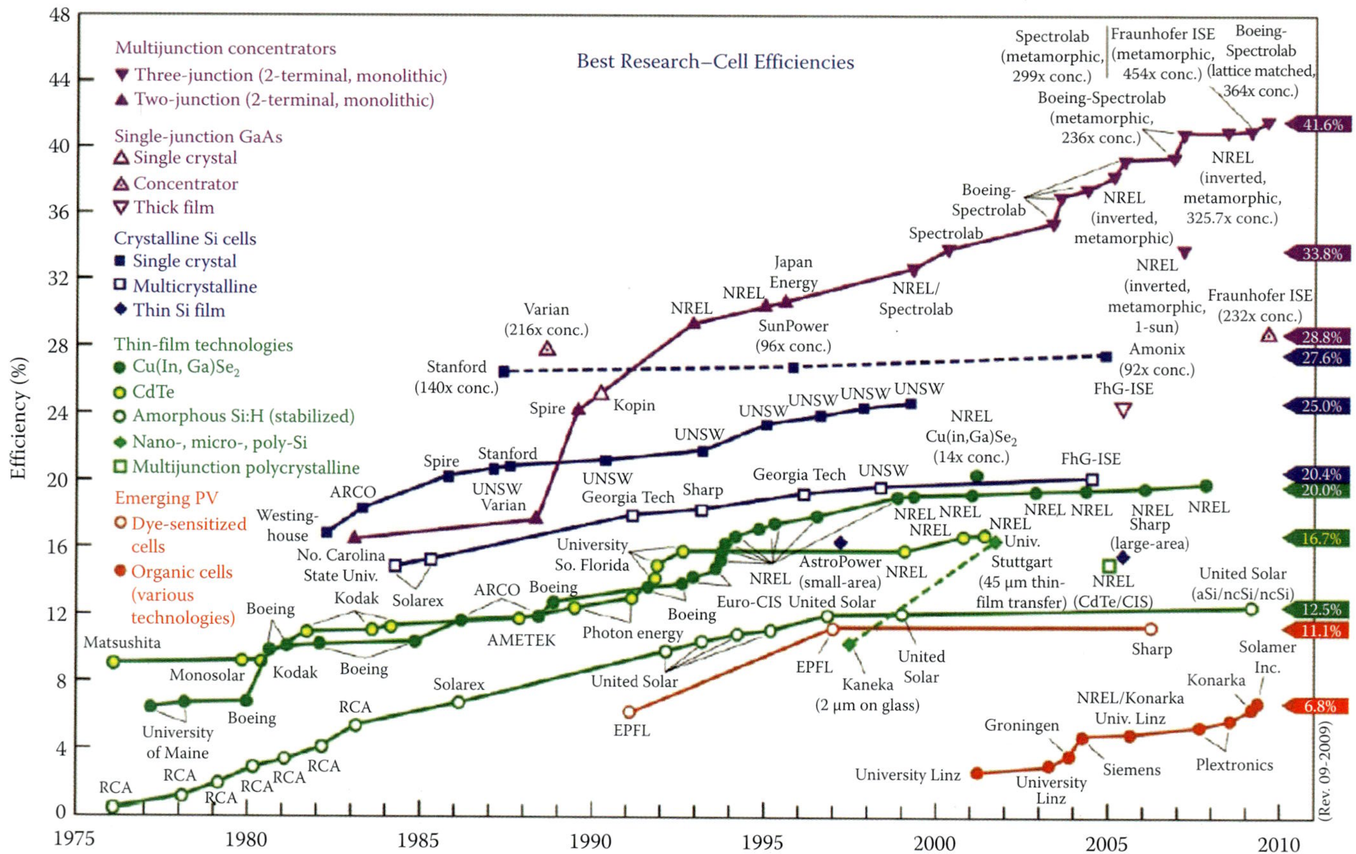

FIGURE 6.26
Best research–cell efficiencies. (From Jayarama Reddy, P., Cell efficiencies, in *Science and Technology of Photovoltaics*, 2nd edn., CRC Press, Boca Raton, FL; Kurtz, S., Opportunities and challenges for development of a mature concentrating photovoltaic power industry, NREL, NREL/TP-520-43208, revised November, 2009.)

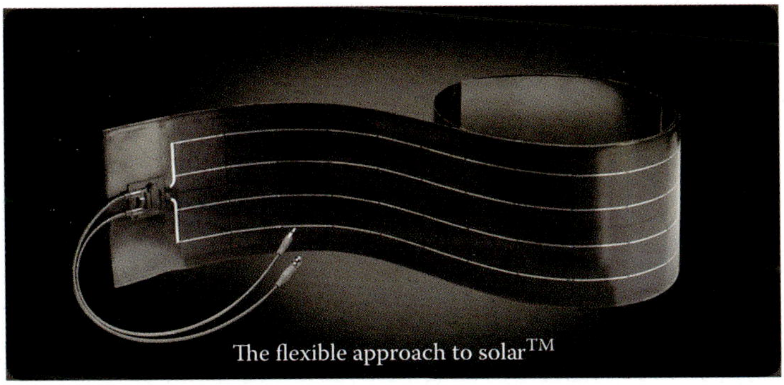

FIGURE 6.28
Thin film flexible photovoltaic CIGS module. (Courtesy of SoloPower, Inc., San Jose, CA.)

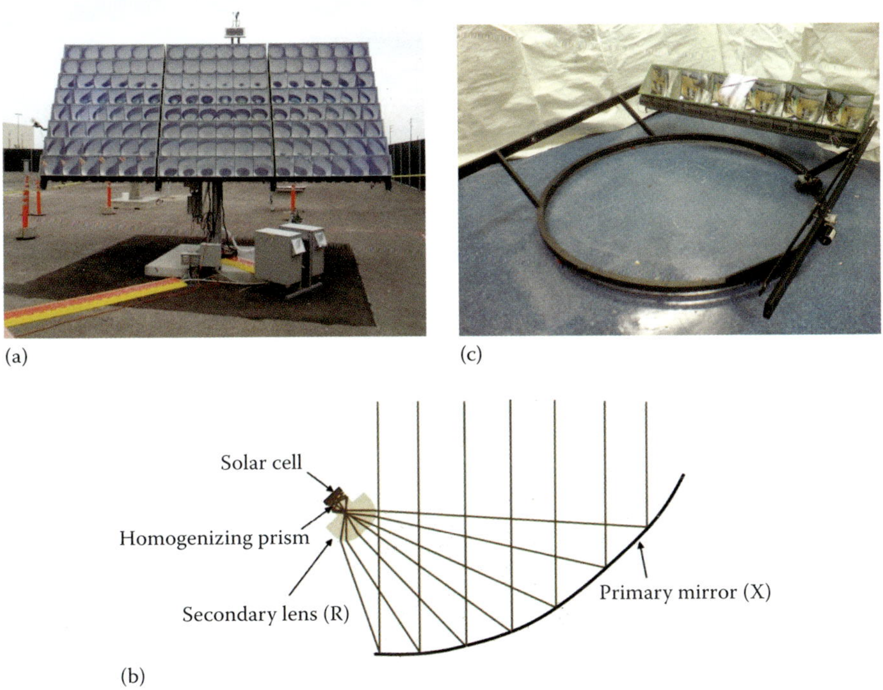

FIGURE 6.35
Dual-axis rooftop tracker for Boeing high-concentration PV system developed at the University of Colorado, Boulder, Colorado. (a) Solar array mounted on conventional pedestal. (Courtesy of Adam Plesniak, Guy Martins, John Hall, Andy Messina and other members of the Boeing CPV Engineering Team.) (b) Ray trace pattern of the off-axis SMS3D non-imaging optical system in the Boeing CPV design. (Courtesy of Pablo Benitez, Professor, Universidad Polytechnic de Madrid and Light Prescription Innovators, LLC.) (c) Advanced dual axis tracker with four of the six string Boeing high concentration modules installed. Tracker was developed by Senior Design Teams at the University of Colorado, Boulder, Colorado, Dr. Frank Kreith, faculty advisor. (Photo courtesy of Kane Chinnel and Nate Bailey).

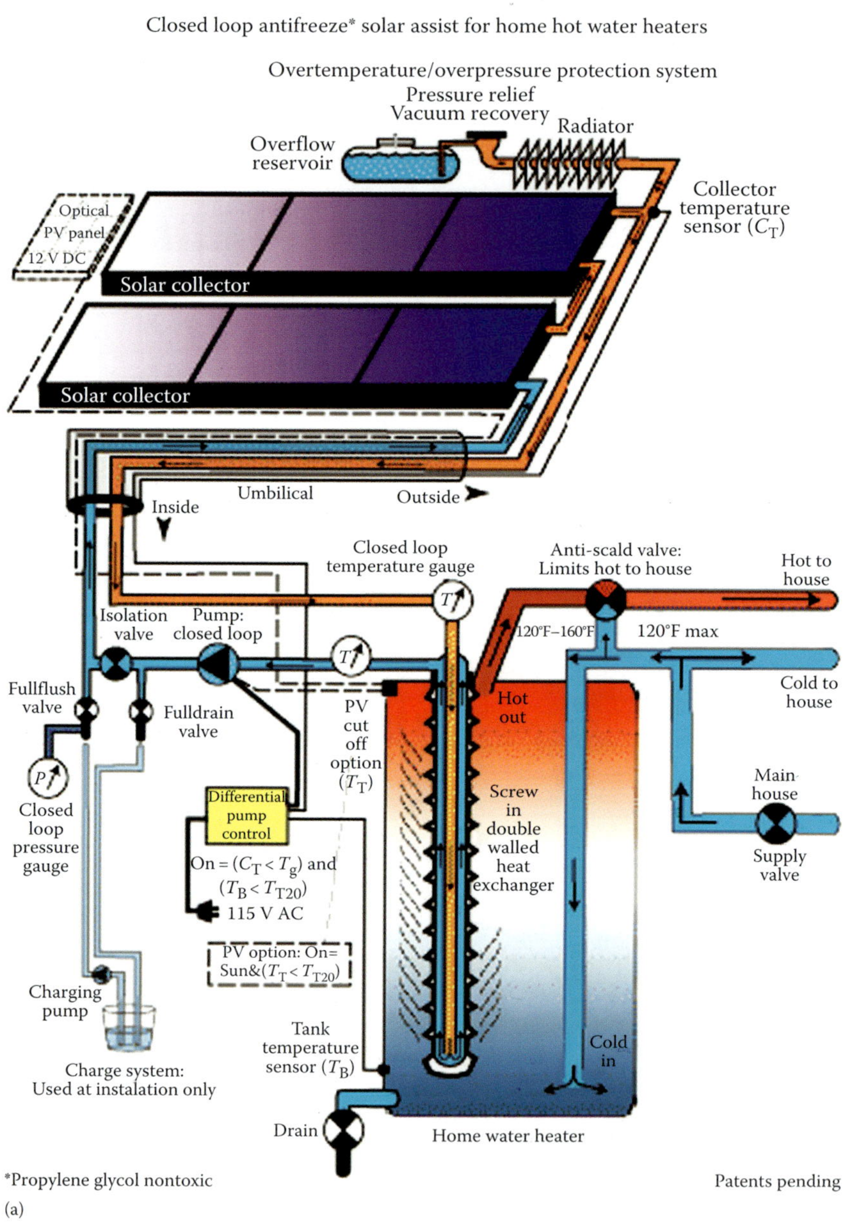

FIGURE 7.1
(a) Schematic of an active flat-plate solar collector with freeze protection.

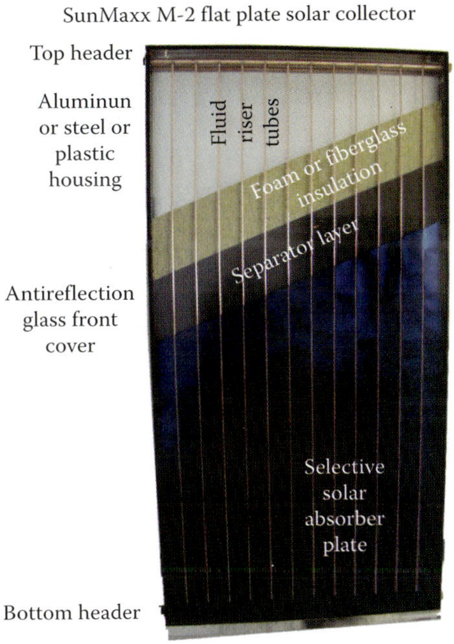

(b)

FIGURE 7.1 (continued)
(b) Photograph of an active flat-plate solar collector. (Courtesy of Barry Butler, PhD, Butler Sun Solutions, Inc., Solana Beach, CA.)

FIGURE 7.2
Photo of active flat-plate solar collector on Kreith residence.

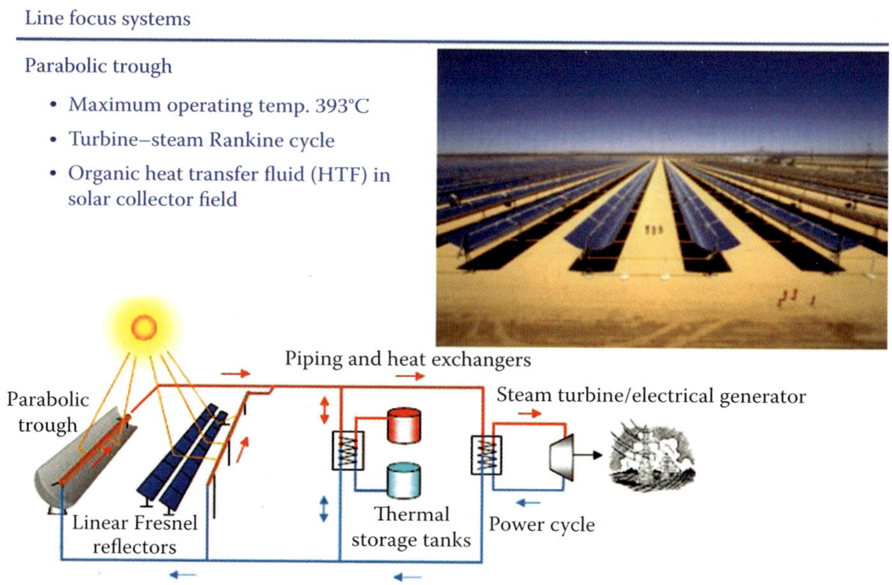

FIGURE 8.27
Line focus solar thermal power system. (Courtesy of Clifford Ho, Sandia National Laboratories.)

FIGURE 8.28
Sky Fuel solar sky trough near Golden, Colorado, viewed by professor Frank Kreith (right) and National Institute of Standards and Technology researcher, Dr. Isaac Garaway (left). (Courtesy of CU Engineering, 2009, Casey A. Cass Photographer.)

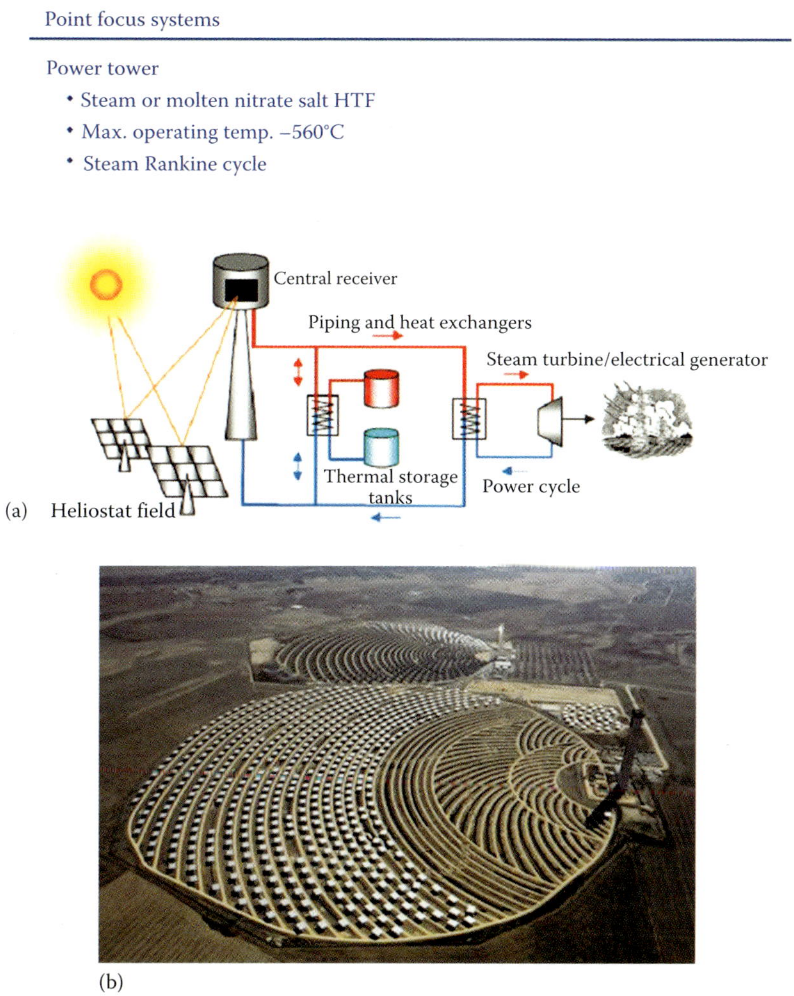

FIGURE 8.31
(a) Schematic diagram of point focus solar thermal power tower system. (b) Photograph of two Spanish Central Receiver Systems. (Courtesy of Clifford Ho, Sandia National Laboratories.)

FIGURE 8.32
Heliostats for power tower.

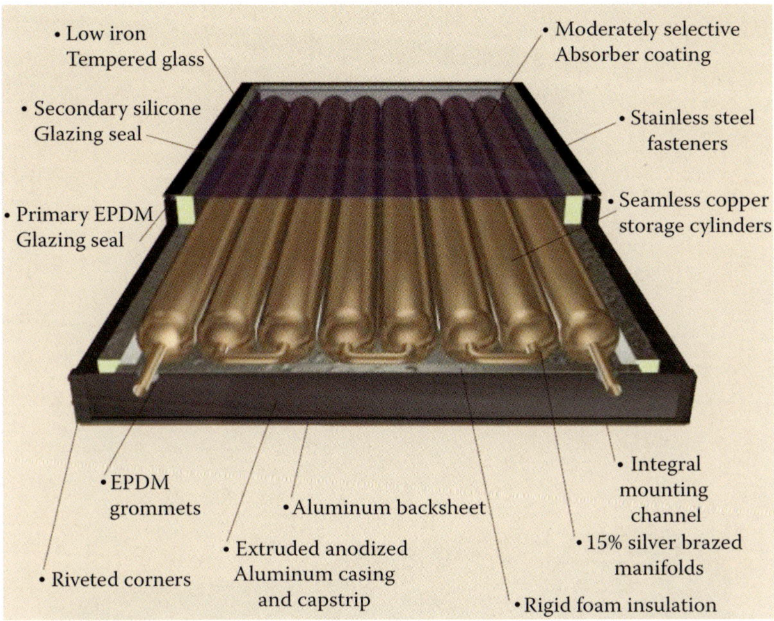

(a)

(b)

FIGURE 9.1
CopperHeart passive solar collector with integrated storage. (Courtesy of SunEarth, Inc., Fontana, CA.) (a) Photograph of installed module on residence in Arizona. (b) Schematic of CopperHeart integral collector storage system.

FIGURE 11.20
La Rance tidal power station. (From Khaligh, A. and Onar, O.C., *Energy Harvesting,* Figure 3.6, CRC Press, Boca Raton, FL, 2010.)

FIGURE 11.31
Pelamis WEC device at sea. (From Khaligh, A. and Onar, O.C., *Energy Harvesting*, Figure 4.72, CRC Press, Boca Raton, FL, 2010.)

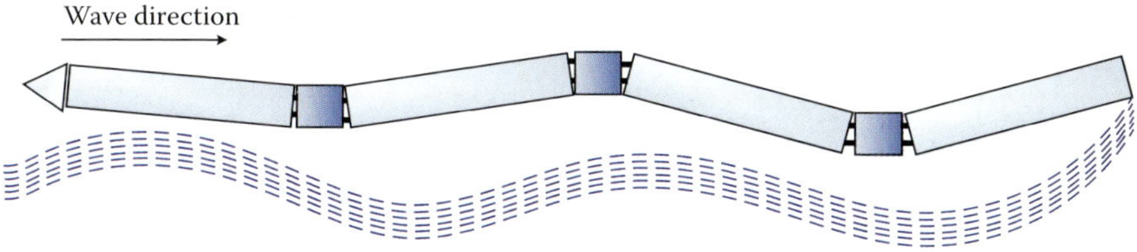

FIGURE 11.32
General layout of the Pelamis device. (From Khaligh, A. and Onar, O.C., *Energy Harvesting*, Figure 4.73, CRC Press, Boca Raton, FL, 2010.)

(a)

(b)

FIGURE 12.5
Fuel-efficient cars. (a) 2010 Toyota Prius: Toyota's third-generation Prius is a "parallel" hybrid; the 80 hp (60 kW) electric motor and gas engine work together to produce an average 50 mpg. (b) 2010 Ford Fusion: The Fusion can travel up to 47 mph (75 kph) on electric power alone. The motor is powered by a nickel-metal hydride battery pack, and the car's engine is a 2.5 L Atkinson Cycle I-4 gasoline engine. The Fusion gets 41 mpg in the city and 36 mpg highway. (Courtesy of *Solar Today*, November/December, 2009.)

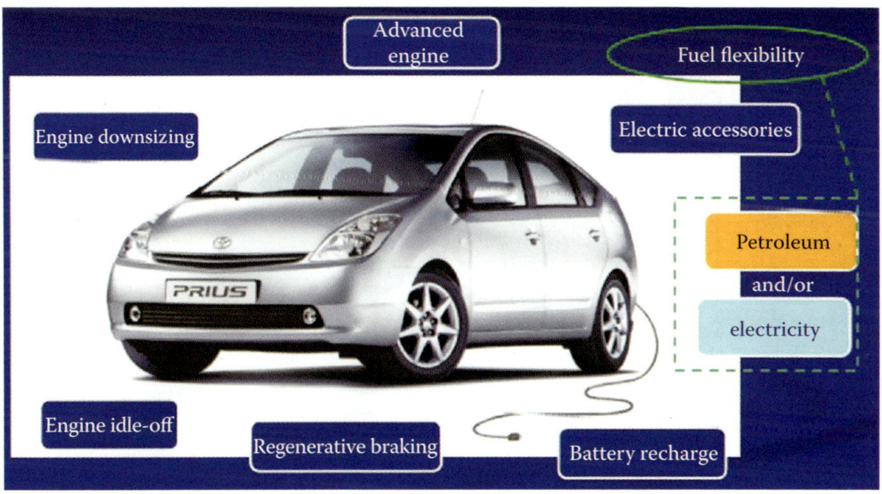

FIGURE 12.6
Plug-in hybrid electric vehicle. (Courtesy of NREL, Golden, CO.)

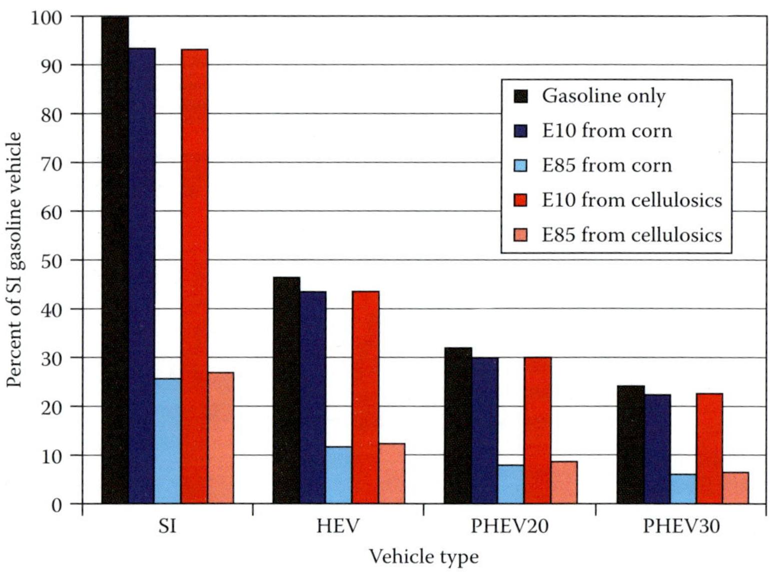

FIGURE 12.7
Petroleum requirement as a percentage of that for SI gasoline vehicle (calculated by West and Kreith, See Ref. [8]).

the sun. Kutcher and Christensen [3] have given a detailed thermal analysis of unglazed transpired solar collectors.

The most important components, whose properties determine the efficiency of solar thermal collectors, are glazings and absorbers.

7.2.3 Glazings

The purpose of a glazing or transparent cover is to transmit the shorter wavelength solar radiation but block the longer wavelength reradiation from the absorber plate, and to reduce the heat loss by convection from the top of the absorber plate. Consequently, an understanding of the process and laws that govern the transmission of radiation through a transparent medium is important. Section 7.1 describes in brief the transmission of radiation through materials.

Glass is the most common glazing material. Figure 7.10 shows transmittance of glass as a function of wavelength. Transparent plastics, such as polycarbonates and acrylics are also used as glazings for flat-plate collectors. The main disadvantage of plastics is that their transmittance in the longer wavelength is also high, therefore, they are not as good a trap as glass. Other disadvantages include deterioration over a period of time due to ultraviolet solar radiation. Their main advantage is resistance to breakage. Although glass can break easily, this disadvantage can be minimized by using tempered glass.

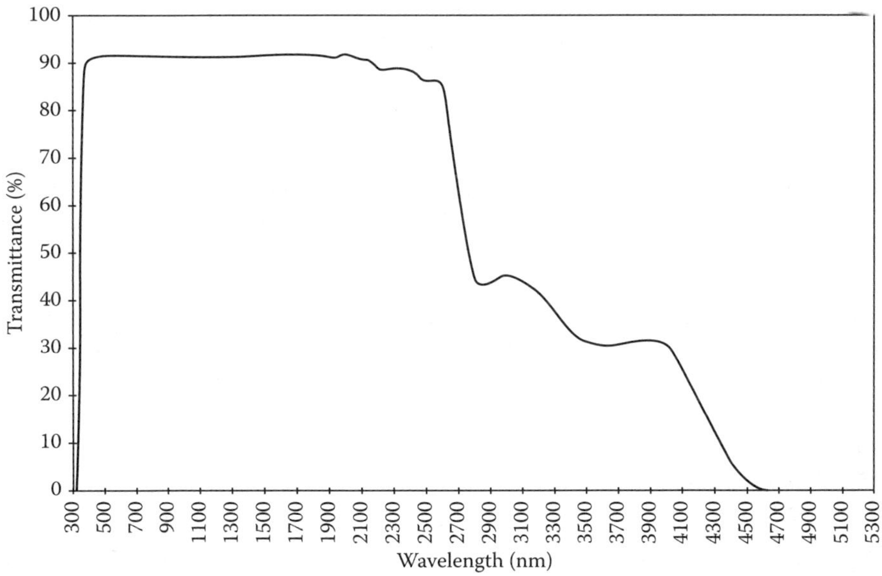

FIGURE 7.10
Spectral transmittance of 3 mm low iron float glass.

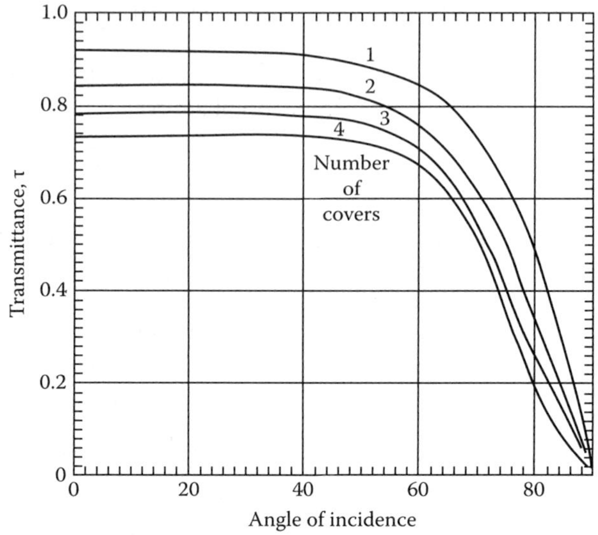

FIGURE 7.11
Transmittance of multiple glass covers having an index of refraction of 1.526 in.

In order to minimize the upward heat loss from the collector, more than one transparent glazing may be used. However, with the increase in the number of cover plates, transmittance is decreased. Figure 7.11 shows the effect of number of glass cover plates on transmittance.

7.2.3.1 Absorbers

The purpose of the absorber is to absorb as much of the incident solar radiation as possible, reemit as little as possible, and allow efficient transfer of heat to a working fluid. The most common forms of absorber plates in use are shown in Figure 7.12. The materials used for absorber plates include copper, aluminum, stainless steel, galvanized steel, plastics, and rubbers. Copper seems to be the most common material used for absorber plates and tubes because of its high thermal conductivity and high corrosion resistance. However, copper is quite expensive. For low-temperature applications (up to about 50°C or 120°F) a plastic material called ethylene-propylene polymer (trade names EPDM, HCP, etc.) can be used to provide inexpensive absorber material. To compensate for the low thermal conductivity, a large surface area is provided for heat transfer.

In order to increase the absorption of solar radiation and to reduce the emission from the absorber, the metallic absorber surfaces are painted or coated with flat black paint or some selective coating. A selective

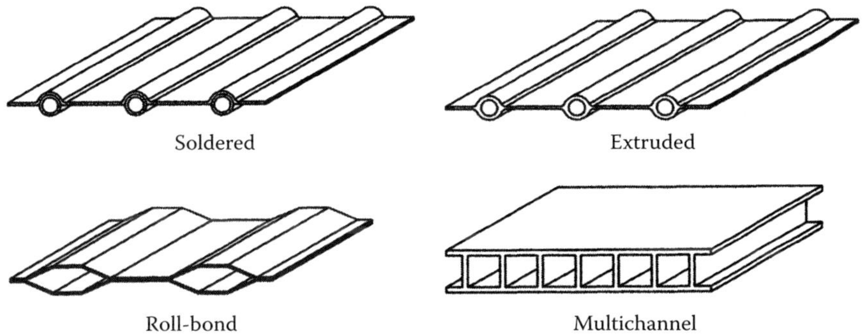

FIGURE 7.12
Common types of absorber plates.

coating has high absorptivity in the solar wavelength range (0.3–3.0 μm). Absorptivities and emissivities of some common selective surfaces are given in Table 7.3.

A simple and inexpensive collector consists of a black painted corrugated metal absorber on which water flows down open, rather than enclosed in tubes. This type of collector is called a *trickle collector* and is usually built on site. Although such a collector is simple and inexpensive, it has the disadvantages of condensation on the glazing and a higher pumping power requirement.

7.2.4 Energy Balance for a Flat-Plate Collector

The thermal performance of any type of solar thermal collector can be evaluated by an energy balance that determines the portion of the incoming radiation delivered as useful energy to the working fluid. For a flat-plate collector of an area A_c this energy balance on the absorber plate is

$$I_c A_c \tau_s \alpha_s = q_u + q_{loss} + \frac{de_c}{dt} \tag{7.8}$$

where

I_c is the solar irradiation on a collector surface
τ_s is the effective solar transmittance of the collector cover(s)
α_s is the solar absorptance of the collector–absorber plate surface
q_u is the rate of heat transfer from the collector–absorber plate to the working fluid
q_{loss} is the rate of heat transfer (or heat loss) from the collector–absorber plate to the surroundings
de_c/dt is the rate of internal energy storage in the collector

The instantaneous efficiency of a collector η_c is simply the ratio of the useful energy delivered to the total incoming solar energy, or

$$\eta_c = \frac{q_u}{A_c I_c}, \tag{7.9}$$

In practice, the efficiency must be measured over a finite time period. In a standard performance test, this period is on the order of 15 or 20 min, whereas for design, the performance over a day or over some longer period t is important. Then we have for the average efficiency

$$\eta_c = \frac{\int_0^1 q_u dt}{\int_0^1 A_c I_c dt}, \tag{7.10}$$

where t is the time period over which the performance is averaged.

A detailed and precise analysis of the efficiency of a solar collector is complicated by the nonlinear behavior of radiation heat transfer. However, a simple linearized analysis is usually sufficiently accurate in practice. In addition, the simplified analytical procedure is very important because it illustrates the parameters of significance for a solar collector and how these parameters interact. For a proper analysis and interpretation of these test results, an understanding of the thermal analysis is imperative, although for design and economic evaluation, the results of standardized performance tests are generally used.

7.2.4.1 Collector Heat-Loss Conductance

In order to obtain an understanding of the parameters determining the thermal efficiency of a solar collector, it is important to develop the concept of *collector heat-loss conductance*. Once the collector heat-loss conductance U_c is known, and when the collector plate is at an average temperature T_c, the collector heat loss can be written in the simple form

$$q_{loss} = U_c A_c (T_c - T_a) \tag{7.11}$$

The simplicity of this relation is somewhat misleading because the collector heat-loss conductance cannot be specified without a detailed analysis of all the heat losses. Figure 7.13 shows a schematic diagram of a single-glazed collector, while Figure 7.14a shows the thermal circuit with all the elements that must be analyzed before they can be combined into a single conductance element shown in Figure 7.14b. The analysis below shows an example of how this combination is accomplished.

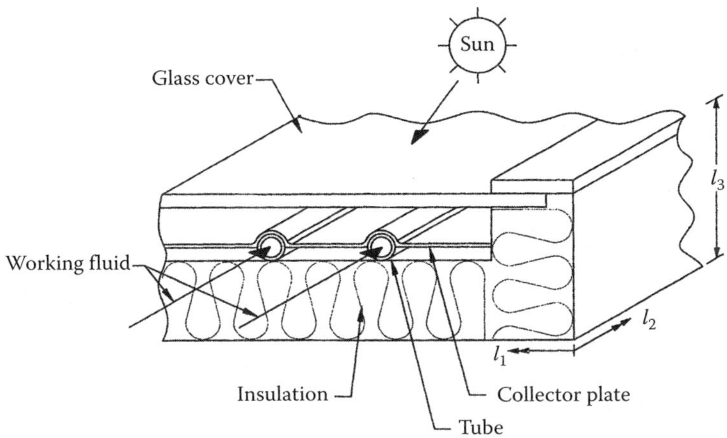

FIGURE 7.13
Schematic diagram of solar collector.

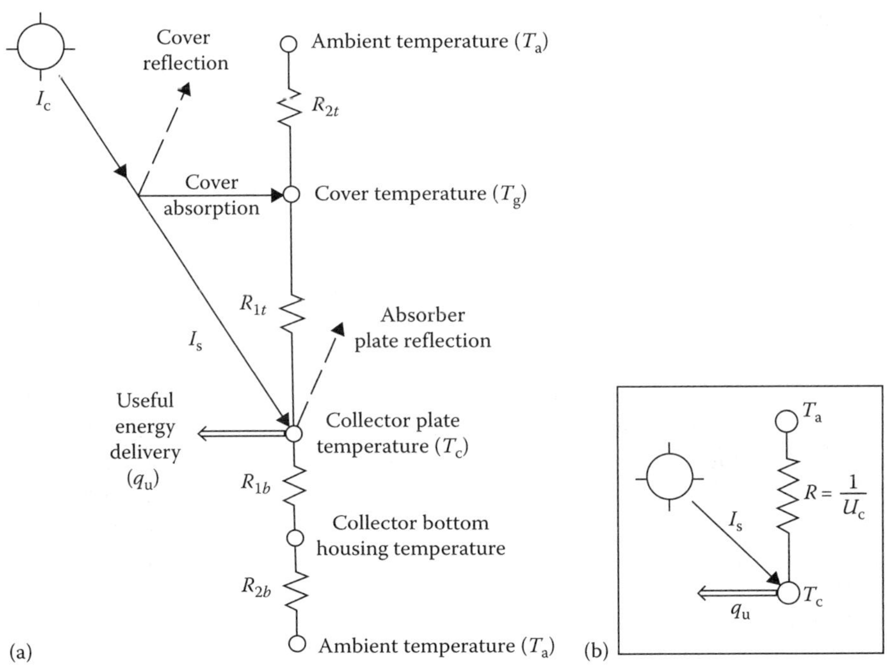

FIGURE 7.14
Thermal circuits for a flat-plate collector shown in Figure 7.13: (a) detailed circuit; (b) approximate, equivalent circuit to (a). In both circuits, the absorber energy is equal to $\alpha_s I_s$, where $I_s = \tau_s I_c$. Collector assumed to be at uniform temperature T_c.

In order to construct a model suitable for a thermal analysis of a flat-plate collector, the following simplifying assumptions will be made:

1. The collector is thermally in steady state.
2. The temperature drop between the top and bottom of the absorber plate is negligible.
3. Heat flow is one-dimensional through the cover as well as through the back insulation.
4. The headers connecting the tubes cover only a small area of the collector and provide uniform flow to the tubes.
5. The sky can be treated as though it were a black-body source for infrared radiation at an equivalent sky temperature.
6. The irradiation on the collector plate is uniform.

For a quantitative analysis, let the plate temperature be T_c and assume solar energy is absorbed at the rate $I_s\alpha_s$. Part of this energy is then transferred as heat to the working fluid, and if the collector is in the steady state, the other part is lost as heat to the ambient air if $T_c > T_a$. Some of the heat loss occurs through the bottom of the collector. It passes first through the back to the environment. Since the collector is in steady state, according to Equation 7.8

$$q_u = I_c A_c \tau_s \alpha_s - q_{loss}, \tag{7.12}$$

where q_{loss} can be determined using the equivalent thermal circuit, as shown in Figure 7.14:

$$q_{loss} = U_c A_c (T_c - T_a) = \frac{A_c (T_c - T_a)}{R}. \tag{7.13}$$

There are three parallel paths to heat loss from the hot collector absorber plate at T_c to the ambient at T_a: top, bottom, and edges. Because the edge losses are quite small compared to the top and the bottom losses, they are quite often neglected. However, they can be estimated easily if the insulation around the edges is of the same thickness as the back. The edge loss can be accounted for by simply adding the areas of the back (A_c) and the edges (A_e) for back heat loss.

Therefore, the overall heat loss coefficient is

$$U_c A_c = \frac{A_c}{R} = \frac{A_c}{R_{1t} + R_{2t}} + \frac{A_c + A_e}{R_{1b} + R_{2b}}. \tag{7.14}$$

The thermal resistances can be found easily from the definition. For example,

$$R_{1b} = \frac{l_i}{k_i} \quad \text{and} \quad R_{2b} = \frac{1}{h_{c,bottom}}, \tag{7.15}$$

where

k_i and l_i are, respectively, the thermal conductivity and thickness of the insulation

$h_{c,bottom}$ is the convective heat transfer coefficient between the collector and the air below the collector

In a well-insulated collector, R_{2b} is much smaller than R_{1b} and usually neglected. Referring to Figure 7.13

$$A_e = 2(l_1 + l_2)l_3 \tag{7.16}$$

Since the heat loss from the top is by convection and radiation, it is more complicated than the bottom heat loss. Convection and radiation provide two parallel paths for heat loss from the absorber plate at T_c to the glass cover at T_g, and from the glass cover to the ambient. That is, the series resistance of R_{1t} and R_{2t} consists of

Therefore,

$$\frac{1}{R_{1t}} = \frac{1}{R_{r,1}} + \frac{1}{R_{c,1}} = h_{r,1} + h_{c,1} \tag{7.17}$$

and

$$\frac{1}{R_{2t}} = \frac{1}{R_{r,\infty}} + \frac{1}{R_{c,\infty}} = h_{r,\infty} + h_{c,\infty} \tag{7.18}$$

Since thermal radiative heat transfer is proportional to the fourth power of the temperature, R_r and h_r are found as follows:

Radiative heat transfer from the plate to the glass cover

$$q_{r_{c \to R}} = \sigma A_c \frac{\left(T_c^4 - T_g^4\right)}{(1/\varepsilon_{p,i} + 1/\varepsilon_{s,i} - 1)} = h_{r,1} A_c (T_c - T_g), \tag{7.19}$$

where

$\varepsilon_{p,i}$ is the infrared emittance of the plate

$\varepsilon_{g,i}$ is the infrared emittance of the glass cover

Therefore

$$h_{r,1} = \frac{\sigma(T_c + T_g)\left(T_c^2 + T_g^2\right)}{(1/\varepsilon_{p,i} + 1/\varepsilon_{g,i} - 1)}. \tag{7.20}$$

Similarly, from the radiative heat transfer between the glass plate (at T_g) and the sky (at T_{sky}) we can find that

$$q_{r_{g®sky}} = \varepsilon_{g,i}\sigma A_c\left(T_g^4 - T_{sky}^4\right) = h_{r,\infty}A_c(T_g - T_a), \tag{7.21}$$

or

$$h_{r,\infty} = \varepsilon_{g,i}\sigma\frac{\left(T_g^4 - T_{sky}^4\right)}{(T_g - T_a)}. \tag{7.22}$$

Evaluation of the collector heat-loss conductance defined by Equation 7.14 requires iterative solution of Equations 7.19 and 7.21, because the unit radiation conductances are functions of the cover and plate temperatures, which are not known a priori. A simplified procedure for calculating U_c for collectors with all covers of the same material, which is often sufficiently accurate and more convenient to use, has been suggested by Hottel and Woertz [4] and Klein [5]. It is also suitable for application to collectors with selective surfaces. For this approach, the collector top loss in watts is written in the form [6]

$$q_{\text{top loss}} = \frac{(T_c - T_a)A_c}{N/(C/T_c)[(T_c - T_a)/(N+f)]^{0.33} + 1/h_{c,\infty}} + \frac{\sigma\left(T_c^4 - T_a^4\right)A_c}{1/[\varepsilon_{p,i} + 0.05N(1-\varepsilon_{p,i})] + (2N + f - 1)/\varepsilon_{g,i} - N}, \tag{7.23}$$

where

$f = (1 - 0.04h_{c,\infty} + 0.005h_{c,\infty}^2)(1 + 0.091N)$
$C = 250[1–0.0044(\beta - 90)]$
N = number of covers
$h_{c,\infty} = 5.7 + 3.8V$
$\varepsilon_{g,i}$ = infrared emittance of the covers
V = wind speed in m/s

The values of $q_{\text{top loss}}$ calculated from Equation 7.23 agreed closely with the values obtained from Equation 7.22 for 972 different observations encompassing the following conditions:

$320 < T_c < 420\,K$
$260 < T_a < 310\,K$
$0.1 < \varepsilon_{p,i} < 0.95$
$0 \le V \le 10\,m/s$
$1 \le N \le 3$
$0 \le \beta \le 90$

The standard deviation of the differences in $U_c = q_{top\ loss}/A_c(T_c - T_a)$ was $0.14\,W/m^2 \cdot K$ for these comparisons.

7.2.5 Thermal Analysis of Flat-Plate Collector–Absorber Plate

In order to determine the efficiency of a solar collector, the rate of heat transfer to the working fluid must be calculated. If transient effects are neglected [5,7,8], the rate of heat transfer to the fluid flowing through a collector depends on the temperature of the collector surface from which heat is transferred by convection to the fluid, the temperature of the fluid, and the heat transfer coefficient between the collector and the fluid. To analyze the rate of heat transfer, consider first the condition at a cross section of the collector with flow ducts of rectangular cross sections, as shown in Figure 7.15. Solar radiant energy

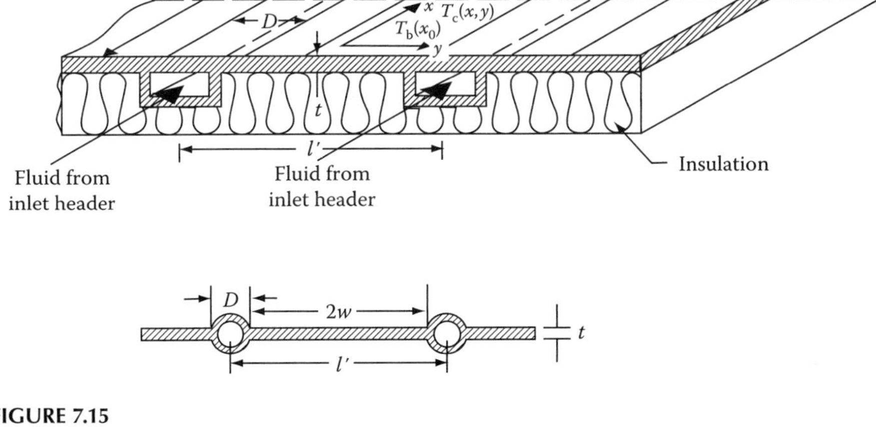

FIGURE 7.15
Sketch showing coordinates and dimensions for collector plate and fluid ducts.

impinges on the upper face of the collector plate. A part of the total solar radiation falls on the upper surface of the flow channels, while another part is incident on the plates connecting any two adjacent flow channels. The latter is conducted in a transverse direction toward the flow channels. The temperature is maximum at any midpoint between adjacent channels, and the collector plate acts as a fin attached to the walls of the flow channel. The thermal performance of a fin can be expressed in terms of its efficiency. The fin efficiency η_f is defined as the ratio of the rate of heat flow through the real fin to the rate of heat flow through a fin of infinite thermal conductivity, that is, a fin at a uniform temperature. We shall now derive a relation to evaluate this efficiency for a flat-plate solar collector.

If U_c is the overall unit conductance from the collector-plate surface to the ambient air, the rate of heat loss from a given segment of the collector plate at x, y in Figure 7.15 is

$$q(x,y) = U_c[T_c(x,y) - T_a]dxdy, \tag{7.24}$$

where

T_c is the local collector-plate temperature ($T_c > T_a$)
T_a is the ambient air temperature
U_c is the overall unit conductance between the plate and the ambient air

U_c includes the effects of radiation and free convection between the plates, the radiative and convective transfer between the top of the cover and the environment, and conduction through the insulation. Its quantitative evaluation has been previously considered.

If conduction in the x direction is negligible, a heat balance at a given distance x_0 for a cross section of the flat-plate collector per unit length in the x direction can be written in the form

$$\alpha_s I_s dy - U_c(T_c - T_a)dy + \left(-kt\left.\frac{dT_c}{dy}\right|_{y,x_0}\right) - \left(-kt\left.\frac{dT_c}{dy}\right|_{y+dy,x_0}\right) = 0. \tag{7.25}$$

If the plate thickness t is uniform and the thermal conductivity of the plate is independent of temperature, the last term in Equation 7.25 is

$$\left.\frac{dT_c}{dy}\right|_{y+dy,x_0} = \left.\frac{dT_c}{dy}\right|_{y,x_0} + \left(\frac{d^2T_c}{dy^2}\right)_{y,x_0} dy,$$

and Equation 7.25 can be cast into the form of a second-order differential equation:

$$\frac{d^2T}{dy^2} = \frac{U_c}{kt}\left[T_c - \left(T_a + \frac{\alpha_s I_s}{U_c}\right)\right] \tag{7.26}$$

The boundary conditions for the system described above at a fixed x_0, are

1. At the center between any two ducts the heat flow is 0, or at y = 0, $dT_c = 0$.
2. At the duct the plate temperature is $T_b(x_0)$, or at $y = w = (l' - D)/2$, $T_c = T_b(x_0)$ where $T_b(x_0)$ is the fin-base temperature.

If we let $m^2 = U_c/kt$ and $\phi = T_c - (T_a + \alpha_s I_s/U_c)$, Equation 7.26 becomes

$$\frac{d^2\phi}{dy^2} = m^2\phi, \tag{7.27}$$

subject to the boundary conditions

$$\frac{d\phi}{dy} = 0 \quad \text{at} \quad y = 0,$$

and

$$\phi = T_b(x_0) - \left(T_a + \frac{\alpha_s I_s}{U_c}\right) \quad \text{at} \quad y = w.$$

The general solution of Equation 7.27 is

$$\phi = C_1 \sinh my + C_2 \cosh my. \tag{7.28}$$

The constants C_1 and C_2 can be determined by substituting the two boundary conditions and solving the two resulting equations for C_1 and C_2. This gives

$$\frac{T_c - (T_a + \alpha_s I_s/U_c)}{T_b(x_0) - (T_a + \alpha_s I_s/U_c)} = \frac{\cosh my}{\cosh mw}, \tag{7.29}$$

From the preceding equation, the rate of heat transfer to the conduit from the portion of the plate between two conduits can be determined by evaluating the temperature gradient at the base of the fin, or

$$q_{\text{fin}} = -kt\frac{dT_c}{dy}\bigg|_{y=w} = \frac{1}{m}\left\{\alpha_s I_s - U_c\left[T_b(x_0) - T_a\right]\tanh mw\right\}. \tag{7.30}$$

Since the conduit is connected to fins on both sides, the total rate of heat transfer is

$$q_{\text{total}}(x_0) = 2w\left\{\alpha_s I_s - U_c\left[T_b(x_0) - T_a\right]\right\}\frac{\tanh mw}{mw}. \tag{7.31}$$

If the entire fin were at the temperature $T_b(x_0)$, a situation corresponding physically to a plate of infinitely large thermal conductivity, the rate of heat transfer would be a maximum, $q_{\text{total, max}}$. As mentioned previously, the ratio of the rate of heat transfer with a real fin to the maximum rate obtainable is the fin efficiency η_f. With this definition, Equation 7.31 can be written in the form

$$q_{\text{total}}(x_0) = 2w\eta_f\left\{\alpha_s I_s - U_c\left[T_b(x_0) - T_a\right]\right\}, \tag{7.32}$$

where $\eta_f \equiv \tanh mw/mw$.

The fin efficiency η_f is plotted as a function of the dimensionless parameter $w(U_c/kt)^{1/2}$ in Figure 7.16. When the fin efficiency approaches unity, the maximum portion of the radiant energy impinging on the fin becomes available for heating the fluid.

In addition to the heat transferred through the fin, the energy impinging on the portion of the plate above the flow passage is also useful. The rate of useful energy from this region available to heat the working fluid is

$$q_{\text{duct}}(x_0) = D\left\{a_s I_s - U_c\left[T_b(x_0) - T_a\right]\right\}. \tag{7.33}$$

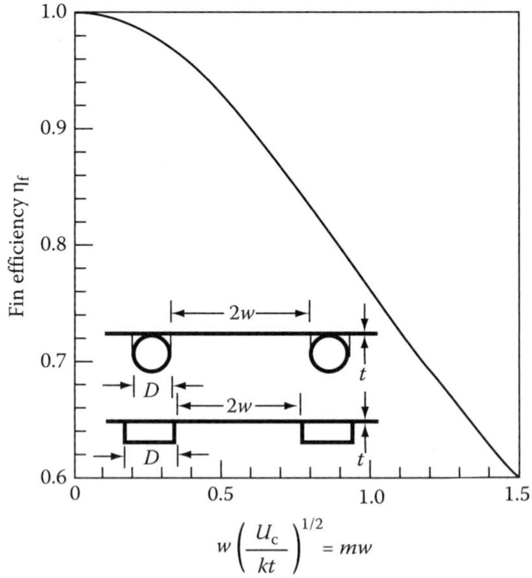

FIGURE 7.16
Fin efficiency for tube and sheet flat-plate solar collectors.

Thus, the useful energy per unit length in the flow direction becomes

$$q_u(x_0) = (D + 2w\eta)\{\alpha_s I_s - U_c[T_b(x_0) - T_a]\}. \quad (7.34)$$

The energy $q_u(x_0)$ must be transferred as heat to the working fluid. If the thermal resistance of the metal wall of the flow duct is negligibly small and there is no contact resistance between the duct and the plate, the rate of heat transfer to the fluid is

$$q_u(x_0) = P\bar{h}_{c,i}[T_b(x_0) - T_f(x_0)], \quad (7.35)$$

where P is the perimeter of the flow duct, which is $2(D + d)$ for a rectangular duct. Contact resistance may become important in poorly manufactured collectors in which the flow duct is clamped or glued to the collector plate. Collectors manufactured by such methods are usually not satisfactory.

7.2.6 Collector Efficiency Factor

To obtain a relation for the useful energy delivered by a collector in terms of known physical parameters, the fluid temperature, and the ambient temperature, the collector temperature must be eliminated from Equations 7.34 and 7.35. Solving for $T_b(x_0)$ in Equation 7.35 and substituting this relation in Equation 7.34 gives

$$q_u(x_0) = l'F'\{\alpha_s I_s - U_c[T_f(x_0) - T_a]\}, \quad (7.36)$$

where

F' is called the collector efficiency factor [9]

$l' = (2w + D)$

F' is given by

$$F' = \frac{1/U_c}{l'[1/(U_c(D + 2w\eta_f)) + 1/(\bar{h}_{c,i}P)]}. \quad (7.37a)$$

For a flow duct of circular cross section of diameter D, F' can be written as

$$F' = \frac{1/U_c}{l'[1/(U_c(D + 2w\eta_f) + 1/(\bar{h}_{c,i}\pi D)]} \quad (7.37b)$$

TABLE 7.6

Typical Values for the Parameters That Determine the Collector Efficiency Factor F' for a Flat-Plate Collector in Equation 7.37

		U_c
2 glass covers	$4\,W/m^2 \cdot K$	$0.685\,Btu/h \cdot ft^2 \cdot °F$
1 glass cover	$8\,W/m^2 \cdot K$	$1.37\,Btu/h \cdot ft^2 \cdot °F$
		kt
Copper plate, 1 mm thick	$0.38\,W/K$	$0.72\,Btu/h \cdot °F$
Steel plate, 1 mm thick	$0.045\,W/K$	$0.0866\,Btu/h \cdot °F$
		$\bar{h}_{c,i}$
Water in laminar flow forced convection	$300\,W/m^2 \cdot K$	$52\,Btu/h \cdot ft^2 \cdot °F$
Water in turbulent flow forced convection	$1500\,W/m^2 \cdot K$	$254\,Btu/h \cdot ft^2 \cdot °F$
Air in turbulent forced convection	$100\,W/m^2 \cdot K$	$17.6\,Btu/h \cdot ft^2 \cdot °F$

Physically, the denominator in Equations 7.37 is the thermal resistance between the fluid and the environment, whereas the numerator is the thermal resistance between the collector surface and the ambient air. The collector-plate efficiency factor F' depends on U_c, $\bar{h}_{c,i}$, and η_f. It is only slightly dependent on temperature and can, for all practical purposes, be treated as a design parameter. Typical values for the factors determining the value of F' are given in Table 7.6.

The collector efficiency factor increases with increasing plate thickness and plate thermal conductivity, but it decreases with increasing distance between flow channels. Also, increasing the heat transfer coefficient between the walls of the flow channel and the working fluid increases F', but an increase in the overall conductance U_c will cause F' to decrease.

7.2.7 Collector Heat-Removal Factor

Equation 7.36 yields the rate of heat transfer to the working fluid at a given point x along the plate for specified collector and fluid temperatures. However, in a real collector, the fluid temperature increases in the direction of flow as heat is transferred to it. An energy balance for a section of flow duct dx can be written in the form

$$\dot{m}c_p\left(T_f\big|_{x+dx} - T_f\big|_x\right) = q_u(x)dx. \tag{7.38}$$

Substituting Equation 7.36 for $q_u(x)$ and $T_f(x) + (dT_f(x)/dx)\,dx$ for $T_f\big|_{x+dx}$ in Equation 7.38 gives the differential equation

$$\dot{m}c_p\frac{dT_f(x)}{dx} = l'F'\left\{\alpha_s I_s - U_c\left[T_f(x) - T_a\right]\right\}. \tag{7.39}$$

Separating the variables gives, after some rearranging,

$$\frac{dT_f(x)}{T_f(x)-T_a-\alpha_s I_s/U_c}=\frac{l'F'U_c}{\dot{m}c_p}dx. \tag{7.40}$$

Equation 7.40 can be integrated and solved for the outlet temperature of the fluid $T_{f,out}$ for a duct length L, and for the fluid inlet temperature $T_{f,in}$ if we assume that F' and U_c are constant, or

$$\frac{T_{f,out}-T_a-\alpha_s I_s/U_c}{T_{f,in}-T_a-\alpha_s I_s/U_c}=\exp\left(-\frac{U_c l'F'L}{\dot{m}c_p}\right). \tag{7.41}$$

To compare the performance of a real collector with the thermodynamic optimum, it is convenient to define the heat-removal factor F_R as the ratio between the actual rate of heat transfer to the working fluid and the rate of heat transfer at the minimum temperature difference between the absorber and the environment. The thermodynamic limit corresponds to the condition of the working fluid remaining at the inlet temperature throughout the collector. This can be approached when the fluid velocity is very high. From its definition F_R can be expressed as

$$F_R=\frac{Gc_p(T_{f,out}-T_{f,in})}{\alpha_s I_s-U_c(T_{f,in}-T_a)}, \tag{7.42}$$

where G is the flow rate per unit surface area of collector $\dot{m}/A_c$. By regrouping the right-hand side of Equation 7.42 and combining with Equation 7.41, it can easily be verified that

$$\begin{aligned}F_R&=\frac{Gc_p}{U_c}\left[1-\frac{\alpha_s I_s/U_c-(T_{f,out}-T_a)}{\alpha_s I_s/U_c-(T_{f,in}-T_a)}\right]\\&=\frac{Gc_p}{U_c}\left[1-\exp\left(-\frac{U_cF'}{Gc_p}\right)\right].\end{aligned} \tag{7.43}$$

Inspection of the above relation shows that F_R increases with increasing flow rate and approaches as an upper limit F', the collector efficiency factor. Since the numerator of the right-hand side of Equation 7.42 is q_u, the rate of useful heat transfer can now be expressed in terms of the fluid inlet temperature, or

$$q_u=A_cF_R\left[\alpha_s I_s-U_c(T_{f,in}-T_a)\right]. \tag{7.44}$$

If a glazing above the absorber plate has transmittance τ_s then

$$q_u=A_cF_R\left[\tau_s\alpha_s I_c-U_c(T_{f,in}-T_a)\right], \tag{7.45}$$

and instantaneous efficiency η_c is

$$\eta_c = \frac{q_u}{I_c A_c} = F_R\left[\tau_s\alpha_s - \frac{U_c(T_{f,in} - T_a)}{I_c}\right] \tag{7.46}$$

Equation 7.46 is also known as the *Hottel–Whillier–Bliss* equation. This is a convenient form for design, because the fluid inlet temperature to the collector is usually known or can be specified.

Example 7.1

Calculate the averaged hourly and daily efficiency of a water solar collector on January 15, in Boulder, Colorado. The collector is tilted at an angle of 60° and has an overall conductance of 8.0 W/m² · K on the upper surface. It is made of copper tubes, with a 1 cm ID, 0.05 cm thick, which are connected by a 0.05 cm thick plate at a center-to-center distance of 15 cm. The heat transfer coefficient for the water in the tubes is 1500 W/m² · K, the cover transmittance is 0.9, and the solar absorptance of the copper surface is 0.9. The collector is 1 m wide and 2 m long, the water inlet temperature is 330 K, and the water flow rate is 0.02 kg/s. The horizontal insolation (total) I_h and the environmental temperature are tabulated below. Assume the diffuse radiation accounts for 25% of the total insolation.

Solution

The total radiation received by the collector is calculated from Equation 5.42, and neglecting the ground reflected radiation

$$I_c = I_{d,c} + I_{b,c} = 0.25 I_h \cos^2\left(\frac{60}{2}\right) + (1 - 0.25) I_h R_b.$$

Time (h)	I_h (W/m²)	T_{amb} (K)
7–8	12	270
8–9	80	280
9–10	192	283
10–11	320	286
11–12	460	290
12–13	474	290
13–14	395	288
14–15	287	288
15–16	141	284
16–17	32	280

The tilt factor R_b is obtained from its definition in Chapter 5 (see Equation 5.56):

$$R_b = \frac{\cos i}{\sin\alpha} = \frac{\sin(L-\beta)\sin\delta_s + \cos(L-\beta)\cos\delta_s\cos h_s}{\sin L\sin\delta_s + \cos L\cos\delta_s\cos h_s}$$

where $L = 40°$ $\delta_s = -21.1$ on January 15, and $\beta = 60°$. The hour angle h_s equals 15° for each hour away from noon.

The fin efficiency is obtained from Equation 7.32:

$$\eta_f = \frac{\tanh m(l' - D)/2}{m(l' - D)/2}$$

where

$$m = \left(\frac{U_c}{kt}\right)^{1/2} = \left(\frac{8}{390 \times 5 \times 10^{-4}}\right)^{1/2} = 6.4,$$

and

$$\eta_f = \frac{\tanh 6.4(0.15 - 0.01)/2}{6.4(0.15 - 0.01)/2} = 0.938.$$

The collector efficiency factor F' is, from Equation 7.37

$$F' = \frac{1/U_c}{l'\left[1/\left(U_c(D + 2w\eta_f)\right) + 1/(h_{c,i}\pi D)\right]}$$

$$= \frac{1/8.0}{0.15\left[1/8.0(0.01 + 0.14 \times 0.938) + 1/1500\pi \times 0.01\right]} = 0.92.$$

Then we obtain the heat-removal factor from Equation 7.43:

$$F_R = \frac{Gc_p}{U_c}\left[1 - \exp\left(-\frac{U_c F'}{Gc_p}\right)\right].$$

Time (h)	I_h (W/m²)	R_b	$I_{d,c}$ (W/m²)	$I_{b,c}$ (W/m²)	I_c (W/m²)	q_u (W)	T_{amb} (K)	η_c
7–8	12	10.9	1	98	99	0	270	0
8–9	80	3.22	5	193	198	0	280	0
9–10	192	2.44	12	351	363	0	283	0
10–11	320	2.18	20	523	543	148	286	0.137
11–12	460	2.08	29	718	747	482	290	0.322
12–13	474	2.08	30	739	769	512	290	0.333
13–14	395	2.18	25	646	671	351	288	0.261
14–15	287	2.49	18	525	543	175	288	0.162
15–16	141	3.22	9	341	350	0	284	0
16–17	32	10.9	2	261	263	0	280	0

$$F_R = \frac{0.001 \times 4184}{8.0}\left[1 - \exp\left(-\frac{8.0 \times 0.922}{0.01 \times 4184}\right)\right] = 0.845.$$

From Equation 7.45, the useful heat delivery rate is

$$q_u = 2 \times 0.845\left[I_c \times 0.81 - 8.0(T_{f,in} - T_{amb})\right].$$

The efficiency of the collector is $\eta_c = q_u/A_c I_c$ and the hourly averages are calculated in the table above.

Thus, $\sum I_c = 4546\,W/m^2$ and $\sum q_u = 1668\,W$. The daily average efficiency is obtained by summing the useful energy for those hours during which the collector delivers heat and dividing by the total insolation between sunrise and sunset. This yields

$$\bar{\eta}_{c,day} = \frac{\sum q_u}{\sum A_c I_c} = \frac{1668}{2 \times 4546} = 0.183 \quad \text{or} \quad 18.3\%$$

7.2.8 Transient Effects

The preceding analysis assumed that steady-state conditions exist during the operation of the collector. Under actual operating conditions, the rate of insolation will vary and the ambient temperature and the external wind conditions may change. To determine the effect of changes in these parameters on the performance of a collector, it is necessary to make a transient analysis that takes the thermal capacity of the collector into account.

As shown in [10], the effect of collector thermal capacitance is the sum of two contributions: the *collector storage* effect, resulting from the heat required to bring the collector up to its final operating temperature, and the *transient* effect, resulting from fluctuations in the meteorological conditions. Both effects result in a net loss of energy delivered compared with the predictions from the zero capacity analysis. This loss is particularly important on a cold morning when all of the solar energy absorbed by the collector is used to heat the hardware and the working fluid, thus delaying the delivery of useful energy for some time after the sun has come up.

Transient thermal analyses can be made with a high degree of precision [11], but the analytical predictions are no more accurate than the weather data and the overall collector conductance. For most engineering applications, a simpler approach is therefore satisfactory [12]. For this approach, it will be assumed that the absorber plate, the ducts, the back insulation, and the working fluid are at the same temperature. If back losses are neglected, an energy balance on the collector plate and the working fluid for a single-glazed collector delivering no useful energy can be written in the form

$$(\overline{mc})_{\mathrm{p}} \frac{d\overline{T}_{\mathrm{p}}(t)}{dt} = A_{\mathrm{c}} I_{\mathrm{s}} \alpha_{\mathrm{s}} + A_{\mathrm{c}} U_{\mathrm{p}} \left[\overline{T}_{\mathrm{g}}(t) - \overline{T}_{\mathrm{p}}(t) \right], \tag{7.47}$$

where

$(\overline{mc})_{\mathrm{p}}$ is the sum of the thermal capacities of the plate, the fluid, and the insolation

I_{s} is the insolation on the absorber plate

U_{p} is the conductance between the absorber plate at $\overline{T}_{\mathrm{p}}$ and its cover at $\overline{T}_{\mathrm{g}}$

Similarly, a heat balance on the collector cover gives

$$(mc)_{\mathrm{g}} \frac{d\overline{T}_{\mathrm{g}}(t)}{dt} = A_{\mathrm{c}} U_{\mathrm{p}} \left[\overline{T}_{\mathrm{p}}(t) - \overline{T}_{\mathrm{g}}(t) \right] - A_{\mathrm{c}} U_{\infty} \left[\overline{T}_{\mathrm{g}}(t) - T_{\mathrm{a}} \right], \tag{7.48}$$

where

$U_{\infty} = (h_{\mathrm{c},\infty} + h_{\mathrm{r},\infty})$ (see Equation 7.18)

$(mc)_{\mathrm{g}}$ = thermal capacity of the cover plate

Equations 7.47 and 7.48 can be solved simultaneously and the transient heat loss can then be determined by integrating the instantaneous loss over the time during which transient effects are pronounced. A considerable simplification in the solution is possible if one assumes that at any time, the collector heat loss and the cover heat loss are equal, as in a quasi-steady state, so that

$$U_{\infty} A_{\mathrm{c}} \left[\overline{T}_{\mathrm{g}}(t) - T_{\mathrm{a}} \right] = U_{\mathrm{c}} A_{\mathrm{c}} \left[\overline{T}_{\mathrm{p}}(t) - T_{\mathrm{a}} \right] \tag{7.49}$$

Then, for a given air temperature, the differentiation of Equation 7.49 gives

$$\frac{d\overline{T}_{\mathrm{g}}(t)}{dt} = \frac{U_{\mathrm{e}} d\overline{T}_{\mathrm{p}}(t)}{U_{\infty} dt}. \tag{7.50}$$

Adding Equations 7.48 through 7.50 gives a single differential equation for the plate temperature:

$$\left[\left(\overline{mc} \right)_{\mathrm{p}} + \frac{U_{\mathrm{c}}}{U_{\infty}} (mc)_{\mathrm{R}} \right] \frac{d\overline{T}_{\mathrm{p}}(t)}{dt} = \left[\alpha_{\mathrm{s}} I_{\mathrm{s}} - U_{\mathrm{c}} (\overline{T}_{\mathrm{p}}(t) - T_{\mathrm{a}}) \right] A_{\mathrm{c}} \tag{7.51}$$

Equation 7.51 can be solved directly for given values of I_{s} and T_{a}. The solution to Equation 7.51 then gives the plate temperature as a function of time, for an initial plate temperature $T_{\mathrm{p},0}$, in the form

$$\overline{T}_{\mathrm{P}}(t)-T_{\mathrm{a}}=\frac{\alpha_{\mathrm{s}}I_{\mathrm{s}}}{U_{\mathrm{c}}}-\left[\frac{\alpha_{\mathrm{s}}I_{\mathrm{s}}}{U_{\mathrm{c}}}-(T_{\mathrm{p},0}-T_{\mathrm{a}})\right]\exp\left[-\frac{U_{\mathrm{c}}A_{\mathrm{c}}t}{\left(\overline{mc}\right)_{\mathrm{p}}+(U_{\mathrm{c}}/U_{\infty})_{\mathrm{c}}(mc)_{\mathrm{g}}}\right]. \quad (7.52)$$

Collectors with more than one cover can be treated similarly, as shown in [13].

For a transient analysis, the plate temperature $\overline{T}_{\mathrm{p}}$ can be evaluated at the end of a specified time period if the initial value of $\overline{T}_{\mathrm{p}}$ and the values of α_{s}, I_{s}, U_{c}, and T_{a} during the specified time are known. Repeated applications of Equation 7.52 provide an approximate method of evaluating the transient effects. An estimate of the net decrease in useful energy delivered can be obtained by multiplying the effective heat capacity of the collector, given by $(\overline{mc})_{\mathrm{p}}+(U_{\mathrm{c}}/U_{\infty})(mc)_{\mathrm{g}}$, by the temperature rise necessary to bring the collector to its operating temperature. Note that the parameter $[(\overline{mc})_{\mathrm{p}}+(U_{\mathrm{c}}/U_{\infty})(mc)_{\mathrm{g}}]/U_{\mathrm{c}}A_{\mathrm{c}}$ is the *time constant* of the collector [11,14] and small values of this parameter will reduce losses resulting from transient effects.

Example 7.2

Calculate the temperature rise between 8 and 10 a.m. of a 1 × 2 m single-glazed water collector with a 0.3 cm thick glass cover if the heat capacities of the plate, water, and back insulation, are 5, 3, and 2 kJ/K, respectively. Assume that the unit surface conductance from the cover to ambient air is 18 W/m² · K and the unit surface conductance between the collector and the ambient air is U_{c} = 6 W/m²·K. Assume that the collector is initially at the ambient temperature. The absorbed insolation $\alpha_{\mathrm{s}}I_{\mathrm{s}}$ during the first hour averages 90 W/m² and between 9 and 10 a.m., 180 W/m². The air temperature between 8 and 9 a.m. is 273 K and that between 9 and 10 a.m., 278 K.

Solution

The thermal capacitance of the glass cover is $(mc)_{\mathrm{g}} = (\rho V c_{\mathrm{p}})_{\mathrm{g}}$ = (2500 kg/m³) (1 m × 2 m × 0.003 m) (1 kJ/kg·K) = 15 kJ/K. The combined collector, water and insulation thermal capacity is equal to

$$(\overline{mc})_{\mathrm{p}}+\frac{U_{\mathrm{c}}}{U_{\infty}}(mc)_{\mathrm{g}}=5+3+2+0.3\times 15=15.5\ \mathrm{kJ/K}.$$

From Equation 7.52 the temperature rise of the collector given by

$$T_{\mathrm{p}}-T_{\mathrm{a}}=\frac{\alpha_{\mathrm{s}}I_{\mathrm{s}}}{U_{\mathrm{c}}}-\left[\frac{\alpha_{\mathrm{s}}I_{\mathrm{s}}}{U_{\mathrm{c}}}-(T_{\mathrm{p},0}-T_{\mathrm{a}})\right]\exp\left[-\frac{U_{\mathrm{c}}A_{\mathrm{c}}t}{(\overline{mc})_{\mathrm{p}}+(U_{\mathrm{c}}/U_{\infty})(mc)_{\mathrm{g}}}\right].$$

At 8 a.m., $T_{p,o} = T_a$, therefore the temperature rise, between 8 and 9 a.m., is

$$= \frac{90}{6}\left[1 - \exp\left(-\frac{2 \times 6 \times 3{,}600}{15{,}500}\right)\right] = 15 \times 0.944 = 14.2\,\text{K}.$$

Thus, at 9 a.m., the collector temperature will be 287.2 K. Between 9 and 10 a.m., the collector temperature will rise as shown below:

$$= 278 + \frac{180}{6} - (30 - 9.2)0.056 = 306.3\,\text{K}(91\,°\text{F})$$

Thus, at 10 a.m., the collector temperature has achieved a value sufficient to deliver useful energy at a temperature level of 306 K.

7.2.9 Air-Cooled Flat-Plate Collector

The basic air-cooled flat-plate collector shown in Figure 7.17 differs from the liquid-based collectors described in preceding sections because of the relatively poor heat transfer properties of air. For example, in turbulent flow in a given conduit for a fixed value of Reynolds number, the convection heat transfer coefficient for water is about 50 times greater than that for air. As a result, it is essential to provide the largest heat transfer area possible to remove heat from the absorber surface of an air-heating collector. This makes air collectors expensive and difficult to locate in space heating of buildings.

The most common way to achieve adequate heat transfer in air collectors is to flow air over the entire rear surface of the absorber as shown. The heat transfer analysis of such a collector does not involve the fin effect or the tube-to-plate bond conductance problem, which arises in liquid collectors. The heat transfer process is essentially that of an unsymmetrically heated duct of large aspect ratio (typically 20–40).

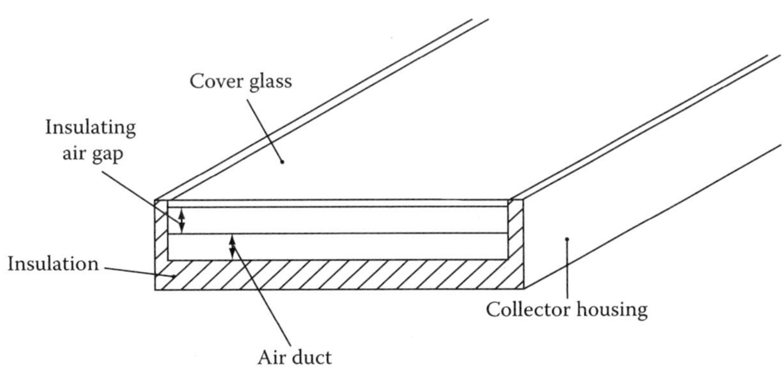

FIGURE 7.17
Schematic diagram of a basic air-heating flat-plate collector with a single glass (or plastic) cover.

7.3 Evacuated Tube Collectors

Two general methods exist for significantly improving the performance of solar collectors above the minimum flat-plate collector level. The first method increases solar flux incident on the receiver. It will be described in the next section on concentrators. The second method involves the reduction of parasitic heat loss from the receiver surface. This can be accomplished by placing a tubular receiver inside another transparent tube and evacuating the annular space between them to about 10–4 mm Hg. This arrangement eliminates convection losses and is commonly referred to as an evacuated tube collector. Figure 7.18 is a schematic cross section of a single tube in an evacuated tube collector.

Evacuated tube devices have been proposed as efficient solar energy collectors since the early twentieth century. In 1909, Emmett [15] proposed several evacuated tube concepts for solar energy collection. Speyer [16] also proposed a tubular evacuated flat-plate design for high-temperature operation. With the recent advances in vacuum technology, evacuated tube collectors can be reliably mass-produced. Their high-temperature effectiveness

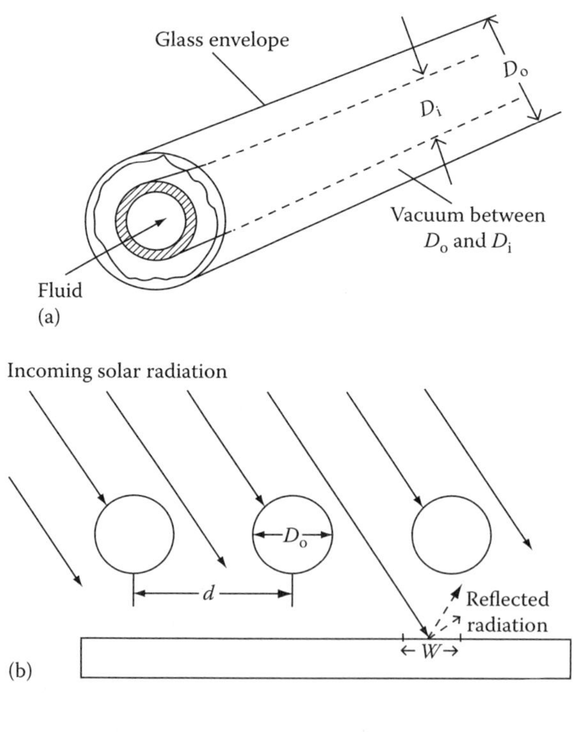

FIGURE 7.18
A schematic cross section of (a) single tube in an evacuated tube collector, and (b) a cross section of an assembly of tubes from a collector. (From Goswami, Y. et al., *Principles of Solar Engineering*, 2nd edn., Taylor & Francis, Philadelphia, 2000. With permission.)

is essential for the efficient operation of proposed solar air-conditioning systems and process heat systems.

Simple tubes with an evacuated annulus are suitable for various concentrating collectors such as parabolic troughs or compound parabolic concentrators to be discussed in Chapter 8. But for flat-plate stationary collectors, an arrangement with several tubes in parallel as shown in Figure 7.18 is required.

Since close packing of concentric-tube collector (CTC) tubes in an array can result in shading losses at any angle other than normal incidence, it is cost-effective to space the tubes apart and to use a back reflector in order to capture radiation passing between the tubes. Figure 7.18 shows the geometry of part of a CTC array with tube spacing *d*; Figure 7.18 is a cutaway drawing of one tubular assembly. Beekley and Mather [17] have analyzed the CTC in detail and have shown that a tube spacing one envelope diameter apart (i.e., $d = 2D$) maximizes daily energy gain. The specular reflector improves performance by 10% or more.

CTC arrays can collect both direct and diffuse radiation. Each radiation component must be analyzed in turn. The optical efficiency η_o may be expressed as

$$\eta_o = \frac{\tau_e \alpha_r \bar{I}_{eff}}{I_{b,c} + I_{d,c}} \tag{7.53}$$

where

I_{eff} is the effective solar radiation both directly intercepted and intercepted after reflection from the back reflector

$I_{b,c}$ and $I_{d,c}$ are, respectively, the beam and diffuse radiation components intercepted per unit collector aperture area

τ_e and α_r are the glass envelope transmittance and receiver absorptance, respectively

The subscript "e" denotes envelope (tube) properties

The total effective insolation I_{eff} can be calculated by summing directly intercepted and reflected radiation, that is,

$$I_{eff} = I_b\left(\cos i + \cos i \rho \Delta \frac{W}{D}\right) + I_{d,c}\left[\pi F_{TS}(1 + \rho \bar{F})\right] \tag{7.54}$$

where

F_{TS} is the radiation shape factor from a tube to the sky dome

$\bar{F}$ is a measure of the shape factors of diffusely illuminated strips to collector tubes

For tubes spaced one diameter apart, $F_{TS} \cong 0.43$ and $\bar{F} \cong 0.34$. The optical efficiency η_o (Equation 7.53) is not, therefore, a simple collector property independent of operating conditions, but rather a function of time through the incidence angles.

7.3.1 Thermal Analysis of an Evacuated Tube Collector

The heat loss, q_L, from an evacuated tubular collector occurs primarily by radiation from the absorber surface and can be expressed as

$$q_L = U_c(T_r - T_a) = \frac{T_r - T_a}{R_1 + R_2 + R_3} \tag{7.55}$$

where
- R_1 is the thermal radiative resistance between the surface of the tube and the inner surface of the glass tube
- R_2 is the thermal resistance of the glass tube
- R_3 is the thermal resistance between the outer surface of the glass tube and the environment by radiation and convection

R_2 is negligible for commercial designs and R_1 and R_3 are approximately given by

$$R_1 = \frac{(1/\varepsilon_r) + (1/\varepsilon_g) - 1}{\sigma(T_r + T_g)\left(T_r^2 + T_g^2\right)} \tag{7.56}$$

$$R_3 = \frac{1}{h_c + \sigma\varepsilon_g(T_g + T_a)(T_g^2 + T_a^2)} \tag{7.57}$$

T_r is the receiver (absorber) temperature, and h_c is the external convection coefficient for the glass envelope. Tests have shown that the loss coefficient U_c for a single tube is between 0.5 and 1.0 W/m$^2 \cdot$°C, which is in agreement with the analysis.

The CTC energy delivery rate q_u on an aperture area basis can be written as

$$q_u = \tau_e\alpha_r I_{eff}\frac{A_t}{A_c} - U_c(T_r - T_a)\frac{A_r}{A_c} \tag{7.58}$$

where
- A_t is the projected area of a tube (its diameter)
- A_r is the receiver or absorber area

The receiver-to-collector aperture area ratio is $\pi D_r/d$, where d is the center-to-center distance between the tubes. Therefore,

$$q_u = \frac{D_r}{d}\left[\tau_e\alpha_r I_{eff} - \pi U_c(T_r - T_a)\right]. \tag{7.59}$$

For optimum performance, a specularly reflecting back surface and a tube spacing of one envelope diameter apart is recommended.

7.4 Experimental Testing of Collectors

The performance of solar thermal systems for heating and cooling depends largely on the performance of solar collectors. Therefore, experimental measurement of thermal performance of solar collectors by standard methods is important and necessary. The experimentally determined performance data is needed for design purposes and for determining the commercial value of the collectors. The thermal performance of a solar collector is determined by establishing an efficiency curve from the measured instantaneous efficiencies for a combination of values of incident solar radiation, ambient temperature and inlet fluid temperature. An instantaneous efficiency of a collector under steady-state conditions can be established by measuring the mass flow rate of the heat transfer fluid, its temperature rise across the collector ($T_{f,out} - T_{f,in}$), and the incident solar radiation intensity (I_c) as

$$\eta_c = \frac{q_u}{A_c I_c} = \frac{\dot{m} C_p (T_{f,out} - T_{f,in})}{A_c I_c}. \tag{7.60}$$

The efficiency, η_c, of a collector under steady state can also be written according to the Hottel–Whillier–Bliss equation (Equation 7.46) as

$$\eta_c = F_R \tau_s \alpha_s - F_R U_c \frac{(T_{f,in} - T_a)}{I_c} \tag{7.61}$$

where $\tau_s\alpha_s$ is the optical efficiency, η_{opt}.

Equation 7.61 suggests that for constant values of F_R and U_c, if η_c is plotted with respect to $(T_{f,in} - T_a)/I_c$, a linear curve will result, with a y intercept of $F_R\tau_s\alpha_s$ and a slope of $-F_R U_c$. Figure 7.19 shows a typical thermal performance curve for a flat-plate collector. Since τ_s and α_s can be measured independently, a thermal performance curve of a flat-plate collector allows us to establish the value of F_R and U_c also. Figure 7.20 shows typical performance curves of various glazed and unglazed flat-plate solar collectors.

In Equation 7.61 the product $\tau_s\alpha_s$ will change with the angle of incidence. Since flat-plate collectors are normally fixed, the angle of incidence changes throughout the day. A relationship can be written between the actual or effective $(\tau_s\alpha_s)$ and $(\tau_s\alpha_s)_n$ for normal incidence as

$$(\tau_s\alpha_s) = (\tau_s\alpha_s)_n K_{\tau\alpha} \tag{7.62}$$

where

$(\tau_s\alpha_s)_n$ is the value of the product for normal angle of incidence

$K_{\tau\alpha}$ is called the incidence angle modifier

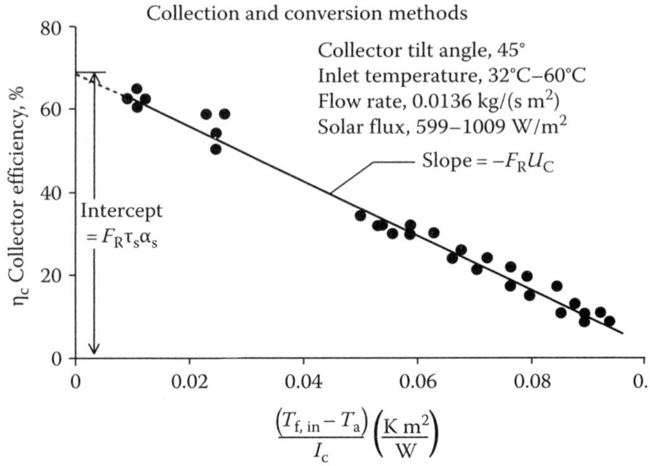

FIGURE 7.19
Experimental thermal efficiency curve for a double-glazed flat-plate liquid-type of solar collector. (From Goswami, Y. et al., *Principles of Solar Engineering,* 2nd edn., Taylor & Francis, Philadelphia, 2000. With permission.)

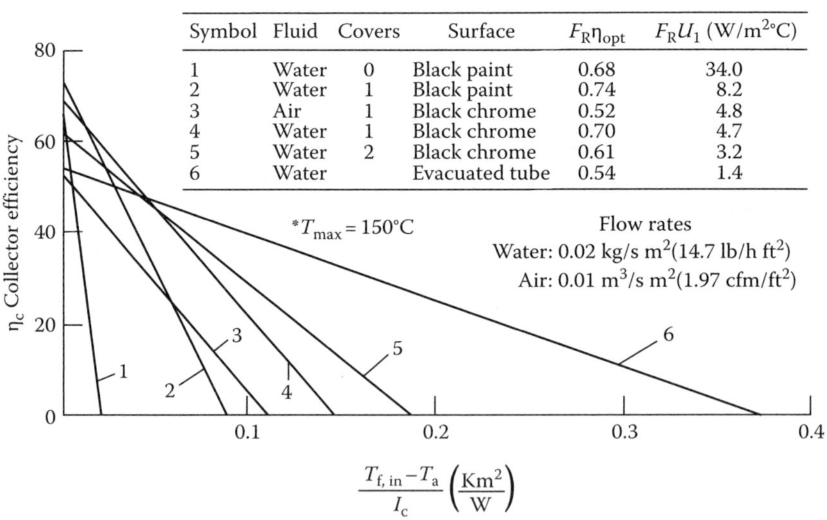

Symbol	Fluid	Covers	Surface	$F_R\eta_{opt}$	$F_R U_l$ (W/m²°C)
1	Water	0	Black paint	0.68	34.0
2	Water	1	Black paint	0.74	8.2
3	Air	1	Black chrome	0.52	4.8
4	Water	1	Black chrome	0.70	4.7
5	Water	2	Black chrome	0.61	3.2
6	Water		Evacuated tube	0.54	1.4

FIGURE 7.20
Typical performance curves for various flat-plate solar collectors.

Therefore, the thermal performance of a flat-plate collector may be written as

$$\eta_c = F_R\left[K_{\tau\alpha}(\tau_s\alpha_s)_n - U_c\frac{(T_{f,in} - T_a)}{I_c}\right]. \tag{7.63}$$

Thermal performance of collectors is usually found experimentally for normal angles of incidence in which case $K_{\tau\alpha} = 1.0$. The incidence angle modifier is then measured separately. It has been established that $K_{\tau\alpha}$ is of the form

$$K_{\tau\alpha} = 1 - b\left(\frac{1}{\cos i} - 1\right), \tag{7.64}$$

where

b is a constant

i is the angle of incidence

Figure 7.21 shows the effect of incidence angle on $K_{\tau\alpha}$ for flat-plate collectors.

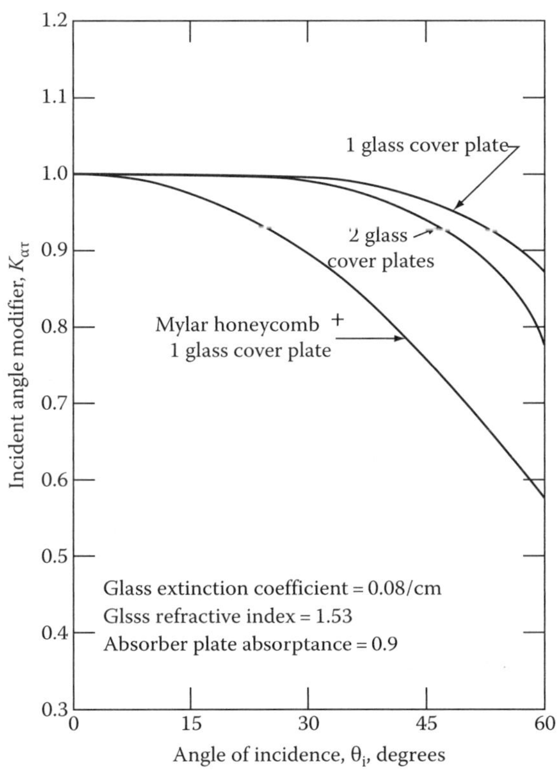

FIGURE 7.21
Incident angle modifier for three flat-plate solar collectors. (Reprinted from ASHRAE Standard 93-77, Methods of testing to determine the thermal performance of solar collectors, Atlanta, GA.)

7.4.1 Testing Standards for Solar Thermal Collectors

Standard testing procedures adopted by regulating agencies in various countries establish ways of comparing the thermal performance of various collectors under the same conditions. In the United States, the thermal performance standards established by the American Society of Heating, Refrigeration and Air-Conditioning Engineers (ASHRAE) have been accepted for thermal performance tests of solar thermal collectors. Similar standards have been used in a number of countries. The standards established by ASHRAE include

ASHRAE Standard 93-77, "Methods of testing to determine the thermal performance of solar collectors."

ASHRAE Standard 96-80, "Methods of testing to determine the thermal performance of unglazed solar collectors."

ASHRAE Standard 93-77 specifies the procedures for determining the time constant, thermal performance, and the incidence angle modifier of solar thermal collectors using a liquid or air as a working fluid. Figure 7.22 shows a schematic of a standard testing configuration for thermal performance testing. The tests are conducted under quasi-steady-state conditions.

7.4.1.1 Time Constant

In order to avoid the transient effects, performance of a collector is measured and integrated over at least the time constant of the collector. The time

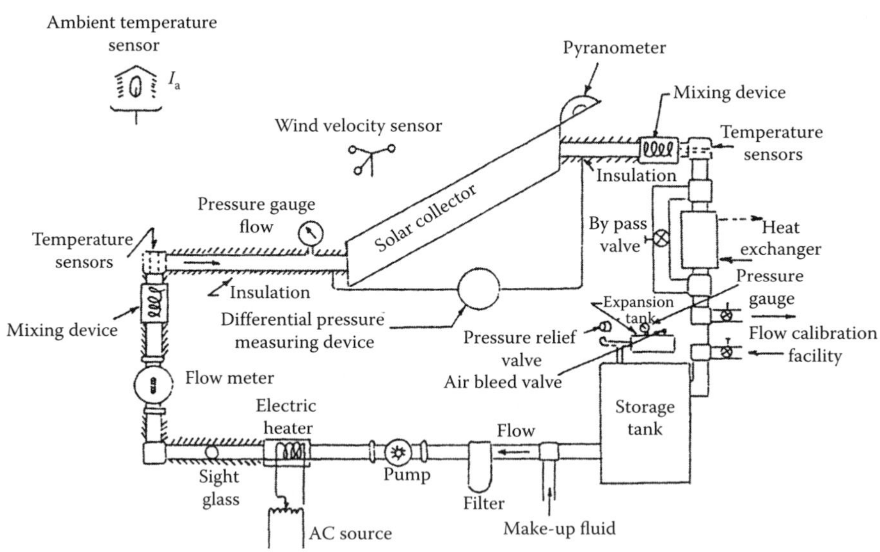

FIGURE 7.22
A testing configuration for a liquid-type solar collector. (Reprinted from ASHRAE Standard 93-77, Methods of testing to determine the thermal performance of solar collectors, Atlanta, GA.)

constant is measured by operating the collector under steady or quasi-steady conditions with a solar flux of at least 790 W/m^2 and then abruptly cutting the incident flux to zero, while continuing to measure the fluid inlet ($T_{f,i}$) and exit ($T_{f,e}$) temperatures. The fluid inlet temperature is maintained at ±1°C of the ambient temperature. The time constant is then determined as the time required to achieve

$$\frac{T_{f,e} - T_{f,i}}{T_{f,e,initial} - T_{f,i}} = 0.30. \tag{7.65}$$

7.4.1.2 Thermal Performance

For determining the thermal performance of a collector, ASHRAE Standard 93-77 specifies test conditions to get at least four data points for the efficiency curve. The test conditions include

1. Near normal incidence ($i \leq 5°$) angle tests close to solar noon time
2. At least four tests for each $T_{f,i}$, two before and two after solar noon
3. At least four different values of $T_{f,i}$ to obtain different values of $\Delta T/I_c$, preferably to obtain ΔT at 10%, 30%, 50% and 70% of stagnation temperature rise under the given conditions of solar intensity and ambient conditions

7.4.1.3 Incidence Angle Modifier

ASHRAE 93-77 specifies that a curve for incidence angle modifier be established by determining the efficiencies of a collector for average angles of incidence of 0°, 30°, 45°, and 60° while maintaining $T_{f,i}$ at ±1°C of the ambient temperature. Since $(T_{f,i} - T_a) \simeq 0$, the incident angle modifier according to Equation 7.63 is

$$K_{\tau\alpha} = \frac{\eta_c}{F_R(\tau_s\alpha_s)_n}. \tag{7.66}$$

The denominator is the y intercept of the efficiency curve.

7.5 Calculations of Heating and Hot Water Loads in Buildings

Energy requirements for space heating or service water heating can be calculated from basic conservation of energy principles. For example, the heat required to maintain the interior of a building at a specific temperature is the total of all heat transmission losses from the structure and heat required to warm and humidify the air exchange with the environment by infiltration and ventilation.

Comfort in buildings has long been a subject of investigation by the ASHRAE. The ASHRAE has developed extensive heat load calculation procedures embodied in the *ASHRAE Handbook of Fundamentals* [18]. The most frequently used load calculation procedures will be summarized in this section; the reader is referred to the ASHRAE handbook for details.

Figure 7.23 shows the combinations of temperature and humidity that are required for human comfort. The shaded area is the standard U.S. comfort

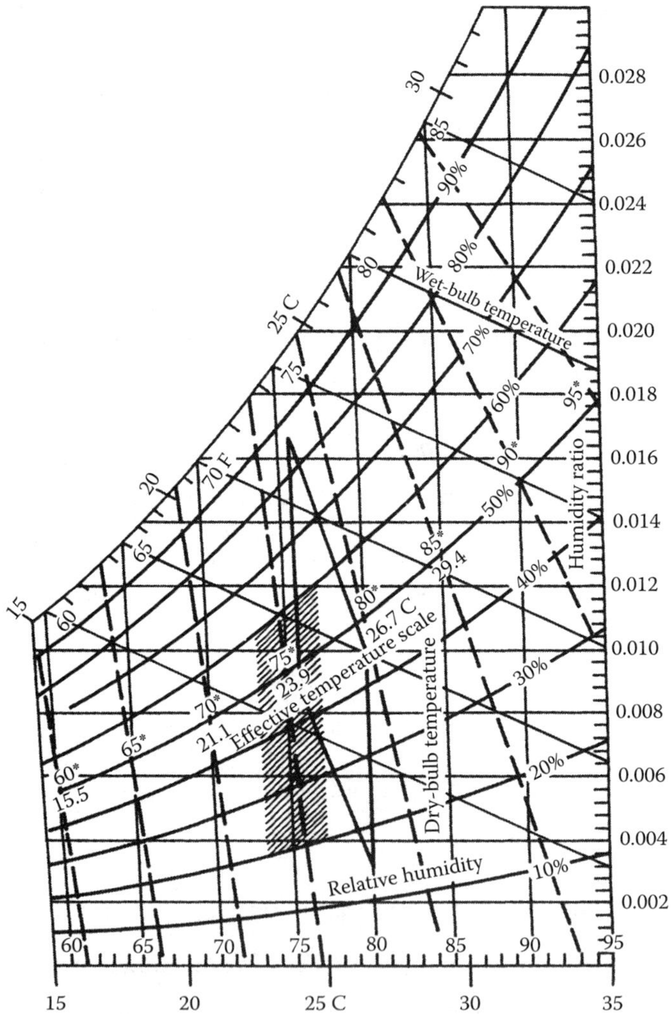

FIGURE 7.23
Heat loss calculations for buildings.

level for sedentary persons. Many European countries have human comfort levels from 3°C to 7°C below U.S. levels. If activity of a continuous nature is anticipated, the comfort zone lies to the left of the shaded area; if extra clothing is worn, the comfort zone is displaced similarly.

7.5.1 Calculation of Heat Loss

It is outside the scope of this book to describe the details of the heat load calculations for buildings. However, the method is described in brief in this section. For details, one should refer to the *ASHRAE Handbook of Fundamentals* [18] or some textbook on heating and air-conditioning. Table 7.7 lists the components of heat loss calculations of a building.

Complete tables of thermal properties of building materials are given in Appendix A. Transmission heat losses through attics, unheated basements, and the like are buffered by the thermal resistance of the unheated space. For example, the temperature of an unheated attic lies between that of the heated space and that of the environment. As a result, the ceiling of a room below an attic is exposed to a smaller temperature difference and consequent lower heat loss than the same ceiling without the attic would be. The

TABLE 7.7

Heating Load Calculations for Buildings

Heating Load Component	Equations 7.67 through 7.70	Descriptions/References
Walls, roof, ceilings, glass	$q = U \cdot A(T_i - T_o)$	T_i, T_o are inside and outside air temperature, respectively. U values of composite section are calculated from the thermal properties of components given in Appendix A
Basement floors and walls below ground level	$q_f = U_f^* A_f$	U_f^* has special units of W/m². Values of U_f^* for various ground water temperatures are given in Appendix A
Concrete floors on ground	$q_{fe} = F_e P_e (T_i - T_o)$	P_e is the perimeter of the slab. F_e, values are given in Appendix A
Infiltration and ventilation air	$q_{sensible} = Q\rho_a Cp_a (T_i - T_o)$ or $= 1200^* Q(T_i - T_o)$ W	Q is the volume of airflow in m³/s. ρ_a and Cp_a are density and specific heat of air
	$q_{latent} = Q\rho_a h_{fg} \Delta W$ or $= 2808^a Q \Delta W$ W	h_{fg} is the latent heat of water at room temperature. ΔW is humidity ratio difference between inside and outside air

[a] Assuming $\rho_a = 1.2\,kg/m^3$; $h_{fg} = 2340\,J/kg$, $Cp_a = 1000\,J/kg\,°C$.

effective conductance of thermal buffer spaces can easily be calculated by forming an energy balance on such spaces.

The example below is an illustration of the heat-loss calculation method described in this section.

Example 7.3

Calculate the heat load on a house for which the wall area is 200 m², the floor area is 600 m², the roof area is 690 m², and the window area totals 100 m². Inside wall height is 3 m. The construction of the wall and the roof is shown in Figure 7.24.

Solution

The thermal resistance of the wall shown in Figure 7.24 can be found by the electrical resistance analogy as

$$R_{wa} = R_{outside\ air} + R_{wood\ siding} + R_{sheathing} + R_{comb} + R_{wall\ board} + R_{inside\ air}$$

Combined thermal resistance for the studs and insulation (R_{comb}) is found as

$$\frac{1}{R_{comb}} = \left(\frac{A_{stud}}{R_{stud}} + \frac{A_{insulation}}{R_{insulation}}\right)\frac{1}{A_{stud} + A_{insulation}}$$

Assuming that the studs occupy 15% of the wall area

$$\frac{1}{R_{comb}} = \frac{0.15}{0.77} + \frac{0.85}{1.94}$$

or

$$R_{comb} = 1.58\ \frac{m^2\,°C}{W}$$

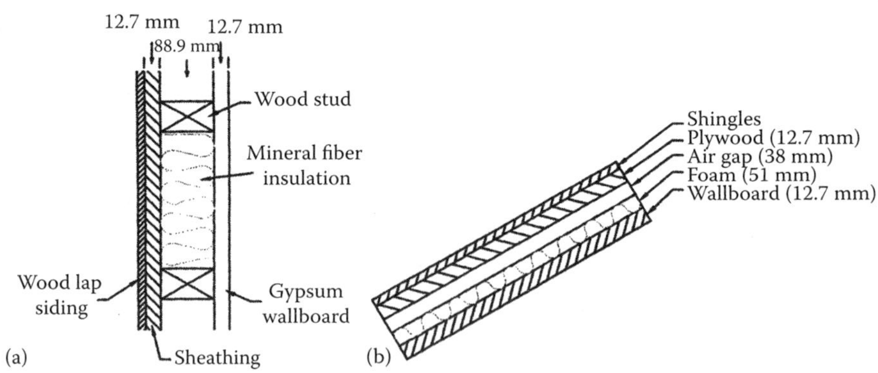

FIGURE 7.24
Cross-sections of (a) the wall and (b) the roof for Example 7.3.

Therefore, the wall thermal resistance, R_{wa}, can be found as

Element	Thermal Resistance (m²°C/W)
Outside air (6.7 m/s wind)	0.030
Wood bevel lap siding	0.14
12.7 mm sheathing	0.23
88.9 mm combined wood stud and mineral fiber insulation	1.58
12.7 mm gypsum wallboard	0.079
Inside air (still)	0.12
	R_{wa} = 2.179

Therefore,

$$U_{wa} = \frac{1}{R_{wa}} = \frac{1}{2.179} = 0.46\ \text{W/m}^2\text{°C}.$$

The heat loss through the windows depends on whether they are single- or double-glazed. In this example, single-glazed windows are installed, and a U factor equal to 4.7 W/m²°C is used. (If double-glazed windows were installed, the U factor would be 2.4 W/m²°C.)

The roof is constructed of 12.7 mm gypsum wall board, 51 mm foam insulation board, 38 mm still air, 12.7 mm plywood, and asphalt shingles (wooden beams and roofing paper are neglected for the simplified calculations here). Therefore,

$$U_{rf} = \frac{1}{\underset{\text{outside air}}{0.030} + \underset{\text{Shingels}}{0.077} + \underset{\text{Plywood}}{0.11} + \underset{\text{Air Gap}}{0.17} + \underset{\text{Foam}}{2.53} + \underset{\text{Wallboard}}{0.079} + \underset{\text{Inside air}}{0.1}} = 0.32\ \text{W/m}^2\text{°C}$$

If the respective areas and U factors are known, the rate of heat loss per hour for the walls, windows, and roof can be calculated:

$$\text{Walls:} \quad q_{wa} = (200\ \text{m}^2) \times 0.46\ \text{W/m}^2\text{°C} = 92\ \text{W/°C}$$

$$\text{Windows:} \quad q_{wi} = (100\ \text{m}^2) \times 0.47\ \text{W/m}^2\text{°C} = 470\ \text{W/°C}$$

$$\text{Roof:} \quad q_{rf} = (690\ \text{m}^2) \times 0.32\ \text{W/m}^2\text{°C} = 220\ \text{W/°C}$$

$$\text{Total} = 782\ \text{W/°C}$$

If double-glazed windows were used, the total heat loss would be reduced to 552 W/°C. As shown in Example 2.5, the installation of double glazed windows is cost effective and repays the energy invested in less than 2 years. Thus, this shows that before installing an active system to supply the heat necessary for maintaining adequate comfort level, energy conservation measures such as using double-glazed rather than single-glazed windows is the preferred option.

The infiltration and ventilation rate Q for this building is assumed to be 0.5 ACH (air changes per hour). The sensible and latent heat loads of the infiltration air may be calculated using the equations given in Table 7.7. Therefore,

$$Q = 0.5 \times (600\ \text{m}^2 \times 3\ \text{m}) = 900\ \text{m}^3/\text{h} = 0.25\ \text{m}^3/\text{s, volume}$$

$$q_{\text{sensible}} = 0.25\ \text{m}^3/\text{s} \times (1.2\ \text{kg/m}^3)(1000\ \text{J/kg °C}) = 300\ \text{W/°C}.$$

In residential buildings, humidification of the infiltration air is rarely done. Neglecting the latent heat, the total rate of heat loss q_{tot} is the sum of q_{sensible} and q_{tr}:

$$q_{\text{tot}} = (782 + 300) = 1082\ \text{W/°C}.$$

This calculation is simplified for purposes of illustration. Heat losses through the slab surface and edges have been neglected, for example.

More refined methods of calculating energy requirements on buildings do not use the steady-state assumption used above [19]. The thermal inertia of buildings may be expressly used as a load-leveling device. If so, the steady-state assumption is not met and the energy capacitance of the structure must be considered for accurate results. Many adobe structures in the U.S. Southwest are built intentionally to use daytime sun absorbed by 1 ft thick walls for nighttime heating, for example.

7.5.2 Internal Heat Sources in Buildings

Heat supplied to a building to offset energy losses is derived from both the heating system and internal heat sources. Table 7.8 lists the common sources of internal heat generation for residences. Commercial buildings such as hospitals, computer facilities, or supermarkets will have large internal gains specific to their function. Internal heat gains tend to offset heat losses from a building but will add to the cooling load of an air-conditioning system. The magnitude of the reduction in heating system operation will be described in the next section.

7.5.3 The Degree-Day Method

The preceding analysis of heat loss from buildings expresses the loss on a per unit temperature difference basis (except for unexposed floor slabs).

TABLE 7.8

Some Common Internal Sensible Heat Gains That Tend to Offset the Heating Requirements of Buildings

Type	Magnitude (W or J/s)
Incandescent lights	Total W
Fluorescent lights	Total W
Electric motors	746 × (hp/efficiency)
Natural gas stove	$8.28 \times m^3/h$
Appliances	Total W
A dog	50–90
People	
Sitting	70
Walking	75
Dancing	90
Working hard	170

Source: ASHRAE, *Handbook of Fundamentals*, American Society of Heating, Refrigerating and Air-Conditioning Engineers, Atlanta, GA, 1997.

In order to calculate the peak load and total annual load for a building, appropriate design temperatures must be defined for each. The outdoor design temperature is usually defined statistically, such that the actual outdoor temperature will exceed the design temperature 97.5% or 99% of the time over a long period. The design temperature difference (ΔT) is then the interior building temperature minus the outdoor design temperature. The design ΔT is used for rating nonsolar heating systems, but is not useful for the selection of solar systems, since solar systems rarely provide 100% of the energy demand of a building at peak conditions.

A more useful index of heating energy demand is the total annual energy requirement for a building. This quantity is somewhat more difficult to calculate than the peak load. It requires a knowledge of day-to-day variations in ambient temperature during the heating season and the corresponding building heat load for each day. Building heat loads vary with ambient temperatures, as shown in Figure 7.25. The environmental temperature T_{nl}, above which no heat need be supplied to the building, is a few degrees below the required interior temperature T_i because of internal heat-generation effects.

The no-load temperature at which internal source generation q_i just balances transmission and infiltration losses can be determined from the energy balance

$$q_i = \overline{UA}(T_i - T_{nl}), \tag{7.71}$$

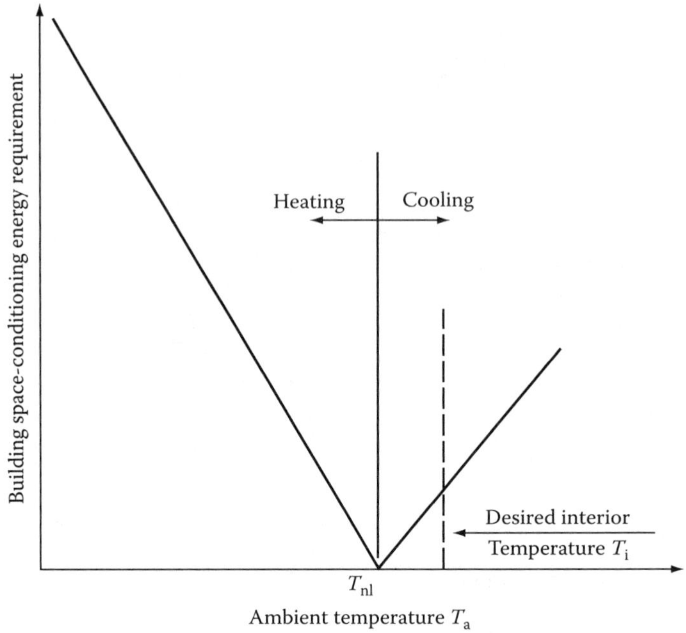

FIGURE 7.25
Building load profile versus ambient temperature showing no-load temperature T_{nl} and desired interior temperature T_i.

where $\overline{UA}$ is the overall loss coefficient for the building (W/°C). Then

$$T_{nl} = T_i - \frac{q_i}{\overline{UA}} \tag{7.72}$$

The total annual heat load on the building, Q_T can be expressed as

$$Q_T = \int_{365\text{ days}} \overline{UA}(T_{nl} - T_a)^+ dt, \tag{7.73}$$

in which all arguments of the integral are functions of time. The superscript $^+$ indicates that only positive values are considered. In practice, it is difficult to evaluate this integral; therefore, three simplifying assumptions are made:

1. $\overline{UA}$ is independent of time.
2. T_{nl} is independent of time.
3. The integral can be expressed by the sum.

$$\overline{UA}\sum_{n=1}^{365}(T_{nl}-\overline{T}_a)_n^+ \tag{7.74}$$

where n is the day number, and the daily average temperature $\overline{T}_a$ can be approximated by ½($T_{\alpha,max} + T_{a,min}$), in which $T_{a,max}$ and $T_{a,min}$ are the daily maximum and minimum temperatures, respectively.

The quantity $(T_{nl} - T_a)^+$ is called the *degree-day unit*. For example, if the average ambient temperature for a day is 5°C and the no-load temperature is 20°C, 15°C-days are said to exist for that day. However, if the ambient temperature is 20°C or higher, 0 degree-days exist, indicating 0 demand for heating that day. Degree-day totals for monthly $\Sigma_{month}(T_{nl} - T_a)^+$ and annual periods can be used directly in Equation 7.74 to calculate the monthly and annual heating energy requirements.

In the past, a single value of temperature has been used throughout the United States as a universal degree-day base, 65.0°F or 18.3°C. This practice is now outdated, since many homeowners and commercial building operators have lowered their thermostat settings in response to increased heating fuel costs, thereby lowering T_{nl}. Likewise, warehouses and factories operate well below the 19°C level. Therefore, a more generalized database of degree-days to several bases (values of T_{nl}) has been created by the U.S. National Weather Service (NWS, http://www.ncdc.noaa.gov/oa/documentlibrary/hcs/hdd.200707–200811.pdf).

Example 7.4

A building located in Denver, Colorado, has a heat-loss coefficient $\overline{UA}$ of 1000 kJ/h °C and internal heat sources of 4440 kJ/h. If the interior temperature is 20°C (68°F), what are the monthly and annual heating energy requirements? A gas furnace with 65% efficiency is used to heat the building.

Solution

In order to determine the monthly degree-day totals, the no-load temperature (degree-day basis) must be evaluated from Equation 7.72.

$$T_{nl} = 20 - \frac{4440}{1000} = 15.6°\text{C}(60°\text{F})$$

The monthly degree C-days for Denver are taken from the U.S. National Weather Service and given in Table 7.9. The energy demand is calculated as

$$\text{Energy demand} = \overline{UA} \times 24\frac{\text{h}}{\text{day}} \times \text{degree C-days.}$$

TABLE 7.9

Monthly and Annual Energy Demands for Example 7.4

Month	Degree C-Days	Energy Demand[a] (GJ)
January	518	12.4
February	423	10.2
March	396	9.5
April	214	5.2
May	68	1.6
June	14	0.3
July	0	0
August	0	0
September	26	0.6
October	148	3.6
November	343	8.2
December	472	11.3
	2622	62.9

[a] Energy demand equals $\overline{UA}$ × degree C-days × 24h/day.

The monthly energy demand is given in Table 7.9.

The annual energy demand of 62.9 GJ is delivered by a 65% efficient gas furnace. Therefore,

$$\text{Average annual purchased energy} = \frac{62.9}{0.65}\,\text{GJ} = 96.8\ \text{GJ}.$$

7.5.4 Service Hot Water Load Calculation

Service hot water loads can be calculated precisely with the knowledge of only a few variables. The data required for calculation of hot water demand are

Water source temperature (T_s)
Water delivery temperature (T_d)
Volumetric demand rate (Q)

The energy requirement for service water heating q_{hw} is given by

$$q_{hw}(t) = \rho_w Q(t) c_{pw}\left[T_d - T_s(t)\right] \tag{7.75}$$

where
ρ_w is the water density
c_{pw} is its specific heat

TABLE 7.10

Approximate Service Hot-Water Demand Rates

	Demand per Person	
Usage Type	**L/Day**	**gal/Day**
Retail store	2.8	0.75
Elementary school	5.7	1.5
Multifamily residence	76.0	20.0
Single-family residence	76.0	20.0
Office building	11.0	3.0

The demand rate, $Q(t)$, varies in general with time of day and time of year; likewise, the source temperature varies seasonally. Source temperature data are not compiled in a single reference; local water authorities are the source of such temperature data.

Few generalized data exist with which to predict the demand rate Q. Table 7.10 indicates some typical usage rates for several common building types. Process water-heating rates are peculiar to each process and can be ascertained by reference to process specifications.

Example 7.5

Calculate the monthly energy required to heat water for a family of four in Nashville, Tennessee. Monthly source temperatures for Nashville are shown in Table 7.11, and the water delivery temperature is 60°C (140°F).

Solution

For a family of four, the demand rate Q may be found using a demand recommended from Table 7.10:

$$Q = 4 \times 76 \text{ L/day} = 0.30 \text{ m}^3/\text{day}$$

The density of water can be taken as 1000 kg/m^3 and the specific heat as 4.18 kJ/kg · °C.

Monthly demands are given by

$$\begin{aligned} q_m &= (Q \times \text{days/month})(\rho_w c_{pw})[T_d - T_s(t)] \\ &= (0.30 \times \text{days/month})(1000 \times 4.18)[60 - T_s(t)]. \end{aligned}$$

The monthly energy demands calculated from the equation above with these data are tabulated in Table 7.11.

TABLE 7.11

Water Heating Energy Demands for Example 7.5

Month	Days/ Month	Demand (m^3/Month)	Source Temperature °C	Energy Requirement (GJ/Month)
January	31	9.3	8	2.0
February	28	8.4	8	1.8
March	31	9.3	12	1.9
April	30	9.0	19	1.5
May	31	9.3	17	1.7
June	30	9.0	21	1.5
July	31	9.3	22	1.5
August	31	9.3	24	1.4
September	30	9.0	24	1.4
October	31	9.3	22	1.5
November	30	9.0	14	1.7
December	31	9.3	12	1.9

7.6 Solar Water-Heating Systems

Solar water-heating systems represent the most common application of solar energy at the present time. Small systems are used for domestic hot water applications while larger systems are used in industrial process heat applications. There are basically two types of water-heating systems: *natural circulation* or passive solar system (thermosyphon) and *forced circulation* or active solar system. Natural circulation solar water heaters are simple in design and of low cost. Their application is usually limited to nonfreezing climates, although they may also be designed with heat exchangers for mild freezing climates. Forced circulation water heaters are used in freezing climates and for commercial and industrial process heat.

7.6.1 Natural Circulation Systems

The natural tendency of a less dense fluid to rise above a denser fluid can be used in a simple solar water heater to cause fluid motion through a collector [20]. The density difference is created within the solar collector where heat is added to the liquid. In the system shown in Figure 7.26, as water gets heated in the collector, it rises to the tank, and the cooler water from the tank moves to the bottom of the collector, setting up a natural circulation loop. It is also called a *thermosyphon loop*. Since this water heater does not use a pump, it is a passive water heater. For the thermosyphon to work, the storage tank must be located higher than the collector.

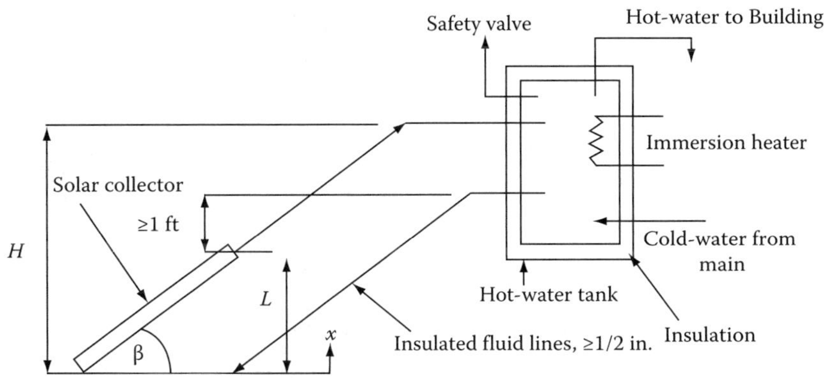

FIGURE 7.26
Schematic diagram of thermosyphon loop used in a natural circulation, service water-heating system. The flow pressure drop in the fluid loop must equal the buoyant force "pressure" $\left[\int_0^L gp(x)dx - \rho_{\text{stor}} gL\right]$, where $\rho(x)$ is the local collector fluid density and ρ_{stor} is the tank fluid density, assumed uniform.

The flow pressure drop in the fluid loop (ΔP_{FLOW}) must equal the buoyant force "pressure difference" ($\Delta P_{\text{BOUYANT}}$) caused by the differing densities in the hot and cold legs of the fluid loop:

$$\Delta P_{\text{FLOW}} = \Delta P_{\text{BOUYANT}}$$

$$= \rho_{\text{stor}} gH - \left[\int_{0}^{L} \rho(x) g dx + \rho_{\text{out}} g(H - L)\right], \tag{7.76}$$

where
  H is the height of the legs
  L is the height of the collector (see Figure 7.26)
  $\rho(x)$ is the local collector fluid density
  ρ_{stor} is the tank fluid density
  ρ_{out} is the collector outlet fluid density, the latter two densities assumed uniform

The flow pressure term ΔP_{FLOW}, is related to the flow loop system head loss, which is in turn directly connected to friction and fitting losses and the loop flow rate:

$$\Delta P_{\text{FLOW}} = \oint_{\text{LOOP}} \rho d(h_{\text{L}}), \tag{7.77}$$

where $h_{\text{L}} = KV^2$, with K being the sum of the component loss velocity factors (see any fluid mechanics text), and V the flow velocity.

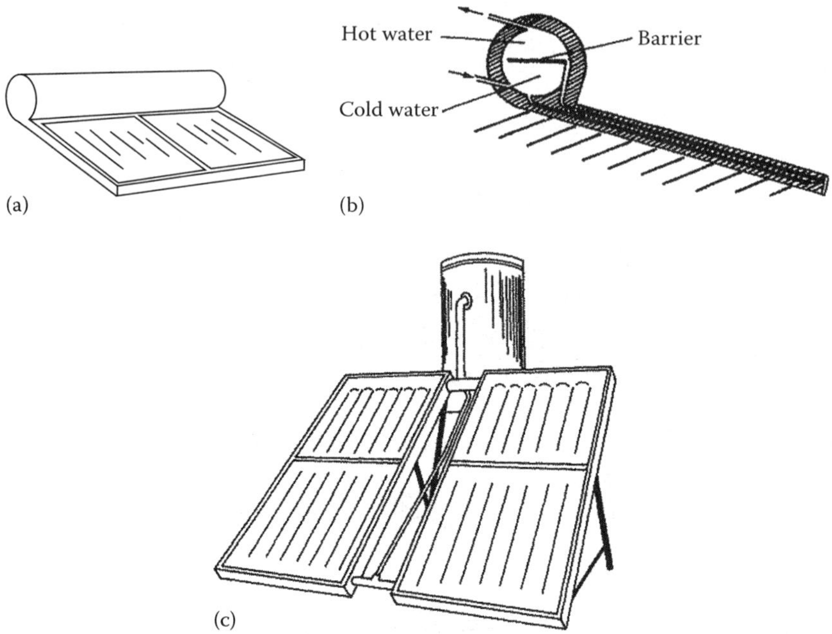

FIGURE 7.27
Passive solar water heaters: (a) compact model using combined collector and storage, (b) section view of the compact model, and (c) tank and collector assembly.

Since the driving force in a thermosyphon system is only a small density difference and not a pump, larger-than-normal plumbing fixtures must be used to reduce pipe friction losses [21]. In general, one pipe size larger than normal would be used with a pump system is satisfactory. Figure 7.27 shows some passive water heaters.

Since the hot water system loads vary little during a year, the angle of tilt is that equal to the latitude, that is, $\beta = L$. The temperature difference between the collector inlet water and the collector outlet water is usually 8°C–11°C during the middle of a sunny day [20]. After sunset, a thermosyphon system can reverse its flow direction and lose heat to the environment during the night. To avoid reverse flow, the top header of the absorber should be at least 30 cm below the cold leg fitting on the storage tank, as shown; otherwise a check valve would be needed.

To provide heat during long cloudy periods, an electrical immersion heater can be used as a backup for the solar system. The immersion heater is located near the top of the tank to enhance stratification and so that the heated fluid is at the required delivery temperature. Tank stratification is desirable in a thermosyphon to maintain flow rates as high as possible. Insulation must be applied over the entire tank surface to control heat loss.

Several features inherent in the thermosyphon design limit its utility. If it is to be operated in a freezing climate, a nonfreezing fluid must be used, which in turn requires a heat exchanger between collector and potable water storage. (If potable water is not required, the collector can be drained during cold periods instead.) Heat exchangers of either the shell-and-tube type or the immersion-coil type require higher flow rates for efficient operation than a thermosyphon can provide. Therefore, the thermosyphon is usually limited to nonfreezing climates. For mild freeze climates, a heat exchanger coil welded to the outer surface of the tank and filled with an antifreeze may work well.

Example 7.6

Determine the "pressure difference" available for a thermosyphon system with 1 m high collector and 2 m high legs. The water temperature input to the collector is 25°C and the collector output temperature is 35°C. If the overall system loss velocity factor (K) is 15.6, estimate the system flow velocity.

Solution

Equation 7.76 is used to calculate the pressure difference, with the water densities being found from the steam tables at http://www.nuc.berkeley.edu/courses/classes/E-115/Reader/steam_table_A-1.Pdf (see Tables A.7.12 and A.7.13).

$$\rho_{stor}(25°C) = 997.009 \text{ kg/m}^3$$

$$\rho_{stor}(35°C) = 994.036 \text{ kg/m}^3$$

$P_{coll.ave.}$ (30°C) = 996.016 kg/m^3 (*note*: average collector temperature used in "integral") and with $H = 2$ and $L = 1$ m:

$$\Delta P_{BOUYANT} = (997.009)9.81(2) - [(996.016)9.81(1) + (994.036)9.81(1)] = 38.9 \text{ N/m}^2 \text{(pa)}$$

The system flow velocity is estimated from the system K given, the pressure difference calculated above, taking the average density of the water around the loop (at 30°C) and substituting into Equation 7.77:

$$\Delta P_{BOUYANT} = (\rho_{loop.ave.})(h_L)_{loop} = (\rho_{loop.ave.})KV^2$$

$$V^2 = \frac{38.9}{(996.016)(15.6)}$$

$$V = 0.05 \text{ m/s}$$

7.6.2 Forced-Circulation Systems

If a thermosyphon system cannot be used for climatic, structural, or architectural reasons, a forced-circulation system is required.

Figure 7.28 shows three configurations of forced-circulation systems: (1) open loop, (2) closed loop, and (3) closed loop with drainback. In an

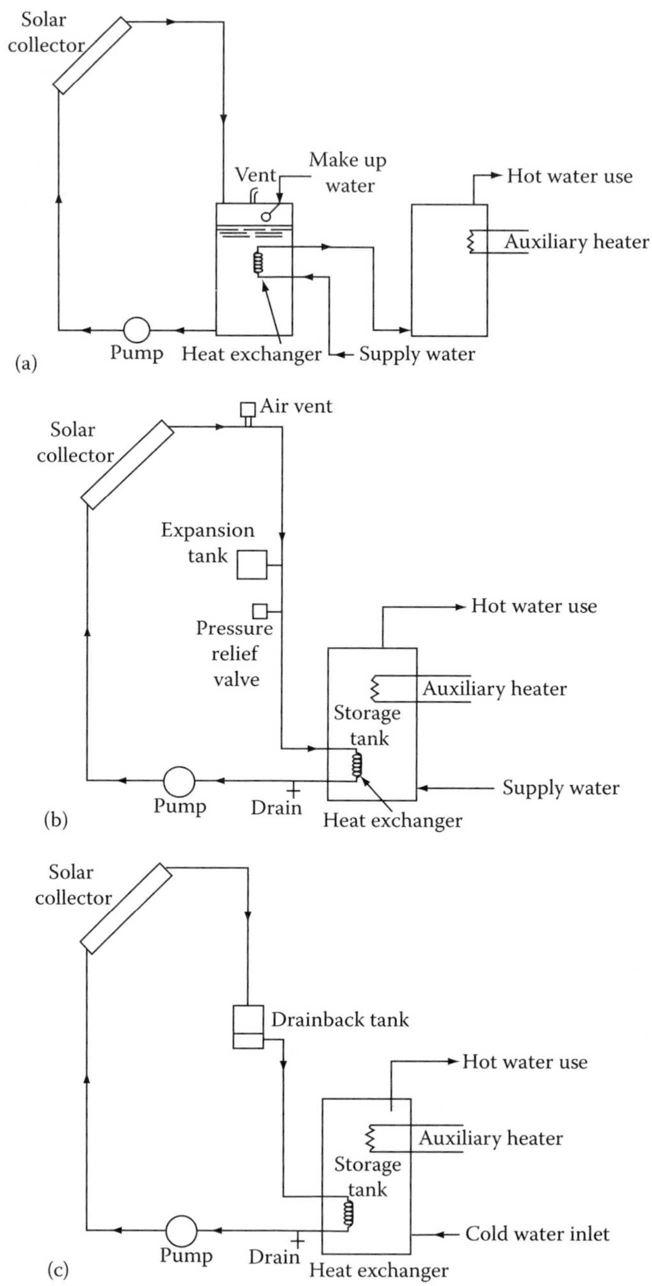

FIGURE 7.28
Typical configurations of solar water-heating systems: (a) open-loop system, (b) closed-loop system, and (c) closed-loop drainback system. (Adapted from Goswami, D.Y., *Alternative Energy in Agriculture*, vol. 1, CRC Press, Boca Raton, FL, 1986.)

open-loop system (Figure 7.28a) the solar loop is at atmospheric pressure, therefore, the collectors are empty when they are not providing useful heat. A disadvantage of this system is the high pumping power required to pump the water to the collectors every time the collectors become hot. This disadvantage is overcome in the pressurized closed loop system (Figure 7.28b) since the pump has to overcome only the resistance of the pipes. In this system, the solar loop remains filled with water under pressure.

In order to accommodate the thermal expansion of water from heating, a small (about 2 gal capacity) expansion tank and a pressure relief valve are provided in the solar loop. Because water always stays in the collectors of this system, antifreeze (propylene glycol or ethylene glycol) is required for locations where freezing conditions can occur. During stagnation conditions (in summer), the temperature in the collector can become very high, causing the pressure in the loop to increase. This can cause leaks in the loop unless some fluid is allowed to escape through a pressure-release valve. Whether as the result of leaks or of draining, air enters the loop, causing the pumps to run dry. This disadvantage can be overcome in a closed loop drainback system that is not pressurized (Figure 7.28c). In this system, when the pump shuts off, the water in the collectors drains back into a small holding tank while the air in the holding tank goes up to fill the collectors. The holding tank can be located where freezing does not occur, but still at a high level to reduce pumping power. In all three configurations, a differential controller measures the temperature differential between the solar collector and the storage, and turns the circulation pump on when the differential is more than a set limit (usually 5°C) and turns it off when the differential goes below a set limit (usually 2°C). Alternatively, a photovoltaic (PV) panel and a DC pump may be used. The PV panel will turn on the pump only when solar radiation is above a minimum level. Therefore, the differential controller and the temperature sensors may be eliminated.

7.6.3 Industrial Process Heat Systems

For temperatures of up to about 100°C, required for many industrial process heat applications, forced circulation flat-plate collector water-heating systems described above can be used. The systems, however, will require a large collector area, storage and pumps, etc. For higher temperatures, evacuated tube collectors or concentrating collectors must be used. Industrial process heat systems are described in more detail in Chapter 8.

7.7 Liquid-Based Solar-Heating Systems for Buildings

Solar space-heating systems can be classified as active or passive depending on the method utilized for heat transfer. A system that uses pumps and/or blowers for fluid flow in order to transfer heat is called an active system.

On the other hand, a system that utilizes natural phenomena for heat transfer is called a passive system. Examples of passive solar space-heating systems include direct gain, attached greenhouse, and storage wall (also called Trombe wall). Passive solar heating systems are described in Chapter 9. In this section, configurations, design methods, and control strategies for active solar-heating systems are described.

7.7.1 Physical Configurations of Active Solar Heating Systems

Figure 7.29 is a schematic diagram of a typical space-heating system. The system consists of three fluid loops—collector, storage, and load. In addition, most space-heating systems are integrated with a domestic water-heating system to improve the year-long solar load factor.

Since space heating is a relatively low-temperature use of solar energy, a thermo-dynamic match of collector to task indicates that an efficient flat-plate collector or low-concentration solar collector is the thermal device of choice.

The collector fluid loop contains fluid manifolds, the collectors, the collector pump, and heat exchanger, an expansion tank, and other subsidiary components. A collector heat-exchanger and antifreeze in the collector loop are normally used in all solar space-heating systems, since the existence of a significant heating demand implies the existence of some subfreezing weather.

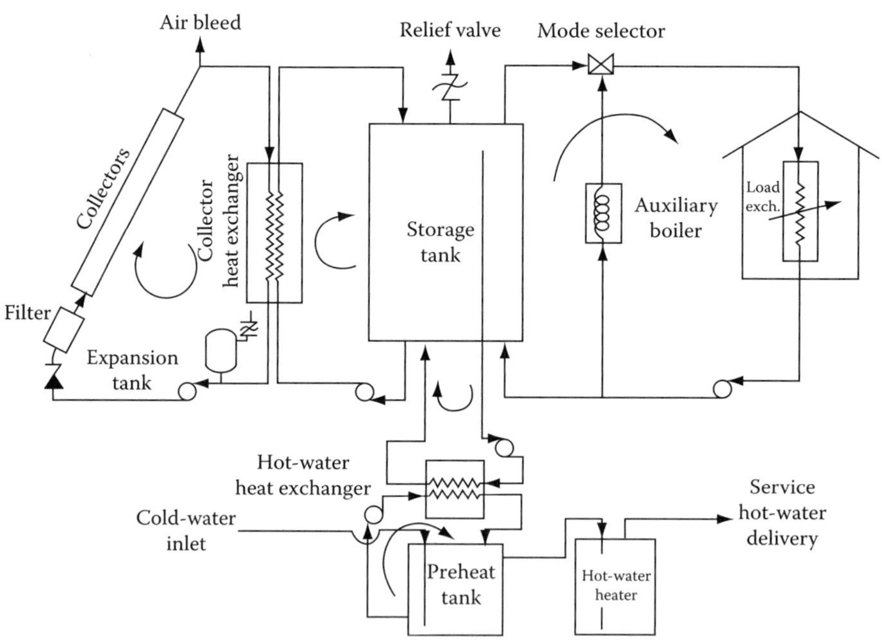

FIGURE 7.29
Typical solar-thermal system for space heating and hot-water heating showing fluid transport loops and pumps.

The storage loop contains the storage tank and pump as well as the tube side of the collector heat exchanger. To capitalize on whatever stratification may exist in the storage tank, fluid entering the collector heat exchanger is generally removed from the bottom of storage. This strategy ensures that the lowest-temperature fluid available in the collector loop is introduced at the collector inlet for high efficiency. The energy delivery-to-load loop contains the load device, baseboard heaters or fin-and-tube coils, and the backup system with a flow control (mode selector) valve.

7.7.2 Solar Collector Orientation

The best solar collector orientation is such that the average solar incidence angle is smallest during the heating season. For tracking collectors, this objective is automatically realized. For fixed collectors in the northern hemisphere, the best orientation is due south (due north in the southern hemisphere), tilted up from the horizon at an angle of about 15° greater than the local latitude.

Although due south is the optimum azimuthal orientation for collectors in the northern hemisphere, variations of 20° east or west have little effect on annual energy delivery [22]. Off-south orientations greater than 20° may be required in some cases because of obstacles in the path of the sun. These effects may be analyzed using sun-path diagrams and shadow-angle protractors as described in Chapter 5.

7.7.3 Fluid Flow Rates

For the maximum energy collection in a solar collector, it is necessary that it operates as closely as possible to the lowest available temperature, which is the collector inlet temperature. Very high fluid flow rates are needed to maintain a collector–absorber surface nearly isothermal at the inlet temperature. Although high flows maximize energy collection, practical and economic constraints put an upper limit on useful flow rates. Very high flows require large pumps and excessive power consumption and lead to fluid conduit erosion.

Figure 7.30 shows the effect of mass flow rate on annual energy delivery from a solar system. It is seen that the law of diminishing returns applies and that flows beyond about $50\ \mathrm{kg/h\ m_c^2}$ ($\sim 10\ \mathrm{lb/h\ ft_c^2}$) have little marginal benefit for collectors with loss coefficients on the order of $6\ \mathrm{W/m_c^2\ °C}$ ($\sim 1\ \mathrm{Btu/h\ ft_c^2 °F}$). In practice, liquid flows in the range of $50-75\ \mathrm{kg/h\ m_c^2}$ ($10-15\ \mathrm{lb/h\ ft_c^2}$) of water equivalent are the best compromise among collect heat-transfer coefficient, fluid pressure drop and energy delivery. However, an infinitely large flow rate will deliver the most energy if pumping power is ignored for a non-stratified storage. If storage stratification is desired, lower flow rates must be used, since high flow destroys stratification. In freezing climates, an antifreeze working fluid is necessary for collectors. Attempts to drain collectors fully for freeze protecting have not been successful.

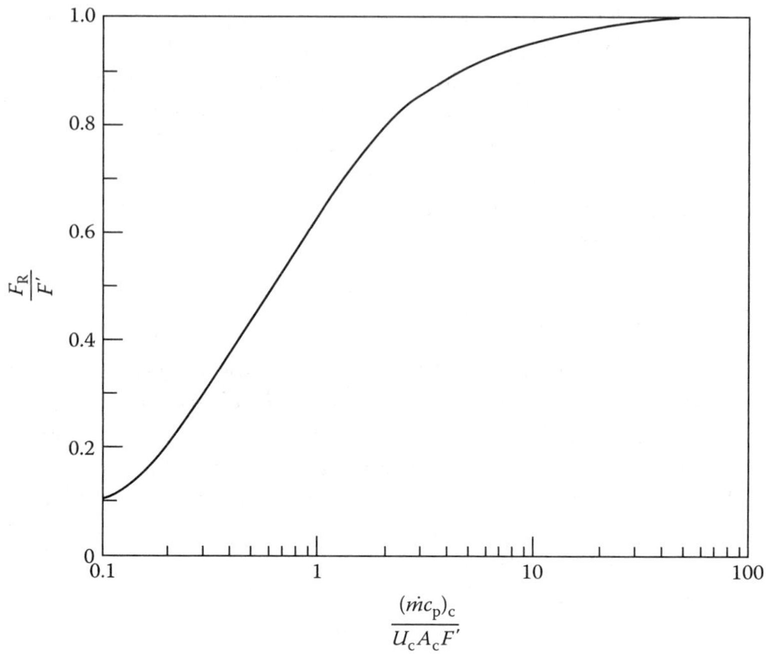

FIGURE 7.30
Effect of fluid flow rate on collector performance as measured by the heat-removal factor F_R; F' is the plate efficiency factor.

7.7.4 Unglazed Transpired Wall System for Air Preheating

Ventilation air preheating systems using wall-mounted unglazed transpired solar air collectors are the only active solar air heating systems that have found market acceptance in commercial and industrial buildings [2]. Such systems preheat the ventilation air in a once-through mode without any storage. Figure 7.31 shows a transpired wall system in which the air is drawn through a perforated absorber plate by the building ventilation fan. Kutcher and Christensen [3] presented a thermal analysis of this system. From a heat balance on the transpired unglazed collector, the useful heat collected is

$$q_u = I_c A_c a_s - U_c A_c (T_{out} - T_a) \tag{7.78}$$

The overall heat loss coefficient U_c, which is due to radiative and convective losses, is given as

$$U_c = \frac{h_r}{\epsilon_{hx}} + h_c \tag{7.79}$$

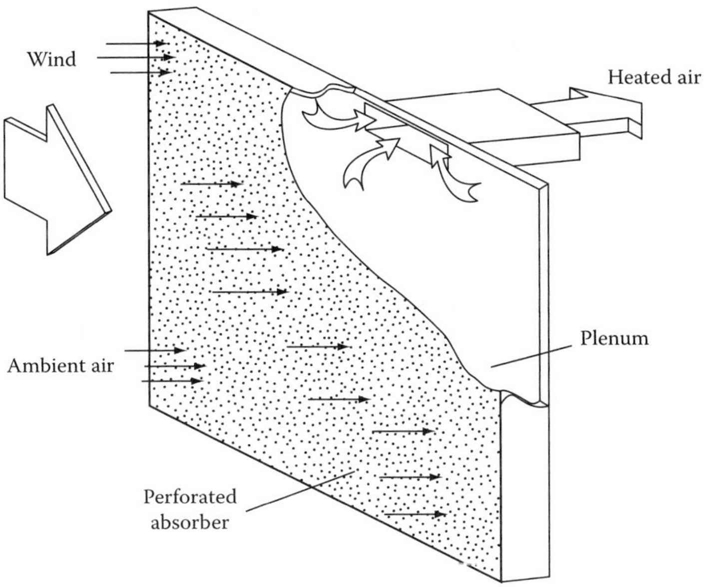

FIGURE 7.31
Unglazed transpired solar collector.

where
- $\in_{hx}$ is absorber heat exchanger effectiveness
- h_r is a linearized radiative heat transfer coefficient
- h_c is the convective heat loss coefficient

The heat exchanger effectiveness for air flowing through the absorber plate is defined as

$$\in_{hx} = \frac{T_{out} - T_{\alpha}}{T_c - T_a} \tag{7.80}$$

The forced convective heat loss coefficient due to a wind velocity of U_∞ is given as

$$h_c = 0.82 \frac{U_\infty \nu c_p}{L \nu_0} \tag{7.81}$$

where
- ν is the kinematic viscosity of air in m^2/s
- c_p the specific heat in J/kg · K
- ν_0 the suction velocity
- L is the height of the collector

Radiation heat loss occurs both to the sky and to the ground. Assuming the absorber is gray and diffuse with an emissivity $\in_c$, the radiative loss coefficient h_r is

$$h_r = \in_c \sigma \frac{(T_c^4 - F_{cs}T_{sky}^4 - F_{cg}T_{gnd}^4)}{T_c - T_a} \quad (7.82)$$

where F_{cs} and F_{cg} are the view factors between the collector and the sky, and collector and the ground, respectively. For a vertical wall with infinite ground in front of it, both F_{cs} and F_{cg} will be 0.5 each. Using the above equations, Kutcher and Christensen [3] showed that the predicted performance matches the measured performances well. Figure 7.32 shows their predicted thermal performances.

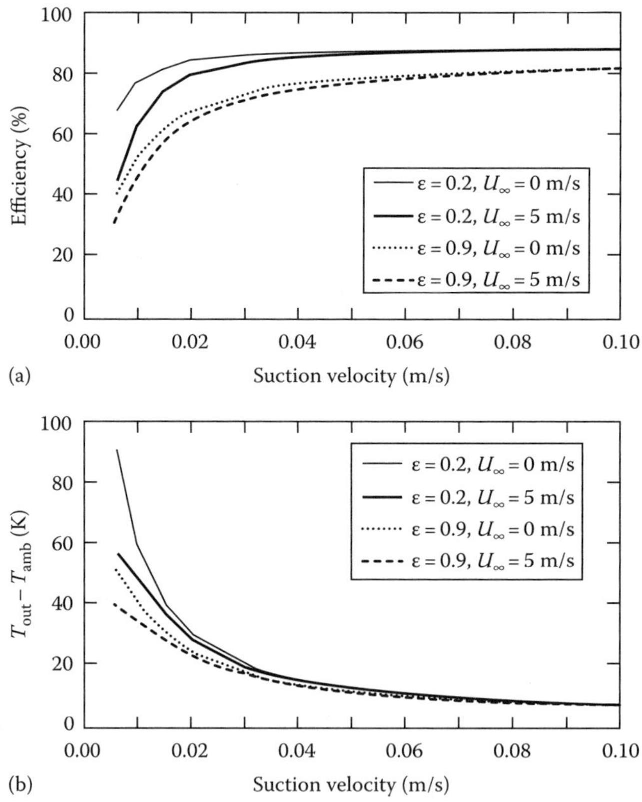

FIGURE 7.32
Predicted performance of unglazed transpired collector. (a) Efficiency vs. suction velocity. (b) Temperature difference vs. suction velocity. (From Kutcher, C.F. and Christensen, C.B., Unglazed transpired solar collectors, in *Advances in Solar Energy*, 7, 283–307, 1992. With permission.)

7.8* Methods of Modeling and Design of Solar Heating Systems

Several methods of modeling and design of solar space and water heating have been developed including *f*-chart, SLR, Utilizability and TRNSYS. The first two of these methods are described briefly in the following sections. Before design of a solar heating system can start, one must know the loads and the availability of solar radiation at the location. The estimation of solar radiation is described in Chapter 5 and the estimation of the loads has been described briefly earlier in this chapter.

7.8.1 Design of a Liquid-Based Solar Heating System by *f*-Chart

Klein and his coworkers [19,23,24–26] developed a method of simplified prediction of the performance of a solar heating system based on a large number of detailed simulations for various system configurations in various locations in the United States. The results from these simulations were then correlated with dimensionless parameters on charts that are general in form and usable anywhere. The charts are called *f*-charts, denoting a parameter f_s, the fraction of monthly load supplied by solar energy. The dimensionless groups used in the *f*-charts are derived from a nondimensionalization of the equations of governing energy flows; the groups therefore have a physical significance as described below. The *f*-chart method has been developed for standard solar heating and hot water system configurations.

A schematic diagram for standard solar heating systems based on liquid heat transfer fluids is shown in Figure 7.29. Certain deviations from this configuration can be handled in the *f*-chart method. For example, the collector-to-storage heat exchanger in the liquid-based system may be eliminated.

In the *f*-chart method, the fraction of load supplied by solar energy, f_s, is correlated with two dimensionless parameters, called the loss parameter P_L and the solar parameter P_s. Parameters P_L and P_s are defined as

$$P_L = \frac{A_c F_{hx} F_R U_c \Delta t (T_R - \bar{T}_a)}{L}, \tag{7.83}$$

and

$$P_s = \frac{A_c F_{hx} F_R \bar{I}_c \overline{(\tau\alpha)}}{L}. \tag{7.84}$$

P_s and P_L are measures, respectively, of the long term insolation gain by the collector and long term thermal loss per unit load. The parameters in Equations 7.83 and 7.84 and in the *f*-chart are described in Table 7.12.

* This section may be omitted in an introductory course.

TABLE 7.12

Definition of Parameters in *f*-Chart

Parameter	Definition
A_c	Net collector aperture area (m^2).
f_s	The solar load fraction: percentage of monthly load carried by solar system.
F_R	Collector heat-removal factor. See Equation 7.43.
F_{hx}	Collector loop heat-exchanger factor.
	$F_{hx} = 1$ if no collector heat exchanger is used.
$\bar{I}_c$	Total monthly collector-plane insolation (J/m^2-month).
L	Total monthly water-heating load (J/month). See Example 7.5.
$\bar{T}_a$	Monthly average ambient temperature (°C).
T_R	Reference temperature with a value of 100°C.
U_c	Collector heat-loss coefficient (W/m^2 °C).
Δt	Number of seconds per month (s/month).
$\overline{\tau\alpha}$	Monthly averaged collector transmittance–absorptance product.

P_s may be rewritten as

$$P_s = \frac{A_c \bar{I}_c F_{hx} F_R (\tau\alpha)_n}{L} \frac{(\overline{\tau\alpha})}{(\tau\alpha)_n} \tag{7.85}$$

As shown earlier in the chapter, $F_R U_c$ and $F_R(\tau\alpha)_n$ in the equations above can be determined directly from the slope and intercept of a collector efficiency curve plotted, as shown in Figure 7.33. In addition, Klein suggests that the time-averaged value $F_R(m)$ is related to the normal incidence value $F_R(\tau\alpha)_n$ as

$$\frac{F_R(\overline{\tau\alpha})}{F_R(\tau\alpha)} = 0.95, \tag{7.86}$$

for a surface tilted within ±20° of the local latitude.

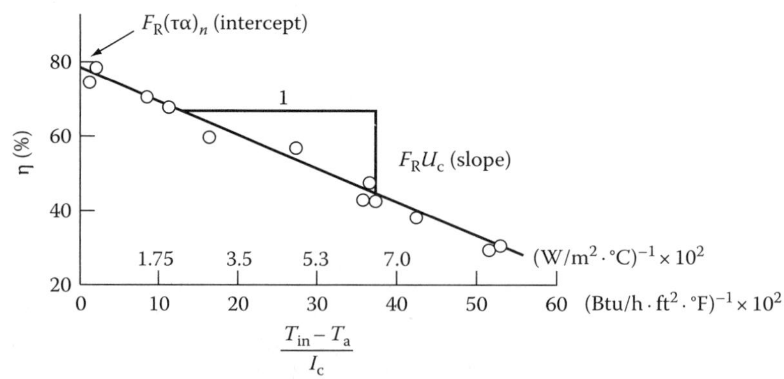

FIGURE 7.33
Typical flat-plate collector efficiency curve showing method of evaluating F_R $(\tau\alpha)_n$ and U_c.

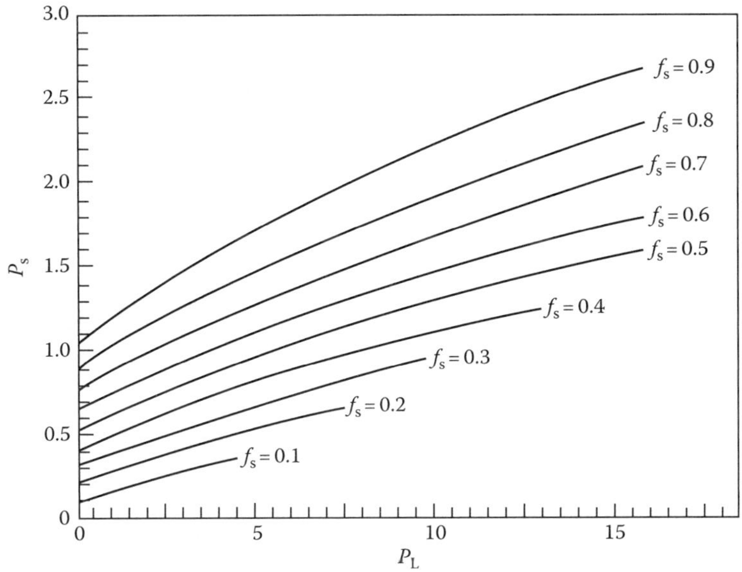

FIGURE 7.34
The *f*-chart for solar water-heating systems. (From Khan, A.Y., *Int. J. Energy Res.*, 18, 643, 1994; Goswami, Y. et al., *Principles of Solar Engineering*, 2nd edn., Taylor & Francis, Philadelphia, 2000.)

Note that Figure 7.33 is a plot of efficiency versus collector *inlet* fluid temperature, as distinguished from *average* fluid temperature. Such curves are usually available from the collector manufacturers. Figure 7.34 shows the *f*-chart for a liquid-based solar heating and hot water system. The results of the computer model used to generate the f_s curves in Figure 7.34 can also be expressed in the form of an empirical equation as [25]

$$f_s = 1.029P_s - 0.065P_L - 0.245P_s^2 + 0.0018P_L^2 + 0.0215P_s^3 \tag{7.87}$$

valid for the range

$$0 \le P_s \le 3.0; \quad 0 \le P_L \le 18.0 \quad \text{and} \quad 0 \le f_s \le 1.0.$$

It is further required that $P_s > PJ'12$ to ensure that the monthly insolation is above the useful threshold and that thermal losses are below an upper bound at which energy absorbed is equal to energy lost from the absorber plate.

The *f*-chart in Figure 7.34 is based on nominal values of collector flow rate, thermal storage mass, and load heat-exchanger effectiveness. It is possible to use the *f*-chart for other values of these important parameters if the loss parameter and a solar parameter are appropriately modified. Table 7.13 contains the

TABLE 7.13

Nominal Values of Physical Parameters and Modifying Groups for *f*-Chart Use[a] for Liquid-Based Systems

Parameter	Nominal Value	Modified Parameter[b]	
Flow rate[c]			
$\frac{(\dot{m}c_p)_c}{A_c}$	$0.0128\,\mathrm{L\ H_2O\ equivalent/s \cdot m_c^2}$	$P_L = P_{L,nom}\frac{F_{hx}F_R}{(F_{hx}F_R)_{nom}}$	(7.88)
		$P_s = P_{s,nom}\frac{F_{hx}F_R}{(F_{hx}F_R)_{nom}}$	(7.89)
Storage volume (water)			
$V_s = \left(\frac{M}{\rho A_C}\right)_s$	$75\,\mathrm{L\ H_2O/m_c^2}$	$P_L = P_{L,nom}\left(\frac{V_s}{75}\right)^{-0.25}$	(7.90)
Load heat exchanger[d]			
$\frac{\varepsilon_L(\dot{m}c_p)_{air}}{Q_L}$	2.0	$P_s = P_{s,nom}\left\{0.393 + 0.651\exp\left[-0.139\frac{Q_L}{\varepsilon_L(\dot{m}c_p)_{air}}\right]\right\}$	(7.91)

[a] Table prepared from data and equations presented in [23,24].

[b] Multiply basic definition of P_s and P_L in Equations 7.83 and 7.84 by factor for non-nominal group values; $(F_{hx}F_R)_{nom}$, refers to values of $F_{hx}F_R$ at collector rating or test conditions.

[c] In liquid systems, the correction for flow rate is small and can usually be ignored if variation is no more than 50% below the nominal value.

[d] $(\dot{m}c_p)_{min}$ is the minimum fluid capacitance rate, usually that of air for the load heat exchanger; Q_L is the heat load per unit temperature difference between inside and outside of the building.

nominal values used in generating the *f*-chart and the dimensionless groups that are used to modify the loss or the solar parameter for other system values. The *f*-chart in Figure 7.34 may be used for a water-heating system only with the following modification for P_L:

$$P_L = \frac{F_{hx}F_R A_c U_c \Delta t(11.6 + 1.18T_{w,o} + 3.86T_{w,i} - 2.32\overline{T}_a)}{L}, \tag{7.92}$$

where $T_{w,i}$ and $T_{w,o}$ are water supply and delivery temperatures, respectively. Klein [24] has prepared *f*-charts for air-based solar heating systems also, as shown in Figure 7.34. Loss and solar parameters, P_L and P_s, are identical to those for liquid systems and are defined in Equations 7.83 and 7.84. Since air systems do not use a collector-to-storage heat exchanger, the heat-exchanger factor is given by $F_{hx} = 1.0$.

The data from which the *f*-chart was created can be expressed by an empirical equation as

$$f_s = 1.040\,P_s - 0.065\,P_L - 0.159\,P_s^2 + 0.00187\,P_L^2 - 0.0095\,P_s^3 \tag{7.93}$$

valid for the ranges

$$0 \le P_s \le 3.0 \quad 0 \le P_L \le 18.0 \quad 0 \le f_s \le 1.0.$$

It is further required that $P_s > 0.07\,P_L$ to ensure that the monthly insolation is above the useful threshold level and that thermal losses are below an upper bound at which energy absorbed is equal to energy lost from the absorber plate.

Example 7.7

Calculate the annual heating energy delivery of a solar space-heating system using a double-glazed, flat-plate collector in Bismarck, North Dakota. The building and solar system specifications are given below:

Building specifications

Location: 47°N latitude (see Tables A5.2 and A5.3)

Space-heating load: 28,000 kJ/°C · day

Solar system specifications

Collector loss coefficient: $F_R U_c = 4.44\,W/m^2 \cdot k$

Collector optical efficiency (average): $F_R(\overline{\tau\alpha}) = 0.70$

Collector tilt: $\beta = L + 15° = 62°$

Collector area: $A_c = 60\,m^2$

Collector fluid flow rate: $\dot{m}_c/A_c = 55\,kg/h\,m^2$

Collector fluid heat capacity: $c_{p_c} = 1.9\,kJ/°C \cdot kg$ (antifreeze)

Storage capacity: 83 kg of H_2O/m^2

TABLE 7.14

Climatic and Solar Data for Bismarck, North Dakota

Month	Average Ambient Temperature (°C)	Heating Degree C-Days	Horizontal Solar Radiation (Langleys/Day)
January	−13.2	978	157
February	−10.3	801	250
March	−3.8	687	356
April	6.1	368	447
May	12.4	188	550
June	17.7	68	590
July	21.6	10	617
August	20.7	19	516
September	14.2	140	390
October	8.2	313	272
November	−1.7	601	161
December	−9.1	851	124

Storage fluid flow rate: $\dot{m}_s/A_c = 100\,\text{kg of } H_2O/\text{h m}^2$

Storage fluid heat capacity: $c_{p_c} = 4.187$ kJ/kg °C (water)

Heat-exchanger effectiveness: 0.75

Load heat exchanger: $\epsilon_L(\dot{m}c_p)_{air}/Q_L = 2.0$

Climatic data

Climatic data from the NWS are tabulated in Table 7.14.

Solution

The *f*-chart method is amenable to a step-by-step application. The following order is suggested:

1. Calculate monthly collector-plane insolation for each month.
2. Calculate solar and loss parameters P_s and P_L (Equations 7.83 and 7.84)) for each month, and heat-exchanger penalty factor F_{hx}.
3. Evaluate f_s from the *f*-chart for each month.

$$F_{hx} = \left\{1 + \left[\frac{F_R U_c A_c}{(\dot{m}c_p)_c}\right]\left[\frac{(\dot{m}c_p)_c}{\epsilon_{hx}\,(\dot{m}c_p)_{min}} - 1\right]\right\}^{-1}$$

4. Calculate total annual energy delivery from monthly totals.

Each of these steps is shown in Table 7.15.

The following are the calculations for the month of January. Results for the other months can be found similarly.

From Table A.5.2 (by interpolation):

$$\overline{H}_{o,h} = 3{,}011\,\text{W}\cdot\text{h/m}^2\text{-day} = 10{,}874\,\text{kJ/m}^2\text{-day}.$$

TABLE 7.15

The *f*-Chart Summary for Example 7.7

Month	Collector-Plane Radiation (kJ/m²-Day)	Monthly Energy Demand (GJ)	P_L[a]	P_s[a]	F_s
January	18,378	27.38	2.81	0.83	0.53
February	20,654	22.43	3.34	1.14	0.69
March	19,327	19.24	3.66	1.24	0.73
April	16,491	10.25	6.22	1.99	0.91
May	15,913	5.26	11.30	3.75	1.00
June	15,356	1.90	29.39	10.01	1.00
July	16,604	.28	190.00	73.43	1.00
August	16,988	.53	101.53	39.69	1.00
September	17,925	3.92	14.85	5.66	1.00
October	18,874	8.76	7.11	2.67	1.00
November	14,926	16.86	4.09	1.10	0.63
December	15,399	23.83	3.11	0.80	0.49
Annual	—	140.64	—		0.68

[a] $P_s > 3.0$ or $P_L > 18.0$ implies $F_s = 1.0$; no correction for storage size and flow rates is required.

From Table 7.14

$$\overline{H_h} = 157 \text{ langleys/day} = 6572 \text{ kJ/m}^2\text{- day}.$$

Using Equation 5.50

$$\bar{K}_T = \frac{\bar{H}_h}{\bar{H}_{o,h}} = 0.604.$$

By Equation 5.52

$$\frac{\bar{D}_h}{\bar{H}_h} = 0.775 + 0.347\left(h_{ss} - \frac{\pi}{2}\right) - \left[0.505 + 0.0261\left(h_{ss} - \frac{\pi}{2}\right)\right]\cos(2\bar{K}_T - 1.8) = 0.2156.$$

Therefore,

$$\bar{D}_h = \frac{\bar{D}_h}{\bar{H}_h} \times \bar{H}_h = 0.2156 \times 6572 \text{ kJ/m}^2\text{-day}$$

$$= 1417 \text{ kJ/m}^2\text{-day}.$$

Using Equation 5.53

$$\bar{B}_h = \bar{H}_h - \bar{D}_h = 4662.3 \text{ kJ/m}^2\text{-day}$$

From Equation 5.23 we can find

$$\delta_s = 23.45° \sin\left[\frac{360(284+15)}{365}\right]^{\circ} = -21.27°.$$

Using Equation 5.30

$$\begin{aligned} h_{sr(\alpha=0)} &= -\cos^{-1}(-\tan L \cdot \tan\delta_s) \\ &= -\cos^{-1}[-\tan(47°)\cdot\tan(-21\cdot 27°)] \\ &= -65.3° \quad \text{or} \quad -1.14 \text{ rad.} \end{aligned}$$

Therefore, using Equation 5.56

$$\begin{aligned} \bar{R}_b &= \frac{\cos(L-\beta)\cos\delta_s \sin h_{sr(\alpha=0)} \sin(L-\beta)\sin\delta_s}{\cos L \cos\delta_S \sin h_{sr(\alpha=0)} + h_{sr(\alpha=0)} \sin L \sin\delta_s} \\ &= \frac{\cos(47°-62°)\cos(-21\cdot 27°)\sin(-65.3°) - 1.14\sin(47°-62°)\sin(-21\cdot 27°)}{\cos 47° \cos(-21\cdot 27°)\sin(-65.3) - 1.14\sin 47° \sin(-21\cdot 27°)} \\ &= 3.36. \end{aligned}$$

Using Equation 5.58

$$\bar{R}_d = \cos^2\left(\frac{\beta}{2}\right) = \cos^2\frac{62°}{2} = 0.7347.$$

Then, using Equation 5.60, neglecting reflection

$$\begin{aligned} \bar{H}_c &= R_b \cdot \bar{B}_h + \bar{R}_d \cdot \bar{D}_h \\ &= 3.363 \times 5155 \text{ kJ/m}^2\text{-day} + 0.7347 \times 1417 \text{ kJ/m}^2\text{-day} \\ &= 18378 \text{ kJ/m}^2\text{-day}. \end{aligned}$$

The monthly energy demand is

$$\begin{aligned} \text{Load} &= \bar{U}_a \times \text{degree-C-days} \\ &= 28{,}000 \text{ kJ/°C day} \times 978\text{°C-day} \\ &= 27.384 \text{ GJ}, \end{aligned}$$

$$\begin{aligned} (\dot{m}c_p)_c &= \left(\frac{\dot{m}_c}{A_c}\right) \times A_c \times c_p \\ &= 55 \text{ kg/h m}^2 \times 60 \text{ m}^2 \times 1.9 \text{ kJ/kg °K} \\ &= 6270 \text{ kJ/h °K} \quad \text{or} \quad 1742 \text{ W/K} \end{aligned}$$

$$(\dot{m}c_p)_s = \left(\frac{\dot{m}_s}{A_c}\right) \times A_c \times c_{ps}$$

$$= 100\ \text{kg/m}^2\ \text{h} \times 60\ \text{m}^2 \times 4.187\ \text{kJ/kg °K}$$

$$= 25{,}122\ \text{kJ/h °K} \quad \text{or} \quad 6{,}978\ \text{W/K}$$

$$(\dot{m}c_p)_{min} = (\dot{m}c_p)_c$$

Therefore, F_{hx} becomes

$$F_{hx} = \left\{1 + \left[\frac{F_R U_c A_c}{(\dot{m}c_p)_c}\right]\frac{1}{\epsilon_{hx}} - 1\right\}^{-1}$$

$$= \left\{1 + 4.44\ \text{W/m}^2\text{-K} \times 60\ \text{m}^2/1742\ \text{W/K}[1/0.75 - 1]\right\}^{-1}$$

$$= 0.951.$$

Using Equation 7.83

$$P_L = \frac{A_c F_{hx} F_R U_c \Delta t (T_R - \bar{T}_a)}{\text{Load}}$$

$$= \frac{60\ \text{m}^2 \times 0.951 \times 4.44\ \text{W/m}^2 - \text{K} \times (31 \times 24 \times 3600)\text{h}}{27.384\ \text{GJ} \times (10^9\ \text{kJ/1 GJ})}(100°\text{C} - (-13.2°\text{C}))$$

$$= 2.81.$$

Using Equation 7.84

$$P_s = \frac{A_c F_{hx} F_R I_c (\overline{\tau\alpha})}{\text{Load}}$$

$$= \frac{60\ \text{m}^2 \times 0.951 \times 0.7 \times (18378\ \text{kJ/m}^2\ \text{day} \times 31\ \text{days})}{27.384\ \text{GJ} \times (10^6\ \text{kJ/1 GJ})}$$

$$= 0.83.$$

From Figure 7.34 or Equation 7.87

$$f_s = 0.53.$$

A simple way to analyze a solar heating system is to utilize the copyrighted online *f*-chart tool by Beckman and Klein [3]. A free trial of the software is available for download at http://www.fchart.com/fchart/fchartdemo1.shtml. With this software, one can choose from nine different systems including active domestic hot water, building storage, pool heating, a general heating system, and five different types of collectors. The user must input several specific design parameters and can choose a specific location. This software can be especially beneficial to design systems for optimizing the solar fraction, f_s, by iterating through different design parameters. In addition to providing the f_s value for the specifications input, a thorough life cycle economic analysis of the system is provided. The software includes weather data for over 300 locations, and in both English and SI units.

7.8.2 The Utilizability Method

The utilizability method is used to predict the long-term performance of a solar thermal system used for space heating, hot water, industrial process heat, or thermal power systems. It is based on finding the long-term utilizability, ϕ, of a thermal collector. ϕ is defined as the fraction of solar flux absorbed by a collector and delivered to the working fluid. The utilizability method is described in detail in Chapter 8 with applications in solar industrial process heat and solar power systems.

7.9* Solar Cooling

In some ways, solar energy is better suited to space cooling than to space heating, but this application of solar energy has so far not found much commercial success. The seasonal variation of solar energy is well suited to the space-cooling requirements of buildings. Since the warmest seasons of the year correspond to periods of high insolation, solar energy is most available when comfort cooling is most needed. Moreover, the efficiency of solar collectors increases with increasing insolation and increasing environmental temperature. Consequently, in the summer, the amount of energy delivered per unit surface area of collector can be larger than that in winter.

There are several approaches that can be taken to solar space cooling and refrigeration. Because of the limited operating experience with solar-cooling systems, their design must be based on basic principles and experience with conventional cooling systems. The material presented in this chapter will therefore stress the fundamental principles of operation of refrigeration cycles and combine them with special features of the components in a solar system. This chapter presents the cooling requirements of buildings and the basics of active solar cooling techniques based on vapor-compression and vapor-absorption refrigeration cycles and desiccant humidification.

* This section may be omitted in an introductory course.

7.9.1 Cooling Requirements for Buildings

The cooling load of a building is the rate at which heat must be removed to maintain the air in a building at a given temperature and humidity. It is usually calculated on the basis of the peak load expected during the cooling season. For a given building, the cooling load depends primarily on

1. Inside and outside dry-bulb temperatures and relative humidities
2. Solar-radiation heat load and wind speed
3. Infiltration and ventilation
4. Internal heat sources

A method of calculating the cooling load is presented in detail in [18].

The steps in calculating the cooling load of a building are as follows:

1. Specify the building characteristics: wall area, type of construction, and surface characteristics; roof area, type of construction, and surface characteristics; window area, setback, and glass type; and building location and orientation.
2. Specify the outside and inside wet- and dry-bulb temperatures.
3. Specify the solar heat load and wind speed.
4. Calculate building cooling load resulting from the following: heat transfer through windows; heat transfer through walls; heat transfer through roof; sensible and latent heat gains resulting from infiltration and ventilation, sensible and latent heat gains (water vapor) from internal sources, such as people, lights, cooking, etc.

Equations 7.94 through 7.100 may be used to calculate the various cooling loads for a building. Cooling loads resulting from lights, building occupants, etc., may be estimated from [18]. For unshaded or partially shaded windows, the load is

$$\dot{Q}_{\text{wi}} = A_{\text{wi}}\left[F_{\text{sh}}\bar{\tau}_{\text{b,wi}}I_{\text{h,b}}\frac{\cos i}{\sin\alpha} + \bar{\tau}_{\text{d,wi}}I_{\text{h,d}} + \bar{\tau}_{\text{r,wi}}I_{\text{r}} + I_{\text{wi}}(T_{\text{out}} - T_{\text{in}})\right]. \tag{7.94}$$

For shaded windows, the load (neglecting sky diffuse and reflected radiation) is

$$\dot{Q}_{\text{wi,sh}} = A_{\text{wi,sh}}U_{\text{wi}}(T_{\text{out}} - T_{\text{in}}). \tag{7.95}$$

For unshaded walls, the load is

$$\dot{Q}_{wa} = A_{wa}\left[\bar{\alpha}_{s,wa}\left(r_r + I_{h,d} + I_{h,b}\frac{\cos i}{\sin\alpha}\right) + U_{wa}(T_{out} - T_{in})\right]. \tag{7.96}$$

For shaded walls, the load (neglecting sky diffuse and reflected radiation) is

$$\dot{Q}_{wa,sh} = A_{wa,sh}\left[U_{wa}(T_{out} - T_{in})\right]. \tag{7.97}$$

For the roof, the load is

$$\dot{Q}_{rf} = A_{rf}\left[\bar{\alpha}_{s,rf}\left(I_{h,d} + I_{h,b}\frac{\cos i}{\sin\alpha}\right) + U_{rf}(T_{out} - T_{in})\right]. \tag{7.98}$$

Sensible-cooling load due to infiltration and ventilation is

$$\dot{Q}_i = \dot{m}_a(h_{out} - h_{in}) = \dot{m}_a Cp_a(T_{out} - T_{in}). \tag{7.99}$$

Latent load due to infiltration and ventilation is

$$\dot{Q}_w = \dot{m}_a\,(W_{out} - W_{in})\lambda_w, \tag{7.100}$$

where

$\dot{Q}_{wi}$ is the heat flow through unshaded windows of area A_{wi}
$\dot{Q}_{wi,sh}$ is the heat flow through shaded windows of area $A_{wi,sh}$
$\dot{Q}_{wu}$ is the heat flow through unshaded walls of area A_{wa}
$\dot{Q}_{wu,sh}$ is the heat flow through shaded walls of area $A_{wa,sh}$
$\dot{Q}_{rf}$ is the heat flow through roof of area A_{rf}
$\dot{Q}_i$ is the heat load resulting from infiltration and ventilation
$\dot{Q}_w$ is the latent heat load
$I_{h,b}$ is the beam component of insolation on horizontal surface
$I_{h,d}$ is the diffuse component of insolation on horizontal surface
I_r is the ground-reflected component of insolation
W_{out}, W_{in} is the outside and inside humidity ratios
U_{wi}, U_{wa}, U_{rf} is the overall heat-transfer coefficients for windows, walls and roof, including radiation
$\dot{m}_a$ is the net infiltration and ventilation mass flow rate of dry air
Cp_a is the specific heat of air (approximately $1.025\,\text{kJ/kg}\cdot\text{K}$ for moist air)
T_{out} is the outside dry-bulb temperature
T_{in} is the indoor dry-bulb temperature
F_{sh} is the shading factor (1.0 = unshaded, 0.0 = fully shaded)

$\bar{\alpha}_{s,wa}$ is the wall solar absorptance
$\bar{\alpha}_{s,rf}$ is the roof solar absorptance
i is the solar-incidence angle on walls, windows, and roof
h_{out}, h_{in} is the outside and inside air enthalpy
α is the solar-altitude angle
λ_w is the latent heat of water vapor
$\bar{\tau}_{b,wi}$ is the window transmittance for beam (direct) insolation
$\bar{\tau}_{d,wi}$ is the window transmittance for diffuse insolation
$\bar{\tau}_{r,wi}$ is the window transmittance for ground-reflected insolation

Recent ASHRAE handbooks recommend the use of the CLTD (cooling load temperature difference) method. For more details of the method, one should refer to [18].

Example 7.8

Determine the cooling load for a building in Phoenix, Arizona, with the specifications tabulated in Table 7.16.

Solution
To determine the cooling load for the building just described, calculate the following factors in the order listed:

1. Incidence angle for the south wall i at solar noon can be written from Equations 5.44 and 5.28 as

$$\cos i = \cos\beta\cos(L-\delta_s)+\sin\beta\sin(L-\delta_s)$$
$$= 0.26. \tag{7.101}$$

2. Solar altitude a at solar noon (from Equation 5.28)

$$\sin\alpha = \sin\delta_s\sin L+\cos\delta_s\cos L\cos h_s = \cos(L-\delta_s) = \cos 15° = 0.966.$$

3. South-facing window load (from Equation 7.94)

$$\dot{Q}_{wi} = 40\left\{\left(0.20\times 0.6\times 185\frac{0.26}{0.966}\right)+(0.81\times 80)+(0.60\times 70)+\left[1.09(100-75)\right]\right\} = 5600\text{ Btu/h}.$$

4. Shaded-window load (from Equation 7.95)

$$\dot{Q}_{wi,sh} = (3\times 40)[1.09(100-75)] = 3270\text{ Btu/h}.$$

TABLE 7.16
Specifications for Example 7.8

Factor	Description or Specification
Building characteristics	
Roof	
Type of roof	Flat, shaded
Area $A_{rf,sh}$, ft^2	1700
Walls (painted white)	
Size, north and south, ft	8 × 60 (two)
Size, east and west, ft	8 × 40 (two)
Area A_{wa}, north and south walls, ft^2	$480 - A_{wi} = 480 - 40 = 440$ (two)
Area A_{wa}, east and west walls, ft^2	$320 - A_{wi} = 320 - 40 = 280$ (two)
Absorptance $\bar{\alpha}_{s,wa}$ of white paint	0.12
Windows	
Size, north and south, ft	4 × 5 (two)
Size, east and west, ft	4 × 5 (two)
Shading factor F_{sh}	0.20
Insolation transmittance	$\bar{\tau}_{b,wi} = 0.60$; $\bar{\tau}_{d,wi} = 0.81$; $\bar{\tau}_{r,wi} = 0.60$
Location and latitude	Phoenix, Arizona; 33°N
Date	August 1
Time and local-solar-hour angle H_s	Noon; $H_s = 0$
Solar declination δ_s, deg	17°–55′
Wall surface tilt from horizontal β	90°
Temperature, outside and inside, °F	$T_{out} = 100$; $T_{in} = 75$
Insolation I, Btu/h · ft^2	$I_{h,b} = 185$; $I_{h,d} = 80$; $I_r = 70$
U factor for walls, windows and roof	$U_{wa} = 0.19$; $U_{wi} = 1.09$; $U_{rf} = 0.061$
Infiltration, lbm dry air/h	Neglect
Ventilation, lbm dry air/h	Neglect
Internal loads	Neglect
Latent heat load Q_w, percent	30% of wall sensible heat load[a]

[a] Approximate rule of thumb for Phoenix.

5. South-facing wall load (from Equation 7.96)

$$\dot{Q}_{wa} = (480 - 40)\left\{0.12\left[70 + 80 + \left(185\frac{0.26}{0.966}\right)\right] + 0.19(100 - 75)\right\}$$

$$= 12{,}610 \text{ Btu/h}.$$

6. Shaded-wall load (from Equation 7.97)

$$\dot{Q}_{wa,sh} = [(480 + 320 + 320) - (3 \times 40)][0.19(100 - 75)] = 4750 \text{ Btu/h}.$$

7. Roof load (from Equation 7.98)

$$\dot{Q}_{rf} = 1700[\bar{\alpha}_{s,rf} \times 0 + 0.061(100 - 75)] = 2600 \text{ Btu/h}.$$

8. Latent-heat load (30% of sensible wall load)

$$\dot{Q}_{w} = 0.3[(480 + 480 + 320 + 320) - (4 \times 40)][0.19(100 - 75)] = 2050 \text{ Btu/h}.$$

9. Infiltration load

$$\dot{Q}_{i} = 0.$$

10. Total cooling load for the building described in the example

$$\dot{Q}_{tot} = \dot{Q}_{wi} + \dot{Q}_{wi,sh} + \dot{Q}_{wa} + \dot{Q}_{wa,sh} + \dot{Q}_{rf} + \dot{Q}_{w} + \dot{Q}_{i},$$

$$\dot{Q}_{tot} = 30,880 \text{ Btu/h}.$$

Looking at the various heat loads, it is apparent that the largest heat load comes from the south-facing walls. The heat load from these walls is 3 times larger than from the shaded walls, and it would therefore behoove the architect or engineer to consider the options for reducing the heat load through the southern walls. Architectural options such as overhangs to shade these windows could be helpful. These options are discussed in Chapter 9.

7.9.2 Vapor-Compression Cycle

The two principal methods of lowering air temperature for comfort cooling are refrigeration with actual removal of thermal energy from the air or evaporation cooling of the air with adiabatic vaporization of moisture into it. Refrigeration can be used under any humidity condition of entering air, whereas evaporative cooling can be used only when the entering air has a comparatively low relative humidity.

The most widely used air-conditioning method employs a vapor-compression refrigeration cycle. Another method uses an absorption refrigeration cycle similar to that of the gas refrigerator. The vapor-compression refrigeration cycle requires energy input into the compressor, which may be provided as electricity from a PV system. Referring to Figure 7.35, the compressor raises the pressure of the refrigerant, which also increases its temperature. The compressed high-temperature refrigerant vapor then transfers thermal energy via a heat exchanger to the ambient environment in the condenser, where it condenses to a high-pressure liquid at a temperature close to, but above, the environmental temperature. The liquid refrigerant is then passed through the expansion valve where its pressure is reduced, resulting in a vapor–liquid mixture at a much lower temperature. The low-temperature refrigerant is then used to cool air or water in the evaporator where the liquid refrigerant evaporates by absorbing heat from the medium being

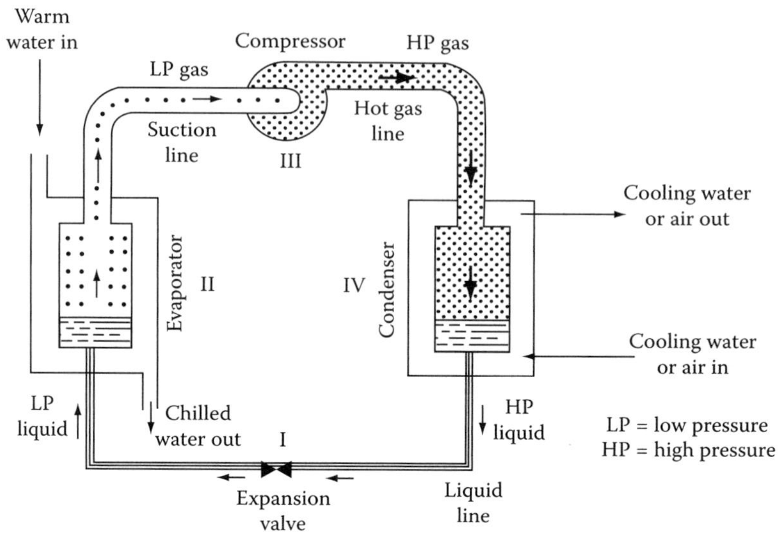

FIGURE 7.35
Schematic diagram illustrating the basic refrigeration vapor-compression cycle. (From Goswami, Y. et al., *Principles of Solar Engineering*, 2nd edn., Taylor & Francis, Philadelphia, 2000.)

cooled. The cycle is completed by the vapor returning to the compressor. If water is cooled by the evaporator, the device is usually called a chiller. The chilled water is then used to cool the air in the building.

The principle of operation of a vapor-compression refrigeration cycle can be illustrated conveniently with the aid of a pressure–enthalpy diagram, as shown in Figure 7.36. The ordinate is the pressure of the refrigerant in N/m^2 absolute, and the abscissa its enthalpy in kJ/kg. The Roman numerals in Figure 7.36 correspond to the physical locations in the schematic diagram of Figure 7.35.

Process I is a throttling process in which hot liquid refrigerant at the condensing pressure p_c passes through the expansion valve, where its pressure is reduced to the evaporator pressure p_e. This is an isenthalpic (constant enthalpy) process, in which the temperature of the refrigerant decreases. In this process, some vapor is produced and the state of the mixture of liquid refrigerant and vapor entering the evaporator is shown by point A. Since the expansion process is isenthalpic, the following relation holds:

$$h_{ve}f + h_{le}(1-f) = h_{lc}, \tag{7.102}$$

where

f is the fraction of mass in vapor state

subscripts v and l refer to vapor and liquid states, respectively

c and e refer to states corresponding to condenser and evaporator pressures, respectively

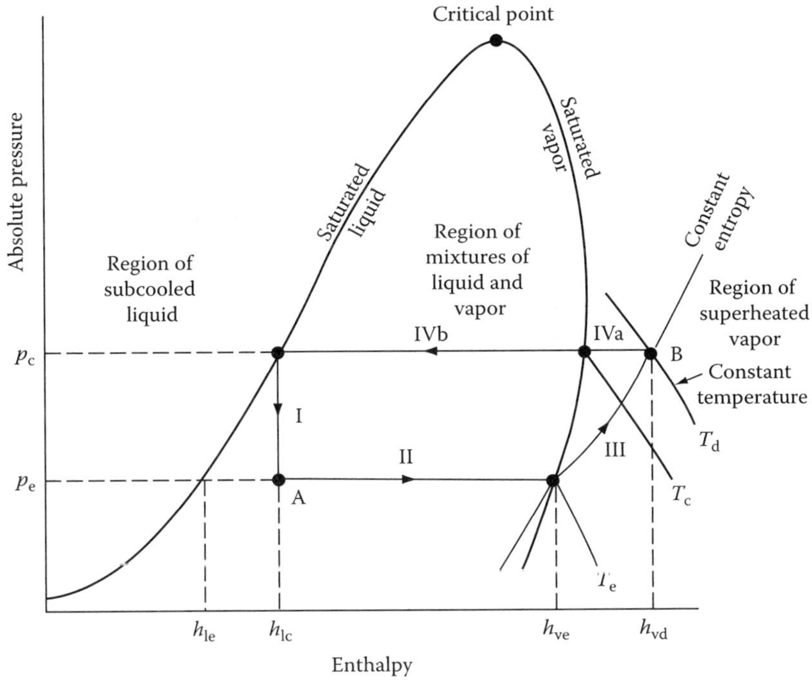

FIGURE 7.36
Simple refrigeration cycle on pressure–enthalpy diagram. (From Goswami, Y. et al., *Principles of Solar Engineering*, 2nd edn., Taylor & Francis, Philadelphia, 2000.)

and

$$f = \frac{h_{lc} - h_{le}}{h_{ve} - h_{le}} \tag{7.103}$$

Process II represents the vaporization of the remaining liquid. This is the process during which heat is removed from the chiller. Thus, the specific refrigeration effect per kilogram of refrigerant q_r is

$$q_r = h_{ve} - h_{lc} \text{ in kJ/kg (Btu/lb)} \tag{7.104}$$

In the United States, it is still common practice to measure refrigeration in terms of tons. One ton is the amount of cooling produced if 1 ton of ice is melted over a period of 24 h. Since 1 ton = 907.2 kg and the latent heat of fusion of water is 334.9 kJ/kg,

$$1 \text{ ton} = \frac{(907.2 \text{ kg}) \times (334.9 \text{ kJ/kg})}{(24 \text{ h}) \times (3{,}600 \text{ s/h})} = 3.516 \text{ kW} = 12{,}000 \text{ Btu/h}. \tag{7.105}$$

If the desired rate of refrigeration requires a heat-transfer rate of $\dot{Q}_r$, the rate of mass flow of refrigerant necessary $\dot{m}_r$ is

$$\dot{m}_r = \frac{\dot{Q}_r}{(h_{ve} - h_{lc})}. \tag{7.106}$$

Process III in Figure 7.36 represents the compression of refrigerant from pressure p_e to p_c. The process requires work input from an external source, which may be obtained from a solar-driven expander-turbine or a solar electrical system. In general, if the heated vapor leaving the compressor is at the condition represented by point B in Figure 7.36, the work of compression W_c is

$$W_c = \dot{m}_r(h_{vd} - h_{ve}). \tag{7.107}$$

In an idealized cycle analysis, the compression process is usually assumed to be isentropic.

Process IV represents the condensation of the refrigerant. Actually, sensible heat is first removed in the subprocess IVa as the vapor is cooled at constant pressure from T_d to T_c and latent heat is removed at the condensation temperature T_c, corresponding to the saturation pressure p_c in the condenser. The heat-transfer rate in the condenser $\dot{Q}_c$ is

$$\dot{Q}_c = \dot{m}_r(h_{vd} - h_{lc}) \tag{7.108}$$

This heat must be rejected into the environment, either to cooling water or to the atmosphere if no water is available.

The overall performance of a refrigeration machine is usually expressed as the ratio of the heat transferred in the evaporator $\dot{Q}_r$ to the shaft work supplied to the compressor. This ratio is called the *coefficient of performance* (COP), defined by

$$\mathrm{COP} = \frac{\dot{Q}_r}{W_c} = \frac{h_{ve} - h_{lc}}{h_{vd} - h_{ve}} \tag{7.109}$$

The highest COP for any given evaporator and condenser temperatures would be obtained if the system were operating on a reversible Carnot cycle. Under these conditions [18]

$$\mathrm{COP(Carnot)} = \frac{T_e}{T_d - T_e}. \tag{7.110}$$

The above cycle has been idealized. In practice, the liquid entering the expansion valve is several degrees below the condensing temperature, while the vapor entering the compressor is several degrees above the evaporation temperature. In addition, pressure drops occur in the suction, discharge, and liquid pipelines, and the compression is not truly isentropic. Finally, the work required to drive the compressor is somewhat larger than W_c above, because of frictional losses. All of these factors reduce the COP below the maximum and must be taken into account in a realistic engineering design.

Example 7.9

Calculate the amount of shaft work to be supplied to a 1 ton (3.52 kW) refrigeration plant operation at evaporator and condenser temperatures of 273 and 309 K, respectively, using Refrigerant 134a (R-134a) as the working fluid. The properties of Refrigerant 134a are tabulated in Table 7.17. Also calculate the COP and the mass flow rate of the refrigerant based on the ideal cycle described above.

Solution

From the property table, the enthalpies for process I are

Saturated vapor at 273 K $h_{ve} = 247.2$ kJ/kg

Saturated liquid at 309 K $h_{lc} = 100.3$ kJ/kg

Saturated liquid at 273 K $h_k = 50.0$ kJ/kg

Therefore, from Equation 7.103

$$f = \frac{100.3 - 50.0}{247.2 - 50.0} = 0.255.$$

The mass flow rate of refrigerant $\dot{m}_r$ is obtained from Equation 7.106 and the enthalpies above, or

$$\dot{m}_r = \frac{3.52\ \text{kW}}{(247.2 - 100.3)\ \text{kJ/kg}} = 0.024\ \text{kg/s}.$$

TABLE 7.17

Properties of Refrigerant 134a for Example 7.9

Temperature (K)	Absolute Pressure (kPa)	Vapor-Specific Volume (m^3/kg)	Liquid Enthalpy (kJ/kg)	Vapor Enthalpy (kJ/kg)	Vapor Entropy (kJ/kg · K)
Saturated					
273	292.8	0.0689	50.02	247.2	0.919
309	911.7	0.0223	100.25	266.4	0.9053
Superheated					
308.5	900	0.0226	—	266.18	0.9054
313	900	0.0233	—	271.3	0.9217
312.4	1000	0.0202	—	268.0	0.9043
313	1000	0.0203	—	268.7	0.9066

The specific shaft-work input required is

$$\frac{W_c}{\dot{m}_r} = h_{vd} - h_{ve}.$$

The entropy s_e of the saturated vapor entering the compressor at 273 K and 292.8 kPa is 0.919 kJ/kg · K. From the property table, superheated vapor at a pressure of 911.7 kPa has an entropy of 0.919 kJ/kg · K at a temperature of 313 K with an enthalpy of 270.8 kJ/kg. Thus, the energy input to the working fluid by the compressor is

$$W_c = 0.024(270.8 - 247.2) = 0.566 \text{ kW}.$$

Finally, the heat-transfer rate from the refrigerant to the sink, or cooling water in the condenser, is from Equation 7.108:

$$\dot{Q}_c = \dot{m}_r(h_{vd} - h_{lc}) = 0.024(270.8 - 100.3) = 4.09 \text{ kW}.$$

The COP of the thermodynamic cycle is

$$\text{COP} = \frac{247.2 - 100.3}{270.8 - 247.2} = 6.2,$$

whereas the Carnot COP is 273/36 or 7.6.

7.9.3 Absorption Air-Conditioning

In an absorption system, the refrigerant is evaporated or distilled from a less volatile liquid absorbent, the vapor is condensed in a water- or air-cooled condenser, and the resulting liquid is passed through a pressure-reducing valve to the cooling section of the unit. There it cools the water as it evaporates, and the resulting vapor flows into a vessel, where it is reabsorbed in the stripped absorbing liquid and pumped back to the heated generator. The heat required to evaporate the refrigerant in the generator can be supplied directly from solar energy, as shown in Figure 7.37.

Absorption air-conditioning is compatible with solar energy since a large fraction of the energy required is thermal energy at temperatures that flat-plate solar collectors such as previously described can provide. Figure 7.43 is a schematic of an absorption refrigeration system. Absorption refrigeration differs from vapor-compression air-conditioning only in the method of compressing the refrigerant. The compression process is shown to the left of the dashed line in Figure 7.38. The pressurization is accomplished by first dissolving the refrigerant in a liquid, called the absorbent, in the absorber

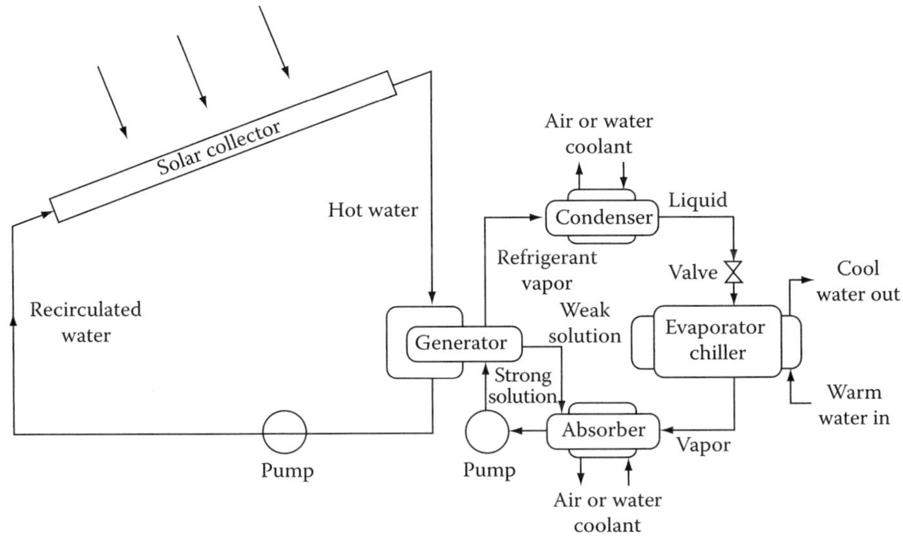

FIGURE 7.37
Schematic diagram of a solar-powered absorption refrigeration system.

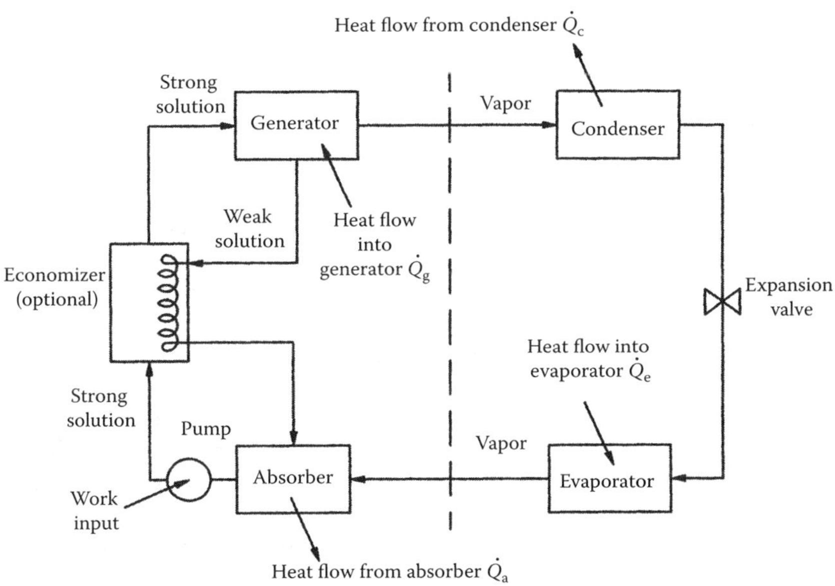

FIGURE 7.38
Diagram of heat and fluid flow of absorption air conditioner, with economizer. (From Klein, S.A. et al., *J Eng. Power*, 96A, 109, 1974.)

section. This liquid or strong solution is then pumped to a high pressure with an ordinary liquid pump. The low boiling-point refrigerant is then driven from solution by the addition of heat in the generator. By this means, the refrigerant vapor entering the condenser is compressed without the large input of high-grade shaft work that the vapor-compression air-conditioning demands.

The effective performance of an absorption cycle depends on the two materials that comprise the refrigerant–absorbent pair. Desirable characteristics for the refrigerant–absorbent pair follow:

1. The absence of a solid-phase absorbent.
2. A refrigerant more volatile than the absorbent so that separation from the absorbent occurs easily in the generator.
3. An absorbent that has a strong affinity for the refrigerant under conditions in which absorption takes place.
4. A high degree of stability for long-term operations.
5. Nontoxic and nonflammable fluids for residential applications. This requirement is less critical in industrial refrigeration.
6. A refrigerant that has a large latent heat so that the circulation rate can be kept low.
7. A low fluid viscosity that improves heat and mass transfer and reduces pumping power.
8. Fluids that must not cause long term environmental effects.

Lithium bromide–water (LiBr–H_2O) and ammonia–water (NH_3–H_2O) are the two pairs that meet most of the requirements. In the LiBr–H_2O system, water is the refrigerant and LiBr is the absorber, while in the NH_3–H_2O system, ammonia is the refrigerant and water is the absorber. Because the LiBr–H_2O system has high volatility ratio, it can operate at lower pressures and, therefore, at the lower generator temperatures achievable by flat-plate collectors. A disadvantage of this system is that the pair tends to form solids. LiBr has a tendency to crystallize when air is cooled, and the system cannot be operated at or below the freezing point of water. Therefore, the LiBr–H_2O system is operated at evaporator temperatures of 5°C or higher. Using a mixture of LiBr with some other salt as the absorbent can overcome the crystallization problem. The NH_3–H_2O system has the advantage that it can be operated down to very low temperatures. However, for temperatures much below 0°C, water vapor must be removed from ammonia as much as possible to prevent ice crystals from forming. This requires a rectifying column after the boiler. Also, ammonia is a safety Code Group B2 fluid (ASHRAE Standard 34-1992), which restricts its use indoors [18].

If the pump work is neglected, the COP of an absorption air conditioner can be calculated from Figure 7.38

$$\text{COP} = \frac{\text{Cooling effect}}{\text{Heat input}} = \frac{\dot{Q}_e}{\dot{Q}_g} \tag{7.111}$$

The COP values for absorption air-conditioning range from 0.5 for a small, single-stage unit to 0.85 for a double-stage, steam-fired unit. These values are about 15% of the COP values that can be achieved by a vapor-compression air conditioner. It is difficult to compare the COP of an absorption air conditioner with that of a vapor-compression air conditioner directly because the efficiency of electric power generation or transmission is not included in the COP of the vapor-compression air-conditioning. The following example illustrates the thermodynamics of an LiBr–H_2O absorption refrigeration system.

Example 7.10

A LiBr–H_2O, absorption refrigeration system such as that shown in Figure 7.39 is to be analyzed for the following requirements:

1. The machine is to provide 352 kW of refrigeration with an evaporator temperature of 5°C, an absorber outlet temperature of 32°C, and a condenser temperature of 43°C.
2. The approach at the low-temperature end of the liquid heat exchanger is to be 6°C.
3. The generator is heated by a flat-plate solar collector capable of providing a temperature level of 90°C.

Determine the COP, absorbent and refrigerant flow rates and heat input.

Solution
For the analytical evaluation of the LiBr–H_2O cycle, the following simplifying assumptions are made:

1. At those points in the cycle for which temperatures are specified, the refrigerant and absorbent phases are in equilibrium.
2. With the exception of pressure reductions across the expansion device between points 2 and 3, and 8 and 9 in Figure 7.39, pressure reductions in the lines and heat exchangers are neglected.
3. Pressures at the evaporator and condenser are equal to the vapor pressure of the refrigerant, that is, water, as found in steam tables.
4. Enthalpies for LiBr–H_2O mixtures are given in Figure 7.40.

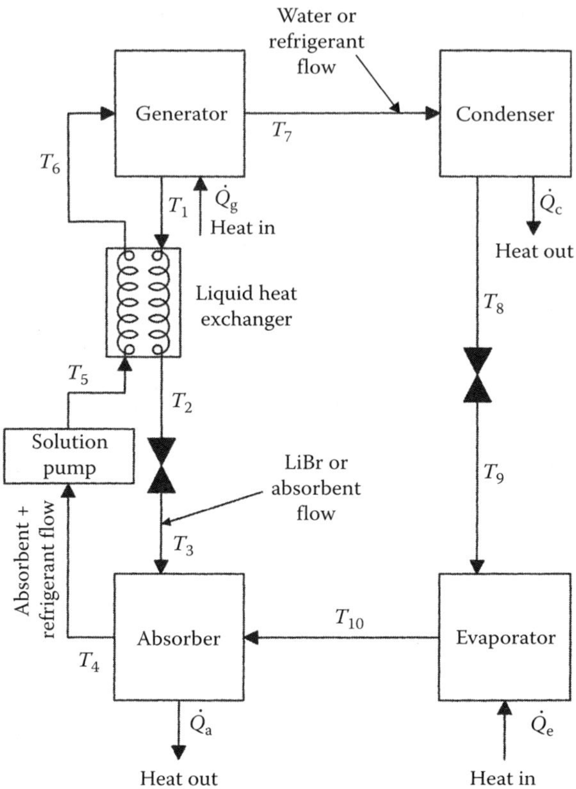

FIGURE 7.39
Li Br–H_2O absorption refrigeration cycle (see Table 7.18). (From Klein, S.A. et al., *J Eng. Power*, 96A, 109, 1974. With permission.)

As a first step in solving the problem, set up a table (Table 7.18) of properties, for example, given

Generator temperature = 90°C = $T_1 = T_7$,

Evaporator temperature = 5°C = $T_9 = T_{10}$,

Condenser temperature = 43°C = T_8,

Absorber temperature = 32°C = T_4.

Neglecting the pump work $T_5 \approx T_4 = 32°C$.

Since the approach at the low temperature end of the heat exchanger is 6°C,

$$T_2 = T_5 + 6°C = 38°C$$

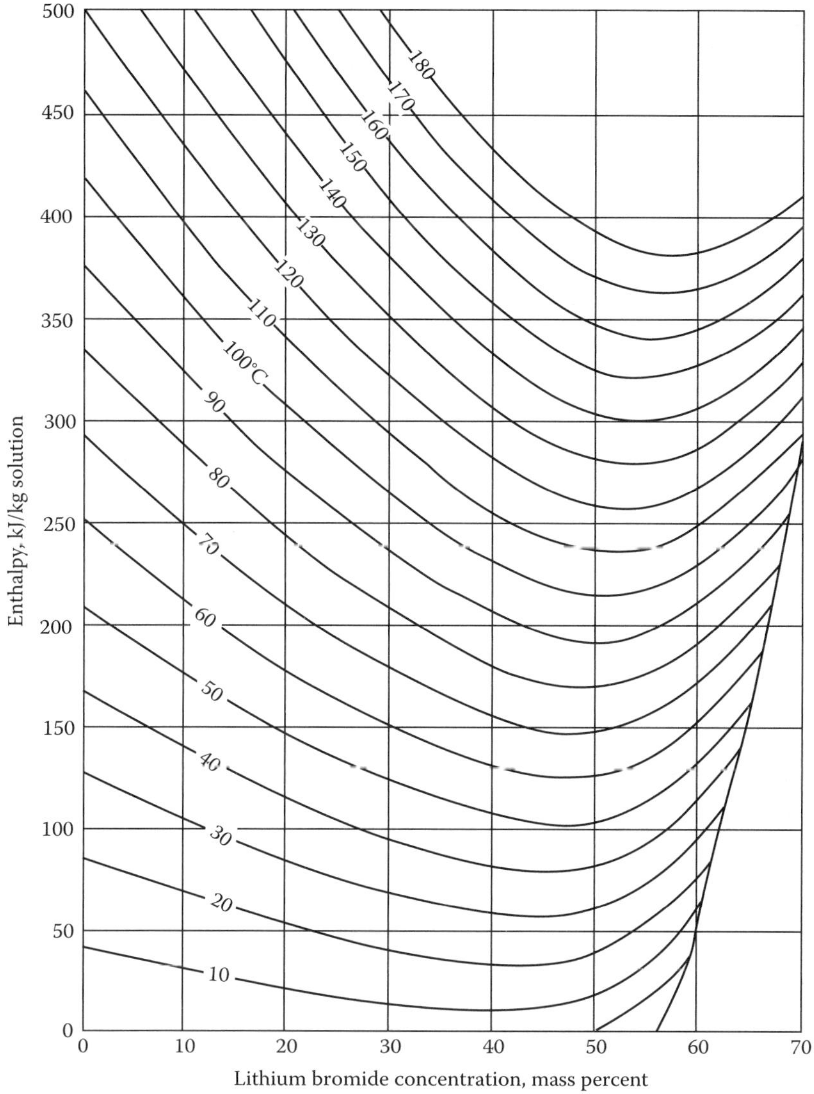

Equations Concentration range $40 < X < 70\%$ LiBr Temperature range $15°C < t < 165°C$

$$h = \Sigma_0^4 A_a X^n + t\, \Sigma_0^4 B_n X^n + t^2\, \Sigma_0^4 C_n X^n \text{ in kJ/kg, where } t = °C \text{ and } X = \%\text{ LiBr}$$

$A_0 = -2024.33$	$B_0 = 18.2829$	$C_0 = -3.7008214$ E−2
$A_1 = 163.309$	$B_1 = -1.1691757$	$C_1 = 2.8877666$ E−3
$A_2 = -4.88161$	$B_2 = 3.248041$ E−2	$C_2 = -8.1313015$ E−5
$A_3 = 6.302948$ E−2	$B_3 = -4.034184$ E−4	$C_3 = 9.9116628$ E−7
$A_4 = -2.913705$ E−4	$B_4 = 1.8520569$ E−6	$C_4 = -4.4441207$ E−9

FIGURE 7.40
Enthalpy–concentration diagram for LiBr–H_2O solutions. (From Goswami, Y. et al., *Principles of Solar Engineering*, 2nd edn., Taylor & Francis, Philadelphia, 2000.)

TABLE 7.18

Thermodynamic Properties of Refrigerant and Absorbent for Figure 7.39

Condition No. in Figure 7.39	Temperature (°C)	Pressure (kPa)	LiBr Weight Fraction	Flow (kg/kg H_2O)	Enthalpy (kJ/kg)
1	90	8.65	0.605	11.2	215
2	38	8.65	0.605	11.2	110
3	38	0.872	0.605	11.2	110
4	32	0.872	0.53	12.2	70
5	32	0.872	0.53	12.2	70
6	74	8.65	0.53	12.2	162
7	90	8.65	0	1.0	2670
8	43	8.65	0	1.0	180
9	5	0.872	0	1.0	180
10	5	0.872	0	1.0	2510

Source: Kreider, J.F. and Kreith, F., *Solar Heating and Cooling*, revised 1st edn., Hemisphere Publishing Corporation, Washington, DC, 1977.

and

$$T_3 \simeq T_2 = 38°\text{C}.$$

Since the fluid at conditions 7, 8, 9, and 10 is pure water, the properties can be found from the steam tables. Therefore,

$$P_7 = P_8 = \text{Saturation pressure of } H_2O \text{ at } 43°\text{C} = 8.65 \text{ kPa},$$

and

$$P_9 = P_{10} = \text{Saturation pressure of } H_2O \text{ at } 5°\text{C} = 0.872 \text{ kPa}.$$

Therefore,

$$P_1 = P_2 = P_5 = P_6 = P_7 = 8.66 \text{ kPa},$$

and

$$P_3 = P_4 = P_{10} = 0.872 \text{ kPa}.$$

Enthalpy

$$h_9 = h_8 = 180 \text{ kJ/kg (saturated liquid at 43°C)},$$

$$h_{10} = 2510 \text{ kJ/kg (saturated vapor enthalpy at 6°C)},$$

and

$$h_7 = 2760 \text{ kJ/kg (superheated vapor at 8.65 kPa, 90°C)}.$$

For the LiBr–H_2O mixture, conditions 1 and 4 may be considered equilibrium saturation conditions, which may be found from Figures 7.40 and 7.41 as

For

$$T_4 = 32°\text{C and } P_4 = 0.872 \text{ kPa}, \quad X_r = 0.53, \quad h_4 = 70 \text{ kJ/kg-sol}.$$

Therefore,

$$h_5 = 70 \text{ kJ/kg-sol}.$$

and for

$$T_1 = 90°\text{C} \quad \text{and} \quad P_1 = 8.65 \text{ kPa}, \quad X_{ab} = 0.605, \quad h_1 = 215 \text{ kJ/kg sol.},$$

for

$$T_3 = 38°\text{C}, \quad X_3 = 0.605, \quad h_3 = 110 \text{ kJ/kg-sol.},$$

$$h_2 = h_3 = 110 \text{ kJ/kg}$$

Relative flow rates for the absorbent (LiBr) and the refrigerant (H_2O) are obtained from material balances. A total material balance on the generator gives

$$\dot{m}_6 = \dot{m}_1 + \dot{m}_7,$$

while a LiBr balance gives

$$\dot{m}_6 X_s = \dot{m}_1 X_{ab},$$

where

X_{ab} is the concentration of LiBr in absorbent of solution
X_r is the concentration of LiBr in refrigerant–absorbent of solution

Substituting $(\dot{m}_1 + \dot{m}_7)$ for $\dot{m}_6$ gives

$$\dot{m}_1 X_s + \dot{m}_7 X_s = \dot{m}_1 X_{ab}.$$

Equations

1. $t = \Sigma_0^3 B_n X^n + I' \Sigma_0^3 A_n X^n$ Solution temp.°C
2. $t' = (t - \Sigma_0^3 B_n X^n) / \Sigma_0^3 A_n X^n$ Refrigerant temp.°C
3. $\text{Log } P = C + D/T' + E/T'^2$ P = kPa: T' = K
4. $T' = \dfrac{-2E}{D + [D^2 - 4E(C - \log P)]^{0.3}}$

$A_0 = 2.0007\ 55$ $B_0 = 124.937$ $C = 7.05$
$A_1 = 0.169\ 76$ $B_1 = -7.716\ 49$ $D = -1596.49$
$A_2 = -3.133\ 362$ E−03 $B_2 = 0.152\ 286$ $E = -104\ 095.5$
$A_3 = 1.976\ 68$ E−05 $B_3 = -7.950\ 90$ E−04

Range: $-15°C < t' < 110°C$
$5°C < t < 175°C$
$45 < X < 70\%$ LiBr

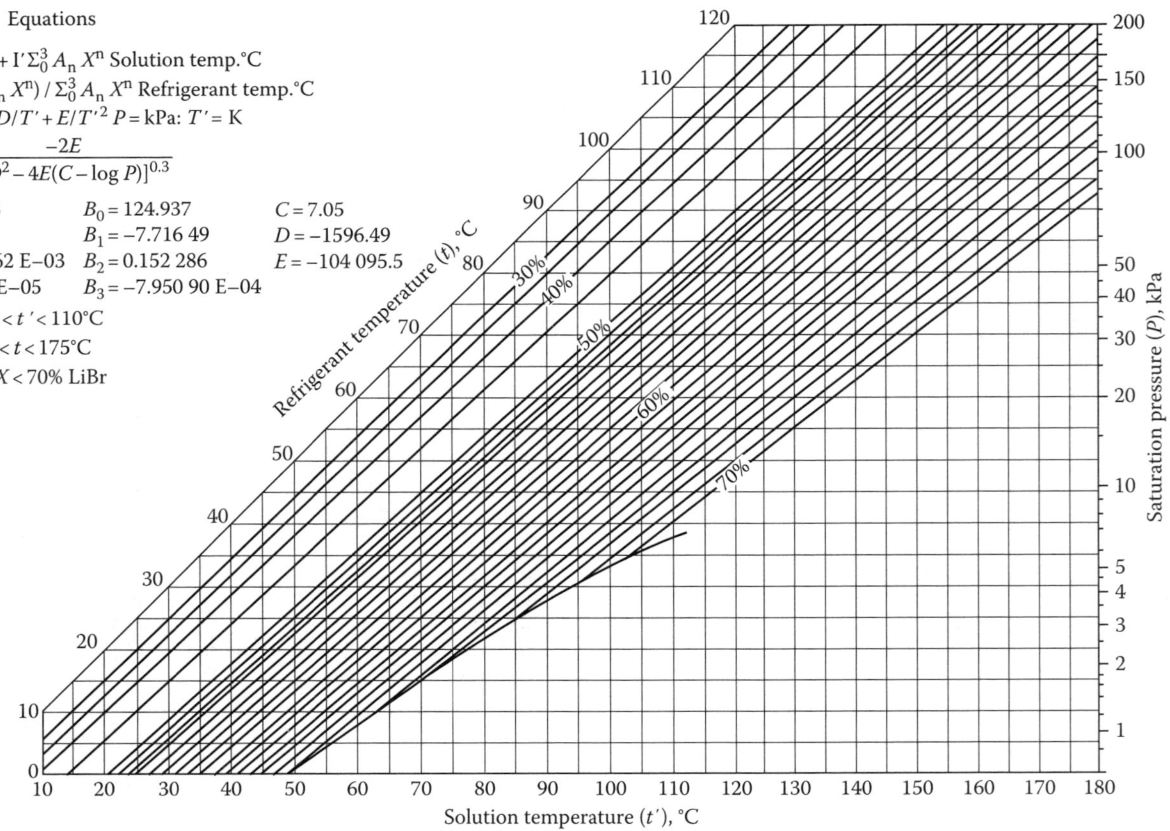

FIGURE 7.41
Equilibrium chart for LiBr–H_2O solutions. (From Goswami, Y. et al., *Principles of Solar Engineering*, 2nd edn., Taylor & Francis, Philadelphia, 2000.)

Since the fluid entering the condenser is pure refrigerant, that is, water, $\dot{m}_7$ is the same as the flow rate of the refrigerant $\dot{m}_r$:

$$\frac{\dot{m}_1}{\dot{m}_7} = \frac{X_s}{X_{ab} - X_s} = \frac{\dot{m}_{ab}}{\dot{m}_r},$$

where

$\dot{m}_{ab}$ is the flow rate of absorbent
$\dot{m}_r$ is the flow rate of refrigerant

Substituting for X_s and X_{ab} from the table gives the ratio of absorbent-to-refrigerant flow rate:

$$\frac{\dot{m}_{ab}}{\dot{m}_r} = \frac{0.53}{0.605 - 0.53} = 7.07.$$

The ratio of the refrigerant–absorbent solution flow rate $\dot{m}_s$ to the refrigerant-solution flow rate $\dot{m}_r$ is

$$\frac{\dot{m}_s}{\dot{m}_r} = \frac{\dot{m}_{ab} + \dot{m}_r}{\dot{m}_r} = 7.07 + 1 = 8.07.$$

Now Table 7.18 is complete except for T_6 and h_6, which may be found from an energy balance at the heat exchanger.

$$\dot{m}_s h_5 + \dot{m}_{ab} h_1 = \dot{m}_{ab} h_2 + \dot{m}_s h_6.$$

Hence

$$h_6 = h_5 + \left[\frac{\dot{m}_{ab}}{\dot{m}_s}(h_1 - h_2)\right] = 70 + \frac{7.07}{8.07}[215 - 110] = 162 \text{ kJ/kg of solution.}$$

The temperature corresponding to this value of enthalpy and a LiBr mass fraction 0.53 is found from Figure 7.41 to be 74°C.

The flow rate of refrigerant required to produce the desired 352 kW of refrigeration is

$$\dot{Q}_e = \dot{m}_r(h_{10} - h_9),$$

where $\dot{Q}_e$ is the cooling effect produced by the refrigeration unit, and

$$\dot{m}_r = \frac{352}{2510 - 180} = 0.15 \text{ kg/s}.$$

The flow rate of the absorbent is

$$\dot{m}_{ab} = \frac{\dot{m}_{ab}}{\dot{m}_r}\dot{m}_r = 7.07 \times 0.15 = 1.06 \text{ kg/s},$$

while the flow rate of the solution is

$$\dot{m}_s = \dot{m}_{ab} + \dot{m}_r = 1.06 + 0.15 = 1.21 \text{kg/s}.$$

The rate at which heat must be supplied to the generator $\dot{Q}_g$ is obtained from the heat balance

$$\begin{aligned}\dot{Q}_g &= \dot{m}_r h_7 + \dot{m}_{ab} h_1 - \dot{m}_s h_6 \\ &= [(0.15 \times 2670) + (1.06 \times 215)] - (1.21 \times 1.62) \\ &= 432 \text{ kW}.\end{aligned}$$

This requirement, which determines the size of the solar collector, probably represents the maximum heat load that the collector unit must supply during the hottest part of the day.

The coefficient of performance COP is

$$\text{COP} = \frac{\dot{Q}_e}{\dot{Q}_g} = \frac{352}{432} = 0.81$$

The rate of heat transfer in the other three heat-exchanger units—the liquid heat exchanger, the water condenser, and the absorber—is obtained from heat balances. For the liquid heat exchanger this gives

$$\dot{Q}_{1-2} = \dot{m}_{ab}(h_1 - h_2) = 1.06[215 - 110] = 111 \text{kW},$$

where $\dot{Q}_{1-2}$ is the rate of heat transferred from the absorbent stream to the refrigerant–absorbent stream. For the water condenser, the rate of heat transfer Q_{7-8} rejected to the environment is

$$\dot{Q}_{7-8} = \dot{m}_r(h_7 - h_8) = 0.15(2670 - 180) = 374 \text{ kW}.$$

The rate of heat removal from the absorber can be calculated from an overall heat balance on this system:

$$\dot{Q}_a = \dot{Q}_{7-8} - \dot{Q}_g - \dot{Q}_e = 374 - 432 - 352 = -410 \text{ kW}.$$

Explicit procedures for the mechanical and thermal design as well as the sizing of the heat exchangers are presented in standard heat-transfer texts. In large commercial units, it may be possible to use higher concentrations of LiBr, operate at a higher absorber temperature, and thus save on heat-exchanger cost. In a solar-driven unit, this approach would require concentrator-type or high-efficiency flat-plate solar collectors.

Ammonia–water systems are quite similar to LiBr–H_2O systems except that a rectifier is needed after the boiler to condense water vapor. They are treated by Goswami et al. in Ref. [27].

7.10* Solar Desiccant Dehumidification

In hot and humid regions of the world experiencing significant latent cooling demand, solar energy may be used for dehumidification using liquid or solid desiccants. Rangarajan et al. [28] compared a number of strategies for ventilation air-conditioning for Miami, Florida, and found that a conventional vapor-compression system could not even meet the increased ventilation requirements of ASHRAE Standard 62-1989. By pretreating the ventilation air with a desiccant system, proper indoor humidity conditions could be maintained and significant electrical energy could be saved. A number of researchers have shown that a combination of a solar desiccant and a vapor-compression system can save from 15% to 80% of the electrical energy requirements in commercial applications, such as supermarkets [29–37].

7.10.1 Solid Desiccant Cooling System

In a desiccant air-conditioning system, moisture is removed from the air by bringing it in contact with the desiccant and followed with sensible cooling of the air by a vapor-compression cooling system, vapor absorption cooling systems, or evaporative cooling system. The driving force for the process is the water vapor pressure. When the vapor pressure in air is higher than on the desiccant surface, moisture is transferred from the air to the desiccant until equilibrium is reached (see Figure 7.42). In order to regenerate the desiccant for reuse, the desiccant is heated, which increases the water vapor pressure on its surface. If air with lower vapor pressure is brought in contact with this desiccant, the moisture passes from the desiccant to the air (Figure 7.42). Two types of desiccants are used: solids, such as silica gel and lithium chloride; or liquids, such as salt solutions and glycols.

The two solid desiccant materials that have been used in solar systems are silica gel and the molecular sieve, a selective absorber. Figure 7.43 shows the equilibrium absorption capacity of several substances.

* This section may be omitted in an introductory course.

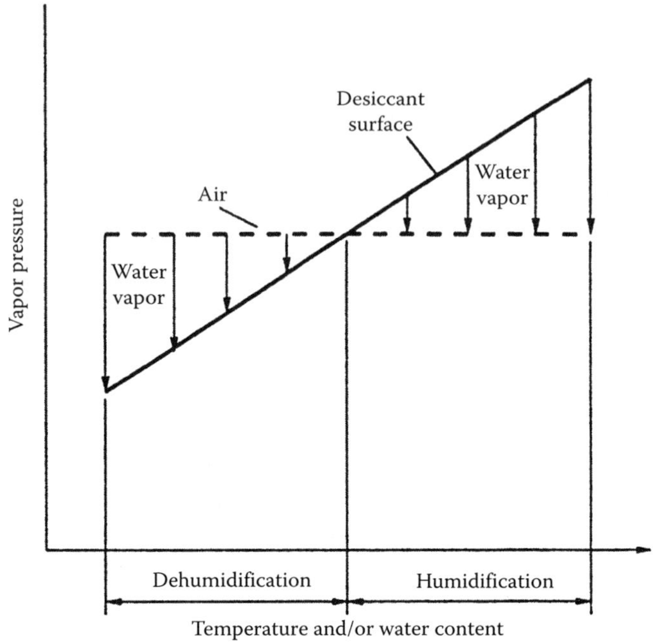

FIGURE 7.42
Vapor pressure versus temperature and water content for desiccant and air.

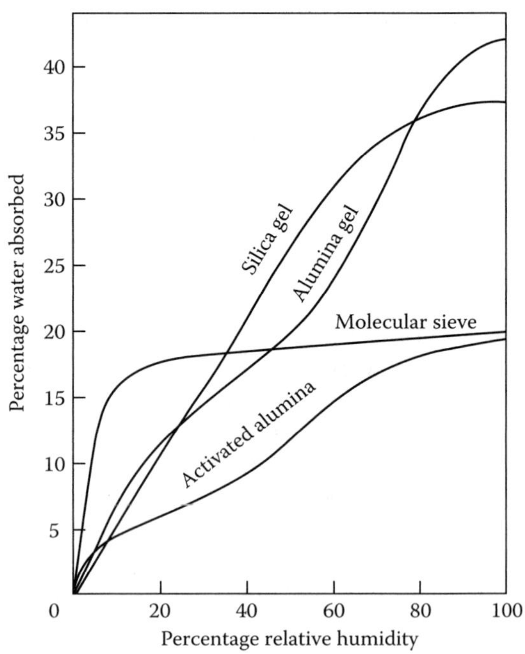

FIGURE 7.43
Equilibrium capacities of common water absorbents.

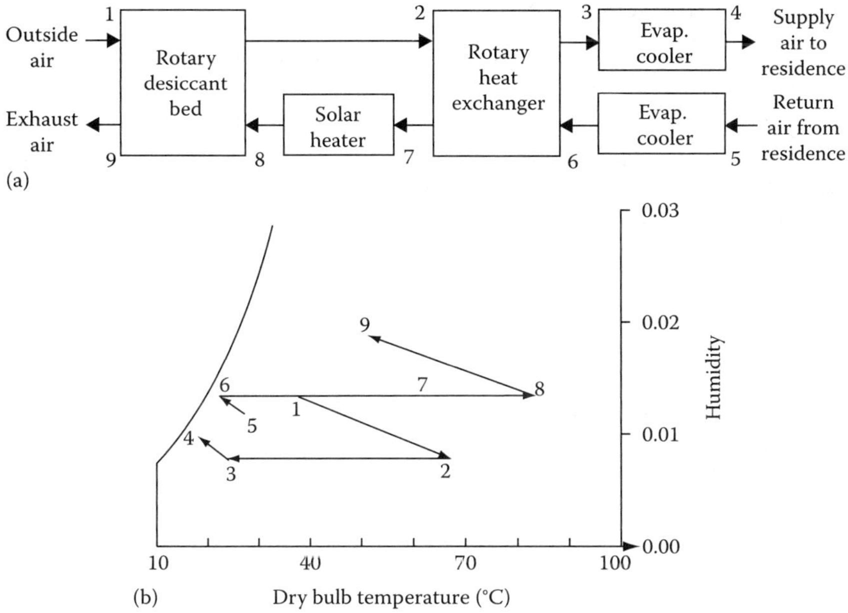

FIGURE 7.44
Schematic of a desiccant cooling ventilation cycle: (a) schematic of airflow and (b) process on a psychromatic chart.

Note that the molecular sieve has the highest capacity up to 30% humidity, and silica gel is optimal between 30% and 75%—the typical humidity range for buildings.

Figure 7.44 is a schematic diagram of a desiccant cooling ventilation cycle (also known as Pennington cycle), which achieves both dehumidification and cooling. The desiccant bed is normally a rotary wheel of a honeycomb-type substrate impregnated with the desiccant. As the air passes through the rotating wheel, it is dehumidified while its temperature increases (processes 1 and 2) due to the latent heat of condensation. Simultaneously, a hot air stream passes through the opposite side of the rotating wheel, which removes moisture from the wheel. The hot and dry air at state 2 is cooled in a heat-exchanger wheel to condition 3 and further cooled by evaporative cooling to condition 4. Air at condition 3 may be further cooled by vapor-compression or vapor-absorption systems instead of evaporative cooling. The return air from the conditioned space is cooled by evaporative cooling (processes 5 and 6), which in turn cools the heat exchanger wheel. This air is then heated to condition 7. Using solar heat, it is further heated to condition 8 before going through the desiccant wheel to regenerate the desiccant. A number of researchers have studied this cycle, or an innovative variation of it, and have found thermal COPs in the range of 0.5–2.58 [38].

7.10.2 Liquid Desiccant Cooling System

Liquid desiccants offer a number of advantages over solid desiccants. The ability to pump a liquid desiccant makes it possible to use solar energy for regeneration more efficiently. It also allows several small dehumidifiers to be connected to a single regeneration unit. Since a liquid desiccant does not require simultaneous regeneration, the liquid may be stored for later regeneration when solar heat is available. A major disadvantage is that the vapor pressure of the desiccant itself may be enough to cause some desiccant vapors to mix with the air. This disadvantage, however, may be overcome by proper choice of the desiccant material.

A schematic of a liquid desiccant system is shown in Figure 7.45. Air is brought in contact with concentrated desiccant in a countercurrent flow in a dehumidifier. The dehumidifier may be a spray column or packed bed. The packings provide a very large area for heat and mass transfer between the air and the desiccant. After dehumidification, the air is sensibly cooled before entering the conditioned space. The dilute desiccant exiting the dehumidifier is regenerated by heating and exposing it to a countercurrent flow of a moisture-scavenging air stream.

Liquid desiccants commonly used are aqueous solutions of LiBr, lithium chloride, calcium chloride, mixtures of these solutions, and triethylene glycol (TEG) [1]. Vapor pressures of these common desiccants are shown in Figure 7.46 as a function of concentration and temperature, based on a number of

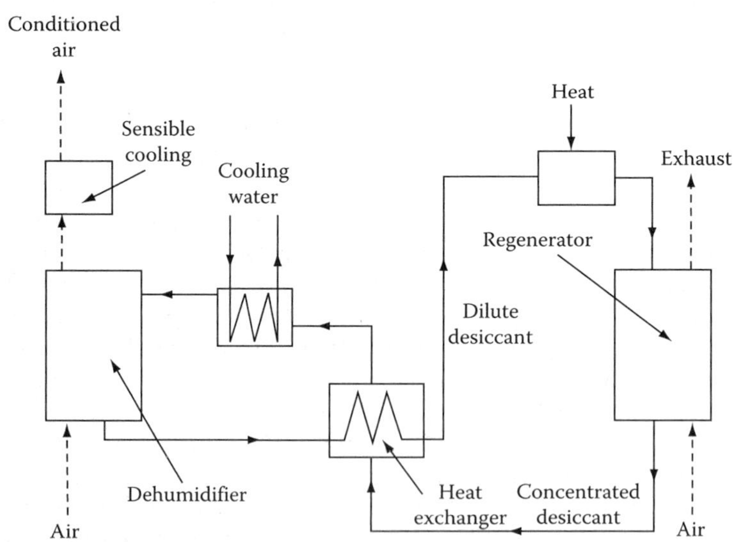

FIGURE 7.45
A conceptual liquid desiccant cooling system.

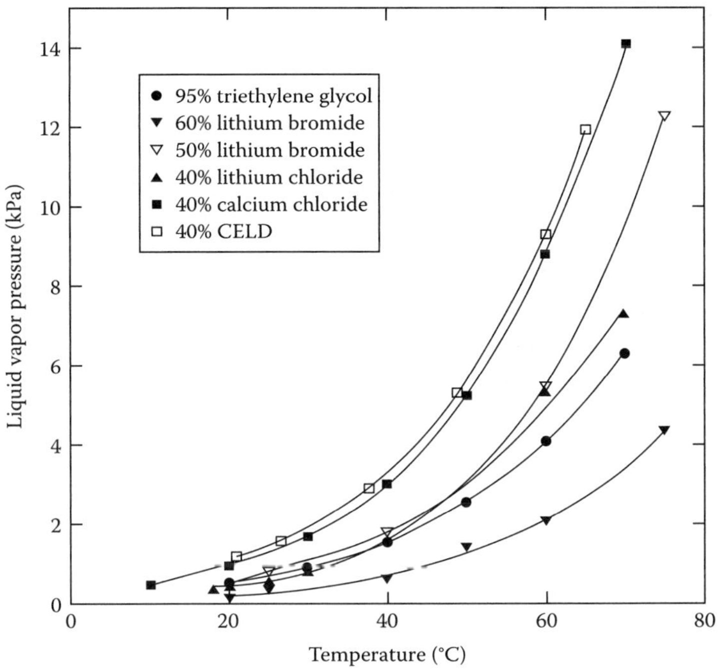

FIGURE 7.46
Vapor pressures of liquid desiccants.

TABLE 7.19
Physical Properties of Liquid Desiccants at 25°C

Desiccant	Density, $\rho \cdot 10^{-3}$ (kg/nV)	Viscosity, $\mu \cdot 10^3$ (N s/m²)	Surface Tension, $\gamma \cdot 10^3$ (N/m)	Specific Heat, c_p (kJ/kg-°C)	References
95% by weight triethylene glycol	1.1	28	46	2.3	[22]
55% by weight LiBr	1.6	6	89	2.1	[43,44]
40% calcium chloride	1.4	7	93	2.5	[43,45,46]
40% by weight lithium chloride	1.2	9	96	2.5	[43]
40% by weight CELD	1.3	5	—	—	[47]

Source: Oberg, V. and Goswami, D.Y., *Sol. Eng.*, 128, 155, 1998.

references [39–43]. Other physical properties important in the selection of desiccant materials are listed in Table 7.19. Although salt solutions and TEG have similar vapor pressures, salt solutions are corrosive and have higher surface tension. The disadvantage of TEG is that it requires higher pumping power because of higher viscosity.

Problems

7.1 Calculate the heat-removal factor for a collector having an overall heat-loss coefficient of 6 W/m² K and constructed of aluminum fins and tubes. Tube-to-tube centered distance is 15 cm, fin thickness is 0.05 cm, tube diameter is 1.2 cm, and fluid tube heat-transfer coefficient is 1200 W/m² · K. The cover transmittance to solar radiation is 0.9 and is independent of direction. The solar absorptance of the absorber plate is 0.9, the collector is 1 m wide and 3 m long, and the water flow rate is 0.02 kg/s. The water temperature is 330 K.

7.2 Calculate the efficiency of the collector described in Problem 7.1 on March 1 at a latitude of 40°N between 11 and 12 a.m. Assume that the total horizontal insolation is 450 W/m², the ambient temperature is 280 K, and the collector is facing south.

7.3 Calculate the plate temperature in Example 7.2 at 10 a.m., if the insolation during the first 3 h is 0, 150, and 270 W/m², and the air temperature is 285 K.

7.4 Calculate the overall heat-transfer coefficient, neglecting edge losses, for a collector with a double glass cover, with the following specifications:

Plate-to-cover spacing	3 cm
Plate emittance	0.9
Ambient temperature	275 K
Wind speed	3 m/s
Glass-to-glass spacing	3 cm
Glass emittance	5 cm
Back insulation thickness	5 cm
Back insulation thermal conductivity	0.04 W/m K
Mean plate temperature	340 K
Collector tilt	45°

7.5 The graph in Figure 7.20 gives the results of an ASHRAE standard performance test for a single-glazed flat-plate collector. If the transmittance for the glass is 0.90 and the absorptance of the surface of the collector plate is 0.92, determine

The collector heat-removal factor F_R

The overall heat-loss conductance of the collector U_c in W/m² K

The rate at which the collector can deliver useful energy in W/m² K when the insolation incident on the collector per unit area is 600 W/m² K, the ambient temperature is 5°C, and inlet water is at 15°C

The maximum flow rate through the collector, with cold water entering at a temperature of 15°C, that will give an outlet temperature of at least 60°C if this collector is to be used to supply heat to a hot-water tank.

7.6 Calculate the overall heat-loss coefficient for a solar collector with a single glass cover having the following specifications:

Spacing between plate and glass cover	5 cm
Plate emittance	0.2
Plate absorptance at equilibrium temperature	0.2
Ambient air temperature	283 K
Wind speed	3 m/s
Back insulation thickness	3 cm
Conductivity of back insulation material	0.04 W/m K
Mean plate temperature	340 K
Collector tilt	45°

Note that the solution to this problem requires trial and error. Start by assuming a glass temperature and then determine whether the heat gained by the glass equals the heat loss from the glass at that temperature. If the heat gain is larger than the heat loss, repeat the calculations with a slightly higher temperature. After you have calculated the heat-loss coefficient, compare your answer with the one obtained from Equation 7.23.

7.7 Standard tests on a commercially available flat plate collector gave a thermal efficiency of

$$\eta = 0.7512 - \frac{0.138(T_{f,in} - T_a)}{I_c}$$

$$K_{ra} - 1 - 0.15(1/\cos(i) - 1)$$

where $(T_{f,in} - T_a)/I_c$ is in °C – m²/W.

Find the useful energy collected from this collector each hour and for the whole day in your city on September 15.

Assume that all the energy collected is transferred to water storage with no losses. Calculate the temperature of the storage for each hour of the day. Assume a reasonable ambient temperature profile for your city. Given

Collector area = 6 m²

Collector tilt = 30° (South facing in northern hemisphere, North facing in southern hemisphere).

Storage volume = 0.3 m³ (water)

Initial storage temperature = 30°C

7.8 The heat-removal factor F_R permits solar collector delivery to be written as a function of collector fluid *inlet* temperature T_f in Equation 7.46. Derive the expression for a factor analogous to F_R, relating collector energy delivery to fluid *outlet* temperature.

7.9 In nearly all practical situations, the argument of the exponential term in Equation 7.43 for F_R is quite small. Use this fact along with Taylor's series expansion to derive an alternate equation for F_R. Determine the range where the alternate equation and Equation 7.43 agree to within 1%.

7.10 What is the operating temperature for an evacuated-tube collector operating at 50% efficiency if the insolation is 800 W/m^2? Use data from Figure 7.20.

7.11 Derive an expression for the heat-loss conductance U_c for a flat-plate collector in which convection and conductance are completely eliminated in the air layers by use of a hard vacuum.

7.12 Calculate the optical efficiency on November 13 of an evacuated-tube collector array at noon and 2 p.m. if the direct normal insolation is 600 W/m^2 and the diffuse insolation is 100 W/m^2. The effective optical transfer function $\tau_e\alpha_r$ is 70 j percent, and the tubes are spaced one diameter apart in front of a white painted surface with reflectance ρ of 60%.

7.13 The no-load temperature of a building with internal heat sources is given by Equation 7.72. How would this equation be modified to account for heat losses through the surface of an unheated slab, the heat losses being independent of ambient temperature?

7.14 An unheated garage is placed on the north wall of a building to act as a thermal buffer zone. If the garage has roof area A_r, window area A_{wi}, door area A_d, and wall area A_{wa}, what is the effective U value for the north wall of the building if its area is A_n? The garage floor is well insulated and has negligible heat loss. Express the effective U value in terms of the U values and areas of the several garage surfaces.

7.15 What is the annual energy demand for a building in Denver, Colorado, if the peak heat load is 150,000 Btu/h based on a design temperature difference of 75°F? Internal heat sources are estimated to be 20,000 Btu/h, and the design building interior temperature is 70°F.

7.16 What is the January solar load fraction for a water-heating system in Washington, District of Columbia, using 100 m^2 of solar collector if the water demand is 4 m^3/day at 65°C with a source temperature of 12°C? No heat exchanger is used and the solar collector efficiency curve is given in Figure 7.33; the solar collector is tilted at an angle equal to the latitude.

7.17 Repeat Problem 7.16 for Albuquerque, New Mexico, in July if the water source temperature is 17°C.

7.18 Explain how the *f*-chart (Figure 7.34) can be used *graphically* to determine the solar load fraction for a range of collector sizes once the solar and loss parameters have been evaluated for only one system size. *Hint*: Consider a straight line passing through the origin and the point (P_s, P_L).

7.19 In an attempt to reduce cost, a solar designer has proposed replacing the shell-and-tube heat exchanger in Figure 7.30 with a tube coil immersed in the storage tank. The shell-and-tube heat exchanger originally specified had a surface area of 10 m^2 and a U value of 2000 W/m^2 K to be used with a 100 m^2 solar collector. Using Equation 7.66 to estimate the U value of the submerged coil, how much length of 0.5 in. (1.27 cm) diameter copper pipe would be needed to achieve the same value of UA product as the shell-and-tube heat exchanger? What percentage of the storage tank volume would be consumed by this coil if 50 kg of water is used per square meter of collector? Use a storage water temperature of 60°C and a collector water outlet temperature of 70°C for the calculations.

7.20 If a solar system delivers 2500 MJ/m^2 · year with a water flow rate of 30 kg/m$_c^2$ and a plate efficiency factor $F' = 0.93$, how much energy will it deliver if the flow rate is doubled? Neglect the effect of flow rate on F'. The collector has a heat-loss conductance of 4 W/m^2 °C.

7.21 How large (MJ/h) should a heat-rejection system be if it must dump the entire heat production of a 1000 m^2 solar collector array in Denver, Colorado on August 21 if the collector is at 100°C and the ambient temperature is 35°C? Use solar collector data in Figure 7.33 and hourly solar radiation data in Appendix A.

7.22 Use the *f*-chart to determine the amount of solar energy that can be delivered in Little Rock, Arkansas, in January for the following solar and building conditions:

Building:
Load: 40 million Btu/month
Latitude: 35°N

Solar system:
Collector tilt: 55°, facing south
Area: 1000 ft^2
Ambient temperature: 40.6°F
Collector efficiency curve: see Figure 7.33
No heat exchanger used
Nominal storage, flow rate, and load heat-exchanger values used

7.23 Using the data in Problem 7.77, calculate system performance for a horizontal and a vertical collector. Assume Equation 7.77 applies to both cases.

7.24 The *f*-chart is based on $F_R(\tau\alpha)$ and F_RU_c values, which can be deduced from a plot of collector efficiency versus $(T_{f\,in} - T_a)/I$. If such a plot is not available but (a) a plot of efficiency versus $(\overline{T}_f - T_a)/I$ is available or (b) a plot of efficiency versus $(T_{f\,out} - T_a)/I$ is available, how can $F_R(\tau\alpha)$ and F_RU_c be calculated from the slope and intercept of these two curves? Express your results in terms of the slopes, intercepts, and fluid capacitance rate $\dot{m}c_p/A_c$.

7.25 The schematic diagrams below illustrate the operation of a solar-assisted heat-pump system and a solar system augmented with a heat pump. Discuss the advantages and disadvantages of each system with respect to different climatic conditions.

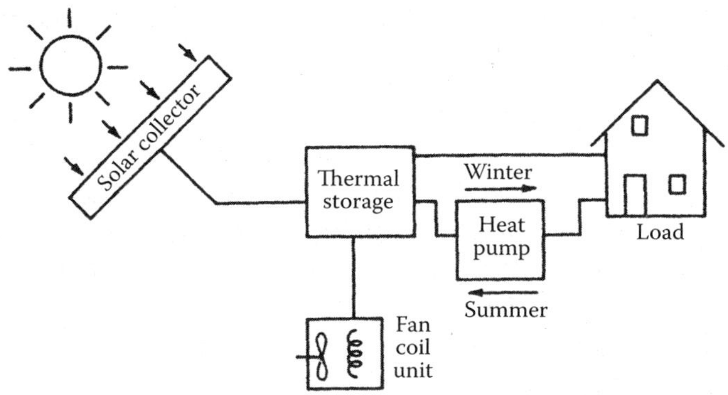

Solar-assisted heat-pump system

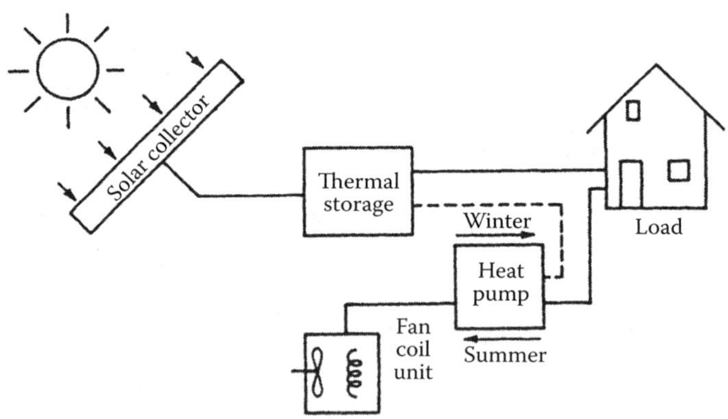

Solar system augmented with heat pump

7.26 The table below gives the characteristics of a building in Houston, Texas. Determine the cooling load for July 30 at solar noon. Any information regarding the load not given may be assumed or neglected.

Factor	Description or Specification
Roof	
Type	Flat, shaded
Area $A_{rf,sh}$, (m²)	250
U factor (W/m² K)	$U_{rf} = 0.35$
Walls	
Type	Vertical, painted white
Orientation, size (m × m)	North, South, 3 × 10; East, West, 3 × 25
U factor (W/m² K)	$U_{wa} = 1.08$
Windows	
Orientation, area (m²)	North, 8; South, 8; East, 20; West, 25
U factor (W/m² K)	$U_{wi} = 6.2$
Insolation transmittance	$\bar{\tau}_{b,wi} = 0.60$; $\bar{\tau}_{d,wi} = 0.80$; $\bar{\tau}_{r,wi} = 0.55$
Temperature	
Inside, outside (°C)	$T_{in} = 24$; $T_{out} = 37$
Insolation	
Beam, diffuse, reflected (W/m²)	$I_{h,b} = 580$; $I_{h,d} = 250$; $I_r = 200$

7.27 An air-conditioning system working in a vapor-compression cycle is used to manage the load for the building in Problem 7.25. If the high and low pressure in the cycle are 915 and 290 kPa, respectively, and the efficiency of the compressor is 90%, find the flow rate of Refrigerant 134a (R-134a) used for the equipment and the COP of the cycle.

7.28 Consider the absorption refrigeration cycle, shown in the line diagram below, that uses lithium bromide as carrier and water as refrigerant to provide 1 kW of cooling. By using steam tables and the chart giving the properties of lithium bromide and water, calculate first

a. Heat removed from the absorber
b. Heat removed from the condenser
c. Heat added to the evaporator
d. COP of the cycle

Then calculate, for a flat-plate collector with a $F_R(\tau\alpha)$ intercept of 0.81 and a F_RU_c of 3 W/m² · K, the area required for operation in Arizona at noon in August for a 3-ton unit. Assume that the enthalpy of the water vapor leaving the condenser can be approximated by the equation

$$h_{vc} = 2463 + 1.9T_c \text{ kJ/kg},$$

and that the enthalpy of the liquid water is

$$h_{\ell c} = 4.2T_c \text{ kJ/kg},$$

where T_c is the temperature of the evaporator in °C. In the analysis assume that evaporation occurs at 1°C and condensation at 32°C.
Answers: (a) 1.85 kW, (b) 1.0 kW, (c) 0.96 kW, and (d) 0.51 kW.

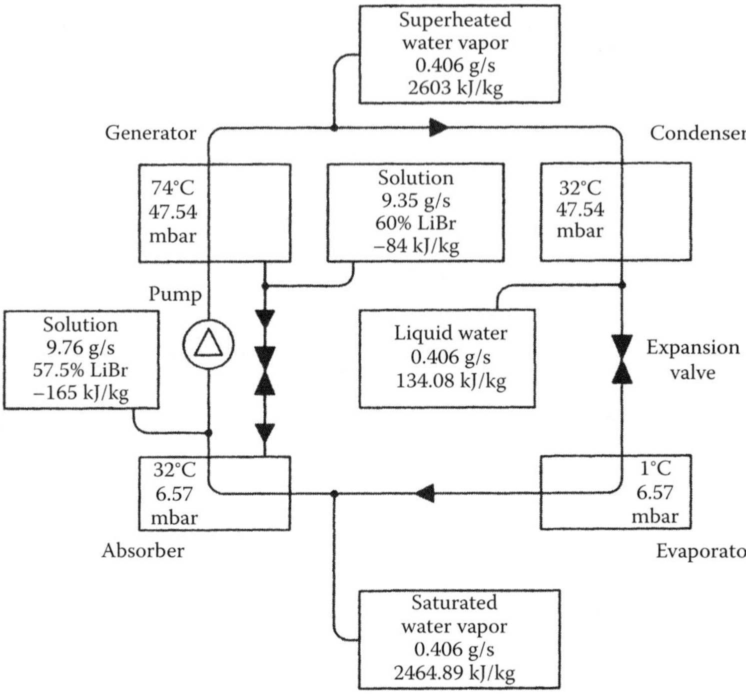
Superheated
water vapor
0.406 g/s
2603 kJ/kg
Generator
Condenser
74°C
47.54
mbar
Solution
9.35 g/s
60% LiBr
−84 kJ/kg
32°C
47.54
mbar
Pump
Solution
9.76 g/s
57.5% LiBr
−165 kJ/kg
Liquid water
0.406 g/s
134.08 kJ/kg
Expansion
valve
32°C
6.57
mbar
1°C
6.57
mbar
Absorber
Evaporator
Saturated
water vapor
0.406 g/s
2464.89 kJ/kg

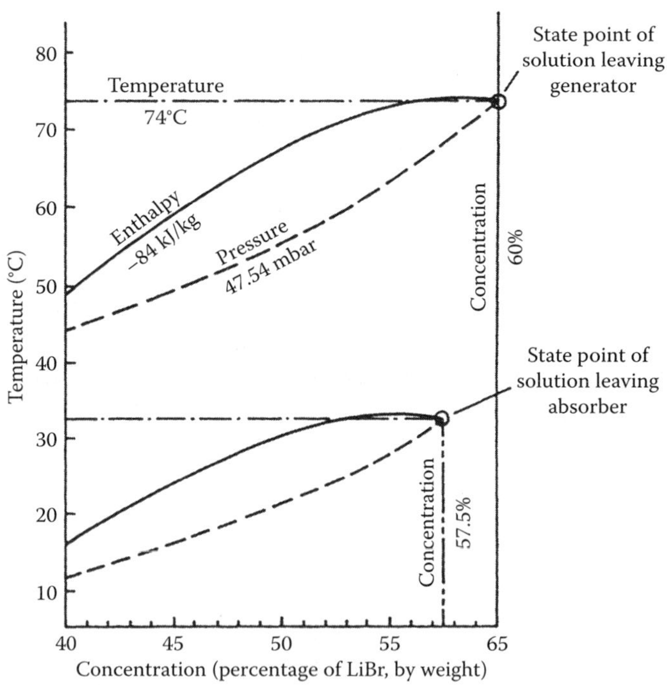
State point of
solution leaving
generator
Temperature
74°C
Enthalpy
−84 kJ/kg
Pressure
47.54 mbar
Concentration
60%
State point of
solution leaving
absorber
Concentration
57.5%
Temperature (°C)
80
70
60
50
40
30
20
10
40
45
50
55
65
Concentration (percentage of LiBr, by weight)

7.29 Make a preliminary design for a solar-driven Rankine refrigeration machine to provide the temperature environment required below the surface of a 20 × 40 m ice skating rink that is to operate all year in the vicinity of Denver, Colorado. State all your assumptions.

7.30 The table below shows data from a dehumidification process using aqueous solution of lithium chloride (LiCl) in a packed tower. The dessicant leaving the dehumidifier is passed through another packed tower for regeneration. For the regeneration process, it is known that the rate of evaporation of water as a function of the concentration $X(kg_{LiCL}/kg_{sol})$ and temperature, T(°C) of the desiccant is given by the following equation:

$$m_{evap} = (a_0 + a_1T + a_2T^2) + (b_0 + b_1T + b_2T^2)X + (c_0 + c_1T + c_2T^2)X^2 (g/s),$$

where

$a_0 = 285077, \quad b_0 = -1658652, \quad c_0 = 2412282,$

$a_1 = -8992, \quad b_1 = 52326, \quad c_1 = -76112,$ and

$a_2 = 70.88, \quad b_2 = -412, \quad c_2 = 600.$

Find the temperature to which a flat-plate collector must raise the temperature of the desiccant for the regeneration process. Assume that the flow rate of liquid desiccant in both processes (dehumidification, regeneration) is the same.

	Air			**Desiccant**	
Variable	**Inlet**	**Outlet**	**Inlet**	**Outlet**	**Variable**
Temperature (°C)	30.4	32.6	30	33.1	Temperature (°C)
RH (%)	66.7	36.7	35	—	Concentration (%)
Mass (kg/h)	260	—	850	—	Mass (kg/h)

7.31 A distant cousin Martha just passed away, peacefully in her sleep at the age of 105, and left you with one of her many real estate properties. The last time you had dinner with Martha, you described this great class you were taking on sustainable energy. She was thrilled to hear your passion and ideas about a sustainable energy future. Martha changed her will to give you an old building on the stipulation that you create an energy efficiency showcase restaurant/café/bar. For this homework, you are going to use eQUEST to determine the energy performance of the building as is (termed "baseline") and explore how specific building improvements will change that baseline energy use. See the eQUEST Background below for more information on this freeware.

Please follow the steps below, and then generate a report documenting the results. Your report should not be a bunch of disjointed bar charts but a few well-explained sample bar charts and a few tables that summarize the results. Follow the typical homework format.

eQUEST is a freeware created by the U.S. Department of Energy. You can go to www.doe2.com to download eQUEST and learn more about this freeware. eQUEST uses DOE-2.2 as a simulation engine and allows the user to model buildings either through the Schematic Design wizard (for preliminary analysis) or the Design Development wizard (for more complex analysis). eQUEST also has a detailed editor to modify the input parameters (you need to be familiar with DOE-2 BDL to use the detailed editor). Parametric analysis can be performed by eQUEST using the Energy Efficiency Measure (EEM) wizard or a more advanced editor called Parametric Runs.

For this homework, you should download eQUEST version 3.6. You will be using the Schematic Design and EEM Wizards.

Step 1. Download eQUEST.

Step 2. Open eQuest.
eQUEST Startup Options. Select "Create a New Project via the Wizard"
Which Wizard? Select "Schematic Design Wizard."

Step 3. Create baseline building using the Schematic Design Wizard

eQUEST Schematic Design Wizard

You will walk through 41 Wizard Screens—the building you have inherited is described by all the Wizard defaults except for data described below.

Wizard Screen 1 (General Information)

Project Name: Up to you—be creative.

Building Type: *Select* "Restaurant, Bar/Lounge"

Building Location, Utilities, and Rates

Coverage: *Select* "All eQUEST Locations"

Region: *Select* "Colorado"

City: *Select* "Boulder" Areas and Floors

Building Area: *Enter* "3000"

Wizard Screen 2 (Building Footprint)

Footprint Shape: *Select* " 'L' Shape"

Building Orientation

Plan North: *Select* "South" Floor Heights

Check Pitched Roof

Wizard Screen 3 (Building Envelope Constructions)

Add'l Insulation for Above Grade Walls: *Select* "R-11 batt"

Wizard Screen 6 (Exterior Doors)
Describe Up To 3 Door Types
Door Type South North West East

1. Glass 1 0 1 0
2. Opaque 1 1 0 0 Door Dimensions and Construction

1. For Construction *select* "Double Clr/Tint(2006Version)"
2. For Construction *select* "Wood, Solid core Flush, 1–3/8 in."

Wizard Screen 7 (Exterior Windows)

Describe Up To 3 Window Types

1. Single Clr/Tint
2. Double Clr/Tint

Window Dimensions, Positions and Quantities South North West East

1. 20 20 20 20 2.. 0 0 0 0
   Click "Finish"

Step 4. Determine baseline building performance.

- On the Actions navigation bar, *click* "Simulate Building Performance."
- Once the simulation is complete, view the summary results/reports. Pretty cool, huh?

Step 5. Use the Energy Efficiency Measure (EEM) Wizard to explore changing the amount of window area and location of those windows. Try the following options:

(i) Baseline
(ii) Option 1: S = 20%, N = 5%, W = 20%, E = 20%
(iii) Option 2: S = 50%, N = 20%, W = 20%, E = 20%
(iv) Option 3: S = 20%, N = 20%, W = 50%, E = 20%
(v) Option 4: S = 50%, N = 20%, W = 50%, E = 20%

To use the EEM Wizard, return to the Building Description Mode. Click on Energy Efficiency Measure Wizard on the Actions navigation bar.

Energy Efficiency Measure Creation Screen

Measure Category: Select "Building Envelope"

Measure Type: Select "Window Area"

Apply Measure To: Select "baseline" (if asked)

Click "OK"

EEM Run Information Screen

One Measure, "Window Area EEM", should appear in the Select Measure to View/Edit window. Highlight that Measure, and then change the name in the EEM Run Name to something like "window area option 1." *Click* on the "EEM Run Details…" button, and then edit the window area characteristics on that screen. You can use the "Create Run" button to add all the options—you will walk through the same process for each. *Click* "Finish" when you have created the options desired.

On the Actions navigation bar, *click* "Simulate Building Performance."

Step 6. Use the EEM Wizard to determine the payback periods of the following measures:

(i) Install better insulation in the walls; use R-19 at a cost of $\$0.25/\text{ft}^2$.
(ii) Install better insulation in the ceiling; use R-30 at a cost of $\$0.35/\text{ft}^2$.
(iii) Install high-efficiency furnace with a seasonal efficiency of AFUE = 0.90 with an incremental cost of $750.
(iv) 65°F when occupied and 55°F when unoccupied using a programmable thermostat at a cost of $100.
(v) Install a daylighting photosensor at a cost of $100.
(vi) Implement all the measures listed above simultaneously.

For the economic analysis, use the simple payback period approach and use an electricity cost of $0.09/kWh and a natural gas cost of $1.25/therm.

7.32 A survey of one million people near Miami, Florida, shows that 80% of them use natural gas and 20% use electricity for their hot water needs in their homes. Estimate the saving in fossil fuel, surface area of the collectors, and annual reduction in CO_2 emissions if flat-plate solar water heaters similar to the one described in the text are used to help with hot water demands (for the one million people surveyed). Assume that Miami is at 24°N latitude. Suggestion…a spreadsheet will make this easier.

Assumptions

- Electric water heating efficiency = 98%
- Natural gas water heating efficiency = 75%
- Coal to electricity efficiency = 35%
- Average person uses 76 L of hot water per day (avg. temp demanded is 140°F)
- Ground reflection is negligible.
- Assume 11 h of daylight

- Assume mass flow rate of 0.01 kg/s
- I_h = average direct irradiation on a horizontal surface during daylight hours = 590 W/m^2
- I_s = average direct irradiation on a 24° incline during daylight hours = 640 W/m^2

Solar water heater design

Front view:

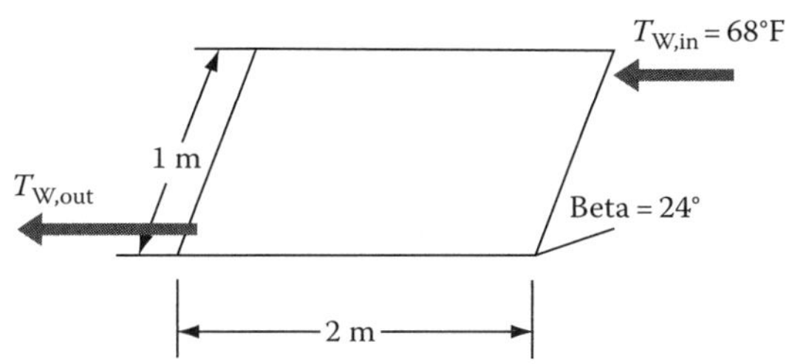

Cross-sectional view:

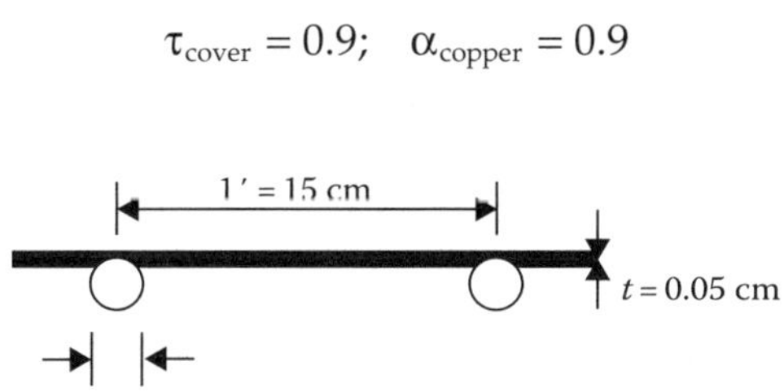

7.33 Estimate the electric power requirement, in kW, of a 1400 ft^2 floor area (three-bedroom home) with three occupants. Using your home power estimate, predict the power requirements for a city of 300,000 people. Use these results to estimate the area of silicon solar cells required to satisfy the community power requirements. Write a short narrative discussing your assumptions and analysis.

Appendix A

TABLE A.7.1

Conductance and Resistance Values for External Air Surfaces

Wind Condition / Position of Surface	Direction of Heat Flow	Type of Surface: Foil		Aluminum-Coated Paper		Nonreflective Building Materials	
		Conductance C, Btu/(h) (ft^2) (°F)	Resistance R, 1/[Btu/(h) (ft^2) (°F)]	Conductance C, Btu/(h) (ft^2) (°F)	Resistance R, 1/[Btu/(h) (ft^2) (°F)]	Conductance C, Btu/(h) (ft^2) (°F)	Resistance R, 1/[Btu/(h) (ft^2) (°F)]
Still air							
Horizontal	Up	0.76	1.32	0.91	1.10	1.63	0.61
45° slope	Up	0.73	1.37	0.88	1.14	1.60	0.62
Vertical	Horizontal	0.59	1.70	0.74	1.35	1.46	0.68
45°slope	Down	0.45	2.22	0.60	1.67	1.32	0.76
Horizontal	Down	0.22	4.55	0.37	2.70	1.08	0.92
7.5-mph wind							
Any position	Any direction (for summer calculations)					4.00	0.25
15-mph wind Any position	Any direction (for summer calculations)					6.00	0.17

Source: Courtesy of Johns-Mansville, Denver, CO.

TABLE A.7.2

Conductance and Resistance Values for Internal Air Surfaces

				Type of Surface					
				Foil and Nonreflective Building Materials		Aluminum-Coated Paper and Nonreflective Building Materials		Both Surfaces Nonreflective Building Materials	
Position of Air Space	Direction of Heat Flow[a]	Thickness (in.)	Temp. Coad.	Conductance *C*, Btu/(h) (ft²) (°F)	Resistance *R*, 1/[Btu/(h) (ft²) (°F)³	Conductance *C*, Btu/(h) (ft²) (°F)	Resistance *R*, 1/[Btu/(h) (ft²) (°F)³	Conductance *C*, Btu/(h) (ft²) (°F)	Resistance, *R*, 1/[Btu/(h) (ft²) (°F)]
Horizontal	Up	3	W	0.45	2.23	0.59	1.71	1.15	0.87
		4							
		3	S	0.44	2.26	0.61	1.63	1.32	0.76
		4							
		4	W	0.37	2.73	0.50	1.99	1.07	0.94
		4	S	0.36	2.75	0.53	1.87	1.24	0.80
45° slope	Up	3	W	0.36	2.78	0.50	2.02	1.06	0.94
		4							
		3	S	0.36	2.81	0.53	1.90	1.24	0.81
		4							
		4	W	0.33	3.00	0.47	2.13	1.04	0.96
		4	S	0.33	3.00	0.51	1.98	1.21	0.82
Vertical	Horizontal	3	W	0.29	3.48	0.42	2.36	0.99	1.01
		4							
		3	S	0.31	3.28	0.48	2.10	1.19	0.84
		4							
		4	W	0.29	3.45	0.43	2.34	0.99	1.01
		4	S	0.29	3.44	0.46	2.16	1.17	0.91

45° slope	Down	3 4	W	0.28	3.57	0.42	2.40	0.98	1.02
		3 4	S	0.31	3.24	0.48	2.09	1.19	0.84
		4		0.23	4.41	0.36	2.75	0.93	1.08
		4		0.23	4.36	0.40	2.50	1.11	0.90
Horizontal	Down	$\frac{3}{1}$	W	0.28	3.55	0.42	2.39	0.98	1.02
		$1\frac{1}{2}$	S	0.17	5.74	0.31	3.21	0.88	1.14
		4	W	0.11	8.94	0.25	4.02	0.81	1.23
		$\frac{3}{4}$	S	0.31	3.25	0.48	2.08	1.19	0.84
		$1\frac{1}{2}$	S	0.19	5.24	0.36	2.76	1.07	0.93
		4	S	0.12	8.08	0.30	3.38	1.01	0.99

Source: Courtesy of Johns-Manville, Denver, CO.

Note: W, winter; S, summer.

[a] Heat flows from hot to cold. For ceiling instillation, the direction of heat flow would normally be "up" for winter and "down" for summer. In a floor the direction of heat flow would be "down" in winter and "up" in summer. Heat flow in walls would be in a horizontal direction.

TABLE A.7.3

Conductance and Resistance Values for Exterior Siding Materials

Material	Description	Conductivity k, Btu/(h) (ft^2) (°F/in.)	Thickness (in.)	Conductance C, Btu/(h) (ft^2) (°F)	Resistance R, 1/[Btu/(h) (ft^2) (°F)]
Brick	Common	5.0	4	1.25	0.80
Brick	Face	9.0	4	2.27	0.44
Stucco		5.0	1	5.0	0.20
Asbestos cement shingles				4.76	0.21
Wood shingles	16–7$\frac{1}{2}$-in. exposure			1.15	0.87
Wood shingles	Double 16–12 in. exposure			0.84	1.19
Wood shingles	Plus $\frac{5}{16}$ in. insulated backerboard			0.71	1.40
Asbestos cement siding	$\frac{1}{4}$ in. lapped			4.76	0.21
Asphalt roll siding				6.50	0.15
Asphalt insulating siding			$\frac{1}{2}$	0.69	1.46
Wood	Drop siding, 1 × 8 in.			1.27	0.79
Wood	Bevel, $\frac{1}{2}$ × 8 in. lapped			1.23	0.81
Wood	Bevel, $\frac{3}{4}$ × 10 in. lapped			0.95	1.05
Wood	Plywood, $\frac{3}{8}$ in. lapped			1.59	0.59
Hardboard	Medium density	0.73	$\frac{1}{4}$	2.94	0.34
	Tempered	1.00	$\frac{1}{4}$	4.00	0.25
Plywood lap siding			$\frac{3}{8}$	1.79	0.56
Plywood flat siding			$\frac{3}{8}$	2.33	0.43

Source: Courtesy of Johns-Mansville, Denver, CO.

TABLE A.7.4

Conductance and Resistance Values for Sheathing and Building Paper

Material	Description	Conductivity k, Bm/(h) (ft²) (°F/in.)	Thickness (in.)	Conductance C, Btu/(h) (ft²) (°F)	Resistance R, 1/[Btu/(h) (ft²) (°F)]
Gypsum		1.11	$\frac{3}{8}$	3.10	0.32
			$\frac{1}{2}$	2.25	0.45
			$\frac{5}{8}$	1.75	0.57
Plywood		0.80	$\frac{1}{4}$	3.20	0.31
			$\frac{3}{8}$	2.13	0.47
			$\frac{1}{2}$	1.60	0.62
			$\frac{5}{8}$	1.28	0.78
			$\frac{3}{4}$	1.07	0.93
Nail-base sheathing		0.44	$\frac{1}{2}$	0.88	1.14
Wood sheathing	Fir or pine	0.80	$\frac{3}{4}$	1.06	0.94
Sheathing paper	Vapor-permeable			16.70	0.06
Vapor barrier	2 layers mopped 15-lb felt			8.35	0.12
	Plastic film			Negl.	Negl.

Source: Courtesy of Johns-Mansvilie, Denver, CO.

TABLE A.7.5

Conductance and Resistance Values for Masonry Materials

Material	Description	Conductivity k, Btu/(h) (ft²) (°F/in.)	Thickness (in.)	Conductance C, Btu/(h) (ft²)(°F)	Resistance R, 1/[Btu/ (h) (ft²)(°F)]
Concrete blocks, three-oval core	Sand and gravel aggregate		4	1.40	0.71
			8	0.90	1.11
			12	0.78	1.28
	Cinder aggregate		4	0.90	1.11
			8	0.58	1.72
			12	0.53	1.89
	Lightweight aggregate		4	0.67	1.50
			8	0.50	2.00
			12	0.44	2.27
Hollow clay tile	1 cell deep		4	0.90	1.11
	2 cells deep		8	0.54	1.85
	3 cells deep		12	0.40	2.50
Gypsum partition tile	3 × 12 × 30 in. solid		3	0.79	1.26
	3 × 12 × 30 in. 4-cell		3	0.74	1.35
	4 × 12 × 30 in. 3-cell		4	0.60	1.67
Cement mortar		5.0	1	5.0	0.20
Stucco		5.0	1	5.0	0.20
Gypsum	Poured	1.66	1	1.66	0.60
	Precast	2.80	2	1.40	0.71
	Sand and gravel or stone	12.0	1	12.0	0.08
Lightweight concrete	Perlite or zonolite mixture				
	1:4 mix, 36 1b/ft³	0.72–0.75	1	0.74	1.35
	1:5 mix, 30 1b/ft³	0.61–0.72	1	0.67	1.49
	1:6 mix, 27 1b/ft³	0.54–0.61	1	0.58	1.72
	1:8 mix, 22 1b/ft³	0.47–0.54	1	0.51	1.96
Stone		12.5	1		0.08

Source: Courtesy of Johns-Mansville, Denver, CO.

TABLE A.7.6

Conductance and Resistance Values for Woods

Material	Description	Conductivity k, Btu/(h) (ft²) (°F/in.)	Thickness (in.)	Conductance C, Btu/(h) (ft²) (°F)	Resistance R, 1/[Btu/(h) (ft²) (°F)]
Maple, oak and similar hardwoods	45 lb/ft³	1.10	$\frac{3}{4}$	1.47	0.68
Fir, pine and similar softwoods	32 1b/ft³	C.80	$\frac{3}{4}$	1.06	0.94
			$1\frac{1}{2}$	0.53	1.89
			$2\frac{1}{2}$	0.32	3.12
			$3\frac{1}{2}$	0.23	4.35

Source: Courtesy of Johns-Mansville, Denver, CO.

TABLE A.7.7

Conductance and Resistance Values for Wall-Insulation Materials

Material	Description	Conductivity k, Btu/(h) (ft²) (°F/in.)	Thickness (in.)	Conductance C, Btu/(h) (ft²) (°F)	Resistance R, 1/[Btu/(h) (ft²) (°F)]
Fiber-glass roof insulation			$\frac{15}{16}$	0.27	3.70
			$1\frac{1}{16}$	0.24	4.17
			$1\frac{3}{16}$	0.19	5.26
			$1\frac{5}{8}$	0.15	6.67
			$1\frac{7}{8}$	0.13	7.69
			$2\frac{1}{4}$	0.11	9.09
Urethane roof insulation	Thickness includes membrane roofing on both sides	0.13	$\frac{4}{5}$	0.19	5.26
			1	0.15	6.67
			$1\frac{1}{5}$	0.12	8.33

(continued)

TABLE A.7.7 (continued)

Conductance and Resistance Values for Wall-Insulation Materials

Material	Description	Conductivity k, Btu/(h) (ft²) (°F/in.)	Thickness (in.)	Conductance C, Btu/(h) (ft²) (°F)	Resistance R, 1/[Btu/(h) (ft²) (°F)]
Styrofoam SM and TG	2.1 lb/ft³	0.19	$\frac{3}{4}$	0.25	3.93
			1	0.19	5.26
			$1\frac{1}{2}$	0.13	7.89
			2	0.95	10.52
Wood shredded	Cemented in preformed slabs	0.60	1	0.60	1.67
Insulating board	Building and service board, decorative ceiling panels	0.38	$\frac{3}{8}$	1.01	0.99
			$\frac{1}{2}$	0.76	1.32
			$\frac{9}{16}$	0.68	1.48
			$\frac{3}{4}$	0.51	1.98
Thermal, acoustical fiber glass		0.39	$2\frac{3}{4}$	0.14	7.00
		0.36	4	0.09	11.00
		0.34	$6\frac{1}{2}$	0.05	19.00
Corkboard	6.4 lb/ft³	0.26	1	0.26	3.85
Expanded polystyrene					
Extruded	1.8 lb/ft³	0.25	1	0.25	4.00
Molded beads	1.0 lb/ft³	0.26	1	0.26	3.85
Urethane foam thurane (dow chemical)	1.9 lb/ft³	0.17	$\frac{3}{4}$	0.23	4.41
			$1\frac{1}{2}$	0.11	8.82
			2	0.09	11.76
Fiberglass perimeter			1	0.23	4.30
Insulation			$1\frac{1}{4}$	0.19	5.40
Fiberglas form board			1	0.25	4.00

Source: Courtesy of Johns-Mansville, Denver, CO.

TABLE A.7.8

Conductance and Resistance Values for Roofing Materials

Material	Description	Conductivity k, Btu/(h) (ft²) (°F/in.)	Thickness (in.)	Conductance C, Btu/(h) (ft²) (°F)	Resistance R, 1/[Btu/(h) (ft²) (°F)]
Asbestos cement shingles	120 lb/ft³			4.76	0.21
Asphalt shingles	70 lb/ft³			2.27	0.44
Wood shingles				1.06	0.94
Slate			$\frac{1}{2}$	20.0	0.05
Asphalt roll roofing	70 lb/ft³			6.50	0.15
Built-up roofing	Smooth or gravel surface		$\frac{3}{8}$	3.00	0.33
Sheet metal				Negl.	Negl.

Source: Courtesy of Johns-Mansville, Denver, CO.

TABLE A.7.9

Conductance and Resistance Values for Flooring Materials

Material	Description	Conductivity k, Btu/(h) (ft²) (°F/in.)	Thickness (in.)	Conductance C, Btu/(h) (ft²) (°F)	Resistance R, 1/[Btu/(h) (ft²) (°F)
Asphalt, vinyl, rubber, or linoleum tile				20.0	0.05
Cork tile		0.45	$\frac{3}{8}$	3.60	0.28
Terrazzo		12.5	1	12.50	0.08
Carpet and fibrous pad				0.48	2.08
Carpet and rubber pad				0.81	1.23
Plywood subfloor		0.80	$\frac{5}{8}$	1.28	0.78
Wood subfloor		0.80	$\frac{3}{4}$	1.06	0.94
Wood, hardwood finish		1.10	$\frac{3}{4}$	1.47	0.68

Source: Courtesy of Johns-Mansville, Denver, CO.

TABLE A.7.10

Conductance and Resistance Values for Interior Finishes

Material	Description	Conductivity k, Btu/(h) (ft²) (°F/in.)	Thickness (in.)	Conductance C, Btu/(h) (ft²) (°F)	Resistance R, 1/[Btu/(h) (ft²) (°F)]
Gypsum board		1.11	$\frac{3}{8}$	3.10	0.32
			$\frac{1}{2}$	2.25	0.45
Cement plaster	Sand aggregate	5.0	$\frac{1}{2}$	10.00	0.10
			$\frac{3}{4}$	6.66	0.15
Gypsum plaster	Sand aggregate	5.6	$\frac{1}{2}$	11.10	0.09
			$\frac{5}{8}$	9.10	0.11
Gypsum plaster	Lightweight aggregate	1.6	$\frac{1}{2}$	3.12	0.32
			$\frac{5}{8}$	2.67	0.39
Gypsum plaster on					
Metal lath	Sand aggregate		$\frac{3}{4}$	7.70	0.13
Metal lath	Lightweight aggregate		$\frac{3}{4}$	2.13	0.47
Gypsum board, $\frac{3}{8}$ in.	Sand aggregate		$\frac{7}{8}$	2.44	0.41
Insulating board		0.38	$\frac{1}{2}$	0.74	1.35
Plywood		0.80	$\frac{3}{8}$	2.13	0.47

Source: Courtesy of Johns-Mansville, Denver, CO.

TABLE A.7.11

Conductance and Resistance Values for Glass

Material	Description	Conductivity k, Btu/(h) (ft^2) (°F/in.)	Thickness (in.)	Conductance C (U), Btu/ (h) (ft^2) (°F)	Resistance R (1/U), 1/[Btu/(h) (ft^2) (°F)]
Single-plate				1.13	0.88
Double-plate	Air space, $\frac{3}{16}$ in.			0.69	1.45
Storm windows	Airspace, 1–4 in.			0.56	1.78
Solid-wood door	Actual thickness, $1\frac{1}{2}$ in.			0.49	2.04
Storm door, wood and glass	With wood and glass storm door			0.27	3.70
Storm door, metal and glass	With metal and glass storm door			0.33	3.00

Source: Courtesy of Johns-Mansville, Denver, CO.

TABLE A.7.12

Saturated Steam and Water—SI Units[a]

Temperature (K)	Pressure (MN/m²)	Specific Volume (m³/kg)		Specific Internal Energy (kJ/kg)		Specific Entropy (kJ/kg)			Specific Entropy (kJ/kg · K)	
		v_f	v_g	u_f	u_g	h_f	h_{fg}	h_g	s_f	s_g
273.15	0.0006109	0.0010002	206.278	−0.03	2375.3	−0.02	2501.4	2501.3	−0.0001	9.1565
273.16	0.0006113	0.0010002	206.136	0	2375.3	+0.01	2501.3	2501.4	0	9.1562
278.15	0.0008721	0.0010001	147.120	+20.97	2382.3	20.98	2489.6	2510.6	+0.0761	9.0257
280.13	0.0010000	0.0010002	129.208	29.30	2385.0	29.30	2484.9	2514.2	0.1059	8.975
283.15	0.0012276	0.0010004	106.379	42.00	2389.2	42.01	2477.7	2519.8	0.1510	8.9008
286.18	0.0015000	0.0010007	87.980	54.71	2393.3	54.71	2470.6	2525.3	0.1957	8.8279
288.15	0.0017051	0.0010009	77.926	62.99	2396.1	62.99	2465.9	2528.9	0.2245	8.7814
290.65	0.0020000	0.0010013	67.004	73.48	2399.5	73.48	2460.0	2533.5	0.2607	8.7237
293.15	0.002339	0.0010018	57.791	83.95	2402.9	83.96	2454.1	2538.1	0.2966	8.6672
297.23	0.0030000	0.0010027	45.665	101.04	2408.5	101.05	2444.5	2545.5	0.3545	8.5776
298.15	0.003169	0.0010029	43.360	104.88	2409.8	104.89	2442.3	2547.2	0.3674	8.5580
302.11	0.004000	0.0010040	34.800	121.45	2415.2	121.46	2432.9	2554.4	0.4226	8.4746
303.15	0.004246	0.0010043	32.894	125.78	2416.6	125.79	2430.5	2556.3	0.4369	8.4533
306.03	0.005000	0.0010053	28.192	137.81	2420.5	137.82	2423.7	2561.5	0.4764	8.3951
308.15	0.005628	0.0010060	25.216	146.67	2423.4	146.68	2418.6	2565.3	0.5053	8.3531
309.31	0.006000	0.0010064	23.739	151.53	2425.0	151.53	2415.9	2567.4	0.5210	8.3304
312.15	0.007000	0.0010074	20.530	163.39	2428.8	163.40	2409.1	2572.5	0.5592	8.2758
313.15	0.007384	0.0010078	19.523	167.56	2430.1	167.57	2406.7	2574.3	0.5725	8.2570
314.66	0.008000	0.0010084	18.103	173.87	2432.2	173.88	2403.1	2577.0	0.5926	8.2287
316.91	0.009000	0.0010094	16.203	183.27	2435.2	183.29	2397.7	2581.0	0.6224	8.1872

318.15	0.009593	0.0010099	15.258	188.44	2436.8	188.45	2394.8	2583.2	0.6387	8.1648
318.96	0.010000	0.0010102	14.674	191.82	2437.9	191.83	2392.8	2584.7	0.6493	8.1502
323.15	0.012349	0.0010121	12.032	209.32	2443.5	209.33	2382.7	2592.1	0.7038	8.0763
327.12	0.015000	0.0010141	10.022	225.92	2448.7	225.94	2373.1	2599.1	0.7549	8.0085
328.15	0.015758	0.0010146	9.568	230.21	2450.1	230.23	2370.7	2600.9	0.7679	7.9913
333.15	0.019940	0.0010172	7.671	251.11	2456.6	251.13	2358.5	2609.6	0.8312	7.9096
333.21	0.020000	0.0010172	7.649	251.38	2456.7	251.40	2358.3	2609.7	0.8320	7.9085
338.15	0.025030	0.0010199	6.197	272.02	2463.1	272.06	2346.2	2618.3	0.8935	7.8310
342.25	0.030000	0.0010223	5.229	289.20	2468.4	289.23	2336.1	2625.3	0.9439	7.7686
343.15	0.031190	0.0010228	5.042	292.95	2469.6	292.98	2933.8	2626.8	0.9549	7.7553
348.15	0.038580	0.0010259	4.131	313.90	2475.9	313.93	2221.4	2635.3	1.0155	7.6824
349.02	0.040000	0.0010265	3.993	317.53	2477.0	317.58	2319.2	2636.8	1.0259	7.6700
353.15	0.047390	0.0010291	3.407	334.86	2482.2	334.91	2308.8	2643.7	1.0753	7.6122
354.48	0.050000	0.0010300	3.240	340.44	2483.9	340.49	2305.4	2645.9	1.0910	7.5939
358.15	0.057830	0.0010325	2.828	355.84	2488.4	355.90	2296.0	2651.9	1.1343	7.5445
359.09	0.060000	0.0010331	2.732	359.79	2489.6	359.86	2293.6	2653.5	1.1453	7.5320
363.10	0.070000	0.0010360	2.365	376.63	2494.5	376.70	2283.3	2660.0	1.1919	7.4797
363.15	0.070140	0.0010360	2.361	376.85	2494.5	376.92	2283.2	2660.1	1.1925	7.4791
366.65	0.080000	0.0010386	2.087	391.58	2498.8	391.66	2274.1	2665.8	1.2329	7.4346
368.15	0.084550	0.0010397	1.9819	397.88	2500.6	397.96	2270.2	2668.1	1.2500	7.4159

Source: Bolz, R.E. and Tuve, G.L., eds., *CRC Handbook of Tables for Applied Engineering Science*, 2nd edn., Chemical Rubber Co., Cleveland, OH, 1973. With permission.

[a] Subscripts: f refers to a property of liquid in equilibrium with vapor; g refers to a property of vapor in equilibrium with liquid; f_g refers to a change by evaporation.

TABLE A.7.13

Superheated Steam—SI Units (Continued)

Pressure (MN/m²) (Saturation Temperature)[a]		Temperature									
		50°C	100°C	150°C	200°C	300°C	400°C	500°C	700°C	1000°C	1300°C
		323.15 K	373.15 K	423.15 K	473.15 K	573.15 K	673.15 K	773.15 K	973.15 K	1273.15 K	1573.15 K
(318.96 K)	*h*	2592.6	2687.5	2783.0	2879.5	3076.5	3279.6	3489.1	3928.7	4640.6	5409.7
	s	8.1749	8.4479	8.6882	8.9038	9.2813	9.6077	9.8978	10.4028	11.0393	11.5811
0.020	*v*	7.412	8.585	9.748	10.907	13.219	15.529	17.838	22.455	29.378	36.301
(60.06°C)	*u*	2442.2	2514.6	2587.3	2660.9	2811.9	2968.8	3132.2	3479.5	4053.0	4683.7
(333.21 K)	*h*	2590.4	2686.2	2782.3	2879.1	3076.3	3279.4	3489.0	3928.6	4640.6	5409.7
	s	7.8498	8.1255	8.3669	8.5831	8.9611	9.2876	9.5778	10.0829	10.7193	11.2612
0.040	*v*	3.683	4.279	4.866	5.448	6.606	7.763	8.918	11.227	14.689	18.151
(75.87°C)	*u*	2438.8	2512.6	2586.2	2660.2	2811.5	2968.6	3132.1	3479.4	4052.9	4683.6
(349.02 K)	*h*	2586.1	2683.8	2780.8	2878.1	3075.8	3279.1	3488.8	3928.5	4640.5	5409.6
	s	7.5192	7.8003	8.0444	8.2617	8.6406	8.9674	9.2577	9.7629	10.3994	10.9412
0.060	*v*	2.440	2.844	3.238	3.628	4.402	5.174	5.944	7.484	9.792	12.100
(85.94°C)	*u*	2435.3	2510.6	2585.1	2659.5	2811.2	2968.4	3131.9	3479.4	4052.9	4683.6
(359.09 K)	*h*	2581.7	2681.3	2779.4	2877.2	3075.3	3278.8	3488.6	3928.4	4640.4	5409.6
	s	7.3212	7.6079	7.8546	8.0731	8.4528	8.7799	9.0704	9.5757	10.2122	10.7541
0.080	*v*	1.8183	2.127	2.425	2.718	3.300	3.879	4.458	5.613	7.344	9.075
(93.50°C)	*u*	2431.7	2508.7	2583.9	2658.8	2810.8	2968.1	3131.7	3479.3	4052.8	4683.5
(366.65 K)	*h*	2577.2	2678.8	2777.9	2876.2	3074.8	3278.5	3488.3	3928.3	4640.4	5409.5
	s	7.1775	7.4698	7.7191	7.9388	8.3194	8.6468	8.9374	9.4428	10.0794	10.6213

0.100	v	1.4450	1.6958	1.9364	2.172	2.639	3.103	3.565	4.490	5.875	7.260
(99.63°C)	u	2428.2	2506.7	2582.8	2658.1	2810.4	2967.9	3131.6	3479.2	4052.8	4683.5
(372.78 K)	h	2572.7	2676.2	2776.4	2875.3	3074.3	3278.2	3488.1	3928.2	4640.3	5409.5
	s	7.0633	7.3614	7.6134	7.8343	8.2158	8.5435	8.8342	9.3398	9.9764	10.5183
0.200	v	0.6969	0.8340	0.9596	1.0803	1.3162	1.5493	1.7814	2.244	2.937	3.630
(120.23°C)	u	2409.5	2496.3	2576.9	2654.4	2808.6	2966.7	3130.8	3478.8	4052.5	4683.2
(393.38 K)	h	2548.9	2663.1	2768.8	2870.5	3071.8	3276.6	3487.1	3927.6	4640.0	5409.3
	s	6.6844	7.0135	7.2795	7.5066	7.8926	8.2218	8.5133	9.0194	9.6563	10.1982
0.300	v	0.4455	0.5461	0.6339	0.7163	0.8753	1.0315	1.1867	1.4957	1.9581	2.4201
(133.55°C)	u	2389.1	2485.4	2570.8	2650.7	2806.7	2965.6	3130.0	3478.4	4052.3	4683.0
(406.70 K)	h	2522.7	2649.2	2761.0	2865.6	3069.3	3275.0	3486.0	3927.1	4639.7	5409.0
	s	6.4319	6.7965	7.0778	7.3115	7.7022	8.0330	8.3251	8.8319	9.4690	10.0110
0.400	v	0.3177	0.4017	0.4708	0.5342	0.6548	0.7726	0.8893	1.1215	1.4685	1.8151
(143.63°C)	u	2366.3	2473.8	2564.5	2646.8	2804.8	2964.4	3129.2	3477.9	4052.0	4682.8
(416.78 K)	h	2493.4	2634.5	2752.8	2860.5	3066.8	3273.4	3484.9	3926.5	4639.4	5408.08
	s	6.2248	6.6319	6.9299	7.1706	7.5662	7.8985	8.1913	8.6987	9.3360	9.8780
0.500	v		0.3146	0.3729	0.4249	0.5226	0.6173	0.7109	0.8969	1.1747	1.4521
(151.86°C)	u		2461.5	2557.9	2642.9	2802.9	2963.2	3128.4	3477.5	4051.8	4682.5
(425.01 K)	h		2618.7	2744.4	2855.4	3064.2	3271.9	3483.9	3925.9	4639.1	5408.6
	s		6.4945	6.8111	7.0592	7.4599	7.7938	8.0873	8.5952	9.2328	9.7749

Source: Bolz, R.E. and Tuve, G.L., eds., *CRC Handbook of Tables for Applied Engineering Science*, 2nd edn., Chemical Rubber Co., Cleveland, OH, 1973. With permisssion.

[a] Symbols: v = specific volume, m^3/kg; u = specific internal energy, kJ/kg; h = specific enthalpy, kJ/kg; s = specific entropy, kJ/K · kg.

References

1. Löf, G.O.G. and Tybout, R.A. (1972) Model for optimizing solar heating design, ASME Paper 72-WA/SOL-8.
2. Kutcher, C.F. (1996) Transpired solar collector systems: A major advance in solar heating, in: Paper presented at the *World Engineering Congress*, November 6–8, 1996, Atlanta, GA.
3. Kutcher, C.F. and Christensen, C.B. (1992) Unglazed transpired solar collectors, *Advances in Solar Energy* 7, 283–307.
4. Hottel, H.C. and Woertz, B.B. (1942) Performance of flat-plate solar-heat collectors, *Trans. Am. Soc. Mech. Eng*. 64, 91.
5. Klein, S.A. (1975) Calculation of flat-plate collector loss coefficients, *Sol. Energy* 17, 79–80.
6. Agarwal, V.K. and Larson, D.C. (1981) Calculation of the top loss coefficient of a flat plate collector, *Solar Energy* 27, 69–71.
7. Hottel, H.C. and Whillier, A. (1958) Evaluation of flat-plate collector performance, *Trans. Conf. Use Sol. Energy* 2(1), 74.
8. Vant-Hull, L.L. (1976) *Proc. SPIE Solar Energy Utilization Conf. II* 85, 104. See also Vant-Hull, L.L. 1976, 85, 111.
9. Bliss, R.W. (1959) The derivations of several plate efficiency factors useful in the design of flat-plate solar-heat collectors, *Sol. Energy* 3, 55.
10. Klein, S.A., Duffie, J.A., and Beckman, W.A. (1974) Transient considerations of flat-plate solar collectors, *J Eng. Power* 96A, 109–114.
11. Kreith, F. and Bohn, M.S. (1997) *Principles of Heat Transfer*, 5th edn., PWS Publishers, Boston, MA.
12. Duffie, J.A. and Beckman, W.A. (1980) *Solar Energy Thermal Processes*, John Wiley & Sons, New York.
13. Collares-Pereira, M., Rabl, A., and Winston, R. (1977) Lens mirror combinations with maximal concentration, Enrico Fermi Institute Rept., No. EFI 77-20.
14. Duff, W.S. (1976) Optical and thermal performance of three line focus collectors, ASME Paper 76-WA/HT-15.
15. Emmett, W.L.R. (1911) Apparatus for utilizing solar heat, U.S. Patent 980, 505.
16. Speyer, F. (1965) Solar energy collection with evacuated tubes, *J. Eng. Power* 87, 270.
17. Beekley, D.C. and Mather, G.R. (1975) *Analysis and Experimental Tests of a High Performance, Evacuated Tube Collector*, Owens-Illinois, Toledo, OH.
18. ASHRAE. (1997) *Handbook of Fundamentals*, American Society of Heating, Refrigerating and Air-Conditioning Engineers, Atlanta, GA.
19. Klein, S.A. et al. (1975) A method for simulation of solar processes and its application, *Sol. Energy* 17, 29–37.
20. Close, D.J. (1962) The performance of solar water heaters with natural circulation, *Sol. Energy* 6, 33.
21. Phillips, W.F. and Cook, R.D. (1975) Natural circulation from a flat plate collector to a hot liquid storage tank, ASME Paper 75-HT-53.
22. Kreider, J.F. and Kreith, F. (1977) *Solar Heating and Cooling*, revised 1st edn., Hemisphere Publ. Corp., Washington, DC.

23. Klein, S.A. (1976) A design procedure for solar heating systems, PhD dissertation, University of Wisconsin, Madison. For an approach similar to the *f*-chart for other solar-thermal systems operating above a minimum temperature above that for space-heating (~ 20°C), see Klein, S.A. and Beckman, W.A. (1977) A general design method for closed loop solar energy systems, in: *Proc. 1977 ISES Meeting*.
24. Beckman, W.A., Klein, S.A., and Duffie, J.A. (1977) *Solar Heating Design by the F-Chart Method*, John Wiley & Sons, New York.
25. Klein, S.A., Beckman, W.A., and Duffie, J.A. (1976) A design procedure for solar heating systems, *Sol. Energy* 18, 113.
26. Klein, S.A., Beckman, W.A., and Duffie, J.A. (1976) A design procedure for solar air heating systems, in: Paper presented at the *1976 ISES Conference*, American Section, Winnipeg, Manitoba, August 15–20.
27. Goswami, Y., Kreith, F., and Kreider, J. (2000) *Principles of Solar Engineering*, 2nd edn., Taylor & Francis, Philadelphia.
28. Rangarajan, K., Shirley, D.B. III, and Raustad, R.A. (1989) Cost-effective HVAC technologies to meet ASHRAE Standard 62-1989 in hot and human climates, *ASHRAE Trans*. pt. 1, 166–182.
29. Meckler, H. (1994) Desiccant-assisted air conditioner improves IAQ and comfort, *Heat. Piping Air Cond*. 66(10), 75–84.
30. Meckler, M. (1995) Desiccant outdoor air preconditioners maximize heat recovery ventilation potentials, *ASHRAE Trans*. 101(2), 992–1000.
31. Meckler, M. (1988) Off-peak desiccant cooling and cogeneration combine to maximize gas utilization, *ASHRAE Trans*. 94(1), 575–596.
32. Meckler, M., Parent, Y.O., and Pesaran, A.A. (1993) Evaluation of dehumidifiers with polymeric desiccants, Gas Institute Report, Contract No. 5091-246-2247, Gas Research Institute, Chicago, IL.
33. Oberg, V. and Goswami, D.Y. (1998) Experimental study of heat and mass transfer in a packed bed liquid desiccant air dehumidifier, In: J.H. Morehouse and R.E. Hogan (eds.), *Solar Engineering*, ASME, New York, pp. 155–166.
34. Oberg, V. and Goswami, D.Y. (1998) A review of liquid desiccant cooling, *Adv. Sol. Energy* ASES 12, 431–470.
35. Spears, J. W. and Judge, J. (1997) Gas-fired desiccant system for retail super center. *ASHRAE J*. 39, 65–69.
36. Thombloom, M. and Nimmo, B. (1994) Modification of the absorption cycle for low generator firing temperatures, in: *Joint Solar Engineering Conf.*, ASME, New York.
37. Thombloom, M. and Nimmo, B. (1995) An economic analysis of a solar open cycle desiccant dehumidification system, in: *Solar Engineering—1995, Proc. of the 13th Annual ASME Conference*, Hawaii, vol. 1, pp. 705–709.
38. Pesaran, A.A., Penney, T.R., and Czanderna, A.W. (1992) Desiccant cooling: State-of-the art assessment, National Renewable Laboratory, Golden Colorado, NREL, Report No. NREL/TP-254-4147.
39. Cyprus Foote Mineral Company. (1995) Technical data on lithium bromide and lithium chloride, Bulletins 145 and 151, Cyprus Foote Mineral Company, Kings Mountain, NC.
40. Dow Chemical Company. (1996) *Calcium Chloride Handbook*, Dow Chemical Company, Midland, MI.
41. Dow Chemical Company. (1992) *A Guide to Glycols*, Dow Chemical Company, Midland, MI.

42. Ertas, A., Anderson, E.E., and Kiris, I. (1992) Properties of a new liquid desiccant solution—Lithium chloride and calcium chloride mixture, *Sol. Energy* 49, 205–212.
43. Zaytsev, I.O. and Aseyev, G.G. (1992) *Properties of Aqueous Solutions of Electrolytes*, CRC Press, Boca Raton, FL.
44. Kettleborough, C.F. and Waugaman, D.G. (1995) An alternative desiccant cooling cycle, *J. Sol. Energy Eng.* 117, 251–255.
45. Close, D.J. (1965) Rock pile thermal storage for comfort air conditioning, *Mech. Chem. Eng. Trans. Inst. Eng. (Australia)* MC-1, 11.
46. Khan, A.Y. (1994) Sensitivity analysis and component modeling of a packed-type liquid desiccant system at partial load operating conditions, *Int. J. Energy Res.* 18, 643–655.
47. Löf, G.O.G. and Tybout, R.A. (1974) Design and cost of optimal systems for residential heating and cooling by solar energy, *Sol. Energy* 16, 9.

8

Solar Process Heat and Thermal Power

> The illusion of unlimited powers, nourished by astonishing scientific and technological achievements, has produced the concurrent illusion of having solved the problem of production.
>
> —**E. F. Schumacher**

8.1 Historical Perspective

Attempts to harness the sun's energy for power production date back to at least 1774 [1], when the French chemist Lavoisier and the English scientist Joseph Priestley discovered oxygen and developed the theory of combustion by concentrating the rays of the sun on mercuric oxide in a test tube, collecting the gas produced with the aid of solar energy, and burning a candle in the gas. Also, during the same year an impressive picture of Lavoisier was published in which he stands on a platform near the focus of a large glass lens and is carrying out other experiments with focused sunlight (Figure 8.1).

A century later, in 1878, a small solar power plant was exhibited at the World Fair in Paris (Figure 8.2). To drive this solar steam engine, sunlight was concentrated from a parabolic reflector onto a steam boiler located at its focus; this produced the steam that operated a small reciprocating steam engine that ran a printing press. In 1901, a 10 hp solar steam engine was operated by A.G. Eneas in Pasadena, California [2]. It used a 700 ft^2 focusing collector in the shape of a truncated cone, as shown in Figure 8.3. Between 1907 and 1913, the American engineer F. Shuman developed solar-driven hydraulic pumps; in 1913 he built, jointly with C.V. Boys, a 50 hp solar engine for pumping irrigation water from the Nile near Cairo in Egypt (Figure 8.4). This device used long parabolic troughs that focused solar radiation onto a central pipe with a concentration ratio of 4.5:1.

With the increasing availability of low-cost oil and natural gas, interest in solar energy for power production waned. Except for C.G. Abbott, who exhibited in 1936 a ½-hp solar-powered engine at an International Power Conference in Washington, District of Columbia and in 1938 in Florida, an improved, somewhat smaller version with a flash boiler, there was very little activity in the field of solar power between 1915 and 1950. Interest in solar power revived

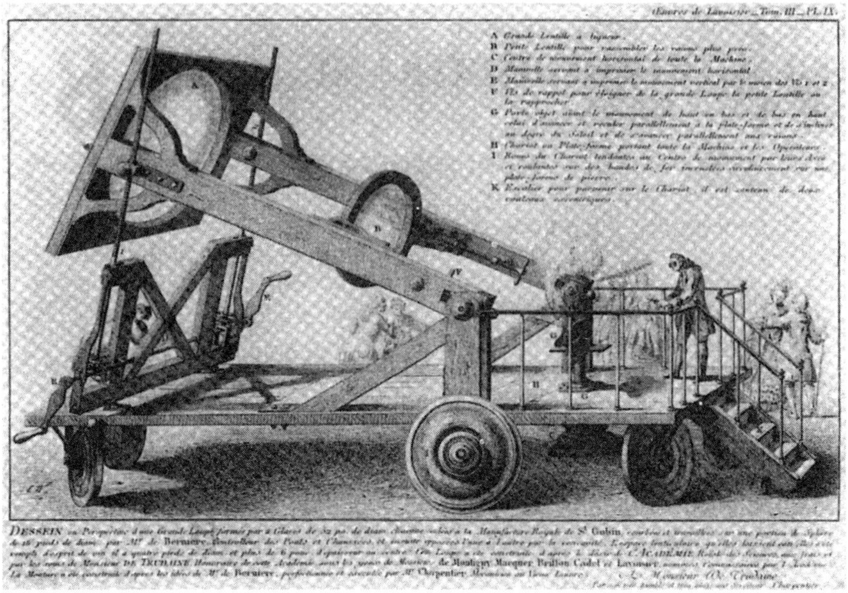

FIGURE 8.1
Solar furnace used by Lavoisier in 1774. (Photo courtesy of Bibliotheque Nationale de Paris. Lavoisier, *Oeuvres*, vol. 3.)

FIGURE 8.2
Parabolic collector powered a printing press at the 1878 Paris Exposition. (From Goswami, Y. et al., *Principles of Solar Engineering*, 2nd edn., Taylor & Francis, Philadelphia, 2000.)

FIGURE 8.3
Irrigation pumps were run by a solar-powered steam engine in Arizona in the early 1900s. The system consisted of an inverted cone that focused rays of the sun on the boiler. (From Goswami, Y. et al., *Principles of Solar Engineering,* 2nd edn., Taylor & Francis, Philadelphia, 2000.)

FIGURE 8.4
Solar irrigation pump (50 hp) operating in 1913 in Egypt. (From Goswami, Y. et al., *Principles of Solar Engineering*, 2nd edn., Taylor & Francis, Philadelphia, 2000.)

in 1949 when, at the centennial meeting of the American Association for the Advancement of Science in Washington, District of Columbia, one session was devoted to future energy sources. At that time, the potentials as well as the economic problems of solar energy utilization were clearly presented by Daniels [3]. Some important conferences that considered solar power generation were held by UNESCO in 1954, the Association for Applied Solar Energy in 1955, the U.S. National Academy of Sciences in 1961, and the United Nations in 1961. In addition, a research and development program supported by the National Aeronautics and Space Administration to build a solar electric power system capable of supplying electricity for the U.S. space program was undertaken in the 1960s. However, widespread interest developed only after research funds became available for the development of earth-bound solar electric power and process heat after the Arab oil embargo in 1973.

8.2 Solar Industrial Process Heat

Industrial processes consumed over 30 quadrillion (10^{15}) Btus of energy in 2001 [4] and this amount is increasing. A study conducted for the Energy Research and Development Administration by the InterTechnology Corporation [5] indicated that solar energy has the potential of providing about 20% of this energy. The economic outlook for industrial solar heat appears to be extremely favorable because process heat solar collectors could be used throughout the year and each system can be designed to fit the temperature level required for its specific applications, which is particularly important in the use of process heat. Table 8.1 shows the amount of heat

TABLE 8.1

Summary of U.S. Industrial Heat Usage by SIC Category for 1971 and 1994

	Quantities in 10^{12} kJ	
SIC Group	**1971**	**1994**
20 Food and kindred products	779	1,254
21 Tobacco products	14	W[a]
22 Textile mills	268	327
23 Apparel	22	W[a]
24 Lumber and wood products	188	518
25 Furniture	37	73
26 Paper and allied products	2,006	2,812
27 Printing and publishing	16	118
28 Chemicals	2,536	5,621
29 Petroleum products	2,576	6,688
30 Rubber	158	303
31 Leather	19	W[a]
32 Stone, clay, and glass	1,541	996
33 Primary metals	3,468	2,597
34 Fabricated metal products	295	387
35 Machinery	283	260
36 Electrical equipment	213	256
37 Transportation	310	383
38 Instruments	53	113
39 Miscellaneous	72	W[a]
Subtotal	14,854	22,854

Source: Intertechnology Corporation, Analysis of the economic potential of solar thermal energy to provide industrial process heat, ERDA Rep. No. C00/2829, 1997.

W[a] = withheld to avoid disclosing data for individual establishments.

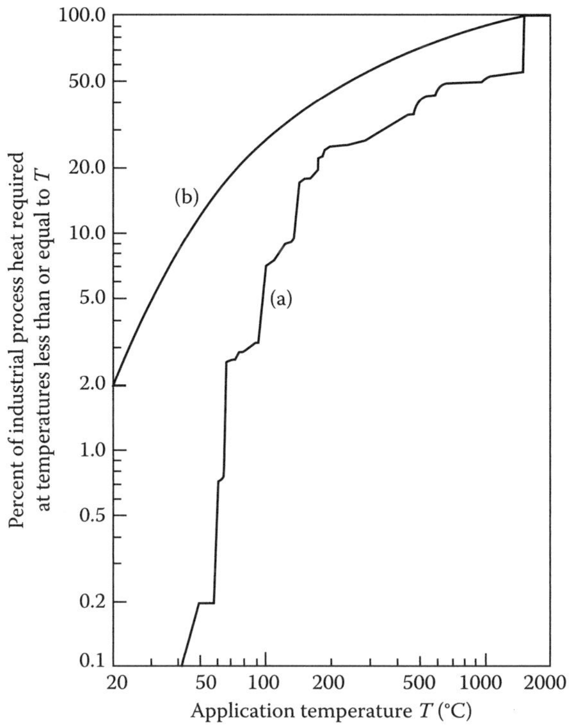

FIGURE 8.5
Distribution of U.S. process heat use by required temperature level: (a) heat requirements; (b) IPH requirements plus preheat from 15°C. (From Intertechnology Corporation, Analysis of the economic potential of solar thermal energy to provide industrial process heat, ERDA Rep. No. C00/2829, 1997.)

used by selected industries in the United States. A majority of the heat is used in mining, food, textiles, lumber, paper, chemicals, petroleum products, stone–clay–glass, and primary metals [6]. The breakdown of industrial energy usage is as follows [7]: process steam, 41%; direct process heat, 28%; shaft drive, 19%; feedstock, 9%; and other 3%. In addition to the quantity of heat, quality (i.e., temperature) is also very important to match the proper solar collection system to the application. Figure 8.5 shows the cumulative process heat use by temperature requirement [7]. It is seen that about 25% of the heat is used at temperatures below 100°C, which may be provided by flat-plate collectors or compound parabolic concentrators (CPCs), about 50% of the heat is used at temperatures below 260°C, and 60% below 370°C. Therefore, a large percent of all the process heat could be delivered by parabolic trough collectors (PTC).

The selection of the type of solar collectors depends on the process temperature requirements. Table 8.2 gives the temperature requirements and the type of solar collectors suitable for the process. The type of storage depends on the temperature requirement, the storage duration, the required

TABLE 8.2

Estimated Date of Technical Readiness for Various Solar IPH Systems

	Energy Form[a]	Temperature (°C)	Simple Air Heaters	Flat Plates	Fixed Compound Surfaces	Single-Tracking Troughs	Central Receivers
Aluminum							
Bayer process digestion	Steam	216				X	
Automobile and truck manufacturing							
Heating solutions	Steam (water)	49–82	X	X			
Heating makeup air in paint booths	Air	21–29	X				
Drying and baking	Air	163–218			X	X	
Concrete block and brick							
Curing product	Steam	74–177			X	X	
Gypsum							
Calcining	Air	160			X	X	
Curing platerboard	Steam (air)	299				X	X
Chemicals							
Borax, dissolving and thickening	Steam	82–99		X	X		
Borax, drying	Air	60–77	X	X			
Bromine, blowing brine/distillation	Steam	107			X		
Chlorine, brine heating	Steam (water)	66–93	X	X			
Chlorine, caustic evaporation	Steam	143–149			X	X	
Phosphoric acid, drying	Air	121			X		
Phosphoric acid, evaporation	Steam	160				X	
Potassium chloride, leaching	Steam	93		X	X		
Potassium chloride, drying	Air	121			X	X	
Sodium metal, salt purification	Steam	135			X	X	
Sodium metal, drying	Steam (air)	116				X	

Food						
Washing	Water	49–71	X	X		
Concentration	Steam (water)	38–43	X	X	X	
Cooking	Steam	121–188			X	X
Drying	Steam (air)	121–232			X	X
Glass						
Washing and rinsing	Water	71–93	X	X		
Laminating	Air	100–177			X	X
Drying glass fiber	Air	135–141			X	X
Decorating	Air	21–93	X	X	X	
Lumber						
Kiln drying	Air	66–99	X	X	X	
Glue preparation/plywood	Steam	99–177			X	X
Hot pressing/fiberboard	Steam	199				X
Log conditioning	Water	82		X		
Mining (Frasch sulfur)	Pressurized					
Extraction	Water	160–166			X	X
Paper and pulp						
Kraft pulping	Steam	182–188				X
Kraft liquor evaporation	Steam	138–143			X	X
Kraft bleaching	Steam	138–143			X	X
Papermaking (drying)	Steam	177				X
Plastics						
Initiation	Steam	121–146			X	X
Steam distillation	Steam	146			X	X
Flash separation	Steam	216				X

(continued)

TABLE 8.2 (continued)

Estimated Date of Technical Readiness for Various Solar IPH Systems

	Energy Form[a]	Temperature (°C)	Simple Air Heaters	Flat Plates	Fixed Compound Surfaces	Single-Tracking Troughs	Central Receivers
Extrusion	Steam	146			X	X	
Drying	Steam	188				X	
Blending	Steam	121			X	X	
Synthetic rubber							
Initiation	Steam (water)	121			X		
Monomer recovery	Steam	121			X		
Drying	Steam (air)	121			X		X
Steel							
Pickling	Steam	66–104	X	X	X		
Cleaning	Steam	82–93		X	X		
Textiles							
Washing	Water	71–82	X	X			
Preparation	Steam	49–113	X	X	X		
Mercerizing	Steam	21–99	X	X	X		
Drying	Steam	60–135	X	X	X		
Finishing	Steam	60–149		X	X	X	

Source: Kreider, J.F., *Medium and High Temperature Solar Processes*, Academic Press, New York, 1979.

[a] Preferred form (secondary form).

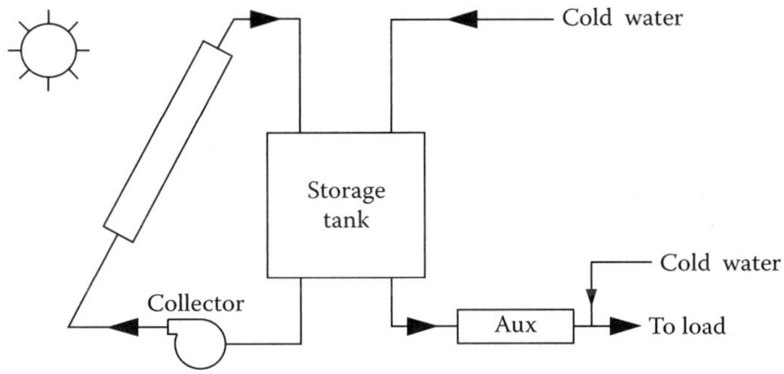

FIGURE 8.6
Schematic of a low-temperature solar industrial process heat system.

energy density (space constraints), and the charging and discharging characteristics. These topics are discussed in detail in Chapter 10. The storage duration for solar industrial process heat (SIPH) systems is usually short since the solar systems are designed to displace part of the fossil fuel requirements. Land availability can be critical for SIPH for existing industries. However, in many cases roofs of industrial buildings can be utilized for this purpose.

Since SIPH systems use components and systems already described, this section will give some examples of SIPH systems. Most of the industrial process heat systems below 200°C require hot water, steam, or hot air. A typical low-temperature SIPH system is shown schematically in Figure 8.6. If the heat is needed in process air, water-to-air heat exchangers may be used or air heating collectors and rock storage is used. Hot air is needed typically in agricultural drying, which may be provided by passive solar air heaters [8].

8.2.1 SIPH for Textile Industries

The textile industry is one of the 10 largest energy consuming industries. Of all the energy used in the textile industry, 60%–65% is used in wet processing, including dyeing, finishing, drying, and curing. The energy for wet processing is used as hot water and steam. The textile industry in the United States uses approximately 500 billion liters of water per day and approximately 25% of this water is used at an average temperature of 60°C. Tables 8.3 and 8.4 show typical calculations for determining energy consumption for jet dyeing [9] and tenter frame drying [10], respectively, for 100% textured polyester circular knit fabric [11]. In the analysis in Table 8.3, Wagner assumed 40% moisture content in the fabric. In carpet dyeing, moisture may be as much as 300% [12]. Drying involves the use of high-pressure (6 atm) steam so that the condensed water may be used for other processes. There are no known examples of SIPH for textile drying, but there are a number of examples of SIPH for textile dyeing. One of those is for a dyeing

TABLE 8.3
Sample Calculations[a]: Heat for Pressure Jet Dyeing

Scour	Energy Consumption GJ/100 kg Fabric
Heat bath, 21°C–60°C	163
Heat cloth, 21°C–60°C	9
Raise bath to 129°C	293
Raise cloth to 129°C	14
Replace heat loss from radiation	
During cycle 21°C–129°C	23
During dyeing at 129°C	65
Scour at 60°C	172
Total	739

Sources: Wagner, R., *Chem. Colorists*, 9, 52, 1977; Goswami, B.C. and Langley, J., A review of the potential of solar energy in the textile industry, in *Progress in Solar Engineering*, Hemisphere Publishing Corporation, Washington, DC, 1987.
[a] Fabric load = 227 kg, bath ratio 10 of 1 cycle time = 2.75 h.

operation at the Riegel Textile Corp. plant in LaFrance, South Carolina [11,13]. Figure 8.7 shows a schematic of the system.

The original system used evacuated tube collectors made by General Electric Corp. That system failed because of repeated tube breakage. Since the dyeing process used water at a temperature of 70°C, the evacuated tube collectors were replaced with flat-plate collectors. The flat-plate collectors at this plant have copper absorbers and tubes painted flat black, low iron textured and tempered glass cover, and bronze-enameled steel frames. The system has 621 m^2 of collector area.

The dyeing process at this plant is an atmospheric Dye Beck batch process at a maximum temperature of 90°C, which is also typical of the textile industry in the United States [11]. The batch dyeing process involves heating of approximately 4500 L of inlet water from 10°C–25°C to 90°C. The typical dye cycle is shown in Table 8.5. The refurbished SIPH system at Riegel Textile plant has been operating successfully. The system was simulated using TRNSYS computer model [15] and there was good agreement between the actual performance of the system and that predicted by TRNSYS. According to measured performance of the system from August 14 to October 8, 1983, the system operated with an efficiency of 46% [11].

8.2.2 SIPH System for Milk Processing

Food systems use approximately 17% of the U.S. energy [16], almost 50% of which is for food processing as hot water (<100°C) or hot air. Proctor and Morse [13] show that over 40% of the energy demand in the beverage industry in Australia was in the form of hot water between 60°C and 80°C.

TABLE 8.4

Energy Consumption during Tenter Frame Drying

A. Steps	**Energy Requirements (MJ/100 kg H_2O)**
Evaporate water	
100 kg × 4.18 kJ/kg°C (100°C–21°C)	33
Latent heat of vaporization	226
Raise steam of 121°C	
100 kg × 1.9 kJ/kg °C (121°C–100°C)	4
Heat air (24 lb air/lb H_2O)	
2400 kg × 1.015 kJ/kg°C (121°C–21°C)	244
Heat fabric to 250	
100 kg × 100/40 × 2.08 × (121°C–21°C)	52
Dryer run at 121°C	559
Dryer run at 149°C	647 (+15%)

Effect of Fabric Moisture Content on Tenter Frame Energy Demand

	Energy Requirements (MJ/100 kg Fabric)	
B. Steps	**30%**	**80%**
Evaporate water		
Raise temperature (21°C–100°C)	10.0	26.5
Latent heat of vaporization	67.7	180.5
Raise steam temperature to 121°C	1.4	3.5
Raise air to 121°C	73.3	195.2
Total	152.4	405.7

Sources: Hebrank, W.H., *Am. Dyestuff Rep.*, 63, 34, 1975; Goswami, B.C. and Langley, J., A review of the potential of solar energy in the textile industry, in *Progress in Solar Engineering*, Hemisphere Publishing Corporation, Washington, DC, 1987.

Considering the temperature and heat requirements of food processing, one would be tempted to conclude that SIPH would be ideal. However, many food processing requirements are seasonal, which may not be economical considering the present price of the conventional fuels unless the SIPH system can be used for a majority of the year. One application that is year-round is milk processing. Singh et al. [16] simulated a SIPH system for milk processing in the United States using TRNSYS. The unit operations (and their respective temperatures) for this plant compatible with solar thermal energy are

1. Boiler feed makeup water (100°C)
2. Pasteurizer makeup water (21°C)
3. Case washer and rinsing (49°C)
4. Clean up (71°C)
5. High temperature short-time cleanup (79°C)
6. Bottle water (93°C)

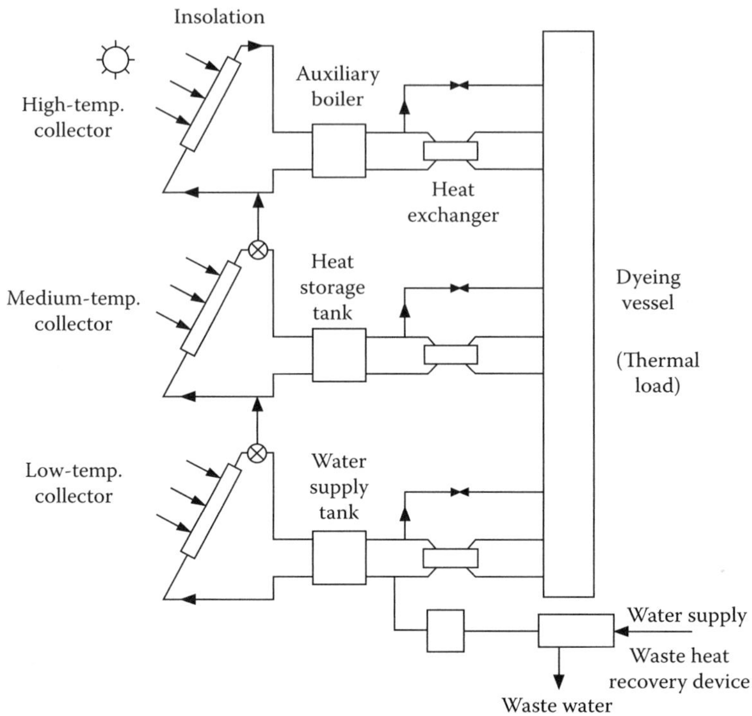

FIGURE 8.7
Schematic of solar energy system at Riegel Textile Corporation plant at LaFrance, South Carolina.

TABLE 8.5
Typical Dye Beck Heat Requirements

Dye Process Operation	Temperature of Dye Beck (°C)	Time Interval of Cycle (h)	Percent of Total Process Energy (%)
Heat the initial load	16–32	0.5	7
Dyeing preparation	32–88	0.5	22
Dye period	88	7–9	68
Cool and reheat for dye fixation	32–43	1.0	3

They estimated that for a plant producing 170,000 kg/week of milk and 98,000 kg/week of orange juice, a total of 621.3 GJ (or 80%) of energy demand was compatible with solar energy in summer and 724.9 GJ (or 93%) of energy demand was compatible in winter. Simulating a solar system similar to the one in Figure 8.6, they found that a 4000 m^2 collector area could provide about 30%–35% solar fraction for the milk-processing plant in Madison, Wisconsin, Fresno, California or Charleston, South Carolina [16].

8.3 Parabolic Collectors

In order to provide thermal energy at temperatures above 100°C, it is necessary to use concentrating collectors. Concentration of solar radiation is achieved by reflecting or refracting the flux incident on an aperture area A_a onto a smaller receiver/absorber area A_r. An optical concentration ratio, CR_o, is defined as the ratio of the solar flux I_r on the receiver to the flux, I_a, on the aperture, or

$$CR_o = \frac{I_r}{I_a}, \tag{8.1}$$

while a geometric concentration ratio CR is based on the areas shown in Figure 8.8, or

$$CR = \frac{A_a}{A_r}. \tag{8.2}$$

CR_o gives a true concentration ratio because it accounts for the optical losses from the reflecting and refracting elements. However, since it has no relationship to the receiver area, it does not give an insight into the thermal losses which are proportional to the receiver area. In the analyses in this book, only geometric concentration ratio CR will be used.

Concentrators are inherently more efficient at a given temperature than are flat-plate collectors, since the area from which heat is lost is smaller than the aperture area. In the flat-plate device, both areas are equal in size. A simple energy balance illustrates this principle. The useful energy delivered by a collector q_u is given by

$$q_u = \eta_o I_c A_a - U_c (T_c - T_a) A_r, \tag{8.3}$$

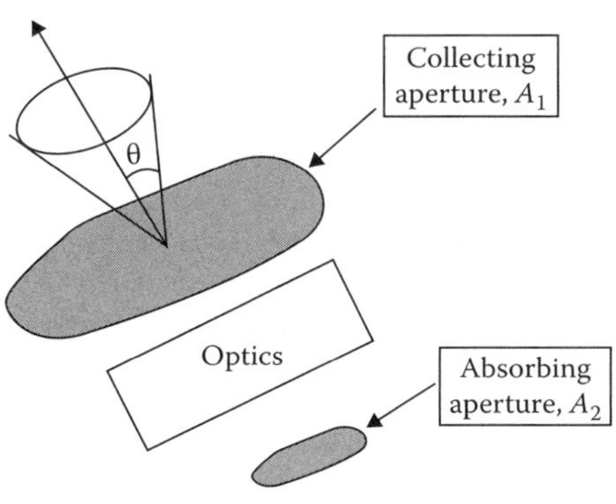

FIGURE 8.8
Definition of geometric concentration.

in which η_o is the optical efficiency and other terms are as defined previously. The instantaneous collector efficiency is given by

$$\eta_c = \frac{q_n}{I_c A_a}, \tag{8.4}$$

from which, using Equation 8.3,

$$\eta_c = \eta_o - \frac{U_c (T_c - T_a)}{I_c} \frac{1}{CR} \tag{8.5}$$

For the flat plate CR $\cong$ 1 and for concentrators CR > 1. As a result, the loss term (second term) in Equation 8.5 is smaller for a concentrator and the efficiency is higher. This analysis is necessarily simplified and does not reflect the reduction in optical efficiency that frequently, but not always, occurs because of the use of imperfect mirrors or lenses in concentrators. The evaluation of U_c in Equation 8.5 in closed form is quite difficult for high-temperature concentrators, because radiation heat loss is usually quite important and introduces nonlinearities because radiation is proportional to T^4. One disadvantage of concentrators is that they can collect only a small fraction of the diffuse energy incident at their aperture. This property is an important criterion in defining the geographic limits to the successful use of concentrators and is described shortly.

From basic thermodynamics, it can be shown [14] that there exists a maximum geometric concentration allowed by physical conservation laws. This maximum concentration, C_{max}, is related to the angular field of view shown in Figure 8.8 [17] according to

$$C_{max} = \frac{n}{\sin\theta} \quad \text{for a two-dimensional concentrator such as a parabolic trough} \tag{8.6}$$

$$C_{max} = \frac{n^2}{\sin^2\theta} \quad \text{for a three-dimensional concentrator such as a parabaloid dish} \tag{8.7}$$

In the above equation, n is the index of refraction at the absorber surface and θ is the half-angle of acceptance seen from the collector aperture. A system that can obtain this concentration limit is referred to as "ideal." All conventional imaging systems fall short of this limit by a factor of at least 2–4.

The second law of thermodynamics prescribes not only the geometric limits of concentration as shown above, but also the operating temperature limits of a concentrator. The radiation emitted by the sun and absorbed by the receiver of a concentrator, q_{abs}, is

$$q_{abs} = \tau\alpha_s A_s F_{sa} \sigma T_s^4, \tag{8.8}$$

where

T_s is the effective temperature of the sun

τ is the overall transmittance function for the concentrator, including the effects of any lenses, mirrors, or glass covers

A_s is the area of the sun

F_{sa} is the fraction of radiation emitted by A_s that reaches the area A_a

If the acceptance half-angle θ_{max} is selected to just accept the sun's disk of angular measure $\theta_s(\theta_s \sim 1/4°)$, we have, by reciprocity ($F_{as}A_a = F_{sa}A_s$), for an ideal concentrator with compound curvature,

$$q_{abs} = \tau\alpha_s A_a \sin^2\theta_s \sigma T_s^4. \tag{8.9}$$

If convection and conduction could be eliminated, all heat loss q_L is by radiation, and

$$q_L = \varepsilon_{ir} A_r \sigma T_r^4, \tag{8.10}$$

where ε_{ir} is the infrared emittance of the receiver surface. Radiation inputs to the receiver from a glass cover or the environment can be ignored for this upper limit analysis.

An energy balance on the receiver is then

$$q_{abs} = q_L + \eta_c q_{abs}, \tag{8.11}$$

where η_c is the fraction of energy absorbed at the receiver that is delivered to the working fluid. Substituting for Equations 8.9 and 8.10 in Equation 8.11, we have

$$(1-\eta_c)\tau\alpha_s A_a \sin^2\theta_s \sigma T_s^4 = \varepsilon_{ir} A_r \sigma T_r^4, \tag{8.12}$$

Because $CR = A_a/A_r$ and $CR_{max} = 1/\sin^2\theta_s$, the maximum temperature of the receiver is

$$T_r = T_s\left[(1-\eta_c)\tau\frac{\alpha_s CR}{\varepsilon_{ir} CR_{max}}\right]^{1/4}. \tag{8.13}$$

8.3.1 Compound Parabolic Concentrators

Until recently it was believed that no useful concentration could be achieved without tracking. However, the invention of Roland Winston [18] showed that the techniques of so-called non-imaging optics provides designs that can deliver moderate levels of concentration with completely stationary concentrators. Figure 8.9 shows a schematic cross section of

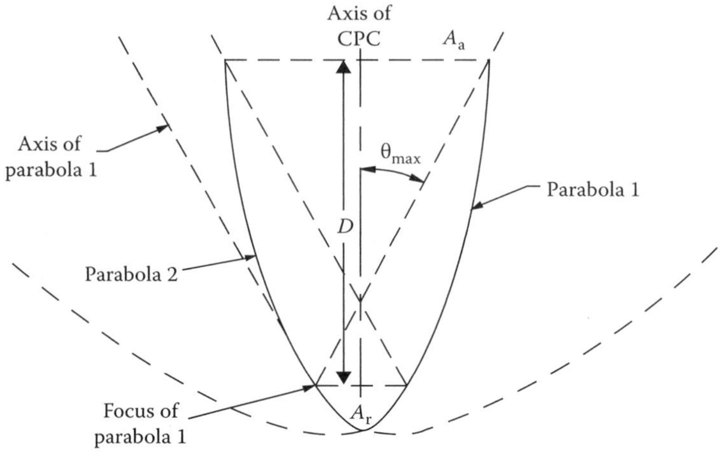

FIGURE 8.9
Schematic cross section of a CPC showing parabolic segments, aperture, and receiver. (From Goswami, Y. et al., *Principles of Solar Engineering*, 2nd edn., Taylor & Francis, Philadelphia, 2000. With permission.)

FIGURE 8.10
Commercial CPC trough collector module with evacuated tube receiver. (Courtesy of Professor Roland Winston, University of California, Merced, CA.)

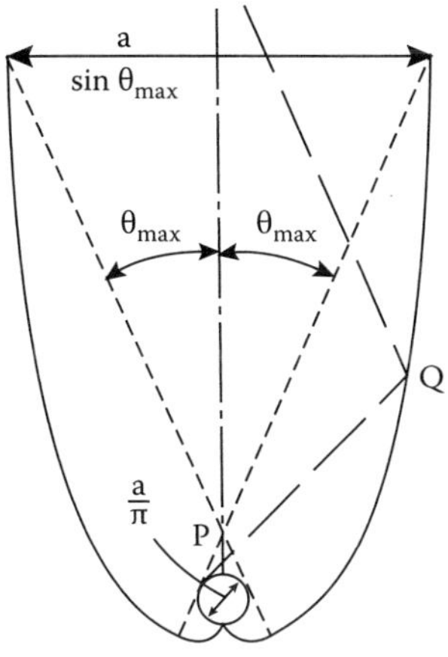

FIGURE 8.11
Cross-section of CPC collector concept with tubular receiver.

the original CPC concept. The concentrator is seen to be formed from two distinct parabolic segments, the foci of which are located at the opposing receiver surface end points. The axes of the parabolic segments are oriented away from the CPC axis by the acceptance angle θ_{max}. The slope of the reflector surfaces at the aperture is parallel to the CPC optical axis. Figure 8.10 is a photograph of a CPC collector. Different types of planar and tubular receivers have been proposed for CPCs. Of most interest in this book is the tubular-type receiver shown in Figure 8.11 since high-pressure heat-transfer fluid can flow through the tube.

The most common way to utilize a CPC collector is shown schematically in Figure 8.12 [17]. The long axis of the CPC trough is aligned in an east west direction and that of the normal through the trough apertures is tilted downward from the zenith by an angle equal to the latitude. The angular acceptance is a wedge of half-angle $\pm\theta_c$. Whenever the sun's path lies within this orange slice, all of the direct solar radiation is collected, concentrated, and delivered to the absorber. In addition, some of the diffuse radiation entering through the aperture is also directed toward the absorber.

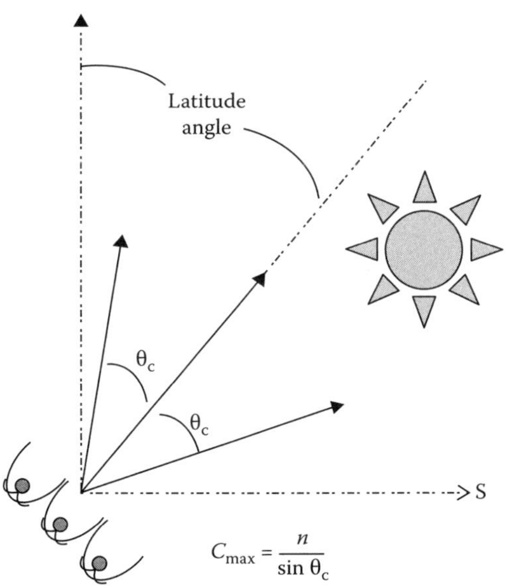

FIGURE 8.12
Schematic of an east-west aligned CPC.

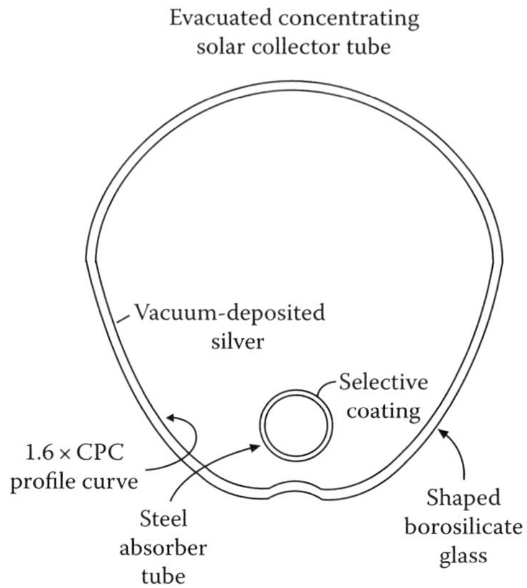

FIGURE 8.13
Schematic cross-section of the advanced CPC integrated optics with evacuated tube receiver. (From O'Gallagher, J.J., *Nonimaging Optics in Solar Energy*, Morgan & Claypool Publishers, Ft. Collins, Co, 2008. With permission.)

Advanced CPC collectors integrate the CPC optics into the design, as shown in Figure 8.13 [17]. The integrated CPC shown in Figure 8.13 has an acceptance angle of 70° and a concentration ratio of 1.6.

A comparison of the predicted and experimental performance is shown in Figure 8.14. From this data, it appears that this type of non-tracking CPC

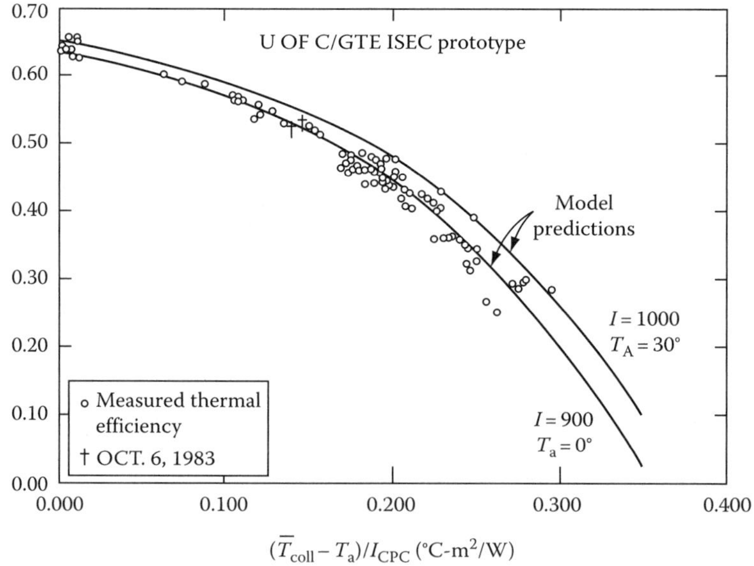

FIGURE 8.14
Comparison of the predicted and experimental performance of a non-tracking CPC. (From O'Gallagher, J.J., *Nonimaging Optics in Solar Energy*, Morgan & Claypool Publishers, Ft. Collins, Co, 2008. With permission.)

collector is a simple and effective method for delivering solar thermal energy in the temperature range between 100°C and 300°C. This type of collector has great potential in the field of solar industrial process heat as well as some solar cooling systems described in Chapter 7.

8.3.2 Single-Axis Tracking Parabolic Trough Collectors

Parabolic trough solar concentrators (PTSCs) have been found to be cost-effective solar conversion systems and are growing in importance. Figure 8.15a is a schematic diagram of this type of single-axis tracking collector and Figure 8.15b is a photograph of a commercial PTSC.

PTSCs have been built in many sizes up to the 6 m width apertures system constructed by Sky Fuel in Colorado. Parabolic trough collectors have been

(a)

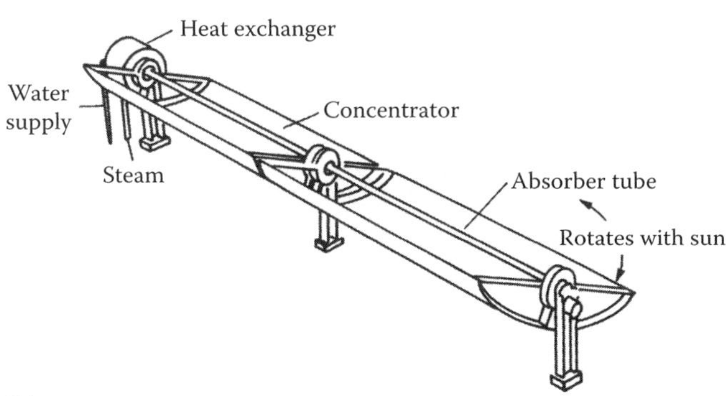

(b)

FIGURE 8.15
(a) Photograph and (b) schematic diagram of single axis tracking parabolic collector.

used in one of the early solar hybrid power plants installed by LUZ in Barstow, California. This 355 MWe plant has operated successfully more than 30 years. At present, PTSCs are being installed as fuel-saving devices in existing natural gas power plants, as well as in new power plant construction. The following will illustrate the design, energy return on energy investment, and cost of energy delivered for PTSCs. Details on the mechanical aspects of design are presented in [19].

The geometric relations for a parabolic trough concentrator are shown schematically in Figure 8.16.

The surface of the reflector/concentrator has a parabolic cross section so that all rays of sunlight parallel to its axis are reflected from the surface to a focal line. The basic geometric relation that defines the parabolic shape for a point source is

$$y = \frac{x^2}{4f} \tag{8.14}$$

where f is the focal length of the concentrator. If D is the width of the concentrator aperture, d the absorber tube diameter, then the concentrator rim angle, ϕ, is

$$\tan\left(\frac{\phi}{2}\right) = \frac{D}{4f} \tag{8.15}$$

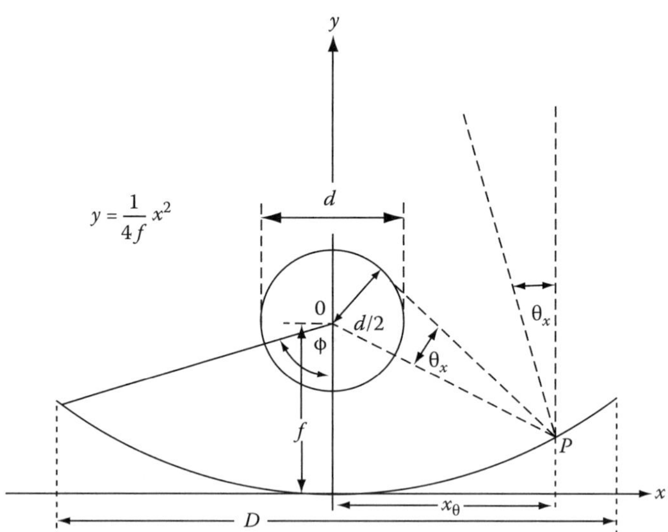

FIGURE 8.16
Geometric relations for a PTSC.

and the geometric concentration ratio, C, is

$$C = \frac{D}{\pi d} \tag{8.16}$$

In practice, parabolic troughs use concentrators with rim angles between 70° and 110°, as this range provides a balance between optical and structural requirements of the concentrator.

In addition to the optical error introduced by the finite specularity of the reflective surface, a number of other optical errors contribute to widening of the reflected beam at the receiver, such as concentrator slope error, tracking error, and receiver mislocation errors. Generally, these errors can be characterized by Gaussian distributions, so the total amount of beam spreading can be calculated as the root mean square (rms) value of all the optical errors.

The sun is not a point source, but the sun's finite size can be approximated with an rms (root mean square) width of $\sigma_{sun} = 2.8$ mrad (milli-radians). However, for line focus optics, the apparent width of the sun increases with incidence angle and a good average of the all-day rms width is 5.0 mrad. The total beam spreading, σ_{total}, due to optical errors and the sun shape can be expressed as

$$\sigma_{total} = \left(\sigma_{optical}^2 + \sigma_{sun}^2\right)^{\frac{1}{2}} \tag{8.17}$$

The optical error is given by

$$\sigma_{opt} = \left[4\sigma_{con}^2 + \sigma_{track}^2 + \lambda(\theta)\left(4\sigma_{con}^2 + \sigma_{spec}^2\right) + \sigma_{disp}^2 + \sigma_{spec}^2\right]^{\frac{1}{2}} \tag{8.18}$$

where

σ_{con} = rms angular deviation of concentrator from perfect parabola (slope error)

σ_{track} = rms angular spread due to sun tracking error

σ_{spec} = rms angular spread of reflected beam due to imperfect specularity of reflector

σ_{disp} = equivalent rms angular spread, which accounts for the imperfect placement of the receiver

$\lambda(\theta)$ = coefficient that accounts for the rim-angle-dependent contribution of longitudinal mirror errors to transverse beam spreading (*Note*: A value of 0.1 is recommended for $\lambda(\theta)$ for concentrators with rim angles between 80° and 100°.)

The first step in a PTSC design is to calculate the intercept factor, γ, defined as the fraction of the reflected beam radiation that is intercepted by the absorber tube. Bendt et al. [20] have shown that the intercept factor γ depends on the product ($\sigma_{total}C$) as well as the concentrator rim angle, ϕ, as shown in Figure 8.17.

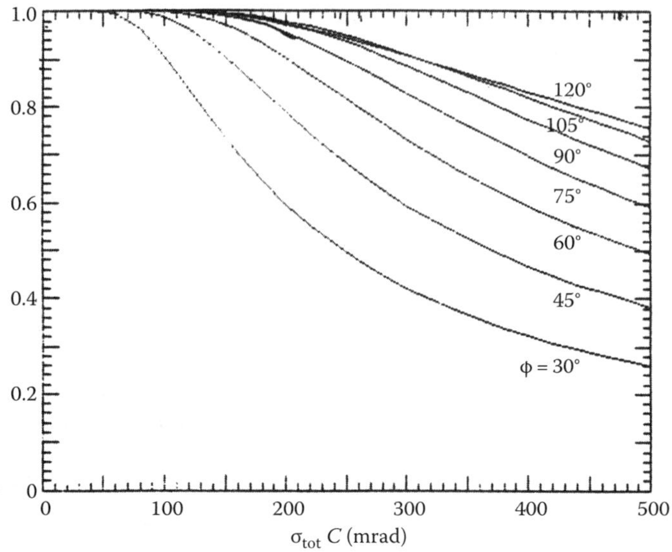

FIGURE 8.17
Intercept factor vs. σ_{tot} for different rim angles. (From Gee, R., *Long Term Average Performance Benefits of Parabolic Trough Improvements*, SERI, Golden, CO, SERI/TR 632-439, March 1980. With permission.)

Once σ_{total} has been estimated, various concentration ratios (and thus receiver diameters) can be evaluated for their resulting impact on the intercept factor. A reasonable value for the intercept factor from an engineering perspective is 95%, i.e. 5% spill over losses.

The absorber/receiver for the parabolic trough system is usually a metal tube surrounded by a glass tube. The heat loss coefficient for this type of receiver/absorber is

$$U_L = \frac{\text{Rate of heat loss}}{(T_{absorber} - T_{amb})A_{absorber}} \tag{8.19}$$

Figure 8.18 shows experimentally measured absorber heat loss coefficients as a function of absorber temperature for various designs.

The reference trough receiver is the simplest construction with air in the annulus. The other curves show the improvement, that is the reduction in the heat loss coefficient, with other designs. The annulus between these two tubes can be evacuated (case E) to reduce conduction and convection losses to a minimum. This is called an evacuated tube collector.

To determine the collector efficiency equation, we can use the results of extensive tests conducted by Sandia National Laboratories. The results of these tests, as summarized by Gee and May [22], show that the collector efficiency, η_c, can be expressed by an equation of the form

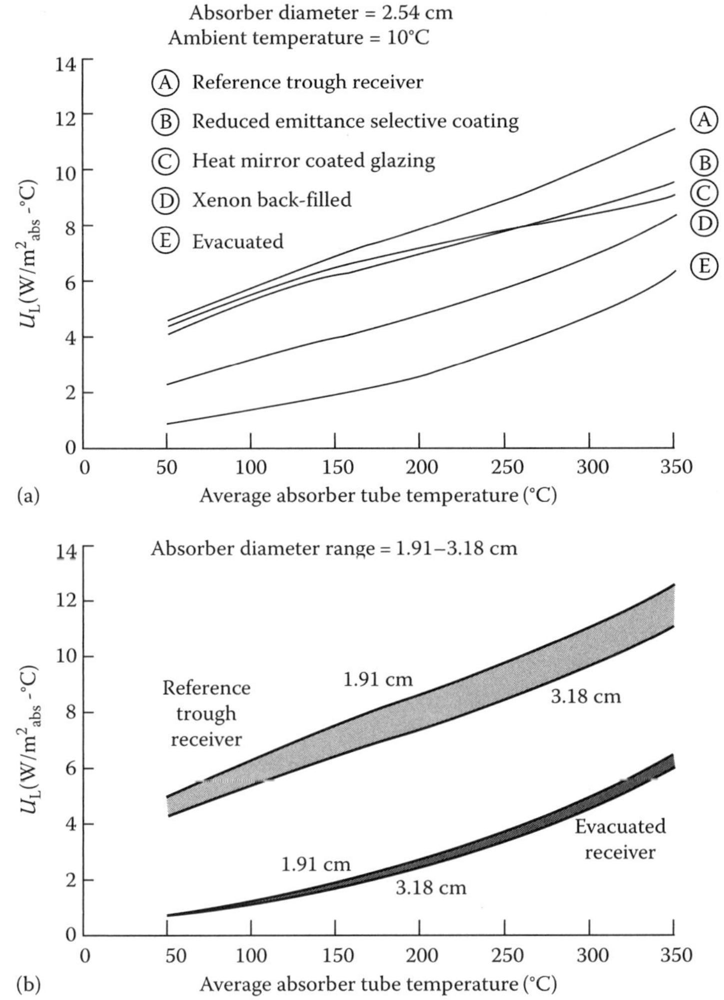

FIGURE 8.18
(a) Absorber heat loss coefficients vs. temperature and (b) heat loss coefficient variance with absorber diameter. (From Gee, R., *Long Term Average Performance Benefits of Parabolic Trough Improvements*, SERI, Golden, CO, SERI/TR 632-439, March 1980. With permission.)

$$\eta_c = \eta_{opt} - \frac{U_L(T_{absorber} - T_{ambient})}{I} \tag{8.20}$$

where
I is the insolation on the aperture
η_{opt} is the product of the reflectance of the concentrator reflector surface, ρ, the transmittance of the receiver glazing, τ, and the absorbtance of the coating on the receiver tube, α

The heat loss coefficient from the absorber tube, U_L, can be related empirically to the absorber temperature, T_a, by the expression

$$U_L = A + B(T_a - T_\infty) \tag{8.21}$$

where
T_∞ is the ambient temperature
A and B are empirical coefficients to describe heat loss from the absorber

These coefficients are usually obtained from test data, as shown in Figure 8.18.

Example 8.1

Derive a relationship for the efficiency of a parabolic collector with a receiver/absorber tube corresponding to the reference case in Figure 8.18 as a function of insolation and temperature difference between the collector surface and ambient. The reflectance of the concentrator surface equals 0.94, the transmittance of the glass tube surrounding the inner tube is 0.95, and the absorbtance of the coating on the receiver tube equals 0.95. Assume that the average direct normal solar irradiance is 560 W/m^2, σ_{total} = 10.5 mrad and the axis of the collector has an East–West orientation.

Solution
The first step in the solution is to select a concentration ratio and rim angle for the collector that results in an intercept factor (that is, the portion of reflected sunlight that strikes the absorber tube) of at least 95%. For this selection, we use the results shown in Figure 8.17, which plots the intercept factor (γ) at a function of $\sigma_{tot} \times C$, where σ_{tot} is the total rms beam spread including the sun's angular width, which is taken as 0.6° or 10.5 mrad in accordance with the data. If we select a rim angle of 90° and an intercept factor of 0.96, the product $\sigma_{total} \times C$ is about 200. Hence, the concentration ratio equals

$$C = \frac{200}{10.5} = 19.$$

Using the information in the problem statement, the optical efficiency equals

$$\eta_{opt} = \rho\tau\alpha = 0.94 \cdot 0.95 \cdot 0.95 = 0.79$$

For the reference case shown in Figure 8.17, a linear curve fit for the relation between the heat loss coefficient, U_L, and the average absorber tube temperature yields the relationship

$$U_L = 3 + 0.025(T_{absorber} - T_{ambient}) = 3 + 0.25\Delta T$$

Hence, the efficiency for this collector, η_c, can be approximated by the equation

$$\eta_c = \eta_{opt} - \frac{(A + B\Delta T)\Delta T}{I} = 0.79 - \frac{(3 + 0.025\Delta T)\Delta T}{I}$$

where ΔT equals $(T_{absorber} - T_{ambient})$.

Note that the relation for the linear fit is based on a unit receiver area, whereas the efficiency equation is based on a unit aperture area. Since we need a relationship based on an aperture area, U_L needs to be divided by the concentration ratio which yields the following relation for the efficiency curve:

$$\eta_c = 0.79 - \frac{\left(\frac{3}{19} - \frac{0.025}{19}\Delta T\right)\Delta T}{I} = 0.79 - \frac{(0.16 - 0.0013\Delta T)\Delta T}{I}$$

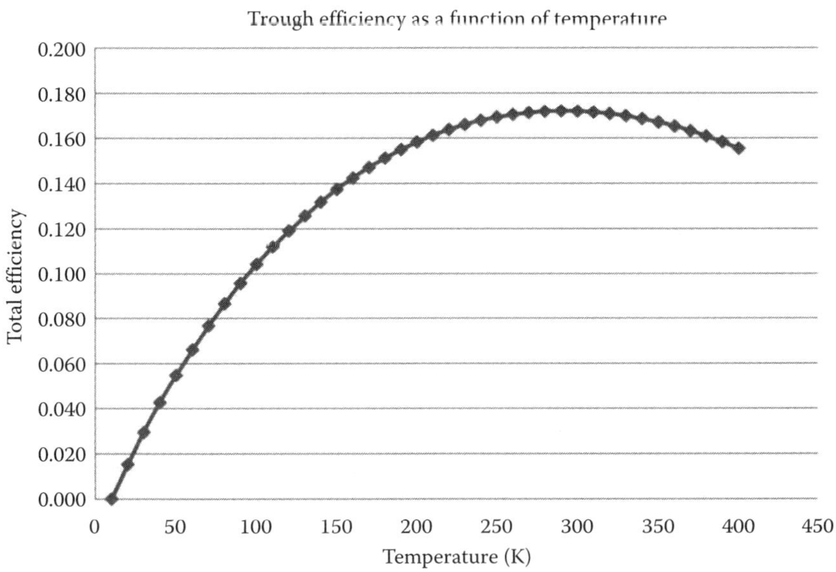

The next example illustrates the economics and the EROI for a tracking parabolic trough solar thermal system.

Example 8.2

In 1986, a parabolic trough-type solar water heater was installed for a correctional facility in Denver, Colorado. The system was designed to deliver 50°C water and a performance model predicted that if energy for O&M is subtracted from the gross output, a net of 3.85 TJ/year (3.666 × 10^6 Btu/year) will be delivered from solar heater. The solar energy displaced natural gas at \$5.00/GJ. The other pertinent information is listed below:

Total investment in first quarter (1986 dollars)	\$115,000
Value of fuel displaced in first quarter (1986 dollars)	\$19,250
Discount rate for present worth	10%

For this problem, omit effects of taxes, depreciation, and decommissioning to simplify analysis and emphasize the main points. Then calculate

a. Simple rate of return
b. Discounted payback time
c. Discounted rate of return
d. Energy payback time
e. Energy return on energy investment (EROI) using the cost breakdown in 1986 dollars in the second table below
f. The levelized cost of energy

Solution

The simple rate of return on investment approach does not take into account the time value of money. For this example, the simple rate of return is

$$\frac{\text{Savings per year (1986\$)} \times 100}{\text{Initial investment (1986\$)}} = \frac{19,250}{115,000} \times 100 = 16.7\%/\text{year}$$

The inverse of the fractional rate of return gives the simple economic payback time as 6 years.

The present worth of the future savings is calculated by taking into account the time value of money and discounting future savings to dollars at the beginning of the project in 1986. Discount factors are taken from Table 8.6. For simplicity,

TABLE 8.6

Present Worth Calculations for Example 8.2

		Present Worth of Fuel Savings in 1986 \$	
Year	**Discount Factor**[a]	**Annual**	**Cumulative**
1	0.909	17,500	17,500
2	0.826	15,900	33,400
3	0.751	14,460	47,860
4	0.683	13,150	61,010
5	0.621	11,950	72,960
6	0.564	10,850	83,810
7	0.513	9,880	93,690
8	0.466	8,980	102,670
9	0.424	8,160	110,830
10	0.386	7,420	118,250

[a] A discount rate of 10% per year is used. The discount factor is given by $1/(1 + i)^N$, where i is the fractional discount rate, 0.10 per year, and N is the number of years.

annual discount factors are used, as though the fuel bill was paid only once at the end of each year. The calculations are shown in Table 8.6.

When the cumulative present worth in 1986 dollars of the energy savings equals the initial investment of $115,000, then the original investment has been paid back in discounted dollars. From Table 8.6, this occurs at 9.6 years and that is the discounted payback time. The discounted return is the average annual present worth of cumulative savings over the payback time, i.e., $115,000/9.6 = $11,980/year and the discounted rate of return is

$$\frac{\$11,980/\text{year}}{\$115,000} \times 100 = 10.4\% \text{ per year.}$$

We next present an energy analysis of the same parabolic trough solar water heating system. The estimated 1986 costs of materials and labor to construct the system are shown in the second column of Table 8.7. Also shown are the annual operating and maintenance (O&M) cost and energy input. O&M costs ranging from $0.13* to $1.34 per ft^2/year (CGSSC, 1984), which is 6.4% per year of the total capital investment, have been used here. This value is conservative because it is considerably higher than the 3% per year often assumed for O&M. The energy requirement for O&M, estimated to be 234 million Btu/year, was subtracted from the estimated solar system output of 3900 million Btu/year to give the net energy output of 3666 million Btu/year. The energy embodied in all the materials and in the labor, per dollar of cost, is given in the third column. These values come from input/output energy analysis of the corresponding sector of the economy and are in 1977 dollars [8]. They must be converted to 1986 dollars by the inflation factor of 0.593. The 1986 product of cost and energy intensity gives the energy incorporated into the system with each material or labor item used, and is shown in the fourth column. The sum of these individual items gives the total energy embodied in this system. Hence, the energy investment is equal to 5.43 × 10^9 Btu/year. The rate of energy return on the energy investment is thus

$$\frac{3.66\times 10^9 \text{ Btu/year}}{5.43\times 10^9 \text{ Btu}} \times 100 = 67.5\%$$

which corresponds to an energy payback time of 1.48 years when the net energy produced equals that invested. The EROI is the fractional rate of energy return times the project lifetime in years. Thus, for a 15 year life, the EROI is 10:1. The levelized cost of energy delivered by the system over 15 years obtained from Equation 2.19 is

$$\text{LCOE} = \$5.85/\text{GJ} = \$0.02/\text{kWh}$$

The system on which this example is based [23] was built in 1986 and continued operation until 2006, when a road was built through the land on which

TABLE 8.7

Parabolic Trough System Costs for Example 8.2

Item	Dollar Cost ($86)	Energy Intensity (Btu/$77)	Embodied Energy (10^6 Btu)
Collector materials			
Aluminum	16,250	158,201	1524
Steel	7,530	115,724	517
Borosilicate tubing	5,445	58,724	190
Adhesive	1,422	95,178	80
Wire rope	673	115,724	46
Fasteners	1,449	115,724	99
Solvent	335	171,572	34
Reflective film	18,103	99,470	1067
Nylon washers	135	32,765	3
Selective film	642	66,761	25
Electric motors	311	33,556	6
Pillow block bearings	2,320	42,638	59
Power jack	1,959	28,793	33
Shaft couplings	30	44,247	1
Silicone O-rings	270	53,271	9
Rebar	620	115,724	43
Bolts, nuts, washers	640	44,427	17
Concrete	1,310	216,631	168
Shipping	1,830	29,198	32
Fabrication	14,790	39,556	347
Installation	10,420	39,556	244
Collector total	86,484		4544
Balance of plant			
Steel piping	3,091	115,724	212
Pipe insulation	3,050	79,474	144
Flexhoses	2,610	27,116	42
Glycol	820	95,178	46
Pump and motor	532	27,731	9
Batteries	220	69,024	9
Expansion tank	110	42,621	3
Valves	1,580	36,093	34
Heat exchanger	1,280	42,621	32
Controls	2,300	23,412	32
Security fence	4,120	47,163	115
Plant installation	8,890	39,556	209
Plant total	28,603		887
Grand total	115,087		5431
Operation and maintenance (annual)	7,480	52,770	234

Source: Courtesy of Kenneth May, Private Communication.

Notes: Dollar and energy costs for a 10,000 ft^2 parabolic trough system with north–south orientation and no thermal storage in Denver, Colorado. (A price deflator ratio of 0.593 was used to convert $86 to $77.)

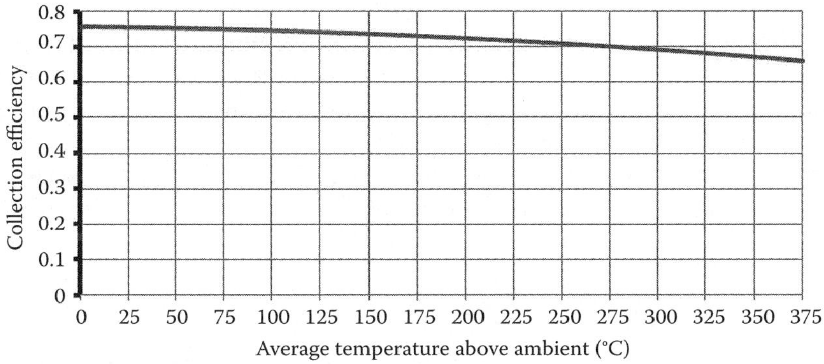

FIGURE 8.19
Projected efficiency of advanced parabolic trough collector. (From Farr, A. and Gee, R., The SkyTrough™ parabolic trough solar collector, in Paper *90090 ASME 3rd International Conference on Energy Sustainability*, San Francisco, CA, 2009; Courtesy of R. Gee.)

the collector was located and the system had to be dismantled. Hence, the real life of the system was 20 years and could easily have been extended another 10 years. With a 30 year life, the EROI would have been 20:1. This shows that the claim often heard in some circles that solar energy systems take more energy to build than they deliver is patently false!

A significant advance in parabolic reflector technology has recently been reported by [24]. In this innovative parabolic trough technology, the traditional reflecting surface of silvered glass is replaced by a highly reflective and shatterproof silver polymer film, called ReflecTech® Mirror Film, that is laminated to a thin aluminum substrate. This film allows for larger and fewer panel segments than in previous trough designs and reduces the weight, the cost, as well as the amount of material required for the reflective surface. The technology is based upon a patent by NREL and ReflecTech [25]. The trough is currently undergoing extensive tests at the NREL test site, and a preliminary estimate of its efficiency is shown in Figure 8.19.

8.4 Long-Term Performance of SIPH Systems

In order to assess the economic viability of any solar process, its cumulative energy delivery over its economic life (in years or decades) must be known. It is very difficult to calculate this number accurately since (1) solar systems and their energy delivery are subject to the vagaries of local microclimate, which can change on a timescale on the order of hours and (2) future weather cannot be predicted at a high level of detail. The standard approach used to estimate future performance of a solar system is to use a typical year of past weather data and assume that it will represent the future on the average, to engineering accuracy. Two common methods used to predict long-term performance of SIPH systems are the utilizability and the TRNSYS methods [15,27]. These methods can also be used for the design of SIPH systems. The following sections describe both methods.

8.4.1 The Utilizability Method

Utilizability, ø, has been used to describe the fraction of solar flux absorbed by a collector, which is delivered to the working fluid. On a monthly timescale

$$\bar{\phi} = \frac{Q_u}{F\bar{\eta}_o}\bar{I}_c < 1.0, \tag{8.22}$$

where

Overbars denote monthly means

Q_u is the monthly averaged daily total useful energy delivery

$\bar{\phi}$ is the fraction of the absorbed solar flux, which is delivered to the fluid in a collector operating at a fixed temperature T_c

The $\bar{\phi}$ concept does not apply to a system comprised of collectors, storage, and other components wherein the value of T_c varies continuously. The fixed temperature mode will occur if the collector is a boiler, if very high flow rates are used, if the fluid flow rate is modulated in response to flux variations to maintain a uniform T_c value, or if the collector provides only a minor fraction of the thermal demand. However, if the flow is modulated, note that the value of F (i.e., F', F_R) may not remain constant to engineering accuracy.

When T_c is not constant in time as in the case of a collector coupled to storage, the $\bar{\phi}$ concept cannot be applied directly. However, for most concentrators with CR > 10, the value of U_c is small and the collector is relatively insensitive to a *small* change of operating temperatures. To check this assumption for a particular process, values of $\bar{\phi}$ at the extremes of the expected temperature excursion can be compared.

The value of $\bar{\phi}$ depends upon many system and climatic parameters. However, Collares-Pereira and Rabl [28] have shown that only three are of first order—the clearness index $\overline{K}_T$ (see Chapter 5), the critical intensity ratio $\bar{X}$ (discussed in Section 8.4.2), and the ratio r_d/r_T (discussed in Section 8.4.3). The first is related to insolation statistics, the second to collector parameters and operating conditions, and the last to collector tracking and solar geometry.

8.4.2 Critical Solar Intensity Ratio *X*

The instantaneous efficiency equation for many solar collectors has been shown to be of the form

$$\eta_c = F\left(\eta_o - \frac{U_c \Delta T^+}{I_c}\right), \quad (\eta_c > 0) \tag{8.23}$$

where
- ΔT^+ is the value of a collector to ambient temperature difference if positive
- η_c is zero otherwise
- F is a heat exchanger factor (F', F_R), the expression for which depends on the definition of ΔT^+

It is possible but not always economical to operate the solar collector system if $\eta_c > 0$. In practice, $\eta_c \geq \eta_{min} > 0$ is usually the system turn-on criterion since it is not worthwhile to operate collector loops for cases where η_c is very small.

Equation 8.23 can be used to determine the solar intensity level above which useful energy collection can take place. Solving Equation 8.23 for I_c, collector output begins when

$$I_c \geq \frac{U_c \Delta T^+}{(\eta_o - \eta_{min}/F)}. \tag{8.24}$$

Since $\eta_{min} \ll 1$ and F is close to 1, for convenience the second term in the denominator can be dropped and a dimensionless critical intensity ratio X defined as

$$X \equiv \frac{U_c \Delta T^+}{\eta_o I_c} \leq 1.0. \tag{8.25}$$

X is seen to be the ratio of collector heat loss to absorbed solar flux at $\eta_c = 0$, i.e., at the no-net-energy-delivery condition. In many cases, the daily or monthly averaged daily critical intensity ratio $\bar{X}$ is of more interest and is defined as

$$\bar{X} \equiv \frac{U_c \overline{\Delta T^+ \Delta t_c}}{\overline{\eta_o} \bar{I}}, \tag{8.26}$$

where
- $\overline{\eta_o}$ is the daily averaged optical efficiency
- $\overline{\Delta T^+}$ is the daily mean temperature difference *during collection*

These can also be expressed as

$$\overline{\Delta T} = \frac{1}{\Delta t_c} \int_{t_o}^{t_o+\Delta t_c} (T_c - T)\,dt, \tag{8.27}$$

$$\overline{\eta_o} = \frac{\int_{t_o}^{t_o \Delta t} \eta_o I_c dt}{\int_{t_o}^{t_o+\Delta t_c} I_c dt}, \tag{8.28}$$

and

$$\overline{I_c} = \frac{1}{\Delta t_c} \int_{t_o}^{t_o=\Delta t_c} I_c dt. \tag{8.29}$$

The collector cut-in time t_o and cut-off time $t_o + \Delta t_c$ are described shortly. The time t = [0, 24] h and is related to the solar hour angle h_s by $t = (180 + h_s)/15$; Δt_c is the collection period in hours. In Equation 8.27 T_c can be collector surface, average fluid, inlet fluid or outlet fluid temperature, depending upon the efficiency data basis.

8.4.3 Collection Period (Δt_c)

The collection period Δt_c can be dictated either by optical or thermal constraints. For example, with a fixed collector, the sun may pass beyond the acceptance limit or be blocked by another collector, and collection would then cease. Alternately, a high-efficiency, solar-tracking concentrator operating at relatively low temperature might be able to collect from sunrise to sunset. A third scenario would be for a relatively low concentration device operating at high temperature to cease to have a positive efficiency during daylight at the time that heat losses are equal to absorbed flux. In this case, the cutoff time is dictated by thermal properties of the collector and the operating conditions.

Collares-Pereira and Rabl [28] have suggested a simple procedure to find the proper value of Δt_c. Useful collection Q_c is calculated using the optical time limit first, i.e., $\Delta t_c = 2\min\{[h_{sr}(\alpha = 0), h_{sr}(i = 90)]/15\}$. Second, Q_u is calculated for a time period slightly shorter, say by one-half hour, than the optical limit. If this value of Q_u is larger than that for the first, optically limited case, the collection period is shorter than the optical limit. The time period is then further reduced until the maximum Q_u is reached.

The above method assumes that collection time is symmetric about solar noon. This is almost never the case in practice since the heat collected for an hour or so in the morning is required to warm the fluid and other masses to operating temperature. A symmetric phenomenon does not occur in the afternoon. If the time constant of the thermal mass in the collector loop is known, the collection period may be assumed to begin at t_o, $\Delta t_c/2\,\text{h}$ (from above symmetric calculation) before noon decreased by two or three time constants. Another asymmetry can occur if solar flux is obstructed during low sun angle periods in winter. It is suggested that r_T and r_d from Table 8.8 under asymmetric collection conditions be calculated from

$$r_T = \frac{\left[r_T(h_{s,\text{stop}}) + r_T(h_{s,\text{start}})\right]}{2}, \tag{8.30}$$

$$r_d = \frac{\left[r_d(h_{s,\text{stop}}) + r_d(h_{s,\text{start}})\right]}{2} \tag{8.31}$$

where the collection starting and stopping hour angles account for transients, shading, etc., as described above:

$$h_{s,\text{start}} = 180 - 15t_o, \tag{8.32}$$

$$h_{s,\text{stop}} = 180 - 15(t_o + \Delta t_c). \tag{8.33}$$

8.4.4 Empirical Expressions for Utilizability

Empirical expressions for $\bar{\phi}$ have been developed for several collector types [28].

For nontracking collectors,

$$\bar{\phi} = \exp\{-\bar{X} - (0.337 - 1.76\bar{K}_T + 0.55 r_d/r_T)\bar{X}^2]\} \tag{8.34}$$

$$\text{for} \quad \bar{\phi} > 0.4, \quad \bar{K}_T = [0.3, 0.5], \quad \text{and} \quad \bar{X} = [0, 1.2].$$

and

$$\bar{\phi} = 1 - \bar{X} + \left(0.50 - 0.67\bar{K}_T + \frac{0.25 r_d}{r_T}\right)\bar{X}^2 \tag{8.35}$$

$$\text{for} \quad \bar{\phi} > 0.4, \quad \bar{K}_T = [0.5, 0.75], \quad \text{and} \quad \bar{X} = [0, 1.2].$$

The $\bar{\phi}$ expression for tracking collectors (CR > 10) is

$$\bar{\phi} = 1.0 - (0.049 + 1.44\bar{K}_T)\bar{X} + 0.341\bar{K}_T\bar{X}^2 \tag{8.36}$$

$$\text{for} \quad \bar{\phi} > 0.4, \quad \bar{K}_T = [0, 0.75], \quad \text{and} \quad \bar{X} = [0, 1.2].$$

TABLE 8.8

Parameters r_T and r_d Used to Calculate Monthly Solar Flux Incident on Various Collector Types[a]

Collector Type	r_T[b,c,d]	r_d[e]
Fixed aperture concentrators which do not view the foreground	$\left[\frac{\cos(L-\beta)}{(d\cos L)}\right]\left\{\begin{matrix}-ah_{coll}\cos h_{sr}(i=90)+[a-b\cos h_{sr}(i=90)]\sin h_{coll}\\ +\left(\frac{b}{2}\right)\sin h_{coll}\cos h_{coll}+h_{coll}\end{matrix}\right\}$	$\left(\frac{\sin h_{coll}}{d}\right)\{[\cos(L+\beta)/\cos L)-[1/CR)])+\left(\frac{h_{coll}}{d}\right)$ $\{[\cos h_{sr}(\alpha=0)/CR)]-[\cos(L-\beta)/\cos h_{st}(i=90)]\}$
East–west axis tracking[f]	$\frac{1}{d}\int_0^{h_{cell}}\left\{\left[\frac{(a+b\cos x)}{\cos L}\right]\times\sqrt{\cos^2 x+\tan^2\delta_s}\right\}dx$	$\frac{1}{d}\int_0^{h_{cell}}\left\{\left(\frac{1}{\cos L}\right)\sqrt{\cos^2 x+\tan^2\delta_s}-\left[\frac{1}{(CR)}\right][\cos x-\cos h_{sr}(\alpha=0)]\right\}dx$
Polar tracking	$(ah_{coll}+b\sin h_{coll})/(d\cos L)$	$(h_{coll}/d)\left\{\left(\frac{1}{\cos L}\right)+[\cos h_{sr}(\alpha=0)/(CR)]\right)-\sin h_{coll}/[d(CR)$
Two-axis tracking	$(ah_{coll}+b\sin h_{coll})/(d\cos\delta_s\cos L)$	$(h_{coll}/d)\{(1/\cos d_s\cos L)+[\cos h_{sr}(a=0)/(CR)]\}-\sin h_{coll}/[d(CR)$

[a] The collection hour angle value h_{coll} not used as the argument of trigonometric functions is expressed in radians. Note that the total interval $2h_{coll}$ is assumed to be centered about solar noon.

[b] $a = 0.409 + 0.5016\sin[h_{sr}(\alpha = 0) - 60°]$.

[c] $b = 0.6609 - 0.4767\sin[h_{sr}(\alpha = 0) - 60°]$.

[d] $d = \sin h_{st}(\alpha = 0) - h_{sr}(\alpha = 0)\cos h_{sr}(\alpha = 0)$.

[e] CR is the collector concentration ratio.

[f] Elliptic integral tables to evaluate terms of the form of $\int_0^h \sqrt{\cos^2 x+\tan^2\delta_s}\,dx$ contained in r_T and r_d can be found in standard handbooks. Use the identity $\cos\delta_s = \sin(90° - \delta_s)$ and multiply the integral by $\cos\delta_s$, a constant. For computer implementation a numerical method can be used. For hand calculations, use Weddle's rule or Cote's formula.

Also,

$$\bar{\phi} = 1.0 - \bar{X} \quad (8.37)$$

for $\bar{\phi} > 0.4$, $\overline{K_T} > 0.75$ (very sunny climate), and $\bar{X} = [0, 1.0]$ for any collector type.

Equations 8.34 through 8.37 were developed using curve-fitting techniques emphasizing large $\bar{\phi}$ values since this is the region of interest for most practical designs. Hence, they should be considered accurate to ±5% only for $\bar{\phi} > 0.4$.

In order to use this method to find monthly average daily solar radiation on a tracking solar concentrator, the tilt factor must be continuously changed due to tracking. However, Collares-Pereira and Rabl [28] have given an expression for various modes of tracking that can be used to calculate the average insolation $\bar{H}_c$ given below:

$$\bar{H}_c = \bar{r}_T \bar{H}_h - \bar{r}_d \bar{D}_h \quad (8.38)$$

Equations for $\bar{r}_T$ and $\bar{r}_d$ are given in Table 8.8.

An example calculation to illustrate the use of the long-term method will be worked in stepwise fashion. The sequential steps are as follows:

1. Evaluate $\bar{K}_T$ from terrestrial $\bar{H}_h$ data and extraterrestrial $\bar{H}_{o,h}$ data.
2. Calculate r_d/r_T for the concentration ratio and tracking mode for the collector.
3. Calculate the critical intensity ratio $\bar{X}$ from Equation 8.26 using a long-term optical efficiency value $\bar{\eta}_o$ and monthly average collector-plane insolation:

$$\bar{I}_c = \left(r_T - \frac{r_d \bar{D}_h}{\bar{H}_h} \right) \bar{H}_h. \quad (8.39)$$

The collection time Δt_c may need to be determined in some cases for non-tracking, low-concentration collectors by an iterative method, as described in the next section.

Example 8.3

Find the energy delivery of a polar-mounted, parabolic trough collector operated for 8 h per day ($\Delta t_c = 8$) during March in Albuquerque, New Mexico ($L = 35°$N). The collector has an optical efficiency $\bar{\eta}_o$ of 60%, a heat loss coefficient $U_c = 0.5\,W/m^2\,°C$, CR = 20, and heat removal factor $F_R = 0.95$. The collector is to be operated at 150°C. The ambient temperature is 10°C.

Solution

Following the three-step procedure above, the clearness index is calculated:

$$\bar{H}_{o,h} = 8.15 \frac{\text{kWh}}{\text{m}^2 \cdot \text{day}}$$

$$\bar{H}_h = 19.31 \frac{\text{MJ}}{\text{m}^2 \cdot \text{day}} = 5.36 \frac{\text{kWh}}{\text{m}^2 \cdot \text{day}}$$

$$K_T = \frac{\bar{H}_h}{\bar{H}_{o,h}} = \frac{5.36}{8.15} = 0.66.$$

The geometric factors r_d and r_T are calculated from expressions in Table 8.8:

$$r_T = \frac{(ah_{coll} + b \sin h_{coll})}{d \cos L}$$

$$r_d = \left(\frac{h_{coll}}{d}\right) \times \left(\frac{1}{\cos L} + \frac{\cos h_{sr}(\alpha = 0)}{\text{CR}}\right) - \left(\frac{\sin h_{coll}}{d \times \text{CR}}\right),$$

where

$$h_{coll} = \frac{\Delta t_c}{2} \times 15° = 60° = 1.047 \text{ rad},$$

If the collection period is centered about solar noon,

$$h_{sr}(\alpha = 0) = 90° = 1.571 \text{ rad},$$

$$a = 0.409 + 0.5016 \sin 30° = 0.66,$$

$$b = 0.6609 - .4767 \sin 30° = 0.42,$$

$$d = \sin 90° - 1.571 \cos 90° = 1.0,$$

in which case

$$r_T = \frac{(0.66 \times 1.047 + 0.42 \times \sin 60°)}{1.0 \cos 35°} = 1.29,$$

$$r_d = \left(\frac{1.047}{1.0}\right) \times \left(\frac{1}{\cos 35°} + \frac{\cos 90°}{20}\right) - \left(\frac{\sin 60°}{1.0 \times 20}\right) = 1.23$$

Finally, the critical intensity ratio is

$$\bar{X} = \frac{U_c \overline{\Delta T^+} \Delta t_c}{\bar{\eta}_o \bar{I}_c},$$

and the collector plane insolation $\bar{I}_c$ from Equation 8.39 is

$$\bar{I}_c = (1.29 - 1.23 \times 0.27) \times 5.36 = 5.13 \frac{\text{kWh}}{\text{m}^2},$$

so

$$\bar{X} = \frac{0.5 \times (150 - 10) \times 8}{0.6 \times 5130} = 0.182$$

The utilizability $\bar{\phi}$ from Equation 8.36 is

$$\bar{\phi} = 1.0 - (0.049 + 1.44 \times 0.66)(0.182) + 0.341 \times 0.66 \times 0.182^2 = 0.83.$$

Finally, the useful energy is

$$Q_u = F_R \bar{\eta}_o \bar{I}_c \bar{\phi} = 0.95 \times 0.6 \times 5.13 \times 0.83 = 2.43 \frac{\text{kWh}}{\text{m}^2 \cdot \text{day}}$$

for the month of March on the average.

Example 8.4

Calculations in Example 8.3 were based on $\Delta t_c = 8\,\text{h}$. Repeat for 10 h to see the effect of collection time if a symmetric collection period about noon is used.

Solution

The values of r_T and r_d for $h_{coll} = 75° = 1.31$ rad are

$$r_T = \frac{(0.66 \times 1.31 + 0.42 \times \sin 75°)}{1.0 \cos 35°} = 1.55$$

$$r_d = \left(\frac{1.31}{1.0}\right) \times \left(\frac{1}{\cos 35°} + \frac{\cos 90°}{20}\right) - \left(\frac{\sin 75°}{1.0 \times 20}\right) = 1.55$$

The collector plane insolation is then

$$\bar{I}_c = (1.55 - 1.55 \times 0.27) \times 5.36 = 6.06 \frac{\text{kWh}}{\text{m}^2},$$

and

$$\bar{X} = \frac{0.5 \times (150 - 10) \times 10}{0.6 \times 6065} = 0.192$$

Then $\bar{\phi}$ is 0.82 from Equation 8.36 and the useful energy delivery is 2.83 kWh/m² day. Hence, it is worthwhile operating the collector for at least 10 h. The calculation can be repeated by the reader for an asymmetric case 4 h before noon and 6 h after to determine the effect of warm up.

8.4.5 Yearly Collector Performance

Procedures for determining the daily and monthly thermal collector output have been discussed in Section 8.4.4. However, in many cases, it is the yearly useful energy delivered by a thermal collector that is of interest for economic assessment and evaluation of sustainability.

The yearly total useful energy (Q_{CY}) delivered by a solar collector can be determined if the inlet temperature to the collector is approximately constant during the year. Using results of hour by hour simulations for 26 stations in the United States combined with typical mean year radiation data [29], Rabl [30] proposed empirical correlations to determine Q_{CY} for different types of solar collectors. Rabl found that, for a given collector type (flat plate, CPC one-axis tracking, two-axis tracking), it is only necessary to know the latitude L and the critical radiation level, I_c, the average clearness index, K to estimate the yearly collectible solar energy with relative ease. The critical radiation level (or minimum threshold radiation) can be calculated from the HWB equation by setting $q_C = 0$. This yields

$$I_C = \frac{U_L(T_{ci} - T_{amb})}{\eta_o} \tag{8.40}$$

where

T_{ci} is the inlet temperature to the collector
T_{amb} is the yearly average ambient temperature
U_L is the collector/receiver heat loss coefficient
η_o is the optical efficiency

For equator-facing flat-plate collectors with tilt equal to latitude, the average beam radiation at normal incidence, I_{bn}, is related to the average annual clearness index, $\bar{K}$, by the empirical correlation

$$I_{bn} = 137\bar{K} - 0.34, \tag{8.41}$$

where I_{bn} is in kW/m². For this type of flat plate collector, Q_{CY} is given by the following empirical relation

$$Q_{CY} = A_C F_R \eta_n \times \begin{pmatrix} \left[5.215 + 6.973\tilde{I}_{bn}) + (-5.412 + 4.293\tilde{I}_{bn})L + (1.403 - 0.899\tilde{I}_{bn})L^2\right] \\ +\left[(-18.596 - 5.931\tilde{I}_{bn}) + (15.468 + 18.845\tilde{I}_{bn})L + (-0.164 - 35.510\tilde{I}_{bn})L^2\right]I_C \\ +\left[(14.601 - 3.570\tilde{I}_{bn}) + (-13.675 - 15.549\tilde{I}_{bn})L + (1.620 + 30.564\tilde{I}_{bn})L^2\right]I_C^2 \end{pmatrix}, \tag{8.42}$$

where Q_{CY} is in GJ/m² y, I_c and I_{bn} in kW/m², and the latitude, L, in radians.

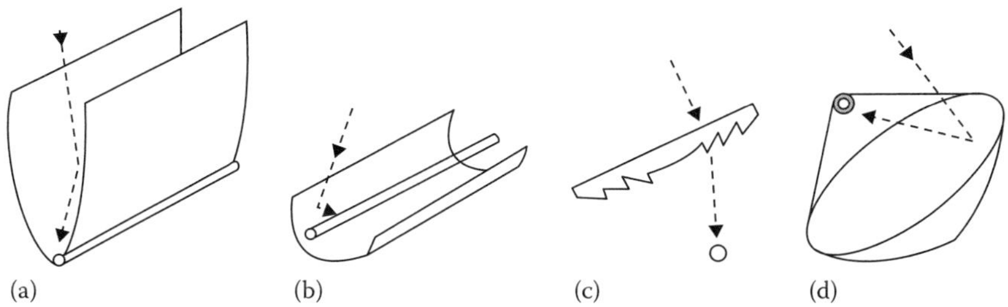

FIGURE 8.20
Commonly used concentrating collector designs: (a) compound parabolic collectors (C = 11–12); (b) parabolic trough (C = 2–40); (c) Fresnel lens (C = 50–500); and (d) parabaloids (C = 100–10,000).

Similar correlations for other collector types shown in Figure 8.20 are presented in Table 8.9.

It should be noted that, for a given location, Equation 8.42 reduces to a simple quadratic equation of the form

$$Q_{CY} = A_C F_R \eta_n(\tilde{a} + \tilde{b} I_C + \tilde{c} I_C) \tag{8.43}$$

Example 8.5

Obtain a quadratic expression for the yearly total energy delivered by a solar collector located in Athens, Greece, (L = 38° which is equal to 0.66 rad) with an average clearness index, $\bar{K} = 0.626$.

Solution
From the data given, we obtain

$$\tilde{I}_{bn} = 1.37 \times 0.626 - 0.34 = 0.518 \text{ kW/m}^2.$$

Using the above values and L and I_{bn}, Equation 8.43 reduces to

$$\frac{Q_{CY}}{A_C F_R \eta_n} = 7.122 - 13.095 I_C + 6.017 I_C^2$$

where
Q_{CY} is in GJ/(m^2 y)
I_C is in kW/m^2

Similar quadratic relations can be obtained for other collectors at a given location from the equations in Table 8.9. It should be noted that this procedure is empirical in nature and its accuracy is limited to approximately ±10%. If higher accuracy is required, the reader is referred to [31] or to obtain daily or monthly average values for the collectible energy.

TABLE 8.9

Correlations of the Yearly Collectible Energy for Different Collector Types

Collector Type	$\frac{Q_{CY}}{A_a F_R \eta_n}$
CPC (geometric concentration ratio $C = 1.5$, acceptance half angle 35°, tilt equal to latitude, due equator-facing)	$[(1.738 + 11.758\tilde{I}_{bn}) + (1.990 - 8.875\tilde{I}_{bn})L + (-3.236 + 7.617\tilde{I}_{bn})L^2]$ $+[(-13.240 - 14.688\tilde{I}_{bn}) + (3.979 + 43.653\tilde{I}_{bn})L + (7.345 - 52.556\tilde{I}_{bn})L^2]I_c$ $+[(14.015 - 1.437\tilde{I}_{bn}) + (-11.884 - 25.852\tilde{I}_{bn})L + (-0.0079 + 39.538\tilde{I}_{bn})L^2]$
Collector with aperture tracking about horizontal east–west axis	$[-0.098 + 11.944\tilde{I}_{bn} -] + [-0.599 - 30.363\tilde{I}_{bn} +]I_c + [1.093 + 17.606\tilde{I}_{bn} - 17.290\tilde{I}_{bn}^2]\tilde{I}_{bn}^2$
Collector with aperture tracking about horizontal north–south axis	$[(0.640 + 11.981\tilde{I}_{bn}) + (-2.365 + 7.979\tilde{I}_{bn})L + (2.380-)L^2]$ $+[(-6.021 - 19.086\tilde{I}_{bn}) + (-4.592 + 20.298\tilde{I}_{bn})L + (10.570 - 22.978\tilde{I}_{bn})L^2]I_c$ $+[(6.440 + 5.219\tilde{I}_{bn}) + (6.986 - 30.500\tilde{I}_{bn})L + (-14.095 + 40.089\tilde{I}_{bn})L^2]I_c^2$
Collector with aperture tracking about polar axis	$[-0.075 + 14.432\tilde{I}_{bn} - 0.592\tilde{I}_{bn}^2] + [-0.780 - 35.670\tilde{I}_{bn} +]I_c + [1.373 + 19.604\tilde{I}_{bn} - 23.965\tilde{I}_{bn}]I_c^2]$
Two-axis tracker	$[-0.147 + 16.084\tilde{I}_{bn} - 0.792\tilde{I}_{bn}^2 + [-0.886 - 37.659\tilde{I}_{bn} + 26.983\tilde{I}_{bn}^2]I_c$ $+[1.700 + 18.883\tilde{I}_{bn} - 24.887\tilde{I}_{bn}^2 I_c^2$

Source: Gordon, J.M. and Rabl, A., *Appl. Opt.*, 31, 7332, 1992.

Notes: Q_{CY} = Yearly collectible energy (GJ/(m² y)).

A_a = Collector net aperture area (m²).

F_R = Collector heat removal factor.

η_n = Collector optical efficiency at normal solar incidence.

I_{bn} = Annual average beam radiation at normal solar incidence (kW/m²).

L = Latitude (rad).

I_c = Critical radiation level defined by Equation 8.39 (kW/m²).

The corresponding correlation of the yearly collectible energy for flat plate collectors in given by Equation 8.43.

8.4.6 TRNSYS—Computer Simulation Program

TRNSYS (transient system simulation) is a sequential-modular transient simulation program developed at the Solar Energy Laboratory of the University of Wisconsin [32]. It is a widely used, detailed, design tool involving hourly simulations of a solar energy system over an entire year. TRNSYS computer program contains FORTRAN subroutines of almost all the components that are necessary to build a solar energy system. The component models, which are either empirical or analytical, describe the component performance with algebraic and/or differential equations. A system simulation model is created by interconnecting the models of individual components. The resulting set of simultaneous algebraic and/or differential equations is solved by TRNSYS. Table 8.10 gives a list of standard TRNSYS component subroutines available in the TRNSYS library. This public domain software is constantly being upgraded and is backed by technical support.

8.5 Solar Thermal Power

Electricity is fast becoming the energy form of choice all over the world even, unfortunately, for space and water-heating applications. There are two basic approaches to solar electric power generation. One is by photovoltaic process, a direct energy conversion. The other approach is to convert sunlight to heat and then heat to mechanical energy by a thermodynamic power cycle and, finally, convert the mechanical energy to electricity. This indirect approach, called solar thermal power, is based on well-established principles of thermal power. A vast majority of electricity in the world is produced by thermal power conversion. Most of the thermal power production in the world is based on Rankine cycle and to a smaller extent Brayton cycle. Both of these are applicable to solar thermal power conversion, with Rankine cycle being the most popular. Stirling cycle has also shown great potential and solar thermal power systems based on this cycle are under development.

8.5.1 Rankine Cycle

Most existing thermal power plants are based on the Rankine cycle. The basic ideal Rankine cycle is shown in Figure 8.21, which also shows a

TABLE 8.10

List of the Standard TRNSYS Component Subroutines

- Utility components
  - Data reader
  - Time-dependent forcing function
  - Algebraic operator
  - Solar radiation processor
  - Quantity integrator
  - Collector array shading
  - Psychometrics
  - Load profile sequencer
  - Convergence promoter
  - Weather data generator
- Solar collectors
  - Linear efficiency data
  - Detailed performance map
  - Single/biaxial incidence angle modifier
  - Theoretical flat plate
  - Theoretical CPC
- Thermal storage
  - Stratified liquid storage
  - Rock bed thermal storage
  - Algebraic tank (plug-flow)
  - Variable volume tank
- Controllers
  - On/off differential controller
  - Four-stage room thermostat
  - Microprocessor controller
- Auxiliary heating and cooling
  - On/off auxiliary heater
  - Absorption air-conditioner
  - Dual-source heat pump
  - Cooling coil
  - Conditioning equipment
  - Part load performance
  - Cooling tower
  - Parallel chillers
- Heat exchangers
  - Heat exchanger
  - Waste heat recovery
- Building loads and structures
  - Energy/degree-hour space heating/cooling load
  - Pitched roof and attic
  - Detailed zone (transfer function)
  - Overhand and wingwall shading
  - Window with variable insulation
  - Thermal storage wall
  - Attached sunspace
  - Multi-zone building and BID
- Combine subsystems
  - Liquid collector-storage subsystem
  - Air collector-storage subsystem
  - Domestic water heating subsystem
  - Thermosyphon collector-storage subsystem
- Outputs
  - Printer
  - Plotter
  - Histogram plotter
  - Simulation summary
  - Economic analysis
- User contributed components
  - Electric storage battery
  - Regulator/invertor
  - Combined PV subsystem
  - Flat-plate solar collector supplement
- Hydronics
  - Pump or fan
  - Flow diverter/tee piece/mixing valve
  - Pipe or duct

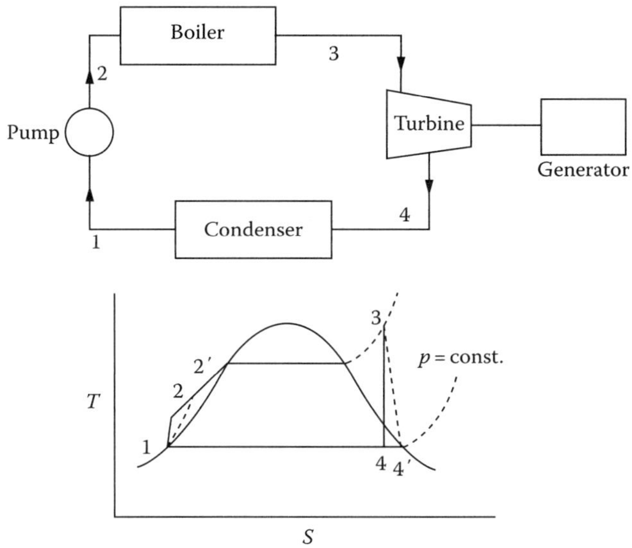

FIGURE 8.21
Basic Rankine power cycle.

temperature–entropy (*T*–*S*) diagram for steam as a working fluid. The ideal cycle consists of the following processes:

Process

1-2 Saturated liquid from the condenser at state 1 is pumped to the boiler at state 2 isentropically.

2-3 Liquid is heated in the boiler at constant pressure. The temperature of the liquid rises until it becomes a saturated liquid. Further addition of heat vaporizes the liquid at constant temperature until all of the liquid turns into saturated vapor. Any additional heat superheats the working fluid to state 3.

3-4 Steam expands isentropically through a turbine to state 4.

4-1 Steam exiting the turbine is condensed at constant pressure until it returns to state 1 as saturated liquid.

In an actual Rankine cycle, the pumping and the turbine expansion processes are not ideal. The actual processes are 1-2′ and 3-4′ respectively. For the above cycle

$$\text{Turbine efficiency } \eta_{\text{turbine}} = \frac{h_3 - h_{4'}}{h_3 - h_4} \tag{8.44}$$

$$\text{Pump efficiency } \eta_{\text{pump}} = \frac{h_1 - h_2}{h_1 - h_{2'}} \tag{8.45}$$

$$\text{Net work output } = (h_3 - h_{4'}) - (h_{2'} - h_1) \tag{8.46}$$

$$\text{Heat input} = h_3 - h_{2'} \tag{8.47}$$

$$\text{Pump work} = h_{2'} - h_1 = \frac{\upsilon(P_2 - P_1)}{\eta_{\text{pump}}} \tag{8.48}$$

$$\text{Cycle efficiency } = \frac{\text{Net work output}}{\text{Heat input}} = \frac{(h_3 - h_{4'}) - (h_{2'} - h_1)}{h_3 - h_{2'}} \tag{8.49}$$

where h represents enthalpy and υv is the specific volume at state 1.

Example 8.6

In a simple steam Rankine cycle, shown in Figure 8.21, steam exits the boiler at 7.0 MPa and 540°C. The condenser operates at 10 kPa and rejects heat to the atmosphere at 40°C. Find the Rankine cycle efficiency, η_R, and compare it to the Carnot cycle efficiency. Both pump and turbine operate at 85% efficiencies.

Solution

The Rankine cycle efficiency is obtained as follows:

$$\eta_R = \frac{(h_3 - h_{4'}) - (h_{2'} - h_1)}{h_3 - h_{2'}},$$

$$\eta_{\text{turbine}} = \frac{h_3 - h_{4'}}{h_3 - h_4},$$

and

$$\eta_{\text{pump}} = \frac{h_2 - h_1}{h_{2'} - h_1} = \frac{\upsilon_1(P_2 - P_1)}{h_{2'} - h_1}$$

The enthalpies at the state points are found as follows:

$$h_1 = h_f(10\text{ kPa}) = 191.8\text{ kJ/kg}.$$

Pump work from 1 to 2′

$$ {}_1W_{2'} = \frac{\upsilon_1(P_2 - P_1)}{\eta_{\text{pump}}} = h_{2'} - h_1, $$

$$ h_{2'} = h_1 + \frac{\upsilon_1(P_2 - P_1)}{\eta_{\text{pump}}} $$

$$ = 191.8\ \text{kJ/kg} + \frac{(0.00101\ \text{m}^3/\text{kg})(7000 - 10)\ \text{kPa}}{0.85} $$

$$ = 200.1\ \text{kJ/kg} $$

$$ h_3(540°\text{C}, 7\ \text{MPa}) = 3506.9\ \text{kJ/kg} $$

States 3 and 4 have the same entropy.

Therefore,

$$ s_3 = 6.9193\ \text{kJ/kg K} = s_4. $$

Saturated vapor entropy at state 4 (10 kPa)

$$ s_{4g} = 8.1502\ \text{kJ/kg K} $$

Since s_4 is less than s_{4g}, 4 is a wet state

$$ s_4 = s_f + Xs_{fg} \quad \text{or} \quad s_g - Ms_{fg}, $$

where X is the vapor quality and M is the moisture.

$$ M_4 = \frac{s_g - s_4}{s_{fg}} = \frac{8.1502 - 6.9193}{7.5009} = 0.1641 \text{ or } 16.41\%, $$

$$ h_4 = h_g - Mh_{fg} = 2584.7 - 0.1641(2392.8) = 2192\ \text{kJ/kg} $$

Therefore,

$$ h_{4'} = h_3 - \eta_{\text{turbine}}(h_3 - h_4) $$

$$ = 3507 - 0.85(3507 - 2192) = 2389\ \text{kJ/kg} $$

Actual moisture at the turbine exhaust is

$$h_{4'} = h_g - Mh_{fg},$$

$$M_{4'} = \frac{h_g - h_{4'}}{h_{fg}} = \frac{2584.7 - 2389}{2391}$$

$$= 0.0817 \text{ or } 8.17\%.$$

$$\text{Net work} = (h_3 - h_{4'}) - (h_2 - h_1) = 1109 \text{ kJ/kg}$$

$$\eta = \frac{\text{Net work}}{h_3 - h_{2'}} = \frac{1109}{3507 - 200} = 0.3354 \text{ or } 33.54\%.$$

$$\eta_{\text{Carnot}} = \frac{813\,\text{K} - 313\,\text{K}}{813\,\text{K}} = 0.615 \text{ or } 61.5\%$$

Several improvements can be made to the basic Rankine cycle in order to increase the cycle efficiency. For example, the pressure in the boiler can be increased. However, that will result in increased moisture in the steam exiting the turbine. In order to avoid this problem, the steam is expanded to an intermediate pressure and then reheated in the boiler. The reheated steam is expanded in the turbine until the exhaust pressure is reached.

Another improvement to the basic Rankine cycle is the regenerative cycle in which expanded steam is extracted at various points in the turbine and mixed with the condensed water to preheat it in the feedwater heaters. These modifications to improve the performance of the basic Rankine cycle are treated in standard thermodynamics texts such as [33] or *Mechanical Engineering Handbooks*.

8.5.2 Components of a Rankine Power Plant

Major components of a Rankine power plant include boiler, turbine, condenser, pumps (condensate pump, feedwater booster, boiler feed pump), and heat exchangers (open heaters and closed heaters). All of the components of a solar thermal power plant are the same as those in a conventional thermal power plant except the boiler. The boiler in a solar thermal power plant includes a solar collection system, a storage system, an auxiliary fuel heater and heat exchangers. Figure 8.22 shows a schematic representation of a solar boiler. The maximum temperature from the solar system depends on the type of solar collection system (parabolic trough collectors, central receiver with heliostat field, parabolic dishes, etc.). If the temperature of the fluid from the solar system/storage is less than the required temperature for the turbine, an auxiliary fuel is used to boost the temperature. A fossil fuel or a biomass fuel may be used as the auxiliary fuel.

A condenser is a large heat exchanger that condenses the exhaust vapor from the turbine. Steam turbines employ surface-type condensers, mainly shell and tube heat exchangers operating under vacuum. The vacuum

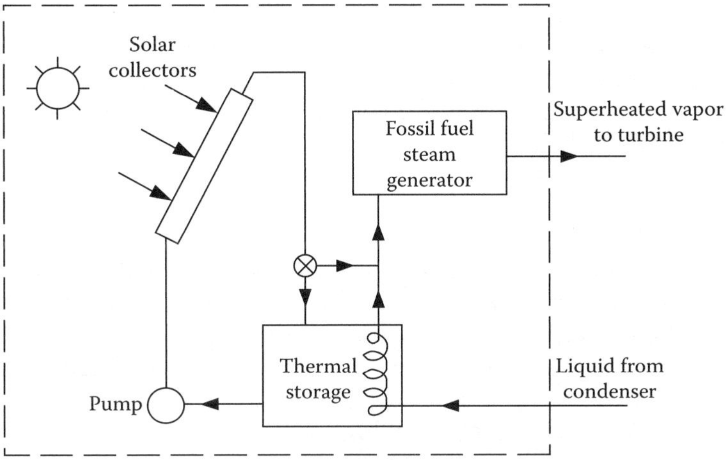

FIGURE 8.22
Schematic of a solar boiler.

in the condenser reduces the exhaust pressure at the turbine blade exit to maximize the work in the turbine. The cooling water from either a large body of water such as a river or a lake, or from cooling tower, circulates through the condenser tubes. The cooling water is cooled in the cooling tower by evaporation. The air flow in the cooling tower is either natural draft (hyperbolic towers) or forced draft (see Figure 8.23).

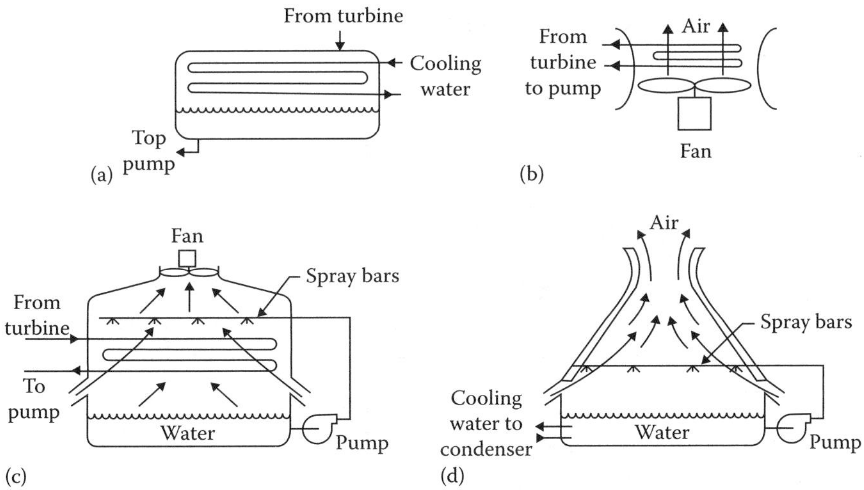

FIGURE 8.23
Types of condenser and/or heat rejection used in Rankine cycle solar power systems: (a) tube-and-shell condenser; (b) dry cooling tower; (c) wet cooling tower; and (d) natural-draft cooling tower. (Adapted from Vant-Hull, L.L., *Solar Today*, November/December issue, pp. 13–16, 1992. With permission.)

The condensate and feed water pumps are motor-driven centrifugal pumps, while the boiler feed pumps may be motor- or turbine-driven centrifugal pumps.

Steam turbines and generators are described in detail in a number of thermodynamics books [34,35] and is not discussed here.

8.5.3 Choice of Working Fluid

Working fluid in a solar Rankine cycle is chosen based on the temperature from the solar collection system. The working fluid must be such that it optimizes the cycle efficiency based on the expected temperature from the source. Steam is the most common working fluid in a Rankine cycle. Its critical temperature and pressure are 374°C and 22.1 MPa. Therefore, it can be used for systems operating at fairly high temperatures. Systems employing parabolic trough, parabolic dish, or central receiver collection systems can use steam as a working fluid. Other major advantages of steam are that it is nontoxic, environmentally safe, and inexpensive. Its major disadvantage is its low molecular weight, which requires very high turbine speeds in order to get high turbine efficiencies.

Steam is a wetting fluid, that is, as it is expanded in a turbine, once it reaches saturation, any further expansion increases the moisture content. In other words, steam becomes wetter as it expands, as shown in Figure 8.24a. On the other hand, a fluid that has a *T–S* (temperature–entropy) diagram similar to that shown in Figure 8.24b is called a drying fluid. As seen from this figure, even though the working fluid passes through the two-phase region, it may exit the turbine as superheated. Normally, the turbine

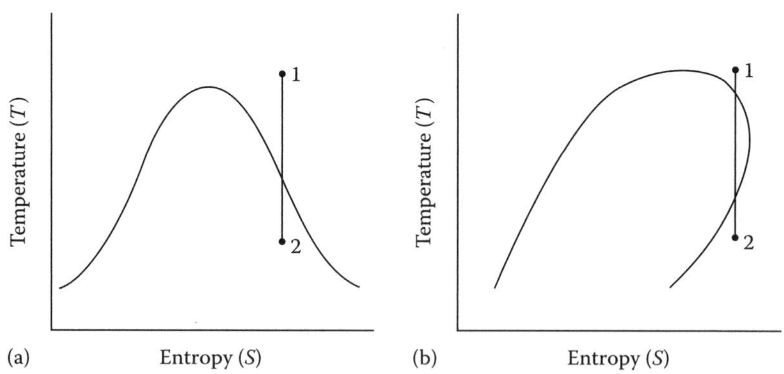

FIGURE 8.24
T–S characteristics of (a) wetting and (b) drying types of working fluids.

speed is so high that, in such a case, there is no condensation in the turbine. Examples of drying fluids include hydrocarbons (toluene, methanol, isobutane, pentane, hexane), chlorofluorocarbons (CFCs such as R-11, R-113), and ammonia. Since a drying fluid does not get wetter on expansion from a saturated vapor condition in an ideal or real process, it does not have to be superheated. Therefore, a Rankine cycle using a drying fluid may be more efficient than the cycle using a wetting fluid. In fact, a drying fluid may be heated above its critical point so that that upon expansion it may pass through the two-phase dome. Because of the *T–S* characteristics, the fluid may pass through the two-phase region and still exit from the turbine as superheated. These characteristics can be used to increase the resource effectiveness of a cycle by as much as 8% [36].

Figure 8.25 and Table 8.11 give some characteristics of candidate working fluids for Rankine cycles.

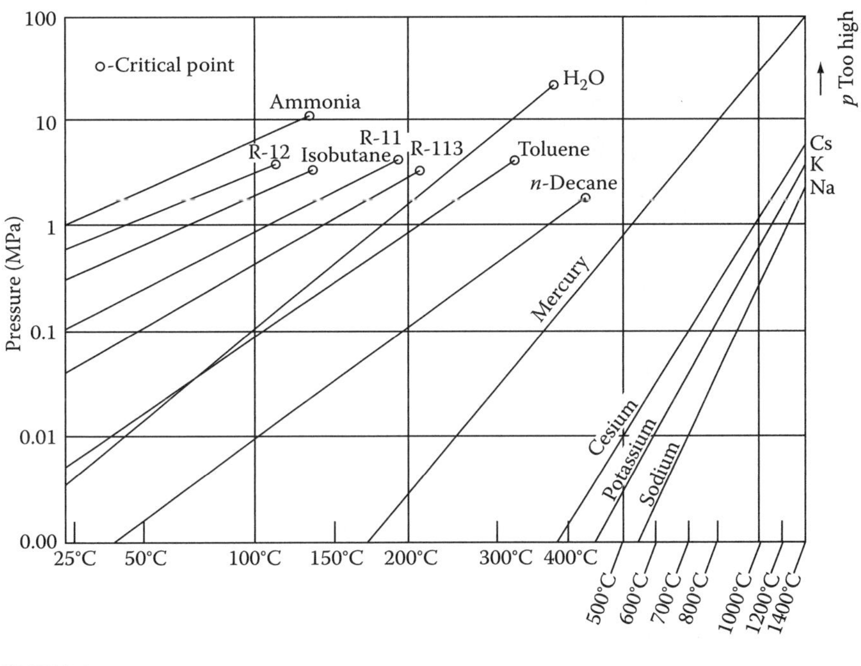

FIGURE 8.25
Saturation pressure–temperature relationships for potential Rankine cycle working fluids. (Adapted from Vant-Hull, L.L., *Solar Today*, November/December issue, pp. 13–16, 1992. With permission.)

TABLE 8.11

Physical and Thermodynamic Properties of Prime Candidate Rankine Cycle Working Fluids

Property	Water	Methanol	2-Methyl Pyridine, H_2O	Fluorinol 85	Toluene	R-113	Ammonia	Isobutane
Molecular weight	18	32	33	88	92	187	17	58
Boiling point (1 atm) (°C)	100	64	93	75	110	48	−330	−12
Liquid density (kg/m^3)	999.5	749.6	934	1370	856.9	2565	682	594
Specific volume (saturated vapor at boiling point) (m^3/kg)	1.69	0.80	0.87	0.31	0.34	0.14	1.124	0.35
Maximum stability temperature (°C)	—	175–230	370–400	290–330	400–425	175–230	300[a]	>200
Wetting-drying	W	W	W	W	D	D	D	D
Heat of vaporization at 1 atm (kJ/kg)	2256	1098	879	442	365	1370	1370	367
Isentropic enthalpy drop across turbine (kJ/kg)	348–1160	162–302	186–354	70–186	116–232	23–46	200–600	120–380

Source: Adapted from Kolb, G.J., *Sol. Energy*, 62(1), 51, 1998.

[a] Anhydrous ammonia in the presence of iron. Small trace of water increases this limit.

8.6 Examples of Solar Thermal Power Plants

There are several successful solar thermal power plants working on Rankine and Stirling cycles, using parabolic troughs, parabolic dishes, and central receiver towers. An example of each type is treated below.

8.6.1 Parabolic Trough–Based Power Plant

Luz Corporation developed components and commercialized parabolic trough collector-based solar thermal power by constructing a series of hybrid solar power plants from 1984 to 1991. Starting with their first 14 MW_e, solar electric generating station (SEGS I) in Southern California, Luz added a series of SEGS power plants with total generating capacity of 354 MW_e. Figure 8.26 shows a schematic of the SEGS VIII and IX plants. Table 8.12 provides information about the solar field, power block, and other general information about SEGS I and SEGS IX plants. All of these plants use solar energy to heat the working fluid and natural gas as the auxiliary fuel to achieve the desired temperature. On average, 75% of the energy is obtained from the sun and 25% from natural gas. With power plant electrical conversion efficiencies of the order of 40% and the solar field efficiencies of 40%–50%, overall efficiencies for solar-to-electricity conversion of order of 15% are being achieved

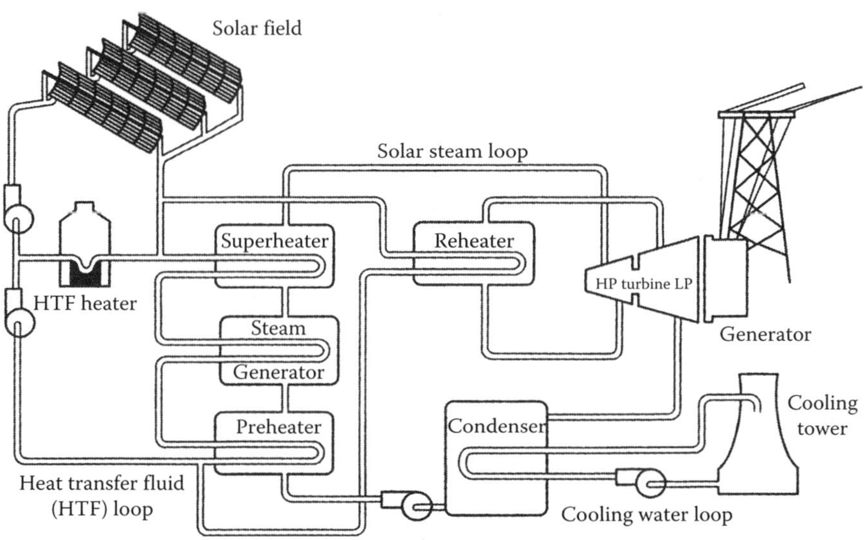

FIGURE 8.26
Flow of heat-transfer fluid through the SEGS VIII and IX plants. (Adapted from de Laquil, P. et al., *Solar Thermal Electric Technology in Renewable Energy*, Island Press, Washington, DC, 1993.)

TABLE 8.12

Important Characteristics of SEGS I to SEGS IX Plants

	Units	I	II	III	IV	V	VI	VII	VIII	IX
Power block										
Turbine-generator output	Gross MW_e	14.7	33	33	33	33	33	33	88	88
Output to utility	Net MW_e	13.8	30	30	30	30	30	30	80	80
Turbine-generator set										
Solar steam conditions										
Inlet pressure	Bar	35.3	27.2	43.5	43.5	43.5	100	100	100	100
Reheat pressure	Bar	0	0	0	0	0	17.2	17.2	17.2	17.2
Inlet temperature	°C	415	360	327	327	327	371	371	371	371
Reheat temperature	°C	N/A	N/A	N/A	N/A	N/A	371	371	371	371
Gas mode steam conditions[a]										
Inlet pressure	Bar	0	105	105	105	105	100	100	100	100
Reheat pressure	Bar	0	0	0	0	0	17.2	17.2	17.2	17.2
Inlet temperature	°C	0	510	510	510	510	510	510	371	371
Reheat temperature	°C	N/A	N/A	N/A	N/A	N/A	371	371	371	371
Electrical conversion efficiency										
Solar mode[b]	Percent	31.5	29.4	30.6	30.6	30.6	37.5	37.5	37.6	37.6
Gas mode[c]	Percent	0	37.3	37.3	37.3	37.3	39.5	39.5	37.6	37.6

Solar field										
Solar collector assemblies										
LS 1 (128 m²)										
LS 2 (235 m²)		560	536	0	0	0	0	0	0	0
LS 3 (545 m²)		48	518	980	980	992	800	400	0	0
Number of mirror segments		0	0	0	0	32	0	184	852	888
Field aperture area		41,600	96,464	117,600	117,600	126,208	96,000	89,216	190,848	198,912
Field inlet temperature	m²	82,960	190,338	230,300	230,300	250,560	188,000	194,280	464,340	483,960
Field outlet temperature	°C	240	231	248	248	248	293	293	293	293
Annual thermal efficiency	°C	307	321	349	349	349	390	390	390	390
Peak optical efficiency	Percent	35	43	43	43	43	42	43	53	50
System thermal losses	Percent of peak	71	71	73	73	73	76	76	80	80
		17	12	14	14	14	15	15	15	15
Heat transfer fluid type										
Inventory		Esso 500	VP-1	VP-1	VP-1	VP-1	VP-1	VP-1	VP-1	VP-1
Thermal storage capacity	m³	3,213	416	403	403	461	416	416	1,289	1,289
	MWh_t	110	0	0	0	0	0	0	0	0
General										
Annual power outlet	Net MWh/year	30,100	80,500	91,311	91,311	99,182	90,850	92,646	252,842	256,125
Annual gas power use	10^9 m³/year	4.76	9.46	9.63	9.63	10.53	8.1	8.1	24.8	25.2

[a] Gas superheating contributes 18% of turbine inlet energy.

[b] Generator gross electrical output divided by solar field thermal input.

[c] Generator gross electrical output divided by thermal input from gas-fired boiler or HTF-heater.

in these plants. It is claimed that the cost of electricity from these plants has decreased from about 30 ¢/KWh to less than 10 ¢/KWh.

Recently, several solar trough power plants similar to the SEGS installation have been built and placed into operation. A notable example is Nevada Solar One. The plant is not 100% renewable, but has, in accordance with Nevada state law, a maximum of 2% fossil fuel backup from a small natural gas heater, which is also used to prevent the water from freezing and provide power on cloudy days or during peak demand periods.

Nevada Solar One went online on June 27, 2007 on the southeast fringe of Boulder City, Nevada. The plant has a nominal capacity of 64 MW and a maximum capacity of 75 MW. The project required an investment of $266 million and electricity production is estimated to be 134 million kWh/year (www.accionna.com/About-Us/Our-Projects/U-S-/Nevada-Solar-One.aspx & www.techno9logynewsdaily.com/node/7150). The time of construction for the plant was about 16 months and it occupies approximately 400 acres (1.6 km^2), while the solar collectors cover about 1.2 km^2. Nevada Solar One uses 760 parabolic troughs that concentrate solar radiation onto evacuated tube collectors that heat the heat-transfer fluid to 391°C (730°F). The heat-transfer fluid is a molten salt, which passes through a heat exchanger to produce steam that drives a conventional 70 MW reheat steam turbine. (www.nevadasolarone.net/themes/zen/pdfs/NSO_FaceSheet_Inert.pdf & www.solartoday.org/2007/mar_apr07/nevada_solar_one.htm).

Example 8.7

The parabolic trough solar collector, whose efficiency was analyzed in Example 8.1, is to be used for electric power production. Assuming that the efficiency of the power generation unit equals one-half of the Carnot efficiency, determine

a. The temperature at which the solar trough power generation system achieves its highest efficiency.
b. The levelized cost of energy from this trough generation system assuming a life of 25 years, an effective discount rate of 3%, and assuming that the total installed system cost is $600 per m^2 of collector aperture area. This includes the cost of the installed solar collector field, as well as the power block and balance of system (turbine, cooling tower, power transformers/interconnection, etc.).

Solution

According to thermodynamic principles, the conversion of thermal energy into electric energy increases in efficiency as the temperature of the thermal source increases. But, Equation 8.20 shows that the efficiency of the collector decreases with increasing operating temperature. In order to obtain the optimum operating temperature, both of these relations need to be taken into account.

If we assume that the high-temperature source in the Carnot efficiency equals the operating temperature of the receiver/absorber, T_a, and the ambient temperature is 20°C, the electric power output, P_e, is equal to

$$P_e = I\eta_c \times 0.5\left(\frac{\Delta T}{T_\infty}\right) = I\left[0.79 - \frac{(0.16 - 0.0013\Delta T)\Delta T}{I}\right] \times 0.5\left(\frac{\Delta T_a}{T_\infty}\right).$$

P_e is shown in the spreadsheet below as a function of operating temperature. We observe that the peak annual electric generation occurs at about 300°C. The annual electricity generation at the optimum operating temperature, as shown in the last column, is approximately 247 (kWh/m²)/year.

To calculate the LCOE, we first assume that annual operation and maintenance costs are insignificant compared to the capital cost of the system. The levelized cost of energy delivered by the system over 25 years obtained from Equation 2.19 is

$$\text{LCOE} = \$0.14/\text{kWh}$$

An advanced version of a solar trough and a line focus solar thermal power system are shown in Figures 8.27 and 8.28.

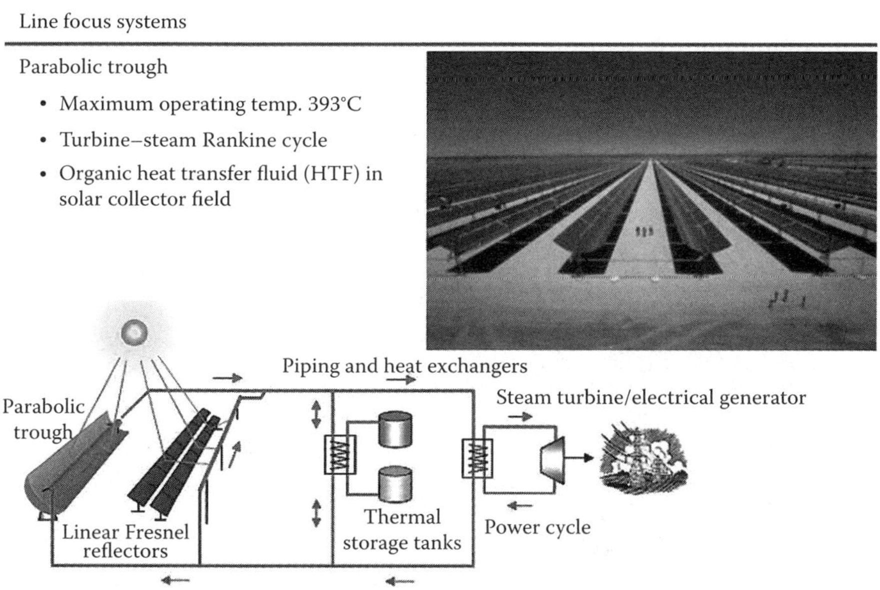

FIGURE 8.27
(See color insert.) Line focus solar thermal power system. (Courtesy of Clifford Ho, Sandia National Laboratories.)

FIGURE 8.28
(See color insert.) Sky Fuel solar sky trough near Golden, Colorado, viewed by professor Frank Kreith (right) and National Institute of Standards and Technology researcher, Dr. Isaac Garaway (left). (Courtesy of CU Engineering, 2009, Casey A. Cass Photographer.)

8.6.2 Central Receiver Systems

In a power tower or central receiver system (CRS), the solar receiver is mounted on top of a tower and solar radiation is concentrated on its surface by means of a large paraboloid consisting of a field of individual collector/reflector modules called heliostats, as shown in Figure 8.29. CRS has a great potential for future electricity production because it can achieve high temperatures in a working fluid, which leads to high thermodynamic efficiencies of power generation. Figure 8.30 is a schematic diagram of a central

FIGURE 8.29
Photograph of Solar One central receiver pilot plant in Barstow, California. (From Goswami, Y. et al., *Principles of Solar Engineering*, 2nd edn., Taylor & Francis, Philadelphia, 2000. With permission.)

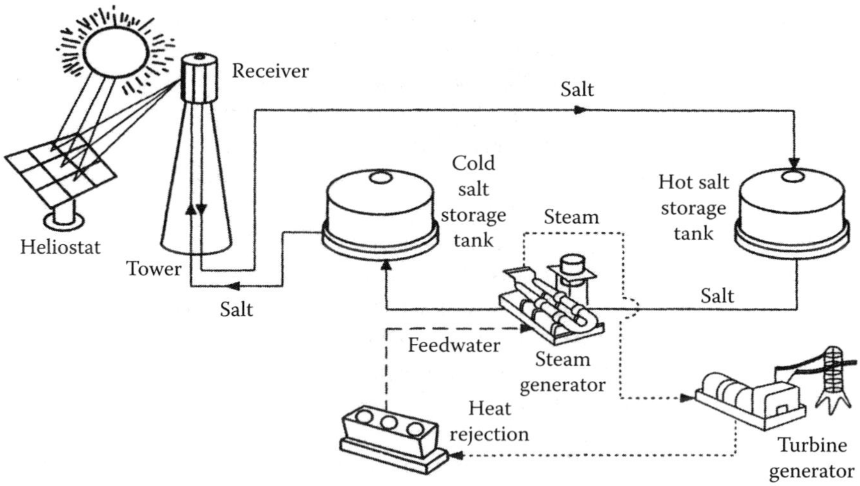

FIGURE 8.30
Schematic of Solar Two central-receiver plant configuration. (Adapted from Hebrank, W.H., *Am. Dyestuff Rep.*, 63, 34, 1975.)

receiver system. Heliostats direct the sunlight to the receiver on top of the tower. A heat-transfer fluid is pumped through a riser to the receiver where it is heated and then via a downcomer to the thermal storage unit and/or a heat engine, which is usually located on the ground near the tower. Typical optical concentration factors range from 200 to 1000 and plant sizes from 10 to 200 MW_e are typical for future systems. The heat flux impinging on the receiver can be between 300 and 1000 kW/m^2 and allows the working fluid to achieve temperatures up to 1000°C. CRS can achieve high annual capacity factors, either by using thermal storage or by operating in a hybrid mode with a fossil fuel supplement [38]. With storage, CRS solar plants have been shown to operate more than 4500 h per year at nominal power [39].

Three promising power tower technologies are under consideration.

1. Molten salt technology
2. Volumetric air technology
3. Saturated steam technology

The technology favored by industry in the United States is based on molten salt technology that utilizes large thermal storage capacity with molten nitrate salts as the working fluid. The decision to use molten salt was based on a three-year long utility study designed to compare efficiencies and costs of various options [40]. The recommendations are based on experience with a 20 MW central receiver power plant, called Solar Two, which operated in California in 1999. Solar Two demonstrated advanced molten salt technology at a scale sufficient to allow commercialization of the technology. The plant operated successfully with solar energy collected efficiently over a broad range of operating conditions and showed that low-cost thermal energy storage systems can operate reliably and efficiently. The unique thermal storage capability allowed solar energy to be collected and stored under high insolation conditions so that electric power could also be generated at night or whenever required by the utility, even when the sun was not shining. Figure 8.30 is a schematic diagram of the molten salt CRS system, with cold and hot storage tanks, and Table 8.13 shows the characteristics of four sizes of molten salt CRS. Solar 100, shown in the fourth column, is based on a conceptual design of a 100 MW_e commercialized central receiver system.

Three types of receiver configurations have been considered for a molten salt power plant. The cavity receiver consists of an insulated enclosure within which a heat absorbing structure is located. Radiation enters through the aperture, is subjected to multiple reflection, and thus enhances the effective absorptivity of the cavity. Radiation losses can occur only through the aperture and the absorbing area of the receiver can be enlarged without substantially increasing the radiation losses. However, convective losses increase as a result of a larger heated area and cavity receivers tend to be large and heavy. Moreover, the aperture size restricts the field of view and spillage can be considerable. In a plain or flat-plate receiver, the aperture of a cavity is replaced by a flat heat exchanging surface. The backside can be insulated, but overall losses are appreciable.

TABLE 8.13

Characteristics of Four Sizes of Molten Salt Power Towers

Parameter	Solar Two (Mature)	Solar Tres	Solar 50/Solar Cuatro	Solar 100
Plant rating	10	15	50	100
Location	Barstow, California	Córdoba, Spain	Southern Spain	Southwestern United States
Annual solar insolation, kWh/m^2	2,700	2,067	2,067	2,700
Capacity factor, %	20	65	69	70
Field area, m^2	8,1400	263,000	971,000	1,466,000
Receiver thermal rating, MW	42	120	466	796
Thermal storage size, MWh	110	610	1,850	3,820
Steam generator rating, MW	35	37	130	254
Annual net energy production, MWh	16,600	75,500	302,000	613,000
Peak net efficiency	0.13	0.19	0.22	0.22
Annual net efficiency	0.08	0.14	0.15	0.16
Levelized energy cost, $/kWh	N/A	0.16	0.12	0.08

Source: Romero, M. et al., *J. Sol. Energy Eng.*, 124, 98, May 2002.

The external receiver consists of a cylinder with a tubular structure containing a working fluid passing through the pipes on the outside of a cylinder. It can receive solar radiation from all directions and is relatively simple and lightweight because no enclosure is required. The design can accommodate heat flux densities as high as 850 kW/m² with molten salt and the tower needs to be only about 70% as tall as the tower for a cavity receiver, which reduces the cost appreciably. The disadvantage of an external receiver is that the reflected fraction of incident radiation is lost and the temperature of the molten salt heat-transfer fluid must be maintained above a certain limit to avoid freezing. This is accomplished, as shown in Figure 8.30, by having hot and cold storage tanks.

The sizing procedure of the receiver from optical and geometric consideration is presented in [42]. However, the details of the optical design for a central receiver system are beyond the scope of this book. But, from the engineering perspective of sustainable energy design, it is possible to estimate the performance of a typical central receiver system thermodynamically, including the size of the receiver, the number of heliostats required, the height of the tower, and the ground area necessary for a given power output as well as the EROI for the system. In any high-temperature concentrating system, the receiver is the focal point of the design. Because radiation and convection losses are proportional to the area of the receiver, it should be as small as possible which requires high concentration. But the receiver must also be large enough to intercept and absorb a large fraction of the solar radiation directed toward it by the heliostats and transfer the heat from the sun to a working fluid. If the designer specifies a more perfect optical system to reduce the receiver size, this will increase the cost of the heliostats and may complicate the thermal design.

The efficiency of the receiver η_{th} equals [1 – (thermal loss/solar input)] and it is given by the following equation:

$$\eta_{th} = 1 - \frac{A_{receiver}\left[\varepsilon\sigma(T^4 - T_a^4) + c(T - T_a)^{1.25}\right]}{G_b A_{collector}\eta_{col}\eta_{int}\alpha} \tag{8.50}$$

or

$$\eta_{th} = 1 - \frac{\sigma(T^4 - T_a^4) + (c/\varepsilon)(T - T_a)^{1.25}}{G_b C\eta_{col}\eta_{int}} \tag{8.51}$$

where

$C = \dfrac{\alpha A_{collector}}{\varepsilon A_{receiver}} = C_g \dfrac{\alpha}{\varepsilon}$

T_a is the ambient temperature (Kelvin)

ε is the emissivity of the receiver at the temperature T

σ is the Stefan–Boltzmann constant, 5.67×10^{-8} W/m² K⁴

c is the temperature-independent convective loss coefficient, typically between 1 and 25 W/m² $K^{1.25}$

Also, α is the absorptivity for direct beam solar radiation (G_b) and η_{col} is the collector efficiency, which includes reflectivity (typically 0.85–0.95), cosine of angle of incidence effects (typically 0.7–0.9), and any factor to account for shading of mirror surfaces or blocking of reflected sunlight (typically 0.9–1.0). It is conventional to use the actual mirrored area of the heliostats for $A_{collector}$. The fraction of reflected light that is intercepted by the receiver is η_{int}, representing scattering and spillage losses. We see that the geometric concentration, C, plays an important role in suppressing the effects of thermal losses.

Based on the experience gained from the operation of Solar Two, Vant-Hull [43] proposed the following procedure for estimating the size and performance of a utility-scale 100 MW central receiver system. This approach circumvents the need for using advanced optical analysis in the calculations, but relies on some empirical constants based on the Solar Two system design parameters and performance. He proposed a nominal allowance for parasitic power of 10% including the cold salt water pumps (6.2 MW_e), hot salt pumps (0.7 MW_e), feedwater pumps (2.2 MW_e), heliostat field operation (0.8 MW_e), and some miscellaneous load estimated at 3.3 MW_e. The analysis also assumed that the ratio of thermal energy output from the receiver to the turbine inlet power rating should be 1.8 to allow filling the thermal storage when generating rated power on a design day. With a reheat turbine efficiency of 43.2%, the receiver design point thermal output power (DPTOP) necessary to power a 100 MW system is given by

$$\begin{aligned} \text{DPTOP} &= \text{turbine output} + \text{parasitic energy} \times 1.8/\text{turbine efficiency} \\ &= (100 + 10) \times 1.8/0.32 = 468\ \text{MW}_t. \end{aligned} \tag{8.52}$$

Some additional design parameters for this system are listed in Table 8.14.

Receiver thermal losses depend on the receiver area and the local surface temperature. A first estimate of the receiver area can be obtained from the allowable flux-density limit. Due to end effects, and the reduced allowable flux at the high-temperature outlet (on the south side of the receiver system), the average allowable flux was reduced from the specified 825 kW/m^2 to about 440 kW/m^2, resulting in a required receiver area, A_e, of A_e = 468 MW/0.44 MW/m^2 = 1.064 m^2.

For the first commercial design, an external receiver of cylindrical shape with an aspect ratio of 1.3 was selected to allow the highly peaked flux distribution to be leveled by selection of a vertical distribution of aim points, resulting in a receiver radius of 16.1 m. To provide the required 1000°F steam temperature at the turbine inlet, the salt storage temperature was set at 1050°F, which must also be the receiver outlet temperature. The working fluid is draw salt that begins to freeze at 430°F. To prevent freezing, a feedwater preheat outlet temperature of 455°F and a cold salt storage tank temperature of 500°F were specified. This defined the receiver inlet temperature at 500°F. Allowing for a flux-dependent 100–200 K temperature rise through the

TABLE 8.14

Solar Two and First Commercial Plant from Utility Study

	Solar Two	First Commercial Utility Study
Design point power (electric)	10 MW	100 MW
Scheduled acceptance test	June 1996	~2,000
Location	15 km E of Barstow, California, 35°N, 117°W	Arizona, Nevada, or Southern California
Site elevation	593 m	400–800 m
Heliostat field	Glass/metal	Glass\metal, or membrane
Configuration	Radial stagger	Radial stager
Reflective area (supplement to Solar One field)	71,000 m^2 at 39 m^2 + 10,000 m^2 at 95 m^2	884,000 m^2 at 150 m^2
Constraint on field (system trade)	South field favored to reduce peak flux on north side of receiver.	
Ground coverage	27% avg., 40%–18%	25% avg., 40%–16%
Collector area (land)	0.3 km^2, ~75 ac	3.54 km^2, ~875 ac
Receiver	External, salt in tube	External, salt in tube
Focal height above heliostats	73.04 m	185 m
Diameter	5.05 m	16.1 m
Length	6.21 m	21.0 m
Heat absorbing area	98.6 m^2	1,068 m^2
Design point power (absorbed thermal)	42.2 MW	468 MW
Absorbed flux density, maximum	850[a] kW/m^2	820[a] kW/m^2
Average	430[a] kW/m^2	440[a] kW/m^2
Thermal storage unit	Hot and cold tank	Hot and cold tank
Medium	Draw salt	Draw salt
Rating (thermal)	3 h at 35 MW	6 h at 260 MW
Turbogenerator rating (nominal, thermal)	35 MW	260 MW
Turbogenerator rating (nominal, electric)	12.5 MW	110 MW
Solar multiple (receiver/turbine)	1.2	1.8
Heat engine efficiency	35% rated	42.3% rated
Power generation efficiency[b]	5.8% annual avg.	14.6% annual avg.

Source: Arizona public service utility solar central receiver study, Report No. DOE/AL/38741-1, November 1988.

[a] Solar Two final receiver design is more aggressive than the preliminary commercial design, which was completed several years earlier.

[b] Estimated as electrical output/(annual insolation · reflective area).

receiver tube walls results in a receiver surface temperature ranging from 750 to 850 K. Using the average of those temperatures for the receiver surface temperature T, 300 K for the ambient (T_a), and 0.88 for ε (typical for rough oxidized steel), and an empirical relation of $c = 5\,\text{W/m}^2\,\text{K}^{1.25}$, the thermal loss, T_L, from the receiver is

$$T_L = A_{\text{receiver}}\left[\varepsilon\sigma\left(T^4 - T_a^t\right) + c(T - T_a)^{1.25}\right]$$

$$T_L = 1{,}056\left[0.88 \cdot 5.67 \times 10^{-8} \cdot (800^4 - 300^4) + 5(800 - 300)^{1.25}\right] \tag{8.53}$$

$$T_L = 1{,}056[20{,}000 + 11{,}600] = 31.6 \text{ MW}$$

The resulting thermal efficiency is

$$\eta_{\text{th}} = 1 - \frac{\text{TL}}{\text{incident power}} = 1 - \frac{\text{TL}}{\text{DPTOP} + \text{TL}}$$

$$\eta_{\text{th}} = 1 - \frac{31.6}{468 + 31.6} = 0.937 = 93.7\% \tag{8.54}$$

The required collector (heliostat) area at the design point is given by

$$\text{Absorbed solar power} = \text{DPTOP} + \text{TL} = G\rho A_c(\text{CSB}\,T\eta)\alpha \tag{8.55}$$

Hence

$$A_c = (468 + 31.6) \cdot 10^6/(950 \cdot 0.92 \cdot (0.83 \cdot 0.92 \cdot 0.93 \cdot 0.99) \cdot 0.92) = 884{,}000 \text{ m}^2$$

Here

G is the design point beam insolation (=950 W/m² reflector area)
ρ is the average heliostat reflectivity
CSB is the combined effect of cosine, shading, and blocking losses
T is the transmission loss from the heliostat to the receiver
η is the receiver interception factor
α is the receiver absorptivity

Rightfully, the term in parenthesis should be computed for each heliostat and then averaged, but using field average values for each item is a reasonable approximation.

If the ground coverage, F, is about 25% of the heliostat/collector area, A_c, with a nominal 3% addition for roads and the power block, the land area for the field, A_f, is

$$A_f = (1.03)\frac{A_c}{F} = (1.03)\frac{884{,}000}{0.25} = 3.64 \text{ km}^2 = 1.4 \text{ mi}^2 = 900 \text{ ac.}$$

Although the field is biased toward the south to minimize north to south variations in the receiver flux, the south radius is still only about 70% of the north field radius. This is approximated by replacing π by 2.5.

$$A_f = \pi R^2 - \left(0.3R \cdot \frac{\pi R}{2}\right) \cong 2.5R^2 \tag{8.56}$$

So

$$R = \left(\frac{A_f}{2.5}\right)^{0.5} = \left(\frac{3.640,000}{2.5}\right)^{0.5} = 1,210 \text{ m}$$

In this case, the range of receiver elevation angles from the edge of the field ($\theta_{rim} = 90° - \psi_{rim}$) is

In the north, $\theta_{rim} = \arctan(185/1210) = 8.7°$
Due south, $\theta_{rim} = \arctan(185/(0.7 \times 1210)) = 12.3°$

The optimum rim angle ψ_{rim} will depend on the relative cost of heliostats and will be decreased if an exceedingly high flux density is required. A higher tower allows a higher heliostat density and a smaller field with a higher concentration. Typical values range about ±2.5° from the preceding figures.

Central receiver tower-based solar thermal power plants have been actively pursued in the United States, Germany, Switzerland, Spain, France, Italy, Russia, and Japan since the concept was first proposed by Alvin Hildebrandt and Lorin Vant Hull in the early 1970s [44]. Table 8.15 summarizes the central receiver plants built in these countries. Solar One is a 10 MW_e power plant built in Southern California as a joint venture of the U.S. Department of Energy, Southern California Edison Company, and the State of California. Figure 8.29 shows a photograph of the Solar One plant. The solar field generated superheated steam at 510°C and 10.3 MPa which ran the turbine. The system had a crushed rock and sand storage system which was charged by steam through a heat transfer oil. Solar One was operated successfully for 6 years and then shut down for a redesign. The redesigned plant is called Solar Two. A schematic of Solar Two is shown in Figure 8.30. Solar Two uses a nitrate salt as the working fluid in the solar loop and as the storage medium.

In April, 2009, Abengoa Solar began commercial operation of a 20 MW solar power tower plant at the Solucar Plateform, near Seville, Spain. PS20 features a higher efficiency receiver, improved control systems, and a better thermal energy storage system than the first tower, PS10. The second system brings the power capacity to 20 MW, double that of the original PS10 tower. A photograph of both of these is shown in Figures 8.31 and 8.32.)

TABLE 8.15

Central-Receiver Pilot Plants Constructed in Several Countries

Name	Location	Size MW_e	Receiver Fluid	Start-Up	Sponsors
Eurelios	Adrano, Sicily	1	Water/steam	1981	European Community
SSPS/CRS	Tabernas, Spain	0.5	Sodium	1981	Austria, Belgium, Italy, Greece, Spain, Sweden, Germany, Switzerland, United States
Sunshine	Nio, Japan	1	Water/steam	1981	Japan
Solar One	Daggett, California	10	Water/steam	1982	U.S. DOE, SCE[a], CEC[a], LADWP[a]
Themis	Targasonne, France	2.5	Hitec salt	1982	France
CESA-1	Tabernas, Spain	1	Water/steam	1983	Spain
MSEE	Albuquerque, New Mexico	0.75	Nitrate salt	1984	U.S. DOE, EPRI [a], U.S. Industry and Utilities
C3C-5	Crimea, former Soviet Union	5	Water/steam	1985	Former Soviet Union

Source: De Laquil, P. et al., Solar-thermal electric technology, in *Renewable Energy*, Island Press, Washington, DC, 1993.

[a] SCE, Southern California Edison Company; CEC, California Energy Commission; LADWP, Los Angeles Department of Water and Power; EPRI, Electric Power Research Institute.

Detailed EROI estimates for solar central receiver power plants have been made by Lorrin Vant-Hull in a series of peer reviewed articles [37,45–47]. The EROI estimates were based on the detailed commercial central receiver plant design, which served as the reference for the 10 MW electric Solar One plant (incorporating 6 h of oil thermocline storage) and a similar analysis for the second generation dispatchable plant incorporating 2-tank molten salt storage (Solar Two). In both cases, the EROI was about 20 for electric power output to thermal (or fossil) energy input assuming a 30 year life and 6 h of storage. The heliostats represented 75% of the energy input. The discussion of the EROI included also consideration for recycling some of the components and it was estimated that up to 41% of the initial capital energy may be recoverable.

An up-to-date description of simulation tools for designing utility-scale central receiver system power plants is presented in [9]. The tools included the DELSOL code of Kistler [48], the TRNSYS of Klein [32], as well as an interface program PTGEN. With the aid of these codes, it is possible to predict the performance of solar central receiver power plants. Continuous development and improvement of PTGEN is anticipated in the National Renewable Energy Laboratory/DOE solar advisory model.

Point focus systems

Power tower

- Steam or molten nitrate salt HTF
- Max. operating temp. –560°C
- Steam Rankine cycle

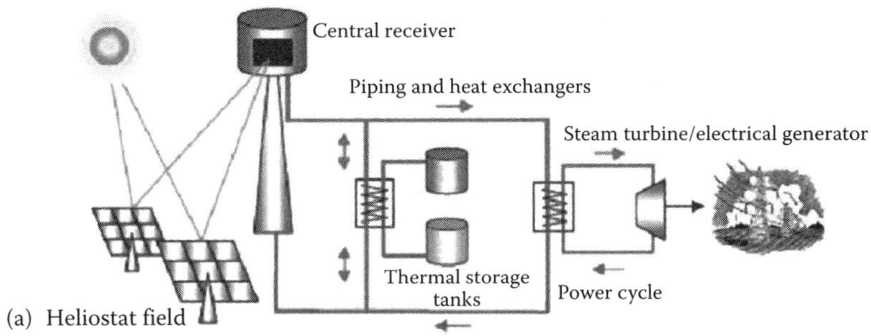

(b)

FIGURE 8.31
(See color insert.) (a) Schematic diagram of point focus solar thermal power tower system. (b) Photograph of two Spanish Central Receiver Systems. (Courtesy of Clifford Ho, Sandia National Laboratories.)

8.7 Parabolic Dish Systems and Stirling Engines

A parabolic dish–based solar thermal power plant was built in Shenandoah, Georgia. This plant was designed to operate at a maximum temperature of 382°C and to provide electricity (450 kW_e), air-conditioning, and process steam (at 173°C) for an industrial complex. The system consisted of 114 parabolic dish concentrators and was decommissioned in 1990.

FIGURE 8.32
(See color insert.) Heliostats for power tower.

Stirling engines have recently received increased attention because they can utilize the high temperature produced by concentrated solar energy, which can be achieved with parabolic dish concentrators. Stirling engines are a promising technology for solar applications because they have the potential of operating at very high efficiency. In fact, the efficiency of the ideal Stirling cycle is the same as that of a Carnot cycle operating between the same temperatures. Since a Stirling engine is an external combustion engine, it can use any fuel or concentrated radiation. Thus, it could be operated at night with an auxiliary source such as natural gas or methane from a landfill. Stirling engines can operate at high temperatures, typically between 600°C and 800°C, resulting in conversion efficiency of 30%–40%. Experience with Stirling engines that have recently been developed for cryogenic applications can be applied to solar-driven Stirling systems [50].

8.7.1 Thermodynamics of a Stirling Cycle

Figure 8.33 shows the thermodynamic process of an ideal Stirling cycle with a perfect gas as the working fluid on pressure–volume and temperature–entropy diagrams. The gas is compressed isothermally (at constant temperature) from state 1 to state 2 by means of heat rejection at the low temperature of the cycle, T_L. The gas is then heated at constant volume from state 2 to state 3, followed by an isothermal expansion from state 3 to state 4. During this expansion process, heat is added at the high temperature in the cycle, T_H. Finally, the gas is cooled at constant volume from T_H to T_L during the process from state 4 to state 1. The cross-hatched areas in the temperature–entropy diagram during the constant volume process between states 2 and 3 represents the heat addition to the working gas while raising its temperature

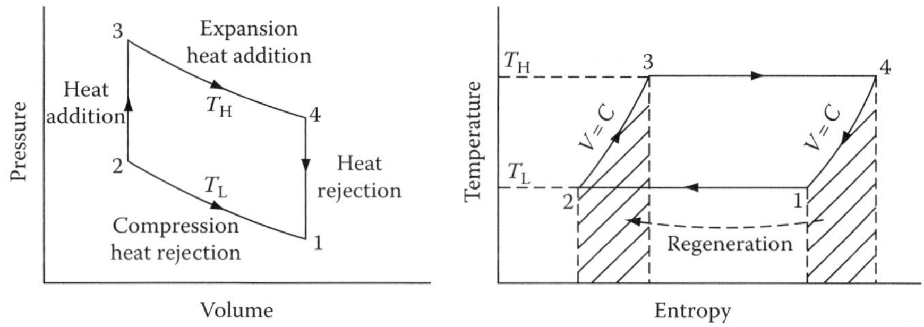

FIGURE 8.33
Thermodynamic diagrams of an ideal Stirling engine.

from T_L to T_H. Similarly, the crosshatched area between states 4 and 1 represents the heat rejection as the gas is cooled from T_H to T_L. Note that the heat addition from state 2 to 3 is equal to the heat rejection from state 4 to 1 and that the processes occur between the same temperature limits. In the ideal cycle, the heat rejected between 4 and 1 is stored and transferred by perfect regeneration to the gas in process 2 to 3. Hence, the only external heat addition in the cycle occurs in the process between states 3 and 4, and is given by

$$Q_{3-4} = -W_{3-4} = \int_3^4 p dV = mRT_H \ln\frac{V_4}{V_3}. \tag{8.57}$$

The work input for compression from state 1 to 2 is

$$W_{1-2} = -\int_1^2 p dV = mRT_L \ln\frac{V_1}{V_2}$$

$$= -mRT_L \ln\frac{V_2}{V_1}, \tag{8.58}$$

Noting that the ratio $V_2/V_1 = V_3/V_4$, and combining the above two equations, the net work output is

$$mR \ln\frac{V_3}{V_4}(T_H - T_L). \tag{8.59}$$

Therefore, the cycle efficiency

$$\eta = \frac{\text{Net work out}}{\text{Heat input}} = \frac{T_H - T_L}{T_H}. \tag{8.60}$$

This efficiency, which is equal to the Carnot cycle efficiency, is based on the assumption that regeneration is perfect, which is not possible in practice.

Therefore, the cycle efficiency would be lower than that indicated by the above equation. For a regeneration effectiveness e as defined below, the efficiency is given by [38]

$$\eta = \frac{T_H - T_L}{T_H + [(1-e)/(k-1)][(T_H - T_L/\ln(V_1/V_2)]} \tag{8.61}$$

where $e = \frac{T_R - T_L}{T_H - T_L}$ and T_R = regenerator temperature, and $k = C_p/C_v$ for the gas.

For perfect regeneration ($e = 1$), the above expression reduces to the Carnot efficiency. It is also seen from the above equation that regeneration is not necessary for the cycle to work because even for $e = 0$ the cycle efficiency is not zero.

Example 8.8

A Stirling engine with air as the working fluid operates at a source temperature of 400°C and a sink temperature of 80°C. The compression ratio is 5.

Assuming perfect regeneration, determine the following:

1. Expansion work.
2. Heat input.
3. Compression work.
4. Efficiency of the machine.

If the regenerator temperature is 230°C, determine

5. The regenerator effectiveness.
6. Efficiency of the machine.
7. If the regeneration effectiveness is zero, what is the efficiency of the machine?

Solution

1. Expansion work per unit mass of the working fluid. Assuming air as an ideal gas,

$$w_{34} = -\int_3^4 P dv = RT_H \ln\frac{\upsilon_4}{\upsilon_3} = (0.287)(400 + 273)\ln 5 = -310.9 \text{ kJ/kg}.$$

   Minus sign shows work output.

2. Heat input per unit mass of the working fluid:

$$q_{34} = w_{34} = 310.9 \text{ kJ/kg}.$$

3. Compression work per unit mass of the working fluid:

$$w_{12} = -\int_1^2 P d\upsilon = RT_L \ln\frac{\upsilon_2}{\upsilon_1} = -(0.287)(80+273)\ln\left(\frac{1}{5}\right) = 163.1\,\text{kJ/kg}.$$

4. Efficiency of the machine:

$$\eta = \frac{T_H - T_L}{T_H} = \frac{400-80}{(400+273)} = 0.475 = 47.5\%.$$

5. The regenerator effectiveness:

$$e = \frac{T_R - T_L}{T_H - T_L} = \frac{230-80}{400-80} = 0.469.$$

6. Efficiency of the machine:

$$\eta = \frac{T_H - T_L}{T_H + \dfrac{(1-e)(T_H - T_L)}{(k-1)\ln(\upsilon_1/\upsilon_2)}} = \frac{400-800}{(400+273) + \dfrac{(1-0)}{(1.4-1)}\dfrac{(400-80)}{\ln(5)}} = 0.341,$$

$$\eta = 34.1\%.$$

7. If the regeneration effectiveness is zero, the efficiency of the machine is

$$\eta = \frac{T_H - T_L}{T_H + \dfrac{(1-e)(T_H - T_L)}{(k-1)\ln(\upsilon_1/\upsilon_2)}} = \frac{400-80}{(400+273) + \dfrac{(1-0)}{(1.4-1)}\dfrac{(400-80)}{\ln(5)}} = 0.273,$$

$$\eta = 27.3\%.$$

8.7.2 Examples of Solar Stirling Power Systems

In order to understand how the Stirling cycle shown in Figure 8.33 may operate, in practice, the simple arrangement and sequence of processes shown in Figure 8.34 is helpful. In the proposed arrangement, two cylinders with pistons are connected via a porous media, which allows gas to pass through from one cylinder to the other. As the gas passes through the porous media, it exchanges heat with the media. The porous media, therefore, serves as the regenerator. In practice, this arrangement can be realized in three ways, as shown in Figure 8.35, by alpha, beta, and gamma types.

The choice of a working fluid for Stirling engine depends mainly on the thermal conductivity of the gas in order to achieve high heat transfer rates.

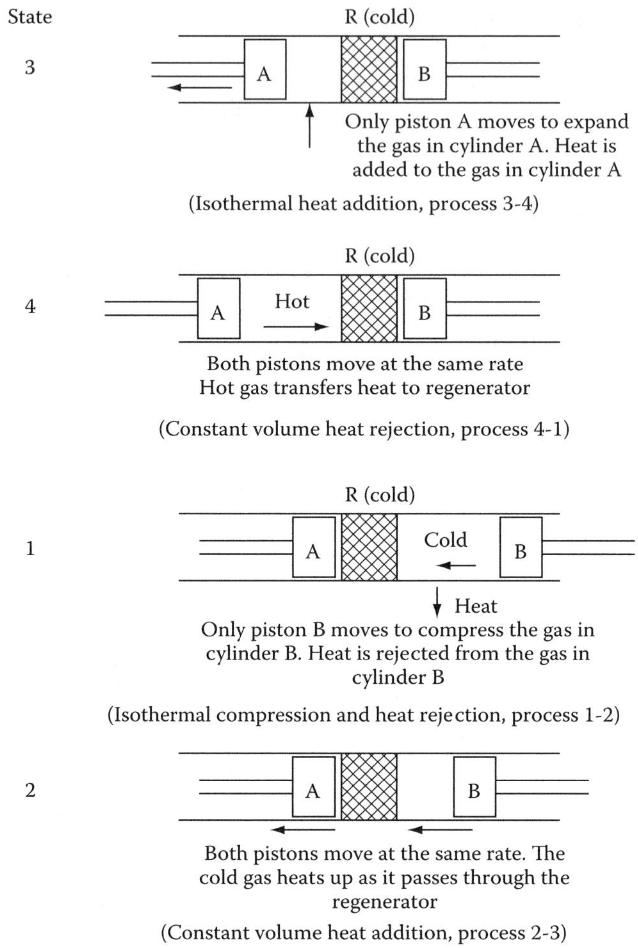

FIGURE 8.34
Stirling cycle states and processes with reference to Figure 8.33.

Air has traditionally been used as the working fluid. Helium has a higher ratio of specific heats (k), which lessens the impact of imperfect regeneration.

In the alpha configuration, there are two cylinders and pistons on either side of a regenerator. Heat is supplied to one cylinder and cooling is provided to the other. The pistons move at the same speed to provide constant volume processes. When all the gas has moved to one cylinder, the piston of that cylinder moves with the other remaining fixed to provide expansion or compression. Compression is done in the cold cylinder and expansion in the hot cylinder. The Stirling Power Systems V-160 engine (Figure 8.36) is based on an alpha configuration.

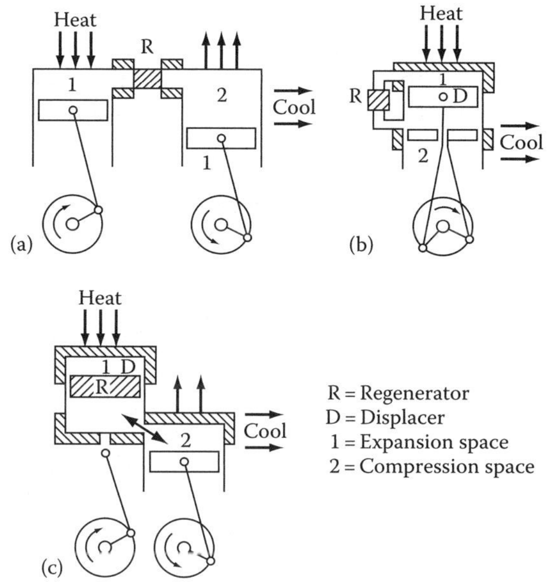

FIGURE 8.35
Three basic types of Stirling engine arrangements. (a) Alpha-type, (b) beta-type, and (c) gamma-type.

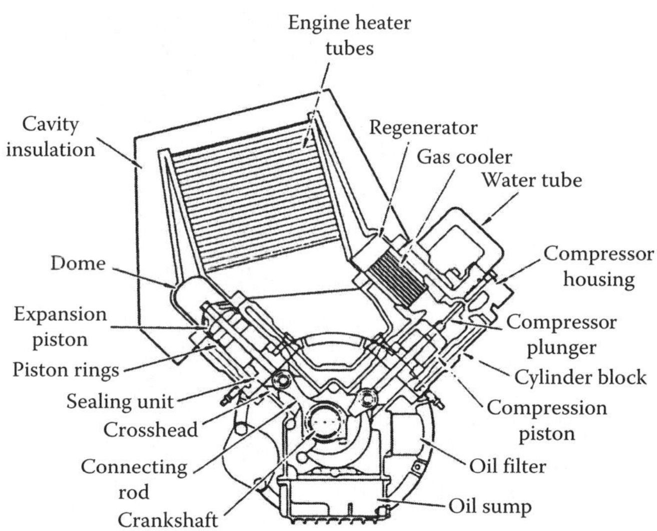

FIGURE 8.36
Stirling power systems/Solo Kleinmotoren V-160 alpha-configuration Stirling engine. (Adapted from Kistler, B.L., *A User's Manual for DELSOL3: A Computer Code for Calculating the Optical Performance and Optimal System Design for Solar Thermal Central Receiver Plants*, Sandia National Labs, Albuquerque, NM, 1986, SAND86-8018.)

The beta configuration has a power piston and a displacer piston, which divides the cylinder into hot and cold sections. The power piston is located on the cold side and compresses the gas when the gas is in the cold side and expands it when it is in the hot side. The original patent of Robert Stirling was based on beta configuration, as are free piston engines.

The gamma configuration also uses a displacer and a power piston. In this case, the displacer is also the regenerator, which moves gas between the hot and cold ends. In this configuration, the power piston is in a separate cylinder.

In a piston/displacer drive, the power and displacer pistons are designed to move according to a simple harmonic motion to approximate the Stirling cycle. This is done by a crankshaft or a bouncing spring/mass second-order mechanical system [51].

In a kinematic engine, the power piston is connected to the output shaft by a connecting rod crankshaft arrangement. Free piston arrangement is an innovative way to realize the Stirling cycle. In this arrangement, the power piston is not connected physically to an output shaft. The piston bounces between the working gas space and a spring (usually a gas spring). The displacer is also usually free to bounce. This configuration is called the Beale free piston Stirling engine after its inventor, William Beale [51]. Since a free piston Stirling engine has only two moving parts,

FIGURE 8.37
The McDonnell Douglas/United Stirling dish-Stirling 25 kW_e module. (From Goswami, Y. et al., *Principles of Solar Engineering*, 2nd edn., Taylor & Francis, Philadelphia, 2000. With permission.)

it offers the potential of simplicity, low cost, and reliability. Moreover, if the power piston is made magnetic, it can generate current in the stationary conducting coil around the engine as it moves. This is the principle of the free piston/linear alternator in which the output from the engine is electricity.

Stirling engines can provide very high efficiencies with high concentration solar collectors. Since practical considerations limit the Stirling engines to relatively small sizes, a Stirling engine fixed at the focal point of a dual tracking parabolic dish provides an optimum match, as shown in Figure 8.37. Therefore, all of the commercial developments to date have been in parabolic dish–Stirling engine combination. The differences in the commercial systems have been in the construction of the dish and the type of Stirling engine. A thorough description of past and current dish-Stirling technologies is presented in Ref. [52].

Problems

8.1 Repeat Example 8.1 for a parabolic collector with an evacuated tube receiver (case E in Figure 8.18a).

8.2 Repeat Example 8.3, but for a collector located in Washington, District of Columbia.

8.3 Consider a Stirling cycle from Example 8.8 with imperfect regenerator ($T_R = 230°C$). Assume ideal gas.

a. Compute the efficiency from

$$\eta = \frac{\text{Work}_{\text{net}}}{\text{Heat}_{\text{input}}},$$

where

$$\text{Heat}_{\text{input}} = q_{34} + mC_v(T_H - T_R).$$

Verify that the efficiency you get is the same as the efficiency obtained by Equation 8.61.

b. Start from the efficiency equation above. Derive the efficiency in the Equation 8.61

8.4 A solar electric engine operating between 5°C and 95°C has an efficiency equal to one-half the Carnot efficiency. This engine is to drive a 4 kW pump. If the collector has an efficiency of 50%, calculate the area needed for operation in Egypt where the insolation averages about 2800 kJ/m^2 h during the day. State any additional assumptions.

8.5 The graph below shows the Rankine cycle efficiency as a function of maximum cycle temperature for different working fluids. Prepare a thermal analysis matching a line focusing collector, a paraboloid dish collector, and a CPC collector with a concentration ratio of 3:1 to a suitable working fluid in Washington, District of Columbia, and Albuquerque, New Mexico. Comment on storage needs in both locations. Low-temperature organic and steam efficiencies were computed with the following assumptions: expander efficiency, 80%; pump efficiency, 50%; mechanical efficiency, 95%; condensing temperature, 95°F; regeneration efficiency, 80%; high side pressure loss, 5%; low side pressure loss, 8%.

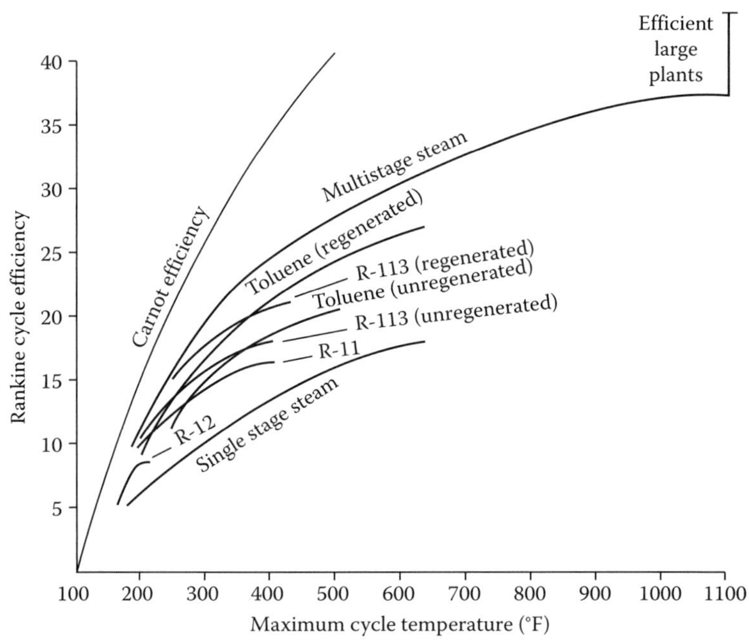

8.6 The schematic diagram shown below for a solar-driven irrigation pump was developed by Battelle Memorial Institute and uses tracking parabolic trough collectors. Solar energy is used to heat the water in the collectors to 423 K, which then vaporizes an organic working fluid that powers the pump-turbine. Calculate the surface area needed to power a 50 hp pump capable of delivering up to 38,000 L/min of water at noon in Albuquerque, New Mexico.

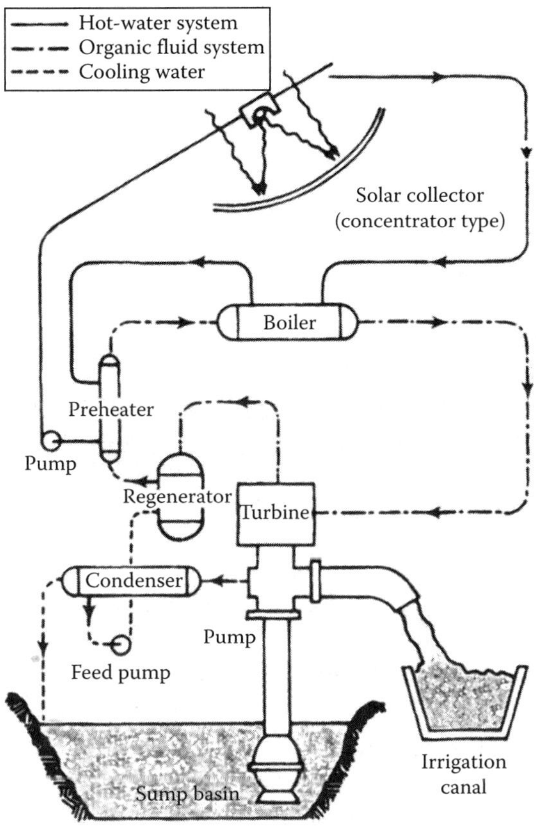

8.7 Write a closed-form expression for the work output of a solar-powered heat engine if the energy delivery of the solar collector at high temperature is given approximately by the expression

$$q_u = (\tau\alpha)_{\text{eff}} I_c - \frac{\sigma\varepsilon\bar{T}_f^4}{CR},$$

(convection and conduction losses are neglected) where $\bar{T}_f$ is the average fluid temperature and CR is the concentration ratio. Two heat engines are to be evaluated:

(1) Carnot cycle:

$$\text{Cycle efficiency } \eta_c = 1 - \frac{T_\infty}{T_f},$$

(2) Brayton cycle:

$$\text{Cycle efficiency } \eta_B = 1 - \frac{C_B T_\infty}{T_f} \quad C_B \geq 1,$$

where

$C_B = (r_p)^{(k-1)/k}$

r_p is the compressor pressure ratio

k is the specific heat ratio of the working fluid

Write an equation with $\bar{T}_f$ as the independent variable that, when solved, will specify the value of $\bar{T}_f$ to be used for maximum work output as a function of concentration ratio, surface emittance and $(\tau\alpha)_{eff}$ product, insolation level, and C_B. Optional: solve the equation derived above for CR = 100, $\epsilon = 0.5$, and $(\tau\alpha)_{eff} = 0.70$, at an insolation level of $1\,kW/m^2$. What is the efficiency of a solar-powered Carnot cycle and a Brayton cycle for which $C_B = 2$?

8.8 Calculate the maximum power and LCOE for a parabolic collector similar to that in Example 8.1, but assume that an evacuated tube receiver is used (case E in Figure 8.18a).

8.9 Prepare a preliminary design for a central receiver system with a power capacity of 50 MW similar to Solar 2 for a location in Nevada.

8.10 It has been estimated that a 400 MW expansion of the existing SEGS solar plant in California would cost $2.4 billion. Estimate the expected capital costs of the generation facility in dollars/kW capacity and the minimum cost of electricity per kWh generated by the addition, if they are operated at full capacity for 6 h/day and have a lifetime of 30 years.

References

1. Trombe, F. (1956) High temperature furnaces, in: *Proceedings of the World Symposium on Applied Solar Energy*, Phoenix, AZ, 1955, Stanford Research Institute, Menlo Park, CA.
2. Daniels, F. (1964) *Direct Use of the Sun's Energy*, Yale University Press, New Haven, CT.
3. Daniels, F. (1949) Solar energy, *Science* 109, 51–57.
4. EIA. (2007) Industrial sector energy demand: Revisions for non-energy-intensive manufacturing. Accessed at http://www.eia.doe.gov/oiaf/aeo/otheranalysis/neim.html.
5. Fraser, M.D. (1976) InterTechnology Corporation assesses industrial heat potential, *Solar Engineering* 1, pp. 11–12, September 1976.
6. Kreider, J.F. (1979) *Medium and High Temperature Solar Processes*, Academic Press, New York.

7. Intertechnology Corporation. (1977) Analysis of the economic potential of solar thermal energy to provide industrial process heat, ERDA Report No. COO/2829-1, 3 vols.
8. Hannon, B., Costanza, R., and Herendeen, R. (1986) Measures of energy cost and value in ecosystems. *Journal of Environmental Economics and Management* 13, 391–401.
9. Wagner, R. (1977) Energy conservation in dyeing and finishing, *Textile Chemists and Colorists* 9, 52.
10. Hebrank, W.H. (1975) Options for reducing fuel usage in textile finishing tenter dryers, *American Dyestuff Reporter* 63, 34.
11. Goswami, B.C. and Langley, J. (1987) A review of the potential of solar energy in the textile industry, in: *Progress in Solar Engineering*, Hemisphere Publishing Corporation, Washington, DC.
12. Lowery, J.F. et al. (1977) Energy conservation in the textile industry, Technical Reports of Department of Energy Project No. Ey-76S-05-5099, School of Textile Engineering, Georgia Institute of Technology, Atlanta, GA, 1977, 1978, and 1979.
13. Proctor, D. and Morse, R.N. (1975) *Solar Energy for Australian Food Processing Industry*, CSIRO Solar Energy Studies, East Melbourne, Victoria, Australia.
14. Goswami, D.Y., Kreith, F., and Kreider, J.F. (2000) *Principles of Solar Engineering*, 2nd edn, Taylor & Francis, Philadelphia.
15. TRNSYS: A transient simulation program, version 11.1, Engineering Experiment Station, Report 38, Solar Energy Laboratory, University of Wisconsin, Madison, WI.
16. Singh, R.K., Lund, D.B., and Buelow, F.H. (1986) Compatibility of solar energy with food processing energy demands, in: *Alternative Energy in Agriculture*, vol. I, CRC Press, Boca Raton, FL.
17. O'Gallagher, J.J. 2008. *Nonimaging Optics in Solar Energy*, Morgan & Claypool Publishers, San Rafael, CA.
18. Winston, R. (1974) Principles of solar concentrators of a novel design, *Solar Energy*, 16, 89.
19. Kreith, F. and West, R.E. (eds.) (1997) *CRC Handbook of Energy Efficiency*, CRC Press, Boca Raton, FL.
20. Bendt, P., Rabl, A., Gaul, H., and Reid, K. *Optical Analysis and Optimization of Line-Focus Solar Collectors*, SERI, Golden, CO, SERI/TR 34-092, September 1979.
21. Gee, R. (1980) Long term average performance benefits of parabolic trough improvements, SERI, Golden, CO, SERI/TR 632-439, March 1980.
22. Gee, R. and May, K. (1997) Parabolic trough concentrating collectors component and system design, in: Kreith, F. and West, R.E. (eds.) *CRC Handbook of Energy Efficiency*, CRC Press, Boca Raton, FL.
23. Kreith, F.K., West, R.E., and Cleveland, C.J. (1987) Energy analysis for renewable energy technologies, *ASHRAE Transactions*, ASHRE 1987, Vol. 93, Part 1.
24. DiGrazia, M., Gee, R., and Jorgensen, G. (2009) ReflecTech mirror film: Attributes and durability for CSP applications, in: *Proceedings of the ASME 3rd International Conference on Energy Sustainability*, San Francisco, CA, July 19–23, 2009.
25. G.J. Jorgensen, Gee, R., and King, D.E. (2006) Durable corrosion and ultraviolet-resistant silver mirror, U.S. Patent No. 6,989,924. Issued January 24, 2006.
26. Farr, A. and Gee, R. (2009) The SkyTrough™ parabolic trough solar collector, in: Paper *90090 ASME 3rd International Conference on Energy Sustainability*, San Francisco, CA, 2009.

27. Reddy, T.A. (1987) *The Design and Sizing of Active Solar Thermal Systems*, Oxford University Press, Oxford, U.K.
28. Collares-Pereira, M. and Ralb, A. (1979) Simple procedure for predicting long term average performance of nonconcentrating and of concentrating solar collectors, *Solar Energy* 23, 235–253.
29. Hall, I., Prairie, R., Anderson, H., and Boes, E. (1978) *Generation of Typical Meteorological Years for 26 SOLMET Stations*, SAND78-1601, Sandia National Laboratories, Albuquerque, NM.
30. Rabl, A. (1981) *Active Solar Collectors and Their Applications*, University Press, Oxford, U.K.
31. Gordon, J.M. and Rabl, A. (1992) Nonimaging compound parabolic concentrator-type reflectors with variable extreme direction, *Applied Optics* 31, 7332–7338.
32. Klein, S.A., Beckman, W.A., Mitchell, J.W., and Duffie, J.A. (2004) *TRNSYS 16—A Transient System Simulation Program, User Manual*, University of Wisconsin, Solar Energy Laboratory, Madison, WI.
33. Cengel, Y.A. and Boles, M.A. (2006) *Thermodynamics: An Engineering Approach*, 6th edn., McGraw-Hill, New York.
34. Fitzgerald, A.E., Kingsley, C.F., and Kusko, A. (1971) *Electric Machinery*, 3rd edn., McGraw-Hill, New York.
35. Kreith, F. and Goswami, D.Y. (eds.) (1997) *The CRC Handbook of Mechanical Engineering*, CRC Press, Boca Raton, FL.
36. Goswami, D.Y., Hingorani, S., and Mines, G. (1991) A laser-based technique for particle sizing to study two-phase expansion in turbines, *Journal of Solar Energy Engineering, Transactions of the ASME* 113(3), 211–218.
37. Vant-Hull, L.L. (1992) Solar thermal central receivers, *Solar Today*, Nov/Dec issue, pp. 13–16.
38. Stine, W.B. and Harrigan, R.W. (1985) *Solar Energy Fundamentals and Design: With Computer Applications*, Wiley, New York.
39. Kolb, G.J. (1998) Economic evaluation of Solar One and hybrid power towers using molten salt technology, *Solar Energy* 62(1), 51–61.
40. Hillesland, T. and Weber, E.R. (1990) Utilities study of solar central receivers, in: Gupta, B.P. and Traugott, W.H. (eds.), *Proceedings of the 4th International Symposium Solar Thermal Technology: Research, Development and Applications*, Santa Fe, NM, Hemisphere Publishing Corp., New York, pp. 165–176.
41. Romero, M., Buck, R., and Pacheco, J.E. (2002) An update on solar central receiver systems, projects, and technologies, *Journal of Solar Energy Engineering* 124, 98–108, May 2002.
42. Wagner, M.J., Klein, S.A., and Reindi, D.T. (2009) Simulation of utility-scale central receiver system power plants, in: *Proceedings of the 2009 ASME 3rd International Conference on Energy Sustainability*, San Francisco, CA, ES2009-90132.
43. Vant-Hull, L.L. (1991) Solar radiation conversion, in: Winter, C.J., Sizmann, R.L., and Vant-Hull, L.L. (eds.), *Solar Radiation Conversion*, Springer, Berlin, Germany, pp. 21–27.
44. Hildebrandt, A.F. and Vant-Hull, L.L. (1974) Tower-top focus solar energy collector, *Journal of Mechanical Engineering* 96(9), 23–27.
45. Vant-Hull, L.L. (1985) Solar thermal power generation, *Natural Resources Journal* 25, 1099–1111.

46. Vant-Hull, L.L. (1992–1993) Solar thermal electricity: An environmentally benign and viable alternative, *Perspectives in Energy* 2, 157–166.
47. Vant-Hull, L.L. (1992) Solar thermal receivers: Current status and future promise, *American Solar Energy Society* 7, 13–16.
48. Kistler, B.L. (1986) *A User's Manual for DELSOL3: A Computer Code for Calculating the Optical Performance and Optimal System Design for Solar Thermal Central Receiver Plants*, Sandia National Labs, Albuquerque, NM, SAND86-8018.
49. De Laquil, P., Kearney, D., Geyer, M., and Diver, R. (1993) Solar-thermal electric technology, in: *Renewable Energy*, Island Press, Washington, DC.
50. Garaway, I. (2009) Personal communication.
51. Stine, W.B. (1998) Stirling engines, in: *The CRC Handbook of Mechanical Engineering*, CRC Press, Boca Raton, FL.
52. Mancini, T., Heller, P., Butler, B., Osborn, B., Schiel, W., Goldberg, V. et al. (2003) Dish-stirling systems: An overview of development and status, *JSEE* 125, 135–151, May 2003.
53. Intertechnology Corporation. (1997) Analysis of the economic potential of solar thermal energy to provide industrial process heat, ERDA Rep. No. C00/2829.
54. O'Gallagher, J.J. (2008) *Nonimaging Optics in Solar Energy*, Morgan & Claypool Publishers, Ft. Collins, CO.
55. Gee, R. (1980) *Long Term Average Performance Benefits of Parabolic Trough Improvements*, SERI, Golden, CO, SERI/TR 632–439, March.
56. Arizona public service utility solar solar central receiver study, Report No. DOE/AL/38741-1, November 1988.

9

Passive Solar Heating, Cooling, and Daylighting

> For the well-being and health...the homesteads should be airy in summer and sunny in winter. A homestead promising these qualities will be longer than it is deep, and the main front would face south.
>
> **—Aristotle**

9.1 Introduction

Passive systems are defined, quite generally, as systems in which the thermal energy flow is by natural means: by conduction, radiation, and natural convection. A *passive heating system* is one in which the sun's radiant energy is converted to heat upon absorption by the building. The absorbed heat can be transferred to thermal storage by natural means or used to directly heat the building. *Passive cooling systems* use natural energy flows to transfer heat to the environmental sinks: the ground, air, and sky.

If one of the major heat transfer paths employs a pump or fan to force flow of a heat transfer fluid, then the system is referred to as having an *active* component or subsystem. *Hybrid systems*—either for heating or cooling—are ones in which there are both passive and active energy flows. The use of the sun's radiant energy for the natural illumination of a building's interior spaces is called *daylighting*. Daylighting design approaches use both solar beam radiation (referred to as *sunlight*) and the diffuse radiation scattered by the atmosphere (referred to as *skylight*) as sources for interior lighting, with historical design emphasis on utilizing skylight.

9.1.1 Distinction between a Passive System and Energy Conservation

A distinction is made between energy conservation techniques and passive solar measures. *Energy conservation features* are designed to reduce the heating and cooling energies required to thermally condition a building: the use of insulation to reduce either heating and cooling loads; and similarly, the use of window shading or window placement to reduce solar gains, reducing summer cooling loads. *Passive features* are designed to increase the use of solar energy to meet heating and lighting loads, plus the use of ambient "coolth" for cooling. For example, window placement to enhance solar gains to meet winter heating loads and/or to provide daylighting is passive solar use, and the use of a thermal chimney to draw air through the building to provide cooling is a passive cooling feature.

9.2 Key Elements of Economic Consideration

The distinction between passive systems, active systems, or energy conservation is not critical for economic calculations, as they are the same in all cases: a trade-off between the life cycle cost of the energy saved (*performance*) and the life-cycle cost of the initial investment, operating, and maintenance costs (*cost*).

9.2.1 Performance: Net Energy Savings

The key performance parameter to be determined is the net annual energy saved by the installation of the passive system. The basis for calculating the economics of any solar energy system is to compare it against a "normal" building; thus, the actual difference in the annual cost of fuel is the difference in auxiliary energy that would be used with and without solar energy. Thus, the energy saved rather than energy delivered, energy collected, useful energy, or some other energy measure, must be determined.

9.2.2 Cost: Over and Above "Normal" Construction

The other significant part of the economic trade-off involves determining the difference between the cost of construction of the passive building and of the "normal" building against which it is to be compared. The convention, adopted from the economics used for active solar systems, is to define a "solar add-on cost." Again, this may be a difficult definition in the case of passive designs because the building can be significantly altered compared to typical construction, since in many cases it is not just a one-to-one replacement of a wall by a different wall, but is more complex and involves assumptions and simulations concerning the "normal" building.

9.2.3 General System Application Status and Costs

Almost one-half million buildings in the United States were constructed or retrofitted with passive features in the 20 years after 1980. Passive heating applications are primarily in single-family dwellings and secondarily in small commercial buildings. Daylighting features, which reduce lighting loads and the associated cooling loads, are usually more appropriate for large office buildings.

A typical passive heating design in a favorable climate might supply up to one-third of a home's original load at a cost of \$5–\$10 per million Btu net energy saved. An appropriately designed daylighting system can supply lighting at a cost of 2.5 to 5¢ per kWh [1].

9.3 Solar Thermosyphon Water Heating

Solar hot water heating systems are composed of a collector and a storage tank. A passive solar collector with integrated storage is shown in Figure 9.1. When the flow between the collector and tank is by natural circulation, these passive solar hot water systems are referred to as thermosyphon systems. This ability of thermosyphon systems to heat water without an externally powered pump has spurred its use in both regions where power is unavailable and where power is very expensive.

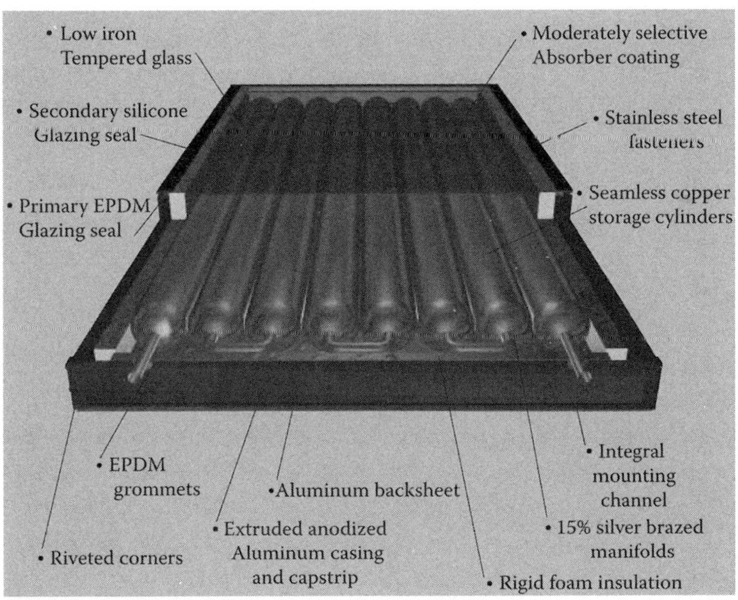

(a)

FIGURE 9.1
(See color insert.) CopperHeart passive solar collector with integrated storage. (Courtesy of SunEarth, Inc., Fontana, CA.) (a) Photograph of installed module on residence in Arizona.

(continued)

(b)

FIGURE 9.1 (continued)
(See color insert.) (b) Schematic of CopperHeart integral collector storage system.

9.3.1 Thermosyphon Concept

The natural tendency of a less dense fluid to rise above a more dense fluid can be used in a simple solar water heater to cause fluid motion through a collector. The density difference is created within the solar collector where heat is added to increase the temperature and decrease the density of the liquid. This collection concept is called a thermosyphon, and Figure 9.2 schematically illustrates the major components of such a system.

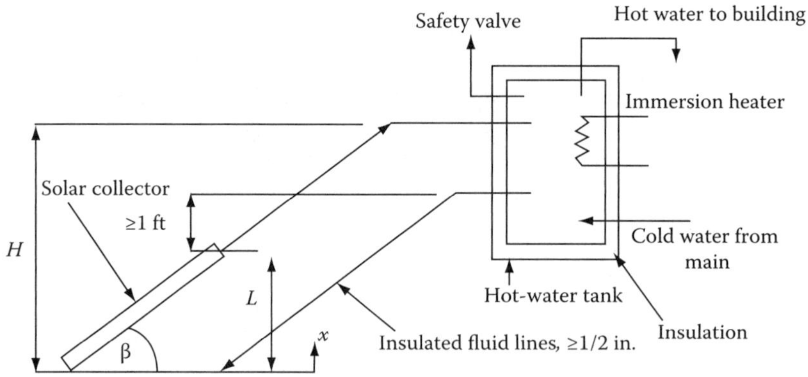

FIGURE 9.2
Schematic diagram of thermosyphon loop used in a natural circulation, service water-heating system. The flow pressure drop in the fluid loop must equal the buoyant force "pressure" $\left[\int_0^L g\rho(x)g\,dx + \rho_{stor} gL)\right]$, where $\rho(x)$ is the local collector fluid density and ρ_{stor} is the tank fluid density, assumed uniform. (From Kreith, F. and West, R., *CRC Handbook of Energy Efficiency and Renewable Energy*, CRC Press, Boca Raton, FL, 2007. With permission.)

The flow pressure drop in the fluid loop (ΔP_{FLOW}) must equal the buoyant force "pressure difference" ($\Delta P_{BUOYANT}$) caused by the differing densities in the "hot" and "cold" legs of the fluid loop:

$$\begin{aligned}\Delta P_{FLOW} &= \Delta P_{BUOYANT} \\ &= \rho_{stor} g H - \left[\int_0^L \rho(x) g dx + \rho_{out} g (H - L)\right]\end{aligned} \tag{9.1}$$

where

H is the height of the "legs" and L the height of the collector (see Figure 9.2)
$\rho(x)$ is the local collector fluid density
ρ_{stor} is the tank fluid density and
ρ_{out} is the collector outlet fluid density, the latter two densities assumed uniform

The flow pressure term ΔP_{FLOW}, is related to the flow loop system headloss, which is in turn directly connected to friction and fitting losses and the loop flow rate:

$$\Delta P_{FLOW} = \oint_{LOOP} \rho d(h_L) \tag{9.2}$$

where $h_L = KV^2$, with K being the sum of the component loss "velocity" factors (see any fluid mechanics text) and V the flow velocity.

9.3.2 Thermo-Fluid System Design Considerations

Since the driving force in a thermosyphon system is only a small density difference and not a pump, larger-than-normal plumbing fixtures must be used to reduce pipe friction losses. In general, one pipe size larger than would be used with a pump system is satisfactory. Under no conditions should piping smaller than ½ in. (12 mm) national pipe thread (NPT) be used. Most commercial thermosyphons use 1 in. (25 mm) NPT pipe. The flow rate through a thermosyphon system is about 1 gal/ft^2 · h (40 L/m^2 · h) in bright sun, based on collector area.

Since the hot-water system loads vary little during a year, the best angle to tilt the collector is that equal to the local latitude. The temperature difference between the collector inlet water and the collector outlet water is usually 15°F–20°F (8°C–11°C) during the middle of a sunny day [3]. After sunset, a thermosyphon system can reverse its flow direction and lose heat to the environment during the night. To avoid reverse flow, the top header of the absorber should be at least 1 ft (30 cm) below the cold leg fitting on the storage tank, as shown.

To provide heat during long cloudy periods, an electrical immersion heater can be used as a backup for the solar system. The immersion heater is located near the top of the tank to enhance stratification so that the heated fluid is at the required delivery temperature at the delivery point. Tank stratification is desirable in a thermosyphon to maintain flow rates as high as possible. Insulation must be applied over the entire tank surface to control heat loss. Figure 9.3 illustrates two common thermosyphon system designs.

Several features inherent in the thermosyphon design limit its utility. If it is to be operated in a freezing climate, a nonfreezing fluid must be used, which in turn requires a heat exchanger between the collector and potable water storage. (If potable water is not required, the collector can be drained during cold periods instead.) Heat exchangers of either the shell-and-tube type or the immersion-coil type require higher flow rates for efficient operation than a thermosyphon can provide. Therefore, the thermosyphon is generally limited to nonfreezing climates. A further restriction on thermosyphon use is the requirement for an elevated tank. In many cases, structural or architectural constraints prohibit raised-tank locations. In residences, collectors

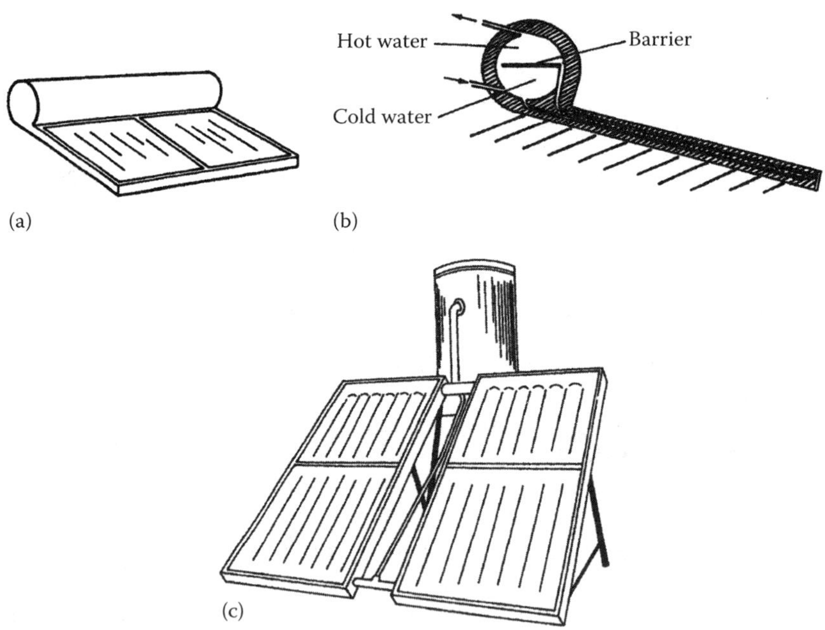

FIGURE 9.3
Passive solar water heaters: (a) compact model using combined collector and storage, (b) section view of the compact model, and (c) tank and collector assembly. (From Kreith, F. and West, R., *CRC Handbook of Energy Efficiency and Renewable Energy*, CRC Press, Boca Raton, FL, 2007. With permission.)

are normally mounted on the roof, and tanks mounted above the high point of the collector can easily become the highest point in a building. Practical considerations often do not permit this application.

Example 9.1

Determine the "pressure difference" available for a thermosyphon system with 1 m high collector and 2 m high "legs." The water temperature input to the collector is 25°C and the collector output temperature is 35°C. If the overall system loss "velocity" factor (K) is 15.6, estimate the system flow velocity.

Solution

Equation 9.1 is used to calculate the pressure difference, with the water densities being found from the temperatures (in steam tables)

$$\rho_{\text{stor}}(25°\text{C}) = 997.009 \text{ kg/m}^3; \quad \rho_{\text{out}}(35°) = 994.036 \text{ kg/m}^3;$$

$$\rho_{\text{coll.ave.}}(30°\text{C}) = 996.016 \text{ kg/m}^3$$ (*note*: average collector temperature used in "integral")

and with $H = 2$ m and $L = 1$ m:

$$\Delta P_{\text{BUOYANT}} = (997.009)9.81(2) - [(996.016)9.81(1) + (994.036)9.81(1)]$$
$$= 38.9 \text{ N/m}^2(\text{Pa}).$$

The system flow velocity is estimated from the "system K" given, the pressure difference calculated above, taking the average density of the water around the loop (at 30°C), and substituting into Equation 9.2:

$$\Delta P_{\text{BUOYANT}} = (\rho_{\text{loop.ave.}})(h_{\text{L}})_{\text{loop}} = (\rho_{\text{loop.ave.}})KV^2$$

$$V^2 = \frac{38.9}{(996.016)(15.6)}$$

$$V = 0.05 \text{ m/s}.$$

9.4 Passive Solar Heating Design Fundamentals

Passive heating systems contain the five basic components of all solar systems, as described in Chapter 7. Typical passive realizations of these components are

1. Collector → windows, walls and floors.
2. Storage → walls and floors, large interior masses (often these are integrated with the collector absorption function).

3. Distribution system → radiation, free convection, simple circulation fans.
4. Controls → moveable window insulation, vents both to other inside spaces or to ambient.
5. Backup system → any nonsolar heating system.

The design of passive systems requires the strategic placement of windows, storage masses, and the occupied spaces themselves. The fundamental principles of solar radiation geometry and availability are instrumental in the proper location and sizing of the system's "collectors" (windows). Storage devices are usually more massive than those used in active systems and are frequently an integral part of the collection and distribution system.

9.4.1 Types of Passive Heating Systems

A commonly used method of cataloging the various passive system concepts is to distinguish three general categories: direct, indirect, and isolated gain. Most of the physical configurations of passive heating systems are seen to fit within one of these three categories.

For direct gain (Figure 9.4), sunlight enters the heated space and is converted to heat at absorbing surfaces. This heat is then distributed throughout the space and to the various enclosing surfaces and room contents.

For indirect gain category systems, sunlight is absorbed and stored by a mass interposed between the glazing and the conditioned space. The conditioned space is partially enclosed and bounded by this thermal storage mass, so a natural thermal coupling is achieved. Examples of the indirect approach are the thermal storage wall, the thermal storage roof, and the northerly room of an attached sunspace.

In the thermal storage wall (Figure 9.5), sunlight penetrates the glazing and is absorbed and converted to heat at a wall surface interposed between the glazing and the heated space. The wall is usually masonry (Trombe wall) or containers filled with water (water wall), although it might contain phase-change material. The attached sunspace (Figure 9.6) is actually a two-zone combination

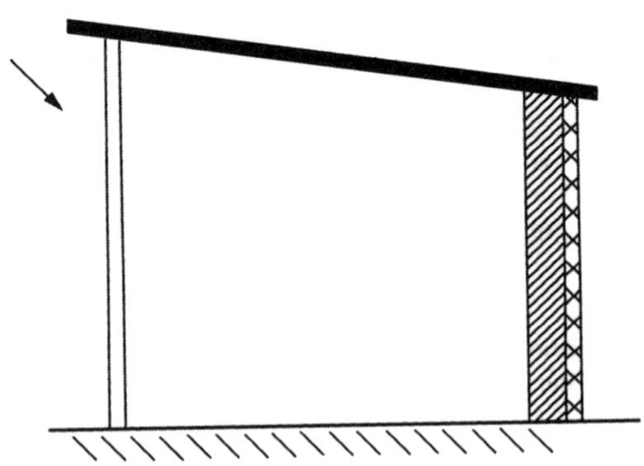

FIGURE 9.4
Direct gain. (From Kreith, F. and West, R., *CRC Handbook of Energy Efficiency and Renewable Energy*, CRC Press, Boca Raton, FL, 2007. With permission.)

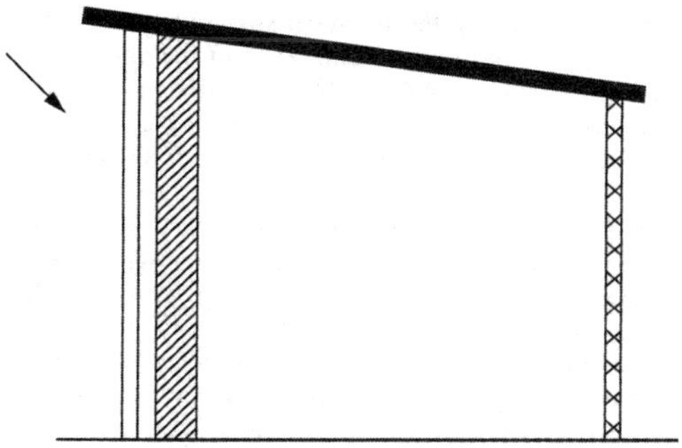

FIGURE 9.5
Thermal storage wall. (From Kreith, F. and West, R., *CRC Handbook of Energy Efficiency and Renewable Energy*, CRC Press, Boca Raton, FL, 2007. With permission.)

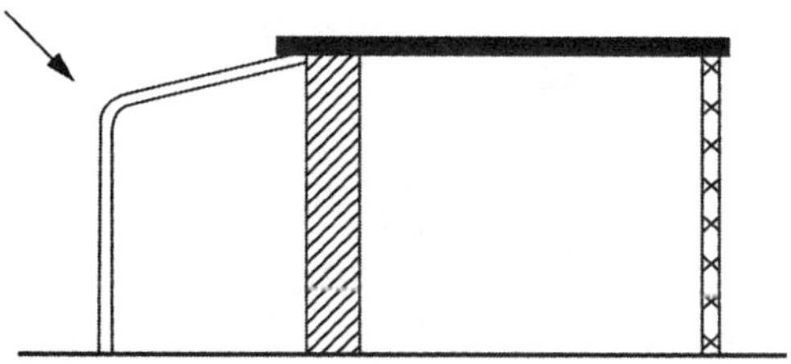

FIGURE 9.6
Attached sunspace. (From Kreith, F. and West, R., *CRC Handbook of Energy Efficiency and Renewable Energy*, CRC Press, Boca Raton, FL, 2007. With permission.)

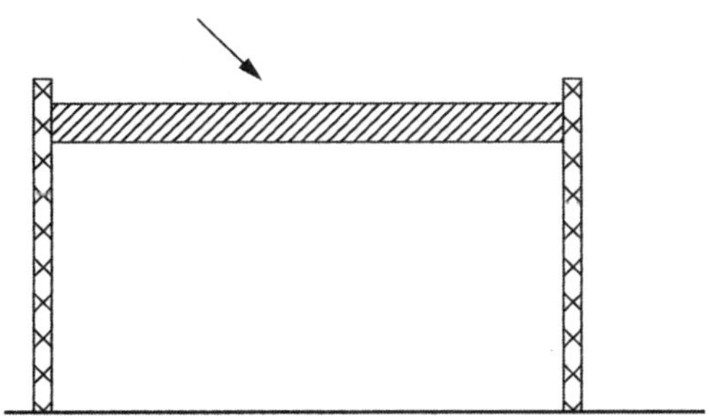

FIGURE 9.7
Thermal storage roof. (From Kreith, F. and West, R., *CRC Handbook of Energy Efficiency and Renewable Energy*, CRC Press, Boca Raton, FL, 2007. With permission.)

of direct gain and thermal storage wall. Sunlight enters and heats a direct gain southerly "sunspace" and also heats a mass wall separating the northerly buffered space, which is heated indirectly. The "sunspace" is frequently used as a greenhouse, in which case the system is called an "attached greenhouse." The thermal storage roof (Figure 9.7) is similar to the thermal storage wall except that the interposed thermal storage mass is located on the building roof.

The isolated gain category concept is an indirect system except that there is a distinct thermal separation (by means of either insulation or physical separation) between the thermal storage and the heated space. The convective (thermosyphon) loop, as depicted in Figure 9.2, is in this category and, while often used to heat domestic water, is also used for building heating. It is most akin to conventional active systems in that there is a separate collector and separate thermal storage. The thermal storage wall, thermal storage roof, and attached sunspace approaches can also be made into isolated systems by insulating between the thermal storage and the heated space.

9.4.2 Fundamental Concepts for Passive Heating Design

Figure 9.8 is an equivalent thermal circuit for the building illustrated in Figure 9.5, the Trombe wall type system. For the heat transfer analysis of the building, three temperature nodes can be identified—room temperature, storage wall temperature, and the ambient temperature. The circuit responds to climatic variables represented by a current injection I_s (solar radiation) and by the ambient temperature T_a. The storage temperature T_s and room temperature T_r are determined by current flows in the equivalent circuit. By using seasonal and annual climatic data, the performance of a passive structure can be simulated, and the results of many such simulations correlated to give the design approaches described below.

9.5 Passive Design Approaches

Design of a passive heating system involves selection and sizing of the passive feature type(s), determination of thermal performance, and cost estimation. Ideally, a cost/performance optimization would be performed by the designer. Owner and architect ideas usually establish the passive feature type, with general size and cost estimation available. However, the thermal performance of a passive heating system has to be calculated.

There are several "levels" of methods that can be used to estimate the thermal performance of passive designs. First-level methods involve a rule of thumb

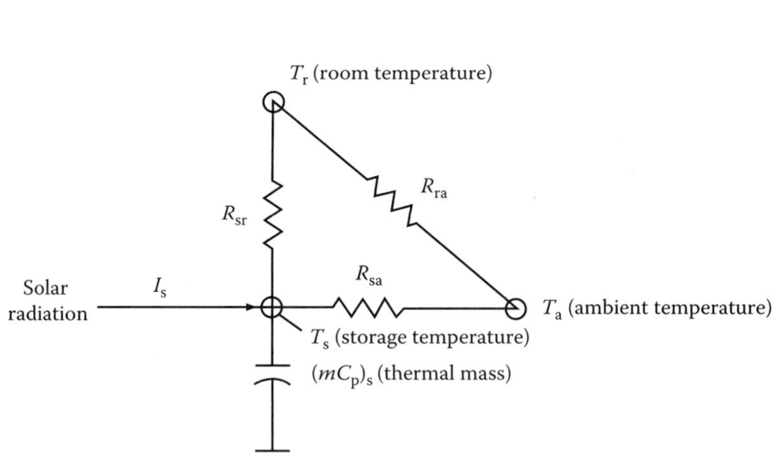

FIGURE 9.8
Equivalent thermal circuit for passively heated solar building in Figure 9.5. (From Kreith, F. and West, R., *CRC Handbook of Energy Efficiency and Renewable Energy*, CRC Press, Boca Raton, FL, 2007. With permission.)

and/or generalized calculation to get a starting estimate for size and/or annual performance. A second-level method involves climate, building, and passive system details, which allow annual performance determination, plus some sensitivity to passive system design changes. Third-level methods involve periodic calculations (hourly, monthly) of performance and permit more detailed variations of climatic, building, and passive solar system design parameters.

These three levels of design methods have a common basis in that they all are derived from correlations of a multitude of computer simulations of passive systems [4,5]. As a result, a similar set of defined terms is used in many passive design approaches:

- A_p, solar projected area, m^2 (ft^2): The net south-facing passive solar glazing area projected onto a vertical plane.
- NLC, net building load coefficient, kJ/CDD (Btu/FDD): Net load of the nonsolar portion of the building per day per degree of indoor–outdoor temperature difference. The CDD and FDD terms refer to centigrade and Fahrenheit degree days, respectively.
- Q_{net}, net reference load, Wh (Btu): Heat loss from nonsolar portion of building is calculated by

$$Q_{net} = \text{NLC} \times (\text{No. of degree days}) \tag{9.3}$$

- LCR, load collector ratio, kJ/m^2 CDD (Btu/ft^2 FDD): Ratio of NLC to A_p,

$$\text{LCR} = \frac{\text{NLC}}{A_p} \tag{9.4}$$

- SSF, solar savings fraction, %: Percentage reduction in required auxiliary heating relative to net reference load,

$$\text{SSF} = 1 - \frac{\text{Auxiliary heat required } (Q_{aux})}{\text{Net reference load } (Q_{net})} \tag{9.5}$$

So, using Equation 9.3,

$$\text{Auxiliary heat required, } Q_{aux} = (1 - \text{SSF}) \times \text{NLC} \times (\text{No. of degree days}). \tag{9.6}$$

The amount of auxiliary heat required is often a basis of comparison between possible solar designs as well as being the basis for determining building energy operating costs. Thus, many of the passive design methods are based on determining SSF, NLC, and the number of degree days in order to calculate the auxiliary heat required for a particular passive system by using Equation 9.6.

9.5.1 The First Level: Generalized Methods

A first estimate or starting value is needed to begin the overall passive system design process. Generalized methods and rules of thumb have been developed to generate initial values for solar aperture size, storage size, solar savings fraction, auxiliary heat required, and other size and performance characteristics. The following rules of thumb are meant to be used with the defined terms presented above.

9.5.1.1 Load

A rule of thumb used in conventional building design is that a design heating load of 120–160 kJ/CDD per m^2 of floor area (6–8 Btu/FDD ft^2) is considered an energy-conservative design. Reducing these nonsolar values by 20% to solarize the proposed south-facing solar wall gives rule-of-thumb NLC values per unit of floor area:

$$\frac{\text{NLC}}{\text{Floor area}} = 100 - 130\ \text{kJ/CDD m}^2(4.8 - 6.4\ \text{Btu/FDD ft}^2) \qquad (9.7)$$

9.5.1.2 Solar Savings Fraction

A method of getting starting-point values for the solar savings fraction is presented in Figure 9.9 [5]. The map values represent optimum SSF in % for

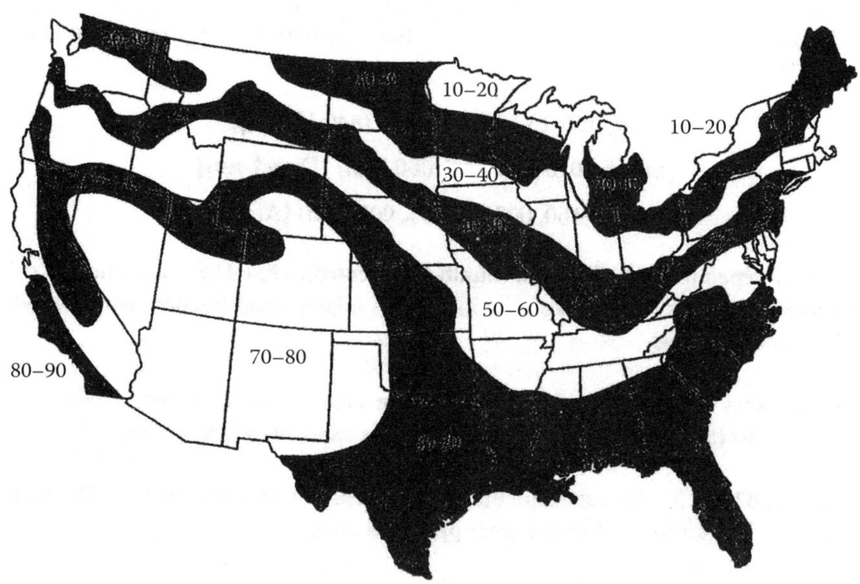

FIGURE 9.9
Starting-point values of SSF in percent. (From PSDH, *Passive Solar Design Handbook, Part One*: Total Environmental Action Inc., *Part Two*: Los Alamos Scientific Laboratory, *Part Three*: Los Alamos National Laboratory. Van Nostrand Reinhold, New York, 1984.)

a particular set of conservation and passive-solar costs for different climates across the United States. With the Q_{net} generated from the NLC rule of thumb above and the SSF read from the map, the Q_{aux} can be determined.

9.5.1.3 LCR

The A_p can be determined using the NLC from above, if the LCR is known. The rule of thumb associated with "good" values of LCR [5] differs depending on whether the design is for a "cold" or "warm" climate:

$$\text{"Good" LCR} = \begin{cases} \text{For cold climate: } 410\ \text{kJ/m}^2\ \text{CDD } (20\ \text{Btu/ft}^2\ \text{FDD}) \\ \text{For warm climate: } 610\ \text{kJ/m}^2\ \text{CDD } (30\ \text{Btu/ft}^2\ \text{FDD}) \end{cases} \tag{9.8}$$

9.5.1.4 Storage

Rules of thumb for thermal mass storage relate storage material total heat capacity to the solar projected area [5]. The use of the storage mass is to provide for heating on cloudy days and to regulate sunny day room air temperature swing. When the thermal mass directly absorbs the solar radiation, each square meter of the projected glazing area requires enough mass to store 613 kJ/C. If the storage material is not in direct sunlight, but heated from room air only, then four times as much mass is needed. In a room with a direct sunlight heated storage mass, the room air temperature swing will be approximately one-half the storage mass temperature swing. For room air–heated storage, the air temperature swing is twice that of the storage mass.

Example 9.2

A Denver, Colorado, building is to have a floor area of 195 m² (2100 ft²). Determine rule-of-thumb size and performance characteristics.

Solution

From Equation 9.5 the NLC is estimated as

$$\begin{aligned} \text{NLC} &= (115\ \text{kJ/CDDm}^2) \times (195\ \text{m}^2) \\ &= 22,400\ \text{kJ/CDD} (11,800\ \text{Btu/FDD}) \end{aligned}$$

Using the "cold" LCR value and Equation 9.4, the passive solar projected area is

$$\begin{aligned} A_p &= \frac{\text{NLC}}{\text{LCR}} = \frac{(22,400\ \text{kJ/CDD})}{(410\ \text{kJ/m}^2\text{CDD})} \\ &= 54.7\ \text{m}^2 (588\ \text{ft}^2) \end{aligned}$$

Locating Denver on the map of Figure 9.9 gives an SSF in the 70%–80% range (use 75%). An annual °C-degree-day value can be found in city climate tables [5,6], and is 3491 CDD (6283 FDD) for Denver. Thus, the auxiliary heat required, Q_{aux} is found using Equation 9.6:

$$Q_{aux} = (1-.75)(22,400 \text{ kJ/CDD})(3,491 \text{ CDD})$$

$$= 19,600 \text{ MJ } (18.5\times10^6 \text{ Btu}) \text{ annually}$$

The thermal storage can be sized using directly solar heated and/or room air heated mass by using the projected area. Assuming brick with a specific heat capacity of 840 J/kg °C; the storage mass is found by

$$A_p \times (613 \text{ kJ/C}) = m \times (840 \text{ J/kg °C})$$

$$m_d = 40,000 \text{ kg } (88,000 \text{ lbm}) \text{ Direct sun}$$

or

$$m_a = 160,000 \text{ kg } (351,000 \text{ lbm}) \text{ Air heated}$$

A more location-dependent set of rules of thumb is presented in [4]. The first rule of thumb relates solar projected area as a percentage of floor area to solar savings fraction, with and without night insulation of the solar glazing:

> A solar projected area of (*B*1)% to (*B*2)% of the floor area can be expected to produce an SSF in (location) of (*S*1)% to (*S*2)%, or, if R9 night insulation is used, of (*S*3)% to (*S*4)%.

where the values of *B*1, *B*2, *S*1, *S*2, *S*3, and *S*4 are found using Table 9.1 for the location. The thermal storage mass rule of thumb is again related to the solar projected area:

> A thermal storage wall should have 14 kg × SSF(%) of water or 71 kg × SSF (%) of masonry for each square meter of solar projected area. For a direct gain space, the mass above should be used with a surface area of at least three times the solar projected area, and masonry no thicker than 10–15 cm. If the mass is located in back rooms, then four times the above mass is needed.

Example 9.3

Determine size and performance passive solar characteristics with the location-dependent set of rules of thumb for the house of the previous example.

Solution

Using Table 9.1 with the 195 m² house in Denver yields

$$\text{Solar projected area} = 12\%-23\% \text{ of floor area}$$

$$= 23.4-44.9 \text{ m}^2$$

$$\text{SSF(no night insulation)} = 27\%-43\%$$

$$\text{SSF(R9 night insulation)} = 47\%-74\%$$

TABLE 9.1

Values to Be Used in the Glazing Area and SSF Relations Rules of Thumb

City	*B*1	*B*2	*S*1	*S*2	*S*3	*S*4
Birmingham, Alabama	0.09	0.18	22	37	34	58
Mobile, Alabama	0.06	0.12	26	44	34	60
Montgomery, Alabama	0.07	0.15	24	41	34	59
Phoenix, Arizona	0.06	0.12	37	60	48	75
Prescott, Arizona	0.10	0.20	29	48	44	72
Tucson, Arizona	0.06	0.12	35	57	45	73
Winslow, Arizona	0.12	0.24	30	47	48	74
Yuma, Arizona	0.04	0.09	43	66	51	78
Fort Smith, Arkansas	0.10	0.20	24	39	38	64
Little Rock, Arkansas	0.10	0.19	23	38	37	62
Bakersfield, California	0.08	0.15	31	50	42	67
Baggett, California	0.07	0.15	35	56	46	73
Fresno, California	0.09	0.17	29	46	41	65
Long Beach, California	0.05	0.10	35	58	44	72
Los Angeles, California	0.05	0.09	36	58	44	72
Mount Shasta, California	0.11	0.21	24	38	42	67
Needles, California	0.06	0.12	39	61	49	76
Oakland, California	0.07	0.15	35	55	46	72
Red Bluff, California	0.09	0.18	29	46	41	65
Sacramento, California	0.09	0.18	29	47	41	66
San Diego, California	0.04	0.09	37	61	46	74
San Francisco, California	0.06	0.13	34	54	45	71
Santa Maria, California	0.05	0.11	31	53	42	69
Colorado Springs, Colorado	0.12	0.24	27	42	47	74
Denver, Colorado	0.12	0.23	27	43	47	74
Eagle, Colorado	0.14	0.29	25	35	53	77
Grand Junction, Colorado	0.13	0.27	29	43	50	76
Pueblo, Colorado	0.11	0.23	29	45	48	75
Hartford, Connecticut	0.17	0.35	14	19	40	64
Wilmington, Delaware	0.15	0.29	19	30	39	63
Washington, District of Columbia	0.12	0.23	18	28	37	61
Apalachicola, Florida	0.05	0.10	28	47	36	61
Daytona Beach, Florida	0.04	0.08	30	51	36	63
Jacksonville, Florida	0.05	0.09	27	47	35	62
Miami, Florida	0.01	0.02	27	48	31	54
Orlando, Florida	0.03	0.06	30	52	37	63
Tallahassee, Florida	0.05	0.11	26	45	35	60
Tampa, Florida	0.03	0.06	30	52	36	63
West Palm Beach, Florida	0.01	0.03	30	51	34	59

(*continued*)

TABLE 9.1 (continued)

Values to Be Used in the Glazing Area and SSF Relations Rules of Thumb

City	*B*1	*B*2	*S*1	*S*2	*S*3	*S*4
Atlanta, Georgia	0.06	0.17	22	36	34	58
Augusta, Georgia	0.06	0.16	24	40	35	60
Macon, Georgia	0.07	0.15	25	41	35	59
Savannah, Georgia	0.06	0.13	25	43	35	60
Boise, Idaho	0.14	0.28	27	38	48	71
Lewiston, Idaho	0.15	0.29	22	29	44	65
Pocatello, Idaho	0.13	0.26	25	35	51	74
Chicago, Illinois	0.17	0.35	17	23	43	67
Moline, Illinois	0.20	0.39	17	22	46	70
Springfield; Illinois	0.15	0.30	19	26	42	67
Evansville, Indiana	0.14	0.27	19	29	37	61
Fort Wayne, Indiana	0.16	0.33	13	17	37	60
Indianapolis, Indiana	0.14	0.28	15	21	37	60
South Bend, Indiana	0.18	0.35	12	15	39	61
Burlington, Iowa	0.18	0.36	20	27	47	71
Des Moines, Iowa	0.21	0.43	19	25	58	75
Mason City, Iowa	0.22	0.44	18	19	56	79
Sioux City, Iowa	0.23	0.46	20	24	53	76
Dodge City, Kansas	0.12	0.23	27	42	46	73
Goodland, Kansas	0.13	0.27	26	39	47	74
Topeka, Kansas	0.14	0.26	24	35	45	71
Wichita, Kansas	0.14	0.26	26	41	45	72
Lexington, Kentucky	0.13	0.27	17	26	35	58
Louisville, Kentucky	0.13	0.27	18	27	35	59
Baton Rouge, Louisiana	0.06	0.12	26	43	34	59
Lake Charles, Louisiana	0.06	0.11	24	41	32	57
New Orleans, Louisiana	0.05	0.11	27	46	35	61
Shreveport, Louisiana	0.08	0.15	26	43	36	61
Caribou, Maine	0.25	0.30	NR		53	74
Portland, Maine	0.17	0.34	14	17	45	69
Baltimore, Maryland	0.14	0.27	19	30	38	62
Boston, Massachusetts	0.15	0.29	17	25	40	64
Alpena, Michigan	0.21	0.42	NR		47	69
Detroit, Michigan	0.17	0.34	13	17	39	61
Flint, Michigan	0.15	0.31	11	12	40	62
Grand Rapids, Michigan	0.19	0.38	12	13	39	61
Sault Ste. Marie, Michigan	0.25	0.50	NR		50	70
Traverse City, Michigan	0.18	0.36	NR		42	62
Duluth, Minnesota	0.25	0.50	NR		50	70
International Falls, Minnesota	0.25	0.50	NR		47	66

TABLE 9.1 (continued)

Values to Be Used in the Glazing Area and SSF Relations Rules of Thumb

City	*B*1	*B*2	*S*1	*S*2	*S*3	*S*4
Minneapolis-St. Paul, Minnesota	0.25	0.50	NR		55	76
Rochester, Minnesota	0.24	0.49	NR		54	76
Jackson, Mississippi	0.06	0.15	24	48	34	59
Meridian, Mississippi	0.08	0.15	23	39	34	58
Columbia, Missouri	0.13	0.26	20	30	41	66
Kansas City, Missouri	0.14	0.29	22	32	44	70
Saint Louis, Missouri	0.15	0.29	21	33	41	65
Springfield, Missouri	0.13	0.26	22	34	40	65
Billings, Montana	0.16	0.32	24	31	53	76
Cut Bank, Montana	0.24	0.49	22	23	62	81
Dillon, Montana	0.16	0.32	24	32	54	77
Glasgow, Montana	0.25	0.50	NR		55	75
Great Falls, Montana	0.18	0.37	23	26	56	77
Helena, Montana	0.20	0.39	21	25	55	77
Lewistown, Montana	0.19	0.38	21	25	54	76
Miles City, Montana	0.23	0.47	21	23	60	80
Missoula, Montana	0.18	0.36	15	16	47	68
Grand Island, Nebraska	0.18	0.36	24	33	51	76
North Omaha, Nebraska	0.20	0.48	21	29	51	76
North Platte, Nebraska	0.17	0.34	25	36	50	76
Scottsbluff, Nebraska	0.16	0.31	24	36	49	74
Elko, Nevada	0.12	0.25	27	39	52	76
Ely, Nevada	0.12	0.23	27	41	50	77
Las Vegas, Nevada	0.09	0.18	35	56	48	75
Lovelock, Nevada	0.13	0.25	32	48	53	78
Reno, Nevada	0.11	0.22	31	48	49	76
Tonopah, Nevada	0.11	0.23	31	48	51	77
Winnemucca, Nevada	0.13	0.26	28	42	49	75
Concord, New Hampshire	0.17	0.34	13	15	45	68
Newark, New Jersey	0.13	0.25	19	29	39	64
Albuquerque, New Mexico	0.11	0.22	29	47	46	73
Clayton, New Mexico	0.10	0.20	28	45	45	73
Farmington, New Mexico	0.12	0.24	29	45	49	76
Los Alamos, New Mexico	0.11	0.22	25	40	44	72
Roswell, New Mexico	0.10	0.19	30	49	45	73
Truth or Consequences, New Mexico	0.09	0.17	32	51	46	73
Tucumcari, New Mexico	0.10	0.20	30	48	45	73
Zuni, New Mexico	0.11	0.21	27	43	45	73
Albany, New York	0.21	0.41	13	15	43	66

(continued)

TABLE 9.1 (continued)

Values to Be Used in the Glazing Area and SSF Relations Rules of Thumb

City	*B*1	*B*2	*S*1	*S*2	*S*3	*S*4
Binghamton, New York	0.15	0.30	NR		35	56
Buffalo, New York	0.19	0.37	NR		36	57
Massena, New York	0.25	0.50	NR		50	71
New York (Central Park), New York	0.15	0.30	16	25	36	59
Rochester, New York	0.18	0.37	NR		37	58
Syracuse, New York	0.19	0.38	NR		37	59
Asheville, North Carolina	0.10	0.20	21	35	36	61
Cape Hatteras, North Carolina	0.09	0.17	24	40	36	60
Charlotte, North Carolina	0.08	0.17	23	38	36	60
Greensboro, North Carolina	0.10	0.20	23	37	37	63
Raleigh-Durham, North Carolina	0.09	0.19	22	37	36	61
Bismarck, North Dakota	0.25	0.50	NR		56	77
Fargo, North Dakota	0.25	0.50	NR		51	72
Minot, North Dakota	0.25	0.50	NR		52	72
Akron-Canton, Ohio	0.15	0.31	12	16	35	57
Cincinnati, Ohio	0.12	0.24	15	23	35	57
Cleveland, Ohio	0.15	0.31	11	14	34	55
Columbus, Ohio	0.14	0.28	13	18	35	57
Dayton, Ohio	0.14	0.28	14	20	36	59
Toledo, Ohio	0.17	0.34	13	17	38	61
Youngstown, Ohio	0.16	0.32	NR		34	54
Oklahoma City, Oklahoma	0.11	0.22	25	41	41	67
Tulsa, Oklahoma	0.11	0.22	24	38	40	65
Astoria, Oregon	0.09	0.19	21	34	37	60
Burns, Oregon	0.13	0.25	23	32	47	71
Medford, Oregon	0.12	0.24	21	32	38	60
North Bend, Oregon	0.09	0.17	25	42	38	64
Pendleton, Oregon	0.14	0.27	22	30	43	64
Portland, Oregon	0.13	0.26	21	31	38	60
Redmond, Oregon	0.13	0.27	26	38	47	71
Salem, Oregon	0.12	0.24	21	32	37	59
Allentown, Pennsylvania	0.15	0.29	16	24	39	63
Erie, Pennsylvania	0.17	0.34	NR		35	55
Harrisburg, Pennsylvania	0.13	0.26	17	26	38	62
Philadelphia, Pennsylvania	0.15	0.29	19	29	38	62
Pittsburgh, Pennsylvania	0.14	0.28	12	16	33	55
Wilkes-Barre-Scranton, Pennsylvania	0.16	0.32	13	18	37	60
Providence, Rhode Island	0.15	0.30	17	24	40	64
Charleston, South Carolina	0.07	0.14	25	41	34	59
Columbia, South Carolina	0.08	0.17	25	41	36	61

TABLE 9.1 (continued)

Values to Be Used in the Glazing Area and SSF Relations Rules of Thumb

City	*B1*	*B2*	*S1*	*S2*	*S3*	*S4*
Greenville-Spartanburg, South Carolina	0.08	0.17	23	38	36	60
Huron, South Dakota	0.25	0.50	NR		58	79
Pierre, South Dakota	0.22	0.43	21	23	58	80
Rapid City, South Dakota	0.15	0.30	23	32	51	76
Sioux Falls, South Dakota	0.22	0.45	18	19	57	79
Chattanooga, Tennessee	0.09	0.19	19	32	33	56
Knoxville, Tennessee	0.09	0.18	20	33	33	56
Memphis, Tennessee	0.09	0.19	22	36	36	60
Nashville, Tennessee	0.10	0.21	19	30	33	55
Abilene, Texas	0.09	0.18	29	47	41	68
Amarillo, Texas	0.11	0.22	29	46	45	72
Austin, Texas	0.06	0.13	27	46	37	63
Brownsville, Texas	0.03	0.06	27	46	32	57
Corpus Christi, Texas	0.05	0.09	29	49	36	63
Dallas, Texas	0.08	0.17	27	44	38	64
Del Rio, Texas	0.06	0.12	30	50	39	66
El Paso, Texas	0.09	0.17	32	53	45	72
Forth Worth, Texas	0.09	0.17	26	44	38	64
Houston, Texas	0.06	0.11	25	43	34	59
Laredo, Texas	0.05	0.09	31	52	39	64
Lubbock, Texas	0.09	0.19	30	49	44	72
Lufkin, Texas	0.07	0.14	26	43	35	61
Midland-Odessa, Texas	0.09	0.18	32	52	44	72
Port Arthur, Texas	0.06	0.11	26	44	34	60
San Angelo, Texas	0.08	0.15	29	48	40	67
San Antonio, Texas	0.06	0.12	28	48	38	64
Sherman, Texas	0.10	0.20	25	41	38	64
Waco, Texas	0.06	0.15	27	45	38	64
Wichita Falls, Texas	0.10	0.20	27	45	41	67
Bryce Canyon, Utah	0.13	0.25	26	39	52	78
Cedar City, Utah	0.12	0.24	28	43	48	75
Salt Lake City, Utah	0.13	0.26	27	39	48	72
Burlington, Vermont	0.22	0.43	NR		46	68
Norfolk, Virginia	0.09	0.19	23	38	37	62
Richmond, Virginia	0.11	0.22	21	34	37	61
Roanoke, Virginia	0.11	0.23	21	34	37	61
Olympia, Washington	0.12	0.23	20	29	38	59
Seattle-Tacoma, Washington	0.11	0.22	21	30	39	59
Spokane, Washington	0.20	0.39	20	24	48	68

(*continued*)

TABLE 9.1 (continued)

Values to Be Used in the Glazing Area and SSF Relations Rules of Thumb

City	*B*1	*B*2	*S*1	*S*2	*S*3	*S*4
Yakima, Washington	0.18	0.36	24	31	49	70
Charleston, West Virginia	0.13	0.25	16	24	32	54
Huntington, West Virginia	0.13	0.25	17	27	34	57
Eau Claire, Wisconsin	0.25	0.50	NR		53	75
Green Bay, Wisconsin	0.23	0.46	NR		53	75
La Crosse, Wisconsin	0.21	0.43	NR		52	75
Madison, Wisconsin	0.20	0.40	15	17	51	74
Milwaukee, Wisconsin	0.18	0.35	15	18	48	71
Casper Wyoming	0.13	0.26	27	39	53	78
Cheyenne, Wyoming	0.11	0.21	25	38	47	74
Rock Springs, Wyoming	0.14	0.28	26	30	54	79
Sheridan, Wyoming	0.16	0.31	22		52	75
Canada						
Edmonton, Alberta	0.25	0.50	NR		54	72
Suffield, Alberta	0.25	0.50	28	30	67	85
Nanaimo, British Columbia	0.13	0.26	26	35	45	66
Vancouver, British Columbia	0.13	0.26	20	28	48	60
Winnipeg, Manitoba	0.25	0.50	NR		54	74
Dartmouth, Nova Scotia	0.14	0.28	17	24	45	70
Moosonee, Ontario	0.25	0.50	NR		48	67
Ottawa, Ontario	0.25	0.50	NR		59	80
Toronto, Ontario	0.18	0.36	17	23	44	68
Normandie, Quebec	0.25	0.50	NR		54	74

Sources: PSDH, *Passive Solar Design Handbook*. Volume One: *Passive Solar Design Concepts*, DOE/CS-127/1, March 1980. Prepared by Total Environmental Action, Inc. (B. Anderson, C. Michal, P. Temple, and D. Lewis); *Volume Two: Passive Solar Design Analysis*, DOE/CS-0127/2, January 1980. Prepared by Los Alamos Scientific Laboratory (J.D. Balcomb, D. Barley, R McFarland, J. Perry, W. Wray and S. Noll). U.S. Department of Energy, Washington, DC, 1980.

Note: NR, not recommended.

Using the rule of thumb for the thermal storage mass:

$$m = 17\ \text{kg} \times 43\% \times 44.9\ \text{m}^2$$

$$= 33{,}000\ \text{kg}\ (72{,}000\ \text{lbm}) \quad \text{Thermal wall or direct gain}$$

Comparing the results of this example to those of the previous example, the two rules of thumb are seen to produce "roughly" similar answers. General system cost and performance information can be generated with results from rule-of-thumb calculations, but a more detailed level of information is needed to determine design-ready passive system type (direct gain, thermal wall, sunspace), size, performance, and costs.

9.5.2 The Second Level: LCR Method

The LCR method is useful for making estimates of the annual performance of specific types of passive system(s) combinations. The LCR method was developed by calculating the annual SSF for 94 reference passive solar systems for 219 U.S. and Canadian locations over a range of LCR values. Table 9.2 includes the description of these 94 reference systems for use both with the LCR method and with the SLR method described below. Tables were constructed for each city with LCR versus SSF listed for each of the 94 reference passive systems. (*Note*: The solar load ratio (SLR) method was used to make the LCR calculations, and this SLR method is described in the next section as the third-level method.) While the complete LCR tables [5] includes 219 locations, Table 9.3 only includes six "representative" cities (Albuquerque, Boston, Madison, Medford, Nashville, Santa Maria), purely due to space restrictions. The LCR method consists of the following steps [5]:

1. Determine the building parameters:
   a. Building load coefficient, NLC
   b. Solar projected area, A_p
   c. Load collector ratio, LCR = NLC/A_p
2. Find the short designation of the reference system closest to the passive system design (Table 9.2).
3. Enter the LCR Tables (Table 9.3):
   a. Find the city
   b. Find the reference system listing
   c. Determine annual SSF by interpolation using the LCR value from above
   d. Note the annual heating degree days (No. of degree of days)
4. Calculate the annual auxiliary heat required:

$$\text{Auxiliary heat required} = (1 - \text{SSF}) \times \text{NLC} \times (\text{No. of degree days})$$

If more than one reference solar system is being used, then find the "aperture area weighted" SSF for the combination. Determine each individual reference system SSF using the total aperture area LCR, then take the "area weighted" average of the individual SSFs.

TABLE 9.2

Designations and Characteristics for 94 Reference Systems

(a) Overall system characteristics	
Masonry properties	
Thermal conductivity (k)	
Sunspace floor	0.5 Btu/h/ft/°F
All other masonry	1.0 Btu/h/ft/°F
Density (Q)	150 lb/ft^3
Specific heat (c)	0.2 Btu/lb/°F
Infrared emittance of normal surface	0.9
Infrared emittance of selective surface	0.1
Solar absorptances	
Waterwall	1.0
Masonry, Trombe wall	1.0
Direct gain and sunspace	0.8
Sunspace: water containers	0.9
Lightweight common wall	0.7
Other lightweight surfaces	0.3
Glazing properties	
Transmission characteristics	Diffuse
Orientation	Due south
Index of refraction	1.526
Extinction of coefficient	0.5 in.$^{-1}$
Thickness of each pane	One-eighth inch
Gap between panes	One-half inch
Infrared emittance	0.9
Control range	
Room temperature	65°F–75°F
Sunspace temperature	45°F–95°F
Internal heat generation	0
Thermocirculation vents (when used)	
Vent area/projected area (sum of both upper and lower vents)	0.06
Height between vents	8 ft
Reverse flow	None
Nighttime insulation (when used)	
Thermal resistance	R9
In place, solar time	5:30 PM to 7:30 AM
Solar radiation assumptions	
Shading	None
Ground diffuse reflectance	0.3

TABLE 9.2 (continued)

Designations and Characteristics for 94 Reference Systems

(b) Direct-gain (DG) system types

Designation	Thermal Storage Capacity[a] (Btu/ft²/°F)	Mass Thickness[a] (in.)	Mass Area-to-Glazing Area Ratio	No. of Glazings	Nighttime Insulation
A1	30	2	6	2	No
A2	30	2	6	3	No
A3	30	2	6	2	Yes
B1	45	6	3	2	No
B2	45	6	3	3	No
B3	45	6	3	2	Yes
C1	60	4	6	2	No
C2	60	4	6	3	No
C3	60	4	6	2	Yes

(c) Vented trombe-wall (TW) system types

Designation	Thermal Storage Capacity[a] (Btu/ft²/°F)	Wall Thickness[a] (in.)	ρck (Btu²/h/ft⁴/°F²)	No. of Glazings	Wall Surface	Nighttime Insulation
A1	15	6	30	2	Normal	No
A2	22.5	9	30	2	Normal	No
A3	30	12	30	2	Normal	No
A4	45	18	30	2	Normal	No
B1	15	6	15	2	Normal	No
B2	22.5	9	15	2	Normal	No
B3	30	12	15	2	Normal	No
B4	45	18	15	2	Normal	No
C1	15	6	7.5	2	Normal	No
C2	22.5	9	7.5	2	Normal	No
C3	30	12	7.5	2	Normal	No
C4	45	18	7.5	2	Normal	No
D1	30	12	30	1	Normal	No
D2	30	12	30	3	Normal	No
D3	30	12	30	1	Normal	Yes
D4	30	12	30	2	Normal	Yes
D5	30	12	30	3	Normal	Yes
E1	30	12	30	1	Selective	No
E2	30	12	30	2	Selective	No
E3	30	12	30	1	Selective	Yes
E4	30	12	30	2	Selective	Yes

(continued)

TABLE 9.2 (continued)

Designations and Characteristics for 94 Reference Systems

(d) Unvented trombe-wall (TW) system types

Designation	Thermal Storage Capacity[a] (Btu/ft²/°F)	Wall Thickness[a] (in.)	ρck (Btu²/h/ft⁴/°F²)	No. of Glazings	Wall Surface	Nighttime Insulation
F1	15	6	30	2	Normal	No
F2	22.5	9	30	2	Normal	No
F3	30	12	30	2	Normal	No
F4	45	18	30	2	Normal	No
G1	15	6	15	2	Normal	No
G2	22.5	9	15	2	Normal	No
G3	30	12	15	2	Normal	No
G4	45	18	15	2	Normal	No
H1	15	6	7.5	2	Normal	No
H2	22.5	9	7.5	2	Normal	No
H3	30	12	7.5	2	Normal	No
H4	45	18	7.5	2	Normal	No
I1	30	12	30	1	Normal	No
I2	30	12	30	3	Normal	No
I3	30	12	30	1	Normal	Yes
I4	30	12	30	2	Normal	Yes
I5	30	12	30	3	Normal	Yes
J1	30	12	30	1	Selective	No
J2	30	12	30	2	Selective	No
J3	30	12	30	1	Selective	Yes
J4	30	12	30	2	Selective	Yes

(e) Waterwall (WW) system types

Designation	Thermal Storage Capacity[a] (Btu/ft²/°F)	Wall Thickness[a] (in.)	No. of Glazings	Wall Surface	Nighttime Insulation
A1	15.6	3	2	Normal	No
A2	31.2	6	2	Normal	No
A3	46.8	9	2	Normal	No
A4	62.4	12	2	Normal	No
A5	93.6	18	2	Normal	No
A6	124.8	24	2	Normal	No
B1	46.8	9	1	Normal	No
B2	46.8	9	3	Normal	No

TABLE 9.2 (continued)

Designations and Characteristics for 94 Reference Systems

Designation	Thermal Storage Capacity[a] (Btu/ft²/°F)	Wall Thickness[a] (in.)	No. of Glazings	Wall Surface	Nighttime Insulation
B3	46.8	9	1	Normal	Yes
B4	46.8	9	2	Normal	Yes
B5	46.8	9	3	Normal	Yes
C1	46.8	9	1	Selective	No
C2	46.8	9	2	Selective	No
C3	46.8	9	1	Selective	Yes
C4	46.8	9	2	Selective	Yes

(f) Sunspace (SS) system types

Designation	Type	Tilt (°)	Common Wall	End Walls	Nighttime Insulation
A1	Attached	50	Masonry	Opaque	No
A2	Attached	50	Masonry	Opaque	Yes
A3	Attached	50	Masonry	Glazed	No
A4	Attached	50	Masonry	Glazed	Yes
A5	Attached	50	Insulated	Opaque	No
A6	Attached	50	Insulated	Opaque	Yes
A7	Attached	50	Insulated	Glazed	No
A8	Attached	50	Insulated	Glazed	Yes
B1	Attached	90/3	Masonry	Opaque	No
B2	Attached	90/3	Masonry	Opaque	Yes
B3	Attached	90/3	Masonry	Glazed	No
B4	Attached	90/3	Masonry	Glazed	Yes
B5	Attached	90/3	Insulated	Opaque	No
B6	Attached	90/3	Insulated	Opaque	Yes
B7	Attached	90/3	Insulated	Glazed	No
B8	Attached	90/3	Insulated	Glazed	Yes
C1	Semienclosed	90	Masonry	Common	No
C2	Semienclosed	90	Masonry	Common	Yes
C3	Semienclosed	90	Insulated	Common	No
C4	Semienclosed	90	Insulated	Common	Yes
D1	Semienclosed	50	Masonry	Common	No
D2	Semienclosed	50	Masonry	Common	Yes
D3	Semienclosed	50	Insulated	Common	No
D4	Semienclosed	50	Insulated	Common	Yes

(continued)

TABLE 9.2 (continued)

Designations and Characteristics for 94 Reference Systems

Designation	Type	Tilt (°)	Common Wall	End Walls	Nighttime Insulation
E1	Semienclosed	90/3	Masonry	Common	No
E2	Semienclosed	90/3	Masonry	Common	Yes
E3	Semienclosed	90/3	Insulated	Common	No
E4	Semienclosed	90/3	Insulated	Common	Yes

Source: PSDH, *Passive Solar Design Handbook. Part One:* Total Environmental Action, Inc., *Part Two:* Los Alamos Scientific Laboratory, *Part Three:* Los Alamos National Laboratory. Van Nostrand Reinhold, New York, 1984.

[a] The thermal storage capacity is per unit of projected area, or, equivalently, the quantity ρck. The wall thickness is listed only as an appropriate guide by assuming ρc = 30 Btu/ft³/°F.

TABLE 9.3

LCR Tables for Six Representative Cities (Albuquerque, Boston, Madison, Medford, Nashville, Santa Maria)

SSF	0.10	0.20	0.30	0.40	0.50	0.60	0.70	0.80	0.90
Santa Maria, California								3053	DD
WW A1	1776	240	119	73	50	35	25	18	12
WW A2	617	259	154	103	74	54	39	28	19
WW A3	523	261	164	114	82	61	45	33	22
WW A4	482	260	169	119	87	65	48	35	24
WW A5	461	263	175	125	92	69	52	38	26
WW A6	447	263	177	128	95	72	54	40	27
WW B1	556	220	128	85	60	43	32	23	15
WW B2	462	256	168	119	88	66	49	36	25
WW B3	542	315	211	151	112	85	64	47	32
WW B4	455	283	197	144	109	83	63	47	32
WW B5	414	263	184	136	103	79	60	45	31
WW C1	569	330	221	159	118	89	67	49	33
WW C2	478	288	197	143	107	81	61	45	31
WW C3	483	318	228	170	130	100	77	57	40
WW C4	426	280	200	149	114	88	68	51	35
TW A1	1515	227	113	70	48	34	24	17	11
TW A2	625	234	134	89	63	46	33	24	16
TW A3	508	231	140	95	68	50	37	27	18
TW A4	431	217	137	95	69	51	38	28	19
TW B1	859	212	112	71	49	35	25	18	12
TW B2	502	209	124	83	59	43	32	23	15
TW B3	438	201	123	84	60	44	33	24	16
TW B4	400	184	112	76	55	40	30	22	14
TW C1	568	188	105	69	48	35	25	18	12

TABLE 9.3 (continued)

LCR Tables for Six Representative Cities (Albuquerque, Boston, Madison, Medford, Nashville, Santa Maria)

SSF	0.10	0.20	0.30	0.40	0.50	0.60	0.70	0.80	0.90
TW C2	435	178	105	70	50	36	27	19	13
TW C3	413	165	97	64	46	33	25	18	12
TW C4	426	146	82	54	38	27	20	14	10
TW D1	403	170	101	67	48	35	25	18	12
TW D2	488	242	152	105	76	57	42	31	21
TW D3	509	271	175	123	90	67	50	36	25
TW D4	464	266	177	127	94	71	53	39	27
TW D5	425	250	169	122	91	69	52	38	26
TW E1	581	309	199	140	102	76	57	42	28
TW E2	512	283	186	132	97	73	55	40	27
TW E3	537	328	225	164	123	94	71	53	36
TW E4	466	287	199	145	109	83	63	47	32
TW F1	713	198	107	68	47	34	25	18	12
TW F2	455	199	120	81	58	42	31	22	15
TW F3	378	190	120	83	60	45	33	24	16
TW F4	311	169	110	77	57	42	32	23	16
TW G1	450	170	98	65	46	33	24	17	12
TW G2	331	163	102	70	51	38	28	20	14
TW G3	278	147	94	66	48	36	27	20	13
TW G4	222	120	78	55	40	30	22	16	11
TW H1	295	137	84	57	41	30	22	16	11
TW H2	226	118	75	52	38	28	21	15	10
TW H3	187	99	64	44	33	24	18	13	9
TW H4	143	75	48	33	24	18	14	10	7
TW I1	318	144	88	59	42	31	23	16	11
TW I2	377	203	132	93	68	51	38	28	19
TW I3	404	226	149	106	78	58	44	32	22
TW I4	387	230	156	113	84	64	48	36	24
TW I5	370	226	155	113	85	65	49	36	25
TW J1	483	271	179	127	94	71	53	39	26
TW J2	422	246	165	119	88	67	50	37	25
TW J3	446	283	199	149	111	85	65	48	33
TW J4	400	254	178	132	100	77	58	43	30
DG A1	392	188	117	79	55	38	26	16	7
DG A2	389	190	121	85	61	45	32	22	14
DG A3	443	220	142	102	77	58	44	31	19
DG B1	384	191	122	86	64	48	35	24	13
DG B2	394	196	127	91	69	53	40	29	19
DG B3	445	222	145	105	80	62	49	37	25
DG C1	451	225	146	104	78	61	47	34	21

(continued)

TABLE 9.3 (continued)

LCR Tables for Six Representative Cities (Albuquerque, Boston, Madison, Medford, Nashville, Santa Maria)

SSF	0.10	0.20	0.30	0.40	0.50	0.60	0.70	0.80	0.90
DG C2	453	226	148	106	80	63	49	37	25
DG C3	509	254	167	121	92	73	58	45	31
SS A1	1171	396	220	142	98	69	49	34	22
SS A2	1028	468	283	190	135	98	71	50	33
SS A3	1174	380	209	133	91	64	45	31	20
SS A4	1077	481	289	193	136	98	71	50	32
SS A5	1896	400	204	127	86	60	42	29	18
SS A6	1030	468	283	190	135	97	71	50	32
SS A7	2199	359	178	109	72	50	35	24	15
SS A8	1089	478	285	190	133	96	69	48	31
SS B1	802	298	170	111	77	55	40	28	18
SS B2	785	366	224	152	108	79	57	41	27
SS B3	770	287	163	106	74	52	37	26	17
SS B4	790	368	224	152	108	78	57	40	26
SS B5	1022	271	144	91	62	44	31	22	14
SS B6	750	356	219	149	106	77	56	40	26
SS B7	937	242	127	80	54	38	27	19	12
SS B8	750	352	215	146	103	75	55	39	25
SS C1	481	232	144	99	71	52	39	28	19
SS C2	482	262	170	120	88	66	49	36	24
SS C3	487	185	107	71	50	36	27	19	13
SS C4	473	235	147	102	74	55	41	30	20
SS D1	1107	477	282	188	132	95	68	48	31
SS D2	928	511	332	232	169	125	92	66	43
SS D3	1353	449	248	160	110	78	56	39	25
SS D4	946	500	319	222	160	117	86	61	40
SS E1	838	378	227	153	108	78	56	40	26
SS E2	766	419	272	190	138	102	75	54	36
SS E3	973	322	178	115	79	56	40	28	18
SS E4	780	393	247	170	122	89	65	47	31
Albuquerque, New Mexico								4292	DD
WW A1	1052	130	62	38	25	18	13	9	6
WW A2	354	144	84	56	39	29	21	15	10
WW A3	300	146	90	62	45	33	24	18	12
WW A4	276	146	93	65	47	35	26	19	13
WW A5	264	148	97	69	50	38	28	21	14
WW A6	256	148	99	70	52	39	30	22	15
WW B1	293	111	63	41	28	20	15	11	7
WW B2	270	147	96	67	49	37	28	20	14
WW B3	314	179	119	84	62	47	35	26	18

TABLE 9.3 (continued)

LCR Tables for Six Representative Cities (Albuquerque, Boston, Madison, Medford, Nashville, Santa Maria)

SSF	0.10	0.20	0.30	0.40	0.50	0.60	0.70	0.80	0.90
WW B4	275	169	116	85	64	49	37	28	19
WW B5	252	159	110	81	61	47	36	27	19
WW C1	333	190	126	89	66	50	38	28	19
WW C2	287	171	115	83	62	47	36	27	18
WW C3	293	191	136	101	77	59	46	34	24
WW C4	264	172	122	91	69	54	41	31	22
TW A1	900	124	60	37	25	17	12	9	6
TW A2	361	130	73	48	33	24	18	13	8
TW A3	293	129	77	52	37	27	20	15	10
TW A4	249	123	76	52	38	28	21	15	10
TW B1	502	117	60	38	26	18	13	9	6
TW B2	291	118	68	45	32	23	17	12	8
TW B3	254	114	68	46	33	24	18	13	9
TW B4	233	104	63	42	30	22	16	12	8
TW C1	332	106	58	37	26	19	14	10	6
TW C2	255	101	58	39	27	20	15	11	7
TW C3	243	94	54	36	25	18	13	10	7
TW C4	254	84	46	30	21	15	11	8	5
TW D1	213	86	50	33	23	17	12	9	6
TW D2	287	139	86	59	43	32	24	17	12
TW D3	294	153	97	68	49	37	27	20	14
TW D4	281	158	104	74	55	41	31	23	16
TW D5	260	151	101	73	57	41	31	23	16
TW E1	339	177	113	78	54	43	32	23	16
TW E2	308	168	109	77	56	42	32	23	16
TW E3	323	195	133	96	72	55	42	31	21
TW E4	287	175	120	88	66	50	38	28	20
TW F1	409	108	57	36	24	17	13	9	6
TW F2	260	110	65	43	31	22	17	12	8
TW F3	216	106	66	45	33	24	10	13	9
TW F4	178	95	61	42	31	23	17	13	9
TW G1	256	93	53	34	24	17	13	9	6
TW G2	189	91	56	38	27	20	15	11	7
TW G3	159	82	52	36	26	20	15	11	7
TW G4	128	68	43	30	22	16	12	9	6
TW H1	168	76	45	31	22	16	12	9	6
TW H2	130	66	41	29	21	15	11	8	6
TW H3	108	56	35	25	8	13	10	7	5
TW H4	83	42	27	19	13	10	7	5	4

(continued)

TABLE 9.3 (continued)

LCR Tables for Six Representative Cities (Albuquerque, Boston, Madison, Medford, Nashville, Santa Maria)

SSF	0.10	0.20	0.30	0.40	0.50	0.60	0.70	0.80	0.90
TW I1	166	73	43	29	20	15	11	8	5
TW I2	221	117	75	52	30	28	21	16	11
TW I3	234	128	83	59	43	32	24	10	12
TW I4	234	137	92	66	49	37	28	21	14
TW I5	226	136	93	67	50	38	29	22	15
TW J1	282	156	102	72	53	40	30	22	15
TW J2	254	146	97	69	51	39	29	22	15
TW J3	269	169	118	86	65	50	38	29	20
TW J4	247	155	106	80	60	46	35	26	18
DG A1	211	97	57	36	22	13	5	—	—
DG A2	227	107	67	46	32	23	16	10	5
DG A3	274	131	83	59	44	34	25	18	10
DG B1	210	97	60	42	30	21	13	6	—
DG B2	232	110	69	49	37	28	21	14	8
DG B3	277	134	85	61	47	37	28	21	14
DG C1	253	120	74	53	39	30	22	14	—
DG C2	271	130	82	59	45	35	26	19	12
DG C3	318	155	96	71	54	43	34	26	18
SS A1	591	187	101	64	44	31	22	16	10
SS A2	531	232	137	92	65	47	34	25	16
SS A3	566	170	90	56	38	27	19	13	8
SS A4	537	230	135	89	63	45	33	23	15
SS A5	980	187	92	56	37	26	18	13	8
SS A6	529	231	136	91	64	47	34	24	16
SS A7	1103	158	74	44	29	20	14	10	6
SS A8	540	226	131	87	61	44	32	23	15
SS B1	403	141	78	50	35	25	18	13	8
SS B2	412	189	111	75	53	39	28	20	14
SS B3	372	130	71	46	31	22	16	11	7
SS B4	403	181	106	72	51	37	27	20	13
SS B5	518	127	65	40	27	19	13	9	6
SS B6	390	179	106	73	52	38	28	20	13
SS B7	457	108	54	33	22	16	11	8	5
SS B8	379	171	102	69	49	35	26	19	12
SS C1	270	126	77	52	37	27	20	15	10
SS C2	282	150	97	68	49	37	28	20	14
SS C3	276	101	57	37	26	19	14	10	7
SS C4	277	135	83	57	41	31	23	17	11

TABLE 9.3 (continued)

LCR Tables for Six Representative Cities (Albuquerque, Boston, Madison, Medford, Nashville, Santa Maria)

SSF	0.10	0.20	0.30	0.40	0.50	0.60	0.70	0.80	0.90
SS D1	548	225	130	85	59	43	31	22	14
SS D2	474	253	162	113	82	61	45	33	22
SS D3	683	212	113	72	49	35	25	17	11
SS D4	484	248	156	107	77	57	42	30	20
SS E1	410	176	103	68	48	35	25	18	12
SS E2	390	208	133	92	67	50	37	27	18
SS E3	487	151	80	51	35	25	18	12	8
SS E4	400	195	120	82	59	43	32	23	15
Nashville, Tennessee								3696	DD
WW A1	588	60	24	13	8	5	3	2	1
WW A2	192	70	38	23	15	11	7	5	3
WW A3	161	72	42	27	18	13	9	6	4
WW A4	148	72	43	29	20	14	10	7	5
WW A5	141	74	46	31	22	16	11	8	5
WW A6	137	74	47	32	22	16	12	8	5
WW B1	135	41	19	10	6	3	2	—	—
WW B2	152	78	48	33	23	17	12	9	6
WW B3	179	97	61	42	30	22	16	12	8
WW B4	164	97	65	46	34	25	19	14	9
WW B5	153	93	63	45	33	25	19	14	9
WW C1	193	105	67	46	33	24	18	13	8
WW C2	169	97	63	44	32	24	18	13	8
WW C3	181	115	79	58	43	33	25	18	12
WW C4	164	104	72	53	39	30	23	17	11
TW A1	509	59	25	13	8	5	3	2	1
TW A2	199	64	33	20	13	9	6	4	3
TW A3	160	65	36	23	15	11	8	5	3
TW A4	136	62	36	23	16	11	8	6	4
TW B1	282	57	26	15	9	6	4	3	2
TW B2	161	59	32	20	13	9	6	4	3
TW B3	141	58	32	21	14	10	7	5	3
TW B4	131	54	30	19	13	9	7	5	3
TW C1	188	53	27	16	10	7	5	3	2
TW C2	144	52	28	18	12	8	6	4	2
TW C3	139	49	27	17	11	8	5	4	2
TW C4	149	45	23	14	9	7	5	3	2
TW D1	99	33	16	9	5	3	2	1	—
TW D2	164	75	44	29	20	14	10	7	5
TW D3	167	82	49	33	23	17	12	8	5

(*continued*)

TABLE 9.3 (continued)

LCR Tables for Six Representative Cities (Albuquerque, Boston, Madison, Medford, Nashville, Santa Maria)

SSF	0.10	0.20	0.30	0.40	0.50	0.60	0.70	0.80	0.90
TW D4	168	91	58	40	29	21	15	11	7
TW D5	160	89	58	40	29	22	16	12	8
TW E1	198	98	59	40	28	20	15	10	7
TW E2	182	95	59	40	29	21	15	11	7
TW E3	197	115	76	54	39	29	22	16	11
TW E4	178	105	70	50	37	27	20	15	10
TW F1	221	50	23	13	8	5	4	2	1
TW F2	139	53	29	18	12	8	6	4	2
TW F3	116	52	30	19	13	9	7	5	3
TW F4	96	47	28	19	13	9	7	5	3
TW G1	137	44	22	13	9	6	4	3	2
TW G2	101	44	25	16	11	8	5	4	2
TW G3	86	41	24	16	11	8	6	4	2
TW G4	69	34	21	14	10	7	5	3	2
TW H1	89	36	20	13	8	6	4	3	2
TW H2	69	33	19	12	9	6	4	3	2
TW H3	59	28	17	11	8	5	4	3	2
TW H4	46	22	13	9	6	4	3	2	1
TW I1	74	26	13	7	4	2	1	—	—
TW I2	125	62	38	25	18	13	9	7	4
TW I3	133	69	43	29	20	15	11	8	5
TW I4	139	78	51	35	26	19	14	10	7
TW I5	137	80	53	37	27	20	15	11	7
TW J1	164	86	54	36	26	19	14	10	6
TW J2	150	82	53	36	26	19	14	10	7
TW J3	165	101	68	49	36	27	20	15	10
TW J4	153	93	63	46	34	25	19	14	10
DG A1	98	34	—	—	—	—	—	—	—
DG A2	130	55	31	19	11	6	—	—	—
DG A3	173	78	47	32	23	16	11	7	2
DG B1	100	36	17	—	—	—	—	—	—
DG B2	134	58	33	22	15	10	6	—	—
DG B3	177	81	49	33	24	18	14	10	6
DG C1	131	52	28	17	9	—	—	—	—
DG C2	161	71	42	28	20	14	10	6	—
DG C3	205	94	57	39	29	22	17	12	8
SS A1	351	100	50	29	19	13	9	6	4
SS A2	328	135	76	49	33	24	17	12	8
SS A3	330	87	41	24	15	10	6	4	2

TABLE 9.3 (continued)

LCR Tables for Six Representative Cities (Albuquerque, Boston, Madison, Medford, Nashville, Santa Maria)

SSF	0.10	0.20	0.30	0.40	0.50	0.60	0.70	0.80	0.90
SS A4	331	133	74	47	32	22	16	11	7
SS A5	595	98	43	24	15	10	7	4	2
SS A6	324	132	75	48	32	23	16	11	7
SS A7	668	79	32	17	10	6	4	2	1
SS A8	330	129	71	45	30	21	15	10	6
SS B1	236	74	38	23	15	10	7	5	3
SS B2	258	110	63	41	28	20	14	10	6
SS B3	212	65	32	19	12	8	5	3	2
SS B4	251	105	60	39	27	19	13	9	6
SS B5	307	65	30	17	10	7	4	3	2
SS B6	241	104	60	39	27	19	14	10	6
SS B7	264	52	23	12	7	5	3	2	—
SS B8	133	98	56	36	25	17	12	9	5
SS C1	141	60	33	21	14	10	7	5	3
SS C2	161	81	50	33	23	17	12	9	6
SS C3	149	48	25	15	10	7	4	3	2
SS C4	160	73	43	28	19	14	10	7	5
SS D1	317	119	64	39	26	18	13	8	5
SS D2	287	147	90	61	43	31	23	16	10
SS D3	405	113	55	33	21	14	10	6	4
SS D4	295	144	87	58	40	29	21	15	10
SS E1	229	89	48	29	19	13	9	6	4
SS E2	233	118	72	48	34	24	18	12	8
SS E3	283	77	37	22	14	9	6	4	2
SS E4	242	111	65	43	29	21	15	11	7
Medford, Oregon								4930	DD
WW A1	708	64	24	11	—	—	—	—	—
WW A2	212	73	38	22	13	7	3	—	—
WW A3	174	75	41	25	16	9	5	2	—
WW A4	158	74	43	27	17	11	6	3	1
WW A5	149	75	45	29	19	12	7	4	2
WW A6	144	75	46	30	20	13	8	4	2
WW B1	154	43	16	—	—	—	—	—	—
WW B2	162	80	48	31	21	14	9	6	3
WW B3	190	100	62	41	28	19	13	8	5
WW B4	171	99	65	45	32	23	16	11	7
WW B5	160	95	63	45	32	23	17	12	7
WW C1	205	108	67	45	31	21	15	10	6
WW C2	178	99	63	43	30	22	15	10	6

(continued)

TABLE 9.3 (continued)

LCR Tables for Six Representative Cities (Albuquerque, Boston, Madison, Medford, Nashville, Santa Maria)

SSF	0.10	0.20	0.30	0.40	0.50	0.60	0.70	0.80	0.90
WW C3	189	117	80	57	42	31	23	16	10
WW C4	170	106	72	52	38	28	21	15	9
TW A1	607	63	25	12	5	—	—	—	—
TW A2	222	68	33	19	11	6	2	—	—
TW A3	175	67	36	21	13	8	4	2	—
TW A4	147	64	36	22	14	9	5	3	1
TW B1	327	61	27	14	7	3	—	—	—
TW B2	178	62	32	19	12	7	4	2	—
TW B3	154	60	33	20	12	8	4	2	1
TW B4	143	56	31	19	12	8	5	2	1
TW C1	212	56	27	15	9	5	2	—	—
TW C2	159	55	28	17	11	7	4	2	—
TW C3	154	52	27	16	10	6	4	2	1
TW C4	167	48	24	14	9	5	3	2	—
TW D1	112	34	14	—	—	—	—	—	—
TW D2	177	77	44	28	18	12	8	5	3
TW D3	180	85	50	32	21	14	9	6	3
TW D4	177	93	58	39	27	19	13	9	5
TW D5	168	92	58	40	28	20	14	10	6
TW E1	213	101	60	39	26	18	12	8	4
TW E2	194	98	59	39	27	19	13	9	5
TW E3	208	118	77	53	38	27	20	13	8
TW E4	186	108	71	49	36	26	19	13	8
TW F1	256	53	23	12	5	—	—	—	—
TW F2	153	56	29	17	10	5	2	—	—
TW F3	125	54	30	18	11	7	3	1	—
TW F4	102	48	28	18	11	7	4	2	1
TW G1	153	46	22	12	7	—	—	—	—
TW G2	109	46	25	15	9	5	3	1	—
TW G3	92	42	24	15	9	6	3	2	—
TW G4	74	35	20	13	8	5	3	2	—
TW H1	97	38	20	12	7	4	1	—	—
TW H2	75	34	19	12	7	5	3	1	—
TW H3	63	29	17	10	7	4	3	1	—
TW H4	49	23	13	8	5	3	2	1	—
TW I1	83	27	10	—	—	—	—	—	—
TW I2	133	64	38	24	16	11	7	4	2
TW I3	142	71	43	28	19	13	9	5	3
TW I4	146	80	51	35	25	17	12	8	5
TW I5	144	82	53	37	26	19	13	9	6
TW J1	175	89	54	36	24	17	11	7	4

TABLE 9.3 (continued)

LCR Tables for Six Representative Cities (Albuquerque, Boston, Madison, Medford, Nashville, Santa Maria)

SSF	0.10	0.20	0.30	0.40	0.50	0.60	0.70	0.80	0.90
TW J2	158	85	53	36	25	18	12	8	5
TW J3	173	103	69	48	35	26	18	13	8
TW J4	160	96	64	45	33	24	17	12	8
DG A1	110	35	—	—	—	—	—	—	—
DG A2	142	58	32	18	9	—	—	—	—
DG A3	187	82	48	32	22	15	9	5	—
DG B1	110	40	15	—	—	—	—	—	—
DG B2	146	61	35	21	13	7	—	—	—
DG B3	193	84	51	34	24	17	12	7	3
DG C1	144	57	29	13	—	—	—	—	—
DG C2	177	75	44	28	19	12	6	—	—
DG C3	224	98	60	41	29	21	14	10	5
SS A1	415	110	51	28	16	9	4	2	—
SS A2	372	146	79	48	31	21	14	8	5
SS A3	397	96	42	21	10	—	—	—	—
SS A4	379	144	76	46	29	19	12	7	4
SS A5	732	111	45	23	12	5	—	—	—
SS A6	368	143	77	47	30	20	13	8	4
SS A7	846	90	33	14	—	—	—	—	—
SS A8	379	140	73	44	27	17	11	6	3
SS B1	274	81	38	21	12	6	3	—	—
SS B2	288	117	65	40	26	18	12	7	4
SS B3	249	71	33	17	8	—	—	—	—
SS B4	282	113	62	38	25	16	11	7	4
SS B5	368	72	30	15	7	—	—	—	—
SS B6	269	111	62	30	25	17	11	7	4
SS B7	323	58	23	10	—	—	—	—	—
SS B8	262	106	57	35	23	15	9	6	3
SS C1	153	62	33	19	11	5	—	—	—
SS C2	172	83	50	32	22	15	10	6	3
SS C3	166	51	24	13	7	3	—	—	—
SS C4	173	76	43	27	18	12	8	5	3
SS D1	367	129	65	37	22	13	7	3	1
SS D2	318	159	92	60	40	27	18	12	7
SS D3	480	124	57	31	18	10	5	2	—
SS D4	328	153	89	57	38	26	17	11	6
SS E1	262	95	48	27	15	7	—	—	—
SS E2	257	124	73	47	31	21	14	9	5
SS E3	334	84	38	20	10	4	—	—	—
SS E4	269	118	67	42	27	18	12	7	4

(continued)

TABLE 9.3 (continued)

LCR Tables for Six Representative Cities (Albuquerque, Boston, Madison, Medford, Nashville, Santa Maria)

SSF	0.10	0.20	0.30	0.40	0.50	0.60	0.70	0.80	0.90
Boston, Massachusetts								5621	DD
WW A1	368	28	9	—	—	—	—	—	—
WW A2	119	41	20	12	7	5	3	2	—
WW A3	101	43	24	15	10	6	4	3	1
WW A4	93	44	26	16	11	7	5	3	2
WW A5	89	45	27	18	12	8	6	4	2
WW A6	87	46	28	19	13	9	6	4	3
WW B1	59	—	—	—	—	—	—	—	—
WW B2	103	52	31	21	15	10	7	5	3
WW B3	123	66	41	28	20	14	10	7	5
WW B4	118	70	46	33	24	18	13	9	6
WW B5	113	69	46	33	25	18	14	10	7
WW C1	135	72	46	31	22	16	12	8	5
WW C2	121	68	44	31	22	16	12	9	6
WW C3	136	86	60	44	33	25	19	14	9
WW C4	124	78	54	40	30	23	17	12	8
TW A1	324	30	11	4	—	—	—	—	—
TW A2	126	37	18	10	6	4	2	1	—
TW A3	102	39	21	13	8	5	3	2	1
TW A4	88	38	22	14	9	6	4	3	2
TW B1	180	32	13	7	4	2	—	—	—
TW B2	104	36	19	11	7	5	3	2	1
TW B3	92	36	19	12	8	5	3	2	1
TW B4	86	34	19	12	8	5	4	2	1
TW C1	122	32	15	9	5	3	2	1	—
TW C2	95	33	17	10	7	4	3	2	1
TW C3	93	31	16	10	6	4	3	2	1
TW C4	102	29	15	9	6	4	3	2	1
TW D1	45	—	—	—	—	—	—	—	—
TW D2	112	49	28	18	12	9	6	4	3
TW D3	113	54	32	21	15	10	7	5	3
TW D4	121	64	41	28	20	15	11	8	5
TW D5	118	66	42	30	21	16	12	8	6
TW E1	138	67	40	27	18	13	9	7	4
TW E2	130	66	41	28	20	14	10	7	5
TW E3	146	84	56	39	29	21	16	11	8
TW E4	133	78	52	37	27	20	15	11	7
TW F1	134	25	10	4	—	—	—	—	—
TW F2	86	30	16	9	5	3	2	1	—
TW F3	72	31	17	11	7	4	3	2	1

TABLE 9.3 (continued)

LCR Tables for Six Representative Cities (Albuquerque, Boston, Madison, Medford, Nashville, Santa Maria)

SSF	0.10	0.20	0.30	0.40	0.50	0.60	0.70	0.80	0.90
TW F4	61	29	17	11	7	5	3	2	1
TW G1	83	24	11	6	3	2	—	—	—
TW G2	63	26	14	9	5	4	2	1	—
TW G3	54	25	14	9	6	4	3	2	1
TW G4	45	21	12	8	5	4	3	2	1
TW H1	54	21	11	6	4	2	1	—	—
TW H2	44	20	11	7	5	3	2	1	—
TW H3	38	17	10	6	4	3	2	1	—
TW H4	30	14	8	5	3	2	2	1	—
TW I1	30	—	—	—	—	—	—	—	—
TW I2	84	41	24	16	11	8	6	4	2
TW I3	91	46	28	19	13	9	7	5	3
TW I4	100	56	36	25	18	13	10	7	5
TW I5	101	58	38	27	20	15	11	8	5
TW J1	114	59	37	25	17	12	9	6	4
TW J2	107	58	37	25	18	13	10	7	4
TW J3	123	75	51	36	27	20	15	11	7
TW J4	115	70	47	34	25	19	14	10	7
DG A1	43	—	—	—	—	—	—	—	—
DG A2	85	34	18	9	—	—	—	—	—
DG A3	125	56	33	22	16	11	7	4	—
DG B1	44	—	—	—	—	—	—	—	—
DG B2	87	36	20	12	7	—	—	—	—
DG B3	129	58	35	24	17	13	9	6	3
DG C1	71	23	—	—	—	—	—	—	—
DG C2	109	47	27	17	12	8	4	—	—
DG C3	151	68	41	28	21	16	12	8	5
SS A1	230	61	29	16	10	6	4	2	1
SS A2	231	93	52	33	22	15	11	7	5
SS A3	205	48	20	10	4	—	—	—	—
SS A4	229	90	49	31	20	14	9	6	4
SS A5	389	58	23	11	6	3	—	—	—
SS A6	226	91	50	32	21	15	10	7	4
SS A7	420	40	12	—	—	—	—	—	—
SS A8	226	86	46	28	19	12	8	6	3
SS B1	151	44	21	12	7	4	2	1	—
SS B2	183	77	43	28	19	13	9	6	4
SS B3	129	36	16	8	3	—	—	—	—
SS B4	176	73	41	26	17	12	8	6	4
SS B5	193	36	15	7	3	—	—	—	—

(continued)

TABLE 9.3 (continued)

LCR Tables for Six Representative Cities (Albuquerque, Boston, Madison, Medford, Nashville, Santa Maria)

SSF	0.10	0.20	0.30	0.40	0.50	0.60	0.70	0.80	0.90
SS B6	169	72	41	26	18	12	9	6	4
SS B7	157	25	7	—	—	—	—	—	—
SS B8	160	66	37	23	16	11	7	5	3
SS C1	84	33	17	10	6	4	2	1	—
SS C2	110	54	33	22	15	11	8	5	3
SS C3	91	26	12	7	4	2	—	—	—
SS C4	109	48	28	18	12	9	6	4	3
SS D1	206	73	38	22	14	9	5	3	2
SS D2	203	103	63	42	29	21	15	10	6
SS D3	264	69	32	18	10	6	4	2	1
SS D4	208	100	60	39	27	19	14	9	6
SS E1	140	51	25	14	8	4	2	—	—
SS E2	161	80	48	32	22	15	11	7	5
SS E3	177	44	19	10	5	2	—	—	—
SS E4	166	75	43	28	19	13	9	6	4
Madison, Wisconsin								7730	DD
WW A1	278	—	—	—	—	—	—	—	—
WW A2	91	27	12	—	—	—	—	—	—
WW A3	77	30	15	8	3	—	—	—	—
WW A4	72	32	17	10	5	—	—	—	—
WW A5	69	33	19	11	7	4	—	—	—
WW A6	67	34	19	12	7	4	2		
WW B1	—	—	—	—	—	—	—	—	—
WW B2	84	41	24	15	10	7	5	3	2
WW B3	102	53	32	21	15	10	7	5	3
WW B4	101	59	39	27	19	14	10	7	5
WW B5	98	59	39	28	20	15	11	8	5
WW C1	113	59	37	25	17	12	8	6	3
WW C2	103	57	37	25	18	13	9	6	4
WW C3	119	75	51	37	28	21	15	11	7
WW C4	109	68	47	34	25	19	14	10	7
TW A1	249	16	—	—	—	—	—	—	—
TW A2	97	26	11	4	—	—	—	—	—
TW A3	79	28	13	7	3	—	—	—	—
TW A4	69	28	15	9	5	3	—	—	—
TW B1	139	20	5	—	—	—	—	—	—
TW B2	81	26	12	6	3	—	—	—	—
TW B3	72	27	13	7	4	2	—	—	—
TW B4	69	26	13	8	5	3	1	—	—
TW C1	96	23	10	4	—	—	—	—	—

TABLE 9.3 (continued)

LCR Tables for Six Representative Cities (Albuquerque, Boston, Madison, Medford, Nashville, Santa Maria)

SSF	0.10	0.20	0.30	0.40	0.50	0.60	0.70	0.80	0.90
TW C2	76	25	12	7	4	2	—	—	—
TW C3	75	24	12	7	4	2	1	—	—
TW C4	84	23	11	6	4	2	1	—	—
TW D1	—	—	—	—	—	—	—	—	—
TW D2	91	39	22	13	9	6	4	2	1
TW D3	93	43	25	16	10	7	5	3	1
TW D4	103	54	34	23	16	12	8	6	4
TW D5	102	56	36	25	18	13	10	7	4
TW E1	115	54	32	21	14	10	7	4	3
TW E2	110	55	34	22	16	11	8	5	3
TW E3	126	72	47	33	24	18	13	9	6
TW E4	116	68	45	32	23	17	13	9	6
TW F1	99	13				—	—	—	—
TW F2	65	20	8	—	—	—	—	—	—
TW F3	55	22	11	5	—	—	—	—	—
TW F4	47	21	11	7	4	2	—	—	—
TW G1	61	14	—	—	—	—	—	—	—
TW G2	47	18	8	4	—	—	—	—	—
TW G3	42	18	9	5	3	—	—	—	—
TW G4	35	16	9	5	3	2			
TW H1	41	13	6	—	—	—	—	—	—
TW H2	34	14	7	4	2	—	—	—	—
TW H3	29	13	7	4	2	1	—	—	—
Tw H4	24	10	6	3	2	1	—	—	—
TW I1	—	—	—	—	—	—	—	—	—
TW I2	68	32	18	12	8	5	3	2	1
TW I3	75	37	22	14	10	7	4	3	2
Tw I4	85	47	30	21	15	11	8	5	3
TW 15	87	50	33	23	16	12	9	6	4
TW J1	95	48	29	19	13	9	6	4	3
TW J2	91	48	30	21	14	10	7	5	3
TW J3	106	65	43	31	23	17	12	9	6
TW J4	100	61	41	29	21	16	12	9	6
DG A1	—	—	—	—	—	—	—	—	—
DG A2	68	25	11	—	—	—	—	—	—
DG A3	109	47	28	18	12	8	5	—	—
DG B1	—	—	—	—	—	—	—	—	—
DG B2	70	27	14	6	—	—	—	—	—
DG B3	114	50	30	20	14	10	7	4	—

(continued)

TABLE 9.3 (continued)

LCR Tables for Six Representative Cities (Albuquerque, Boston, Madison, Medford, Nashville, Santa Maria)

SSF	0.10	0.20	0.30	0.40	0.50	0.60	0.70	0.80	0.90
DG C1	47	—	—	—	—	—	—	—	—
DG C2	91	37	21	13	7	—	—	—	—
DG C3	133	59	35	24	17	13	9	6	3
SS A1	192	47	20	9	3	—	—	—	—
SS A2	200	78	42	26	17	12	8	5	3
SS A3	166	32	—	—	—	—	—	—	—
SS A4	197	74	39	23	15	10	6	4	2
SS A5	329	42	13	—	—	—	—	—	—
SS A6	195	75	40	25	16	11	7	5	3
SS A7	349	22	—	—	—	—	—	—	—
SS A8	192	69	36	21	13	8	5	3	2
SS B1	122	32	13	5	—	—	—	—	—
SS B2	158	64	36	22	15	10	7	5	3
SS B3	100	22	—	—	—	—	—	—	—
SS B4	150	60	33	29	13	9	6	4	2
SS B5	156	24	—	—	—	—	—	—	—
SS B6	145	59	33	20	13	9	6	4	2
SS B7	122	—	—	—	—	—	—	—	—
SS B8	136	54	29	18	11	7	5	3	2
SS C1	61	20	7	—	—	—	—	—	—
SS C2	90	43	25	16	11	7	5	3	2
SS C3	67	16	—	—	—	—	—	—	—
SS C4	90	38	22	13	9	6	4	2	1
SS D1	169	56	26	13	6	—	—	—	—
SS D2	175	86	51	34	23	16	11	7	5
SS D3	221	52	21	10	—	—	—	—	—
SS D4	179	84	49	32	21	15	10	7	4
SS E1	108	34	12	—	—	—	—	—	—
SS E2	135	65	38	24	16	11	7	5	3
SS E3	141	29	8	—	—	—	—	—	—
SS E4	140	61	34	21	14	9	6	4	2

Sources: PSDH, *Passive Solar Design Handbook*. Volume One: *Passive Solar Design Concepts*, DOE/CS-127/1, March 1980. Prepared by Total Environmental Action, Inc. (B. Anderson, C. Michal, P. Temple, and D. Lewis); *Volume Two: Passive Solar Design Analysis*, DOE/CS-0127/2, January 1980. Prepared by Los Alamos Scientific Laboratory (J.D. Balcomb, D. Barley, R McFarland, J. Perry, W. Wray and S. Noll). U.S. Department of Energy, Washington, DC, 1980.

The LCR method allows no variation from the 94 reference passive designs. To treat off-reference designs, sensitivity curves have been produced that illustrate the effect on SSF of varying one or two design variables. These curves were produced for the six "representative" cities, chosen for their wide geographical and climatological ranges. Several of these SSF "sensitivity curves" are presented in Figure 9.10 for storage wall (a, b, c) and sunspace (d) design variations.

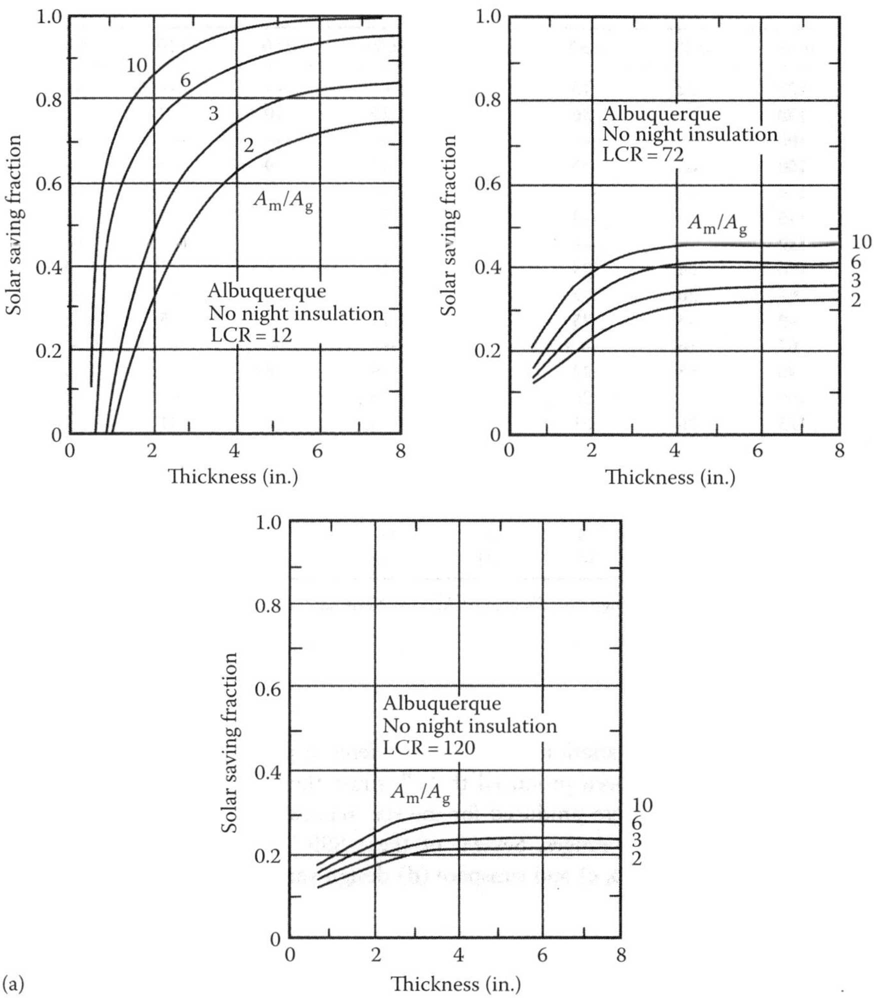

(a)

FIGURE 9.10
(a) Storage wall: Mass thickness. Sensitivity of SSF to off-reference conditions.

(continued)

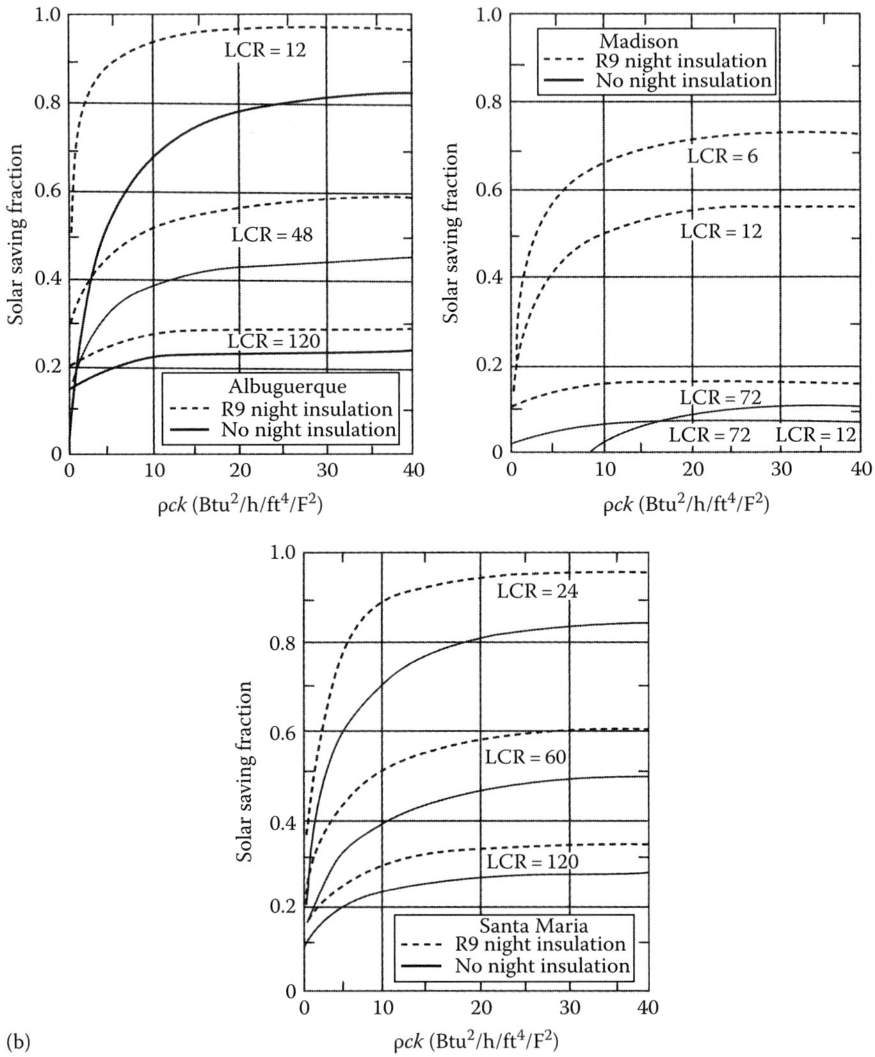

FIGURE 9.10 (continued)
(b) Storage wall: ρ*ck* product.

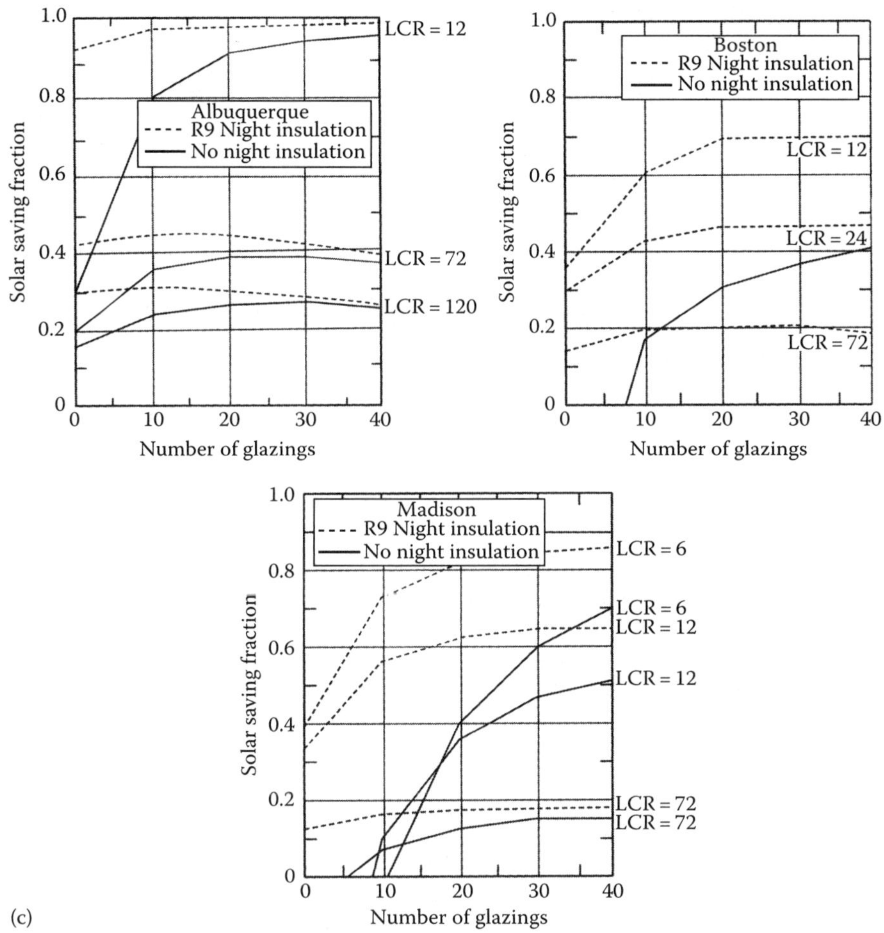

FIGURE 9.10 (continued)
(c) Storage wall: Number of glazings.

(continued)

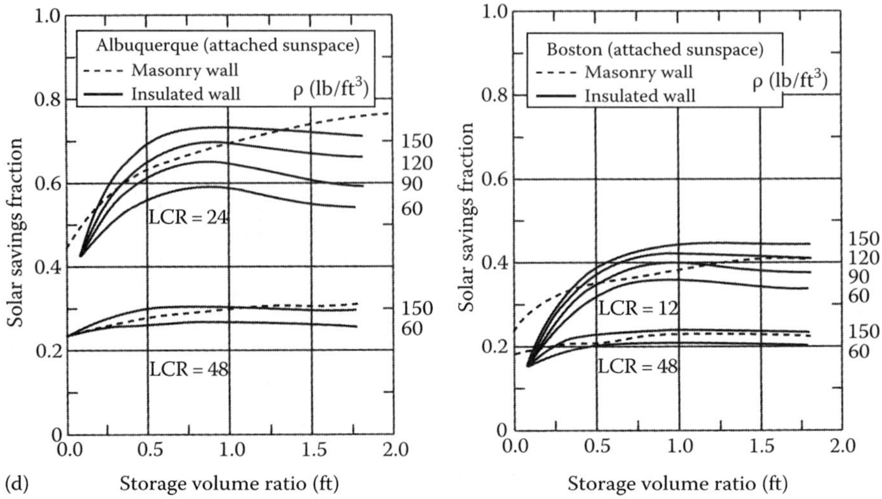

FIGURE 9.10 (continued)
(d) Sunspace: Storage volume to projected area ratio. (From PSDH, *Passive Solar Design Handbook, Part One*: Total Environmental Action Inc., *Part Two*: Los Alamos Scientific Laboratory, *Part Three*: Los Alamos National Laboratory. Van Nostrand Reinhold, New York, 1984.)

Example 9.4

The previously used 2,100 ft^2 building with NLC = 11,800 Btu/FDD is preliminarily designed to be located in Medford, Oregon, with 180 ft^2 of 12 in. thick vented Trombe wall and 130 ft^2 of direct gain, both systems having double glazing, nighttime insulation, and 30 Btu/ft^2 thermal storage capacity. Determine the annual auxiliary energy needed by this design.

Solution
Step 1 yields:

$$\text{NLC} = 11{,}800 \text{ Btu/FDD}$$

$$A_p = 180 + 130 = 20 \text{ ft}^2$$

$$\text{LCR} = \frac{11{,}800}{320} = 36.8 \text{ Btu/FDD ft}^2$$

Step 2 yields: From Table 9.2, the short designations for the appropriate systems are

TWD4 (Trombe wall)

DGA3 (Direct gain)

Step 3 yields: From Table 9.3 for Medford, Oregon, with LCR = 36.8

TWD4: SSF(TW) = 0.42

DGA3: SSF(DG) = 0.37

Determine the "weighted area" average SSF:

$$\text{SSF} = \frac{180(0.42) + 130(0.37)}{320} = 0.39$$

Step 4 yields: Using Equation 9.6 and reading 4930 FDD from Table 9.3

$$Q_{aux} = (1 - 0.39) \times 11{,}800 \text{ Btu} \times 4{,}930 \text{ FDD} = 35.5 \times 10^6 \text{ Btu annually}$$

Using the reference system characteristics yields the thermal storage size:

- Trombe wall ($\rho ck = 30$, concrete properties from Table 9.2c):

$$\begin{aligned} m(\text{TW}) &= \text{density} \times \text{area} \times \text{thickness} \\ &= 150 \text{ lbm/ft}^3 \times 180 \text{ ft}^2 \times 1 \text{ ft} \\ &= 27{,}000 \text{ lbm} \end{aligned}$$

- Direct gain ($\rho ck = 30$, concrete properties) using mass area to glazing area ratio of 6:

$$\text{mass area} = 6 \times 130 = 780 \text{ ft}^2 \text{ of 2 in. thick concrete}$$

$$\begin{aligned} m(\text{DG}) &= 150 \text{ lbm/ft}^3 \times 780 \text{ ft}^2 \times 1/6 \text{ ft} \\ &= 19{,}500 \text{ lbm} \end{aligned}$$

Using the LCR method allows a basic design of passive system types for the 94 reference systems and the resulting annual performance. A bit more design variation can be obtained by using the sensitivity curves of Figure 9.10 to modify the SSF of a particular reference system. For instance, a direct gain system SSF of 0.37 would increase by approximately 0.03 if the mass area-to-glazing area ratio (assumed 6) were increased to 10, and would decrease by about 0.04 if the mass area-to-glazing area ratio were decreased to 3. This information provides a designer with quantitative information for making trade-offs.

9.5.3 The Third Level: SLR Method

The solar-load ratio (SLR) method calculates monthly performance, and the terms and values used are monthly based. The method allows the use of specific location weather data and the 94 reference design passive systems (Table 9.2). In addition, the sensitivity curves (Figure 9.10) can again be used to define performance outside the reference design systems. The result of the SLR method is the determination of the monthly heating auxiliary energy required, which is then summed to give the annual requirement for auxiliary heating energy. Generally, the SLR method gives annual values within ±3% of detailed simulation results, but the monthly values may vary more [5,7]. Thus, the monthly SLR method is more "accurate" than the rule-of-thumb methods, plus providing the designer with system performance on a month-by-month basis.

The SLR method uses equations and correlation parameters for each of the 94 reference systems combined with the insolation absorbed by the system, the monthly degree days, and the system's LCR to determine the monthly SSF. These correlation parameters are listed in Table 9.4 as *A*, *B*, *C*, *D*, *R*, *G*, *H*, and LCRs for each reference system [5]. The correlation equations are

$$\text{SSF} = 1 - K(1 - F) \tag{9.9}$$

where

$$K = 1 + \frac{G}{\text{LCR}} \tag{9.10}$$

$$F = \begin{cases} AX, & \text{when } X < R \\ B - C\exp(-DX), & \text{when } X > R \end{cases} \tag{9.11}$$

$$X = \frac{S/\text{DD} - (\text{LCRs})H}{(\text{LCR})K} \tag{9.12}$$

and *X* is called the generalized solar load ratio. The term *S* is the monthly insolation absorbed by the system per unit of solar projected area. Monthly average daily insolation data on a vertical south-facing surface can be found and/or calculated using various sources [5,8] and the *S* term can be determined by multiplying by a transmission and an absorption factor and the number of days in the month. Absorption factors for all systems are close to 0.96 [5], whereas the transmission is approximately 0.9 for single glazing, 0.8 for double glazing, and 0.7 for triple glazing.

Example 9.5

For a vented, 180 ft^2, double-glazed with night insulation, 12 in. thick Trombe wall system (TWD4) in a NLC = 11,800 Btu/FDD house in Medford, Oregon, determine the auxiliary energy required in January.

Solution

Weather data for Medford, Oregon [5] yields for January (N = 31, days): daily vertical surface insolation = 565 Btu/ft^2 and 880 FDD, so S = (31)(565)(0.8)(0.96) = 13,452 Btu/ft^2-month.

TABLE 9.4
SLR Correlation Parameters for the 94 Reference Systems

Type	*A*	*B*	*C*	*D*	*R*	*G*	*H*	LCR_s	STDV
WW A1	0.0000	1.0000	0.9172	0.4841	−9.0000	0.00	1.17	13.0	0.053
WW A2	0.0000	1.0000	0.9833	0.7603	−9.0000	0.00	0.92	13.0	0.046
WW A3	0.0000	1.0000	1.0171	0.8852	−9.0000	0.00	0.85	13.0	0.040
WW A4	0.0000	1.0000	1.0395	0.9569	−9.0000	0.00	0.81	13.0	0.037
WW A5	0.0000	1.0000	1.0604	1.0387	−9.0000	0.00	0.78	13.0	0.034
WW A6	0.0000	1.0000	1.0735	1.0827	−9.0000	0.00	0.76	13.0	0.033
WW B1	0.0000	1.0000	0.9754	0.5518	−9.0000	0.00	0.92	22.0	0.051
WW B2	0.0000	1.0000	1.0487	1.0851	−9.0000	0.00	0.78	9.2	0.036
WW B3	0.0000	1.0000	1.0673	1.0087	−9.0000	0.00	0.95	8.9	0.038
WW B4	0.0000	1.0000	1.1028	1.1811	−9.0000	0.00	0.74	5.8	0.034
WW B5	0.0000	1.0000	1.1146	1.2771	−9.0000	0.00	0.56	4.5	0.032
WW C1	0.0000	1.0000	1.0667	1.0437	−9.0000	0.00	0.62	12.0	0.038
WW C2	0.0000	1.0000	1.0846	1.1482	−9.0000	0.00	0.59	8.7	0.035
WW C3	0.0000	1.0000	1.1419	1.1756	−9.0000	0.00	0.28	5.5	0.033
WW C4	0.0000	1.0000	1.1401	1.2378	−9.0000	0.00	0.23	4.3	0.032
TW A1	0.0000	1.0000	0.9194	0.4601	−9.0000	0.00	1.11	13.0	0.048
TW A2	0.0000	1.0000	0.9680	0.6318	−9.0000	0.00	0.92	13.0	0.043
TW A3	0.0000	1.0000	0.9964	0.7123	−9.0000	0.00	0.85	13.0	0.038
TW A4	0.0000	1.0000	1.0190	0.7332	−9.0000	0.00	0.79	13.0	0.032
TW B1	0.0000	1.0000	0.9364	0.4777	−9.0000	0.00	1.01	13.0	0.045
TW B2	0.0000	1.0000	0.9821	0.6020	−9.0000	0.00	0.85	13.0	0.038
TW B3	0.0000	1.0000	0.9980	0.6191	−9.0000	0.00	0.80	13.0	0.033
TW B4	0.0000	1.0000	0.9981	0.5615	−9.0000	0.00	0.76	13.0	0.028
TW C1	0.0000	1.0000	0.9558	0.4709	−9.0000	0.00	0.89	13.0	0.039
TW C2	0.0000	1.0000	0.9788	0.4964	−9.0000	0.00	0.79	13.0	0.033
TW C3	0.0000	1.0000	0.9760	0.4519	−9.0000	0.00	0.76	13.0	0.029
TW C4	0.0000	1.0000	0.9588	0.3612	−9.0000	0.00	0.73	13.0	0.026
TW D1	0.0000	1.0000	0.9842	0.4418	−9.0000	0.00	0.89	22.0	0.040
TW D2	0.0000	1.0000	1.0150	0.8994	−9.0000	0.00	0.80	9.2	0.036
TW D3	0.0000	1.0000	1.0346	0.7810	−9.0000	0.00	1.08	8.9	0.036
TW D4	0.0000	1.0000	1.0606	0.9770	−9.0000	0.00	0.85	5.8	0.035
TW D5	0.0000	1.0000	1.0721	1.0718	−9.0000	0.00	0.61	4.5	0.033
TW E1	0.0000	1.0000	1.0345	0.8753	−9.0000	0.00	0.68	12.0	0.037
TW E2	0.0000	1.0000	1.0476	1.0050	−9.0000	0.00	0.66	8.7	0.035
TW E3	0.0000	1.0000	1.0919	1.0739	−9.0000	0.00	0.61	5.5	0.034
TW E4	0.0000	1.0000	1.0971	1.1429	−9.0000	0.00	0.47	4.3	0.033
TW F1	0.0000	1.0000	0.9430	0.4744	−9.0000	0.00	1.09	13.0	0.047
TW F2	0.0000	1.0000	0.9900	0.6053	−9.0000	0.00	0.93	13.0	0.041
TW F3	0.0000	1.0000	1.0189	0.6502	−9.0000	0.00	0.86	13.0	0.036
TW F4	0.0000	1.0000	1.0419	0.6258	−9.0000	0.00	0.80	13.0	0.032

(continued)

TABLE 9.4 (continued)

SLR Correlation Parameters for the 94 Reference Systems

Type	*A*	*B*	*C*	*D*	*R*	*G*	*H*	LCR_s	STDV
TW G1	0.0000	1.0000	0.9693	0.4714	−9.0000	0.00	1.01	13.0	0.042
TW G2	0.0000	1.0000	1.0133	0.5462	−9.0000	0.00	0.88	13.0	0.035
TW G3	0.0000	1.0000	1.0325	0.5269	−9.0000	0.00	0.82	13.0	0.031
TW G4	0.0000	1.0000	1.0401	0.4400	−9.0000	0.00	0.77	13.0	0.030
TW H1	0.0000	1.0000	1.0002	0.4356	−9.0000	0.00	0.93	13.0	0.034
TW H2	0.0000	1.0000	1.0280	0.4151	−9.0000	0.00	0.83	13.0	0.030
TW H3	0.0000	1.0000	1.0327	0.3522	−9.0000	0.00	0.78	13.0	0.029
TW H4	0.0000	1.0000	1.0287	0.2600	−9.0000	0.00	0.74	13.0	0.024
TW I1	0.0000	1.0000	0.9974	0.4036	−9.0000	0.00	0.91	22.0	0.038
TW I2	0.0000	1.0000	1.0386	0.8313	−9.0000	0.00	0.80	9.2	0.034
TW I3	0.0000	1.0000	1.0514	0.6886	−9.0000	0.00	1.01	8.9	0.034
TW I4	0.0000	1.0000	1.0781	0.8952	−9.0000	0.00	0.82	5.8	0.032
TW I5	0.0000	1.0000	1.0902	1.0284	−9.0000	0.00	0.65	4.5	0.032
TW J1	0.0000	1.0000	1.0537	0.8227	−9.0000	0.00	0.65	12.0	0.037
TW J2	0.0000	1.0000	1.0677	0.9312	−9.0000	0.00	0.62	8.7	0.035
TW J3	0.0000	1.0000	1.1153	0.9831	−9.0000	0.00	0.44	5.5	0.034
TW J4	0.0000	1.0000	1.1154	1.0607	−9.0000	0.00	0.38	4.3	0.033
DG A1	0.5650	1.0090	1.0440	0.7175	0.3931	9.36	0.00	0.0	0.046
DG A2	0.5906	1.0060	1.0650	0.8099	0.4681	5.28	0.00	0.00	0.039
DG A3	0.5442	0.9715	1.1300	0.9273	0.7068	2.64	0.00	0.00	0.036
DG B1	0.5739	0.9948	1.2510	1.0610	0.7905	9.60	0.00	0.00	0.042
DG B2	0.6180	1.0000	1.2760	1.1560	0.7528	5.52	0.00	0.00	0.035
DG B3	0.5601	0.9839	1.3520	1.1510	0.8879	2.38	0.00	0.00	0.032
DG C1	0.6344	0.9887	1.5270	1.4380	0.8632	9.60	0.00	0.00	0.039
DG C2	0.6763	0.9994	1.4000	1.3940	0.7604	5.28	0.00	0.00	0.033
DG C3	0.6182	0.9859	1.5660	1.4370	0.8990	2.40	0.00	0.00	0.031
SS A1	0.0000	1.0000	0.9587	0.4770	−9.0000	0.00	0.83	18.6	0.027
SS A2	0.0000	1.0000	0.9982	0.6614	−9.0000	0.00	0.77	10.4	0.026
SS A3	0.0000	1.0000	0.9552	0.4230	−9.0000	0.00	0.83	23.6	0.030
SS A4	0.0000	1.0000	0.9956	0.6277	−9.0000	0.00	0.80	12.4	0.026
SS A5	0.0000	1.0000	0.9300	0.4041	−9.0000	0.00	0.96	18.6	0.031
SS A6	0.0000	1.0000	0.9981	0.6660	−9.0000	0.00	0.86	10.4	0.028
SS A7	0.0000	1.0000	0.9219	0.3225	−9.0000	0.00	0.96	23.6	0.035
SS A8	0.0000	1.0000	0.9922	0.6173	−9.0000	0.00	0.90	12.4	0.028
SS B1	0.0000	1.0000	0.9683	0.4954	−9.0000	0.00	0.84	16.3	0.028
SS B2	0.0000	1.0000	1.0029	0.6802	−9.0000	0.00	0.74	8.5	0.026
SS B3	0.0000	1.0000	0.9689	0.4685	−9.0000	0.00	0.82	19.3	0.029
SS B4	0.0000	1.0000	1.0029	0.6641	−9.0000	0.00	0.76	9.7	0.026
SS B5	0.0000	1.0000	0.9408	0.3866	−9.0000	0.00	0.97	16.3	0.030
SS B6	0.0000	1.0000	1.0068	0.6778	−9.0000	0.00	0.84	8.5	0.028

TABLE 9.4 (continued)

SLR Correlation Parameters for the 94 Reference Systems

Type	*A*	*B*	*C*	*D*	*R*	*G*	*H*	LCR_s	STDV
SS B7	0.0000	1.0000	0.9395	0.3363	−9.0000	0.00	0.95	19.3	0.032
SS B8	0.0000	1.0000	1.0047	0.6469	−9.0000	0.00	0.87	9.7	0.027
SS C1	0.0000	1.0000	1.0087	0.7683	−9.0000	0.00	0.76	16.3	0.025
SS C2	0.0000	1.0000	1.0412	0.9281	−9.0000	0.00	0.78	10.0	0.027
SS C3	0.0000	1.0000	0.9699	0.5106	−9.0000	0.00	0.79	16.3	0.024
SS C4	0.0000	1.0000	1.0152	0.7523	−9.0000	0.00	0.81	10.0	0.025
SS D1	0.0000	1.0000	0.9889	0.6643	−9.0000	0.00	0.84	17.8	0.028
SS D2	0.0000	1.0000	1.0493	0.8753	−9.0000	0.00	0.70	9.9	0.028
SS D3	0.0000	1.0000	0.9570	0.5285	−9.0000	0.00	0.90	17.8	0.029
SS D4	0.0000	1.0000	1.0356	0.8142	−9.0000	0.00	0.73	9.9	0.028
SS E1	0.0000	1.0000	0.9968	0.7004	−9.0000	0.00	0.77	19.6	0.027
SS E2	0.0000	1.0000	1.0468	0.9054	−9.0000	0.00	0.76	10.8	0.027
SS E3	0.0000	1.0000	0.9565	0.4827	−9.0000	0.00	0.81	19.6	0.028
SS E4	0.0000	1.0000	1.0214	0.7694	−9.0000	0.00	0.79	10.8	0.027

Source: PSDH, *Passive Solar Design Handbook. Part One*: Total Environmental Action, Inc., *Part Two*: Los Alamos Scientific Laboratory, *Part Three*: Los Alamos National Laboratory. Van Nostrand Reinhold, New York, 1984.

$$\mathrm{LCR} = \frac{\mathrm{NLC}}{A_\mathrm{p}} = \frac{11,800}{180} = 65.6\ \mathrm{Btu/FDD\ ft^2}$$

From Table 9.4 at TWD4: $A = 0$, $B = 1$, $C = 1.0606$, $D = 0.977$, $R = -9$, $G = 0$, $H = 0.85$, LCRs = 5.8 Btu/FDD ft^2.

Substituting into Equation 9.10 gives

$$K = 1 + \frac{0}{65.6} = 1$$

Equation 9.12 gives

$$X = \frac{13,452/880 - (5.8 \times 0.85)}{65.6 \times 1} = 0.16$$

Equation 9.11 gives

$$F = 1 - 1.0606e^{-0.977 \times 0.16} = 0.09$$

and Equation 9.9 gives

$$\mathrm{SSF} = 1 - 1(1 - 0.09) = 0.09$$

The January auxiliary energy required can be calculated using Equation 9.6:

$$Q_{aux}(\text{January}) = (1 - \text{SSF}) \times \text{NLC} \times (\text{No. of degree days})$$
$$= (1 - 0.09) \times 11{,}800 \times 880$$
$$= 9{,}450{,}000 \text{ Btu}$$

As mentioned, the use of sensitivity curves [5] as in Figure 9.10 will allow SSF to be determined for many off-reference system design conditions involving storage mass, number of glazings, and other more esoteric parameters. Also, the use of multiple passive system types within one building would be approached by calculating the SSF for each system type individually using a "combined area" LCR, and then a weighted-area (aperture) average SSF would be determined for the building.

9.6 Passive Space Cooling Design Fundamentals

Passive cooling systems are designed to use natural means to transfer heat from buildings, including convection/ventilation, evaporation, radiation, and conduction. However, the most important element in both passive and conventional cooling design is to prevent heat from entering the building in the first place. Cooling conservation techniques involve building surface colors, insulation, special window glazings, overhangs and orientation, and numerous other architectural/engineering features.

9.6.1 Solar Control

Controlling the solar energy input to reduce the cooling load is usually considered a passive (versus conservation) design concern because solar input may be needed for other purposes, such as daylighting throughout the year and/or heating during the winter. Basic architectural solar control is normally "designed in" via the shading of the solar windows, where direct radiation is desired for winter heating and needs to be excluded during the cooling season.

The shading control of the windows can be of various types and "controllability," ranging from drapes and blinds, use of deciduous trees, to the commonly used overhangs and vertical louvers. A rule-of-thumb design for determining proper south-facing window overhang for both winter heating and summer shading is presented in Table 9.5. Technical details on calculating shading from various devices and orientations are found in [9,10].

9.6.2 Natural Convection/Ventilation

Air movement provides cooling comfort through convection and evaporation from human skin. [9] places the comfort limit at 79°F for an air velocity of 50 ft/min (fpm), 82°F for 160 fpm, and 85°F for 200 fpm. To determine

TABLE 9.5

South-Facing Window Overhang Rule of Thumb

$$\text{Length of the Overhang} = \frac{\text{Window Height}}{F}$$

(a) Overhang Factors

North Latitude	F[a]
28	5.6–11.1
32	4.0–6.3
36	3.0–4.5
40	2.5–3.4
44	2.0–2.7
48	1.7–2.2
52	1.5–1.8
56	1.3–1.5

(b) Roof Overhang Geometry

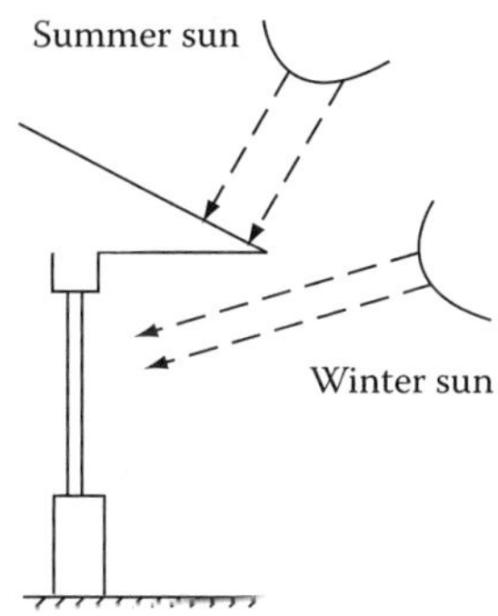

Properly sized overhangs shade out hot summer sun but allow winter sun (which is lower in the sky) to penetrate windows

Source: Halacy, D.S., *Home Energy*, Rodale Press, Emmaus, PA, 1984.

[a] Select a factor according to your latitude. High values provide complete shading at noon on June 21: lower values, until August 1.

whether or not comfort conditions can be obtained, a designer must calculate the volumetric flow rate, Q, which is passing through the occupied space. Using the cross-sectional area, A_x, of the space and the room air velocity, V_a, required, the flow is determined by

$$Q = A_x V_a. \tag{9.13}$$

The proper placement of windows, "narrow" building shape, and open landscaping can enhance natural wind flow to provide ventilation. The air flow rate through open windows for wind-driven ventilation is given by [9]:

$$Q = C_v V_w A_w \tag{9.14}$$

where

- Q is the air flow rate, m^3/s
- A_w is the free area of inlet opening, m^2
- V_w is the wind velocity, m/s
- C_v is the effectiveness of opening = 0.5–0.6 for wind perpendicular to opening, and 0.25–0.35 for wind diagonal to opening

The stack effect can induce ventilation when warm air rises to the top of a structure and exhausts outside, while cooler outside air enters the structure

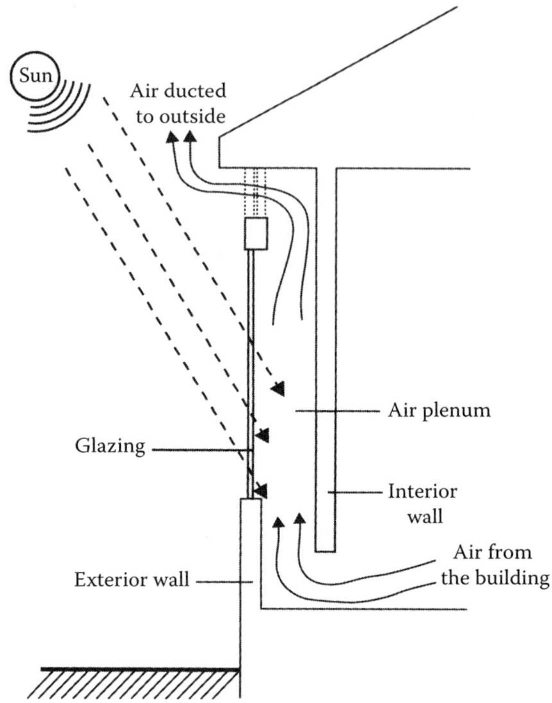

FIGURE 9.11
The stack-effect/solar chimney concept to induce convection/ventilation. (From PSDH. (1980) *Passive Solar Design Handbook*, Volume One: *Passive Solar Design Concepts*, DOE/CS-0127/1, March 1980. Prepared by Total Environmental Action, Inc. (B. Anderson, C. Michal, P. Temple, and D. Lewis); Volume Two: *Passive Solar Design Analysis*, DOE/CS-0127/2, January 1980. Prepared by Los Alamos Scientific Laboratory (J.D. Balcomb, D. Barley, R McFarland, J. Perry, W. Wray, and S. Noll). U.S. Department of Energy, Washington, DC.)

to replace it. Figure 9.11 illustrates the solar chimney concept, which can easily be adapted to a thermal storage wall system. The greatest stack-effect flow rate is produced by maximizing the stack height and the air temperature in the stack, as given by

$$Q = 0.116 A_j \sqrt{h(T_s - T_o)} \tag{9.15}$$

where

Q is the stack flow rate, m^3/s
A_j is the area of inlets or outlets (whichever is smaller), m^2
h is the inlet to outlet height, m
T_s is the average temperature in stack, °C
T_o is the outdoor air temperature, °C

If inlet or outlet area is twice the other, the flow rate will increase by 25%, and by 35% if the areas' ratio is 3:1 or larger.

Example 9.6

A two-story (5 m) solar chimney is being designed to produce a flow of 0.25 m^3/s through a space. The preliminary design features include a 25 cm × 1.5 m inlet, a 50 cm × 1.5 m outlet, and an estimated 35°C average stack temperature on a sunny 30°C day. Can this design produce the desired flow?

Solution

Substituting the design data into Equation 9.15

$$Q = 0.116\,(0.25 \times 1.5)[5(5)]^{1/2}$$
$$= 0.2\ m^3/s$$

Since the outlet area is twice the inlet area, the 25% flow increase can be used:

$$Q = 0.2\,(1.25) = 0.25 m^3/s \quad \text{(answer: Yes, the proper flow rate is obtained)}$$

9.6.3 Evaporative Cooling

When air with less than 100% relative humidity moves over a water surface, the evaporation of water causes both the air and the water to cool. The lowest temperature that can be reached by this direct evaporative cooling effect is the wet-bulb temperature of the air, which is directly related to the relative humidity, with lower wet-bulb temperature associated with lower relative humidity. Thus, dry air (low relative humidity) has a low wet-bulb temperature and will undergo a large temperature drop with evaporative cooling, while humid air (high relative humidity) can only be slightly cooled evaporatively. The wet-bulb temperature for various relative humidity and air temperature conditions can be found via the "psychrometric chart" available in most thermodynamic texts. Normally, an evaporative cooling process cools the air only part of the way down to the wet-bulb temperature. To get the maximum temperature decrease, it is necessary to have a large water surface area in contact with the air for a long time, and interior ponds and fountain sprays are often used to provide this air–water contact area.

The use of water sprays and open ponds on roofs provides cooling primarily via evaporation. The hybrid system involving a fan and wetted mat, the "swamp cooler," is by far the most widely used evaporative cooling technology. Direct, indirect, and combined evaporative cooling system design features are described in [9,11].

9.6.4 Nocturnal and Radiative Cooling Systems

Another approach to passive convective/ventilative cooling involves using cooler night air to reduce the temperature of the building and/or a storage mass. Thus, the building/storage mass is prepared to accept part of the heat load during the hotter daytime. This type of convective system can also be combined with evaporative and radiative modes of heat transfer, utilizing air and/or water as the convective fluid. Work in Australia [3] investigated rock storage beds that were chilled using evaporatively cooled night air. Room air was then circulated through the bed during the day to provide space cooling. The use of encapsulated roof ponds as a thermal cooling mass has been tried by several investigators [12–14] and is often linked with nighttime radiative cooling.

All warm objects emit thermal infrared radiation; the hotter the body, the more energy it emits. A passive cooling scheme is to use the cooler night sky as a sink for thermal radiation emitted by a warm storage mass, thus chilling the mass for cooling use the next day. The net radiative cooling rate, Q_r, for a horizontal unit surface [9] is

$$Q_r = \varepsilon\sigma\left(T_{body}^4 - T_{sky}^4\right) \tag{9.16}$$

where

Q_r is the net radiative cooling rate, W/m^2 (Btu/h ft^2)
ε is the surface emissivity fraction (usually 0.9 for water)
$\sigma = 5.67 \times 10^{-8}$ W/m^2 K^4 (1.714×20^{-9} Btu/h ft^2 R^4)
T_{body} is the warm body temperature, Kelvin (Rankine)
T_{sky} is the effective sky temperature, Kelvin (Rankine)

The monthly average air–sky temperature difference has been determined [15] and Figure 9.12 presents these values for July (in °F) for the United States.

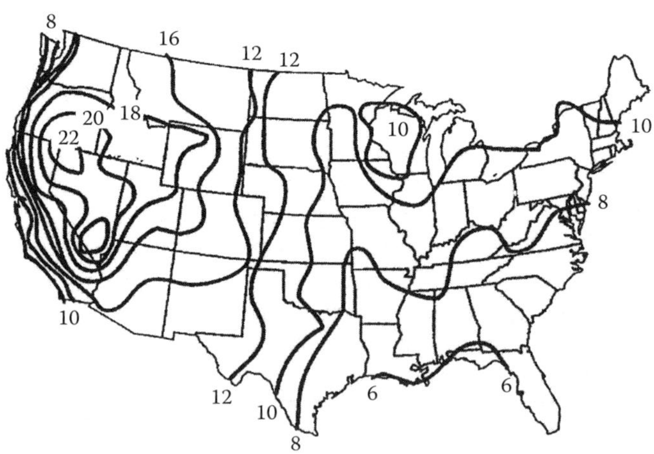

FIGURE 9.12
Average monthly sky temperature depression ($T_{air} - T_{sky}$) for July in °F. (From Martin, M. and Kreider, J.F. *Solar Energy*, 33, 321, 1984.)

Example 9.7

Estimate the overnight cooling possible for a 10 m², 85°F water thermal storage roof during July in Los Angeles.

Solution

Assume the roof storage unit is black with $\varepsilon = 0.9$. From Figure 9.12, $T_{air} - T_{sky}$ is approximately 10°F for Los Angeles. From weather data for Los Angeles airport [5,9], the July average temperature is 69°F with a range of 15°F. Assuming night temperatures vary from the average (69°F) down to half the daily range (15 1/2), the average nighttime temperature is chosen as 69 − (1/2)(15/2) = 65°F. So, T_{sky} = 65 − 10 = 55°F. From Equation 9.16,

$$Q_r = 0.9(1.714 \times 10^{-9})[(460+85)^4 - (460+55)^4]$$

$$= 27.6 \text{ Btu/h ft}^2$$

For a 10 h night and 10 m² (107.6 ft²) roof area,

$$\text{Total radiative cooling} = 27.6(10)(107.6)$$

$$= 29{,}700 \text{ Btu}$$

Note: This does not include the convective cooling possible, which can be approximated (at its maximum rate) for still air [5] by

$$\text{Maximum total } Q_{conv} = hA(T_{roof} - T_{air})(\text{Time})$$

$$= 5(129)(85-55)(10)$$

$$= 161{,}000 \text{ Btu}$$

This is a maximum since the 85°F storage temperature will drop as it cools—which is also the case for the radiative cooling calculation. However, convection is seen to usually be the more dominant mode of nighttime cooling.

9.6.5 Earth Contact Cooling (or Heating)

Earth contact cooling or heating is a passive summer cooling and winter heating technique that utilizes underground soil as the heat sink or source. By installing a pipe underground and passing air through the pipe, the air will be cooled or warmed depending on the season. A schematic of an open loop system and a closed loop air-conditioning system are presented in Figures 9.13 and 9.14, respectively [16].

The use of this technique can be traced back to 3000 BC when Iranian architects designed some buildings to be cooled by natural resources only. In the nineteenth century, Wilkinson [17] designed a barn for 148 cows where a 500 ft long underground passage was used for cooling during the

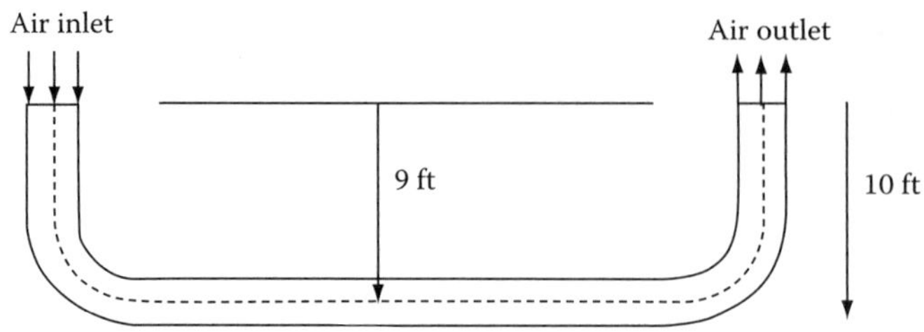

FIGURE 9.13
Open loop underground air tunnel system. (From Goswami, D.Y. and Biseli, K.M. Use of underground air tunnels for heating and cooling agricultural and residential buildings. Report EES-78, Florida Energy Extension Service, University of Florida, Gainesville, FL, August 1994.)

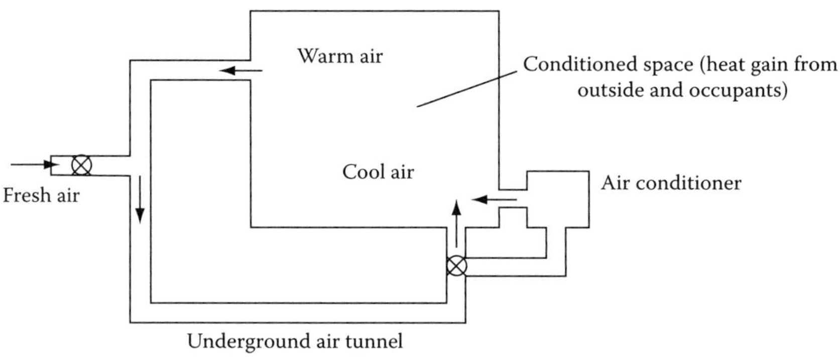

FIGURE 9.14
Schematic of closed-loop air-conditioning system using air tunnel. (From Goswami, D.Y. and Biseli, K.M. Use of underground air tunnels for heating and cooling agricultural and residential buildings. Report EES-78, Florida Energy Extension Service, University of Florida, Gainesville, FL, August 1994.)

summertime. Since that time, a number of experimental and analytical studies of this technique have continued to appear in the literature [18–20]. Goswami and Dhaliwal [21] have given a brief review of the literature, as well as presented an analytical solution to the problem of transient heat transfer between the air and the surrounding soil as the air is made to pass through a pipe buried underground.

9.6.5.1 Heat Transfer Analysis

The transient thermal analysis of the air and soil temperature fields [21] is conducted using finite elements with the convective heat transfer between the air and the pipe and using semi-infinite cylindrical conductive heat transfer to the soil from the pipe. It should be noted that the thermal resistance of the pipe (whether of metal, plastic or ceramic) is negligible relative to the surrounding soil.

9.6.5.1.1 Air and Pipe Heat Transfer

The pipe is divided into a large number of elements and a psychrometric energy balance written for each, depending on whether the air leaves the element (a) unsaturated or (b) saturated.

(a) If the air leaves an element as unsaturated, the energy balance on the element is

$$mC_p(T_1 - T_2) = hA_p(T_{air} - T_{pipe}) \tag{9.17}$$

T_{air} can be taken as $(T_1 + T_2)/2$

Substituting and simplifying we get

$$T_2 = \frac{[(1-U/2)T_1 + UT_{Pipe}]}{(1+U/2)} \tag{9.18}$$

where U is defined as

$$U = \frac{A_p h}{mC_p}$$

(b) If the air leaving the element is saturated, the energy balance is

$$mC_pT_1 + m(W_1 - W_2)H_{fg} = mC_pT_2 + hA_p(T_{air} - T_{pipe}) \tag{9.19}$$

Simplifying, we get

$$T_2 = \frac{\left(1 - \frac{U}{2}T_1\right) + \frac{W_1 - W_2}{C_p}H_{fg} + UT_{Pipe}}{(1+U/2)} \tag{9.20}$$

The convective heat transfer coefficient h in the preceding equations depends on Reynolds number, the shape, and roughness of the pipe. Using the exit temperature from the first element as the inlet temperature for the next element, the exit temperature for the element can be calculated in a similar way. Continuing this way from one element to the next, the temperature of air at the exit from the pipe can be calculated.

9.6.5.1.2 Soil Heat Transfer

The heat transfer from the pipe to the soil is analyzed by considering the heat flux at the internal radius of a semi-infinite cylinder formed by the soil around the pipe. For a small element, the problem can be formulated as

$$\frac{\partial^2 T(r,t)}{\partial r^2} + \frac{1}{r}\frac{\partial T(r,t)}{\partial r} = \frac{1}{\alpha}\frac{\partial T(r,t)}{\partial t} \tag{9.21}$$

with initial and boundary conditions as

$$T(r,0) = T_e$$

$$T(\infty,t) = T_e$$

$$-K\frac{\partial T}{\partial r}(r,t) = q''$$

where

T_e is the bulk earth temperature

q'' is also given by the amount of heat transferred to the pipe from the air by convection, i.e., $q'' = h(T_{air} - T_{pipe})$

9.6.5.2 Soil Temperatures and Properties

Kusuda and Achenback [22] and Labs [23] studied the earth temperatures in the United States. According to both of these studies, temperature swings in the soil during the year are dampened with depth below the ground. There is also a phase lag between the soil temperature and the ambient air temperature, and this phase lag increases with depth below the surface. For example, the soil temperature for light dry soil at a depth of about 10 ft (3.05 m) varies by approximately ±5°F (2.8°C) from the mean temperature (approximately equal to mean annual air temperature) and has a phase lag of approximately 75 days behind ambient air temperature [23].

The thermal properties of the soil are difficult to determine. The thermal conductivity and diffusivity both change with the moisture content of the soil itself, which is directly affected by the temperature of and heat flux from and to the buried pipe. Most researchers have found that using constant property values for soil taken from standard references gives reasonable predictive results [24].

9.6.5.3 Generalized Results from Experiments

Figure 9.15 presents data from [16] for an open system, 100 ft long, 12 in. diameter pipe, buried 9 ft deep. The figure shows the relationship between pipe inlet-to-outlet temperature reduction ($T_{in} - T_{out}$) and the initial soil temperature with ambient air inlet conditions of 90°F, 55% relative humidity for various pipe flow rates.

Other relations from this same report which can be used with the Figure 9.15 data include (1) the effect of increasing pipe/tunnel length on increasing the inlet-to-outlet air temperature difference is fairly linear up to 250 ft and

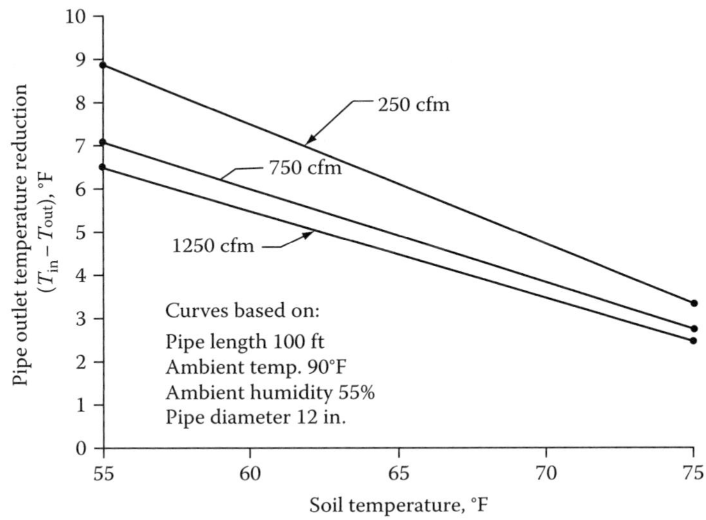

FIGURE 9.15
Air temperature drop through a 100 ft long, 12 in. diameter pipe buried 9 ft underground. (From Goswami, D.Y. and Biseli, K.M. Use of underground air tunnels for heating and cooling agricultural and residential buildings. Report EES-78, Florida Energy Extension Service, University of Florida, Gainesville, FL, August 1994.)

(2) the effect of decreasing pipe diameter on lowering the outlet air temperature is slight, and only marginally effective for pipes less than 12 in. diameter.

Example 9.8

Provide the necessary 12 in. diameter pipe length(s) which will deliver 1500 cfm of 75°F air if the ambient temperature is 85°F and the soil at 9 ft is 65°F.

Solution
From Figure 9.15, for 100 ft of pipe at 65°F soil temperature, the pipe temperature reduction is

$$\begin{aligned} T_{in} - T_{out} &= 6° \text{ (at 250 cfm)} \\ &= 5°\text{F (at 750 cfm)} \\ &= 4.5°\text{F (at 1250 cfm).} \end{aligned}$$

Since the "length versus temperature reduction" is linear (see text above), the 10°F reduction required (85 down to 75) would be met by the 750 cfm case (5°F for 100 ft) if 200 ft of pipe is used. Then, two 12 in. diameter pipes would be required to meet the 1500 cfm requirement.

Answer
Two 12 in. diameter pipes, each 200 ft long. (*Note*: See what would be needed if the 250 cfm or the 1250 cfm cases had been chosen. Which of the three flow rate cases leads to the "cheapest" installation?)

9.7 Daylighting Design Fundamentals

Daylighting is the use of the sun's radiant energy to illuminate the interior spaces in a building. In the last century, electric lighting was considered an alternative technology to daylighting. Today the situation is reversed, primarily due to the economics of energy use and conservation. However, there are good physiological reasons for using daylight as an illuminant. The quality of daylight matches the human eye's response, thus permitting lower light levels for task comfort, better color rendering, and clearer object discrimination [25–27].

9.7.1 Lighting Terms and Units

Measurement of lighting level is based on the "standard candle," where the *lumen* (lm), the unit of *luminous flux* (ϕ), is defined as the rate of luminous energy passing through a 1 m^2 area located 1 m from the candle. Thus, a standard candle generates 4π lumens which radiate away in all directions. The *illuminance* (E) on a surface is defined as the luminous flux on the surface divided by the surface area, $E = \phi/A$. Illuminance is measured in either *lux* (lx), as lumens per square meter, or *footcandles* (fc), as lumens per square foot.

Determination of the daylighting available at a given location in a building space at a given time is important to evaluate the reduction possible in electric lighting and the associated impact on heating and cooling loads. Daylight provides about 110 lm/W of solar radiation, fluorescent lamps about 75 lm/W of electrical input, and incandescent lamps about 20 lm/W; thus daylighting generates only 1/2 to 1/5 the heating that equivalent electric lighting does, significantly reducing the building cooling load.

9.7.2 Approach to Daylighting Design

Aperture controls such as blinds and drapes are used to moderate the amount of daylight entering the space, as are the architectural features of the building itself (glazing type, area, and orientation; overhangs and wingwalls; lightshelves; etc.). Many passive and "active" reflective, concentrating and diffusing devices are available to specifically gather and direct both the direct and diffuse components of daylight to areas within the space [28]. Electric lighting dimming controls are used to adjust the electric light level based on the quantity of the daylighting. With these two types of controls (aperture and lighting), the electric lighting and cooling energy use and demand, as well as cooling system sizing, can be reduced. However, the determination of the daylighting position and time illuminance value within the space is required before energy usage and demand reduction calculations can be made.

Daylighting design approaches use both solar beam radiation (referred to as *sunlight*) and the diffuse radiation scattered by the atmosphere (referred to as *skylight*) as sources for interior lighting, with historical design emphasis being on utilizing skylight. Daylighting is provided through a variety of glazing features, which can be grouped as *sidelighting* (light enters via the side of the space) and *toplighting* (light enters from the ceiling area). Figure 9.16 illustrates several architectural forms producing sidelighting and toplighting, with the dashed lines representing the illuminance distribution

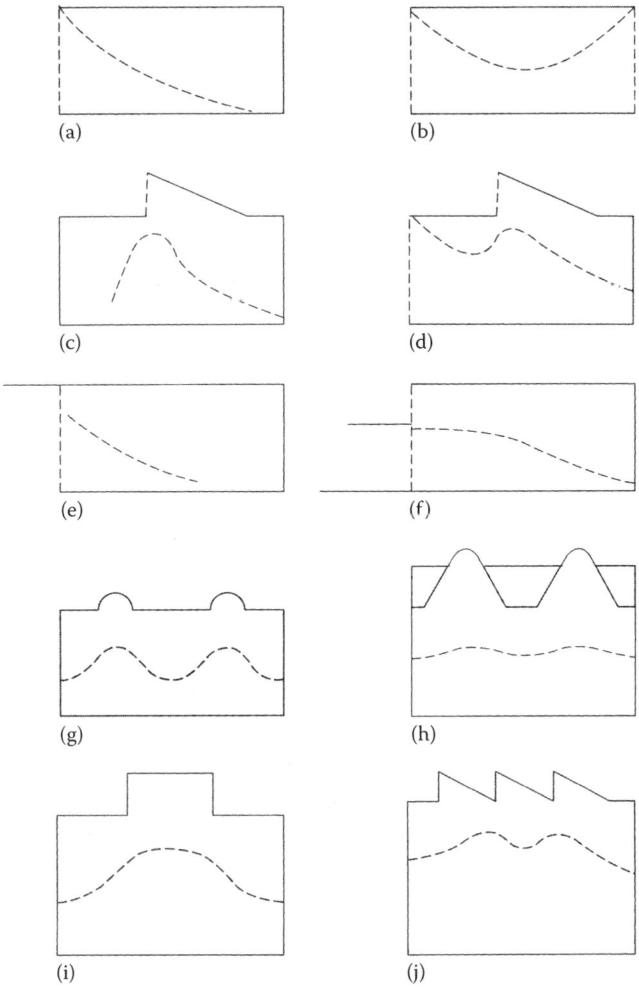

FIGURE 9.16
Examples of sidelighting and toplighting architectural features (dashed lines represent illuminance distributions). (a) Unilateral, (b) bilateral, (c) clerestory, (d) clerestory + unilateral, (e) overhang, (f) overhang + ground reflection, (g) skylight, (h) skylight + well, (i) roof monitor, and (j) sawtooth. (From Murdoch, J.B. *Illumination Engineering—From Edison's Lamp to the Laser*, Macmillan, New York, 1985.)

within the space. The calculation of workplane illuminance depends on whether sidelighting and/or toplighting features are used, and the combined illuminance values are additive.

9.7.3 Sun–Window Geometry

The solar illuminance on a vertical or horizontal window depends on the position of the sun relative to that window. In the method described here, the sun and sky illuminance values are determined using the sun's altitude angle (α) and the sun-window azimuth angle difference (Φ). These angles need to be determined for the particular time of day, day of year, and window placement under investigation.

9.7.3.1 Solar Altitude Angle (α)

The solar altitude angle is the angle swept out by a person's arm when pointing to the horizon directly below the sun and then raising the arm to point at the sun. The equation to calculate solar altitude, α, is

$$\sin\alpha = \cos L \cos\delta \cos H + \sin L \sin\delta \tag{9.22}$$

where

L = local latitude (degrees)
δ = earth-sun declination (degrees) given by $\delta = 23.45 \sin[360\,(n-81)/365]$
n = day number of the year
H = hour angle (degrees) given by

$$H = \frac{(12\text{ noon} - \text{time})(\text{in min})}{4}; \quad (+\text{morning}, -\text{afternoon}) \tag{9.23}$$

9.7.3.2 Sun–Window Azimuth Angle Difference (Φ)

The difference between the sun's azimuth and the window's azimuth needs to be calculated for vertical window illuminance. The window's azimuth angle, γ_w, is determined by which way it faces, as measured from south (east of south is positive, westward is negative). The solar azimuth angle, γ_s, is calculated as

$$\sin\gamma_s = \frac{\cos\delta \sin H}{\cos\alpha} \tag{9.24}$$

The sun–window azimuth angle difference, Φ, is given by the absolute value of the difference between γ_s and γ_w:

$$\Phi = |\gamma_s - \gamma_w| \tag{9.25}$$

9.7.4 Daylighting Design Methods

To determine the annual lighting energy saved (ES_L), calculations using the lumen method described below should be performed on a monthly basis for both clear and overcast days for the space under investigation. Monthly weather data for the site would then be used to prorate clear and overcast lighting energy demands monthly. Subtracting the calculated daylighting illuminance from the design illuminance leaves the supplementary lighting needed, which determines the lighting energy required.

The approach in the method below is to calculate the "sidelighting" and the "skylighting" of the space separately, and then combine the results. This procedure has been computerized (Lumen II/Lumen Micro) and includes many details of controls, daylighting technologies, and weather. ASHRAE [9] lists many of the methods and simulation techniques currently used with daylighting and its associated energy effects.

9.7.4.1 Lumen Method of Sidelighting (Vertical Windows)

The lumen method of sidelighting calculates interior horizontal illuminance at three points, as shown in Figure 9.17, at the 30 in. (0.76 m) work plane on the room-and-window centerline. A vertical window is assumed to extend from 36 in. (0.91 m) above the floor to the ceiling. The method accounts for both direct and ground-reflected sunlight and skylight, so both horizontal and vertical illuminances from sun and sky are needed. The steps in the lumen method of sidelighting are presented next.

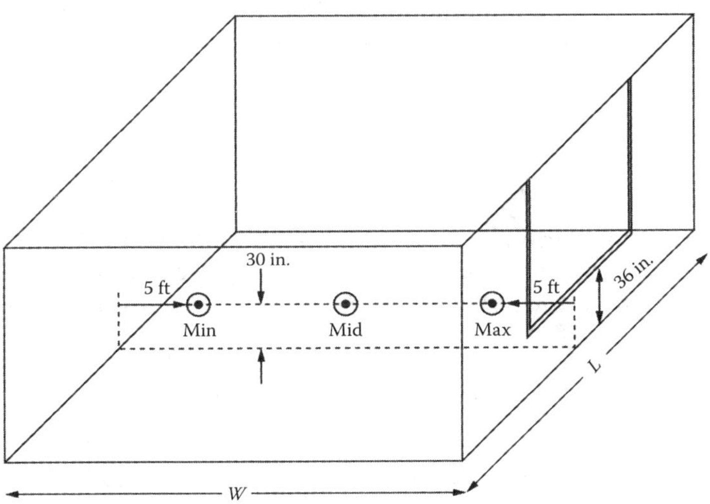

FIGURE 9.17
Location of illumination points within the room (along centerline of window) determined by lumen method of sidelighting. (From Kreith, F. and West, R., *CRC Handbook of Energy Efficiency and Renewable Energy*, CRC Press, Boca Raton, FL, 2007. With permission.)

As mentioned, the incident direct and ground-reflected window illuminance are normally calculated for both a cloudy and a clear day for representative days during the year (various months), as well as for clear or cloudy times during a given day. Thus, the interior illumination due to sidelighting and skylighting can then be examined for effectiveness throughout the year.

Step 1: Incident direct sky and sun illuminances
The solar altitude and sun-window azimuth angle difference are calculated for the desired latitude, date, and time using Equations 9.22 and 9.25, respectively. Using these two angles, the total illuminance on the window (E_{sw}) can be determined by summing the direct sun illuminance (E_{uw}) and the direct sky illuminance (E_{kw}), each determined from the appropriate graph in Figure 9.18.

Step 2: Incident ground-reflected illuminance
The sun illuminance on the ground (E_{ug}), plus the overcast or clear sky illuminance (E_{kg}) on the ground, make up the total horizontal illuminance on the ground surface (E_{sg}). A fraction of the ground surface illuminance is then considered diffusely reflected onto the vertical window surface (E_{gw}), where "gw" indicates from the ground to the window.

The horizontal ground illuminances can be determined using Figure 9.19, where the clear sky plus sun case and the overcast sky case are functions of solar altitude. The fractions of the ground illuminance diffusely reflected onto the window depends on the reflectivity (ρ) of the ground surface (see Table 9.6) and the window-to-ground surface geometry.

If the ground surface is considered uniformly reflective from the window outward to the horizon, then the illuminance on the window from ground reflection is

$$E_{gw} = \frac{\rho E_{sg}}{2} \tag{9.26}$$

A more complicated ground-reflection case is illustrated in Figure 9.20, with multiple "strips" of differently reflecting ground being handled using the angles to the window, where a strip's illuminance on a window is calculated:

$$E_{gw(strip)} = \frac{\rho_{strip} E_{sg}}{2}(\cos\theta_1 - \cos\theta_2) \tag{9.27}$$

And the total reflected onto the window is the sum of the strip illuminances:

$$E_{gw} = \frac{E_{sg}}{2}\left[\rho_1(\cos 0 - \cos\theta_1) + \rho_2(\cos\theta_1 - \cos\theta_2) + \cdots + \rho_n(\cos\theta_{n-1} - \cos 90)\right] \tag{9.28}$$

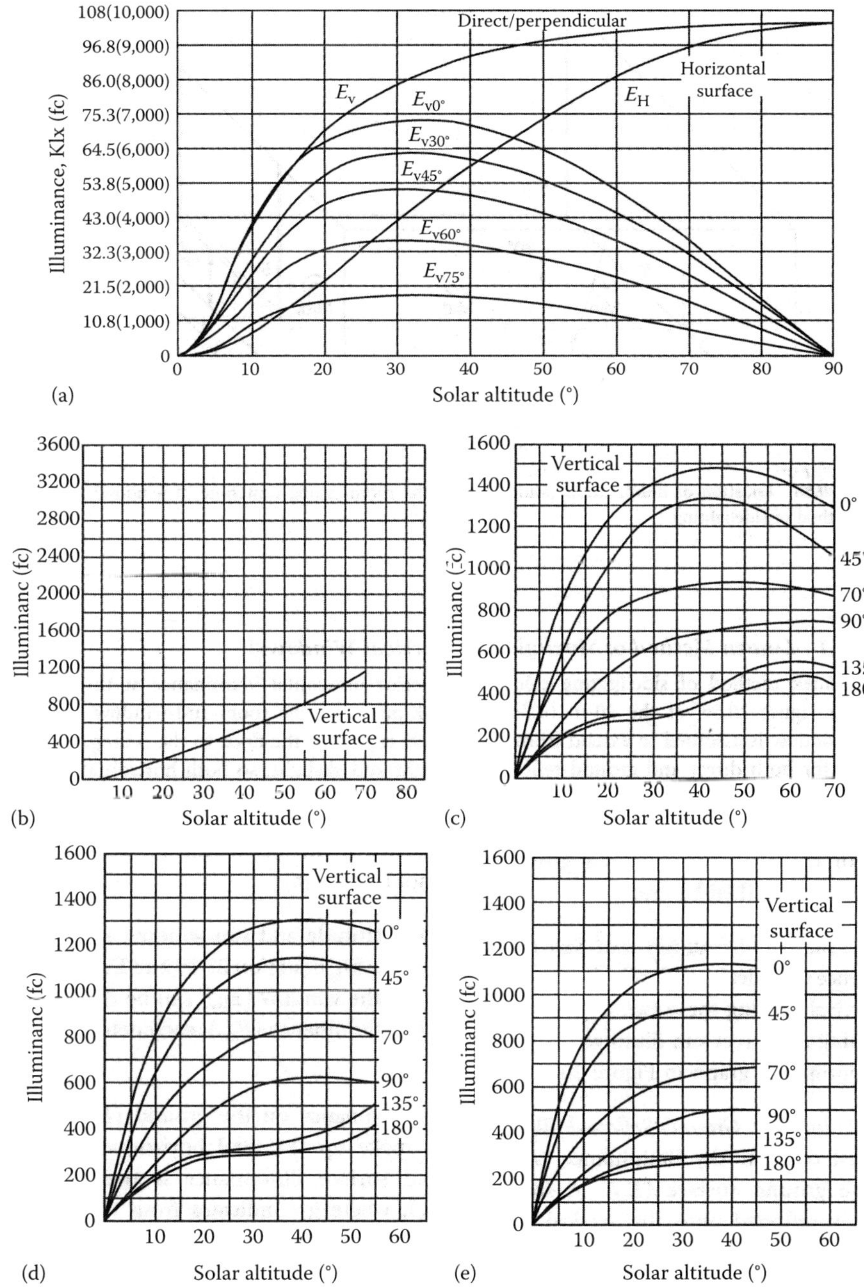

FIGURE 9.18
Vertical illuminance from (a) direct sunlight, (b) overcast skylight, (c) clear summer skylight (d) clear autumn skylight, and (e) clear winter skylight, for various sun–window azimuth angle differences. (Modified from Kaufman, J.E., ed., *The IES Lighting Handbook: Reference Volume*, IESNA, New York, 1984. With permission from the Illuminating Engineering Society of North America.)

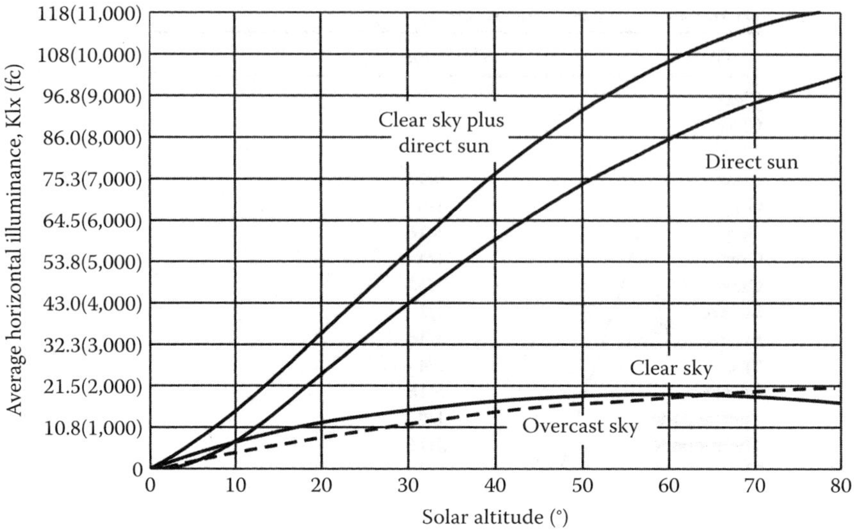

FIGURE 9.19
Horizontal illuminance for overcast sky, direct sun, and clear sky direct sun. (From Murdoch, J.B., *Illumination Engineering—From Edison's Lamp to the Laser*, Macmillan, New York, 1985.)

TABLE 9.6
Ground Reflectivities

Material	ρ (%)
Cement	27
Concrete	20–40
Asphalt	7–14
Earth	10
Grass	6–20
Vegetation	25
Snow	70
Red brick	30
Gravel	15
White paint	55–75

Source: Murdoch, J.B., *Illumination Engineering-From Edison's Lamp to the Laser*, Macmillan, New York, 1985.

Step 3: Luminous flux entering space
The direct sky-sun and ground-reflected luminous fluxes entering the building are attenuated by the transmissivity of the window. Table 9.7 presents the transmittance fraction (τ) of several window glasses. The fluxes entering the space are calculated from the total sun-sky and the ground-reflected illuminances by using the area of the glass, A_w:

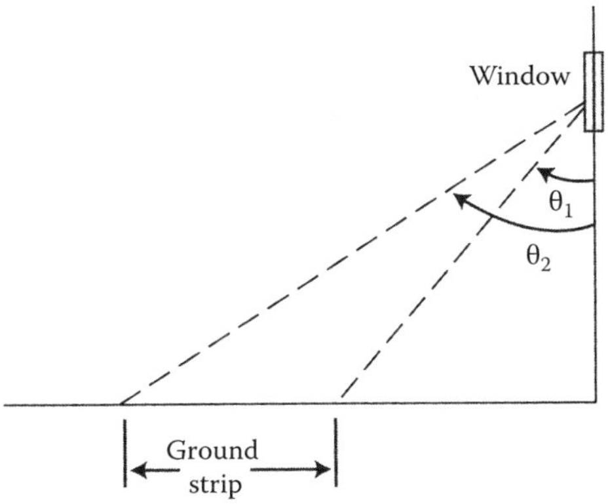

FIGURE 9.20
Geometry for ground "strips." (From Murdoch, J.B., *Illumination Engineering—From Edison's Lamp to the Laser,* Macmillan, New York, 1985.)

$$\phi_{sw} = E_{sw}\tau A_w$$
$$\phi_{gw} = E_{gw}\tau A_w \tag{9.29}$$

Step 4: Light loss factor
The light loss factor (K_m) accounts for the attenuation of luminous flux due to dirt on the window (WDD, window dirt depreciation) and on the room surfaces (RSDD, room surface dirt depreciation). WDD depends on how often the window is cleaned, but a 6-month average for offices is 0.83 and for factories is 0.71 [29].

The RSDD is a more complex calculation involving time between cleanings, the direct–indirect flux distribution, and room proportions. However, for rooms cleaned regularly, RSDD is around 0.94 and for once-a-year-cleaned dirty rooms, the RSDD would be around 0.84.

The light loss factor is the product of the preceding two fractions:

$$K_m = (\text{WDD})(\text{RSDD}) \tag{9.30}$$

Step 5: Work-plane illuminances
As discussed earlier, Figure 9.17 illustrates the location of the work-plane illuminances determined with this lumen method of sidelighting. The three illuminances (max, mid, min) are determined using two coefficients of utilization, the *C* factor and *K* factor. The *C* factor depends on room length, width, and wall reflectance. The *K* factor depends on ceiling-floor height, room width, and wall reflectance. Table 9.8 presents *C* and *K* values for the three cases of incoming fluxes: sun plus clear sky, overcast sky, and ground reflected. Assumed ceiling and floor reflectances are given for this case with no window controls (shades, blinds, overhangs, etc.). These further window control complexities can be found in IES [30], LOF [31], and others. A reflectance of 70% represents light-colored walls, with 30% representing darker walls.

TABLE 9.7

Glass Transmittances

Glass	Thickness (in.)	τ (%)
Clear	$\frac{1}{8}$	89
Clear	$\frac{3}{16}$	88
Clear	$\frac{1}{4}$	87
Clear	$\frac{5}{16}$	86
Gray	$\frac{1}{8}$	61
Gray	$\frac{3}{16}$	51
Gray	$\frac{1}{4}$	44
Gray	$\frac{5}{16}$	35
Bronze	$\frac{1}{8}$	68
Bronze	$\frac{3}{16}$	59
Bronze	$\frac{1}{4}$	52
Bronze	$\frac{5}{16}$	44
Thermopane	$\frac{1}{8}$	80
Thermopane	$\frac{3}{16}$	79
Thermopane	$\frac{1}{4}$	77

Source: Murdoch, J.B., *Illumination Engineering—From Edison's Lamp to the Laser*, Macmillan, New York, 1985.

The work-plane max, mid, and min illuminance are each calculated by adding the sun-sky and ground reflected illuminances, which are given by

$$\begin{aligned} E_{sp} &= \phi_{sw} C_s K_s K_m \\ E_{gp} &= \phi_{gw} C_g K_g K_m \end{aligned} \tag{9.31}$$

where the "sp" and "gp" refer to the sky-to-work-plane and ground-to-work-plane illuminances.

TABLE 9.8

C and *K* Factors for No Window Controls for (a) Overcast Sky, (b) Clear Sky, and (c) Ground Illumination (Ceiling Reflectance, 80%; Floor Reflectance, 20%)

		C: Coefficient of Utilization								K: Coefficient of Utilization							
Room Length		20 ft		30 ft		40 ft		Ceiling Height		8 ft		10 ft		12 ft		14 ft	
Wall Reflectance		70%	30%	70%	30%	70%	30%	Wall Reflectance		70%	30%	70%	30%	70%	30%	70%	30%
	Room Width (ft)								Room Width (ft)								
(a) Illumination by overcast sky																	
Max	20	0.0276	0.0251	0.0191	0.0173	0.0143	0.0137	Max	20	0.125	0.129	0.121	0.123	0.111	0.111	0.0991	0.0973
	30	0.0272	0.0248	0.0188	0.0172	0.0137	0.0131		30	0.122	0.131	0.122	0.121	0.111	0.111	0.0945	0.0973
	40	0.269	.0246	.0182	.0171	.0133	.0130		40	.145	.133	.131	.126	.111	.111	.0973	.0982
Mid	20	0.0159	0.0177	0.0101	0.0087	0.0081	0.0071	Mid	20	0.0908	0.0982	0.107	0.115	0.111	0.111	0.105	0.122
	30	0.0058	0.0050	0.0054	0.0040	0.0034	0.0033		30	0.156	0.102	0.0939	0.113	0.111	0.111	0.121	0.134
	40	0.0039	0.0027	0.0030	0.0023	0.0022	0.0019		40	0.106	0.0948	0.123	0.107	0.111	0.111	0.135	0.127
Min	20	0.0087	0.0053	0.0033	0.0043	0.0050	0.0037	Min	20	0.0908	0.102	0.0951	0.114	0.111	0.111	0.118	0.134
	30	0.0032	0.0019	0.0029	0.0017	0.0020	0.0014		30	0.0924	0.119	0.101	0.114	0.111	0.111	0.125	0.126
	40	0.0019	0.0009	0.0016	0.0009	0.0012	0.0008		40	0.111	0.0926	0.125	0.109	0.111	0.111	0.133	0.130
(b) Illumination by clear sky																	
Max	20	0.0206	0.0173	0.0143	0.0123	0.0110	0.0098	Max	20	0.145	0.155	0.129	0.132	0.111	0.111	0.101	0.0982
	30	0.0203	0.0173	0.0137	0.0120	0.0098	0.0092		30	0.141	0.149	0.125	0.130	0.111	0.111	0.0954	0.101
	40	0.0200	0.0168	0.0131	0.0119	0.0096	0.0091		40	0.157	0.157	0.135	0.134	0.111	0.111	0.0964	0.0991

(continued)

TABLE 9.8 (continued)

C and *K* Factors for No Window Controls for (a) Overcast Sky, (b) Clear Sky, and (c) Ground Illumination (Ceiling Reflectance, 80%; Floor Reflectance, 20%)

		C: Coefficient of Utilization								K: Coefficient of Utilization							
Room Length		20 ft		30 ft		40 ft		Ceiling Height		8 ft		10 ft		12 ft		14 ft	
Wall Reflectance		70%	30%	70%	30%	70%	30%	Wall Reflectance		70%	30%	70%	30%	70%	30%	70%	30%
	Room Width (ft)								Room Width (ft)								
Mid	20	0.0153	0.0104	0.0100	0.0079	0.0083	0.0067	Mid	20	0.110	0.128	0.116	0.126	0.111	0.111	0.103	0.108
	30	0.0082	0.0054	0.0062	0.0043	0.0046	0.0037		30	0.106	0.125	0.110	0.129	0.111	0.111	0.112	0.120
	40	0.0052	0.0032	0.0040	0.0028	0.0029	0.0023		40	0.117	0.118	0.122	0.118	0.111	0.111	0.123	0.122
Min	20	0.0106	0.0060	0.0079	0.0049	0.0067	0.0043	Min	20	0.105	0.129	0.112	0.130	0.111	0.111	0.111	0.116
	30	0.0054	0.0028	0.0047	0.0023	0.0032	0.0021		30	0.0994	0.144	0.107	0.126	0.111	0.111	0.107	0.124
	40	0.0031	0.0014	0.0027	0.0013	0.0021	0.0012		40	0.119	0.116	0.130	0.118	0.111	0.111	0.120	0.118
(c) Ground illumination (ceiling reflectance, 80%; floor reflectance, 20%)																	
Max	20	0.0147	0.0112	0.0102	0.0088	0.0081	0.0071	Max	20	0.124	0.206	0.140	0.135	0.111	0.111	0.0909	0.0859
	30	0.0141	0.0012	0.0098	0.0088	0.0077	0.0070		30	0.182	0.188	0.140	0.143	0.111	0.111	0.0918	0.0878
	40	0.0137	0.0112	0.0093	0.0086	0.0072	0.0069		40	0.124	0.182	0.140	0.142	0.111	0.111	0.0936	0.0879
Mid	20	0.0128	0.0090	0.0094	0.0071	0.0073	0.0060	Mid	20	0.123	0.145	0.122	0.129	0.111	0.111	0.100	0.0945
	30	0.0083	0.0057	0.0062	0.0048	0.0050	0.0041		30	0.0966	0.104	0.107	0.112	0.111	0.111	0.110	0.105
	40	0.0055	0.0037	0.0044	0.0033	0.0042	0.0026		40	0.0790	0.0786	0.0999	0.106	0.111	0.111	0.118	0.118
Min	20	0.0106	0.0071	0.0082	0.0054	0.0067	0.0044	Min	20	0.0994	0.108	0.110	0.114	0.111	0.111	0.107	0.104
	30	0.0051	0.0026	0.0041	0.0023	0.0033	0.0021		30	0.0816	0.0822	0.0984	0.105	0.111	0.111	0.121	0.116
	40	0.0029	0.0018	0.0026	0.0012	0.0022	0.0011		40	0.0700	0.0656	0.0946	0.0986	0.111	0.111	0.125	0.132

Source: IES, *Lighting Handbook, Applications Volume*, Illumination Engineering Society, New York, 1987.

Example 9.9

Determine the clear sky illuminances for a 30-ft-long, 30-ft-wide, 10-ft-high room with a 20-ft-long window with a 3-ft sill. The window faces 10°E of South, the building is at 32° latitude, and it is January 15 at 2 P.M. The ground cover outside is grass, the glass is l/4 in. clear, and the walls light colored.

Solution

Following the steps in the "sidelighting" method:

Step 1: With $L = 32$, $n = 15$, $H = (12 - 14)60/4 = -30$,

$$\delta = 23.45\ \sin[360(15-81)/365] = -21.3°$$

Then, Equation 9.22 yields $\alpha = 41.7°$, Equation 9.24 yields $\gamma_s = -38.7°$, and Equation 9.25 yields

$$\Phi = |-38.7-(+10)| = 48.7°.$$

From Figure 9.18 with $\alpha = 41.7°$ and $\Phi = 48.7°$:

(a) For clear sky (winter, no sun): $E_{kw} = 875$ fc.
(b) For direct sun: $E_{uw} = 4100$ fc.
(c) Total clear sky plus direct: $E_{sw} = 4975$ fc.

(**Note**: A high E_{uw} value probably indicates a glare situation!)

Step 2: Horizontal illuminances from Figure 9.19: $E_{sg} = 7400$ fc.
Then Equation 9.26 yields, with $\rho_{grass} = 0.06$, $E_{gw} = 222$ fc.

Step 3: From Equation 9.29, with $\tau = 0.87$ and $A_w = 140\ \text{ft}^2$,

$$\Phi_{sw} = 4,975(0.87)(140) = 605,955\ \text{lm}$$

$$\Phi_{gw} = 222(0.87)(140) = 27,040\ \text{lm}$$

Step 4: For a clean office room:

$$K_m = (0.83)(0.94) = 0.78$$

Step 5: From Table 9.8, for 30 ft width, 30 ft length, 10 ft ceiling, and wall reflectivity 70%,

(a) Clear sky:

$$C_s,\text{max} = 0.0137; \quad K_s,\text{max} = 0.125$$

$$C_s,\text{mid} = 0.0062; \quad K_s,\text{mid} = 0.110$$

$$C_s,\text{min} = 0.0047; \quad K_s,\text{min} = 0.107$$

(b) Ground reflected:

$$C_g,\text{max} = 0.0098; \quad K_g,\text{max} = 0.140$$

$$C_g,\text{mid} = 0.0062; \quad K_g,\text{mid} = 0.107$$

$$C_g,\text{min} = 0.0041; \quad K_g,\text{min} = 0.0984$$

Then using Equation 9.31,

$$E_{sp},\text{max} = 605{,}955(0.0137)(0.125)(0.78) = 809\ \text{fc}$$

$$E_{sp},\text{mid} = 605{,}955(0.0062)(0.110)(0.78) = 322\ \text{fc}$$

$$E_{sp},\text{min} = 605{,}955(0.0047)(0.107)(0.78) = 238\ \text{fc}$$

$$E_{gp},\text{max} = 27{,}040(0.0098)(0.140)(0.78) = 29\ \text{fc}$$

$$E_{gp},\text{mid} = 27{,}040(0.0062)(0.107)(0.78) = 14\ \text{fc}$$

$$E_{gp},\text{min} = 27{,}040(0.0041)(0.0984)(0.78) = 9\ \text{fc}$$

Thus,

$$E_{\text{max}} = 838\ \text{fc}$$

$$E_{\text{mid}} = 336\ \text{fc}$$

$$E_{\text{min}} = 247\ \text{fc}$$

9.7.4.2 Lumen Method of Skylighting

The lumen method of skylighting calculates the average illuminance at the interior work plane provided by horizontal skylights mounted on the roof. The procedure for skylighting is generally the same as that described above for sidelighting. As with windows, the illuminance from both overcast sky and clear sky plus sun cases are determined for specific days in different seasons and for different times of the day, and a judgment is then made as to the number and size of skylights and any controls needed.

The procedure is presented in four steps: (1) finding the horizontal illuminance on the outside of the skylight; (2) calculating the effective transmittance through the skylight and its well; (3) figuring the interior space light loss factor and the utilization coefficient; and, finally, (4) calculating illuminance on the work plane.

Step 1: Horizontal sky and sun illuminances
The horizontal illuminance value for an overcast sky or a clear sky plus sun situation can be determined from Figure 9.19 knowing only the solar altitude.

Step 2: Net skylight transmittance
The transmittance of the skylight is determined by the transmittance of the skylight cover(s), the reflective efficiency of the skylight well, the net-to-gross skylight area, and the transmittance of any light-control devices (lenses, louvers, etc.).

The transmittance for several flat-sheet plastic materials used in skylight domes is presented in Table 9.9. To get the effective dome transmittance (T_D) from the flat-plate transmittance (T_F) value [32], use

$$T_D = 1.25T_F(1.18 - 0.416T_F) \tag{9.32}$$

If a double-domed skylight is used, then the single-dome transmittances are combined as follows [33]:

$$T_D = \frac{T_{D_1}T_{D_2}}{T_{D_1}T_{D_2} - T_{D_1}T_{D_2}} \tag{9.33}$$

TABLE 9.9
Flat-Plate Plastic Material Transmittance for Skylights

Type	Thickness (in.)	Transmittance (%)
Transparent	$\frac{1}{8}-\frac{3}{16}$	92
Dense translucent	$\frac{1}{8}$	32
Dense translucent	$\frac{3}{16}$	24
Medium translucent	$\frac{1}{8}$	56
Medium translucent	$\frac{3}{16}$	52
Light translucent	$\frac{1}{8}$	72
Light translucent	$\frac{3}{16}$	68

Source: Murdoch, J.B., *Illumination Engineering—From Edison's Lamp to the Laser*, Macmillan, New York, 1985.

If the diffuse and direct transmittances for solar radiation are available for the skylight glazing material, it is possible to follow this procedure and determine diffuse and direct dome transmittances separately. However, this difference is usually not a significant factor in the overall calculations.

The efficiency of the skylight well (N_w) is the fraction of the luminous flux from the dome that enters the room from the well. The well index (WI) is a geometric index (height, h; length, l; width, w) given by

$$\mathrm{WI} = \frac{h(w+l)}{2wl} \tag{9.34}$$

and WI is used with the well–wall reflectance value in Figure 9.21 to determine well efficiency, N_w.

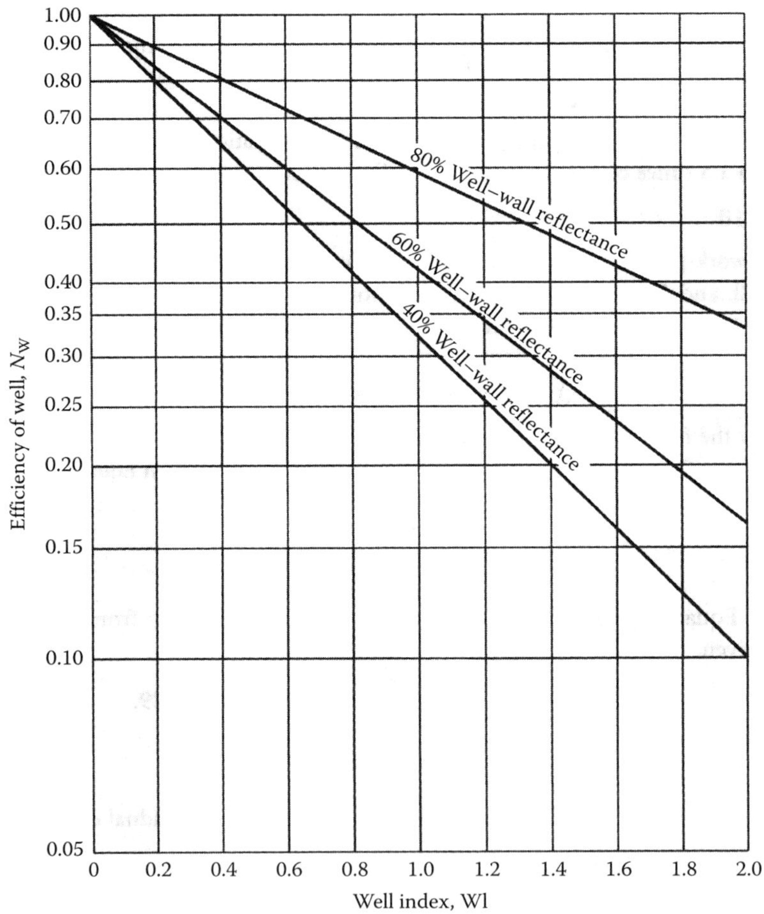

FIGURE 9.21
Efficiency of well versus well index. (Modified from Kaufman, J.E., ed., *The IES Lighting Handbook: Reference Volume*, IESNA, New York, 1984. With permission from the Illuminating Engineering Society of North America.)

With T_D and N_w determined, the net skylight transmittance for the skylight and well is given by

$$T_n = T_D N_w R_A T_C \tag{9.35}$$

where
R_A is the ratio of net to gross skylight areas
T_C is the transmittance of any light-controlling devices

Step 3: Light loss factor and utilization coefficient
The light loss factor (K_m) is again defined as the product of the room surface dirt depreciation (RSDD) and the skylight direct depreciation (SDD) fractions, similar to Equation 9.30. Following the reasoning for the sidelighting case, the RSDD value for clean rooms is around 0.94, and 0.84 for dirty rooms. Without specific data indicating otherwise, the SDD fraction is often taken as 0.75 for office buildings and 0.65 for industrial areas.

The fraction of the luminous flux on the skylight that reaches the work plane (K_u) is the product of the net transmittance (T_n) and the room coefficient of utilization (RCU). Dietz et al. [34] developed RCU equations for office and warehouse interiors with ceiling, wall, and floor reflectances of 75%, 50%, and 30%, and 50%, 30%, and 20%, respectively.

$$\text{RCU} = \frac{1}{1 + A(\text{RCR})^B}, \quad \text{if RCR} < 8 \tag{9.36}$$

where
$A = 0.0288$ and $B = 1.560$ (offices)
$A = 0.0995$ and $B = 1.087$ (warehouses)

and room cavity ratio (RCR) is given by

$$\text{RCR} = \frac{5h_c(l + w)}{lw} \tag{9.37}$$

with h_c the ceiling height above the work plane and l and w being room length and width, respectively.

The RCU is then multiplied by the previously determined T_n to give the fraction of the external luminous flux passing through the skylight and incident on the workplace:

$$K_u = T_n(\text{RCU}) \tag{9.38}$$

Step 4: Work-plane illuminance
The illuminance at the work plane (E_{TWP}) is given by

$$E_{TWP} = E_H\left(\frac{A_T}{A_{WP}}\right)K_u K_m \tag{9.39}$$

where
E_H is the horizontal overcast or clear sky plus sun illuminance from Step 1
A_T is total gross area of the skylights (number of skylights times skylight gross area)
A_{WP} is the work-plane area (generally room length times width)

Note that in Equation 9.39, it is also possible to fix the E_{TWP} at some desired value and determine the skylight area required.

Rules of thumb for skylight placement for uniform illumination include 4%–8% of roof area and spacing less than 1 1/2 times ceiling-to-work-plane distance between skylights [29].

Example 9.10

Determine the work-plane "clear sky plus sun" illuminance for a 30′ × 30′ × 10′ office with 75% ceiling, 50% wall, and 30% floor reflectance with four 4′ × 4′ double-domed skylights at 2 PM on January 15th at 32° latitude. The skylight well is 1′ deep at with 60% reflectance walls, and the outer and inner dome flat-plastic transmittances are 0.85 and 0.45, respectively. The net skylight area is 90%.

Solution
Follow the four steps in the lumen method for skylighting.

Step 1: Use Figure 9.19 with the solar altitude of 41.7° (calculated from Equation 9.26) for the clear sky plus sun curve to get horizontal illuminance:

$$E_H = 7400 \text{ fc}$$

Step 2: Use Equation 9.32 to get domed transmittances from the flat plate plastic transmittances given,

$$T_{D_1} = 1.25(0.85)[1.18 - 0.416(0.85)] = 0.89$$

$$T_{D_2}(T_F = 0.45) = 0.56$$

and Equation 9.33 to get total dome transmittance from the individual dome transmittances:

$$T_D = \frac{(0.89)(0.56)}{(0.89) + (0.56) - (0.89)(0.56)} = 0.52$$

To get well efficiency, use WI = 0.25 from Equation 9.34 with 60% wall reflectance in Figure 9.21 to give N_w = 0.80. With R_A = 0.90, use Equation 9.35 to calculate net transmittance:

$$T_n = (0.52)(0.80)(0.90)(1.0) = 0.37.$$

Step 3: The light loss factor is assumed to be from "typical" values in Equation 9.30: $K_m = (0.75)(0.94) = 0.70$. The room utilization coefficient is determined using Equations 9.36 and 9.37:

$$\text{RCR} = \frac{5(7.5)(30+30)}{(30)(30)} = 2.5$$

$$\text{RCU} = \left[1 + 0.0288(2.5)^{1.560}\right]^{-1} = 0.89$$

and Equation 9.37 yields $K_u = (0.37)(0.89) = 0.33$.

Step 4: The work-plane illuminance is calculated by substituting the above values into Equation 9.39.

$$E_{\text{TWP}} = 122 \text{ fc}$$

9.7.5 Daylighting Controls and Economics

The economic benefit of daylighting is directly tied to the reduction in lighting electrical energy operating costs. Also, lower cooling system operating costs are possible due to the reduction in heating caused by the reduced electrical lighting load. The reduction in lighting and cooling system electrical power during peak demand periods could also beneficially affect demand charges.

The reduction of the design cooling load through the use of daylighting can also lead to the reduction of installed or first-cost cooling system dollars. Normally, economics dictate that an automatic lighting control system must take advantage of the reduced lighting/cooling effect, and the control system cost minus any cooling system cost savings should be expressed as a "net" first cost. A payback time for the lighting control system ("net" or not) can be calculated from the ratio of first costs to yearly operating savings. In some cases, these paybacks for daylighting controls have been found to be in the range of 1–5 years for office building spaces [35].

Controls, both aperture and lighting, directly affect the efficacy of the daylighting system. As shown in Figure 9.22, aperture controls can be architectural (overhangs, light shelves, etc.) and/or window shading devices (blinds, automated louvers, etc.). The aperture controls generally moderate the sunlight entering the space to maximize/minimize solar thermal gain, permit the proper amount of light for visibility, and prevent glare and beam

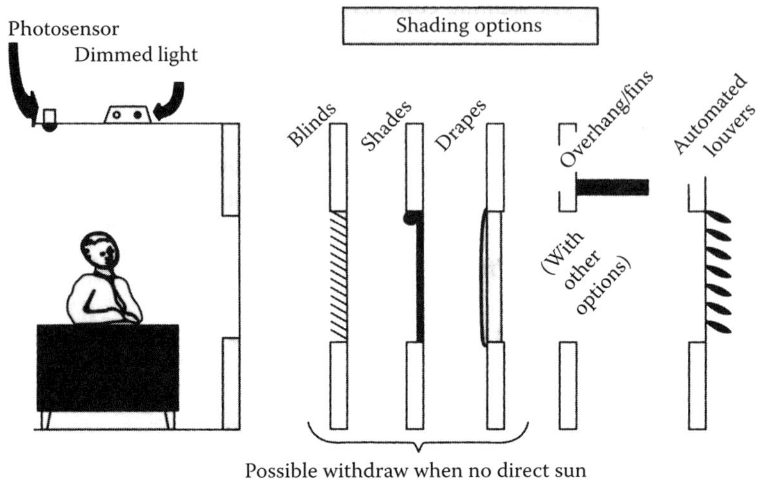

FIGURE 9.22
Daylighting system controls. (From Rundquist, R.A. *ASHRAE Journal*, 11, 30, 1991.)

radiation onto the workplace. Photosensor control of electric lighting allows the dimming (or shutting off) of the lights in proportion to the amount of available daylighting illuminance.

In most cases, increasing the solar gain for daylighting purposes, with daylighting controls, saves more in electrical lighting energy and the cooling energy associated with the lighting than is incurred with the added solar gain [35]. In determining the annual energy savings total from daylighting (ES_T), the annual lighting energy saved from daylighting (ES_L) is added with the reduction in cooling system energy (ΔES_C) and with the negative of the heating system energy increase (ΔES_H):

$$ES_T = ES_L + \Delta ES_C - \Delta ES_H \tag{9.40}$$

A simple approach to estimating the heating and cooling energy changes associated with the lighting energy reduction is by using the fraction of the year associated with the cooling or heating season (f_C, f_H) and the seasonal COP of the cooling or heating equipment. Thus, Equation 9.40 can be expressed as

$$ES_T = ES_L + \frac{f_C ES_L}{COP_C} - \frac{f_H ES_L}{COP_H}$$

$$ES_T = ES_L\left(1 + \frac{f_C}{COP_C} - \frac{f_H}{COP_H}\right) \tag{9.41}$$

It should be noted that the increased solar gain due to daylighting has not been included here but would reduce summer savings and increase winter savings. If it is assumed that the increased wintertime daylighting solar gain approximately offsets the reduced lighting heat gain, then the last term in Equation 9.41 becomes negligible.

To determine the annual lighting energy saved (ES_L), calculations using the lumen method described earlier should be performed on a monthly basis for both clear and overcast days for the space under investigation. Monthly weather data for the site would then be used to prorate clear and overcast lighting energy demands monthly. Subtracting the calculated (controlled) daylighting illuminance from the design illuminance leaves the supplementary lighting needed, which determines the lighting energy required.

This procedure has been computerized and includes many details of controls, daylighting methods, weather, and heating and cooling load calculations. ASHRAE [9] lists many of the methods and simulation techniques currently used with daylighting and its associated energy effects.

Example 9.11

A 30′ × 20′ space has a photosensor dimmer control with installed lighting density of 2.0 W/ft^2. The required workplace illuminance is 60 fc and the available daylighting illuminance is calculated as 40 fc on the summer peak afternoon. Determine the effect on the cooling system (adapted from [35]).

Solution

The lighting power reduction is (2.0 W/ft^2) (30 × 20) ft^2 × (40 fc/60 fc) = 800 W. The space cooling load would also be reduced by this amount (assuming CLF = 1.0):

$$\frac{800\text{ W} \times 3.413\text{ Btuh/W}}{12{,}000\text{ Btuh/ton}} = 0.23\text{ ton}$$

Assuming 1 1/2 tons nominally installed for 600 ft^2 of space at \$2200/ton, the 0.23 ton reduction is "worth" 0.23 × \$2,200/ton = \$506. The lighting controls cost about \$1/ft^2 of controlled area, so the net installed first cost is

$$\text{Net first cost} = \$600\,\text{controls} - \$500\text{ A/C savings} = \$100$$

Assuming the day-to-monthly-to-annual illuminance calculations gave a 30% reduction in annual lighting, the associated operating savings can be determined. Lighting energy savings are

$$ES_L = 0.30 \times 2.0\text{ W/ft}^2 \times 600\,\text{ft}^2 \times 2500\text{ h/year} = 900\text{ kWh}$$

Using Equation 9.41 to also include cooling energy saved due to lighting reduction (with $COP_C = 2.5$, $f_C = 0.5$, and neglecting heating) gives

$$ES_T = 900(1 + 0.5/2.5 - 0) = 1080\text{ kWh}$$

At \$0.10 per kWh, the operating costs savings are \$0.10/kWh × 1080 kWh = \$108 per year.

Thus, the simple payback is approximately 1 year (100/108) for the "net" situation, and a little over 5 1/2 years (600/108) against the "controls" cost alone. It should also be noted that the 800 W lighting electrical reduction at peak hours, with an associated cooling energy reduction of 800 W/2.5 COP = 320 W, provides a peak demand reduction for the space of 1.1 kW, which can be used as a "first-cost savings" to offset control system costs.

Problems

9.1 Explain how window placement in a building could be defined as (a) a passive solar feature, (b) an energy conservation technique, and (c) both of the above.

9.2 Write an equation for calculating the cost of savings life-cycle economics of a proposed passive solar system. Explain why it is important to be able to determine the auxiliary energy required for any given passive (or active) system design.

9.3 Referring to the thermal circuit diagram of Figure 9.7 for the thermal storage (Trombe) wall building, construct appropriate thermal circuits for (a) attached sunspace, (b) thermal storage, and (c) direct gain buildings.

9.4 Using rules of thumb for a 200 m^2 floor area Denver residence, determine (a) the auxiliary heating energy required, (b) the solar projected area, and (c) the concrete storage mass needed for a maximum 10°C daily temperature swing.

9.5 A 2000 ft^2 house in Boston is being designed with NLC—12,000 Btu/F-day and 150 ft^2 of direct gain. The direct gain system includes double glazing, nighttime insulation, and 30 Btu/ft^2 F thermal storage capacity. Using the LCR method, determine (a) the annual auxiliary heating energy needed by this design and (b) the storage mass and dimensions required.

9.6 Compare the annual SSF for 150 ft^2 of the following passive systems for the house in Problem 9.5: (a) direct gain (DGA3), (b) vented Trombe wall (TWD4), (c) unvented Trombe wall (TW14), (d) waterwall (WWB4), and (e) sunspace (SSB4).

9.7 A design modification to the house in Problem 9.5 is desired. A 200 ft^2, vented, 12 in. thick Trombe wall is to be added to the direct gain system. Assuming the same types of glazing and storage as described above, determine (a) the annual heating auxiliary energy needed and (b) the Trombe wall mass.

9.8 Using the SLR method, calculate the auxiliary energy required in March for a 2000 ft^2, NLC 12,000 Btu/F-day house in Boston with a 150 ft^2, night-insulated double-glazed direct gain system with 6 in. thick storage floors of 45 Btu/ft^2 F capacity.

9.9 Calculate the heating season auxiliary energy required for the Boston house in Problem 9.8.

9.10 Determine the length of the overhang needed to shade a south-facing 2 m high window in Dallas, TX (latitude 32°51′), to allow for both winter heating and summer shading.

9.11 A 10 mph wind is blowing directly into an open 3 ft × 5 ft window, which is mounted in a room's 8 ft high by 12 ft wide wall. If the wind's temperature is 80°F, are the room's occupants thermally comfortable?

9.12 Design a stack effect/solar chimney (vented Trombe wall) to produce an average velocity of 03 m/s within a 4 m wide by 5 m long by 3 m high room. Justify your assumptions.

9.13 Estimate the overnight radiant cooling possible from an open, 30°C, 8 m diameter water tank during July in Chicago. What would you expect for convective and evaporative cooling values?

9.14 For the buried pipe example (9.8) in the text, determine which of the three flow rate cases leads to the least expensive installation.

9.15 Using data from Figure 9.14, design a 9 ft deep ground-pipe system for Dallas in June to deliver 1000 cfm at 75°F when the outside air temperature is 90°F.

9.16 A 30 ft × 20 ft office space has a photosensor dimmer control working with installed lighting of 2 W/ft^2. The required workplace illuminance is 60 fc and the available daylighting is calculated as 40 fc on the summer peak afternoon. Determine the payback period for the dimmer control system assuming the following: 1–1/2 ton cooling installed for 600 ft^2 at $2200/ton, lighting control system cost at $1/ft^2, 30% reduction in annual lighting due to daylighting, $0.10/kWh electricity cost, and cooling for 6 months at a $COP_c = 2.5$.

9.17 Determine the illuminances (sun, sky, and ground-reflected) on a vertical, south-facing window at solar noon at 36°N latitude on June 21 and December 21 for (a) a clear day and (b) an overcast day.

9.18 Determine the sidelighting workplace illuminances for a 20 ft long, 15 ft wide (deep), 8 ft high light-colored room with a 15 ft long by 5 ft high window. Assume the direct sun plus clear sky illuminance is 3000 fc and the ground-reflected illuminance is 200 fc.

9.19 Determine the clear sky day and the cloudy day work-plane illuminances for a 30 ft long, 30 ft wide, 10 ft high light-colored room. A 20 ft long by 7 ft high window with ¼-in. clear glass faces 10°E of South, the building is at 32°N latitude, and it is January 15 at 2 PM solar. The ground outside is covered by dead grass.

9.20 Determine the clear day and cloudy day illuminances on a horizontal skylight at noon on June 21 and December 21 in (a) Miami, (b) Los Angeles, (c) Denver, (d) Boston, and (e) Seattle.

9.21 A 3 ft × 5 ft double-domed skylight has outer and inner flat-plate plastic transmittances of 0.8 and 0.7, respectively; a 2 ft deep well with 80% reflectance walls; and a 90% net skylight area. Calculate the net transmittance of the skylight.

9.22 Determine the number and roof placement of 10 ft × 4 ft skylights needed for a 50 ft × 50 ft × 10 ft high office when the horizontal illuminance is 6000 fc, the skylight has 45% net transmittance, and the required workplace illuminance is 100 fc.

9.23 What would be the procedure for producing uniform workplace illuminance when both sidelighting and skylighting are used simultaneously?

Defining Terms

Active system: A system employing a forced (pump or fan) convection heat transfer fluid flow.

Daylighting: The use of the sun's radiant energy for illumination of a building's interior space.

Hybrid system: A system with parallel passive and active flow systems or one using forced convection flow to distribute from thermal storage.

Illuminance: The density of luminous flux incident on a unit surface. Illuminance is calculated by dividing the luminous flux (in lumens) by the surface area (m^2, ft^2). Units are lux (lx) (lumens/m^2) in SI and footcandles (fc) (lumens/ft^2) in English systems.

Luminous flux: The time rate of flow of luminous energy (lumens). A lumen (1 m) is the rate which luminous energy from a 1 candela (cd) intensity source is incident on a 1 m^2 surface 1 m from the source.

Passive cooling system: A system using natural energy flows to transfer heat to the environmental sinks (ground, air, and sky).

Passive heating system: A system in which the sun's radiant energy is converted to heat by absorption in the system, and the heat is distributed by naturally occurring processes.

Sidelighting: Daylighting by light entering through the wall/side of a space.

Skylight: The diffuse solar radiation from a clear or overcast sky, excluding the direct radiation from the sun.

Sunlight: The direct solar radiation from the sun.

Toplighting: Daylighting by light entering through the ceiling area of a space.

References

1. Larson, R., Vignola, F., and West, R. (eds.) (1992) *Economics of Solar Energy Technologies*, American Solar Energy Society (ASES), Boulder, CO, December 1992.
2. Kreith, F. and West, R. (2007) *CRC Handbook of Energy Efficiency and Renewable Energy*, CRC Press, Boca Raton, FL.
3. Close, D.J., Dunkle, R.V., and Robeson, K.A. (1968) Design and performance of a thermal storage air conditioning system, *Mechanical and Chemical Engineering Transactions*, Institute Eng. Australia, MC4, 45.
4. PSDH. (1980) *Passive Solar Design Handbook*, Volume One: *Passive Solar Design Concepts*, DOE/CS-0127/1, March 1980. Prepared by Total Environmental Action, Inc. (B. Anderson, C. Michal, P. Temple, and D. Lewis); Volume Two: *Passive Solar Design Analysis*, DOE/CS-0127/2, January 1980. Prepared by Los Alamos Scientific Laboratory (J.D. Balcomb, D. Barley, R McFarland, J. Perry, W. Wray, and S. Noll). U.S. Department of Energy, Washington, DC.
5. PSDH. (1984) *Passive Solar Design Handbook, Part One:* Total Environmental Action, Inc., *Part Two:* Los Alamos Scientific Laboratory, *Part Three:* Los Alamos National Laboratory. Van Nostrand Reinhold, New York.
6. NCDC. (1992) National Climactic Data Center, *Climatography of the U.S. #81*, Federal Building, Asheville, NC.
7. Duffie, J.A. and Beckman, W.A. (1991) *Solar Engineering of Thermal Processes*, 2nd edn., Wiley, New York.
8. McQuiston, P.C. and Parker, J.D. (1994) *Heating, Ventilating, and Air Conditioning*, 4th edn. Wiley, New York.
9. ASHRAE Handbook. (1989, 1993, 1997). *Fundamentals*, I-P and S-I editions, American Society of Heating, Refrigerating and Air-Conditioning Engineers, Atlanta, GA.
10. Olgyay, A. and Olgyay, V. (1967) *Solar Control and Shading Devices*, Princeton University Press, Princeton, NJ.
11. ASHRAE Handbook. (1991, 1995) *Heating, Ventilating, and Air-Conditioning Applications*, I-P and S-I editions, American Society of Heating, Refrigerating and Air-Conditioning Engineers, Atlanta, GA.
12. Hay, H. and Yellott, J. (1969) Natural air conditioning with roof ponds and movable insulation, *ASHRAE Transactions* 75(1), 165–177.
13. Marlatt, W., Murray, C., and Squire, S. (1984) Roof Pond Systems Energy Technology Engineering Center. Rockwell International Report No. ETEC 6, April.
14. Givoni, B. (1994) *Passive and Low Energy Cooling of Buildings*, Van Nostrand Reinhold, New York.
15. Martin, M. and Berdahl, P. (1984) Characteristics of infrared sky radiation in the United States, *Solar Energy*, 33(314), 321–336.
16. Goswami, D.Y. and Biseli, K.M. (1994) Use of underground air tunnels for heating and cooling agricultural and residential buildings. Report EES-78, Florida Energy Extension Service, University of Florida, Gainesville, FL.
17. U.S.D.A. (1960) Power to produce, *1960 Yearbook of Agriculture*, U.S. Department of Agriculture, Washington, DC.

18. Krarti, M. and Kreider, J.F. 1996. Analytical model for heat transfer in an underground air tunnel, *Energy Conversion Management* 37(10), 1561–1574.
19. Hollmuller, P. and Lachal, B. (2001) Cooling and preheating with buried pipe systems: Monitoring, simulation and economic aspects, *Energy and Buildings* 33, 509–518.
20. De Paepe, M. and Janssens, A. (2003) Thermo-hydraulic design of earth-air heat exchanger, *Energy and Buildings* 35, 389–397.
21. Goswami, D.Y. and Dhaliwal, A.S. (1985) Heat transfer analysis in environmental control using and underground air tunnel, *Journal of the Solar Energy Engineering* 107(May), 141–145.
22. Kusuda, T. and Achenbach, P.R. (1965) Earth temperature and thermal diffusivity at selected stations in the United States, *ASHRAE* Transactions, 71(1), 965.
23. Labs, K. (1981) Regional analysis of ground and above ground climate, Report ORNAL/Sub-81/40451/1, Oak Ridge National Laboratory, Oak Ridge, TN.
24. Goswami, D.Y. and Ileslamlou, S. (1990) Performance analysis of a closed-loop climate control system using underground air tunnel, Journal of the Solar Energy Engineering 112(May), 76–81.
25. Robbins, C.L. (1986) *Daylighting—Design and Analysis*, Van Nostrand, New York.
26. McCluney, R. (1998) Advanced fenestration daylighting systems, in: *International Daylighting Conference'98*, Natural Resources Canada/CETC, Ottawa, Ontario, Canada.
27. Clay, R.A. (2001) Green is good for you, *Monitor on Psychology* 32(4), 40–42.
28. Kinney, L., McCluney, R., Cler, G., and Hutson, J. (2005) New designs in active daylighting: Good ideas whose time has (finally) come, in: *Proceedings of the 2005 Solar World Congress*, Orlando, FL.
29. Murdoch, J.B. (1985) *Illumination Engineering—From Edison's Lamp* to *the Laser*, Macmillan, New York.
30. IES. (1987) *Lighting Handbook, Applications Volume*, Illumination Engineering Society, New York.
31. LOF. (1976) *How to Predict Interior Daylight Illumination*, Libbey-Owens-Ford Co., Toledo, OH.
32. AAMA. (1977) Voluntary standard procedure for calculating skylight annual energy balance, Architectural Aluminum Manufacturers Association Publication 1602.1.1977, Chicago, IL.
33. Pierson, O. (1962) *Acrylics for the Architectural Control of Solar Energy*, Rohm and Haas, Philadelphia, PA.
34. Dietz, P., Murdoch, J., Pokoski, J., and Boyle, J. (1981) A skylight energy balance analysis procedure, *Journal of the Illuminating Engineering Society* 11(1), 27–34, October 1981.
35. Rundquist, R.A. (1991) Daylighting controls: Orphan of HVAC design, *ASHRAE Journal* 11, 30–34, November 1991.

Suggested Readings

Anderson, E.E. (1983) *Fundamentals of Solar Energy Conversion*, Addison-Wesley, Reading, MA.

Anderson, B. and Wells, M. (1981) *Passive Solar Energy*, Brick House Publishing, Andover, MA.

Design Guidelines for the Passive Solar Home. (1980) Pacific Power and Light Pamphlet 6163 9/80 10M ML.

Halacy, D.S. (1984) *Home Energy*, Rodale Press, Emmaus, PA.

Lebens, R.M. (1980) *Passive Solar Heating Design*, Applied Science Publishers Ltd., London, England.

Phillips, R.O. (1980) Making the best use of daylight in buildings, in: Cowen, H.J. (ed.), *Solar Energy Applications in the Design of Buildings*, Applied Science Publishers London, England, pp. 95–120.

For Further Information

The most complete basic reference for passive system heating design is still the *Passive Solar Design Handbook*, all three parts. *Solar Today* magazine, published by the American Solar Energy Society, is the most available source for current practice designs and economics, as well as a source for passive system equipment suppliers.

The *ASHRAE Handbook of Fundamentals* is a general introduction to passive cooling techniques and calculations, with an emphasis on evaporative cooling. *Passive Solar Buildings* and *Passive Cooling*, both published by MIT Press, contain a large variety of techniques and details concerning passive system designs and economics. All the major building energy simulation codes (DOE-2, EnergyPlus, TRNSYS, TSB13, etc.) now include passive heating and cooling technologies.

The Illumination Engineering Society's Lighting Handbook presents the basis for and details of daylighting and artificial lighting design techniques. However, most texts on illumination present simplified format daylighting procedures. Currently used daylighting computer programs include various versions of *Lumen Micro*, *Lightscape*, and *Radiance*.

Passive Solar Design Strategies: Guidelines for Homebuilders (Passive Solar Industries Council, Washington, DC, 1989) presents a user-friendly approach to passive solar design.

10

Energy Storage

10.1 Overview of Storage Technology

In June of 2008, the American Institute of Chemical Engineers published a white paper entitled, "Massive electricity storage" [1]. This paper states that electricity is generated and consumed instantaneously and that the electric power grid essentially has little or no storage capacity to smooth out variations in demand. However, coal-fired and nuclear power plants have turbine generators that can deliver continuous and dispatchable power required by consumers, and these power plants can meet shifting demands by utilizing peaking generators or so-called spinning reserves. On the other hand, renewable power generated by wind or solar radiation is intermittent, and massive electric storage (MES) is the critical technology needed if renewable power is to become a major sustainable source. For system stability and load leveling, large-scale stored power is needed and without sufficient MES, renewable energy cannot serve as a base load energy supplier.

In 2006, the U.S. grid had approximately 1000 GW of installed capacity and delivered 4254 TWh of electricity. Of that total, the energy sources were 49% coal, 22% natural gas (NG), 19% nuclear, 7% hydroelectric, and less than 2.5% came from non-hydro renewable sources. Based on these data, the white paper made an estimate of the upper bound of the U.S. MES size and economic cost for various levels of grid penetration by renewable power. These estimates are shown in Table 10.1.

An inspection of these figures clearly indicates that, in addition to the installation of renewable energy sources, the investment required for storage is enormous and has to be included in planning a sustainable energy future.

Energy storage will play a critical role in any future renewable energy system. There are two principal reasons why energy storage will be of increasing importance to renewable energy:

- Many of the most significant renewable energy sources are intermittent and generate power when the weather or the movement of the sun allows it, rather than when energy is needed.

TABLE 10.1

Upper Bound Estimate of U.S. MES Size and Cost

Grid penetration by renewable power (%)	20	50	75
Firm renewable demand (GW)	200	500	750
Nameplate renewable installed capacity (GW)	570	1430	2150
Capital investment for installed capacity ($ billion)	860	2150	3220
MES power capacity (GW)	114	285	428
MES storage capacity (GWh)	912	2280	3424
MES capital investment ($ billion)	342	855	1284

- A ground transportation system using cars and buses requires energy to be carried by the vehicles. As gasoline becomes more expensive and less available, other forms of fuel and energy storage for cars and buses will be needed.

Energy can be stored in many forms: as mechanical energy in rotating, compressed, or elevated substances; as thermal or electrical energy that can be released as needed; or as electrical charges ready to travel from positive to negative poles as demanded. Storage technologies that can absorb and release electricity have great value because electricity can be converted easily into mechanical or thermal energy. Electricity is also the output of the most promising renewable energy technologies: wind turbines, solar thermal power plants, and photovoltaic cells. There are, however, some applications that can benefit from thermal or mechanical storage technology. For example, thermal energy storage (TES) is needed for the continuous operation of solar thermal power plants and overnight heat storage for heating or cooling of buildings.

The storage methods treated in this chapter can accept and deliver energy in three ways: electrical, mechanical, and thermal. When energy is stored electrically or mechanically, it can be converted to either of the other two forms quite efficiently. For example, electricity can drive a motor with only about 5% energy loss or can provide heat with a resistive heater with virtually no loss at all. The quality of TES depends mostly on its temperature. The efficiency of conversion of thermal energy into power is limited by thermodynamic considerations. The theoretical maximum quantity of useful work, W_{max} (mechanical energy), that can be extracted from a given quantity of heat, Q, is shown in Chapter 8 as

$$W_{max} = \frac{T_1 - T_2}{T_1} Q$$

where

T_1 is the absolute temperature of the stored heat
T_2 is the ambient absolute temperature

Any energy-storage technology must be carefully chosen to accept and produce a form of energy consistent with both the energy source and the final application. Storage technologies that accept and/or produce thermal energy should only be used with heat energy sources or heat applications. Mechanical and electrical technologies are more versatile, but electrical storage technologies are more convenient than mechanical ones because electricity can be easily transmitted [2].

10.1.1 Applications

Table 10.2 presents an overview of the most important energy storage technologies. In this table, each technology is classified by its relevance to one of the following four principles.

Utility shaping: For this application, large capacity storage devices are required to respond to varying electricity demands. Examples would be a solar thermal power plant, which delivers energy generated during the day to a thermal storage device for use at night time or a wind turbine that compresses air during high winds to power a gas turbine when the wind stops.

Power quality: Storage technologies to smooth power delivery during short periods such as outages or switching events must be capable of providing large changes in power output over very short periods. Ultracapacitors and batteries are in this category.

Distributed grid technology. Storage devices for distributed grid technologies must be able to generate energy and storage at customer locations rather than at a central utility power station. Technologies in this category include batteries for photovoltaic power installed on the roof of a building. Other systems include solar thermal or geothermal power that can be implemented on a distributed scale. The required power generation is small compared to utility shaping. They require capacities in the 1–50 kWh range, while utility shaping requires capacities in the 1000 MWh range.

Transportation technologies: The devices that can be used for storage in automobiles and trucks include all types of batteries. Examples are hybrid electric vehicles such as the Prius, and plug-in electric vehicles such as the Volt. Large fleets of plug-in-hybrid-electric vehicles (PHEVs) could also help in utility shaping by utilizing off-peak power to charge the batteries and feeding the grid system from the batteries during peak demand.

The physical modes of energy storage may involve one or more mechanical, thermal, or electromagnetic forms. An energy storage technology may require both a storage reservoir and a converter and transmission system for moving the power to and from the reservoir to its destination and use. A pumped hydro system, for example, contains both elements using potential energy to store and a turbine to deliver the energy. A molten salt thermal

TABLE 10.2

Overview of Energy Storage Methods and Their Applications

	Utility Shaping	Power Quality	Distributed Grid	Automotive
Direct electric				
Ultracapacitors		✓		✓
SMES		✓		
Electrochemical				
Batteries				
Lead acid	✓	✓	✓	
Lithium-ion	✓	✓	✓	✓
Nickel-cadmium	✓	✓		
Nickel-metal hydride				✓
Zebra				✓
Sodium-sulfur	✓	✓		✓
Flow batteries				
Vanadium redox	✓			
Polysulfide bromide	✓			
Zinc bromide	✓			
Electrolytic hydrogen				✓
Mechanical				
Pumped hydro	✓			
Compressed air	✓			
Flywheels		✓		✓
Direct thermal				
Sensible heat				
Liquids			✓	
Solids			✓	
Latent heat				
Phase change	✓		✓	
Hydration–dehydration	✓			
Chemical reaction	✓		✓	
Thermochemical				
Biomass solids	✓		✓	
Ethanol	✓			✓
Biodiesel				✓
Syngas	✓			✓

system for a power tower, on the other hand, requires only one reservoir for storing and delivering energy.

10.1.2 Technology Characterization

Every energy storage technology can be characterized essentially by three important parameters: self-discharge time, storage size, and efficiency. Within a category, the final selection of a specific storage technology can be made by the consideration of cycle life, specific energy, specific power, energy density, and power density.

Self-discharge time (SDT) is the time required for a fully charged, self-standing storage device to reach a certain depth of discharge (DOD). DOD can be described as a percentage of the storage device's useful capacity. For example, 80% DOD means that 20% of the device's energy capacity is left. It should be noted that the relation between SDT and DOD is, in general, not linear and must therefore be measured. SDTs vary greatly from a few minutes to weeks, as, for example, in battery or thermal storage.

Storage size describes the intrinsic scale of the technology. It is the most difficult to define of the three parameters. Some technologies have a fairly large storage size but cannot provide small-scale energy storage. Storage size is often called energy storage capacity and is measured in kWh.

Efficiency is the ratio of the energy output from the device to the energy input. As for energy density and specific energy, the system boundaries must be carefully specified when determining efficiency. It is also important to note the form of energy required at the input and output interconnections and to include the entire system necessary to attach these interconnections. For example, if the system is used to store energy from a utility wind farm, then both the input and output will be AC electricity. However, when comparing a battery with a fuel cell, it is necessary to include the efficiency of an AC-to-DC rectifier for the battery, whereas an AC-powered hydrogen power generation system is required for the fuel cell. DC-to-AC converters are associated with both. Efficiency is also related to the SDT. For example, if the discharge of a thermal storage system occurs much later than charging, the apparent efficiency will be lower because a significant amount of the thermal energy may be lost if the interim between charge and discharge is large.

Cycle life is the number of consecutive charge/discharge cycle that a storage system can undergo while maintaining the system's specifications. Cycle life specifications are made against a selected DOD, depending on the application of the storage device. For example, a battery used in a hybrid electric vehicle may consume only 10%–20% of the energy stored during most of the discharge cycle. The advantage of a thermal storage system is that it can undergo many charge/discharge cycles without significant waste.

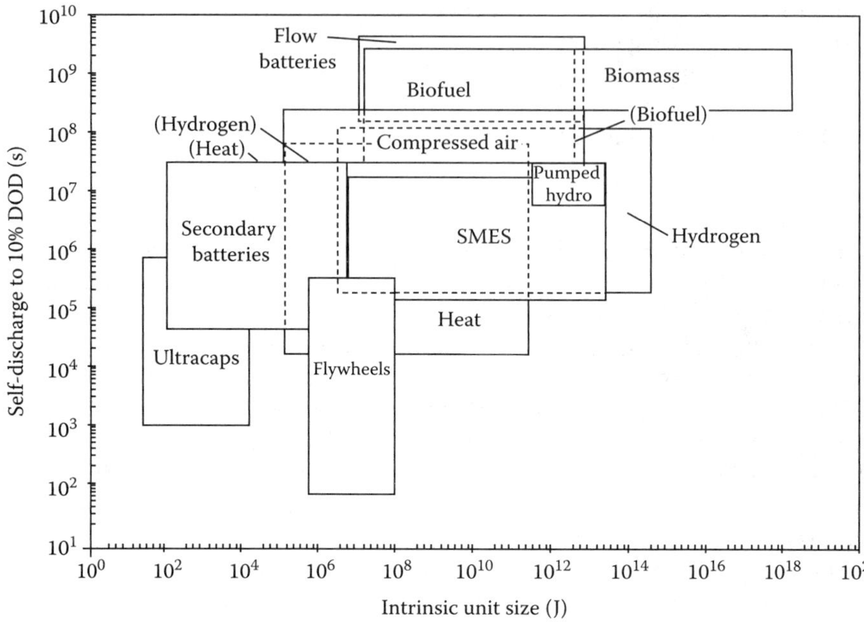

FIGURE 10.1
All storage technologies are mapped by SDT and unit size. Not all hidden lines are shown. Larger SDTs are always more desirable, but importance depends on the application. Intrinsic unit size must be matched to the application.

Specific energy is a measure of how heavy the system is. It is measured in units of energy per mass and here we will represent this quantity in terms of MJ/kg or kWh/kg. The higher the specific energy, the lighter the device. For transportation applications, a high specific energy is necessary, whereas for utility application, specific energy is relatively unimportant.

Energy density is how much space the system occupies. It is measured in units of energy per volume, MJ/L or kWh/m^3.

Storage energy modes depend on the end-use application, but the magnitude of the power load and the time scale are important. Figure 10.1 shows the energy storage technology characteristics of the various storage technologies to be discussed in this chapter according to size and discharge time. For example, to provide electric power for short power outages requiring a large power flux for a short period of time, such as seconds or minutes, capacitors and flywheels are well suited. On the other hand, when large amounts of electric or thermal energy are needed for periods of hours or longer, pumped hydro power, compressed air storage, or thermal storage are the appropriate technologies. Distributed grid, automotives and building applications require a moderate discharge time and a moderate unit size. Figure 10.2 shows the fields of application in terms of discharge time and power rating.

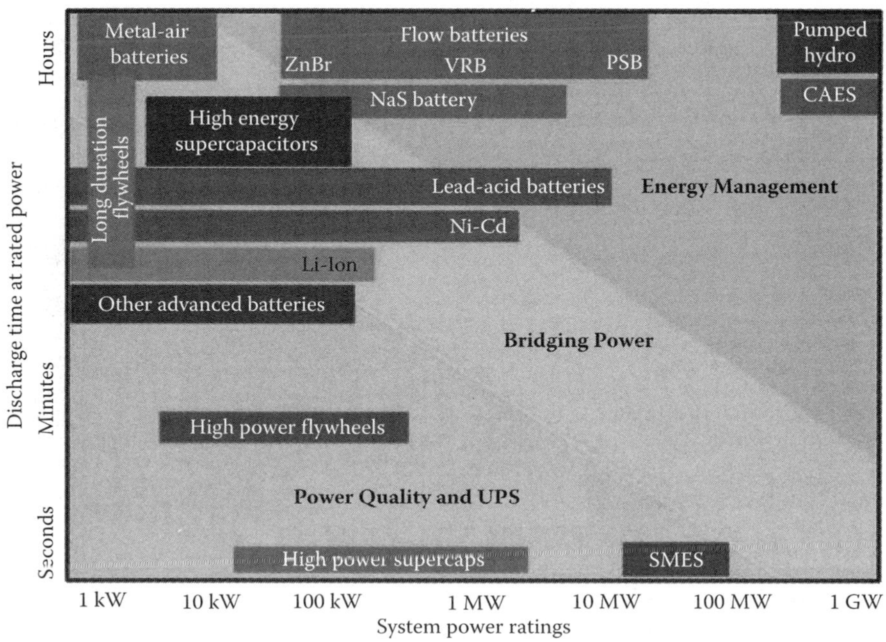

FIGURE 10.2
Discharge time vs. power rating for various energy storage methods. (From Technology comparisons | Ratings, Electricity Storage Association, www.esa.org, 2003. With permission.)

10.2 Mechanical Technologies

Two of the mechanical storage systems discussed below are at this time the only technologies available for large scale and long-term utility size storage, as shown in Figure 10.2. Flywheels are only useful for very short time storage.

10.2.1 Pumped Hydroelectric Energy Storage

Pumped hydroelectric energy storage (PHES) is a mature technology that has been deployed for over a century. Examples as early as the 1890s exist in Italy and Switzerland. PHES uses electricity to pump water uphill into a reservoir where it can be stored. The energy can be recaptured when the water is released and runs back downhill through a turbine. These storage systems allow management of water as well as energy resources. PHES systems are very efficient and, with modern turbine technologies, they can approach 80%–85% round trip efficiencies.

Figure 10.3 is a schematic diagram of a PHES installation at Raccoon Mountain, which is operated by the Tennessee Valley Authority. The Raccoon Mountain pumped storage plant is widely cited as an example of excellent

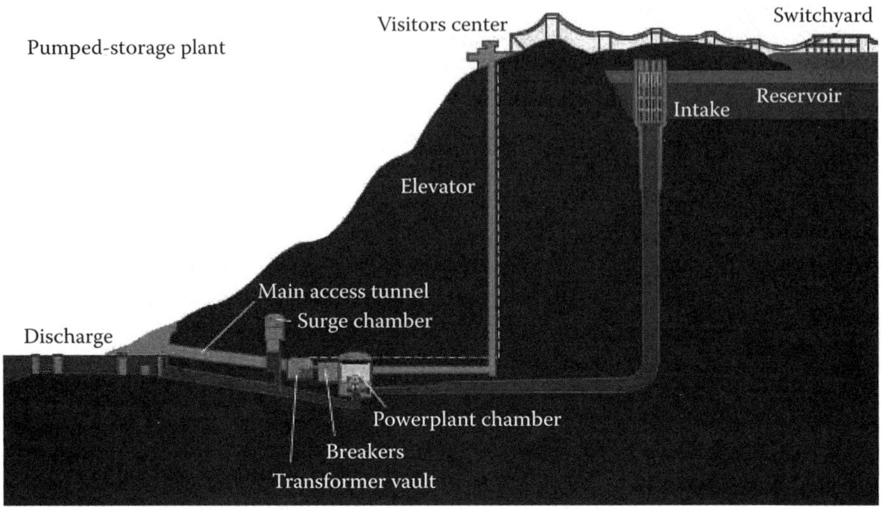

FIGURE 10.3
Schematic of the Raccoon Mountain pumped-storage plant. (From Tennessee Valley Authority, The mountaintop marvel, www.tva.gov/heritage/mountaintop/index.htm. Accessed October 12, 2010.)

engineering for PHES. The plant, which was completed in 1978, has a generating capacity of about 1,600 MW and can run for 22 h to supply about 35,000 MWh of electricity.

The power and capacity of a PHES system is a function of the hydraulic head, the water flow rate, and the efficiency of the pump and the turbine. The head is given by the upper elevation to which the water is pumped and the lower elevation at which the turbine is located. The flow rate can be regulated by valve action. The maximum flow that can be obtained is limited by the reservoir size. The maximum flow rate is dictated by the reservoir size divided by the desired storage time to yield the available flow per unit time, minus a 15% reserve operating condition.

Given the hydraulic head, an upper bound on flow rate and an efficiency of the plant, the power-generating capacity of a pumped hydroelectric installation is given by the equation below:

$$P = QH\rho g\eta \tag{10.1}$$

where

P is the generated output power in Watts (W)
H is the height differential in meters (m)
Q is the fluid flow rate in cubic meters per second (m^3/s)
ρ is the fluid density in kilogram per cubic meter (kg/m^3) L, about 1000 kg/m^3 for water
g is the gravitational constant (9.81 m/s^2 on Earth)
η is the overall efficiency

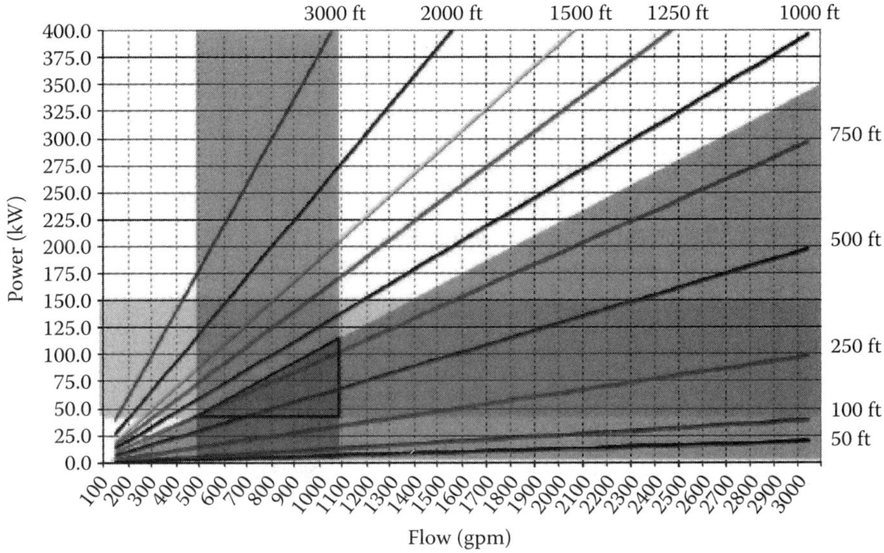

FIGURE 10.4
Pumped hydro turbine power output as a function of flow rate for various hydraulic heads, efficiency = 70%. (From Levine, J.G., Pumped hydroelectric energy storage and spatial diversity of wind resources as methods of improving utilization of renewable energy sources, Thesis, University of Colorado, Boulder, CO, 2007. With permission.)

Given the power capacity of the plant, the limiting value of the upper or lower reservoir dictates the energy capacity. Potential energy generation in kWh is calculated by power output times the operating time. Figure 10.4 shows the power as a function of flow rate and head.

Example 10.1

Calculate the power generated by a PHES with a potential hydraulic height differential between reservoirs of 100 m and a volumetric flow rate of 1000 m^3/s through a hydraulic turbine with an efficiency of 0.8.

Solution
We use Equation 10.1, $P = QH\rho g\eta$. By substituting the appropriate values to find that $P = 1{,}000\,m^3/s \times 100\,m \times 1{,}000\,kg/m^3 \times 9.81 m/s^2 \times 0.8 = 784{,}800{,}000\,We = 785\,MWe$

10.2.1.1 Turbines

Hydro turbines have in the past been used mostly for hydropower generation from large dams. Hydropower produces about 20% of the world's electricity in 150 countries. The technology has evolved for more than 100 years and the literature of hydroelectric power generation can be used as a guide for designing pumped storage facilities.

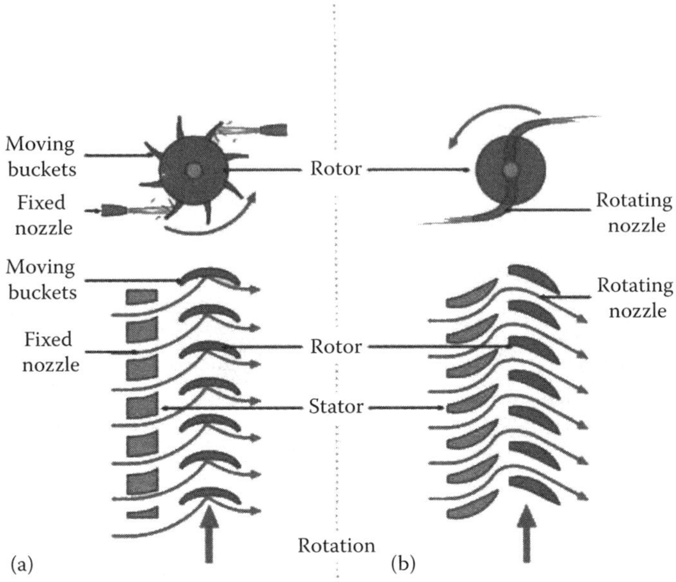

FIGURE 10.5
Schematic diagram of (a) impulse and (b) reaction turbines.

Conventional hydraulic turbines are either impulse or reaction machines (see Figure 10.5). Impulse turbines, often called Pelton turbines, convert the potential energy of a fluid into kinetic energy by expansion in a stationary nozzle to form a jet, which is then directed toward buckets attached to a rotating wheel to create shaft work that can be extracted. Reaction turbines are of the Francis or Kaplan type and utilize both hydraulic pressure and kinetic energy to create rotating shaft work. Each of the turbines has a specific operating range in terms of hydraulic head and power output. The choice of which turbine to use depends on the appropriate characteristics.

Pelton wheels were used in the early days of hydropower. They have a lower efficiency at low hydraulic head in comparison to reaction machine. Pelton wheels are still used, however, for high head resources (about 200 m) and small power outputs (less than 5 MW).

Francis turbines use a set of fixed veins that guide the fluid to the buckets that make up the turbine runner and are mounted on a central shaft. Francis turbines have a large operating range with heads from 40 to 500 m and unit sizes from 10 to up to 1000 MWe. Kaplan turbines are similar to Francis turbines, but the turbine blade angle can be adjusted to improve performance under different flow conditions. They work well at low heads, less than 10 m to about 100 m, with power outputs of the order of 200 MWe.

Once built, hydro power installations have operated successfully for more than a century. It is therefore expected that also pumped storage facilities will have a long life. The main drawback of pumped storage is the limited availability of suitable sites. According to Schoenung et al. [4]:

> There have been a number of studies that have indicated that the availability of storage can be a key element in the successful operation of a robust electric power system. Pumped hydro-storage systems are therefore in widespread use worldwide. In 1996, more than 300 systems with capacities ranging from about 20 MWe to more than 2000 MWe were operating.

10.2.2 Compressed Air Energy Storage

Compressed air energy storage (CAES) is a thermodynamic storage mechanism that resembles a gas turbine power plant. Figure 10.6 schematically shows side by side the operation of both a gas turbine and a CAES system.

The operation of a CAES can be demonstrated by comparing these two thermodynamic systems. In a gas turbine power plant, atmospheric pressure air passes through a compressor, is then delivered to a combustion

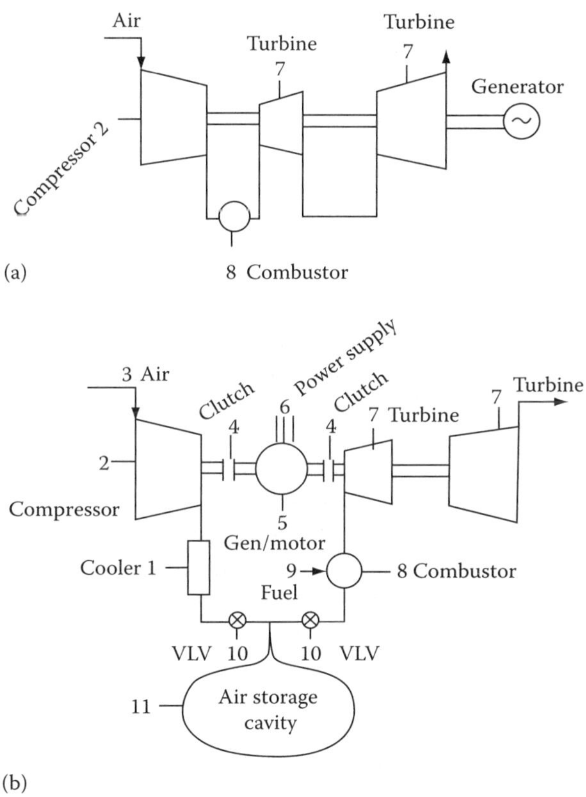

FIGURE 10.6
Comparison of (a) gas turbine and (b) compressed air storage system.

chamber where fuel is injected and the high-pressure (hp) air is heated. The hp heated gas from the combustion chamber then passes through the blades of a turbine whose rotating shaft is connected to an electricity generator. The air from the turbine is exhausted at approximately atmospheric pressure. Approximately two-thirds of the energy generated by the turbine is used to drive the compressor, whereas the remaining one-third is converted to generate electric power.

The components in a CAES are similar to those in a gas turbine, but clutches are added in order that the compressor and the turbine may be connected separately to the generator, which also has to work as a motor. In addition, provision is made to store the hp air from the compressor in a suitable cavity. The air from the cavity is heated in a combustor similarly to the gas turbine power plant, but the timing of discharging the stored hp air can be controlled. Since compression and expansion of the air can take place at different times, the entire output from the turbine is available to generate electricity as needed. In other words, compressed air can be placed in the cavity where it is stored until power is needed. This arrangement is of utmost utility for renewable energy systems such as wind turbines that do not operate continuously. But by using any excess power when available to compress and store energy, the compressed air in the cavity becomes available when winds die down but electricity is needed.

The technology of storing energy in the form of compressed air in underground cavities has been demonstrated to be economical and technically viable. Although the air stored at high pressure in the cavity could be used directly with a low-pressure (lp) turbine, this approach has a low overall thermodynamic efficiency, whereas the hybrid system as described above has an excellent energy utilization factor.

The volume of the air storage is determined by the amount of energy to be stored. In a wind system, for example, the wind energy should be sufficient to supply electric power for an average period consistent with past variation in the available wind power. The rated capacity of the compressor depends on the required length, during which the reservoir is recharged, but the delivery and charging times are not the same. For example, the volume of the reservoir may be sufficient to operate the turbine for 1 h at full load, while the compressor could be designed to recharge the reservoir in 4 h. In this arrangement, the compressor could be sized for one-quarter of the air passing through the turbine and a charging ratio of one to four would be obtained. The design of a CAES system offers considerable flexibility in size and location. Moreover, on a specific cost per rated power capacity, the main part of the equipment costs only a fraction of that for an equivalent gas turbine, since the electrical output of CAES can be several times larger than that of a simple gas turbine alone. There are three basic methods for producing a cavity suitable for CAES air storage, as shown in Figure 10.7.

The best developed of these is "solution mining" of cavities in salt formations. It involves dissolving salts with fresh water and removing the solution after it

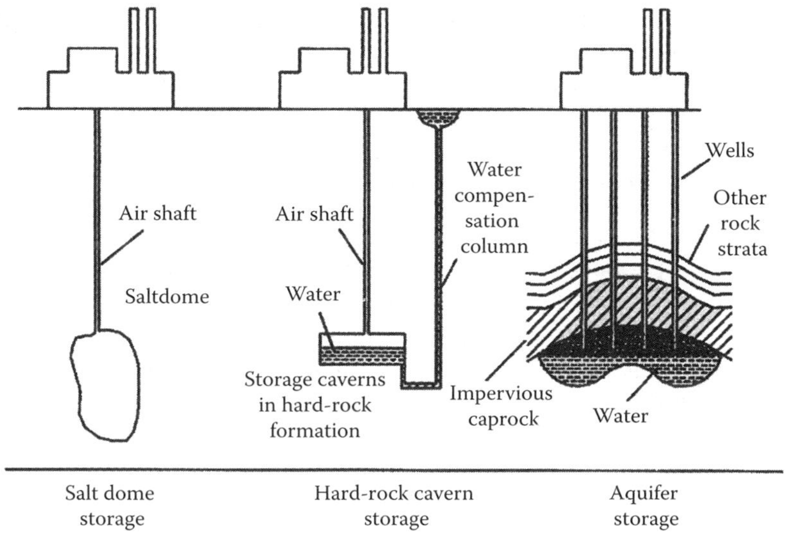

FIGURE 10.7
Methods for producing CAES cavity.

is saturated. The solution mining technology has been used to store gas and oil. Salt caverns are virtually leak tight and suitable sites are widely available.

The second method is to actually produce a cavity in hard rock. This is an expensive operation, but its cost may be offset by using a water compensation leg to make it possible to operate at a constant pressure.

The third option can be used in aquifer regions when there is suitable domed caprock. It requires drilling into the aquifer to develop an air bubble to replace some of the water. This is a slow process, but after the air bubble is created, since air is less viscous than water, it is possible to achieve the required charge and discharge rates. Experience has shown that as many as 50 injection wells may be required to keep the total pressure difference between the charge and discharge cycle to between 10 and 20 bar. This may be the most expensive part of this low-cost aquifer air storage method.

10.2.2.1 Round Trip Efficiency of CAES

CAES is a hybrid generation/storage system because it uses NG as fuel in the round trip process. However, the NG heat input per unit power is much less than for a conventional gas turbine and other heat sources, for example, syngas from biomass, could be used. CAES requires approximately 0.7–0.8 kWh of off-peak or excess electricity in addition to about 4300 Btu (1.26 kWh) of NG heat input to produce 1 kWh of dispatchable electricity. This compares with a heat rate of 11,000 Btu per kWh for a conventional turbine powered by NG.

The round trip efficiency for CAES, defined as [energy out/energy in], can be calculated in two different ways. The first method assumes that all energy inputs, electricity and NG, are of the same quality. In other words, 1 kWh of electricity output requires 0.75 kWh of electricity plus 4300 Btu (1.26 kWh) of NG. In this method, the total energy input is twice the energy output, for an efficiency of 50%. However, the exergy, or quality, of electric energy, is higher than that of NG because, if the NG were used to generate electricity, it would have a conversion efficiency of 30%–40%. Hence, by comparing electricity energy input to electrical energy output using a heat rate of 11,000 Btu per kWh, the gas turbine would produce 0.39 kWh of electricity. Adding that to the 0.75 kWh of off-peak electricity, the total theoretical electricity input is 1.14 kWh to produce 1 kWh of dispatchable electricity. Hence, the round trip efficiency with this perspective, that is, the ratio of electricity in to electricity out, yields an efficiency of about 88%.

10.2.2.2 Comparison between CAES and PHES

Compressed air and pumped hydro are the only storage options for large-scale wind power applications. Each has distinct advantages and disadvantages. PHES has very good round trip efficiency, does not use any auxiliary fuel, can be built for medium and large-sized applications, but finding suitable sites and obtaining water rights may be difficult. In comparison, CAES has a much smaller visible footprint, does not require water rights or dams, is considerably cheaper, and can be built in more places. The major disadvantage of CAES is that it requires a NG booster, which leads to a lower round trip efficiency than PHES unless one compares the efficiency as electricity input to electricity output.

Capital costs for pumped hydro and CAES are shown in Table 10.3. CAES has a low capital investment cost compared to other storage technologies. The capital cost for CAES is between $600 and $700 per kW, excluding the cavern. The PHES cost is somewhat higher due to the extra equipment and controls necessary, as well as the cost of a reservoir. Both methods are superior to battery storage.

TABLE 10.3

Capital Cost Comparison for Pumped Hydro and CAES

	Pumped Hydroelectric (1000 MW)	CAES (300 MW)	Sodium Sulfur Battery (10 MW)
Capital cost: Capacity ($/kW)	600	580	~1200
Capital cost: Energy ($/kWh)	37.5	1.75	~200
Hours of storage	10	40	6–9
Total capital cost ($/kW)	975	650	~3200

Source: Abstracted from Succar, S., Baseload power production from wind turbine arrays coupled to compressed air energy storage, PhD thesis, Princeton University, Princeton, NJ, 2008.

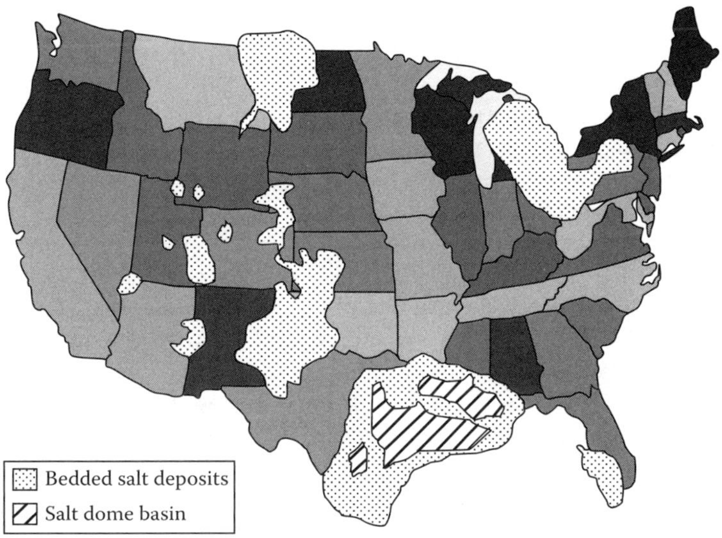

FIGURE 10.8
Map of potential salt cavities for CAES in the United States. (Adapted from Cohn, A. et al., Applications of air saturation to integrated coal gassification CAES power plants, ASME 91-JPGC-GT-2, 1999.)

The primary requirement for siting a CAES plant is a suitable geology with a cavern to hold air at hp. A secondary requirement for the site is vicinity to the generation source, which is likely to be a wind turbine. Typical cavern pressures for CAES range from 500 psi when fully discharged to 1200 psi when fully charged. According to Ref. [5], suitable geological salt rock formations for building a CAES cavity can be found in 85% of the United States. Figure 10.8 shows a map of the United States with suitable sites for potential CAES locations. A simplified estimate of a CAES system in Colorado with a cavern size that allows holding enough air to produce 525 MWh of electricity from a turbine, and an electricity cost of 8.8 cents per kWh showed a simple payback period (without considering the cost of money) of approximately 5 years [3].

10.2.2.3 CAES Volumetric Energy Density

One of the keys to assessing the geologic requirements for CAES is to determine how much electrical energy can be generated per unit volume of storage cavern capacity (E_{GEN}/V_S). The electrical output of the turbine (E_{GEN}) is given by

$$E_{GEN} = \eta_M \eta_G \int_0^t \dot{m}_T W_{CV,TOT} dt \tag{10.2}$$

where

the integral is the mechanical work generated by the expansion of air and fuel in the turbine
$W_{CV,TOT}$ is the total mechanical work per unit mass generated in this process
$\dot{m}_T$ is the air mass flow rate
t is the time required to deplete a full storage reservoir at full output power
η_M is the mechanical efficiency of the turbine
η_G is the electric generator efficiency

All the CAES systems to date are based on two expansion stages, as shown in Figure 10.9. Hence, the work output can be expressed as the sum of the output from the two stages.

The first term in Equation 10.3 represents the work output from the hp turbine that expands the air from the hp turbine inlet pressure (p_1) to the lp turbine inlet pressure (p_2), while the second term is the expansion work of the lp turbine from p_2 to ambient pressure (p_a):

$$W_{CV,TOT} = w_{CV1} + w_{CV2} = -\int_{p_1}^{p_2} v\,dp - \int_{p_2}^{p_a} v\,dp \tag{10.3}$$

Consider first the work output from the first expansion stage. Assuming adiabatic compression and that the working fluid is an ideal gas with a constant specific heat (so that pv^k = a constant, where $k = c_p/c_v$) the work per unit mass for both stages is

$$W_{CV,TOT} = c_p T_2 \left(\frac{c_{p1}T_1}{c_{p2}T_2} \left[1 - \left(\frac{p_2}{p_1} \right)^{\frac{k-1}{k}} \right] + \left[1 - \left(\frac{p_a}{p_2} \right)^{\frac{k-1}{k}} \right] \right) \tag{10.4}$$

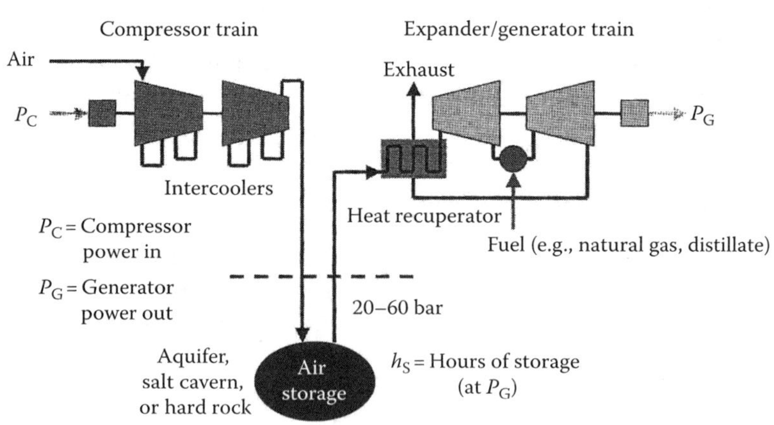

FIGURE 10.9
Typical CAES system.

The total mass flow through the turbine can be expressed as separate air ($\dot{m}_A$) and fuel ($\dot{m}_F$) input terms:

$$\dot{m}_T = \dot{m}_A + \dot{m}_F = \dot{m}_A\left(1 + \frac{\dot{m}_F}{\dot{m}_A}\right) \tag{10.5}$$

Finally, the electric power output per unity cavity volume, V_S, is

$$\frac{E_{GEN}}{V_S} = \frac{\alpha}{V_S}\int_0^t \dot{m}_A\left(\beta + 1 - \left(\frac{p_b}{p_2}\right)^{\frac{k-1}{k}}\right)dt \tag{10.6}$$

where

$$\alpha = \eta_M \eta_G c_{p2} T_2\left(1 + \frac{\dot{m}_F}{\dot{m}_A}\right)$$

and

$$\beta = \frac{c_{p1}T_1}{c_{p2}T_2}\left[1 - \left(\frac{p_2}{p_1}\right)^{\frac{k-1}{k}}\right]$$

The existing CAES plants operate by throttling the reservoir pressure p_s to the hp turbine inlet pressure p_1. If the mass flow and expansion work output are constant, the integral representing the mechanical work in turbine expansion can be reduced to a simple time average, but the net air mass withdrawn from storage is a function of the storage pressure change over the range p_{S2} to p_{S1}:

$$\dot{m}_T = \frac{\Delta m_A}{t}\left(1 + \frac{\dot{m}_F}{\dot{m}_A}\right) \tag{10.7}$$

where

$$\Delta m_A = \frac{V_S p_{S2}}{RT_{S2}} - \frac{V_S p_{S1}}{RT_{S1}} = \frac{V_S p_{S2}}{RT_{S2}}\left(1 - \left[\frac{p_{S1}}{p_{S2}}\right]^{\frac{1}{k}}\right)$$

Substituting the above into Equation 10.5 yields

$$\frac{E_{GEN}}{V_S} = \frac{\alpha M_W p_{S2}}{RT_{S2}}\left(\beta + 1 - \left(\frac{p_b}{p_2}\right)^{\frac{k-1}{k}}\right)\left(1 - \left[\frac{p_{S1}}{p_{S2}}\right]^{\frac{1}{k_S}}\right) \tag{10.8}$$

More detailed information can be found in Ref. [10].

10.2.2.4 Existing CAES Plants

Two CAES systems have previously been designed and operated for many years. One of these is located in Huntorf, Germany and the other one is in McIntosh, Alabama. The experience gained from these existing CAES systems will be valuable in the design and operation of future compressed air storage.

10.2.2.4.1 Huntorf

The Huntorf CAES plant, completed in 1978 near Bremen, Germany was the world's first CAES facility. The 290 MW plant was designed to provide start up during a complete grid outage for nuclear power plants near the North Sea and to provide inexpensive peak power. It has operated successfully for three decades primarily as a peak shaving unit and to supplement hydroelectric storage facilities on the system. Availability and starting reliability for this unit are reported as 90% and 99%, respectively. The plant can provide up to 3 h of storage and has recently been used to help balance the rapidly growing wind output from North Germany [6,7]. The underground portion of the plant consists of two salt caverns (310,000 m^3 total) designed to operate between 48 and 66 bar. The compression and expansion sections draw 108 and 417 kg/s of air, respectively, and are each comprised of two stages. The first turbine stage expands air from 46 to 11 bar. Because gas turbine technology was not compatible with this pressure range, in 1979, a steam turbine was chosen for the hp expansion stage. Although the plant could have operated at a lower heat rate if equipped with heat recuperators to recover exhaust heat from the lp turbine for preheating the gas entering the hp turbine, this feature was omitted in order to minimize system startup time.

10.2.2.4.2 McIntosh

The 110 MW McIntosh plant was built by the Alabama Electric Cooperative on the McIntosh salt dome in southwestern Alabama and has been in operation since 1991. It was designed for 26 h of generation at full power and uses a single salt cavern (560,000 m^3) designed to operate between 45 and 74 bar.

The operational aspects of the plant (inlet temperatures, pressures, etc.) are similar to those of the Huntorf plant. The facility does, however, include a heat recuperator that reduces fuel consumption by approximately 22% at full

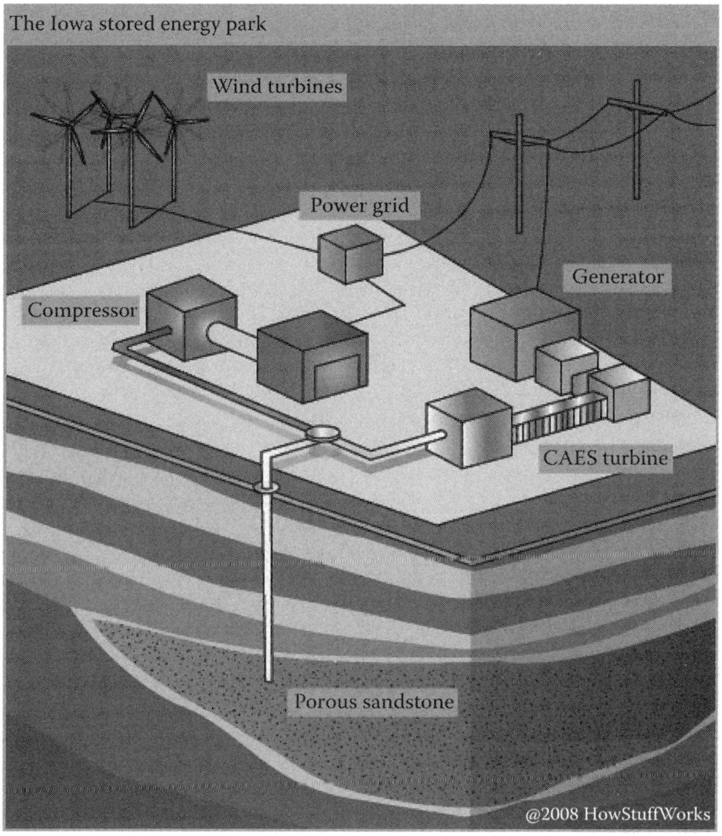

FIGURE 10.10
Schematic diagram of a CAES system proposed for Iowa.

load output and features a dual-fuel combustor capable of burning No. 2 fuel oil in addition to NG [8]. Over 10 years of operation, the plant achieved 92% average starting reliability as well as up to 99.5% average running reliability for the generation cycle and compression cycle, respectively [9].

Figure 10.10 is the schematic of a facility proposed for Iowa, which incorporates all the essential features of a CAES system.

10.2.3 Flywheels

Flywheels have been used for a long time for short-term energy storage. Modern flywheel energy storage systems consist of a massive rotating cylinder on a shaft that is supported on a stator by magnetically levitated bearings that reduce bearing wear and increase system life. By operating a flywheel system in a vacuum environment, drag can be drastically reduced. To complete the system, the flywheel is connected to a motor/

generator that is mounted on the stator that, through power electronics, can interact with the utility grid.

The amount of energy stored in a rotating flywheel is given by

$$E = \frac{I\omega^2}{2}, \tag{10.9}$$

where

I is the flywheel's moment of inertia in kg/m^2
ω is its angular velocity in 1/s
E is in joules

In order to obtain a high specific energy for flywheels, the flywheel speed is increased, since the energy stored increases with the square of the velocity, whereas the specific energy only increases in direct proportion with its mass.

Some of the advantages of flywheels are their low maintenance requirements, long cycle life (better than 10,000 cycles), and long lifetime (in excess of 20 years). High-speed and low-mass flywheels, which are the modern approach to flywheel storage, are made from composites such as carbon fiber. Flywheels can have an energy efficiency of better than 90%, but their energy storage capacity is limited to a short period. Their low energy density and specific energy limit them to voltage regulation and UPS capability.

10.3 Direct Electrical Technologies

10.3.1 Ultracapacitors

An ordinary capacitor stores energy in the electric field between two oppositely charged conductors with a dielectric between them. The dielectric prevents arcing between the plates and permits the plates to hold more charge, thus increasing the maximum potential energy storage. The ultracapacitor, also known as super capacitor, differs from the traditional capacitor in that it uses an extremely thin electrolyte instead of a dielectric. This electrolyte is only a few Angstrom thick, thereby making it possible to increase the energy density of the device. The electrolyte can be made either of an organic or an aqueous material. The aqueous design can operate over a larger temperature range but has a smaller energy density than the organic design. The electrodes are usually made of a porous carbon that increases their surface area as well as the energy density, compared to a traditional capacitor.

Ultracapacitors can effectively equalize voltage variation quickly, which makes them useful for power quality management and for regulating the voltage in automobiles during regular driving. Ultracapacitors can also work in tandem with batteries to relieve peak power needs. They also exhibit very high cycle life of greater than 500,000 cycles and a life span of more than 10 years. The main limitation of ultracapacitors is their ability to maintain charge voltage over any significant period, losing up to 10% of the charge per day.

10.3.2 Superconducting Magnetic Energy Storage

Superconducting magnetic energy storage (SMES) systems can store and discharge energy at high rates. They store energy in the magnetic field created by direct current in a coil of cryogenically cooled, superconducting material. The advantage of a cryogenically cooled, superconducting material over copper is that it reduces electrical resistance to nearly zero. SMES recharge quickly and can repeat the charge-discharge sequence thousands of times without degrading the magnet. They can also achieve full power rather quickly, within 100 ms. Theoretically, a coil of around 150–500 m radius would be able to support a load of 18,000 GJ at 1,000 MW, depending on the peak field and the ratio of the coil's height and diameter. Since no conversion of energy to other forms is involved (e.g., mechanical), the energy is stored directly and round-trip efficiency is expected to be very high. It is believed that mature, commercialized SMES can operate with round-trip efficiency as high as 97% and are a superb technology for providing reactive power on demand. The downside of SMES is the need to cool the coil to cryogenic temperatures, which requires a cryogenic refrigeration system and uses large amount of energy.

10.4 Fundamentals of Batteries and Fuel Cells

Both batteries and fuel cells date back to the early 1800s. Alexander Volta described an electrochemical cell in a letter to the British Royal Society in 1800. This invention and the development of the telegraph in the 1830s led to the deployment and manufacture of batteries, which provided the electric energy for the telegraph, as well as for other inventions. The fuel cell was first demonstrated by Sir William Grove in 1839, but it was not until recently that significant progress toward practical applications occurred.

The operation of batteries and fuel cells are based on similar electrochemical principles and processes. However, batteries are devices for the storage of energy directly, whereas fuel cells are continuous energy converters who by themselves have no inherent storage capability and need hydrogen or some other energy source to operate. The operation of a typical battery is shown schematically in Figure 10.11.

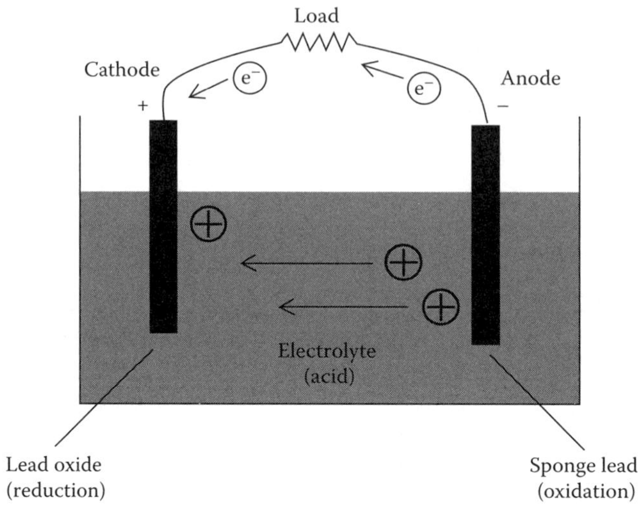

FIGURE 10.11
Schematic diagram of a battery.

The battery has a fixed amount of chemicals that spontaneously produce a flow of electrons if a conducting path is connected to its terminals. Fuel cells require a continuous external supply of chemical reactants or hydrogen and the products are eliminated continuously. The operation of the fuel cells is dependent on a supply of hydrogen, which can be stored externally. Hence, the combination of stored hydrogen and a fuel cell that produces electricity can be considered a storage system, which will be considered later on.

The reaction in both batteries and fuel cells are oxidation-reduction reactions. Both contain electron pairs in contact with an electrolyte, which is the charge-carrying medium, as shown in Figure 10.11. The oxidation reaction takes places at the negative electrode called the *anode* whereas the reduction occurs at the positive electrode called the *cathode*. The anode delivers electrons to the external circuit, which flow from the negative to the positive electrode. However, the conventional current flows in the opposite direction from high to low potential. The electrolyte is usually a liquid solution through which positively charged ions pass from the anode to the cathode. At the cathode, electrons from the external load are neutralized by reacting with positive ions from the electrolyte.

Batteries designed for a single discharge cycle, such as flashlight batteries, are called primary cells, whereas those that can be recharged are called secondary cells. Fuel cells, on the other hand, utilize hydrogen to provide a continuous supply of direct current. There are many different types of batteries and fuel cells available and we will consider the most important of these in the following sections.

10.4.1 Principles of Battery Operation

Batteries find a variety of applications today from tiny button cells in wrist watches to ignition batteries in automobiles and, finally, to large storage batteries in utility power supplies. At present, batteries are considered for storing energy from renewable sources such as solar and wind in order to provide a continuous supply of energy. Batteries are devices for storing an electric charge and automatically delivering an electric current flow on demand. A battery system consists of one or more cells connected in series to provide an open-circuit voltage or electromagnetic force (EMF). The exact value of the EMF depends on the specific cell reactants used. The operation of the battery can be visualized by means of a simple conceptual model shown in Figure 10.12.

When current is drawn from a battery, the voltage drops below its EMF across the terminals, largely due to the internal voltage drop associated with the battery's internal resistance. As shown in Figure 10.12, the terminal voltage in this model is the difference between the EMF and the potential drop across the internal resistance, R_i:

$$V = E - IR_i \tag{10.10}$$

and the current, I, is

$$I = \frac{E}{R_i + R_o} \tag{10.11}$$

The following example illustrates the simplified model for a conventional lead-acid battery.

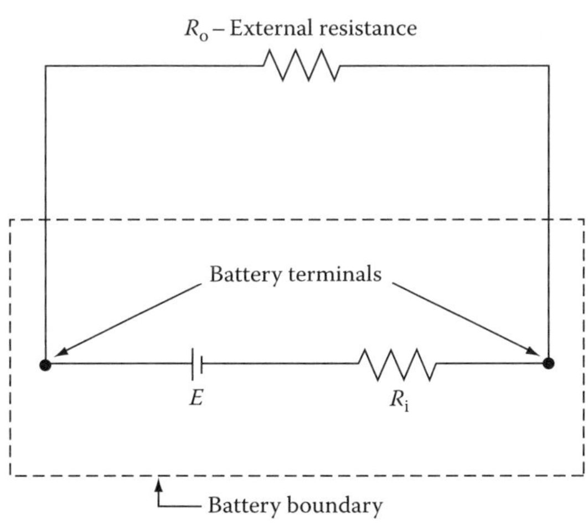

FIGURE 10.12
Simplified battery model.

Example 10.2

Estimate the internal and external resistances for a lead-acid battery with an EMF of 12.7 V. When delivering a current of 50 A through an external resistance of 3 Ω, the voltage measured in operation is 11.1 V. Draw the voltage vs. current characteristics.

Solution

From Equations 10.10 and 10.11, we obtain the internal and external resistances as shown below:

$$R_i = \frac{E-V}{I} = \frac{(12.7\text{ V} - 11.1\text{ V})}{50\text{ A}} = 0.032\ \Omega$$

$$R_o = \frac{E}{I} - R_i = \frac{12.7}{50} - 0.032 = 0.222\,\Omega$$

The current flow through a 3 Ω external resistance is

$$I = \frac{E}{R_i + R_o} = \frac{12.7}{0.032 + 3} = 3.825\text{ A}$$

and the voltage is

$$V = IR_o = (3.825\,\text{A})(3\,\Omega) = 11.5\,\text{V}$$

The voltage vs. current characteristics for the simplified model are shown in Figure 10.13.

The linear model is useful, but represents real battery characteristics only qualitatively. Real batteries can provide current and power to a wide range of resistances with little voltage drop, but only for short periods of time. The linear model fails for longer periods of time, and when large currents are

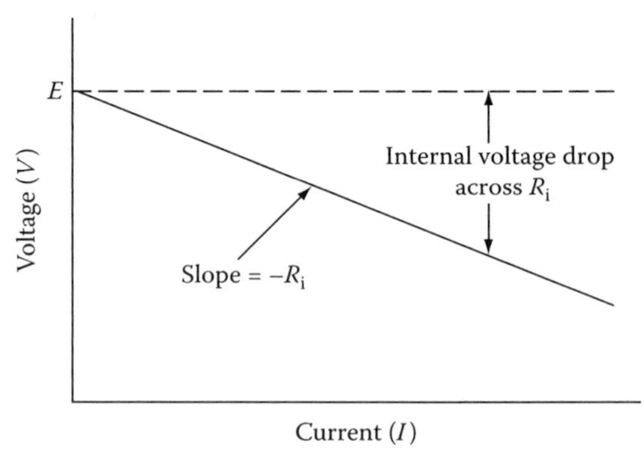

FIGURE 10.13
Current vs. voltage characteristics for simplified battery in problem solution.

drawn because it does not account for the finite amount of stored chemical energy and the rate of reaction limits of real batteries. In real batteries, the terminal voltage and power output drop drastically as the chemicals of the cell are consumed.

For utility scale electric power storage and automotive propulsion, secondary cells offer great opportunities. In secondary cells, the reactions are approximately internally reversible and with an external current source, the internal battery reactions may be reversed during a process called charging. For example, in a solar thermal power plant, the batteries can be charged during periods of sunshine and the energy stored in the battery can then be used for power applications when the sun goes down. In an electric car, the battery can be charged at night and used to power the vehicle in the day.

10.4.2 Cell Physics

When the current is small, the voltage is close to the cell EMF and the power output is small. To analyze the cell physics, assume that the chemical reaction in the battery consumes reactants at an electrode at the rate n_c mol per second. The electrons released in the reaction flow from the electrode through the external load at a rate proportional to the rate of reaction, jn_c, where j is the number of moles of electrons released per mole of reactant. Thus, jn_c is the rate of flow of electrons from the cell in moles of electrons per second. Since there are 6.023×10^{23} electrons per gram-mole and each electron has a charge of 1.602×10^{-19} C, the product is $96{,}488 \times 10^4$ C/g-mol. The charge transported by a gram-mole of electrons is a constant named in honor of the pioneer of electrochemistry, Michael Faraday, Faraday's constant, F,

$$F = (6.023\times10^{23})(1.602\times10^{-19}) = 96,488\times10^{4}\ \text{C/g-mol}$$

With this model, the electric current from a cell, I, can be related to the reaction rate in the cell according to

$$I = jn_cF \tag{10.12}$$

and the instantaneous power, P, delivered by the cell is

$$P = jn_cFV \tag{10.13}$$

It can be seen that the cell electrode, the electrolyte material and the nature and rate of chemical reaction control the maximum cell voltage, cell current, and maximum power output. It is also apparent that the amount of consumable reactants in the battery sets a limit to the battery capacity.

10.5 Rechargeable Batteries

Rechargeable storage batteries offer great opportunities for storing electric power and many different types of batteries are under extensive research at the present time. Although the method of operation for all, except so-called flow batteries, is similar, many different materials and assemblies are being tried. Since at this time it is not clear which battery or groups of batteries will eventually achieve marketability, we will here simply discuss the method of operation of rechargeable batteries and then list the pros and cons of the various types with their potential applications. Figure 10.14 shows ranges of operation for different battery types as specific energy vs. specific power.

Lead-acid batteries have been the workhorse for electric storage for many decades, but they are too heavy and need to be recharged too often for many current applications, especially in the field of transportation. As a result, a worldwide effort is underway to improve and develop new batteries that can meet the demands of various industries including electric utilities and electric automobiles. The field of batteries is changing rapidly and new batteries are constantly entering the market. At this time, it is impossible to pick winners and losers; therefore, the following merely outlines the potential applications, as well as advantages and disadvantages of various typical batteries that are vying for a market share of the future. Only the lead-acid battery will be discussed in detail in order to demonstrate the basic method of operation of rechargeable batteries.

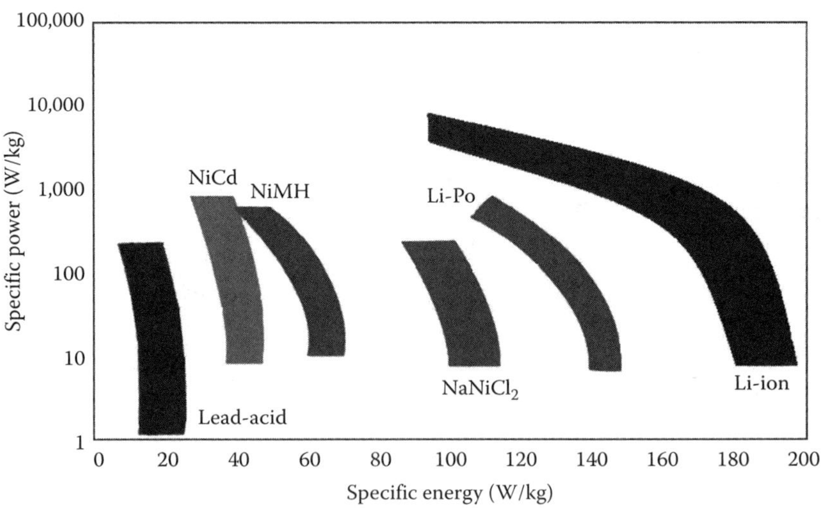

FIGURE 10.14
Approximate operating characteristics of different batteries in terms of specific energy and specific power. (Modified from de Guilbert, A. Batteries and supercapacitors cells for the fully electric vehicle. Saft Groupe SA, Brussels, 2009.)

10.5.1 Lead-Acid Batteries

In order to illustrate the operation of the charging cycle of a secondary battery, we will analyze the operation of a common lead-acid battery shown in Figure 10.15 in some detail.

For such a battery, each cell consists of a porous sponge lead anode and a lead-dioxide cathode, which is usually in the form of a plate. The two electrodes are separated by porous membranes in a sulfuric acid solution. The aqueous electrolyte contains positive hydrogen ion and negative sulfate ions, resulting from disassociation of the sulfuric acid in solution as shown by the chemical equation below:

$$2H_2SO_4 \rightarrow 4H^+ + 2SO_4^{2-} \tag{10.14}$$

As the lead is converted to lead sulfate at the anode, it releases two electrons to the external load as shown below:

$$Pb(s) + SO_4^{2-}(aq) \rightarrow PbSO_4(s) + 2e^- \tag{10.15}$$

Hence, 2 mol of electrons are produced for each mole of sulfate ions and sponge lead reacted, or $j = 2$ for this battery.

In the reduction reaction at the cathode, lead dioxide, using hydrogen and sulfate ions from the electrolyte and electrons from the external circuit, is converted to lead sulfate, as shown by the chemical equation below:

$$PbO_2(s) + 4H^+(aq) + SO_4^{2-}(aq) + 2e^- \leftrightarrow PbSO_4(s) + 2H_2O(l) \tag{10.16}$$

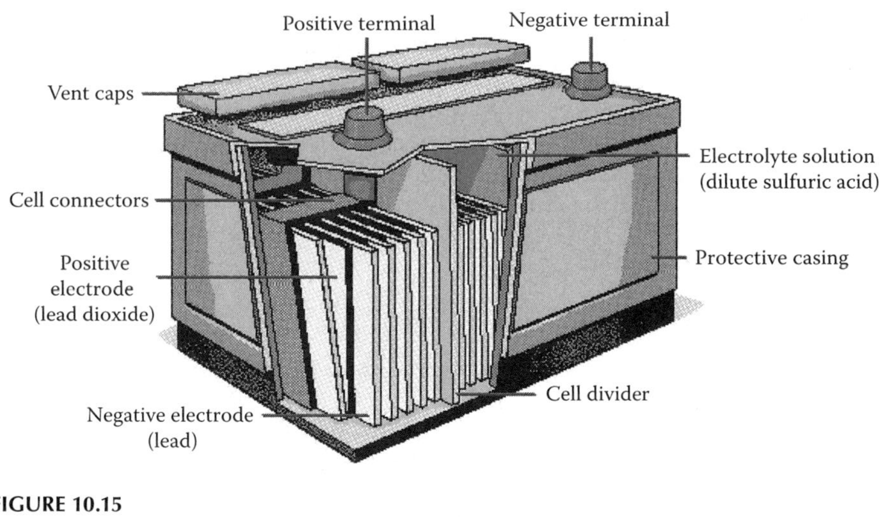

FIGURE 10.15
Schematic diagram of a conventional lead-acid battery.

We observe that the sulfate radical, that is, the sulfuric acid, is consumed at both electrodes and that hydrogen ions combine with oxygen from the cathode material to form water that remains in the cell. The water that is formed as the battery discharges makes the sulfuric acid electrolyte more dilute, as can be seen from the net cell reaction below. The net cell reaction, obtained as the sum of the electrode and electrolyte reactions, is

$$Pb(s) + PbO_2(s) + 2H_2SO_4(aq) \leftrightarrow 2PbSO_4(s) + 2H_2O(l). \quad (10.17)$$

In this and all other rechargeable battery chemistries, left-to-right indicates a battery discharging process and right-to-left indicates recharging. When any of the reactant is depleted, the battery ceases to function. However, the battery can be charged by applying an external power source. This reverses the electron flow and the above reactions are also reversed, as water is consumed to produce sulfate and hydrogen ions from the lead sulfate on the electrodes. Thus, the battery regains its charge and upon demand can again produce electric current.

There are three types of lead-acid batteries on the market: the wet cell, the sealed gel cell, and the sealed absorbed glass mat (AGM). Each of them has certain advantages and disadvantages. The wet cell has a liquid electrolyte, which must be replaced periodically to replenish hydrogen and oxygen escaping during the charge cycle. The sealed gel cell has a silica component added to the electrolyte to stiffen it. The AGM uses a fiberglass-like separator to hold the electrolyte in close proximity to the electrodes. This increases the efficiency. For both the gel and the AGM configuration, the risk of hydrogen explosion is reduced and corrosion is lessened. However, these two types require a slower charging rate. Both the gel cell and the AGM batteries are sealed and pressurized so the oxygen and hydrogen produced during charging are recombined into water.

The advantages of the lead-acid battery technology are low cost and high power density. However, its application for utility storage is very limited because of its short cycle life. A typical installation survives only a maximum of 1500 deep cycles. Therefore, lead-acid batteries have not found much use in commercial and large-scale energy-management application. The largest existing facility is a 140 GJ system in Chino, California, built in 1988. Because of the low specific energy at only 0.8 MJ/kg, lead-acid batteries are not a viable storage option for automobiles.

10.5.2 Nickel Metal (Ni-Cd and Ni-MH)

Description: The nickel cadmium battery is the precursor to the nickel metal hydride battery, but Ni-Cd batteries are losing popularity due to the toxicity of cadmium. Nickel metal-hydride (Ni-MH) batteries operate similar to other secondary batteries. The cathode in a Ni-MH battery uses a hydrogen absorbing alloy whereas cadmium is used in a Ni-Cd battery. During cell discharge, hydrogen ions flow from the anode to the cathode through the separator.

Structure: The cathode of a Ni-MH battery is composed of nickel oxyhydroxide. The anode is generally an alloy of nickel, cobalt, manganese, and/or aluminum combined with a mixture of rare earth metals. The electrolyte is an alkaline, generally potassium hydroxide.

During discharge, the nickel oxyhydroxide is reduced to nickel hydroxide.

$$NiOOH + H_2O + e \rightarrow Ni(OH)_2 + OH^- \quad E'' = 0.52\,V \tag{10.18}$$

And the metal hydride MH is oxidized to the metal alloy M.

$$MH - OH^- \rightarrow M + H_2O + e \quad E'' = 0.83\,V \tag{10.19}$$

The overall reaction on discharge is

$$MH + NiOOH \rightarrow M + Ni(OH)_2 \quad E'' = 0.83\,V \tag{10.20}$$

The process is reversed during charge.

Advantages of Ni-MH vs. Ni-Cd:

- Higher energy density than standard Ni-Cd (30–80 Wh/kg).
- Periodic deep discharge cycling are required less often.
- Much more environmentally friendly than Ni-Cd. Ni-MH is easily recyclable.
- Long shelf life.

Disadvantages:

- Service life limited with repeated deep cycling. More prevalent with high load currents.
- Low discharge current when compared to Li-ion batteries (0.2 C–0.5 C recommended).
- High self-discharge.
- Requires full discharge to prevent memory effect where the capacitance and the voltage drop until the battery is charged to its correct voltage and then discharged fully.

General uses:

- Small electronics
- Hybrid electric and electric cars
- Wireless communications

10.5.3 Lithium Ion

Pure lithium batteries are primary batteries and cannot be recharged due to the instability of lithium. Li-ion batteries are secondary batteries and use lithium compounds instead of pure metallic lithium for their anode. Li-ion batteries, similar to other rechargeable batteries have three primary functional components, the anode, the cathode, and the electrolyte. Lithium ions move from the anode to the cathode during discharging and the opposite way during charging. Most Li-ion batteries use a liquid organic electrolyte, but lithium polymer (Li-Po) batteries use a solid polymer composite that makes them easier to manufacture and more durable. Research on how to increase the power output of Li-ion batteries by decreasing the internal resistance of the battery is currently in progress. Li-ion batteries can be discharged at a continuous rate of 5–20 C with peak currents reaching 40 C. Li-Po batteries can achieve similar rates.

Structure: The cathode is normally made of a metal oxide such as lithium cobalt oxide ($LiCoO_2$) or lithium manganese oxide ($LiMn_2O_4$) attached to a current collector made of aluminum foil. The anode is generally made from layered graphitic carbon with a copper current collector. Recent research and development has shown that silicon nanowire anodes hold more energy and allow greater battery life (http://www.sciencedaily.com/releases/2007/12/071219103105.htm). The electrolyte is a lithium salt such as $LiPF_6$, $LiBF_4$, or $LiClO_4$. These salts are dissolved in an organic solvent, such as ether. In Li-Po batteries, the electrolyte is a composite of a gelled polymer and one of the lithium salts.

The chemical reactions for a typical Li-ion battery are shown below.

$$\text{Positive:}\quad Li_{1-x}CoO_2 \Leftrightarrow LiCoO_2 + xLi^+ + xe^- \tag{10.21}$$

$$\text{Negative:}\quad xLi^+ + e^- + 6C \Leftrightarrow Li_xC_6 \tag{10.22}$$

$$\text{Overall:}\quad LiCoO_2 + 6C \Leftrightarrow Li_{1-x}CoO_2 + Li_xC_6 \tag{10.23}$$

Source: http://www.gpbatteries.com/html/pdf/Li-ion_handbook.pdf

Note that the C in the above equations stands for a very carbon-rich negative material such as graphite. The above equations are written in units of moles, using x as a coefficient.

Advantages of Li-ion batteries compared to Ni-MH:

- Higher energy densities than other battery types (130–200 Wh/kg) compared to Ni-MH, (30–80 Wh/kg)
- Generally much lighter
- Low maintenance (Ni-MH have "memory" effect and have to be fully cycled occasionally)
- Low self-discharge rate—typically 5% per month

Disadvantages of Li-ion batteries:

- Shelf life—the battery's capacity will slowly decay over time regardless of the number of charge/discharge cycles. Depending on conditions, the battery can lose anywhere from 6% to 40% of its capacity over 1 year. (if stored at 40% charge level and 25°C, the battery will lose 4% of its capacity after 1 year.)
- Requires a protective circuit to make sure it is not overcharged or over-discharged as this degrades the battery by causing deposits of material that can short the battery (over-discharge), or cause excessive heat (overcharge).
- If damaged, combustion of the electrolyte, anode, and cathode can occur due to the interaction of lithium and the water vapor in the air.
- Currently is expensive to manufacture although expected to become cheaper as the technology matures.

General uses:

- Electric/hybrid cars (future potential)
- Cell phones and laptops

10.5.4 Flow Batteries

Description: Flow batteries are a novel type of secondary battery in which charged electrolytes are passed through a cell, producing electric charge (see Figure 10.16). The vanadium redox (VR) battery is the most common type of flow battery. The VR battery uses an electrolyte created from electrolytically

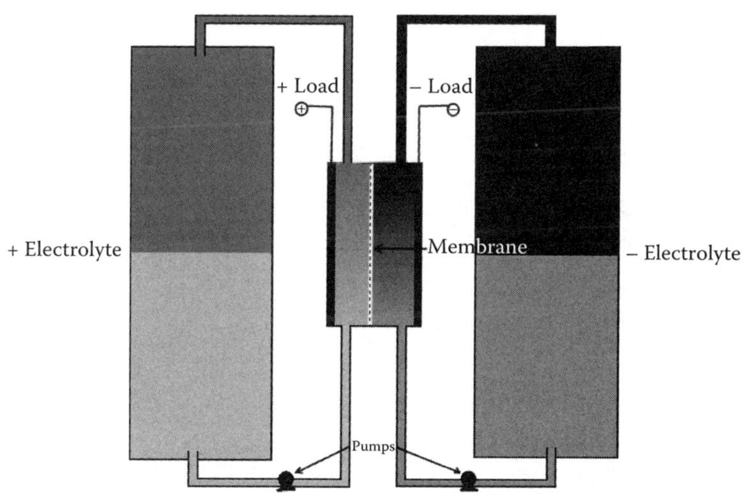

FIGURE 10.16
Schematic diagram of a typical flow battery.

dissolving vanadium pentoxide (V_2O_5) in sulfuric acid (H_2SO_4). The battery consists of two tanks of electrolyte and a membrane connected to a load, as shown in Figure 10.16.

Charging the electrolyte causes the cathode to produce VO^{2+} ions and the anode to produce V^{2+}. During discharge, the charged electrolyte is pumped across a membrane, creating VO_2^+ on the positive tank side and V^{3+} on the negative tank side. This produces an open circuit voltage. Since the chemical structure of the electrolyte does not undergo a change, this battery is very stable.

At the positive electrode

$$VO_2^+ + 2H^+ + e \xleftrightarrow[\text{charge}]{\text{discharge}} VO^{2+} + H_2O \quad E'' = 1.00\,V \tag{10.24}$$

At the negative electrode

$$V^{3+} + e \xleftrightarrow[\text{discharge}]{\text{charge}} V^{2+} \quad E'' = -0.26\,V \tag{10.25}$$

The potential generated by the cathode and anode combine to create the overall potential for the battery.

Advantages:

- Very environmentally friendly—the vanadium electrolyte does not break down and can theoretically last forever thus reducing the waste to near zero for the whole system. Vanadium is a non-toxic element so catastrophic failure of containment is not an issue.
- Capable of very large energy storage. Only limiting factor is how large the electrolyte tanks are. Can reach MWh range (http://www.vrb.unsw.edu.au/).
- Capable of very fast change in power.
- Can be fully charged/discharged without degradation.
- High cycle life (>10,000) after which only the membrane needs to be replaced.
- Very low self-discharge rate—electrolytes are stable at charged levels.
- Cost/kWh decreases as storage size increases.
- Possibility of instant recharge by replacing electrolyte.
- Already scaled to industrial size (http://www.vrb.unsw.edu.au/).
- High efficiency (between 80% and 90% in larger installations).

Disadvantages:

- Energy density only 10–20 Wh/kg and 15–25 Wh/L. Second generation V/Br batteries can achieve 35–50 Wh/kg with higher concentrations of Vanadium in the solution.
- System is complex in larger storage sizes.

Uses:

- Load leveling
- Off-peak production storage for renewable energy
- Possible electric car applications (fill up like gasoline, but no burning of fuel, simply recharged at the station)

10.6 Fuel Cells and Hydrogen

Electricity generation from hydrogen requires a fuel cell as an intermediate step, as discussed in Chapter 1. Details of the principles of operation of the fuel cell and the use of hydrogen as a storage medium are discussed below [10].

10.6.1 Principles of Fuel Cell Operation

The operation of a fuel cell is shown schematically in Figure 10.17.

A fuel cell consists, like a battery, of two electrodes separated by an electrolyte, which transmits ions but is impervious to electrons. In a fuel cell, hydrogen, or another reducing agent, is supplied to the negative electrode, while oxygen (or air) go to the positive electrode. A catalyst on the porous anode promotes dissociation of hydrogen molecules into hydrogen ions and electrons. The H^+

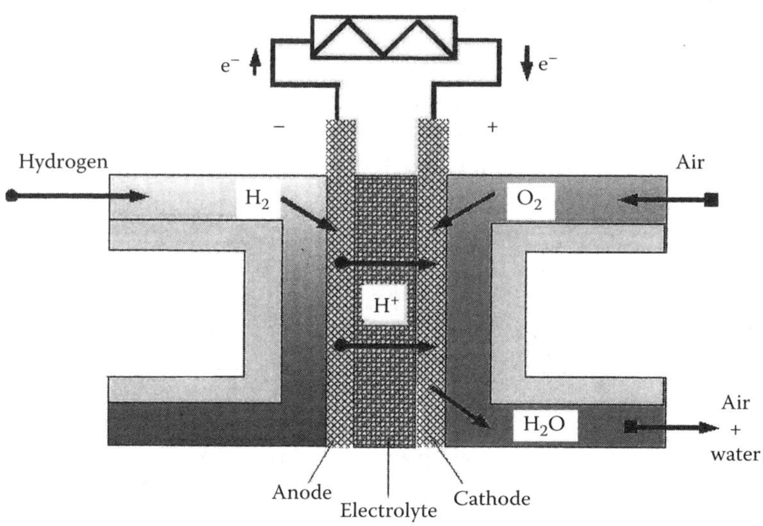

FIGURE 10.17
Schematic diagram of fuel cell operation.

ions migrate through the electrolyte to the cathode where they react with electrons supplied through the external circuit and with oxygen to form water.

Although a fuel cell converts chemical energy into electricity directly, without an intermediate combustion cycle, it does not approach its theoretical efficiency of 100% in practice. Similar to batteries, fuel cells have an efficiency on the order of 40% in practice, but this efficiency is independent of the amount of power generated. Since the efficiency of an assembly of fuel cells is approximately equal to that of a single cell, there are no economies of scale. Therefore, fuel cells are promising for small, localized plants in the range of 1–100 kW capacity. Since fuel cells also reject heat, they can operate as a cogeneration system and could supply a building with both electricity and heat.

10.6.2 Types of Fuel Cells

There are six principal types of fuel cells, as shown in Table 10.4.

- Proton exchange membrane fuel cell (PEMFC)
- Direct methanol fuel cell (DMFC)
- Molten carbonate fuel cell (MCFC)
- Alkaline fuel cell (AFC)
- Phosphoric acid fuel cell (PAFC)
- Solid oxide fuel cell (SOFC)

The above types of fuel cells differ in operating temperatures, efficiency, and cost. Their operating characteristics and some potential applications are summarized in Table 10.4.

PEMFC cells operate at relatively low temperatures (about 80°C), have a high power density, and can change rapidly to meet varying power demand or a fast startup. These features make PEMFC the favorite for automobile transportation. The membrane of a PEMFC is made of thin perfluorosulfonic acid sheets, which acts as an electrolyte and allows the passage of hydrogen ions only. The membrane is coated on both sides with highly dispersed metal alloy particles (currently mostly platinum) that act as catalysts.

DMFC cells also use a polymer membrane as an electrolyte. The most attractive feature of DMFCs is that they can use liquid methanol as a fuel. The anode of a DMFC can draw hydrogen from liquid methanol and thus eliminates the use of a fuel reformer.

MCFC cells use a molten carbonate salt as the electrolyte. They are flexible in their fuel supply and can use coal-derived gases, methane, or NG as fuel. The efficiency of these fuel cells is about 60%, but if operating as a co-generation unit that uses the waste heat, the efficiency overall can be even higher.

TABLE 10.4

Operating Characteristics of Fuel Cells

	PEMFC	DMFC	MCFC	AFC	PAFC	SOFC
Electrolyte	Ion exchange membrane	Polymer membrane	Immobilized liquid molten carbonate	Immobilized liquid phosphoric acid	Potassium hydroxide	Ceramic
Operating temperature (°C)	80	60–130	650	200	60–90	800–1000
Efficiency (%)	40–60	40	45–60	35–40	45–60	50–65
Typical electrical power	Up to 250 kW	<10 kW	>1 MW	>50 kW	Up to 20 kW	>200 kW
Possible applications	Vehicles	Portables	Power stations	Small power stations	Submarines, spacecraft	Power stations

PAFC cells use an anode and a cathode of finely dispersed platinum catalyst on carbon and a silicon-carbide structure that contains phosphoric acid as electrolyte.

AFC cells are commercially available. They operate at low temperatures, below 100°C, and are used in submarines and for space power applications.

SOFC cells work at high temperatures (between 800°C and 1000°C) and utilize a solid ceramic electrolyte such as zirconium oxide stabilized with yttrium oxide instead of a liquid. The efficiency of these cells can reach as high as 60% and they have the potential of being used for large-scale energy storage or generating electricity industrially. They can also be used for cogeneration systems.

The major challenges for widespread application of fuel cells are their cost, durability, and their need for a continuous and an inexpensive supply of hydrogen.

10.6.3 Generation of Hydrogen

There are currently two different methods used for commercial production of hydrogen. The most widely used is NG reforming. In this process, called steam methane reforming, NG is mixed with steam and the resulting mixture is introduced into a catalytic reforming reactor where it is passed through a nickel catalyst and converted at about 900°C into CO and H_2 according to the equation

$$CH_4 + H_2O \rightarrow 3H_2 + CO \quad \Delta H^0 = 206\,\text{kJ/mol} \tag{10.26}$$

The other method is electrolysis, where an electric current is passed through water to separate the hydrogen from the oxygen. Today most commercial water electrolyzers use an alkaline electrolyte system, as shown in Figure 10.18.

In such an electrolytic cell, the following simplified reaction takes place:

In the electrolyte

$$2H_2O \rightarrow 2H_2 + 2OH^- \tag{10.27}$$

At the cathode

$$2H^+ + 2e^- \rightarrow H_2 \tag{10.28}$$

At the anode

$$2OH \rightarrow \frac{1}{2}O_2 + H_2O + 2e^- \tag{10.29}$$

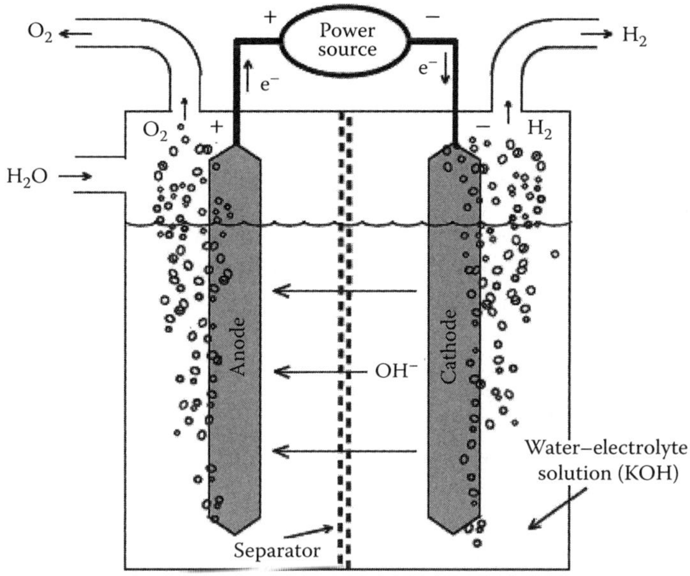

FIGURE 10.18
Schematic diagram of hydrogen generation by electrolysis.

The overall reaction is

$$H_2O \rightarrow \frac{1}{2}O_2 + H_2 \quad \text{where } \Delta H_r = 286\,\text{kJ/mol} \tag{10.30}$$

According to Ref. [11], the estimated cost of hydrogen production by NG reforming or coal gasification is about \$1/kg, whereas the cost via a renewable technology, such as electrolysis with electricity from wind, is between \$6 and \$7/kg. The energy content of 1 kg of H_2 approximately equals the energy content of 1 gal of gasoline.

10.6.4 Storage and Transport

Once the hydrogen is generated, it can be stored and subsequently combusted to provide heat or work, or it can be used to generate electricity with a fuel cell. Gaseous hydrogen has a low specific density and for storage purposes it must be compressed or cryogenically liquefied. If hydrogen is compressed adiabatically to a pressure of 350 bar, a value often assumed suitable for automotive technologies, it consumes about 12% of its higher heating value (HHV). The loss can be reduced appreciably; however, if the compression approaches an isothermal process. If the hydrogen is stored as a liquid at cryogenic temperatures, the process takes about 40% of the HHV with current technology. Moreover, liquid storage is not feasible for automotive applications because mandatory boil-off from the storage container cannot be safely released in closed spaces such as garages or tunnels.

A recent technology proposed for storing hydrogen consists of bonding it into metal hydrides using an absorption process. The energy penalty of this type of storage, which requires only a pressure of about 30 bar, may be less. However, the density of the metal hydride can be between 20 and 100 times the density of the hydrogen stored. Recently, carbon nanotubes have received attention as a potential hydrogen storage medium, but neither of the above processes has been fully developed and commercialized.

10.6.5 Thermodynamics and Economics

As an energy carrier, hydrogen should be compared with the only other widespread and viable alternative: electricity. Once generated and compressed, hydrogen could flow through large pipelines with a loss of approximately 0.8%/100 km [2,12]. According to Refs. [2,13], while the energy loss experienced with high voltage DC (long distance) electric transmission lines is of the order of 0.6%/100 km. Hence, the losses incurred in transmission of energy by hydrogen and electricity are of the same order of magnitude. However, the cost of building a transmission system for hydrogen has been estimated to be on the order of half a trillion dollars, and unless this up-front expenditure is made, transmission of hydrogen would have to be accomplished by truck, which would incur enormous expenses and energy penalties. Moreover, as shown by Kreith and West, the extraction of electricity from hydrogen is associated with irreversible losses. Electrolysis of water to make hydrogen is commercially available at about 70% efficiency. The sequence of steps required of generating electricity for producing hydrogen and then using a fuel cell to generate electricity again is shown in Figure 10.19. Compressing hydrogen to a pressure of 350 bar for transporting it is estimated to be about 88% efficient. And, finally, conversion back into electricity with a fuel cell has an efficiency of approximately 60%, although the theoretical limit is somewhat higher. Taken together, generating power for electrolysis, compression, and generation of electricity in the fuel cell results in, at most, an overall efficiency of only about 50%, compared to the amount of electricity that could have been obtained from the original source. Hence, from a thermodynamic as well as from an economic perspective, the use of hydrogen as a storage and transmission system appears to be inferior to the use of electricity directly. It should also be noted that this loss of energy is the result of basic thermodynamic considerations and would occur irrespective of whether the original generation of electricity came from renewable sources, nuclear power, or fossil fuels. In addition, because two units of electricity are required to produce one unit of useful power at the end use, the amount of CO_2 generation/kW electricity would be double via the hydrogen path compared to any other electric power system.

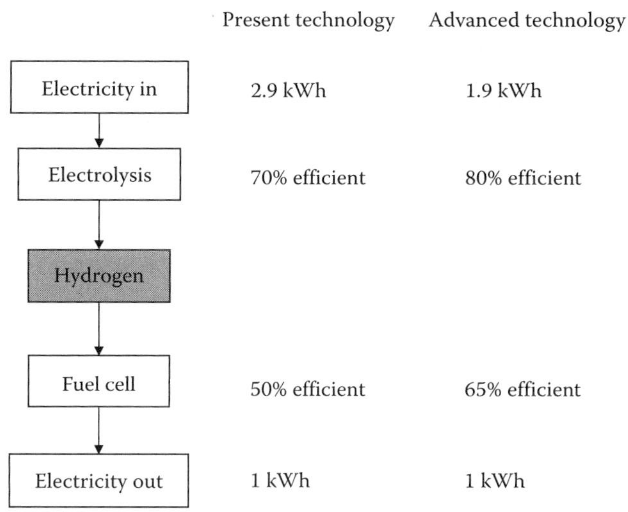

FIGURE 10.19
Generating electricity with a fuel cell using hydrogen produced by electrotysis. (Adapted from Kreith, F. and West, R.E., *J. Energy Resourc. Technol.*, 126, 249, 2004.)

10.7 Thermal Energy Storage

Solar thermal storage is a well-established technology and the principles of design and operation in tanks or in the ground are extensively described in standard textbooks on heat transfer [14,15]. For solar thermal heat and power thermal storage is important in applications such as storing heat overnight for buildings, seasonal heat storage for large-scale heating and/or cooling projects (called seasonal storage), and for storage in solar thermal power plants such as solar troughs or central receivers. Solar storage can also be dispersed through the overall system as, for example, in the thermal mass of a building—an art that has been practiced for centuries. A particularly good example is the Mesa Verde buildings in Colorado, where heat is stored in rock walls, which form a part of the structures.

10.7.1 Sensible Heat

Thermal energy can be stored as sensible heat, as latent heat, or in reversible thermochemical equations. The sensible heat, ΔQ, gained or lost per unit mass by a material changing temperature from T_1 to T_2 is

$$\Delta Q = \int_{T_1}^{T_2} \rho c_p \, dT \tag{10.31}$$

Although both the density, ρ, and the specific heat, c_p, vary with temperature, the variations are usually small enough to use average values for engineering calculations. Equation 10.31 then simplifies so that $\Delta Q = c_p \Delta T$ per unit mass, or $\rho c_p \Delta T$ per unit volume. Commonly used figures of merits are $\$/c_p$ for relative media costs, ρc_p for volumetric heat capacity, using the lowest value of ρ at the highest expected operating temperature for relative container size.

Solid media: Table 10.5 shows the characteristics of some plentiful and economically competitive solid materials for sensible heat storage. Rocks, bricks, and cement are excellent candidates for overnight thermal storage in buildings.

TABLE 10.5

Physical Properties of Some Sensible Heat Storage Materials

Storage Medium	Temperature Range (°C)	Density (ρ) (kg/m³)	Specific Heat (c), (J/kg K)	Energy Density (ρc) (kWh/m³ K)	Thermal Conductivity (W/m K)
Water	0–100	1000	4190	1.16	0.63 at 38°C
Water (10 bar)	0–180	881	4190	1.03	—
50% ethylene glycol-50% water	0–100	1075	3480	0.98	—
Dowtherm A® (Dow Chemical, Co., Midland, Michigan)	12–260	867	2200	0.53	0.122 at 260°C
Therminol 66® (Monsanto Co., St. Louis, Missouri)	9–343	750	2100	0.44	0.106 at 343°C
Draw salt ($50NaNO_3$–$50KNO_3$)[a]	220–540	1733	1550	0.75	0.57
Molten salt ($53KNO_3$/$40NaNO_3$/$7NaNO_3$)[a]	142–540	1680	1560	0.72	0.61
Liquid Sodium	100–760	750	1260	0.26	67.5
Cast iron	m.p. (1150–1300)	7200	540	1.08	42.0
Taconite	—	3200	800	0.71	—
Aluminum	m.p. 660	2700	920	0.69	200
Fireclay	—	2100–2600	1000	0.65	1.0–1.5
Rock	—	1600	880	0.39	—

Note: m.p. = melting point.

[a] Composition in percent by weight.

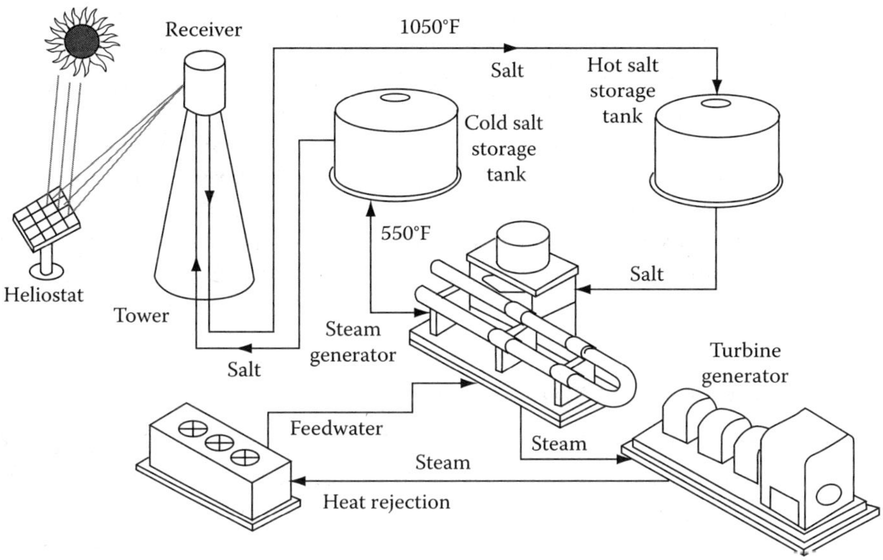

FIGURE 10.20
Two-tank storage for central receiver system.

Liquid media: Table 10.5 shows the characteristics of some useful liquid media for sensible heat storage. Water is clearly superior for temperatures below 100°C, whereas oils, molten salts, or liquid metals are used for solar thermal power applications such as parabolic trough or central receivers.

Dual media: Solids and liquid sensible heat storage materials can be combined in various ways. Rock beds and water tanks have been used in hybrid storage systems for space heating, whereas rocks and oils have been used in single vessels for solar thermal power applications to improve stratification. This approach could reduce the cost compared to relatively expensive dual tank liquid storage systems where two vessels necessary, as shown in Figure 10.20. However, experiences with single tank storage to achieve stratification have not been reliable, and for solar thermal power applications, dual tank storage is preferred.

10.7.2 Phase Change Heat Storage

Latent heat is the thermal energy released or absorbed by a material undergoing a phase transition, such as solid to liquid (melting), or liquid to gas (vaporization). Materials that have a high heat release or absorption during phase change are of interest for thermal storage. Latent heat storage material should have a large heat of transition, high density, appropriate transition temperature, low toxicity, and no gradation with multiple phase change. Materials attractive for storage are paraffin waxes and salt hydrates.

The advantage of using phase change material for storage is that a large amount of heat can be stored in small volumes and the temperature of the material remains constant during phase change. The problem with phase change heat transfer is to find materials that have appropriate temperatures during phase change, and also the design of heat exchangers that transfer heat to or from the material during phase change is exceedingly difficult. Table 10.5 shows the material properties of several common TES materials that may potentially be useful for latent heat storage. The engineering design of latent heat storage systems is difficult and beyond the scope of this book (Table 10.6).

10.7.3 Thermochemical Storage

Thermochemical energy is stored as the bond energy of a chemical compound. In thermochemical energy storage, atomic bonds are broken during a reversible chemical reaction and are catalyzed by an increase in temperature. After thermochemical separation occurs, the constituents are stored apart until the combination reaction is desired. Recombination of the bonds between atoms releases the stored thermochemical energy.

Reversible chemical reactions with reactants and products which can be easily stored as liquids and/or solids are of interest for thermochemical energy storage. Reactions that produce two distinct phases such as a solid and a gas are desirable since the separation of products to prevent back reaction is facilitated. Table 10.7 shows several common thermochemical storage reactions and their standard enthalpy change (ΔH^0) in kJ and their turning temperature (T') in K.

The turning temperature, T', in Table 10.6 is defined as the temperature at which the equilibrium constant is unity and is calculated using the ratio of the standard enthalpy change and the standard entropy change for the reaction. At this temperature, the reactants and products will be present in approximately equal quantities. When $T > T'$, the endothermic storage reaction is favored, meaning heat is necessary for and absorbed during the reaction. Conversely, for $T < T'$, the exothermic reaction dominates, meaning that heat is produced in the reaction.

The primary advantages of thermochemical storage include the high energy density and the long-term, low-temperature, storage capability. However, the thermochemical process is very complex and the thermochemical materials are often expensive and can be hazardous. There is considerable research activity in the field of thermochemical reactions for energy storage, but no large-scale commercial applications exist to date.

10.7.4 Applications

Many factors contribute to choosing an appropriate storage method and material for each TES application. Table 10.8 lists several TES options for

TABLE 10.6

Physical Properties of Latent Heat Storage Materials or PCMs

Storage Medium	Melting Point (°C)	Latent Heat (kJ/kg)	Specific Heat (kJ/kg °C)		Density (kg/m³)		Energy Density (kWh/m³ K)	Thermal Conductivity (W/m K)
			Solid	Liquid	Solid	Liquid		
$LiClO_3 \cdot 3H_2O$	8.1	253	—	—	1720	1530	108	—
$Na_2SO_4 \cdot 10H_2O$ (Glauber's salt)	32.4	25.1	1.76	3.32	1460	1330	92.7	2.25
$Na_2S_2O_3 \cdot 5H_2O$	48	200	1.47	2.39	1730	1665	92.5	0.57
$NaCH_3COO \cdot 3H_2O$	58	180	1.90	2.50	1450	1280	64	0.5
$Ba(OH)_2 \cdot 8H_2O$	78	301	0.67	1.26	2070	1937	162	0.653ℓ
$Mg(NO_3) \cdot 6H_2O$	90	163	1.56	3.68	1636	1550	70	0.611
$LiNO_3$	252	530	2.02	2.041	2310	1776	261	1.35
$LiCO_3/K_2CO_3$, (35:65)[a]	505	345	1.34	1.76	2265	1960	188	—
$LiCO_3/K_2CO_3/Na_2CO_3$ (32:35:33)[a]	397	277	1.68	1.63	2300	2140	165	—
n-Tetradecane	5.5	228	—	—	825	771	48	0.150
n-Octadecane	28	244	2.16	—	814	774	52.5	0.150
HDPE (cross-linked)	126	180	2.88	2.51	960	900	45	0.361
Steric acid	70	203	—	2.35	941	347	48	0.172ℓ

Note: ℓ = liquid.

[a] Composition in percent by weight.

TABLE 10.7
Thermochemical Storage Reactions

Reaction	ΔH^0 (kJ)	T' (K)
$NH_4F(s) \leftrightarrow NH_3(g) + HF(g)$	149.3	499
$Mg(OH)_2(s) \leftrightarrow MgO(s) + H_2O(g)$	81.1	531
$MgCO_3(s) \leftrightarrow MgO(s) + CO_2(g)$	100.6	670
$NH_4HSO_4(l) \leftrightarrow NH_3(g) + H_2O(g) + SO_3(g)$	337.0	740
$Ca(OH)_2(s) \leftrightarrow CaO(s) + H_2O(g)$	109.3	752
$BaO_2(s) \leftrightarrow BaO(s) + \frac{1}{2}O_2(g)$	80.8	1000
$LiOH(l) \leftrightarrow \frac{1}{2}Li_2O(s)\ \frac{1}{2}H_2O(g)$	56.7	1000
$CaCO_3(s) \leftrightarrow CaO(s)\ CO_2(g)$	178.1	1110
$MgSO_4 \leftrightarrow MgO(s) + SO_3(g)$	287.6	1470

Source: Wyman, C. et al., *Solar Energy*, 24(6), 517, 1980.

solar thermal power and their appropriate storage materials. In the table, PCM stands for phase change material, HX means heat exchanger, and VHT means very high temperature.

10.7.5 Thermal Storage for Concentrating Collector Systems

The most important application of thermal storage is in high-temperature concentrator systems such as central receivers or parabolic troughs. Steam is generated in the solar receiver when the sun is up and either passes through the turbine, or, when excess heat is available, it is stored in a tank. The size of the tank will determine the number of hours that the system can operate when the sun is down. For a typical system of this type, steam is raised in the receiver at 40 bar and 250°C, and condensed at 0.6 bars and 5°C. After the steam is condensed, it is returned to the solar receiver as in any typical Rankine cycle. The drawback of this approach is that steam storage requires a large volume, which limits the length of time the turbine can operate when the sun is down.

A more significant application of thermal storage is the two-tank storage system shown in Figure 10.19 that was used in the Solar 2 central receiver. In this approach, a single heat transfer fluid is used in the receiver and for storage. This approach eliminates the need for expensive heat exchangers and the degradation in the availability of the energy that occurs during heat transfer from one fluid to another. The fluid most commonly proposed for use in this kind of application is a molten salt, although other heat transfer fluids such as oils or liquid metals are being considered.

When molten salts are used as the heat transfer fluid, care must be taken to avoid freezing anywhere in the system. Molten salts freeze at a temperature between 120°C and 200°C (250°F–430°F). Thus, the Rankine cycle has to operate between a high temperature achievable in the receiver and a cold temperature that does not drop below the freezing point of the salt. The design of a typical two-tank direct system with approximate temperatures is shown

TABLE 10.8

Options for TES in Solar Power Production

TES Options for Solar Thermal Power	Temperature (°C)	Storage Medium	Type
Small power plants and water pumps			
Organic Rankine	100	Water in thermocline tank or two tanks	Sensible
	300	Petroleum oil in the thermocline tank	Sensible
Steam Rankine with organic fluid receiver	375	Synthetic oil with trickle charge	Sensible
Dish mounted engine generators (buffer storage only)			
Organic Rankine	400	Bulk PCM with indirect HX	Latent
Stirling and air Brayton	800	Bulk PCM with indirect HX	Latent
Advanced air Brayton	1370	Graphite	Sensible
		Encapsulated PCM	Latent
Larger power plants (typically 3 8 h storage)			
Steam Rankine with organic fluid receiver	300	Petroleum oil in thermocline tank or two tanks, evaporation only	Sensible
	300	Petroleum oil/rocks (dual medium) in thermocline tank	Sensible
		Petroleum oil in thermocline tank or two tanks, evaporation only	Sensible
		Petroleum oil/rocks (dual medium) in thermocline tank	Sensible
		Encapsulated PCM with evaporative HX	Latent
Steam Rankine with water-steam receiver		Bulk PCM with indirect HX	Latent
		Bulk PCM with direct HX	Latent
		Pressurized water above ground or underground	Latent
	540	Molten draw salt in themocline tank or two tanks, superheat	Sensible
		Air/rocks	Sensible
		Bulk PCM with direct HX, evaporation stage	Latent
		Solid or liquid decomposition, evaporation stage	Thermochemical

(continued)

TABLE 10.8 (continued)

Options for TES in Solar Power Production

TES Options for Solar Thermal Power	Temperature (°C)	Storage Medium	Type
Steam Rankine with molten draw salt receiver	540	Molten draw salt in thermoline tank or two tanks	Sensible
Steam Rankine with liquid metal receiver	540	Liquid sodium in one tank, mixed, buffer only	Sensible
		Liquid sodium in two tanks	
		Air/rocks	Sensible
		Refractory or cast-iron	Sensible
Brayton with gas-cooled receiver	800	Bulk PCM with indirect HX	Latent
		Solid or liquid decomposition	Thermochemical
		VHT molten salt in two tanks	Sensible
Brayton with liquid-cooled receiver	800	VHT molten salt/refractory (dual medium) in thermocline tank	Sensible
	1100	Bulk glassy slag, liquid and solid bead storage, direct HX	Sensible and Latent

in Figure 10.20. The temperature of the molten salt is raised in the receiver to about 565°C and then passed to the hot salt storage tank. From this tank, salt is extracted to a heat exchanger, which raises the steam for a Rankine cycle power plant, and then returned to the cold salt storage tank at about 290°C. The working fluid is then recirculated to the receiver in a virtually closed loop. A normal Rankine cycle is used for the generation of electricity. The length of time that the system can operate once the sun is down depends on the size of the storage tanks. The advantage of this approach is the simplicity of design, and the disadvantage is that a large amount of heat transfer fluid is needed for the plant to operate.

In systems where the heat transfer fluid is very expensive or cannot be used for some reason as the storage fluid, a two-tank indirect system is used. The two-tank indirect system functions the same way as the two-tank direct system, except that different fluids are used in the receiver heating loop and the storage. Therefore, the heat from the working fluid must be transferred to the storage fluid via a heat exchanger, which the fluid enters at high temperature and leaves at low temperature. Storage fluid from the high temperature tank is then used to generate steam, as in the direct system.

10.7.6 Overnight Storage for Buildings and Domestic Hot Water

TES for domestic hot water or for overnight heating of buildings has been mentioned in previous chapters. These applications require storing a liquid,

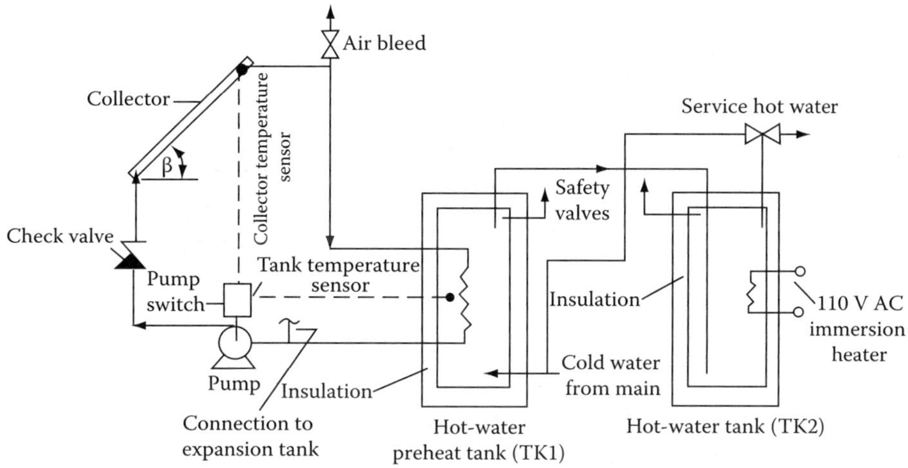

FIGURE 10.21
Schematic diagram of solar heating system with storage.

almost always water, in a tank. A schematic diagram typical of such systems with electric backup is shown in Figure 10.21. Insulation of the storage tank is a simple heat transfer problem treated in standard textbooks. The properties of insulation materials are given in Table 10.9.

The annual average insolation on a surface tilted at an angle equal to the latitude in the midlatitudes of the United States is about 23,000 kJ/m^2 per day (2030 Btu/ft^2 per day). If the average collection efficiency is 30%, 6900 kJ/m^2 per day (610 Btu/ft^2 per day) is delivered by the collector system. The service hot-water demand can be taken as 76 L/day (20 gal/day) per person. If the tap water has a temperature of 283 K (50°F), and the hot water 333 K (140°F), about 210 kJ/L (750 Btu/gal) is required for hot water heating. For a delivery of 230 L/day (60 gal/day) for a family of three, 48.3 MJ/day (46,000 Btu/day) is needed. If the collector delivers 6900 kJ/m^2, a collector with a size of approximately 7 m^2 is needed. This collector is about one-tenth the size of a collector required for heating a typical building in the midlatitudes. A rule of thumb is that 1 m^2 of collector is required for each 30 L of hot water to be delivered. The storage tank should be large enough to hold about a 2 day supply of hot water, or about 500 L for a family of three.

Thermal storage tanks must be insulated to control heat loss. If a storage tank is located within a structure, any losses from the tank tend to offset the active heating demands of the building. However, such storage loss is uncontrolled and may cause overheating in the summer.

The amount of thermal storage used in a solar-heating system is limited by the law of diminishing returns. Although larger storage results in larger annual energy delivery, the increase at the margin is small and hence not cost-effective. Seasonal storage is, therefore, usually uneconomic, although it can be realized in a technical sense. Experience has shown that liquid

TABLE 10.9

Thermal Properties of Some Nonmetals

Material	Average Temperature (°F)	k [Btu/(h·ft·°F)]	c [Btu/(lb$_m$·°F)]	ρ (lb$_m$/ft^3)	α (ft^2/h)
Insulating materials					
Asbestos	32	0.087	0.25	36	~0.01
	392	0.12	—	36	~0.01
Cork	86	0.025	0.04	10	~0.006
Cotton fabric	200	0.046			
Diatomaceous earth, powdered	100	0.030	0.21	14	~0.01
	300	0.036	—		
	600	0.046	—		
Molded pipe covering	400	0.051	—	26	
	1600	0.088	—		
Glass wool					
Fine	20	0.022	—		
	100	0.031	—	1.5	
	200	0.043	—		
Packed	20	0.016	—		
	100	0.022	—	6.0	
	200	0.029	—		
Hair felt	100	0.027	—	8.2	
Kaolin insulating brick	932	0.15	—	27	
	2102	0.26	—		
Kaolin insulating firebrick	392	0.05	—	19	
	1400	0.11	—		
Magnesia, 85%	32	0.032	—	17	
	200	0.037	—	17	
Rock wool	20	0.017	—	8	
	200	0.030	—		
Rubber	32	0.087	0.48	75	0.0024
Building materials					
Brick					
Fire clay	392	0.58	0.20	144	0.02
	1832	0.95			
Masonry	70	0.38	0.20	106	0.018
Zirconia	392	0.84	—	304	
	1832	1.13	—		
Chrome brick	392	0.82	—	246	
	1832	0.96	—		

TABLE 10.9 (continued)

Thermal Properties of Some Nonmetals

Material	Average Temperature (°F)	k [Btu/(h·ft·°F)]	c [Btu/(lb$_m$·°F)]	ρ (lb$_m$/ft^3)	α (ft^2/h)
Concrete					
Stone	~70	0.54	0.20	144	0.019
10% moisture	~70	0.70	—	140	~0.025
Glass, window	~70	~0.45	0.2	170	0.013
Limestone, dry	70	0.40	0.22	105	0.017
Sand					
Dry	68	0.20	—	95	
10% water	68	0.60	—	100	
Soil					
Dry	70	~0.20	0.44	—	~0.01
Wet	70	~1.5	—	—	~0.03
Wood					
Oak ⊥ to grain[a]	70	0.12	0.57	51	0.0041
‖ to grain	70	0.20	0.57	51	0.0069
Pine ⊥ to grain[a]	70	0.06	0.67	31	0.0029
‖ to grain	70	0.14	0.67	31	0.0067
Ice	32	1.28	0.46	57	0.048

Thermal Properties of Metals and Alloys

Material	k, [Btu/(h·ft·°F)] 32°F	212°F	572°F	932°F	c [Btu/(lb$_m$·°F)] 32°F	ρ (lb$_m$/ft^3) 32°F	α (ft^2/h) 32°F
Metals							
Aluminum	117	119	133	155	0.208	169	3.33
Bismuth	4.9	3.9	—	—	0.029	612	0.28
Copper, pure	224	218	212	207	0.091	558	4.42
Gold	169	170	—	—	0.030	1203	4.68
Iron, pure	35.8	36.6	—	—	0.104	491	0.70
Lead	20.1	19	18	—	0.030	705	0.95
Magnesium	91	92	—	—	0.232	109	3.60
Mercury	4.8	—	—	—	0.033	849	0.17
Nickel	34.5	34	32	—	0.103	555	0.60
Silver	242	238	—	—	0.056	655	6.6
Tin	36	34	—	—	0.054	456	1.46
Zinc	65	64	59	—	0.091	446	1.60

(continued)

TABLE 10.9 (continued)

Thermal Properties of Metals and Alloys

Material	k, [Btu/(h·ft·°F)] 32°F	212°F	572°F	932°F	c [Btu/(lb$_m$·°F)] 32°F	ρ (lb$_m$/ft^3) 32°F	α (ft^2/h) 32°F
Alloys							
Admiralty metal	65	64					
Brass, 70% Cu, 30% Zn	56	60	66	—	0.092	532	1.14
Bronze, 75% Cu, 25% Sn	15	—	—	—	0.082	540	0.34
Cast iron							
Plain	33	31.8	27.7	24.8	0.11	474	0.63
Alloy	30	28.3	27	—	0.10	455	0.66
Constantan, 60% Cu, 40% Ni	12.4	12.8	—	—	0.10	557	0.22
18-8 Stainless steel							
Type 304	8.0	9.4	10.9	12.4	0.11	488	0.15
Type 347	8.0	9.3	11.0	12.8	0.11	488	0.15
Steel, mild, 1% C	26.5	26	25	22	0.11	490	0.49

Source: Kreith, F., *Principles of Heat Transfer*, Intext, New York, 1973.

[a] ⊥, perpendicular; ||, parallel.

storage amounts of 50–75 kg H_2O/m^2 (10–15 lb/ft_c^2) are the best compromise between storage tank cost and useful energy delivery.

Since solar energy heating systems operate at temperatures relatively close to the temperatures of the spaces to be heated, storage must be capable of delivering and receiving thermal energy at relatively small temperature differences. The designer must consider the magnitude of these driving forces in sizing heat exchangers, pumps, and air-blowers. The designer must also consider the non-recoverable heat losses from storage—even though storage temperatures are relatively low, surface areas of storage units may be large and heat losses therefore appreciable.

Other mechanical components in solar heating systems include pumps, heat exchangers, air bleed valves, pressure release valves and expansion tanks. Heat exchangers in solar systems are selected based on economic criteria. The best trade-off of energy delivery increase with increasing heat-exchanger size usually results from use of an exchanger with effectiveness in the range of 0.6–0.8. Counterflow heat exchangers are required for this level of effectiveness.

Achievement of the required effectiveness level may dictate fairly high flow rates in the storage tank side of the collector heat exchanger. Flows up to twice that in the collector side can improve exchanger performance significantly in many cases. Since the storage side loop is physically short and has a small pressure drop, increased flow in this loop increases pump energy requirements by a negligible amount. Typical solar heat-exchangers sizes range from 0.05 to 0.10 m^2 of heat-exchanger surface per square meter of net collector area.

In hydronic heating systems, it is essential that all air be pumped from the system. To facilitate this process, air bleed valves located at the high points in a system are used. These are opened during system fill and later if air should collect. Air bleeds are required at points of low velocity in piping systems where air may collect because the local fluid velocity is too low for entrainment.

Control strategies and hardware used in current solar system designs are quite simple and are similar in several respects to those used in conventional systems. The single fundamental difference lies in the requirement for differential temperature measurement instead of simple temperature sensing. In the space heating system shown in Figure 10.21, two temperature signals determine which of three modes is used. The signals used are the collector-storage differential and room temperature. The collector-storage difference is sensed by two thermistors or thermocouples, the difference being determined by a solid-state comparator, which is a part of the control device. Room temperature is sensed by a conventional dual-contact thermostat.

The control system operates as follows. If the first room thermostat contact closes, the mode selector valve and distribution pump are activated in an attempt to deliver the thermal demand from solar-thermal storage. If room temperature continues to drop, indicating inadequate solar availability, the mode selector diverts flow through the backup system instead of the solar system, and the backup is activated until the load is satisfied.

The collector-storage control subsystem operates independently of the heating subsystem described above. If collector temperature, usually sensed by a thermistor thermally bonded to the absorber plate, exceeds the temperature in the bottom of the storage tank by 5°C–10°C (9°F–18°F), the collector pump and heat-exchanger pump (if present) are activated and continue to run until the collector and storage temperature are within about 1°C–2°C (2°F–4°F) of each other. At this point, it is no longer worthwhile to attempt to collect energy, and the pumps are turned off. The collector-storage subsystem also has a high temperature cutout that turns the collector loop pump off when the storage temperature exceeds a set limit.

A heating load device transfers heat from the solar storage to the air in the space. Therefore, a liquid-to-air heat exchanger is sized based on the energy demand of a building. Several generic types of load devices are in common use:

1. Forced-air systems—tube-and-fin coil located in the main distribution duct of a building or zone of a building (see Figure 10.22).
2. Baseboard convection systems—tube-and-fin coils located near the floor on external walls. These operate by natural convection from the convectors to the room air.
3. Heated floors or ceilings—water coils. These transfer heat to large thermal masses that in turn radiate or convect into the space. This heating method is usually called radiant heating.

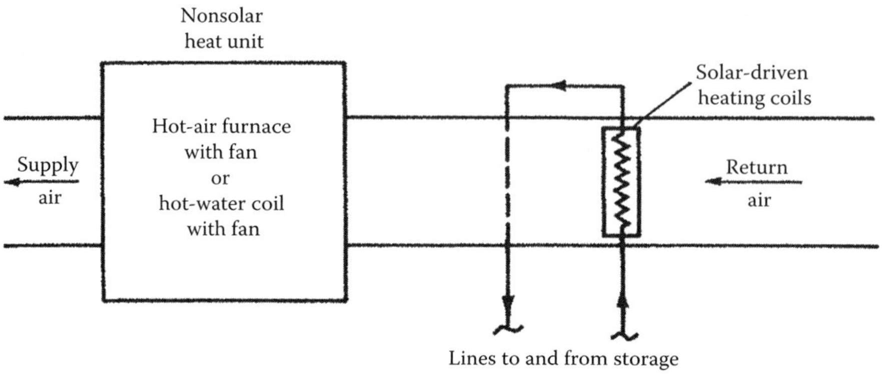

FIGURE 10.22
Forced-air heating system load device location upstream of nonsolar heat exchanger or furnace.

Each load device requires fluid at a different temperature in order to operate under design load conditions, as shown in Figure 10.23. Since baseboard heaters are small in heat-transfer area and rely on the relatively ineffectual mechanism of natural convection, they require the highest fluid temperature. Forced-air systems involve the more efficient forced-convection heat-transfer mode and, hence, are operable at lower fluid temperatures (see Figure 10.23). Radiant heating can use very large heat-transfer areas and is, therefore, operable at relatively low fluid temperatures.

In Figure 10.23, the intersection of the solar fluid temperature line and the load line for a specific configuration is called the *balance point*. At ambient temperatures below the balance point, solar energy cannot provide the entire demand, and some backup is required; above the balance point solar capacity is sufficient to carry the entire load. Note that the load lines are specific to a given building. The solar fluid temperature line is not fixed for a building but depends on solar collector and storage size as well as local solar radiation levels. The line shown in Figure 10.23 is, therefore, an average line.

It is possible to modify load devices to lower the balance point, as shown in Figure 10.23. For example, a forced-air tube-and-fin exchanger can be enlarged by adding one or more additional rows of tubes. This increased heat-transfer area will permit the same energy delivery at a lower fluid temperature. The law of diminishing returns is evident as increasing effectiveness returns progressively less energy.

10.7.7 Long-Term Performance of Solar Collector Systems with Storage

Section 8.4.3 described a method of predicting long-term performance of a solar collector operated at a temporally constant temperature. This situation is a good approximation of the operating conditions experienced by several types of generic thermal systems. Other systems, however, do not operate

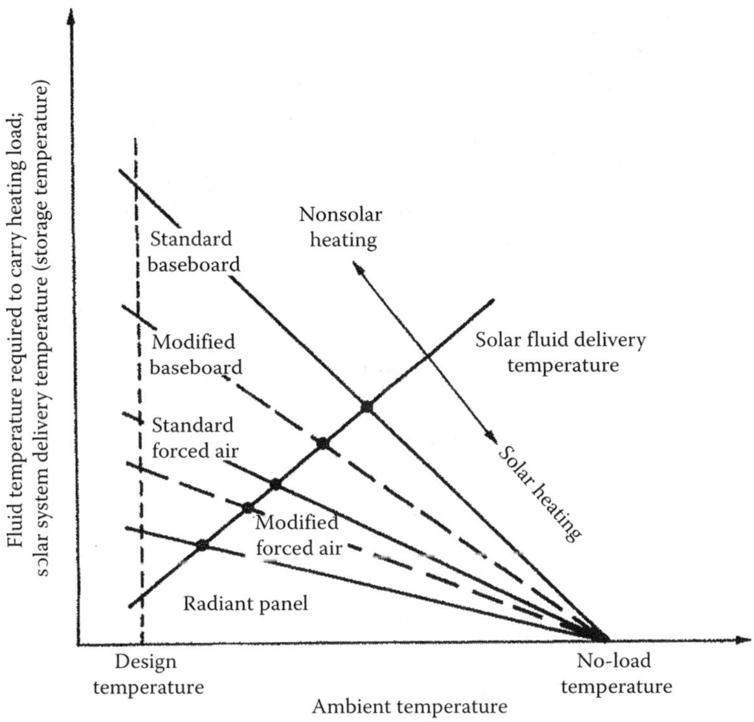

FIGURE 10.23
Heating load diagram for baseboard, forced-air, and radiant systems. Modified baseboard and forced-air systems are oversized in order to carry heating demands at lower temperature. Balance points are indicated by large dots at intersections.

at constant temperature and the $\bar{\phi}$ method cannot be used. Although there is no simplified performance method now extant for varying temperature systems, Klein and Beckman [16] have correlated some modeling results on collector-heat exchanger-storage subsystems coupled to a uniform, process-like load operating above some temperature T_{proc}. The method is called the $\bar{\phi}$, f chart and is described below. Although the method was developed for flat-plate collectors and uses a different $\bar{\phi}$ calculation method than used above, it can be applied equally well to concentrators [17].

The calculation method requires first the determination of the utilizability $\bar{\phi}$ from Chapter 8. This represents the maximum energy deliverable to a load at T_{proc}. When storage is present and collected solar heat is greater than the demand, the temperature of storage, and hence the collector inlet temperature, will rise. (The $\bar{\phi}, f$ method applies only to well-mixed, sensible heat storage with liquid heat transfer fluids and storage media.) Hence, the monthly averaged, daily useful energy collected Q_u will be less than $F\bar{\eta}_o\bar{I}_c\bar{\phi}$, but storage may permit a greater fraction of the demand to be met by solar since the maximum amount of heat $F\bar{\eta}_o\bar{I}_c\bar{\phi}$ collectable may be more than can be used, depending on the demand amount. The $\bar{\phi},f,$ method can be employed to find Q_u in a system with storage.

The technical basis for the $\bar{\phi},f,$ method lies in the nondimensionalization of governing energy equations for a solar thermal system. Dimensionless groups, so identified, are used to correlate monthly thermal energy delivery-to-load for various systems simulated in various climates by an hourly time-scale computer model. The two dimensionless groups identified for use in the $\bar{\phi},f,$ method, in addition to $\bar{\phi}$ ($\bar{\phi}$ is defined in this context relative to the minimum temperature acceptable to the process T_{proc}, not relative to the collector temperature as in the previous section), are a solar parameter P_s, a measure of long-term solar gain by the collector receiver, and a collector heat-loss parameter P_L, a measure of long-term heat loss at a fixed collector-to-ambient temperature difference of 100°C. This 100°C value does not restrict the generality of the results, however. In equation form

$$P_s = \frac{F_R \bar{\eta}_o \bar{I}_c A_c N_d}{L}, \tag{10.32}$$

$$P_L = \frac{F_R U_c A_c N_h 100}{L} \tag{10.33}$$

where

$$F_R = \frac{Gc_p}{U_c}\left[1 - \exp\left(-\frac{U_c F'}{Gc_p}\right)\right]$$

L is the monthly thermal demand

N_d and N_h are the number of days and hours in a month

The $\bar{\phi},f$ chart predicts the monthly solar load fraction $f_s(\bar{\phi}, P_s, P_L) \equiv Q_u/L$ by an empirical equation [16]

$$f_s = \bar{\phi} P_s - a[e^{3.85 f_s} - 1][1 - e^{-0.15 P_L}], \tag{10.34}$$

where $a = 0.015[Mc_p/350A_c]^{-0.76}$ in which m = (kg), c_p = (kJ/kg°C), and A_c = (m²).

Figure 10.24 is one $\bar{\phi},f,$ chart for a standard storage size of 350 kJ/°C-m² and is entered with values of $\bar{\phi}P_s$ and P_L to give a monthly value of f_s. The calculation of $\bar{\phi}P_s$ and P_L is done once for each month of an average year, and the totals are added to give annual performance. An example below shows how the method is used. It is noted that $\bar{\phi}P_s$, the ordinate, is the ratio of maximum possible energy delivered by a collector operating at fixed T_{proc}, ($F\eta_o \bar{I}_c \bar{\phi} A_c$), to the monthly load L. At values of $f_s > 0.4$, the $\bar{\phi}, f,$ curves are not independent of X since at progressively higher load fractions, the average storage and

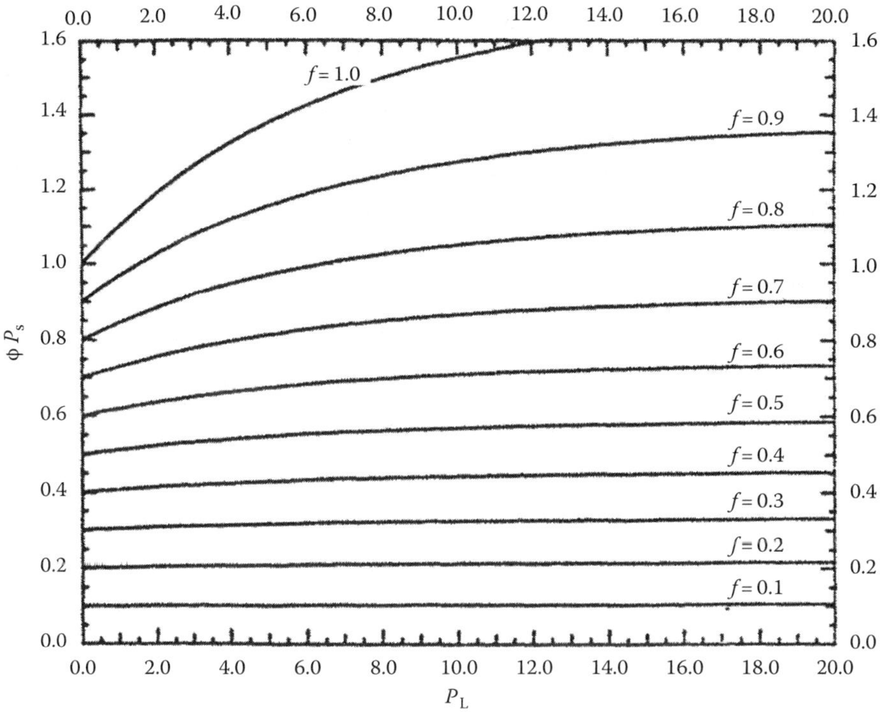

FIGURE 10.24
The $\bar{\phi}, f$ chart used to calculate average, monthly solar fraction $f(f_s)$ of solar-thermal systems. (From Klein, S.A. and Beckman, W.A., *Solar Energy*, 22, 269, 1979. With permission.)

collector temperatures are higher and collected solar heat per unit area is smaller because collector efficiency is lower at higher temperature.
This $\bar{\phi}, f$ chart is based upon several limiting assumptions, which should be noted in interpreting f_s values:

1. The load L is distributed uniformly over the month between the hours of 6:00 and 18:00.
2. Standard storage amount is fixed at 350 kJ/°C m^2 (about 2 gal of H_2O/ft_c^2 or 84 l/m_c^2). For non-standard storage, correction a (see Equation 8.20) is applied.
3. No energy is rejected from storage; therefore, the vessel is assumed to be designed for the peak temperature and pressure expected.
4. Storage is well mixed and no storage-to-load heat exchanger is used.
5. The load device uses solar heat at temperature-independent efficiency to meet the load L. Therefore, the load device cannot be a turbine, for example.
6. No parasitic heat losses from storage occur.

Some of these restrictions can be relaxed using work on the $\bar{\phi}$, f, method conducted by Klein and his coworkers [16]. Users of the $\bar{\phi}$ methods must exercise caution in the proper choice of the $\bar{\phi}$ time scale. In the method presented here, $\bar{\phi}$ and $\bar{I}_c$ are calculated over the collection period Δt_c, not over all daylight hours. The method developed by Klein uses $\bar{\phi}$ and $\bar{I}_c$ for all daylight hours. Although the Collares-Pereira $\bar{\phi}$ value can be used with the $\bar{\phi}$, f, chart, the two methods of finding $\bar{\phi}$ itself must not be confused.

Example 10.3

Repeat Example 8.3 for Albuquerque, New Mexico, for a monthly averaged load of 260 kWh/day using a collector of 100 m². From the previous example, recall that

$$F_R = 0.95$$

$$U_C = 0.5\,\frac{\mathrm{W}}{\mathrm{m^2 C}}$$

$$\bar{\eta}_0 = 0.60$$

$$\bar{I}_c = 5.13\,\frac{\mathrm{kWh}}{\mathrm{m^2}}$$

$$\bar{\phi} = 0.83$$

What is the effect of storage on energy delivery per unit collector area?

Solution

First calculate P_s and P_L, then use the $\bar{\phi}$, f, chart to find the solar fraction.

$$P_s = \frac{F_R\bar{\eta}_0\bar{I}_c A_c N_d}{L} = \frac{0.95\times0.60\times5.13\times100\times31}{260\times31} = 1.12$$

$$P_L = \frac{F_R U_c A_c N_h \times 100}{L} = \frac{0.95\times0.5\times100\times31\times24\times100}{260\times31\times1000} = 0.44$$

The value of f_s from the chart with $P_L = 0.44$ and $\bar{\phi}P_s = 0.93$ is $f_s = 0.90$. Therefore, the energy delivery per unit area is $0.90 \times 260/100 = 2.34$ kWh/m² day, nearly identical to the result using the $\bar{\phi}$ method. This is a result of the low value of U_c for the concentrator and its resulting insensitivity to temperature fluctuations above T_{proc}. The reader may repeat the calculations for a 200 m² collector with $U_c = 2.0$ W/m² °C to show that $\bar{\phi} = 0.392$, $P_L = 3.51$, $P_s = 2.25$, $f_s = 0.78$. The energy delivery per unit area is then 1.01 kWh/m² day compared with 1.15 kWh/m² day predicted by the $\bar{\phi}$ method. Hence, the effect of storage is to reduce the unit energy delivery by 12% for the more efficient collector.

Problems

10.1 Design a compressed air energy storage system for a wind turbine. Assume that a cavity in an abandoned salt mine was available adjacent to the wind turbine. Use all the power generated above 5 kW to drive a compressor, which can compress the ambient air for the storage system. (This is an open-ended problem)

10.2 A hydrogen/oxygen fuel cell stack produces 100 kW of DC power at an efficiency of 60% with water vapor as the product. Determine the hydrogen mass flow rate in g/s and the required cell voltage.

10.3 Derive an expression for the electric energy generation per unit of air volume stored in a cavity for a CAES system. Assume that there are two expansion stages and express the work output as the sum of the output from the two stages. The system operates with a constant cavern pressure, such as a hard rock cavern with a hydraulic compensation.

10.4 Supposing you own some land next to a 50 MW wind farm located on Colorado's Front Range. Your land has a fantastic cliff/ridgeline with an elevation difference of 250 m. As an opportunist, you decide to evaluate whether or not you can make money by building an energy storage system on your land that interfaces with the wind farm. Your plan is to capture excess energy from the wind farm or buy energy from the wind farm when electricity prices are low, and sell that energy back to the grid when electricity prices are high (assume that the utility that owns the grid will grant you a fair contract). You find that a good power rating for your storage system is 50% of the rated wind farm output, or 25 MW. Also, your storage system should be able to generate electricity at rated power for 8 h.

The first step in this problem is to evaluate the technical design of candidate energy storage systems.

a. Pumped hydroelectric. Determine the flow rate and reservoir size needed to accomplish the required power output and energy capacity. Pumped hydroelectric plants generally run at 80%–90% generating efficiency, depending on the size of the machinery. Suggest a reasonable surface area and depth for your two reservoirs. (Hey, wait! You like to fish and plan to stock one of your reservoirs with some northern pike. You cannot leave them high and dry when the plant cycles).

b. Pulley and weight. You cannot kick the feeling that a simpler energy storage system is possible. What if you use a weight and a pulley that stores energy as the weight is lifted (using an electric motor) and energy is released when the weight falls (with an electric generator). You need to dig a shaft. Find the value of weight you would use, and the depth and diameter of the shaft. Is this viable?

c. Propose a new energy storage method. Be creative. Examples may be large springs, thermal storage, steam cycles, batteries, etc. Perform a simple analysis to size the system, and say whether or not your creative idea is viable.

The second step is to evaluate the financial aspects of installing and operating your energy storage system to determine whether you could make money.

a. Pumped hydroelectric costs about \$500 per kW to install your turbomachinery and penstocks, plus \$2 per cubic yard to build reservoirs. Calculate the initial capital cost of your pumped hydroelectric system.
b. You buy energy from the wind farm at \$0.035/kWh between the hours of 10:00 PM and 8:00 AM to charge your storage, and sell energy between 1:00 PM and 9:00 PM back to the grid at \$0.1/kWh. Select a reasonable simple payback period that would motivate you to invest in energy storage. Calculate the maximum capital cost expenditure on your energy storage system that would allow this payback period.
c. Would you decide to build an energy storage system? Why or why not?

10.5 It has been proposed that the United States should strive toward a hydrogen economy. At present, the United States uses about 100 quads (10^{17} Btu) of energy per year. Suppose that 10% should be produced by 2015 with hydrogen fuel cells. The hydrogen is to be made by electrolysis of water in central stations located strategically. The electric power for electrolysis is to be supplied by nuclear power plants with a thermodynamic efficiency of 33% and a capacity factor of 0.90. Each plant should be of 500 MW capacity and operate on a simple Rankine cycle.

Estimate

- The number of plants required.
- The amount of cooling water needed for the condensers (assuming evaporative cooling—2400 kJ cooling for each liter of cooling water).
- The amount of water needed for electrolysis assuming 1500 kJ are used by the fuel cell (i.e., only 750 kJ of energy is provided to the end user for each liter of water fed into the process).
- The total initial capital investment (if the cost of a nuclear plant is estimated at \$6000 per kW and that of a fuel cell at \$2100 per kW) [18].
- Comment on the viability of using hydrogen as an energy carrier on a global scale/niche application [18].

10.6 a. Design a compressed air energy storage system for a wind turbine you analyzed previously. Assume that a cavity in an abandoned salt mine is available adjacent to the wind turbine. Use all the power above 5 kW to drive a compressor, which can compress ambient air. Assume that NG is available for heating the air entering the compressor.

b. Repeat part 1, but assume that the wind turbine is located near a landfill where low Btu gas can be obtained for free. Compare this option with the design in part a, where the gas has to be purchased at $5 per 1000 standard ft^3.

10.7 Derive an expression for the linear battery model power output as a function of the internal-to-external resistance ratio. Nondimensionalize the power output by dividing it by $E^2/R_{i.}$ Tabulate the dimensional power function with a spreadsheet and then plot it. At what point is the maximum nondimensional power achieved?

10.8 Design a battery electric storage system for a power plant with 20 MW peak power delivery for a duration of 4 h and estimate the minimum number of batteries and the current during peak operation. Assume that the plant will use 600 A/h batteries, operating at 400 V DC.

10.9 A hydrogen/oxygen fuel cell operates with a voltage of 0.7 V with water vapor production. Calculate the kW/h of work per kg-mol of hydrogen and determine the cell efficiency.

10.10 You are asked to estimate the amount of heat that can be stored overnight in a solar-heated building, which has a storage space of 6 m^3. The solar system uses water with antifreeze in the collector and there are two options for storing the heat overnight. One option is to use water with an allowable temperature change of 30°C. The other is to use a glauber salt, which undergoes a phase change at approximately 30°C with a heat of fusion of 329 kJ/L. In both instances, a heat exchanger will be necessary in order to meet building code. Discuss the advantages and disadvantages of the two options.

10.11 A neighborhood fuel cell power plant is to be designed for an electric power output of 2000 kW with liquid–water product. Estimate the flow rates of hydrogen and oxygen during peak power production, assuming an 80% efficient power conditioner is used to convert DC to AC power and the fuel cell efficiency is 55%. What is the plant heat rate?

References

1. Lee, B. and Gushee, D. (June 2008) Massive electricity storage, White paper, AICHE, New York.
2. Hammerschlag, R., Pratt, R., and Schaber, C.P. (2004) Energy storage, in: F. Kreith and D.Y. Goswami (eds.), *Handbook of Energy Efficiency and Renewable Energy*, CRC Press, Boca Raton, FL, 2004.
3. Levine, J.G. (2007) Pumped hydroelectric energy storage and spatial diversity of wind resources as methods of improving utilization of renewable energy sources, Thesis, University of Colorado, Boulder, CO.

4. Schoenung, S., Eyer, J.M., Iannucci, J.J., and Horgan, S.A. (1996) Energy storage for a competitive power market, *Annual Review of Energy and the Environment* 21: 347–370.
5. Cohn, A. et al. (1999) Applications of air saturation to integrated coal gassification CAES power plants, ASME 91-JPGC-GT-2.
6. Crotogino, F., Mohmeyer, K.-U., and Scharf, R. (2001) Huntorf CAES: More than 20 years of successful operation, in: *Solution Mining Research Institute Meeting*, Orlando, FL.
7. van der Linden, S. (December 2006) Bulk energy storage potential in the USA, current developments and future prospects, *Energy* 31: 3446–2457.
8. EPRI-DOE (2003) *Handbook of Energy Storage for Transmission and Distribution Applications*, EPRI/DOE, Palo Alto, CA/Washington, DC.
9. Davis, L. and Schainker, R. (2006) Compressed air energy storage (CAES): Alabama electric cooperative McIntosh plant—Overview and operational history, in: *Electricity Storage Association Meeting 2006: Energy Storage in Action*, Knoxville, TN, Energy Storage Association.
10. Succar, S. (September 2008) Baseload power production from wind turbine arrays coupled to compressed air energy storage, PhD thesis, Princeton University, Princeton, NJ.
11. Gupta, R.G. (ed.) (2008) *Hydrogen Fuel—Production, Transport, and Storage*, CRC Press, Boca Raton, FL.
12. Bossel, U. et al. (2003) *The Future of Hydrogen Economy: Bright or Bleak*, Oberrohreorf, Switzerland.
13. Rabinwitz, M. (2000) Power systems of the future, Parts 1–3. *IEEE Power Engineering Review* 20(135): 5–16, 10–15, 21–24.
14. Kreith, F. and Bohn, M.S. (2001) *Principles of Heat Transfer*, 7th edn., Brooks/Cole Pacific Grove, CA.
15. Kreith, F. and West, R.E. (2004) Fallacies of a hydrogen economy: A critical analysis of hydrogen production, *Journal of Energy Resources Technology* 126: 249.
16. Klein, S.A. and Beckman, W.A. (1979) General design method for closed-loop solar energy systems, *Solar Energy* 22: 269.
17. Klein, S.A. (1978) Private communication.
18. National Research Council, *The Hydrogen Economy, Opportunities, Costs, Barriers, and R&D Needs*, U.S. Department of Energy Office of Basic Energy Sciences Committee on Alternatives and Strategies for Future Hydrogen Production and Use, National Academic Press, Washington, DC, Available at http://www.nap.edu/catalog/10922.html
19. Kreith, F. (1973) *Principles of Heat Transfer*, Intext, New York, 1973.
20. Wyman, C., Castle, J., and Kreith, F. (1980) A review of collector and energy storage technology for intermediate temperature applications, *Solar Energy* 24(6): 517–540.

Battery Resources

Battery technology is undergoing intensive research and below are some resources to follow development.

Li-Ion

http://www.sciencedaily.com/releases/2007/12/071219103105.htm
http://www.a123systems.com/products
http://www.sciencedaily.com/releases/2008/11/081120103802.htm
http://electronics.howstuffworks.com/lithium-ion-battery.htm
http://e-articles.info/e/a/title/The-Lithium-Ion-Battery/
http://e-articles.info/e/a/title/Advantages-and-Limitations-of-the-Lithium-Polymer-Battery/
http://www.batteryuniversity.com/parttwo-34.htm
http://www.kokam.com/english/product/battery_main.html
http://www.gpbatteries.com/html/pdf/Li-ion_handbook.pdf

Ni-MH

http://e-articles.info/e/a/title/Advantages-and-disadvantages-of-the-Nickel-Metal-Hydride-(Ni-MH)-Battery/
http://www.batteryuniversity.com/partone-4.htm

Vanadium Redox

http://www.vrb.unsw.edu.au/
http://thefraserdomain.typepad.com/energy/2006/01/vandium_reflux_.html

General

http://www.batteryuniversity.com/partone-9.htm

11

Ocean Energy Conversion

The oceans contain a great deal of energy, but the conversion of this energy into a useful form such as electricity requires advanced technology that is still under development. However, many demonstration prototypes have been designed and built. Valuable information is being obtained from the operation of these plants, and the results should lead to the development of large-scale units that are expected to become economically competitive in the future.

The key conversion technologies of ocean energy are

1. Ocean thermal energy conversion
2. Tidal energy
3. Wave energy

The basic elements of each of these three technologies will be presented in this chapter.

11.1 Ocean Thermal Energy Conversion

Oceans cover 71% of the earth's surface and receive the majority of the solar energy incident upon the earth. In semitropical and tropical oceans, the available temperature difference between the surface, which is warmed by the sun, and the temperature at the lower depths (ΔT_0) is sufficient to operate a Rankine-type power plant using fluids such as ammonia. Since the temperature difference is available 24 h/day virtually throughout the year, ocean thermal energy conversion (OTEC) has vast potential for implementation in many parts of the world. If the energy in a band ±10° latitude near the equator could be extracted with an energy efficiency of 0.5% of the incident solar energy, more than 7×10^5 MW power could be generated.

The main drawback of an OTEC system is that the overall thermal efficiency of the plant is low, and therefore large seawater pumps and large heat exchangers are required to produce adequate power. The maximum efficiency of any type of thermal power plant is the Carnot cycle efficiency, η_c, which for a typical OTEC plant is equal to

$$\eta_c = \frac{T_h - T_c}{T_h} = \frac{\Delta T_0}{T_h}$$

where

T_h is the hot surface water absolute temperature (about 298 K)
T_c is the absolute temperature of the cold water from below (about 278 K)

Thus, the Carnot efficiency is only about 6.7%. This ideal efficiency does not allow for parasitic losses or the temperature differences between the seawater and the working fluid required for evaporation and condensation. A rough rule of thumb is that about 25% of ΔT_0 needs to be allowed for the evaporation, and 25% for the condensing process, leaving only about half of the total temperature difference for the power turbine. For an 80% overall efficiency of the turbine generator, the gross power output efficiency would therefore only be about 3%. Parasitic load takes another 30% or so of the overall thermal efficiency, leaving a net power delivery efficiency of only about 2%–3%.

Figure 11.1 illustrates the four plant types that have been suggested: land-based (or shallow-water), shelf-mounted, floating, and moored plant types relative to the continent/ocean floor profile. The greatest number of potential

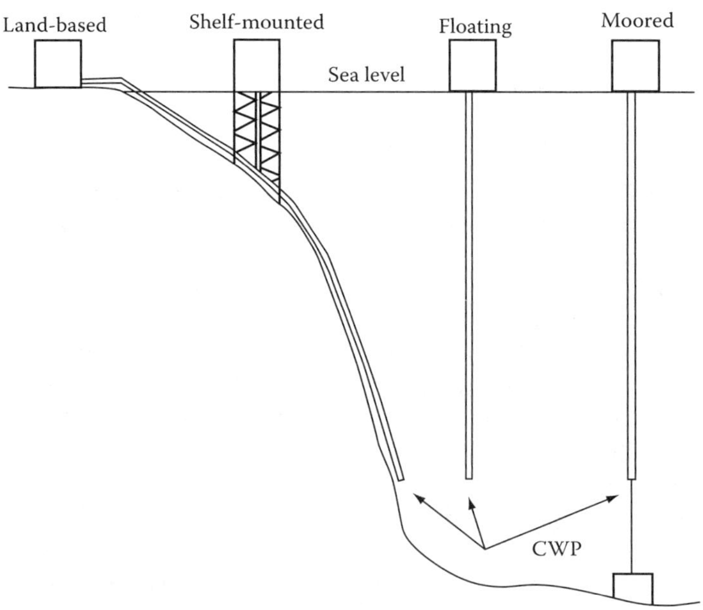

FIGURE 11.1
Basic OTEC plant configurations with cold water pipe (CWP).

sites exists for the floating plant with a suspended CWP. Shelf-mounted or land-based plant types reduce electricity and water transportation, as well as the ocean dynamics/plant operation problems. Except for the CWP, the design and construction of floating and moored types are within current state-of-the-art technology. However, longer pipes are required, which increases costs and reduces the number of potential sites.

Disadvantages of the floating moored plant types are related to electricity and water transportation and the effect of ocean dynamics on plant operation. For example, efficient transportation of electricity (by cable) and/or freshwater (by pipeline or barge) from a plant far offshore may be difficult. An example of the effect of ocean dynamics on plant design and operation is related to the evaporator. The evaporator produces steam as a result of the difference between the operating pressure and the saturation pressure of the warm seawater. Since this difference corresponds to only about 0.1 m of water, it is critical that the warm seawater supply pressure remain essentially constant, in spite of the wave action and platform motion, which may be difficult to achieve.

There are two basic technical approaches to producing power from the ocean thermal resources. Both are based on the Rankine thermodynamic cycle; one uses a closed system with a pressurized working fluid (typically ammonia or a fluorocarbon), and the second is an open system using the seawater itself as the working fluid. The concept of a closed-cycle OTEC (CC-OTEC) system was originally proposed by D'Arsonval [1] in 1881, and in 1930, his student, Claude [2], demonstrated the operation of an open-cycle OTEC (OC-OTEC) plant off the coast of Cuba.

11.1.1 Closed-Cycle Ocean Thermal Energy Conversion

Figure 11.2 is a schematic diagram of a closed-cycle ocean thermal conversion system. Warm surface water is used in the boiler to evaporate a working fluid; the sensible heat liberated by the seawater is the heat input, P_0, for the cycle

$$P_0 = \rho c Q \Delta T_0$$

where

ρ is the density of sea water in kg/m^3
c is the specific heat of seawater in J/kg K
Q is the flow rate in m^3/s

The vapor produced in the boiler passes through a turbine to the condenser and the expansion process generates power for electrical generation. The condenser uses cold water obtained at a depth of about 1000 m. Obviously, a long section of pipe and substantial pumping power are required for the condensation process.

The mechanical features of the closed cycle are very similar to a refrigeration cycle. Except for preventing bio-fueling of the heat exchanger and the

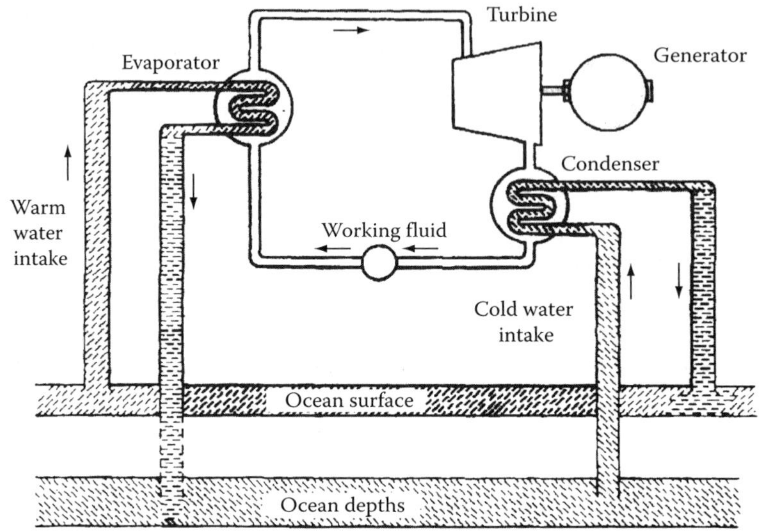

FIGURE 11.2
Schematic diagram of CC-OTEC.

CWP the technology is currently available. It should also be noted that a CC-OTEC system can also be used to produce potable water [3].

Figure 11.3 shows typical temperature vs. depth profiles near Hawaii, Puerto Rico, and Recife, Brazil (designated ATL-1), and Table 11.1 presents seasonal variations in ΔT_0 for the same three sites. The sites near Puerto Rico and Hawaii

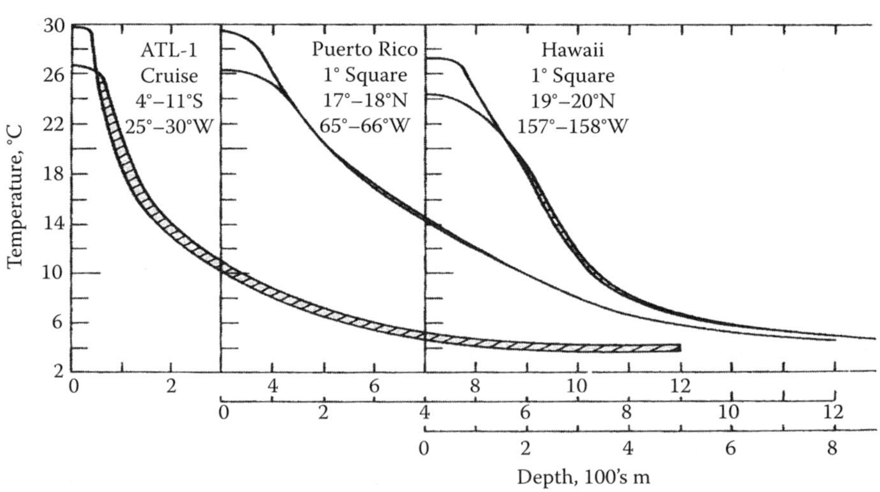

FIGURE 11.3
Temperature profiles vs. depth for Hawaii, Puerto Rico, and Recife, Brazil (ATL-1). (According to Dugger, G. et al., Ocean thermal energy conversion, in Kreider, J.F. and Kreith, F. (eds.), *Solar Energy Handbook*, McGraw-Hill, New York, Chap. 19.)

TABLE 11.1

Monthly Values of ΔT_0 and Power P_{net} (Percentage of Design Value) for Selected Sites

	ATL-1		Puerto Rico		Hawaii	
Month	ΔT_0 (°C)	P_{net} (%)	ΔT_0 (°C)	P_{net} (%)	ΔT_0 (°C)	P_{net} (%)
January	23.5	96	21.3	88	20.6	90
February	24.3	104	20.9	84	20.1	85
March	25.2	115	20.9	84	20.0	84
April	25.2	115	21.2	87	20.6	90
May	24.2	103	22.2	99	21.3	99
June	24.4	105	22.6	104	21.3	99
July	23.6	97	23.0	108	22.4	112
August	24.3	104	23.0	108	22.7	116
September	22.4	84	23.6	116	22.7	116
October	23.3	94	23.9	119	22.5	114
November	22.6	86	23.3	112	21.5	101
December	23.3	94	22.3	100	20.7	92
Average	23.9	99.8	22.3	100.7	21.4	99.8
$P_{net\ avg\ rel}$	100		83.7		74.2	

For details on the power, see [3].

are of particular interest to the United States, because for these islands all of the fuel needs to be imported. If movable or so-called grazing plants were to be installed, they would seek out the highest ΔT based on historical plots of monthly surface temperature contours, as well as daily information from satellites, which would indicate the optimum path for the platform on which the OTEC plant is based. A mini OTEC plant, as shown in Figure 11.4, operated off of Keahole Point, Hawaii in 1977 and generated 50 kW (gross) and 10 kW (net) from a temperature difference of 21.2°C (38.2°F). The system contained all the essential components required to build a full-scale CC-OTEC plant. It used a titanium plate–type heat exchanger and the plant was mounted on a barge moored by a single anchor, which incorporated a 24 in. OD polyethylene CWP as an integral member of the moor. A flexible rubber hose connected the top of the CWP to a moon pool in the center of the barge.

Example 11.1

Estimate the size of heat exchanger and the cold water flow rate required for a 10 MW CC-OTEC plant.

Assume that the cold water comes from a depth of 1000 m at a temperature T_{wc} of 5°C, while the hot water temperature, T_{wh}, is 25°C. Assume that the evaporator as well as the condenser operate at an effective temperature difference of ΔT of 4 K, with an overall heat transfer coefficient U of 6000 W/m^2 K. Use the Carnot efficiency for the power plant as an approximation.

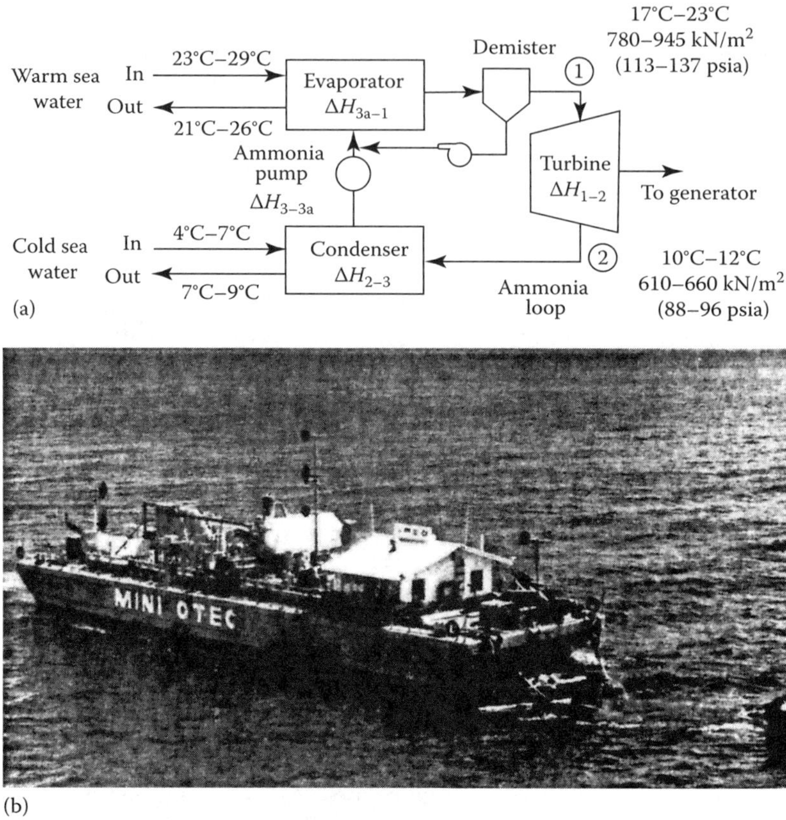

FIGURE 11.4
(a) Simplified diagram for typical CC-OTEC using ammonia as the working fluid and (b) a photo of the CC-OTEC operated in 1979 off Hawaii. (According to Dugger, G. et al., Ocean thermal energy conversion, in Kreider, J.F. and Kreith, F. (eds.), *Solar Energy Handbook*, McGraw-Hill, New York, Chap. 19.)

Solution

The rate of heat transfer to the working fluid in the evaporator of area A is approximately

$$q_{in} = UA\Delta T$$

The rate of enthalpy increase of the water is

$$q_{in} = \dot{Q}\rho c(T_h - T_c) = Q\rho c\left[(T_{wh} - \Delta T) - (T_{wc} + \Delta T)\right] = \dot{Q}\rho c(T_{wh} - T_{wc} - 2\Delta T)$$

where

ρ is the density 1.0 kg/m³

c is the specific heat 4.2×10^3 J/kg K

The Carnot efficiency η_c of the energy conversion process is

$$\eta_c = \frac{T_h - T_c}{T_h} = \frac{T_{wh} - T_{wc} - 2\Delta T}{T_{wh} - \Delta T} = \frac{12}{294} = 0.041$$

The power output, P_0, is given by

$$P_0 = \frac{\dot{Q}\rho c\,(T_{wh} - T_{wc} - 2\Delta T)^2}{(T_{wh} - \Delta T)}$$

and the water flow rate by

$$\dot{Q} = \frac{P_0(T_{wh} - \Delta T)}{\rho c\,(T_{wh} - T_{wc} - 2\Delta T)^2} = \frac{P_0}{\rho c \eta_c}$$

$$= \frac{10 \times 10^3\ \mathrm{Js^{-1}}\ 294\ \mathrm{K}}{10^3\ \mathrm{kg\,m^{-3}} \times 4.2 \times 10^3 (\mathrm{J\,kg^{-1}}) \times (12\ \mathrm{K})^2}$$

$$= 4.86\ \mathrm{m^3/s}$$

This is a very large flow rate. If the CWP were 1 m in diameter, the water velocity would be

$$v = \frac{4Q}{\pi d^2} = \frac{4 \times 4.86}{\pi 1^2} = 6.2\ \mathrm{m/s}$$

For a rate of heat transfer, q, the heat exchanger area is

$$A = \frac{q}{U\Delta T} = \frac{\dot{Q}\rho c\,(T_{wh} - T_{wc} - 2\Delta T)}{U\Delta T}$$

$$= \frac{P_0}{U\Delta T \eta_c}$$

$$= \frac{10 \times 10^6\ \mathrm{W}}{6{,}000\,(\mathrm{W\,m^{-2}\,K})(4\ \mathrm{K}) \times 0.041}$$

$$= 10{,}162\ \mathrm{m^2}$$

This is a huge and expensive heat exchanger surface area.

The heat transfer and power-generation processes are shown schematically in Figure 11.5.

11.1.2 Open-Cycle Ocean Thermal Energy Conversion

OC-OTEC systems can generate electricity and/or freshwater using the temperature difference between surface and deep ocean seawater without the need of an expensive heat exchanger [4]. It is an attractive concept due to the vast amount of ocean energy available and the minimal ecological impact. OC-OTEC is ideally suited for locations, which have a need for alternative electricity and freshwater, and are also close to an ocean

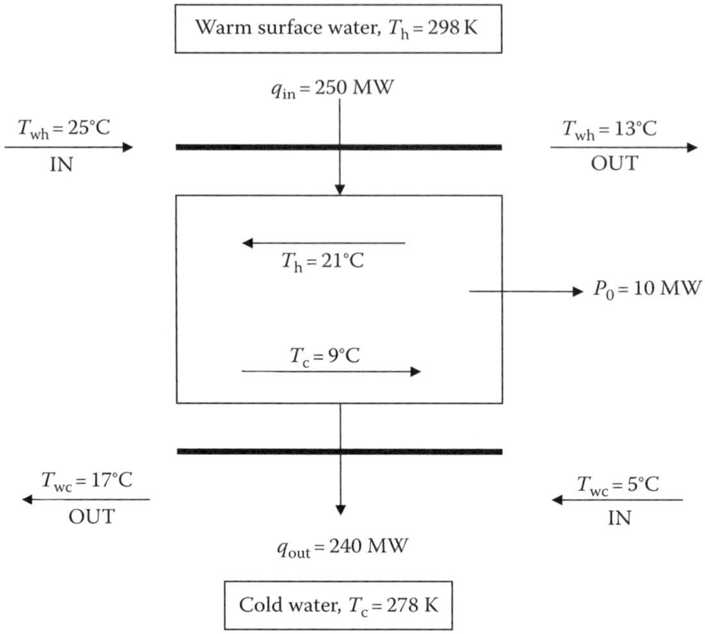

FIGURE 11.5
Schematic diagram for solution of Example 11.1.

TABLE 11.2
Oceanographic Data for Potential OC-OTEC Sites in the United States

Site	Latitude (°N)	Longitude (°W)	Water Depth (m)	Distance from Shore (km)	Temperature Differential[a] (°C)
Puerto Rico	18.0	65.8	1000	3.1	22.3
Miami	25.4	79.9	740[b]	28.4	21.2
Key West	24.2	81.3	914[b]	60.0	21.5
Tampa	27.8	85.2	1000	281.6	20.9
New Orleans	28.8	88.8	1000	80.5	20.8
Island of Hawaii	19.9	156.1	1000	2.3	21.4

[a] Average.
[b] Maximum depth at location.

thermal resource. Table 11.2 provides information about potentially suitable OC-OTEC sites in the United States.

The OC-OTEC concept is shown schematically in Figure 11.6. Warm (25°C typical) surface seawater is introduced into the evaporator, which operates at a pressure slightly below the saturation pressure corresponding to the warm seawater temperature. A small percentage (≈0.5%) of the seawater flashes to produce steam, which is expanded through a turbine connected to an

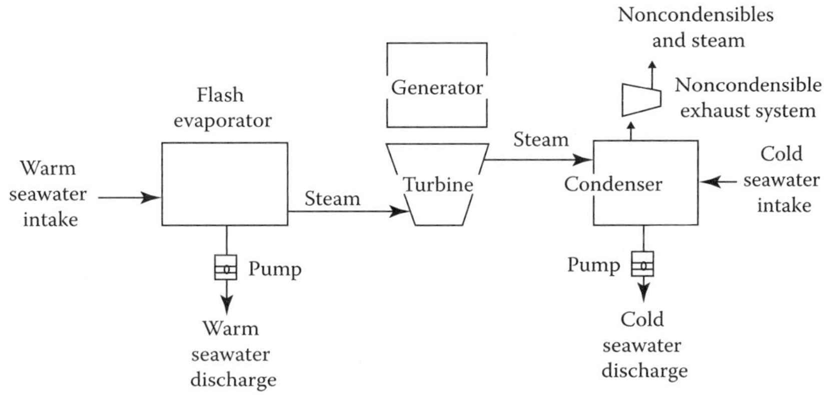

FIGURE 11.6
Conceptual schematic of OC-OTEC system.

electrical generator. Most of the steam is condensed in a direct contact condenser supplied with cold (typically 5°C) seawater from depths of approximately 1000 m. A surface condenser is used only if freshwater production is desired, whereas a more efficient direct-contact condenser is used for electricity production. The remaining steam and noncondensibles (released from the seawater in the evaporator and direct-contact condenser, and from possible leaks in the containment structure) are removed from the condenser by the noncondensible exhaust system. Variations of the basic cycle shown in Figure 11.6 include multiple stages of (evaporator, turbine, condenser) seawater pre-deaeration, and a combination of freshwater and direct-contact condenser with freshwater to seawater heat exchanger.

In an OC-OTEC system, seawater itself is used as the working fluid. A schematic diagram showing a plant cross section for a 100 MW electric open cycle system proposed some time ago by Westinghouse is shown in Figure 11.7. The Westinghouse design proposed turbines with 140 ft tip diameter that were to operate at 225 rpm and deliver up to 100 MW of electric power under optimum conditions (Table 11.3). The hull in this design is located around the vertical axis turbine in an integrated design. An open channel evaporator is located axis-symmetrically around the turbine with warm seawater entering the evaporator through a peripheral submerged sluice. Vapor flashes from the warm water flow along a bent path toward the turbine blade, while the spent vapor leaving the turbine rotor is condensed in a surface condenser centrally located underneath the turbine. More recently, it has been proposed to utilize a direct-contact condenser, which increases the efficiency of the system [5]. The warm seawater leaving the evaporator and the cold seawater leaving the condenser are rejected into the ocean through a common pipe. Depending upon whether the condenser is a surface condenser or a direct-contact condenser, seawater distillation is possible with this approach, and Figure 11.7 illustrates schematically such a dual-purpose system. A detailed description

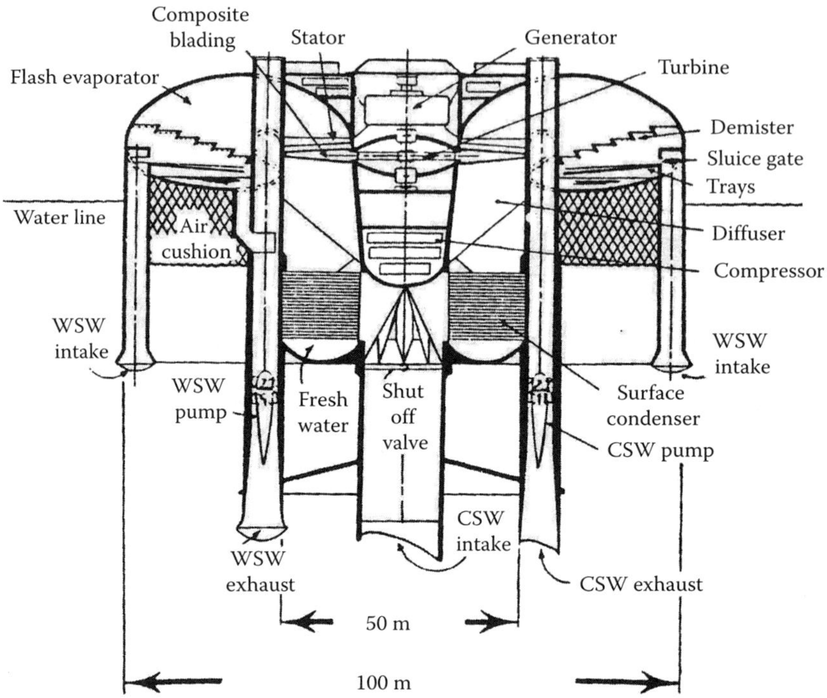

FIGURE 11.7
100 MW_e OC-OTEC plant cross-section (as proposed by Westinghouse). (From Rabas, T.J. et al., OTEC 100-MWe alternate power systems study, Report by Westinghouse Electric Corp. to U.S. DOE under contract EG/77/C/05/1473, March 1979.)

of the OC-OTEC system using a falling jet evaporator and direct-contact condenser, a more advanced version of the Westinghouse concept, has been developed at the Solar Energy Research Institute and is described in [5] (Figure 11.8).

An OC-OTEC plant can be designed for electricity and/or freshwater production. The selection of the seawater flow rates and the temperature drops across various components of the system depends on the desired products. For example, a trade-off must be made between the temperature drop across the evaporator (which affects the amount of steam generated) and the temperature drop across the turbines (which affects the amount of electricity produced). The optimum turbine temperature drop (based on plant economics) is approximately one half of the total available temperature drop for a single-stage electricity-producing plant, but only about one-fourth for a freshwater-producing plant [3].

11.1.2.1 Direct Contact Evaporation and Condensation

The mechanism by which steam is generated in an OC-OTEC system is called direct contact or "flash" evaporation. This process occurs when warm seawater is introduced into the evaporator at a pressure below the vapor pressure

TABLE 11.3

Design and Performance Specifications for a 10 MWe, OC-OTEC Plant

Description	Specification
Warm water temperature	25°C
Warm water flow rate	62,100 kg/s
Warm water pipe diameter	9.27 m
Warm water pump power	1.03 MW
Evaporator size (platform area)	113 m^2
Turbine (number and rotor diameter)	(3) 5.3 m
Gross power	15.77 MW
Cold water temperature	5°C
Cold water flow rate	35,590 kg/s
CWP diameter	4.7 m
Cold water pump power	2.27 MW
Condenser size (platform area)	158 m^2
Purge system flow	1.30 kg/s at 1.14 kPa
Compressor power	0.76 MW
Net power produced	11.7 MW

Source: From Rabas, T.J., et al., OTEC 100-MWe alternate power systems study, Report by Westinghouse Electric Corp. to U.S. DOE under contract EG/77/C/05/1473, March 1979.

corresponding to the liquid inlet temperature, and steam is produced by the combined action of boiling and surface evaporation. Flash evaporation is usually quite violent as a result of the explosive growth of vapor bubbles from nucleation sites in the liquid. The growth of these bubbles shatters the liquid continuum and yields a wide range of droplet sizes. Because of the irregular geometry of the interface, it is practically impossible to define or measure the surface area from which evaporation takes place. Heat transfer in flash evaporation can therefore not be described in terms of a conventional heat transfer coefficient. To quantify the process and present experimental data, another parameter is used "the effectiveness ε." The effectiveness of flash evaporation is defined as the ratio of the temperature difference between the inlet and outlet liquid streams to the temperature difference between the inlet stream and the vapor temperature corresponding to the chamber saturation pressure. This definition is similar to the effectiveness of a conventional heat exchanger, since it is the ratio of the temperature difference actually achieved to the maximum temperature difference thermodynamically available.

Several evaporator options have been proposed and analyzed for OC-OTEC. They include open channel flow, falling films, falling jets, and spout evaporators. These are shown schematically in Figure 11.9. Based upon extensive experiments at the SERI test facility [6], the single-spout evaporator was found to be the most suitable configuration. The effectiveness measured

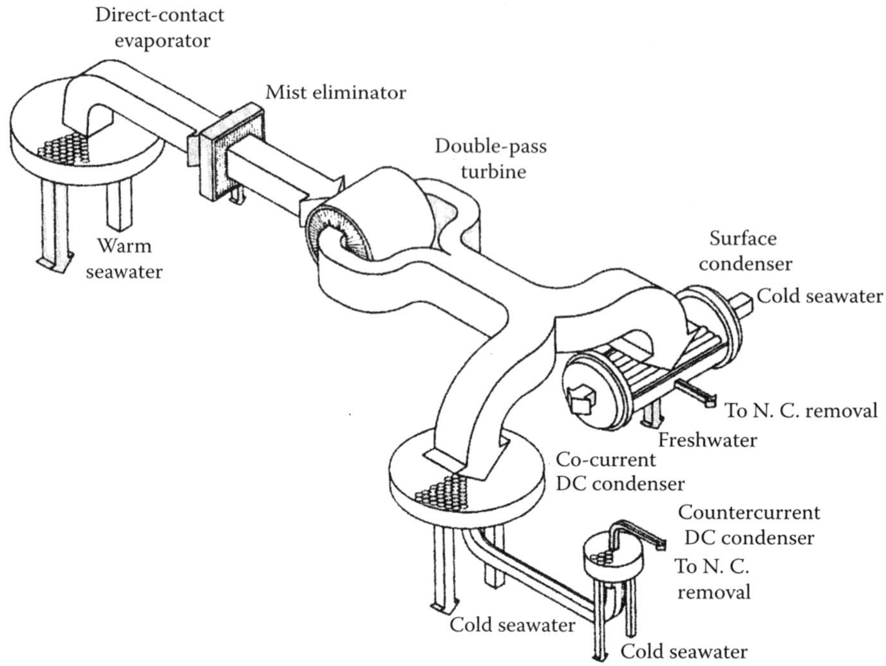

FIGURE 11.8
Schematic of OC-OTEC for power and freshwater production.

in experiments with and without screens placed above the pipe outlet varied between 0.9 and 0.97, as shown in Figure 11.10, with a liquid-side pressure drop of 0.7 m (spout height + kinetic energy loss).

OTEC direct-contact condensation is more complicated than for conventional steam turbine applications, because ocean water contains a large amount of noncondensable gases that may come out of solution and the need to reduce parasitic power to a minimum. Based on extensive experiments, it was found that packed columns with structured packing are the best technology for OC-OTEC. The analysis for these configurations is presented in [4]. The most effective configuration was found to be an initial co-current condenser, followed by a countercurrent section in which noncondensable gases can be removed, as shown in Figure 11.8.

11.1.3 Comparison of Open- and Closed-Cycle OTEC Systems

For a given temperature potential, the OC-OTEC system is thermodynamically superior to the closed-cycle system because a larger portion of the temperature difference between the hot and the cold water is available to produce net power [5]. This is because of the direct-contact heat transfer mode of the evaporator and condenser, as used in the open-cycle system. The steam in the vacuum chamber can attain the temperature of the warm water discharged from the evaporator.

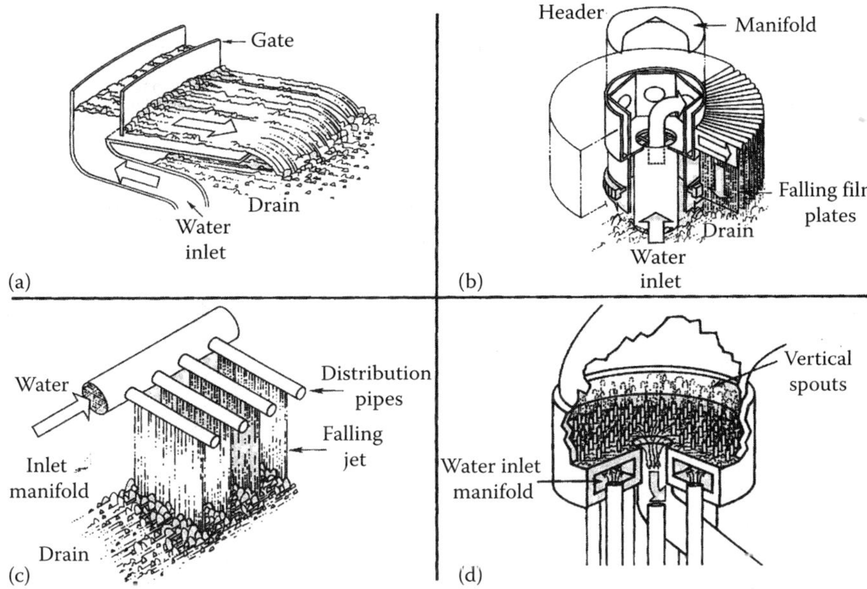

FIGURE 11.9
Contending direct contact heat exchanger geometries for OC-OTEC: (a) open channel flow, (b) falling films, (c) falling jets, and (d) vertical spouts.

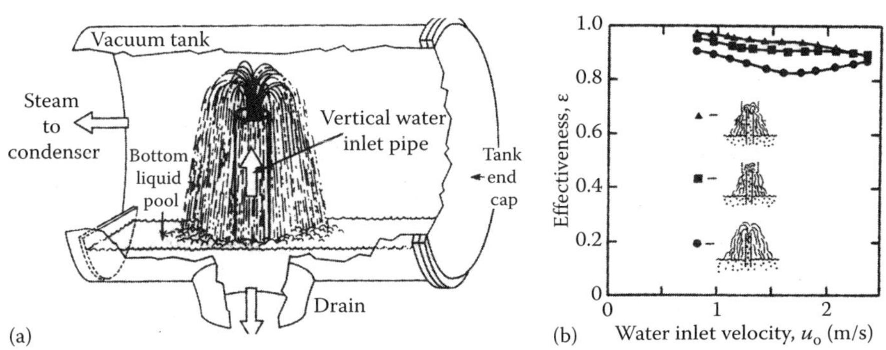

FIGURE 11.10
(a) Sketch of single-spout evaporator in SERI test chamber. (b) Effectiveness of spout evaporator with and without screens vs. water velocity. (From Bharathan, D. and Penney, T., Flash evaporation from falling turbulent jets, SERI-TP, 252-1853, Solar Energy Research Institute, Golden, CO, 1983; Bharathan, D. et al., Measured performance of direct contract jet condensers, SERT-TP, 252-1437, Solar Energy Research Institute, Golden, CO, 1982.)

At the same time, the exhaust steam from the turbine can be condensed at a temperature approaching the condenser cold water exit temperature.

In a closed-cycle system, the working fiuid passing through the turbine is separated from the warm and cold waters by a solid wall. Hence, in both the evaporator and the condenser, a temperature drop is necessary first to transfer the heat from the water through the wall to produce the vapor in the boiler, and then another temperature drop is required to condense the vapor of the working fluid at the outlet from the turbine by cold seawater. Consequently, the heat exchangers for CC-OTEC plants require very large surface areas to minimize the thermodynamically required temperature losses. These exchange surfaces are continuously exposed to corrosion and biofouling in the harsh ocean environment. On the other hand, the turbine in a CC-OTEC system is relatively small and is available as an "off-the-shelf" item. But, because of the low pressure in which the turbine must operate in the open-cycle system, the turbine is very large, and so far, no full-scale turbines for plants 1 MW and greater have been designed and operated.

Any ocean thermal energy conversion system faces the enormous challenge of producing net power with a temperature difference on the order of 20°C. Since the Carnot efficiency of such a plant is only on the order of 6% or 7%, any successful ocean energy conversion system must use as much of the temperature potential as possible and, at the same time, reduce parasitic losses to a minimum. The major parasitic losses in ocean thermal systems are the pumping power required to bring the cold water from a depth of about 1000 m to the surface, the power required to remove noncondensable gases in the open-cycle system, and the degradation of the available thermal energy by temperature drops due to fouling and corrosion across the heat transfer surfaces of a closed-cycle system.

11.1.4 Cold-Water Pipe and Pumping Requirements

The cost of the CWP is the dominant feature in the overall cost for an OTEC system [7]. This pipe is subject to stresses that include drag from ocean current, oscillating forces due to vortex shedding, and possibly forces due to the drift and motion of the platform. Moreover, the weight of the pipe is substantial, and it is not clear whether it should be made of a rigid or flexible material such as polythane. There may also be problems in assembling and positioning the pipe because of its enormous length. A premature failure of the CWP could be a financial blow to the entire OTEC system.

To transport the large quantities of cold and hot water necessary to operate an OTEC system requires a good deal of pumping power. This power will have to be supplied from the gross power output of the OTEC system, and constitutes a loss in net energy. The frictional loss in pumping the water through the CWP can be reduced by increasing the pipe diameter, but this will increase the cost and weight of the pipe. It can also lead to instabilities that could prematurely destroy the cold water support. It should also be noted that the power output estimated in the example below is based on the efficiency of a Carnot engine, whereas a real turbine is less efficient and thereby reduces the net power output.

Example: Pumping Power Requirements

Estimate the power required to move water from a depth of 1000 m for a 10 MW OTEC system described in the previous example. Assume that the diameter of the pipe is 1 m.

Solution
The average water velocity is

$$\bar{u} = \frac{\dot{Q}}{A} = \frac{4.86\ (\text{m}^3/\text{s})}{\pi 0.5^2\ \text{m}^2} = 6.2\ \text{m/s}$$

Hence, the Reynolds number is

$$Re_D = \frac{uD}{\upsilon} = \frac{6.2\ (\text{m/s})1\ \text{m}}{1\times 10^{-6}\ \text{m}^2/\text{s}} = 6.1\times 10^6$$

The flow is turbulent and the friction factor can be estimated from the empirical relation given below:

$$f = \frac{0.184}{Re_D^{0.2}} = \frac{0.184}{(6.1\times 10^6)^{0.2}} = 0.0081$$

The head loss in the pipe equals

$$\Delta H = \frac{2fLu^2}{Dg} = \frac{2\times 0.0081\times 10^3\ \text{m}\times 6.2^2\ \text{m}^2/\text{s}^2}{1.0\ \text{m}\times 9.8\ \text{m/s}^2} = 62.8\ \text{m}$$

Hence, the pumping power required is

$$\begin{aligned} P_{\text{pump}} &= \rho \dot{Q} g \Delta H = 1\times 10^3\ \text{kg/m}^3 \times 4.86\ \text{m}^3/\text{s} \times 9.8\ \text{m/s}^2 \times 62.8\ \text{m} \\ &= 3\times 10^6\ \text{kW} = 3\ \text{MW} \end{aligned}$$

The pumping power requirements could be reduced substantially by using a 2 m diameter pipe.

11.1.5 Economics

Cost estimates have been performed for both open and closed cycle systems [7]. By and large, they have indicated similar results. The main conclusion of one of the most detailed investigations is shown in Figures 11.11 and 11.12. Figure 11.11 shows the capital costs of OC-OTEC plants optimized for power

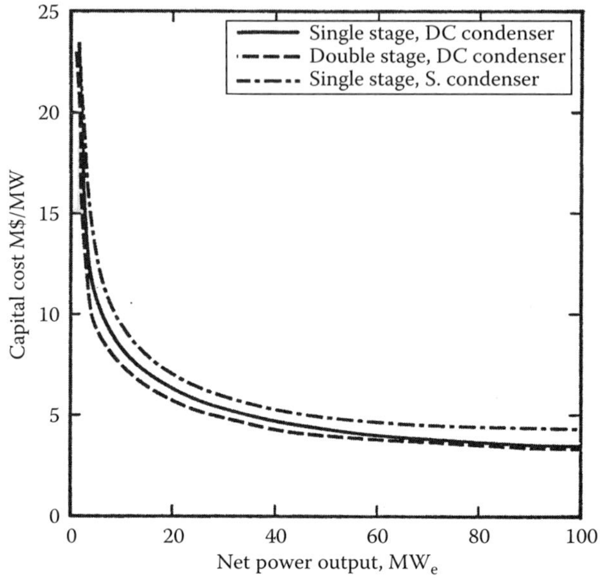

FIGURE 11.11
Capital cost estimates for three types of OC-OTEC plants vs. net power output optimized for power generation (1983$). (From Block, D.L. and Valenzuela, J.A., Thermoeconomic optimization of OC-OTEC electricity and water production plants, SERI/STR-251-2603, U.S. Department of Energy, Washington, DC, May 1985.)

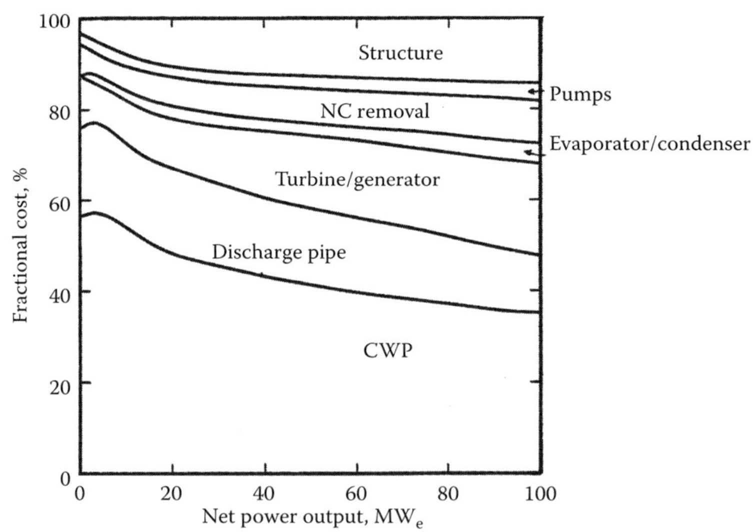

FIGURE 11.12
Fractional component cost of single-stage, direct-contact condenser plant optimized for power output. (From Block, D.L. and Valenzuela, J.A., Thermoeconomic optimization of OC-OTEC electricity and water production plants, SERI/STR-251-2603, U.S. Department of Energy, Washington, DC, May 1985.)

production for a single-stage direct-contact condenser, a double-stage direct-contact condenser, and a single-stage surface condenser. For all three of these configurations, it can be seen that the capital cost decreases rapidly as net power output increases and levels off at net power above 20 MW electric. These results clearly indicate that in order for OTEC to be economically viable, extremely large plants would be necessary. For such large plants, the initial investment is large and this has, so far, discouraged investment in this technology.

Figure 11.12 shows the fractional costs in percentage for the various parts of the OTEC system. These estimates indicate that the largest costs associated with an OTEC plant are the discharge and CWPs, with the CWP costing approximately 40% of the total at a net power output of 20 MW electric, while the discharge pipe would require about half that amount. These cost estimates have been made for several configurations. In the single-stage plants, all of the steam is produced in one flash evaporator, which expands through a single turbine, and is condensed afterward. The difference between the two single-stage plants analyzed is the type of condenser. One plant had a surface condenser allowing for the production of both electricity and freshwater, while the other plant had a direct-contact condenser and could only produce electricity. A direct-contact condenser is cheaper than an equivalent surface condenser, and there are trade-offs between freshwater production capabilities and plant capital costs depending upon the demand. Double-stage plants have potentially slightly higher efficiency and can provide greater flexibility of operation, but they are more complex and also more expensive.

11.2 Tidal Energy

11.2.1 Introduction

Tidal energy has a long history. Small tidal mills were used for grinding corn in Europe in the Middle Ages, but more recently, using tidal energy on a large scale to generate electricity has been proposed for many parts of the world. Tidal energy can be harnessed by building large barrages—essentially long low dams—across the inlet of suitable estuaries. When the tides come in to the shore, they can be trapped by means of sluices in reservoirs called tidal basins behind these barrages. When the tide ebbs, the water behind the barrage can be let out via a turbine somewhat similar to a hydroelectric power plant. However, it is important to distinguish tidal power from hydropower. The former is the result of the interaction of the gravitational pull of the moon and the sun on the oceans, while the latter is derived from the hydrological climate cycle powered by solar energy. A simplified analysis of the earth–moon interaction that produced the tidal behavior is presented in [8].

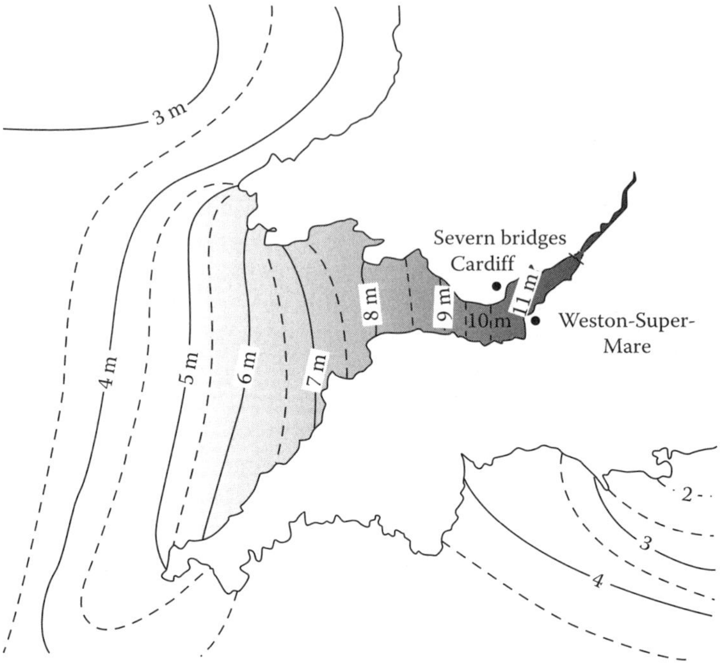

FIGURE 11.13
The effects of concentration of tidal flow in the Severn Estuary (tidal ranges in meters).

Tidal energy technology relies on the twice-daily tides with the concomitant upstream flows and downstream ebbs in estuaries. To extract energy from the rise and fall of the tides, barrages are built across the inlet to the tidal basin and turbines are located in water passages inside the barrages. The potential energy in the difference in water levels across the barrage due to the tidal motion is converted to kinetic energy as fast-moving water passes through the turbine. The rotational kinetic energy of the blades of the turbine can then be used to drive a generator to produce electricity. The average power output from a tidal system is roughly proportional to the square of the difference in water level between high and low tides. A tidal range of at least 5 m is considered the minimum for viable power generation. Sometimes tidal flows can be increased in height using the effect of concentration, as shown schematically [9] in the Severn Estuary near Cardiff in Wales (Figure 11.13).

Although many tidal systems have been proposed, the only major tidal power plant that has been constructed and operated is the La Rance Tidal Power Plant in France (Figure 11.14). The barrage for the plant was built in the 1960s near St. Maillot. A dam, 330 m long, was built in front of a 22 km^2 basin at a location where the tidal range was 8 m. The plant consists of 24 bulb-type turbine generators, 5.35 m in diameter, rated at 10 MW each. Power can be generated irrespective of whether the tide is going in or out.

FIGURE 11.14
A view of La Rance tidal scheme. (Courtesy of http://www.reuk.co.uk/La-Rance-Tidal-Power-Plant.htm)

The plant has now been in operation for 40 years, during which time 20 billion kW hours of electricity has been generated without major incident or mechanical breakdown.

11.2.2 Tidal Power

To estimate the amount of power output from a tidal barrage, assume that in front of the barrage is an infinite ocean, whereas behind it, a rectangular basin delineates the volume into which the water flows. The volume has a surface area A, and the range between high and low tide is H, as shown in Figure 11.15.

The incoming tide flows freely into the basin, but when the tide goes out, the water is held in the basin by closing the sluice gates at a level H above the level of the ocean. Given a rectangular basin, the center of gravity of the mass of water in the basin will be at a height $H/2$ above the sea level. The total volume in the basin will thus be AH and its mass will be ρAH. All of the water stored in the basin could now be made to flow through a turbine from

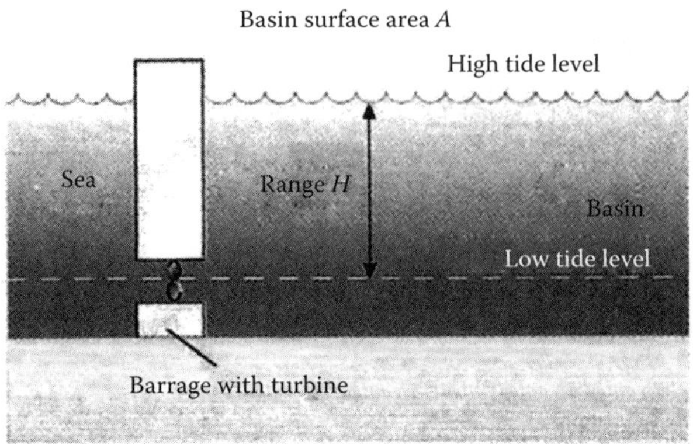

FIGURE 11.15
Nomenclature for tidal power analysis.

the high to the low water level. The maximum potential energy E available per tide, assuming all the water passes through a height $H/2$, is therefore

$$E = \rho AHg\left(\frac{H}{2}\right) \tag{11.1}$$

where g is the gravitational attraction. The main periods of tides are diurnal at about 24 h and semidiurnal at about 12 h 25 min [9].

If the basin is allowed to fill again with the next incoming tide and the cycle repeats with a period t, the average potential energy, P, that could be extracted becomes

$$P = \frac{\rho AH^2 g}{2t} \tag{11.2}$$

For example, if $A = 10\,\text{km}^2$, $H = 5\,\text{m}$, $t = 12\,\text{h}\ 25\,\text{min}$, $P = 26.6\,\text{MW}$.

However, the difference between high and low tidal level, H, varies throughout the year from a maximum H_{max} in the spring, to a minimum, H_{min}, for the neap tides. As shown in [8], this variation in H can be accounted for by using a mean range for all the tides, $\bar{H}$. The mean power production per month, $\bar{P}$, then becomes

$$\bar{P} \sim \frac{\rho Ag}{2t}\left(\frac{H_{max}^2 + H_{min}^2}{2}\right) \tag{11.3}$$

For example, if $H_{max} = 5\,\text{m}$, $H_{min} = 2.5\,\text{m}$, $A = 10\,\text{km}^2$, $\rho = 1030\,\text{kg/m}^3$, and $t = 12\,\text{h}\ 25\,\text{min}$ ($4.47 \times 10^4\,\text{s}$), then $\bar{P} \cong 16.1\,\text{MW}$.

Tidal power availability is dependent on local conditions and there are many potential sites for tidal barrages around the world, as shown in Table 11.4, with

TABLE 11.4

Some World Locations for Potential Tidal Power Projects

Country	Mean Tidal Range (m)	Basin Area (km²)	Installed Capacity (MW)	Approx. Annual Output (TWh per Year)	Annual Plant Load Factor (%)
Argentina					
San José	5.8	778	5,040	9.4	21
Golfo Nuevo	3.7	2,376	6,570	16.8	29
Rio Deseado	3.6	73	180	0.45	28
Santa Cruz	7.5	222	2,420	6.1	29
Rio Gallegos	7.5	177	1,900	4.8	29
Australia					
Secure Bay	7.0	140	1,480	2.9	22
Walcott Inlet	7.0	260	2,800	5.4	22
Canada					
Cobequid	12.4	240	5,338	14.0	30
Cumberland	10.9	90	1,400	3.4	28
Shepody	10.0	115	1,800	4.8	30
India					
Gulf of Kachchh (Kutch)	5.0	170	900	1.6	22
Gulf of Cambay (Khambat)	7.0	1,970	7,000	15.0	24
Korea (Rep)					
Garolim	4.7	100	400	0.836	24
Cheonsu	4.5	—	—	1.2	—
Mexico					
Rio Colorado	6–7	—	—	54	—
United States					
Passamaquoddy	5.5	—	—	—	—
Knik Arm	7.5	—	2,900	7.4	29
Turnagain Arm	7.5	—	6,500	16.6	29
Russian Federation					
Mezeh	6.7	2,640	15,000	45	34
Tugur[a]	6.8	1,080	7,800	16.2	24
Penzhinsk	11.4	20,530	87,400	190	25

Sources: Adapted from World Energy Council Web site, http://worldenergy.org (accessed 23 December 2003); Elliot, D., Tidal power, in *Renewable Energy*, 2nd edn., G. Boyle (ed.), Oxford University Press, Inc., New York, Chap. 6, 2004.

[a] 7000 MW variant also studied.

an estimated total potential of the order of 300 TW hours per year. The total energy dissipated from tides globally is approximately 3000 GW, of which approximately a third is in accessible locations. The realistically recoverable resource has been placed at one-tenth of this amount, or 100 GW, due to practical limitations such as access to major tides [10].

There are essentially three ways to generate power from a tidal system [9]. In the most common method, called ebb generation, the incoming tide passes through the sluice gate in the barrage and the water is trapped in the tidal basin by closing the sluices. The head of water in the basin can then be passed back through the turbine during the outgoing ebb tide, as shown in Figure 11.16. Alternatively, flood generation can use the incoming tide to generate electricity as it passes through the turbines in the barrage (see Figure 11.17). In either of these two cases, two bursts of power are produced during every 24.8 h period. This creates problems because the power production is not uniform. However, the electricity generation can be smoothed out

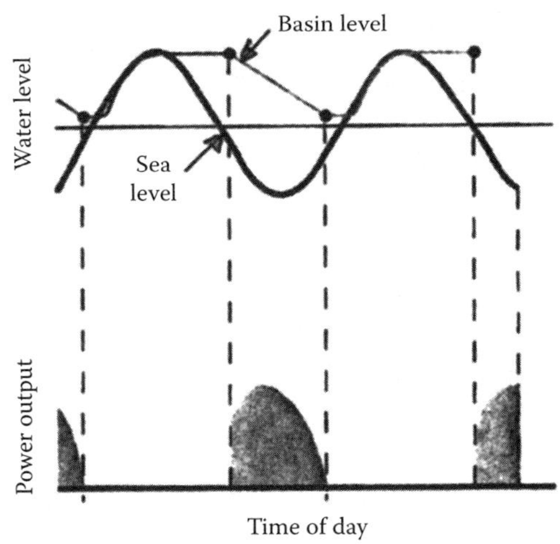

FIGURE 11.16
Schematic diagram of water levels and power outputs for an ebb generation scheme.

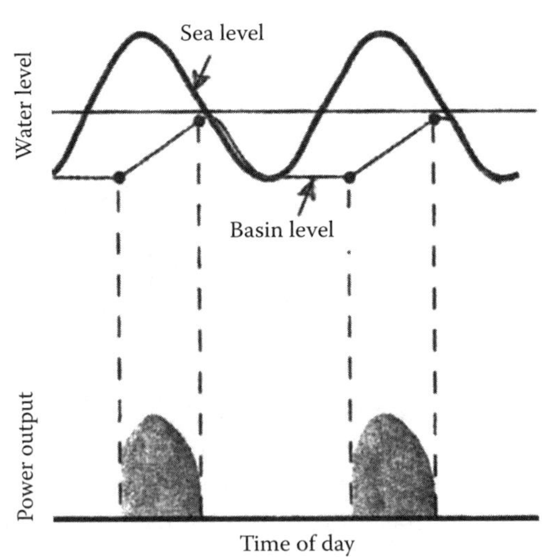

FIGURE 11.17
Schematic diagram of water levels and power outputs for a flood generation scheme.

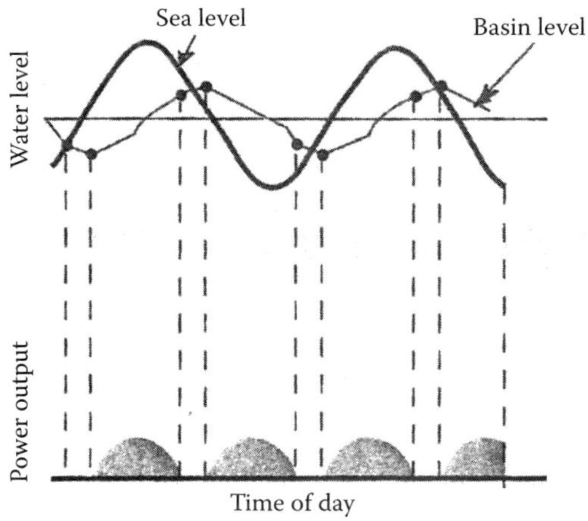

FIGURE 11.18
Schematic diagram of water levels and power outputs for a two-way generation scheme.

by two-way operation, using energy both during the ebb and the flood, as shown in Figure 11.18. Reference [9] presents details of various schemes proposed for two-way operation.

The most popular scheme of power generation is the bulb system, as shown in Figure 11.19. In this system, the turbine generator is sealed in a bulb-shaped enclosure, which is mounted in the flow channel. Water flows around the large bulb and turns the turbine runner or blades, after passing through a distributor. The main disadvantage of this scheme is that to attain access for maintenance of the generator requires cutting off the flow of water. This could be avoided by the use of a rim generator or tubular turbine configurations described in [8].

11.2.2.1 Economics of Tidal Power

The cost of energy from a tidal system will depend on several factors such as location, capital cost, interest rate, turbine performance, length of time for amortization, and difference in water height between high and low tides. The last factor is particularly important because, as shown by Equation 11.2, the power output from a tidal system is a function of the square of the tidal ebb. If amortized over a life of 40 years, the cost of power from the La Rance tidal system will be one of the lowest of any renewable energy schemes (see Figure 11.20).

One of the most promising large-scale tidal systems, which has received considerable interest is the so-called Severn barrage in the United Kingdom [11]. The proposed system would product 8.6 GW, stretching 16 km across the Severn Estuary, and would generate 17 TWh of electric energy per year. The system is shown schematically in Figure 11.21. Several other possible sites for large scale projects exist all around the world, but information about their potential, economic as well as technical, is not generally available.

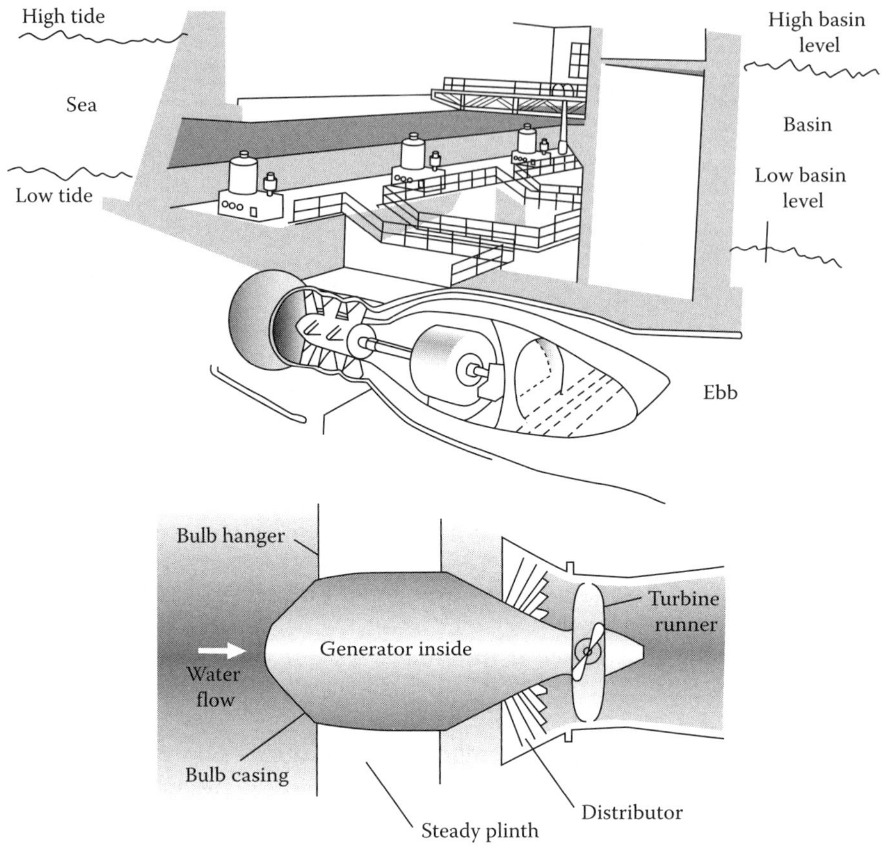

FIGURE 11.19
Conceptual layout of a tidal power generation scheme with details of a bulb turbine as used at La Rance.

A detailed analysis of capital and running costs was made in 1989 for the Severn barrage tidal system in Great Britain. At that time, a detailed study, shown in Table 11.5, indicated that the total capital cost was estimated at £8.28 billion. This estimate was updated to £10–15 billion in 2002. It should be noted that these estimates do not include interest on borrowed capital accrued during construction. This is an important factor, because construction of a tidal power plant may take several years, and during this time, money is expended but no power is produced. It should further be noted that since the system would only operate during tidal cycles, the 8.6 GW turbine capacity of the Severn barrage would only be equivalent to about 2 GW of generating capacity of a conventional base-load power plant because the system functions only intermittently. The precise capacity credit that can be attributed to a tidal system, that is the value-as-replacement for a conventional plant, will depend in practice on the scale and timing of the outputs of the plant.

FIGURE 11.20
(See color insert.) La Rance tidal power station. (From Khaligh, A. and Onar, O.C., *Energy Harvesting*, Figure 3.6, CRC Press, Boca Raton, FL, 2010.)

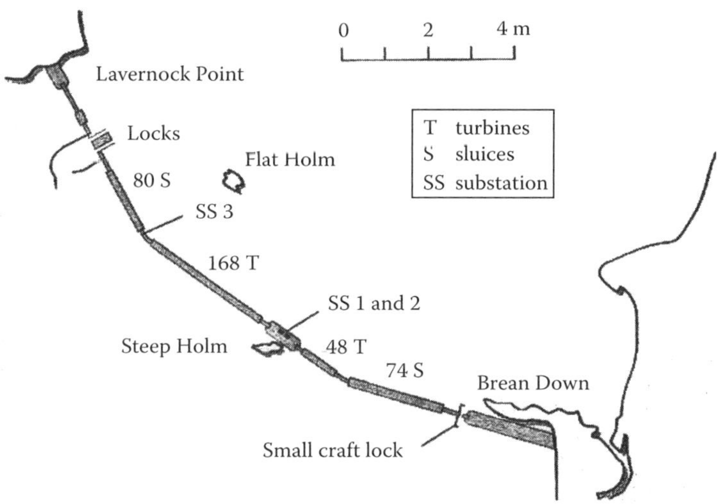

FIGURE 11.21
Layout of the proposed Severn barrage tidal system. (Abstracted from Department of Energy, *The Seven Barrage Project: General Report*, Energy Paper 57, HMSO.)

In other words, the barrage load factor, that is, the percentage of time that the plant can deliver power, was estimated to be about 23% for the Severn barrage. This is about the same as the average output from an intermittent wind power plant, but does not approach that of a nuclear power plant, for which the load factor is about 90%, or of a combined cycle gas turbine with a load

TABLE 11.5

Severn Barrage "Reference Project" Capital and Annual Recurring Costs

	Cost/£ Million[a]
Pre-construction phase	
Feasibility and environmental studies, planning and parliamentary costs	60
Design and engineering	130
Barrage construction	
Civil engineering works[b]	4900
Power generation works	2400
On-barrage transmission and control	380
Management, engineering and supervision	300
Land and urban drainage sea defenses	30
Effluent discharge, port works	80
Barrage capital cost total	8280
Off-barrage transmission and grid reinforcement	
With all transmission line overhead	850
Extra cost for 10% of transmission lines underground	380
Annual costs	
Barrage operation and maintenance	40
Off-barrage costs	30
Annual costs total	70

Sources: Department of Energy, *The Severn Barrage Project: General Report*, Energy Paper 57, HMSO, 1989; Elliot, D., Tidal power, in *Renewable Energy*, 2nd edn., G. Boyle (ed.), Oxford University Press, Inc., New York, Chap. 6, 2004.

[a] The costs presented in Table 6.1, and in Box 6.6 earlier, are in 1988 money terms. In 1991 these estimates were updated, in line with inflation, with the total being put at over £10 billion. More recently, an STPG report, *The Severn Barrage—Definition Study for a New Appraisal of the Project*, commissioned by DTI to feed into the 2002 Energy Review (ETSU, 2002), concluded that the Barrage would cost between £10–15 bn at current prices.

[b] Excluding public road across the barrage (estimated to cost from £135 million to £207 million depending on links provided into road network).

factor of 85% or more. A study conducted for the House of Commons Select Committee on Energy [11] indicated that the capital cost for the Severn Tidal Plant was about £10,000 million, the running cost per annum £86 million, and the electricity price about 5 pence/kWh, assuming a 20 year payback. These figures show that capital intensive projects such as tidal power plants are a good long-term investment, but may require support from government agencies because of their large initial capital requirements and relatively slow payback. However, like many other renewable energy systems, once the capital and interest costs have been paid off, tidal barrages can generate virtually free power without CO_2 generation for the rest of their life, which has been estimated to be at 100 years. Operating costs are relatively small, and since tidal barrages, like other renewable energy systems, require no fuel input, they are excellent prospects for future sustainable energy worldwide. References [12–17] provide additional perspectives for tidal power generation.

11.3 Ocean Wave Energy

Ocean waves are generated by wind blowing over water. Since wind is driven by solar energy, the energy in ocean waves is essentially a stored, moderately high-density form of solar energy. Typical average solar power levels of the order of 100 W m can be transformed into waves with power levels of over 100 kW/m of crest length. The size of the waves generated by wind depends upon the wind speed and its duration. The distance over which wind energy is transferred into the ocean to form waves is called the Fetch. Greater amplitude waves contain more energy per meter of crest length than small waves. In practice, however, the power that can be extracted from the waves rather than their energy content is the useful parameter for engineers.

11.3.1 Deepwater Wave Power

A typical ocean wave can be approximated by a sinusoidal shape of height, H, as shown in Figure 11.22. Defining the average height above the water surface, a (m), the distance between peaks as the wavelength, λ (m), the speed at which the peaks move across the surface of the ocean as the wave velocity, v (m/s), and the time for successive peaks to pass a fixed point as period, P (s), and the wave frequency as $\nu = 1/P$, we can determine important wave characteristics. For example, using this type of simplified and idealized wave motion, Twidell and Weir [18] have proposed a procedure to estimate the power in deep ocean waves, based on theoretical foundations laid in 1977 by Coulson and Jeffrey [19]. The distinctive properties of deepwater waves are summarized below:

1. Surface waves consist of unbroken sine waves of irregular wavelength, phase, and direction.
2. The motion of any particle of water is circular. Whereas the surface forms of a wave show a definite progression, water particles have no progression.
3. Surface water remains on the surface.

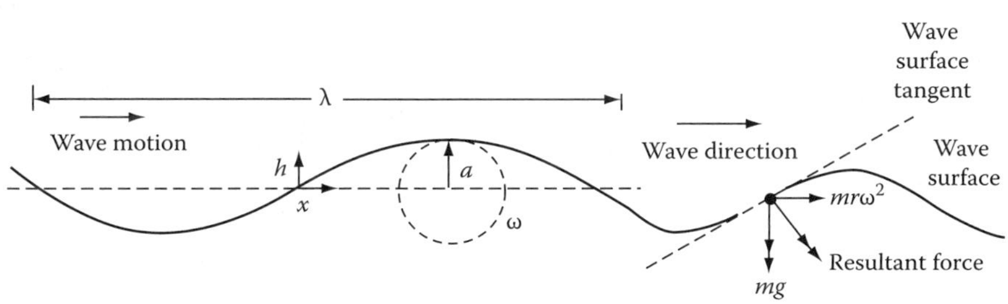

FIGURE 11.22
Characteristics of a sinusoidal wave.

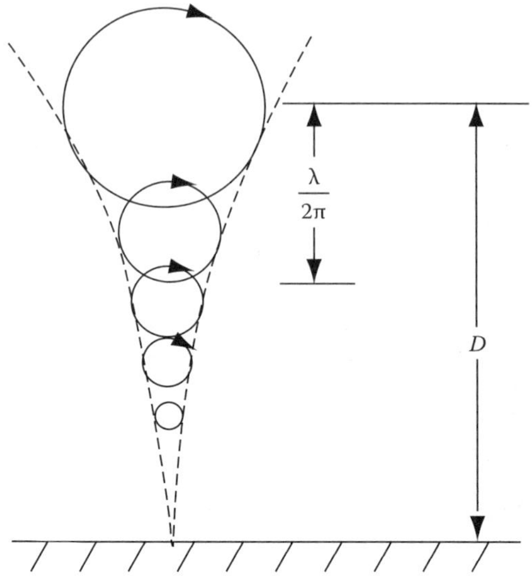

FIGURE 11.23
Motion of a particle in deepwater waves. The motion is circular with angular speed ω (rad/s) and radius *a* (m).

4. The amplitudes of the water particle motions decrease exponentially with depth. At a depth of $\lambda/2\pi$ below the mean surface, the amplitude is reduced to $1/e$ of the surface amplitude.
5. The amplitude, a, of the surface wave is essentially independent of the wavelength, λ, velocity, c, or period T, of the wave and depends on the history of the wind above the surface.

Since most devices proposed for energy extraction are designed for deepwater waves, it may be assumed that the two dominant forces are the result of gravity and circular motion, as shown in Figure 11.23. An average wave for power generation has a wavelength ~100 m, an amplitude ~3 m, and behaves as a deepwater wave of depth greater than ~30 m with a circular motion of water particles as illustrated in Figure 11.23. The circles approximate the motion of a particle in a deepwater wave. This circular motion has an angular velocity, ω (rad/s), and an amplitude, a (m), that decreases exponentially with depth, D, and becomes negligible when $D > (\lambda/2)$. It should be noted that there is no net motion of water in deepwater waves and an object suspended in water will move in circles, as shown in Figure 11.23.

To calculate the power that can be extracted from deepwater waves, we shall simplify the rigorous mathematical treatment. It shows that for a surface wave of amplitude a and wave number k, the radius r of particle motion below the surface is

$$r = ae^{kz} \tag{11.4}$$

where
$k = 2\pi/\lambda$
z is the distance below the surface, as shown in Figure 11.24

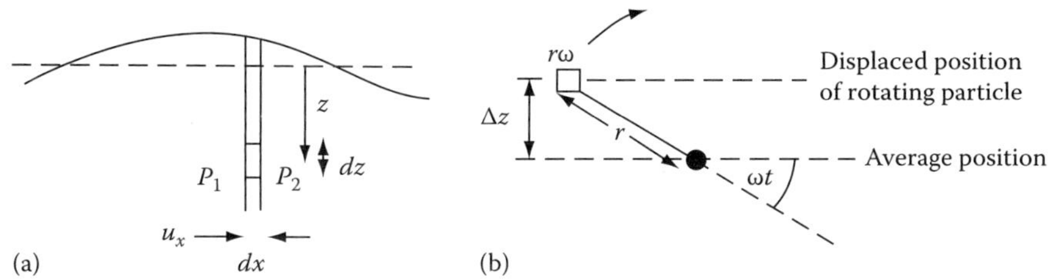

FIGURE 11.24
Local pressure fluctuation in a deepwater wave: (a) pressure in the wave and (b) local displacement of water particle.

The vertical displacement from the average position as a function of time t is

$$\Delta z = r \sin \omega t = ae^{kz} \sin \omega t \tag{11.5}$$

and the horizontal velocity component, u_x, is

$$u_x = r\omega \sin \omega t = \omega ae^{kz} \sin \omega t \tag{11.6}$$

Then, P^1, the power in the wave at x per unit width of wave-front is given by

$$P^1 = \int_{z=-\infty}^{z=0} (p_1 - p_2) u_x dz \tag{11.7}$$

where $(p_1 - p_2)$ is the pressure difference experienced by the element of height dz in the horizontal direction. This pressure difference is equal to the change in potential energy of particles rotating in the circular paths, or

$$p_1 - p_2 = \rho g \Delta z = \rho gae^{kz} \sin \omega t \tag{11.8}$$

Hence, substituting the above relation in Equation 11.7, yields

$$P^1 = \int_{z=-\infty}^{z=0} (\rho gae^{kz} \sin \omega t)(\omega ae^{kz} \sin \omega t) dz = \rho ga^2\omega \int_{z=-\infty}^{z=0} e^{2kz} \sin^2 \omega t dz \tag{11.9}$$

Since the time average over many periods of $\sin^2 \omega t$ can be shown to equal ½

$$P^1 = \left(\frac{\rho ga^2\omega}{2}\right) \int_{z=-\infty}^{z=0} e^{2kz} dz = \left(\frac{\rho ga^2\omega}{2}\right)\left(\frac{1}{2k}\right) \tag{11.10}$$

Introduce the definition of the phase velocity of a wave, *c*, given by

$$c = \left(\frac{\omega}{k}\right) = \left(\frac{gT}{2\pi}\right) = \left(\frac{g\lambda}{2\pi}\right) \tag{11.11}$$

where *T* is the period of a sinusoidal wave given by

$$T = \frac{2\pi}{\omega} = 2\pi\left(\frac{\lambda}{2\pi g}\right)^{1/2} \tag{11.12}$$

With these equations, the power in watts is

$$P' = \left(\frac{\rho g a^2 \omega}{2}\right)\left(\frac{1}{2}\right)\left(\frac{gT}{2\pi}\right) = \frac{\rho g a^2 T}{8\pi}, \tag{11.13}$$

or, as a function of the wavelength, it has the form

$$P' = \left(\frac{\rho g a^2}{8\pi}\right)\left(\frac{2\pi\lambda}{g}\right)^{1/2}$$

We can see that waves of great height and long wavelength are important parameters for effective wave power production.

11.3.2 Surface Wave Power

The waves in the ocean are irregular and are composed of many individual components. The combination of these waves is observed on the surface of the ocean, and the total power per meter of wave-front is the sum of the powers of all the components. It is impossible to measure the height and periods of each of these components, and an averaging process is therefore necessary to estimate the total power.

The variation in surface levels over a period of time can be obtained by means of wave-rider buoys. Although the average height of these waves is zero, a meaningful figure of the wave height is obtained by means of the significant wave height, H_s, which is defined as 4× the root-mean-square of the water elevation. In other words, the instantaneous elevations are squared, which makes the values positive, and then the mean over a number of waves is calculated. Finally, the significant wave height is 4× the square root of the mean. The significant wave height is approximately equal to the average of the highest one-third of the waves.

The second important parameter is the zero-up-crossing period, T_e. It is defined as the average time counted over 10 or more crossings between upward movements of the surface through the mean level. For a typical irregular sea, it can then be shown that the average total power per unit of wave crest, P, is given by

$$P = \frac{\left(H_s^2 T_e\right)}{2}$$

where P is in kW/m.

Figure 11.25a illustrates a typical wave record of the significant wave height and zero-crossing period as a function of time in seconds, whereas Figure 11.25b shows two wave records for the same location but on different days.

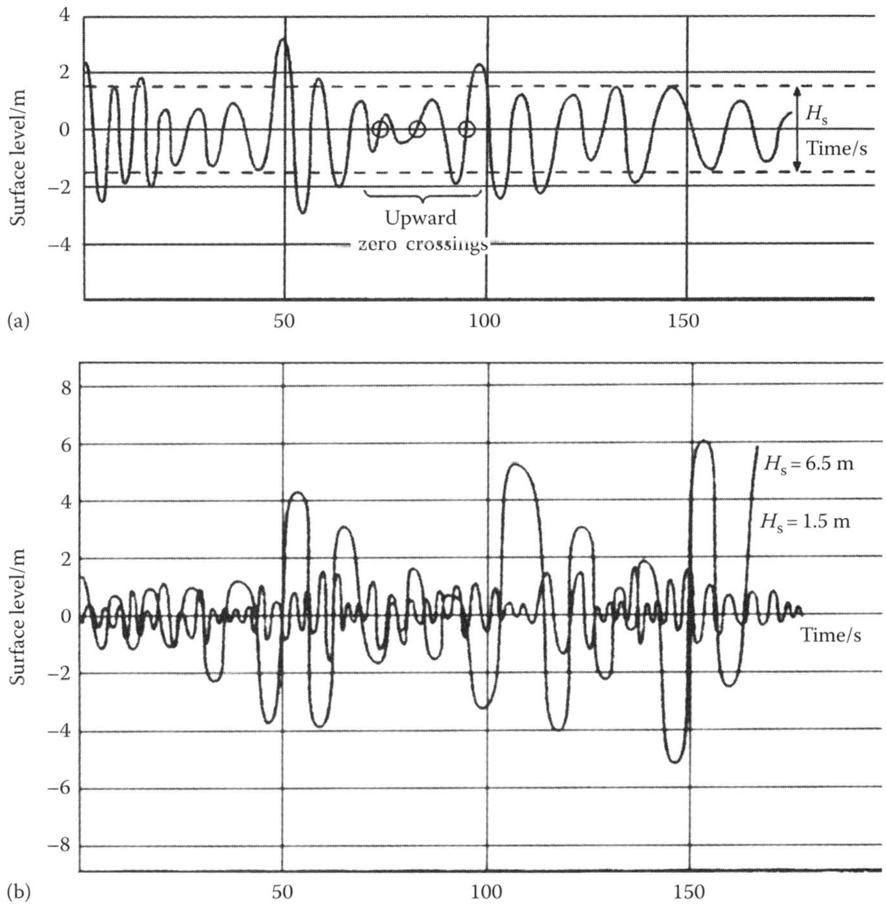

FIGURE 11.25
(a) A typical wave record. In this example, the significant wave height, H_s = 3 m. The successive upward movements of the surface are indicated with small circles. In this case, there are 15 crossings in 150 s, so T_e = 10 s. From this, P(kW/m) = ($3^2 \times 10$)/2 (kW/m) = 45 kW/m. (b) Two wave records are shown here for the same location but represent recordings taken on different days. (From Duckers, L., Wave energy, Chap. 8 in Boyle, G. (ed.), *Renewable Energy*, 2nd edn., Oxford University Press, Oxford, U.K., 2004.)

To obtain the wave power at any location, the ocean states at that location are measured over a whole year. Characterizing the states by the values of H_s and T_e, it is possible to obtain a statistical picture of the distribution of local wave conditions. This picture can be represented as a scatter diagram and gives the relative occurrences in parts per thousands of the contributions of H_s and T_e. An example of such a scatter diagram is shown in Figure 11.26. For this particular location at a zero-crossing period of 10 and a significant wave height of 6 m, the power density is 200 kW/m.

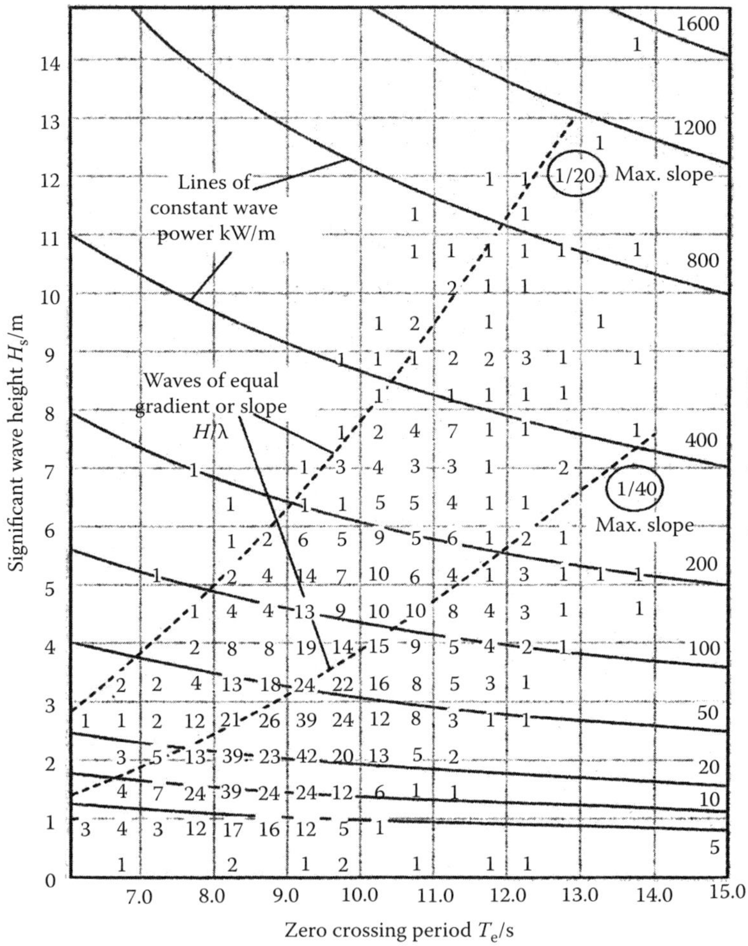

FIGURE 11.26
Scatter diagram of significant wave height (H_s) against zero crossing period (T_e) for 58°N 19°W in the north Atlantic. The numbers on the graph denote the average number of occurrences of each H_s and T_e in each 1000 measurements made over one year. The most frequent occurrences are H_s ~ 2 m, T_e ~ 9 s. (From Boyle, G. [ed.], *Renewable Energy: Power for a Sustainable Future*, 2nd edition, Oxford University Press, 2004. With permission.)

11.3.3 Wave Power Devices

Wave power generation has not yet reached the same state of development as wind power, but there have been many proposals for capturing energy from waves, and quite a few prototypes have been built, tested, and put into operation. So far, however, there have been no large-scale wave energy systems built and operated over any length of time. Consequently, it is not possible to decide which of the various devices have the greatest promise for eventual deployment. This section will therefore restrict itself to describing a few typical devices that have been built and operated.

11.3.3.1 Wave Capture Systems

Wave capture systems are relatively simple. When waves break over a sea wall or dam near the seashore, water is impounded at a height above the average sea level and can be captured in a reservoir. The water from the reservoir can then be returned to the ocean through a conventional low head turbine attached to a generator, similar to a tidal system described in Section 11.2. The flow pattern of a wave capture system, however, is less regular than that of a tidal system.

A 350 kW wave capture system was built and operated successfully in Norway in 1985. A schematic diagram of the system, which was called Tapchan, is shown in Figure 11.27. In this system, waves were funneled through a tapered channel so that larger waves were able to flow over the walls immediately, while the smaller waves increased in height as they moved up in the channel.

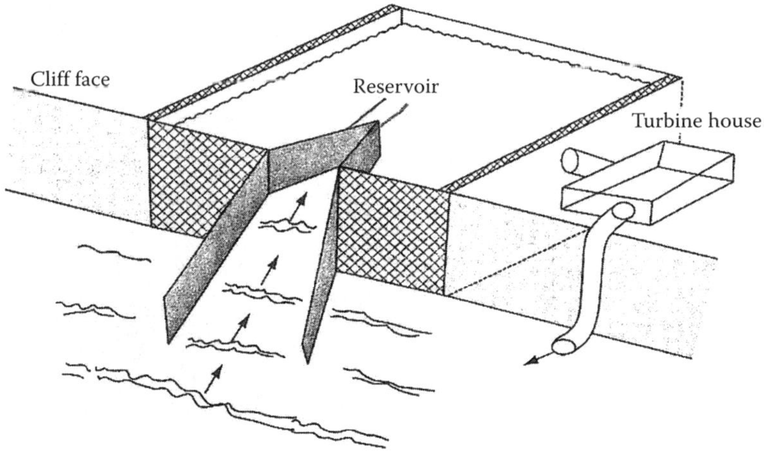

FIGURE 11.27
Schematic diagram of the Tapchan wave capture system in Norway.

A site for a successful wave capture system should have the following characteristics:

1. Deep water close to shore
2. A tidal range of less than 1 m
3. A convenient site for constructing the reservoir, preferably a natural gully in a rock formation near shore
4. Continuous waves with large amounts of energy

Although these types of wave capture systems are conceptually very simple, they suffer in general from a low head, which requires a large amount of water flow for a respectable energy output. On the other hand, the Tapchan concept has very few moving parts, its maintenance costs are low, and its reliability is high. Moreover, the storage reservoir helps to smooth the electrical output.

11.3.3.2 The Oscillating Water Column

Oscillating water column (OWC) systems are simple and robust and have been deployed in a number of countries. They probably represent the most common form of wave energy conversion. A successful OWC system, called the Limpet, was constructed in the year 2000 at the Scottish island Islay. A schematic diagram of this onshore wave power system demonstrating its operation is shown in Figure 11.28. Air is compressed and decompressed by the oscillating water column, and this causes air to be forced through a Wells turbine located above. The Wells turbine is capable of rotating in the same direction, regardless of the direction of the air flow. Thus, it generates power irrespective of the direction

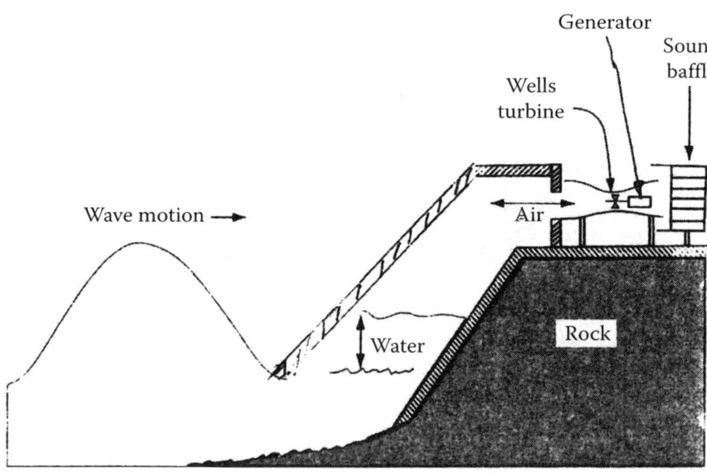

FIGURE 11.28
Schematic diagram of an onshore wave power system using an oscillating water column similar to Limpet.

FIGURE 11.29
Photograph of Limpet. (From Boyle, G. [ed.], *Renewable Energy: Power for a Sustainable Future*, 2nd edition, Oxford University Press, 2004. With permission.)

of movement of the water column. The water column rises with each incoming wave and compresses the air above, which is thus forced to flow through the turbine. As the water column goes down, air is drawn in the opposite direction through the Wells turbine, which generates electricity via a generator. The waves are captured by a reinforced concrete wall set into an excavated rock face on the shore. The existing Limpet system, shown in Figure 11.29, has a capacity of 500 kW, and has been successful in generating electricity for the grid on the island for almost 10 years. The electricity is exported to the utility grid within the Scottish Administration's Renewable Obligation Program.

Power extraction by OWC can be enhanced by reducing the cross-sectional area of the channel approaching the turbine. This couples the slow motion of waves to the faster rotation of the turbine without mechanical gears. Another advantage is that the electrical generator is not in contact with saline water and, therefore, requires relatively little maintenance.

Although a majority of wave power devices are placed onshore, the Japanese have demonstrated a floating device called the Whale, placed off Gokashoa in 1998. The Whale has been tested as a full-scale prototype consisting of a massive structure 500 m in length, 1000 ton in weight, and rated at 110 kW. The wave conditions at Gokashoa Bay were found to be approximately 1 m in wave height, 5–8 s in frequency, and having a mean power density of 4 kW/m. The mean power output was 6–7 kW in an energy period of 6–7 s, yielding an overall efficiency of approximately 15% [21]. Other systems similar in concept to the Whale have been proposed and are described in [18].

11.3.4 Wave Profile Devices

These types of devices float on or near the surface of the ocean and respond to the shape of the waves, rather than the vertical up and down motion. The original idea was conceived by Professor Steven Salter at Edinburgh University in the 1970s, and is often called the "Salter's Duck." The Duck was designed to match the orbital motion of water particles. The concept is theoretically highly efficient, but has not been fully developed.

The most advanced concept of a wave profile device is the Pelamis, or sea snake, shown schematically in Figure 11.30. It consists of a number of cylindrical sections hinged together as an attenuator device. The wave induced motion of the cylinders is resisted at the joints by hydraulic rams that pump

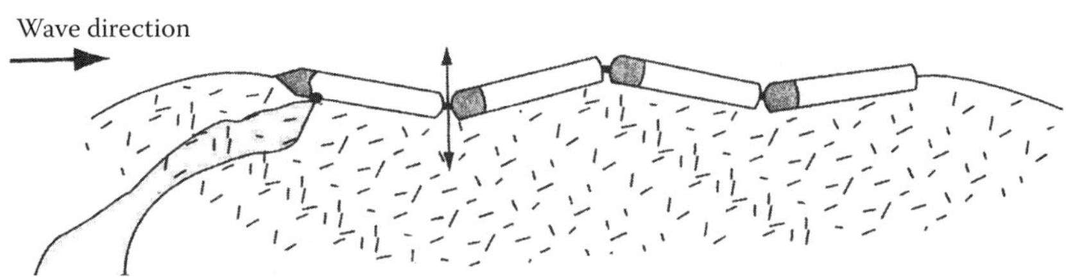

FIGURE 11.30
Schematic side view of a Pelamis wave power device.

FIGURE 11.31
(See color insert.) Pelamis WEC device at sea. (From Khaligh, A. and Onar, O.C., *Energy Harvesting*, Figure 4.72, CRC Press, Boca Raton, FL, 2010.)

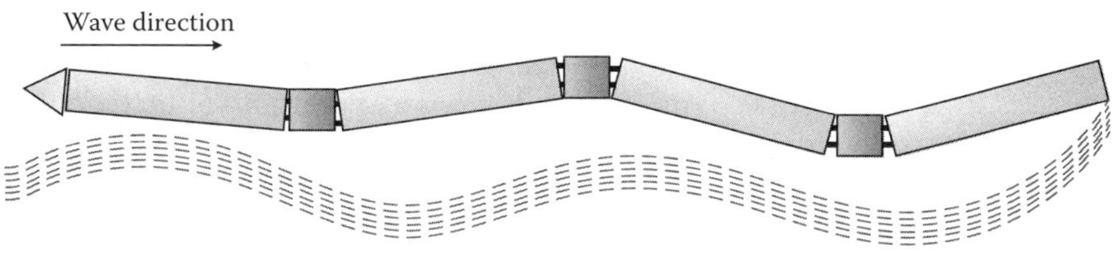

FIGURE 11.32
(See color insert.) General layout of the Pelamis device. (From Khaligh, A. and Onar, O.C., *Energy Harvesting*, Figure 4.73, CRC Press, Boca Raton, FL, 2010.)

high-pressure oil through hydraulic motors via smoothing accumulators. The hydraulic motors then drive electric generators to produce electricity. Several devices can be connected in an array which itself is linked to shore through a seabed cable.

Each Pelamis is held in position by a mooring system, which restrains the device, but at the same time allows the machine to swing head-on to oncoming prevailing waves as the snake spans successive wave crests. A 750 kW electric prototype 120 m long and 3.5 m in diameter was installed in 2005 off the shores of the main island of Orkney in Northern Scotland (see Figures 11.31 and 11.32).

Problems

11.1 Explain in your own words the manner in which the Pelamis wave power device operates. Note that the motion at the hinges produces hydraulic power that is fed into electric generators. A motion picture of the operation of the Pelamis can be viewed at http://www.youtube.com/watch?v=JYzocwUfpNg

11.2 Estimate the period and phase velocity of a deep water wave of 100 m wavelength.

11.3 Using the phase velocity calculated in the previous problem, estimate the power in a deep water wave of 1.5 m amplitude and 100 m wavelength.

11.4 Estimate the mean tidal power produced by a plant with area of 10 km^2, a maximum range of 5 m, and a minimum range of 2.5 m (an average range of 3.7 m).

11.5 A 100 MW ocean thermal gradient power plant is to be designed for a location where the ocean surface temperature is 300 K (80°F) and water at a lower depth is available at 278 K (40°F). If the heat exchangers are sized to operate a power plant using R-22 as the working fluid between

294 and 284 K (70°F and 50°F), calculate the flow rates of R-22 and water required, assuming that the condenser heat exchangers have an overall conductance of 1000 W/m$^2\cdot$K (176 Btu/h$\cdot$ft^2 °F) and an effectiveness of 100%. Also, calculate the plant efficiency and the surface area of the condenser. The saturation pressure of R-22 at 294 K is 9.38×10^5 N m^2 (136.1 psia), the enthalpy is 256.8 kJ/kg (110.4 Btu/lb$_m$), and the entropy is 0.900 kJ/kg K (0.215 Btu/lb$_m$ °R). The enthalpy of the saturated liquid at 284 K is 56.45 kJ/kg (24.27 Btu/lb) and the heat of vaporization is 197.0 kJ/kg (85.68 Btu/lb$_m$). The saturation pressure at 284 K is 6.81×10^5 N m^2 (98.73 psia), while the entropy at 284 K of the saturated liquid is 0.217 kJ/kg K (0.0519 Btu/lb$_m$ °R). Assume efficiencies for the pump and turbine are 80% and 90%, respectively. The specific volume of the saturated R-22 liquid at 284 K is 0.000799 m^3/kg (0.0128 ft^3/lb). A schematic diagram of the system is shown as follows:

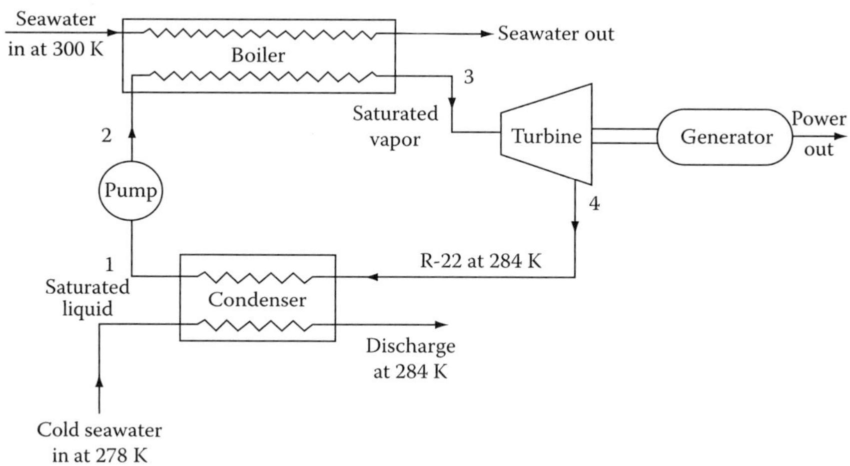

11.6 If the sea power plant described in the previous problem is to deliver power at \$8/10^6 Btu, estimate the maximum permissible cost of the condenser and evaporator heat-exchanger surface in dollars per square foot, assuming a 20-year life, 10% discount rate, and 7% yearly fuel inflation.

The next two problems are open-ended with no solutions:

11.7 It has recently been proposed that ocean thermal energy conversion could be used to produce ammonia, a chemical currently manufactured from natural gas at a total energy consumption of about 2.5% of the U.S. natural gas consumption. Since the manufacture of ammonia could be accomplished anywhere, the plant could be located in a favorable location, for example, off the coast of Brazil, where the difference in temperature between the upper layer of the ocean and the layer at the 600 m depth is approximately 20 K. Using an OTEC power plant of 500 MWe capacity, estimate the number of tons of ammonia that could be produced per year. For a price of $165/ton of ammonia, estimate the time required to regain the capital investment in the OTEC production plant and compare your estimate with the claim that the capital could be repaid in 2.5 years.

11.8 Aluminum sells today at approximately $1850/ton. The production of aluminum is one of the most energy-intensive processes in the metallurgical industry, and it has recently been proposed that an OTEC plant could be used for the production of aluminum. Metallurgical estimates indicate that approximately 18 kWh of energy is required in the refining process for each kilogram of aluminum produced. Assuming a plant load factor of 85%, calculate the size of an OTEC plant required to produce 250,000 ton/year of aluminum, if there is a temperature gradient similar to that in the previous problem. Then calculate the time required to repay an investment cost in the plant and compare your results with the estimate of 3 years obtained in the literature.

References

1. D'Arsonval, A. (1891) Utilisation des forces naturelles, Avenir de l'electricite, *Revue Scientifique*, 17, 370–372.
2. Claude, G. (1930) Power from tropical seas, *Mechanical Engineering*, 52, 1039–1044.
3. Dugger, G., Neff, F., and Snyder, J.E. (1981) Ocean thermal energy conversion, in: Kreider, J.F. and Kreith, F. (eds.), *Solar Energy Handbook*, McGraw-Hill, New York, Chap. 19.
4. Kreith, F. and Barathon, D. (1988) Heat transfer research for ocean thermal energy conversion, *Journal of Heat Transfer*, 110, 5–22, February.
5. Barathon, D., Kreith, F., and Owens, W.L. (1987) An overview of heat and mass transfer in open-cycle OTEC systems, in: Mari, Y. and Yang, W.J. (eds.), *Proceedings of the ASME/JSME Thermal Engineering Joint Conference*, Honolulu, HI, vol. 2, pp. 301–314.
6. Kreith, F. (1981) An overview of SERI solar thermal research facilities, *Heat Transfer Engineering*, 2, 3–4, January–June.
7. Block, D.L. and Valenzuela, J.A. (1985) Thermoeconomic optimization of OC-OTEC electricity and water production plants, SERI/STR-251-2603, U.S. Department of Energy, Washington, DC, May.
8. Twidell, J. and Weir, T. (2006) *Renewable Energy Resources*, 2nd edn., Taylor & Francis, Oxon, U.K., Chap. 13.
9. Elliott, D. (2004) Tidal power, in: Boyle, G. (ed.), *Renewable Energy*, 2nd edn., Oxford University Press, Inc., New York, Chap. 6.
10. Jackson, T. (1992) Renewable energy: Summary paper for the renewable energy series, *Energy Policy*, 20(9), 861–83.

11. Department of Energy. (1989) *The Severn Barrage Project: General Report*, Energy Paper 57, HMSO, cited in Reference [9].
12. Baker, A.C. (1991) *Tidal Power*, Peter Peregrinus, London, U.K.
13. Cavanagh, J.E., Clarke, J.H., and Price, R. (1993) Ocean energy systems, in: Johansson, T.B., Kelly, H., Reddy, A.K.N., Williams, R.H., and Burnham, L. (eds), *Renewable Energy—Sources for Fuels and Electricity*, Earthscan, London, U.K.
14. Charlier, R.C. (2003) Sustainable co-generation from the tides: A review, *Renewable and Sustainable Energy Reviews*, 7, 187–213.
15. Hubbert, M.K. (1971) The energy resources of the Earth, *Scientific American*, 225, 60–87, September.
16. Sorensen, B. (2000) *Renewable Energy*, 2nd edn., Academic Press, London, U.K.
17. Webb, D.J. (1982) Tides and tidal power, *Contemporary Physics*, 23, 419–442.
18. Twidell, J. and Weir, T. (2006) *Renewable Energy Resources*, 2nd edn., Taylor & Francis, Oxon, U.K., Chap. 12.
19. Coulson, C.A. and Jeffrey, A. (1977) *Waves*, Longman, London, U.K.
20. Duckers, L. (2004) Wave energy, in: Boyle, G. (ed.), *Renewable Energy*, 2nd edn., Oxford University Press, Oxford, U.K., Chap. 8.
21. Washyo, Y., Osawa, H., and Ogata, T. (2001) The open sea test of the offshore floating type wave power device "mighty whale": Characteristics of wave energy absorption and power generation, in: *Ocean 2001 Symposium*, cited in [18]).
22. Rabas, T.J., Wittig, J.M., and Finsterwalder, K. (1979) OTEC 100-MWe alternate power systems study, Report by Westinghouse. Electric Corp. to U.S. DOE under contract EG/77/C/05/1473, March.
23. Bharathan, D. and Penney, T. (1983) Flash evaporation from falling turbulent jets, SERI/TP, 252-1853, Solar Energy Research Institute, Golden, CO.
24. Bharathan, D., Olson, D.A., Green, H.G., and Johnson, D.H. (1982) Measured Performance of direct contact jet condensers, SERI-TP, 252-1437, Solar Energy Research Institute, Golden, CO.
25. Khaligh, A. and Onar, O.C. (2010) *Energy Harvesting*, Figure 3.6, CRC Press, Boca Raton, FL.
26. Khaligh, A. and Onar, O.C. (2010) *Energy Harvesting*, Figure 4.72, CRC Press, Boca Raton, FL.
27. Khaligh, A. and Onar, O.C. (2010) *Energy Harvesting*, Figure 4.73, CRC Press, Boca Raton, FL.

12

Transportation

12.1 Introduction

A viable transportation system is a crucial part of a sustainable energy future. Transportation is a complex interdisciplinary topic, which really deserves a book unto itself. However, some of the main issues related to transportation have to do with fuels and energy storage, which are topics covered in this book. This section does not purport to be exhaustive, but examines some of the key issues related to a viable transportation future.

Gasoline and diesel are not only very convenient fuels for ground transportation, but also have a high energy density which permits storage in a relatively small volume—an important asset for automobiles. For example, these liquid fuels have a volumetric specific energy content of about 10,000 kWh/m^3, compared to hydrogen compressed to 100 bar at about 300 kWh/m^3. But known petroleum resources worldwide are being consumed rapidly, and future availability of these resources is bound to decline. At present, more than 97% of the fuel used for ground transportation in the United States is petroleum-based, more than half of it imported, and the percentage of imported petroleum is continuously increasing. Moreover, the increase in cost of gas and oil has become of growing concern to average citizens, and the emission from current transportation systems is a major component of CO_2 pollution that produces global warming.

There is worldwide agreement among oil experts that world oil production will reach a peak sometime between 2020 and 2030. The predictions for the date of peak world oil production according to various estimates [1] are demonstrated in Figure 12.1. The oil production is shown as a function of time for three total amounts of recoverable oil that span the entire range of assumptions by experts. Although new oil fields may be discovered in the future, it is not expected that they will substantially increase the total recoverable amounts. Hence, it is believed that the total amounts lie somewhere between 3000 and 4000 billion barrels of oil (bbl). The obvious conclusion to be drawn from these predictions is that the production

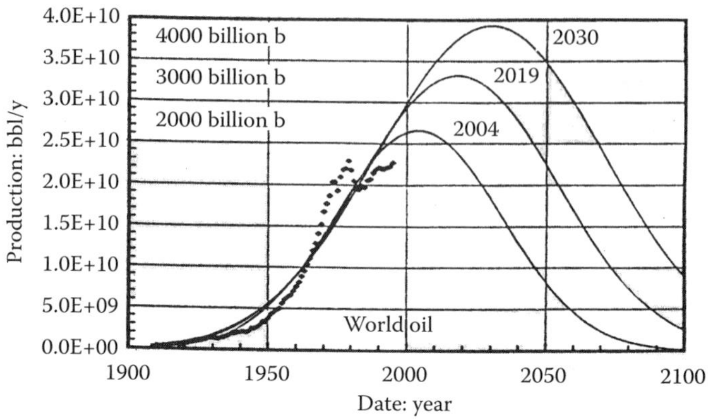

FIGURE 12.1
Oil production vs. time for various recoverable amounts of petroleum. (According to Bartlett, A.A., *Math. Geol.*, 32(1), 2000. With permission.)

peak is imminent and the price of oil will continue to escalate as supplies decline. Thus, planning for a sustainable transportation system that does not depend entirely on petroleum resources is an imperative segment of a sustainable energy future.

12.2 Alternative Fuels

Alternative fuels available to supplement oil as well as their feedstock are shown in Table 12.1 [2]. Inspection of the table shows that the fuels potentially not dependent on a petroleum resource such as oil or natural gas are biodiesel, electricity, ethanol, and hydrogen (via electricity). The potential of producing liquid fuel from biomass has been treated in Chapter 4. There is every reason to believe that biomass will provide a substantial part of the

TABLE 12.1
Feedstocks for Alternative Fuels

Fuel	Feedstock
Propane (LPG)	Natural gas (NG), petroleum
Compressed NG	NG
Hydrogen	NG or (water + electricity)
FT diesel	NG, coal, or algae
Methanol (M85)	NG or coal
Ethanol (E85)	Corn, sugarcane, or cellulosic biomass
Electricity	NG, coal, uranium, or renewables

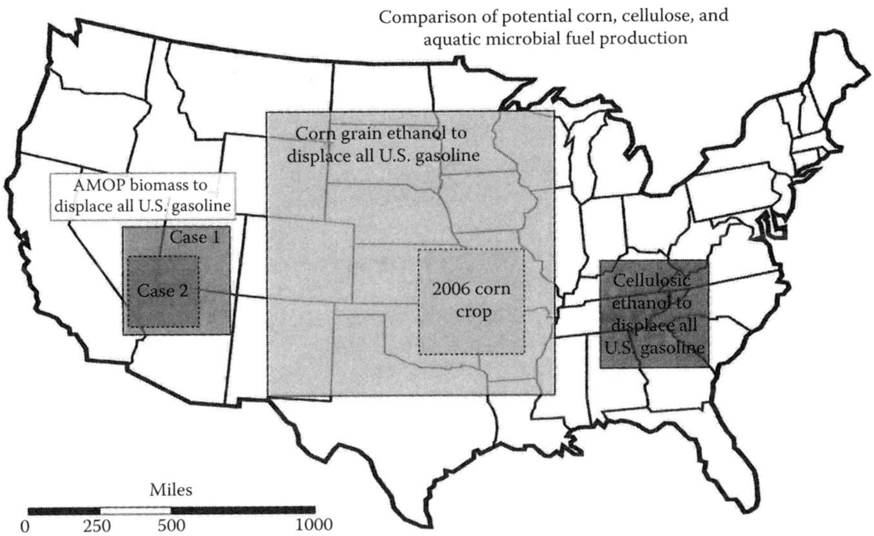

FIGURE 12.2
Relative land area requirement for various liquid fuel biosources. (From Dismukes, C. et al., *Curr. Opin. Biotechnol.*, 19, 235, June 2008. With permission.)

future transportation fuel, especially if ethanol can be produced from cellulosic materials such as switchgrass or bio-waste, or diesel from algae. The reason for this is that ethanol produced by traditional methods from corn kernels has only an energy return on energy investment on the order of 1.25, whereas the EROI for cellulosic ethanol is somewhere between 6 and 8. Diesel from algae may be even better. As shown in Figure 12.2, even cellulosic ethanol would not be able to replace oil, because growing it would require too large a percentage of all the arable lands in the United States, and the production of large amounts of ethanol would compete with production of food, which is of increasing importance for a socially sustainable energy system.

It has recently been proposed [3] to use biofuels derived from aquatic microbial oxygenic photoautotrophs (AMOPs), commonly known as algae. In this study, it was shown that AMOPs are inherently more efficient solar collectors, use less or no land, can be converted to liquid fuels using simpler technologies than cellulose, and offer secondary uses that fossil fuels do not provide. AMOPs have a 6- to 12-fold energy advantage over terrestrial plants because of their inherently higher solar energy conversion efficiency, which is claimed to be between 3% and 9%. Figure 12.2 compares the area needed for three different biomass sources. The data are for corn grain, switchgrass, mixed prairie grasses, and, AMOPs. Each box superimposed on the map of the United States represents the area needed to produce a sufficient amount of biomass to generate enough liquid fuel to displace all the gasoline used in the

TABLE 12.2

Comparison of Some Sources of Biodiesel

Crop	Oil Yield (L/ha)	Land Area Needed (Mha)[a]	Percent of Existing U.S. Cropping Area[a]
Corn	172	1540	846
Soybean	446	594	326
Canola	1,190	223	122
Oil palm	5,950	45	24
Microalgae[b]	136,900	2	1.1
Microalgae[c]	58,700	4.5	2.5

Source: Abstracted from Christi, Y., *Biotechnol. Adv.*, 25, 294–306, 2007.
[a] For meeting 50% of all transport fuel needs of the United States.
[b] 70% oil (by wt) in biomass.
[c] 30% oil (by wt) in biomass.

United States in the year 2007. The two boxes for AMOPs are for 30% and 70% conversion efficiency. The overall solar energy conversion to biofuels works out to about 0.05% for solar-to-ethanol from corn grain and roughly 0.5% from switchgrass-to-ethanol. Comparatively, this value is about 0.5%–1% for AMOPs-to-ethanol or biodiesel.

An even more favorable assessment for the potential of algae to produce biodiesel is presented in [4]. According to this study, microalgae may be a source of biodiesel that has the potential to displace fossil fuel. According to [4], microalgae grow extremely rapidly and are exceedingly rich in oil. Some microalgae double their biomass every day and the oil content of microalgae can exceed 80% by weight. Table 12.2 shows a comparison of some sources of biodiesel that could meet 50% of all of the transportation needs in the United States. According to estimates in [4], only a small percentage of U.S. cropping areas would be necessary to supply 50% of the entire transport fuel needs in the United States. However, no full-scale commercial algae biodiesel production facility has been built and operated for a sufficient time to make reliable predictions regarding the future of algae for a sustainable transportation system.

12.3 Well-to-Wheel Analysis

Rather than simply looking at the efficiency of an engine or a given fuel, a more comprehensive way to determine overall efficiency when evaluating the potential of any new fuel for ground transportation is to use what

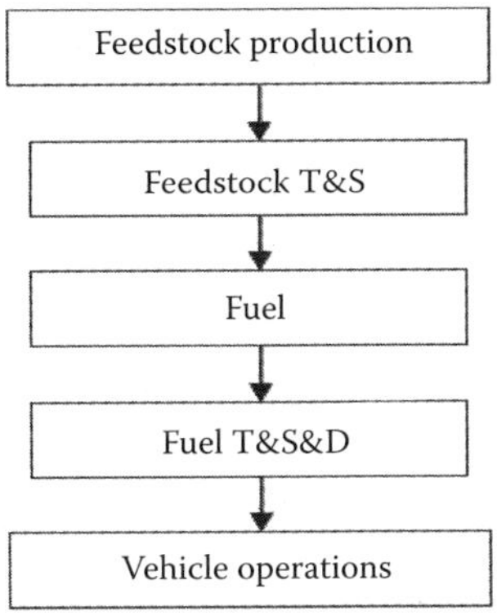

FIGURE 12.3
Steps in a well-to-wheel analysis for ground transportation vehicles.

is called a *well-to-wheel analysis*. The approach to a well-to-wheel analysis is shown schematically in Figure 12.3 [5,15]. The well-to-wheel approach of a fuel cycle includes several sequential steps: feedstock production; feedstock transportation and storage; fuel production; transportation, storage, and distribution (T&S&D) of fuel; and finally vehicle operation. This approach is essential for a fair comparison of different options because each step entails losses. For example, whereas a fuel cell has a much higher efficiency than an internal combustion engine, it depends for its operation on a supply of hydrogen which must be produced by several steps from other sources.

Moreover, there is no infrastructure for transporting and storing hydrogen and this step in the overall well-to-wheel analysis has large losses and contributes to much larger energy requirements compared to a gasoline or electrically driven vehicle.

12.4 Mass Transportation

An opinion circulating widely among state governments and environmentalists is that mass transportation would greatly reduce the total energy consumption for the transportation sector. However, as shown in Table 12.3, based upon data collected by the U.S. Transportation Department in this country, the energy intensity of intercity rail and transit buses, i.e., the energy spent per passenger mile traveled, is virtually the same as the energy intensity of automobiles with current use. This is due to the urban sprawl in major cities that makes it difficult to reach outlying areas by a mass transport network. In other words, unless there are incentives for mass transport or disincentives to use the automobile, thereby achieving higher mass transport load factors—that is, more passengers per mile—on transit buses and light rail, the availability of mass transport systems will not materially change the overall energy use by the transportation sector. Moreover, installing new light rail is very capital-intensive and may not always be worth the energy and/or the money invested in its construction. Although passenger mass transit does not appear to offer large untapped opportunities, using ship or rail instead of air for shipment of freight and commercial goods is a source of enormous fuel reduction potential.

TABLE 12.3

Passenger Travel and Energy Use, 2002

	No. of Vehicles (Thousands)	Vehicle-Miles (Millions)	Passenger-Miles (Millions)	Load Factor (Persons/Vehicle)	Energy Intensities (Btu per Vehicle-Mile)	Energy Intensities (Btu per Passenger-Mile)	Energy Use (Trillion Btu)
Automobiles	135,920.7	1,658,640	2,604,065	1.57	5,623	3,581	9,325.9
Personal trucks	65,268.2	698,324	1,201,117	1.72	6,978	4,057	4,872.7
Motorcycles	5,004.2	9,553	10,508	1.22	2,502	2,274	23.9
Demand response	34.7	803	853	1.1	14,449	13,642	11.6
Vanpool	6.0	77	483	6.3	8,568	1,362	0.7
Buses	[a]	[a]	[a]	[a]	[a]	[a]	191.6
Transit	76.8	2,425	22,029	9.1	37,492	4,127	90.0
Intercity[b]	[a]	[a]	[a]	[a]	[a]	[a]	29.2
School[b]	617.1	[a]	[a]	[a]	[a]	[a]	71.5
Air	[a]	[a]	[a]	[a]	[a]	[a]	2,212.9
Certified route[c]	[a]	5,841	559,374	95.8	354,631	3,703	2071.4
General aviation	211.2	[a]	[a]	[a]	[a]	[a]	141.5
Recreational boats	12,409.7	[a]	[a]	[a]	[a]	[a]	187.2
Rail	18.2	1,345	29,913	22.2	74,944	3,370	100.8
Intercity[d]	0.4	379	5,314	14.0	67,810	4,830	25.7
Transit[e]	12.5	682	15,095	22.1	72,287	3,268	49.3
Commuter	5.3	284	9,504	33.5	90,845	2,714	25.8

Source: Davis, S. and Diegel, S., *Transportation Energy Data Book*, Oak Ridge National Laboratory/U.S. Department of Energy, Oak Ridge, TN, 2004.

[a] Data are not available.

[b] Energy use is estimated.

[c] Includes domestic scheduled services and 1/2 of international scheduled services. These energy intensities may be inflated because all energy use is attributed to passengers; cargo energy use is not taken into account.

[d] Amtrak only.

[e] Light and heavy rail.

12.5 Hybrid Electric Vehicles

Another obvious approach to ameliorating the expected increase in price and lack of availability of petroleum fuel is to increase the mileage of the vehicles. This can be achieved by improving the efficiency of the internal combustion engine, as for instance by using advanced diesel engines which have a higher compression ratio than spark ignition (SI) engines, or by using hybrid electric vehicles (HEVs). Increasing the efficiencies of internal combustion or diesel engines is a highly specialized topic and is not discussed here. However, HEVs offer a near-term option for utilizing improved battery technology, as discussed in Chapter 9.

An HEV is powered by the combination of a battery pack and electric motor—like that of an electric vehicle—and a power generation unit (PGU) which is normally an internal combustion or diesel engine. Unlike electric vehicles however, HEV batteries can be recharged by an onboard PGU, which can be fueled by existing fuel infrastructure.

HEVs can be configured in a parallel or a series design. The parallel design enables the HEV to be powered by both the PGU and the motor, either simultaneously or separately. The series design uses the PGU to generate electricity, which recharges the HEV battery pack and produces power via an electric motor. The key element of either design is that the battery pack, as well as the PGU, can be much smaller than those of a typical electric vehicle or a vehicle powered by an IC engine because the IC engine can be operated at its maximum efficiency nearly all the time.

Currently available HEVs, such as the Toyota Prius, use a parallel configuration, as shown in Figure 12.4. A parallel HEV has two propulsion paths:

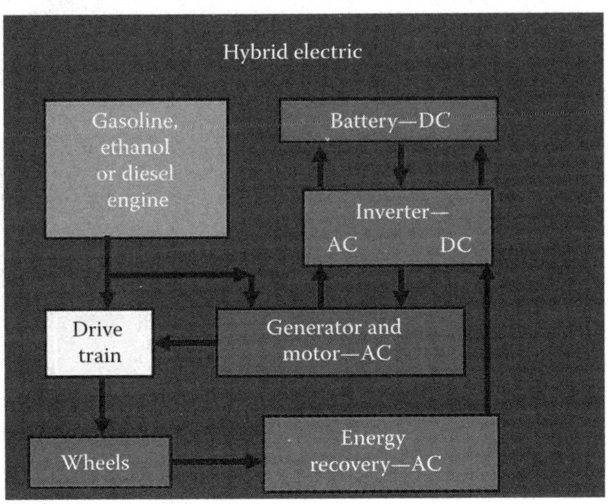

FIGURE 12.4
Schematic of a hybrid electric vehicle.

one from the PGU and one from the motor, while computer chips control the output of each. A parallel configuration HEV has a direct mechanical connection between the PGU and the wheels, as in a conventional vehicle, but also has an electric motor that can drive the wheels. For example, a parallel vehicle could use the electric motor for highway cruising and the power from IC engine for accelerating. The power produced by the PGU also drives the generator, which in turn can charge the battery as needed. The system to transfer electricity from the generator to the battery pack is exactly like that of an EV, with alternating current converted to DC by the inverter. HEV parallel designs also use a regenerative braking feature that converts energy stored in the inertia of the moving vehicle into electric power during deceleration (see Figure 12.5).

(a)

(b)

FIGURE 12.5
(See color insert.) Fuel-efficient cars. (a) 2010 Toyota Prius: Toyota's third-generation Prius is a "parallel" hybrid; the 80 hp (60 kW) electric motor and gas engine work together to produce an average 50 mpg. (b) 2010 Ford Fusion: The Fusion can travel up to 47 mph (75 kph) on electric power alone. The motor is powered by a nickel-metal hydride battery pack, and the car's engine is a 2.5 L Atkinson Cycle I-4 gasoline engine. The Fusion gets 41 mpg in the city and 36 mpg highway. (Courtesy of *Solar Today*, November/December, 2009.)

Some benefits of a parallel configuration vs. a series configuration include the following:

- The vehicle has more power because both the engine and the motor provide power simultaneously.
- Parallel HEVs do not need a separate generator.
- Power is directly coupled to the road, thus operating the vehicle more efficiently.

Energy and cost savings from an HEV depend on many factors, such as the overall car design, cost of fuel, and the cost and efficiency of the batteries. Preliminary estimates indicate that over 5–8 years, reduced size of the motor and savings in gasoline could pay for the additional cost of the batteries. Economically, an HEV would be beneficial when the price per gallon of gasoline or diesel exceeds $3.00/gal, but the lifecycle of the batteries will also have to be considered. If batteries have a lifecycle of 150,000 mi, as claimed by Toyota, they will not need to be replaced during the life of an average vehicle. On the other hand, if the battery life is considerably less and battery replacement is necessary during the expected 10 year life of the car, the operation cost over the life of an HEV would be considerably higher.

12.6 Plug-In Hybrid Electric Vehicles

Given the current state of technology, probably the most promising near-term solution to the ground transportation crisis is the use of plug-in hybrid electric vehicles (PHEVs). PHEVs have the potential of making the leap to the mainstream consumer market because they neither require a new technology nor a new distribution infrastructure. Like hybrids, which are already widely available, PHEVs have a battery and an internal combustion engine for power. But, the difference is that a PHEV has a larger battery capacity and a plug-in charger with which the battery can be recharged whenever the car is parked near a 210 V outlet. A so-called PHEV40 can travel the first 40 mi on grid-supplied electric power when fully charged. When that charge is depleted, the gas or diesel motor kicks in and the vehicle operates like a conventional hybrid. Because most commuter trips are less than 40 mi, it is estimated that a PHEV could reduce gasoline usage by 50% or more for many U.S. drivers. Moreover, using electric energy is cheaper and cleaner than using gasoline in automobile-type ground transportation. Figure 12.6 shows a Prius with plug-in potential added. These plug-in additions are

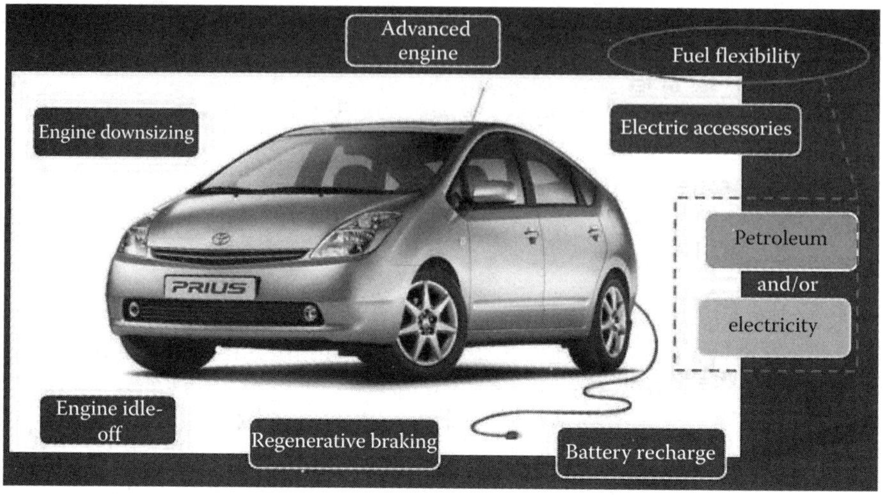

FIGURE 12.6
(See color insert.) Plug-in hybrid electric vehicle. (Courtesy of NREL, Golden, CO.)

available commercially, but because they are not in mass production, their cost is high and Toyota does not honor their warranty when plug-in features are added aftermarket.

An important feature of PHEVs is that their batteries can be charged at night when utilities have excess power available. Utilities have taken notice of the potential energy charging and storage capabilities of PHEVs because off-peak charging would help utilities to use low-cost baseload generation more fully. Furthermore, more advanced vehicle-to-grid concepts would allow utilities to buy back energy from the batteries of vehicle owners during peak demand periods and thus make a fleet of PHEVs into a large, distributed storage-generation network. The arrangement would also enable renewable energy storage by charging PHEVs using solar- or wind-generated excess capacity. As mentioned in the previous section on batteries, lithium-ion and lithium-polymer batteries have the potential to store large charges in a lightweight package, which would make the HEVs even more attractive than the current technology which uses nickel metal hydride batteries.

The efficiency of a PHEV depends on the number of miles the vehicle travels on liquid fuel and electricity, respectively, as well as on the efficiency of the prime movers according to the equation

$$\eta_{\mathrm{PHEV}} = \frac{\text{Energy to wheels}}{\text{Energy from primary source}} = f_1\eta_1\eta_2 + f_2\eta_3\eta_4 \qquad (12.1)$$

where

η_1 is the efficiency from the primary energy source to electricity
η_2 is the efficiency of transmitting energy to the wheels
f_1 is the fraction of energy supplied by electricity
f_2 is the fraction of energy supplied by fuel = $(1 - f_1)$
η_3 is the well-to-wheel efficiency
η_4 is the tank-to-wheel efficiency

PHEVs can be designed with different all-electric ranges. The distance, in miles, that a PHEV can travel on batteries alone is denoted by a number after PHEV. Thus, a PHEV20 can travel 20 mi on fully charged batteries without using the gasoline engine. According to a study by EPRI [7], on average 1/3 of the annual mileage of a PHEV20 is supplied by electricity and 2/3 by gasoline. The percentage depends, of course, on the vehicle design and the capacity of the batteries on the vehicle. A PHEV60 can travel 60 mi on batteries alone and the percentage of electric miles will be greater as will the battery capacity and weight. The tank-to-wheel (more appropriately battery-to-wheel) efficiency for a battery all-electric vehicle according to EPRI [7] is 0.82.

Given the potentials for plug-in hybrid vehicles, the Electric Power Research Institute [7] conducted a large-scale analysis of the cost, the battery requirements, and the economic competitiveness of plug-in vehicles today and within the near-term future. Table 12.4 presents the net present value of life cycle costs over 10 years for a midsized internal combustion engine vehicle such as the Ford Focus [IC], HEVs such as the Prius [HEV], and a

TABLE 12.4

Net Value of Life Cycle Costs over 117,000 mi/10 Years for Conventional Gasoline(IC), HEV and PHEV20 Midsize Vehicles with Gasoline Costs at $1.75/gal

Vehicle Type	IC	HEV	PHEV 20
Battery unit cost ($/kWh)		$385[a]	$316[a]
Incremental vehicle cost ($)		($547)	$224
Battery pack cost ($)	$60	$3,047	$3,893
Fuel costs ($)	$5,401	$3,725	$2,787
Maintenance costs ($)	$5,445	$4,733	$4,044
Battery salvage costs ($)		($54)	($43)
Total lifecycle costs ($)	$10,906	$10,904	$10,905

Source: Extracted from EPRI, Advanced batteries for electric drive vehicles: A technology and cost-effectiveness assessment for battery electric vehicles, power assist hybrid electric vehicles and plug-in hybrid electric vehicles, EPRI Tech. Report 1009299, EPRI, Palo Alto, CA, 2004.

[a] Battery module price at which life cycle parity with CV occurs.

future PHEV20 plug-in electric vehicle. The battery module cost in dollars per kWh is the cost at which the total life cycle costs of all three vehicles would be the same. According to projections for the production of nickel metal hydride battery modules, a production volume of about 10,000 units per year would achieve the necessary cost reduction to make both an HEV and a PHEV20 economically competitive. The EPRI analysis was conducted in 2004 and is, therefore, extremely conservative because it assumed a gasoline cost of $1.75/gal. A reevaluation of the analysis based upon a gasoline cost of $2.50/gal showed that the permitted battery price at which the net present value of conventional internal combustion [IC] vehicles and battery vehicles are equal for battery module costs $1135 for an HEV and $1648 for a PHEV20, respectively.

Table 12.5 shows the electric and plug-in hybrid vehicle battery requirements (module basis for the cost estimates in Table 12.4), and it is apparent that, even with currently available nickel metal hydride (NiMH) batteries, the cost of owning and operating HEVs and PHEVs is competitive with that of an average IC engine vehicle.

A cautionary note in the expectation of future ground transportation systems is the reduction in the rate of petroleum consumption that can be expected as HEVs and PHEVs are introduced into the fleet [8]. In these estimates, a rate of new vehicle sales of 7% of the fleet per year, a retirement rate of 5% per year resulting in a net increase of total vehicles of 2% per year was assumed. This increase is in accordance with previous increase rates between 1966 and 2003. Based upon the existing mileage for IC engine, HEVs and PHEVs, it was estimated that *even if all new cars were HEVs or PHEVs,*

TABLE 12.5

NiMH Battery Cost Assumptions for Table 12.4

	ARB 2000 Report for BEVs		
Assumption	**2003**	**Volume**	**EPRI Assumptions**
Module cost[a]	$300 per kWh	$235 per kWh	Varied[b]
Added cost for pack	$40 per kWh	$20 per kWh	$680 + $13 per kWh[c]
Multiplier for manufacturer and dealer mark-up	1.15	1.15	Varies[c]
Battery life assumptions	6 years	10 years	10 years

[a] Equivalent module costs for an HEV 0 battery is $480 for 2003 and $384 for volume. Equivalent module costs for a PHEV 20 battery is $376 for 2003 and $301 for volume. HEV 0 and PHEV 20 batteries have a higher power to energy ratio and are more costly. These figures are calculated based on Figure 3.3.

[b] Battery module costs were varied in this analysis to determine the effect of battery module cost on life cycle cost.

[c] Manufacturer and dealer mark-up for HEV 0 battery modules estimated at $800, PHEV 20 battery modules $850, pack hardware mark-up assumed to be 1.5. Method documented in 2001 EPRI HEV report.

after 10 years the annual gasoline savings as a percentage of the gasoline usage by an all-gasoline fleet in the same year would only be about 30% for the HEVs and 38% for the PHEVs. These relatively small reductions in the gasoline use are due to the fact that despite introduction of more efficient vehicles, it takes time to replace the existing fleet of cars and the positive effects will not be realized for some years.

12.7 Advanced Ground Transportation with Biomass Fuel

In the previous section, we have analyzed the potential of using batteries combined with traditional engines to reduce the petroleum consumption in the transportation system. However, the scenario used for this analysis can also be extended to determine the combination of PHEVs with biofuels, particularly ethanol made from corn or cellulosic biomass. No similar analysis for using diesel from algae is available at this time.

A scenario for a sustainable transportation system based on fuel from biomass has been presented in [9]. In this analysis, the following four vehicle types combined with various fuel options has been calculated. The preferred mixture in an economy based largely on ethanol (E85) would be 85% ethanol and 15% gasoline that could be used in Flex Fuel automobiles. Currently, the United States is using E10, a mixture of 90% gasoline and 10% ethanol, with ethanol produced from corn. The fuel types used in this analysis are gasoline only, E10 with ethanol made from either corn or cellulosic materials, and E85 with ethanol from either corn or cellulosic materials. The four vehicle combinations are a convention SI engine, an HEV similar to the Toyota Prius, a PHEV20, and a PHEV30. The analysis was based upon an average 2009 performance on the U.S. light vehicle fleet at approximately 20 mpg, for an HEV at 45 mpg, and for a PHEV20 at 65 mpg, according to [7]. For ethanol/gasoline-blended fuels, it was assumed that the gasoline and ethanol are utilized with the same efficiency. That is, the mileage per unit of fuel energy is the same for gasoline and ethanol.

Based on the above assumption, the following parameters were calculated:

1. The miles per gallon of fuel, including the gasoline used to make ethanols (mpg).
2. The petroleum required to drive a particular distance for a case vehicle as a percentage of the petroleum required to drive the same distance by a gasoline-fueled SI vehicle.
3. The carbon dioxide emission rate for case vehicles as a percentage of that for SI gasoline-only.

Based upon the above assumption, one can calculate the miles per gallon of fuel by the following equation:

Miles per gallon of fuel,

$$MF_{ij} = (\text{mi/gal gas}) \times \Big[(\text{gal gas/gal fuel}) + (1 - \text{gal gas/gal fuel}) \times \frac{(\text{energy, LHV/gal ethanol})}{(\text{LHV/gal gasoline})} \Big]$$

$$= MGO_i \times \left[FG_j + (1 - FG_j) \times (\text{LHV ratio}) \right] \qquad (12.2)$$

where

MF_{ij} is the miles per gallon of fuel for vehicles type i and fuel type j
MGO_i is miles per gallon gasoline-only for vehicle i (see Table 12.4)
FG_j is volume fraction of gasoline in fuel type
$j(1 - FG_j)$ is the volume fraction of ethanol in fuel type j
the ratio of lower heating values (LHVs) is LHV ratio = (LHV/gal ethanol)/(LHV/gal gasoline) = 0.6625

The index i indicates the vehicle type as shown in Table 12.4, and the index j denotes the volume fraction of gasoline in an ethanol–gasoline blend as follows: for gasoline only, $FG_1 = 1$; for E10 (i.e., 10 vol% ethanol and 90 vol% gasoline), $FG_2 = 0.90$; and for E85 (85 vol% ethanol, 15 vol% gasoline), $FG_3 = 0.15$. Other ethanol concentrations could be used.

The miles per gallon of gasoline in the fuel, including the petroleum-based fuels used in making of the ethanol in the fuel (by counting the energy of all petroleum-based fuels as gasoline) is given by

Miles per gallon of gasoline in fuel =

$$MG_{ijk} = \frac{MF_{ij}}{\left[FG_j + (1 - FG_j) \times (0.6625) \times (\text{MJ gasoline used in making 1 gal ethanol})/\text{MJ/gal ethanol} \right]}$$

$$= \frac{MF_{ij}}{FG_j + (1 - FG_j) \times (0.6625) \times Rk} \qquad (12.3)$$

where Rk denotes the gallons of gasoline used to make one gallon of ethanol. For corn-based ethanol, $k = 1$, $R1 = 0.06$, while for cellulosic-based ethanol, $k = 2$, $R2 = 0.08$, according to Refs. [10].

12.7.1 Petroleum Requirement

The petroleum required to produce an ethanol–gasoline blend, including the petroleum used to make the ethanol, is expressed as a percentage of the petroleum required for the same miles traveled by the same vehicle type using gasoline-only. For a gasoline-only-fueled vehicle of any type, this percentage is 100%.

A general equation for the percentage petroleum requirement is

$$\text{Petroleum requirement, \%} = 100 \times \frac{(MGO_i)}{(MG_{ijk})} \tag{12.4}$$

12.7.2 Carbon Dioxide Emissions

The CO_2 emissions, including those from making the ethanol and generating the electricity used from the grid by the vehicle, are expressed as a percentage of the emissions produced by the same type of vehicle fueled by gasoline-only traveling the same number of miles.

$$\begin{aligned} &CO_2 \text{ production, \% of that for gasoline-only vehicle} \\ &= 100 \times \left[FE_m \times (\text{kWh/mi}) \times (\text{g-carbon/kWh})\right] + (1/\text{MGO}_i) \times (1 - FE_m) \\ &\quad \times \left[FG_j \times (\text{g-carbon/MJ gasoline}) \times (\text{MJ/gal gasoline})\right] + (1 - FG_j) \\ &\quad \times \frac{\left[(\text{MJ/gal ethanol}) \times (\text{g-carbon/MJ ethanol})\right]}{\left[(\text{g-carbon/MJ gasoline used}) \times (\text{MJ/gal gasoline})/(\text{MGO}_1)\right]} \end{aligned} \tag{12.5}$$

where FE_m is the fraction of the miles driven by electricity from the grid for a plug-in hybrid vehicle. For $m = 0$, any non-PHEV, $FE_0 = 0$; for a PHEV20, m = 1, and $FE_1 = 0.327$ [11]; and for a PHEV30, $m = 2$, and $FE_2 = 0.50$ estimated by analogy with PHEV20. According to Ref. [11], the kWh/mi from the grid = 0.2853, the g-carbon emitted/kWh = 157 (146 average for all electricity generation [12] divided by 0.93, the average transmission efficiency), the g-carbon emitted/MJ gasoline = 94 [11], the MJ/gal gasoline = 121, the MJ/gal ethanol = 80.2, and, CE_k the g-carbon emitted/MJ ethanol. For $k = 1$ (corn), $CE_1 = 87$, and for $k = 2$, cellulosics, $CE_2 = 11$.

So

$$\begin{aligned} &CO_2 \text{ production as \% of that for a gasoline-only vehicle} \\ &= \frac{100 \times \{FE_m \times (0.2853) \times (157) + (1/MF_{ij}) \times (1 - FE_m) \times [FG_j \times (94) \times 121 + (1 - FG_j) \times CE_k \times (80.2)]\}}{(94) \times (121)/21} \end{aligned} \tag{12.6}$$

PROBLEM

Calculate the CO_2 production ratio relative to an IC vehicle for a PHEV30, E85 vehicle using ethanol from cellulosic material.

Solution

The relevant variables are

$$FE_m = 0.5$$

$$MF_{ij} = 85 \times (0.15 + 0.85 \times 0.6625) = 60.61$$

$$Ce_k = 11$$

Using the above parameter, the CO_2 production, as $a\%$ for that of gasoline-only vehicle is

$$\begin{aligned} &= 100 \times \{(0.5 \times 0.2853 \times 157) + (1/60.61) \\ &\quad \times 0.5[0.15 \times 94 \times 121 + 0.85 \times 11 \times 80.2]\}/(94 \times 121/21) \\ &= 4.13 + 2.60 + 1.14 = 7.87\% \end{aligned}$$

Using the results of the above equation, the key conclusions of the analysis are shown in Figures 12.7 and 12.8. Figure 12.7 shows the gasoline requirement of any of the 20 cases considered as a percentage of that for

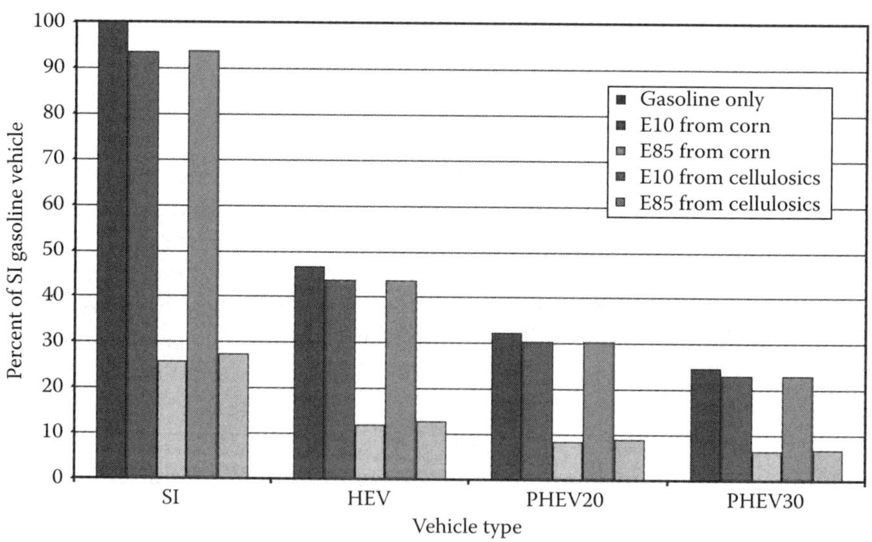

FIGURE 12.7
(See color insert.) Petroleum requirement as a percentage of that for SI gasoline vehicle (calculated by West and Kreith, See Ref. [8]).

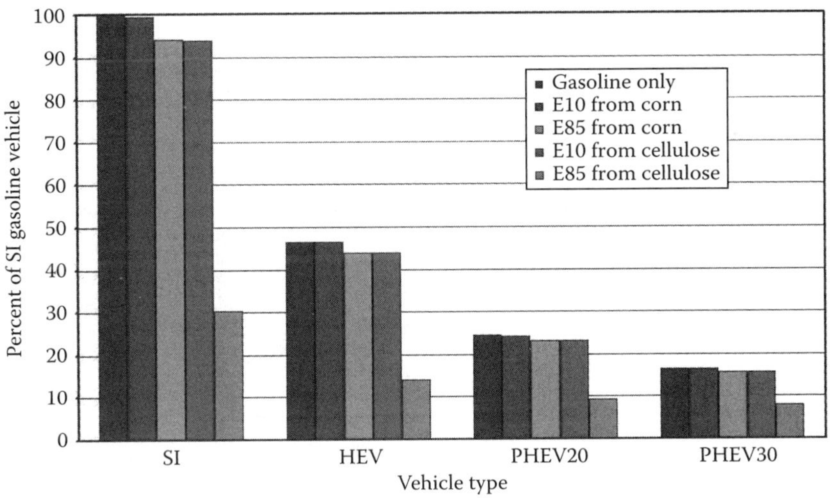

FIGURE 12.8
Carbon dioxide emissions as a percentage of emissions for SI gasoline vehicle.

> an SI gasoline fleet. It can be seen that the gasoline consumption for a PHEV20 using E10 is about 30% of that for the gasoline fleet, whereas it decreases to about 8.8% for a fuel consisting of 85% ethanol from cellulosic with 15% from gasoline (E85 cellulosic). The potential reduction in CO_2 emission shown in Figure 12.8 is equally impressive. The CO_2 emission of a PHEV20 with E85 cellulosic is only about 10% of that for the SI fleet. Since, at present, emission from automobile engines is responsible for almost one-third of the total CO_2 emissions of the country, it is apparent that with existing technology, substantial reduction in CO_2 generation could be achieved by using PHEVs, particularly if the fuel were to be manufactured from cellulosic material.

The final question to be asked in the utilization of PHEVs is whether or not the existing electricity system of the United States could handle the charging of the batteries in a PHEV-based ground transportation system. This question has recently been answered by an analysis in [13]. This analysis clearly showed that with a normal commuting scenario, which was based upon statistical information from a major city, if charging occurred during off-peak hours, no additional generational capacity or transmission requirements would be needed to charge a significant portion of the automotive fleet with PHEVs in the system, as illustrated in Figure 12.9. The analysis also showed that the off-peak charging scenario would also add to the profit of the electric utilities.

An unexpected result of the analysis was that, although the amount of CO_2 emitted by the electric utility will increase from PHEV charging, if this is compared to the corresponding reduction in tailpipe emission to assess the overall environmental impact, the generation/PHEV transportation system

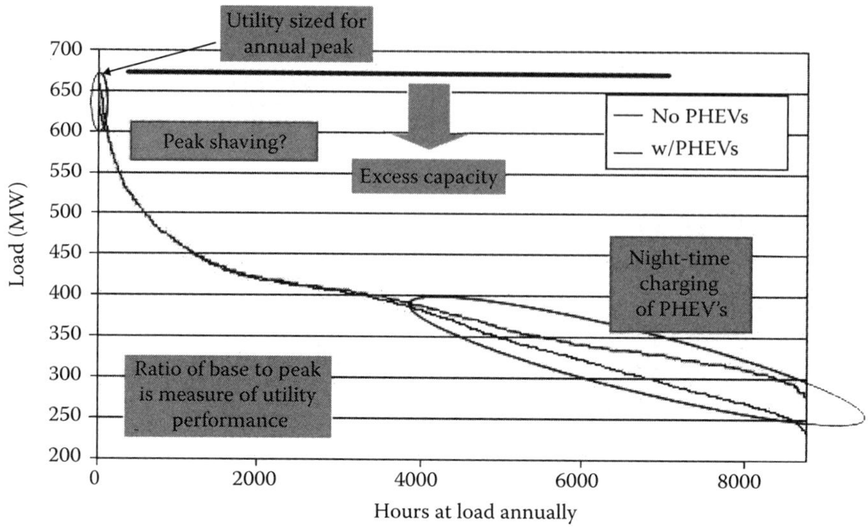

FIGURE 12.9
PHEV impact on utilities' load profile. (Courtesy of NREL, Golden, CO.)

would substantially decrease CO_2 emission even if the current mix of coal, nuclear, and renewable generation facilities were unchanged. This result can be explained, however, because the average efficiency of the electric power system is on the order of 43%, whereas the average efficiency of internal combustion engines is only on the order of 22%. Consequently, the net emission of CO_2 would be reduced by a hybrid system consisting of PHEVs and electric charging during off-peak hours. Moreover, the utility generation profile would be evened out and this would also contribute to reducing CO_2 emissions, as well as the cost of producing electricity.

In summary, in a PHEV transportation system, if charging of batteries is limited to off-peak hours, this hybrid arrangement can

- Reduce the amount of petroleum consumed by the transportation sector
- Reduce the cost of driving
- Reduce CO_2 emission
- Improve the load profile of electric utilities

Although no quantitative analyses are as yet available, it is believed that the availability of electric storage in the batteries of a fleet of PHEVs could also be used to reduce the peak demand on electric utilities by utilizing the vehicle's batteries as a distributed storage system. Details of such an arrangement would have to be worked out by differential charges, incentives, and taxation arranged between utilities and owners of PHEVs.

12.8 Future All-Electric System

The next step in the development of a viable transportation system could be the all-electric car. All-electric vehicles were mandated in California as part of an effort to reduce air pollution about 25 years ago by the California Air Resources Board. Initially, automakers embraced the idea, but it is likely that the acceptance by Detroit was the result of the mandate which required that at least 2% of all the cars sold in California by any one automaker had to be zero emission vehicles. The only zero emission vehicle available at the time was the electric car, and the mandate therefore required selling a certain number of all-electric vehicles. Battery technology at the time was nowhere near ready for commercialization, and in addition there was no infrastructure available for charging vehicle batteries on the road. As a result, the mandate failed to achieve its objective and, as soon as the mandate was lifted, automakers ceased to make electric vehicles. In the meantime, however, battery technology has evolved to where one could potentially envision an all-electric ground transportation system. Some people propose that there should be enough charging stations built to make it possible to charge batteries anywhere in the country, whereas others propose that there should be, instead of gas stations, battery exchange stations that would simply replace batteries as they reach the end of their charge. At present, batteries take too long to be charged during a trip and it may be necessary to combine the two ideas to evolve an all-battery transportation system sometime in the near future.

At this time, it is not possible to assess the future of electric vehicles. But, at the start of 2010, the following information was collected from various sources on the Internet. It is subject to uncertainties, vagrancies, and changes, as the electric vehicle market develops. The student is encouraged to follow development on his or her own. The near-term availability of electric vehicles is difficult to predict. However, President Barack Obama announced in July of 2009 a program with $2.4 billion in federal grants to develop next-generation vehicles and batteries in the United States. In Germany, one of the political party members proposed a €5000 ($7050) subsidy for people who buy electric cars. Japan's Nissan Motor Company unveiled, in July of 2009, the LEAF, an electric car that is claimed to have a range of 100 mi (160 km) on a single battery charge. It is scheduled to go into mass production for a global market in 2012. Mitsubishi Motor Company launched, in June 2009, its own electric vehicle program, and the Chinese automaker, Dong Feng Motor Corporation, has announced that it will team up with a Dutch company to develop and produce electric vehicles. General Motors has announced its Chevrolet Volt, a PHEV which is claimed by its CEO to get 200 mpg of gasoline. Finally, Daimler AG is working with the California-based electric car maker, Tesla Motors, Inc., on developing better batteries and electric drive systems for vehicles.

12.9 Hydrogen for Transportation

Hydrogen was touted as a potential transportation fuel after former President George W. Bush said in his 2001 inaugural address that "a child born today will be driving a pollution-free vehicle… powered by hydrogen." This appeared to be welcome news. However, to analyze whether or not hydrogen is a sustainable technology for transportation one must take into account all the steps necessary to make the hydrogen from a primary fuel source, get it into the fuel tank, and then power the wheels via a prime mover and a drive train. In other words, one must perform a well-to-wheel analysis, as shown in Figure 12.3. There is a loss in each step and to obtain the overall efficiency, one must multiple the efficiencies of all the steps. Using natural gas as the primary energy source, a well-to-wheel analysis [14] showed that a hybrid SI car would have a wheel-to-energy efficiency of 32%; a hydrogen-powered fuel-cell car with hydrogen made by steam reforming of natural gas, 22%; a hydrogen fuel cell car with hydrogen made by electrolysis with electricity from a natural gas–combined cycle power plant with 55% efficiency would have a well-to-wheel efficiency of 12%, as shown in Table 12.6; and, if the hydrogen were produced from photovoltaic cells, the well-to-wheel efficiency of the automobiles would be less than 5%. The estimates in Table 12.6 assume a fuel-cell stack efficiency at peak load of 44.5%, a part load efficiency factor of 1.1, and a transmission efficiency of 90%. For details, see [15].

In addition to the low overall efficiency of a hydrogen transportation system, as well as the high cost compared to other options, it should also be noted that before a hydrogen transportation system could be put into practice, an infrastructure for the distribution and storage of the hydrogen would have to be constructed. An extensive study of the comparative costs of fuel distribution systems conducted at Argonne National Laboratory in 2001 estimated that for a market penetration by the year 2030 40% of hydrogen vehicles with a mileage of 2.5 times that of average conventional vehicles (i.e., about 55 mpg), would minimally cost $320 billion, but could be as high as $600 billion. Based upon these

TABLE 12.6

Well-to-Wheel Efficiency of Fuel-Cell Vehicle with Hydrogen Produced by Electrolysis

NG feedstock production efficiency	95%
Conversion efficiency (NG to electricity)	55%
Electrolysis efficiency (electricity to H_2)	63%
Storage and transmission	97%
Compression efficiency	87%
Overall efficiency of fuel production	28%
Total fuel-cell well-to-wheel efficiency: ($0.28 \times 0.445 \times 1.1 \times 0.9$)	12%

estimates, a national transportation plan based on hydrogen with any currently available technology would be wasteful and inefficient and should not be considered as a pathway to a sustainable energy future [16]. The U.S. Department of Energy, Secretary of Energy and Nobel Laureate, Dr. Steven Chu, concurred with this conclusion and in 2009 cut off funds for development related to hydrogen fuel-cell vehicles in order to focus on other options [17].

Problems

12.1 On February 2006, President George W. Bush said, "All of a sudden, you know, we may be in the energy business by being able to grow grass on the ranch and have it harvested and converted into energy." Analyze the promise of the President for two options:

1. Surveying the technical literature on the topic, we find an article by Schmer et al. [18]. According to this article, the net energy yield of switchgrass grown over 5 years on marginal crop land on farms in the mid-continental United States was 60 GJ/ha. This is equivalent to about 0.2 W/m^2, about 10 times the bioethanol from corn in the United States, and represents a potential source of ethanol sometime in the future.
2. Bioethanol can also be produced from sugarcane in climates such as in Brazil. For sugarcane, the production is 80 ton/ha.
3. Ethanol from corn using technology of 2009.
4. Discuss and compare the above options.

12.2 A BMW hydrogen car uses 2.5 kWh/km. Compare this performance with that of an average U.S. car that is able to get 22 mpg of gasoline.

12.3 From the information supplied below, estimate the percentage of the total land area in the lower 48 states that would have to be used to replace all of the gasoline currently used for transportation with ethanol obtained from switchgrass. Do the same for the arable land in the lower 48 states.

Data:

- Estimated ethanol productivity = 1000 gallons (gal) ethanol per hectare per year (based on switch grass as the crop)—see, http://bioenergy.ornl.gov/paper/misc/switgrs
- 1.51 gal of ethanol are required to obtain the same energy as 1 gal of gasoline, based on their respective lower heating values
- U.S. gasoline consumption in 2008 was 138×10^9 gal/year (information obtained from: http://eia.doe.gov/oil_gas/petroleum/inforglance/consumption/product_supplies/annual-thousand_barrels/finished_motor_gasoline)
- U.S. total land area (lower 48) = 1.92×10^9 ac (information obtained from: http://cia.gov/library/publication/the_world_fact_book)

- U.S. arable land area (suitable for crops) = 0.45×10^9 ac (information obtained from: http://cia.gov/library/publication/the_world_fact_book)
- 1 ha = 2.47 ac

12.4 The combustion of ethanol goes according to the equation:

$$C_2H_5OH + \_\_ \ O_2 \rightarrow \_\_ \ CO_2 + \_\_ \ H_2O \ \text{LHV} = 21 \ \text{MJ/L}$$

where LHV = lower heating value, the symbol "L" designates liter.

1. Calculate the amount of CO_2 produced in g per MJ of energy released (LHV). The density of ethanol at 16°C is 780 g/L.
   The growth of corn and conversion of the corn into ethanol produces 87 g CO_2/(MJ of ethanol) from the fossil fuels required. In addition, 35 g of CO_2/(MJ of ethanol) are produced during the fermentation of corn sugar into ethanol. However, each atom of carbon in the corn sugar, and thus each atom of carbon in the CO_2 released by fermentation of the sugar, originally came from an atom of carbon dioxide in the atmosphere. Therefore, the CO_2 from the fermentation is "carbon neutral," that is, the amount of CO_2 released by the fermentation is balanced by CO_2 removed from the atmosphere in making the corn sugar. [*Note*: The same is true for the fermentation of any biomass.] Likewise, each atom of C in the CO_2 produced by combustion of ethanol is also carbon neutral.
2. Calculate the total amount of CO_2 released to the atmosphere per MJ (LHV) of ethanol made from corn and combusted, and the net amount of CO_2 released per MJ (LHV) of ethanol made from corn and combusted. The net amount is the amount by which the production and combustion of ethanol increases the amount of CO_2 in the atmosphere.
   Gasoline is a mixture of many compounds of carbon and hydrogen, but may be represented by the formula C_8H_{16}, with a LHV of 32.8 MJ/L (L = liter).
3. Calculate the CO_2 emissions in gm CO_2/MJ (LHV) with gasoline as fuel.
4. Estimate the total CO_2 emissions [g CO_2/MJ (LHV)] from "well-to-wheel" by including "well-to-tank" emissions with gasoline as a fuel. The well-to-tank energy efficiency for gasoline is approximately 90% (100 × (MJ energy to tank)/(MJ energy of petroleum "in the well")). Assume that the ratio of CO_2 emissions from tank-to-wheel is also 90% of the total CO_2 emissions.

12.5 A truck engine consumes diesel at a rate of 30 L/h and delivers 65 kW of power to the wheels. If the fuel has a heating value of 43.5 MJ/kg and a density of 800 kg/m^3, determine

1. The thermal efficiency of the engine
2. The waste heat rejected by the engine

12.6 An automobile storage battery with an open circuit voltage of 12.8 V is rated at 260 A/h. If the internal resistance of the battery is 0.2 Ω, estimate the maximum duration of current flow and its value through an external resistance of 1.8 Ω.

12.7 Make a preliminary design for the power train of a hydrogen/oxygen fuel cell–powered automobile. Provide a system description with preliminary design estimates of the power and fuel supply arrangements.

12.8 Explain the difference in operation between hybrid electric vehicles with parallel and series designs.

References

1. Kreith, F. (1999) *Ground Transportation for the 21st Century*, National Conference of State Legislatures, Denver, CO and ASME Press, New York.
2. Kreith, F., West, R.E., and Isler, B. (2002) Legislative and technical perspectives for advanced ground transportation system, *Transportation Quarterly* 96(1), 51–73, Winter 2002.
3. Dismukes, C. et al. (June 2008) Aquatic phototrophs: Efficient alternatives to land-based crops for biofuels, *Current Opinions in Biotechnology* 19, 235–240.
4. Christi, Y. (2007) Biodiesel from microalgae, *Biotechnology Advances* 25, 294–306.
5. Kreith, F. and West, R.E. (2003) Gauging efficiency: Well-to-wheel, *Mechanical Engineering Power*, pp. 20–23.
6. Davis, S. and Diegel, S. (2004) *Transportation Energy Data Book*, Oak Ridge National Laboratory/U.S. Department of Energy, Oak Ridge, TN.
7. EPRI (2004) Advanced batteries for electric drive vehicles: A technology and cost-effectiveness assessment for battery electric vehicles, power assist hybrid electric vehicles and plug-in hybrid electric vehicles, EPRI Tech. Report 1009299, EPRI, Palo Alto, CA.
8. Kreith, F. and West, R.E. (September 2006) A vision for a secure transportation system without hydrogen or oil, *ASME Journal of Energy Resources Technology* 128(9), 236–243.
9. Kreith, F. and West, R.E. (May 2008) A scenario for a secure transportation system based on fuel from biomass, *Journal of Solar Energy Engineering*, 130, 1–6.
10. Farrell, A.E., Plevin, R.J., Turner, B.T., Jones, A.D., O'Hare, M., and Kammen, D.M. (2006) Ethanol can contribute to energy and environmental goals, *Science* 311, 506–508.
11. EPRI (2004) Advanced batteries for electric drive vehicles, EPRI Tech. Report 1009299, EPRI, Palo Alto, CA.
12. EIA (2007) www.cia.doe.gov/fuelelectric/electricity infocard 2005.
13. Himelich, J.B. and Kreith, F. (2008) Potential benefits of plug-in hybrid electric vehicles for consumers and electric power utilities, in: *Proceedings of ASME 2008 IMEC*, Boston, MA, October 31–November 6, 2008.
14. Kreith, F. and West, R.E. (2004) Fallacies of a hydrogen economy: A critical analysis of hydrogen production and utilization, *Journal of Energy Resources Technology* 126, 249–257.

15. Kreith, F., West, R.E., and Isler, B.E. (2002) Efficiency of advanced ground transportation technologies, *Journal of Energy Resources Technology* 24, 173–179.
16. Mince, M. (2001) Infrastructure requirement of advanced technology vehicles, in: *NCSL/TRB Transportation Technology and Policy Symposium*, Argonne National Laboratory, Argonne, IL.
17. M.L. Wald, *New York Times* News Service, 4/8/2009.
18. Schmer, M.R. et al. (2008) Net energy of cellulosic ethanol from switchgrass, *PNAS* 105(2), 464–469, www.pnas.org/cgi/content/full/105/2/464
19. Bartlett, A.A. (2000) An analysis of U.S. and world oil production patterns using Hubbard-style curves, *Mathematical Geology* 32(1).

Index

A

Abbott, C.G., 541
Abengoa Solar, 603
Absorbed glass mat (AGM) batteries, 734
Absorber plates, thermal analysis of for flat-plate collectors, 433–437
Absorptance, 417–418
Absorption air-conditioning, 494–505
Acetone, butanol, and ethanol (ABE) fermentation, 211
Active components of passive systems, 621
Active solar heating systems, physical configurations of, 470–471
Actuator disk theory, 148–152
Advanced ground transportation, analysis of, 821–826
Agricultural residues, use of as biomass feedstocks, 182
Air-cooled flat-plate collectors, 445
Air-heating flat-plate collector, diagram of, 445
Air mass ratio, 281
Air preheating, unglazed transpired wall system for, 472–474
Air surfaces, conductance and resistance values for, 523–525
Air-type flat-plate collectors, 424–425; *see also* Flat-plate collectors
Algal biomass, production of biofuels from, 220–221, 812
Alkali compounds, effect of in direct combustion systems, 235
Alkaline fuel cells (AFC), 740–742
All-electric vehicles, 827
Allowed energy bands, 369
Alloys, thermal properties of, 755–756
Alpha-type Stirling engines, 610–611
Alternative energy sources; *see also* Renewable energy
  levelized and capital cost estimates, 47–49
Alternative fuels, 810–812
  advanced ground transportation analysis of, 821–826
Altitude angle, 260; *see also* Sun angles
Ammonia, use of as transportation fuel, 226
Ammonia–water system, 496
Amorphous silicon thin films, 397
  fabrication of, 388–389
Anaerobic digesters, 193, 196
Anaerobic digestion, 192–195
  microbial phases in, 194
Anaerobic reactor designs, 193
Angle of incidence, 284, 421–422
Angle of refraction, 421–422
Angström–Page estimation, 289–290
Anhydrous ammonia, use of as transportation fuel, 226
Anisotropic diffuse solar radiation, 295–296
Annuities, 69–70
Anode, 728
Appliances, energy conservation example for, 22, 25
Aquatic microbial oxygenic photoautotrophs (AMOPs), 811–812
Aquifer air storage, 719
Arable land production, 9
Array effects, 165
Ash, analysis of in biomass, 189
ASHRAE testing standards for solar collectors, 452–453
Atmospheric boundary layer, 134–135
  surface layer of, 138
Atmospheric stability states, 139
Atoms, electronic structure of, 368
Axial induction factor theory, 151
Axial thrust, 151
Azimuth angle, 260; *see also* Sun angles
  sun–window difference, 682

B

Balance of station (BOS), factors affecting costs, 164
Balance point, 758
Band gap, 367, 369
Barrage load factor, 793
Baseboard convection systems, 757
Batch reactors for anaerobic digestion, 193
Batteries, 709
  cell physics, 731
  fundamentals of, 727–728
  NiMH cost assumptions, 820
  principles of operation of, 729–731
  rechargeable, 732–739
  resources for, 767
  storage capabilities of PHEVs, 818
Beale free piston Stirling engine, 612
Beam solar radiation, 280, 283–284
  long-term measurement models for, 290–298
  measurement of using pyrheliometers, 301–302
Beta-type Stirling engines, 611–612
Betz limit, 150, 152
Biobased transportation fuels, *see* Biofuels
Biocatalysts, 213
Biocrude, fuels from, 219
Biodiesel, 37–39, 811–812
  conversion of triglycerides into, 219–223
Biofuels, 13, 206
  global status and potential of, 37–39
Biogas, 181
  production of by anaerobic digestion, 192–193
Biomass
  advanced ground transportation with fuels from, 821–826
  anaerobic digestion, 192–195
  conversion of to electricity, 230–236
  conversion of to gaseous fuels, 192, 203
  conversion of to hydrogen, 32
  definition of, 181
  energy crops as feedstocks for, 183–185
  feedstocks, 182, 191
  global status and potential of, 36–39
  properties of, 187–190
  size of resource base, 185–187
  societal costs of, 73–74
  thermal gasification of, 195–203
  transportation fuels from, 204–229 (*see also* Transportation fuels; specific fuels)
  waste materials as feedstocks, 182–183
Bio-oil; *see also* Pyrolysis
  ethanol from, 213
  production cost estimates for, 219
Biorenewable resources, *see* Biomass
Bioresources, land area requirement for, 811
Bird and bat kills, wind turbines and, 124–125
Black-body radiation, 250
Blade element momentum (BEM) theory, 148
Boilers, 233
  fluidized bed combustors, 234
Boudouard reaction, 197
Boys, C.V., 541
Building paper, conductance and resistance values for, 527
Buildings
  calculation of heat loss in, 455–458
  calculations of heating and hot water loads in, 453–455
  cooling requirements for, 485–489
  degree-day method for calculating heat loss, 458–462
  energy conservation example for, 22
  internal heat sources in, 458
  liquid-based solar-heating systems for, 469–474
  overnight storage for, 752–753, 756–758
  service hot water load calculation, 462–464
Bulb system of tidal power generation, 791
Bulk density, determination of for biomass, 190
Butanol, 211

C

Cadmium–telluride (CdTe) thin films, 397
Campbell–Stokes sunshine recorder, 303
Capacity factors, 84
  wind turbine performance, 155–156
Capital costs
  estimates for OC-OTEC plants, 783–785
  estimates for renewable energy, 47–49
  Severn barrage, 791–794
Capital recovery factors (CRFs), 112–114
Carbohydrates, use of as an ethanol feedstock, 207–208
Carbon capture systems (CCS), cost estimates for, 50
Carbon–oxygen reaction, 197
Carbon–water reaction, 197
Carrying capacity
  food availability and, 9–10
  water availability and, 5–7
Cash flow, example scenario, 83–87
Cathode, 728
Cellulosic ethanol, 209–210
  EROI for, 811
Central receiver systems, 595–596
  molten salt power tower systems, 596–603
  thermal energy storage in, 750–752
  two-tank storage for, 747
Centrifugal pumps, 395
Char, solid-gas reactions of, 232
China, energy consumption of, 11
CIGS thin films, 398–399
Circumsolar radiation, 295–296
  measurement of using pyrheliometers, 302
Clearness number, 282
Closed-cycle ocean thermal energy conversion (CC-OTEC), 771–775
  comparison with open-cycle systems, 780–782
Closed-loop drainback systems, 467–469
Closed-loop systems, 467–469
Clostridum acetobtylicum, fermentation of butanol from, 211
Cobalt, use of as catalyst in Fischer–Tropsch synthesis, 216
CO_2 emissions, analysis of, 823–826
Co-firing, 235–236
Cold-water pipe (CWP), 782–783
Collection period (Δt_c), 572–573
Collector efficiency equation, 562–563
Collector efficiency factor, 437–438
Collector heat-loss conductance, 428–433
Collector heat-removal factor, 438–442
Collector storage effect, 442
Combustion
  direct, 230–232
  equipment, 233–236
  in thermal gasification, 195, 198, 200–203
Compound parabolic concentrators (CPCs), 555–559
  use of for industrial process heat, 545
Compressed air energy storage (CAES), 717–719
  capital costs, 720
  comparison with pumped hydro, 720–721
  Huntorf plant, 724
  McIntosh plant, 724–725
  methods for producing cavities, 719
  round trip efficiency of, 719–720
  schematic diagram of typical system, 722
  volumetric energy density, 721–724
Compressed gases, use of as transportation fuels, 225–226
Concentrating solar collectors
  concentration limits of, 553–555
  reflecting surfaces for, 419–420
  thermal energy storage for, 750–752
Concentration ratio, 553
Condensation in OC-OTEC, 778–780
Condensers, 586–587
Conductance values
  exterior siding materials, 526
  external air surfaces, 523
  flooring materials, 531
  glass, 533
  interior finishes, 532
  internal air surfaces, 524–525
  masonry materials, 528

roofing materials, 531
sheathing and building paper, 527
wall-insulation materials, 529–530
wood, 529
Conduction band, 367
Conductors, 367
Constant dollars, 70
Constant speed/variable pitch (CSVP) turbines, 156–158
Consumer price index (CPI), 70, 97
Control strategies for thermal energy storage, 757
Cooling load, 485–487
Cooling load temperature difference (CLTD) method, 487
Coolth, 622
CO_2 reduction, example calculations for, 101
Corn
cost of ethanol from, 209
dry milling of, 208
Cost of energy (COE)
example calculation for wind turbines, 162
tidal systems, 791–794
wind turbines, 158–163
Covalent bonds, 367
Critical solar intensity ratio X, 571–572
Cumulative distribution function, 143
Current dollars, 70
Currents in photovoltaic cells, 375–380
Cut-in wind speed, 146
Cut-out wind speed, 147
Cycle life, 711
Czochralski method, 384–385

D

Daily load calculations for photovoltaics, 390
Daily solar radiation on tilted surfaces, 296–297
Dark current, 375
Darrieus wind turbines; *see also* Vertical axis wind turbines
performance of, 130
Daylighting, 621
approach to design, 680–682
controls and economics, 697–700
design fundamentals for, 680
design methods, 683–697
lighting terms and units, 680
sun–window geometry, 682
Declination, 260
Deepwater waves
estimation of power of, 795–798
motion of a particle in, 796
Degree-day method, 458–462
Dehumidification, solar desiccant, 505–509
DELSOL, 604
Dendritic web growth, 386
DePere Wind Project (DPWP), EROI calculation for, 102–105
Deposition techniques, 399–400
Depreciation, 75
Desiccants
liquid, 508–509
solid, 505
Diesel fuel, 206
Diffuse reflection, 414, 419
Diffuse solar radiation, anisotropic, 295–296
Digesters, 193, 196
Dimethyl ether, use of as transportation fuel, 226
Direct contact evaporation in OC-OTEC, 778–780
Direct electrical storage technologies
superconducting magnetic energy storage (SMES), 727
ultracapacitors, 726–727
Direct energy, 87
Direct gain, 628
Direct methanol fuel cells (DMFC), 740–741
Direct solar radiation, 280, 284–288
measurement of using pyrheliometers, 301–302
Distributed grid technology, 709
Domestic hot water, thermal energy storage for, 752–753, 756–758
Double-paned windows
energy analysis example for, 97–102
heat loss through, 457–458

Downdraft gasifiers, 198, 200–201
Dry grind ethanol, 208–209
Drying fluids, 588–589
Dye Beck, 550
typical heat requirements for, 552

E

Earth contact cooling, 675–676
heat transfer analysis of, 676–678
Earth–sun geometric relationship, 259–274
Ebb generation, 790
Economics of hydrogen, 744
Edge-defined film-fed growth (EFG), 386–387
Edge loss, 430
Effective discount rate, 70, 72
Efficiency of energy storage, 711
Efficient Development of Offshore Wind farms (ENDOW) project, 165
Electrical conduction, 369
Electrical energy storage, 708–709
Electric energy, energy conservation example for, 22, 25
Electricity, conversion of biomass to, 230–236
Electric power generation, geothermal energy for, 34–35
Electric vehicles, 827
Electrochemical plating, 399
Electrolysis of water, 31–32
generation of hydrogen by, 742
Electromagnetic energy, 249
Electromagnetic force (EMF), 729
Electromagnetic radiation spectrum, 249
Electronic structure of atoms, 368
Embedded energy, 93–97
EMCORE T1000 cell, 402–403
Emissive power, 252
Encapsulants, 388
Eneas, A.G., 541
Energy
cost for wind, 158–163
definition of, 119–120
major global issues and, 4
societal costs of, 72–74
sustainable development and, 10–18
units and conversion factors, 54–56
Energy analysis methodology, 87–102
tracking parabolic trough collector example, 567–569
Energy balance for a flat-plate collector, 427–428
collector heat-loss conductance, 428–433
Energy conservation
distinction between passive systems and, 621–622
heating and cooling of buildings and, 413
role of, 18–25
Energy consumption
global comparison of, 13
per capita, 10–11
Energy crisis, 3
Energy crops, 183–185
organic components of, 187
Energy density, 46–47, 712
Energy efficiency
economic potential of in North America, 19–21
economic potential of in Western Europe, 23–24
geothermal heat pumps, 33
hydrogen, 32
legislating, 25
sustainable growth and, 2
Energy flow, 14
Energy gap, 373
Energy generation technologies
energy return on energy investment (EROI) of, 16–18
Energy intensities, 93–97
example calculations for windows, 99–101
Energy payback time, 16
Energy return on energy investment (EROI), 16–18
calculation of, 87–102
calculation of for wind energy system, 102–105
cellulosic ethanols, 811
EROI for sample wind energy systems, 106–107
estimates for solar central receiver plants, 604

photovoltaic thin films, 397–398
tracking parabolic trough collector example, 565–567

Energy storage
applications, 709–711
direct electrical technologies, 726–727
importance of to renewable energy, 707–708
mechanical technologies, 713–726
methods for, 708
technology characterization, 711–713

Energy technologies, economics of, 32–33
Energy transitions, 2–3
Energy transport, physics of the sun and, 247–248
Engineered geothermal systems, *see* Enhanced geothermal systems
Engine knock, 206
Enhanced geothermal systems (EGS), 35
Enthalpies of formation, calculations of for biomass, 189–190
Entrained flow gasifiers, 201
Environmental effects, economic costs of, 73–74
Environmental impact, wind energy technology, 123–124
Enzymatic hydrolysis, 209
EROI, *see* Energy return on energy investment
ERSATZ, 305

Ethanol
from bio-oil, 213
cellulosic, 209–210
cost of from corn, 209
global status and potential of, 37–39
grain, 207–209
limitations of as a transportation fuel, 210–211
from syngas, 212–213
volumetric yield of, 210

Ethanol–gasoline blends, petroleum requirement for, 823

Europe
per capita energy use in, 18–19
wind capacity potential in, 169

Evacuated tube collectors, 446–447, 562
thermal analysis of, 448

Evaporative cooling, 489, 492–493, 673
Exterior siding materials, conductance and resistance values for, 526
External air surfaces, conductance and resistance values for, 523
Externalities, 72–73
Extinction coefficient, 258, 421–422
Extraterrestrial radiation, 276–279
Extrinsic semiconductors, 367

F

Faraday's constant, 731
Fast pyrolysis liquid, 217
f-chart, design of liquid-based solar-heating systems by, 475–484

Feedstocks
for alternative fuels, 810
biomass, 182–185, 191
carbohydrates as, 207–208

Fermentation, 193
bio-oil, 213
ethanol, 207
fuels other than ethanol from, 210–212

Fetch, 795
Fin efficiency, 436–437, 441

Fischer–Tropsch liquids, 215–217
methanol synthesis and research in, 214

Fixed carbon, analysis of in biomass, 189
Fixed charge rate (FCR), definition of, 159
Flash evaporation in OC-OTEC, 778–780

Flat-plate collectors, 415–416
air-cooled, 445
air-type, 424–425
collector efficiency factor, 437–438
collector heat-removal factor, 438–442
energy balance for, 427–433
glazings, 425–426
liquid-type, 423–424
thermal analysis of absorber plate, 433–437
thermal circuits for, 429
thermal performance of, 448–451
transient effects, 442–445

types of, 413
use of for industrial process heat, 545
Floating ocean thermal power plants, 770–771
Float zone method, 385
Flood generation, 790
Flooring materials, conductance and resistance values for, 531
Flow batteries, 737–739
Fluidized bed combustors, 234
Fluidized bed gasifiers, 200–201
Flywheels, use of for energy storage, 725–726
Food, sustainability and the availability of, 9–10
Footcandles, 680
Forbidden energy bands, 369
Forbidden gap, 367
Forced-air systems, 757
Forced-circulation systems, 467–469
Fossil fuels
contribution of to total energy consumption, 13
replacing as an energy source, 3
Fractional costs, estimates for OC-OTEC plants, 784–785
Francis turbines, 716
Free piston Stirling engine, 612
Fresnel lens, use of in multijunction PV cells, 400–401
Fuel cells
cost estimates for, 50
fundamentals of, 727–728
generating electricity with, 745
hydrogen, 31–32, 739 (*see also* Hydrogen)
operating characteristics of, 741
principles of operation, 739–740
types of, 740–742
use of methanol to provide hydrogen for, 214
Furans, methylated, 223–225

G

Gamma-type Stirling engines, 611–612
Gas cleaning, 202–203
Gas heating value, 201–202
Gasification, 195–203
Gasifiers
downdraft, 198, 200–201
entrained flow, 201
fluidized bed, 200–201
fluid-plate, 196
low-temperature, 197
performance of, 201–202
producer gas composition from, 200
thermodynamic efficiency of, 203
updraft, 198
Gasoline, 204, 206
Gas turbine power plants, 717–718
Gaussian probability distribution function, 143
GDP, *see* Gross domestic product
Geometric concentration ratio, 553
Geometric correction factors (GCF), 300–301
Geothermal energy, 33–36
Glass
conductance and resistance values for, 533
transmittances, 688
use of in flat-plate solar collectors, 425–426
Grain ethanol, use of as a biofuel, 207–209
Gross domestic product (GDP), consumption of by world countries, 11, 13
Gross national product (GNP), 70
relationship of to energy use, 11
Grove, Sir William, 727

H

Hard rock caverns, 719
Heat exchangers, 756
for closed-cycle OTEC, 782
for open-cycle OTEC, 781
Heat floors or ceilings, 757
Heating fuel, energy conservation example for, 22
Heating load calculations for buildings, 455, 632
Heating load devices, 757–758
diagram for, 759

Heating value, calculations of for biomass, 189
Heat loss
  calculating for flat-plate solar collectors, 430–431
  calculation of for buildings, 454–458
  degree-day method for calculating, 458–462
  example problem for determining, 98–99
Heat loss coefficients, 563–564
Heat pumps, 33
Heat transfer analysis, earth contact cooling, 676–678
Heat transfer, analysis of the rate of in flat-plate collectors, 433–437
Herbaceous energy crops (HEC), 184
Heuristic approach to photovoltaic system sizing, 392–393
Horizontal axis wind turbines (HAWTs), 126, 128; *see also* Wind turbines
  performance of, 130
Horizontal illuminance, 686
Horizontal solar radiation; *see also* Solar radiation
  models based on long-term measured, 290–299
Hottel–Whillier–Bliss equation, 440
Hour angle, 260
Hourly solar radiation on tilted surfaces, 297–299
Human carrying capacity
  food availability and, 9–10
  water availability and, 5–7
Human Development Index (HDI), 10
Huntorf CAES plant, 724–725
Hybrid electric vehicles (HEVs), 815–817
Hybrid systems, 621
Hydrocarbons, synthesis of from microbes, 211
Hydrocracking, 218
Hydroelectric power, 13
  global status and potential of, 36–37, 39
Hydrogen
  cost estimates for use in fuel cells, 50
  fuel cells, 31–32, 739
  generation of, 742–743
  storage and transport of, 743–744
  thermodynamics and economics of, 744
  for transportation, 828–829
  use of as transportation fuel, 225
  utilization of as an energy carrier, 29–31
Hydrogenation reaction, 197, 222
Hydrogen economy, 29
Hydrolysis, 193
  enzymatic, 209
Hydropower *vs.* tidal power, 785
Hydrothermal processing (HTP), 219
Hydrotreating, 218

I

Illuminance, 305, 680
  calculation of for sidelighting design, 683–686
  calculation of for skylighting design, 693
Impulse turbines, 716
Incentives *vs.* mandates, 25
Incidence angle modifier, 453
Index of refraction, 421–422
Indirect gain, 628
Industrial process heat systems, 469
  estimated date of technical readiness for, 546–548
  long-term performance of, 569–570
  solar, 544–549 (*see also* Solar industrial process heat)
  use of TRNSYS to assess performance of, 581
  utilizability method for assessing performance of, 570
Inferred resources (IR), 26
Inflation, 70–72
  indices, 71
Initial capital cost (ICC), definition of, 159
Input–output (I/O) analysis, 89–92
Insolation, 275
  percent of possible, 298
Insulators, 367
Integrated gas combined cycle (IGCC)
  cost estimates for, 50
Intercept factor, 561
Interior finishes, conductance and resistance values for, 532

Internal air surfaces, conductance and resistance values for, 524–525
Internal rate of return (IRR), 76–82
Internet resources, 60–61
Intrinsic semiconductors, 367
Inulin, 207–208
Ionic bonds, 367
Iron, use of as catalyst in Fischer–Tropsch synthesis, 216
Isolated gain, 629
Isoprene synthesis, 212

J

Jack pumps, 395
Jet fuel, 206
Junction current, 375

K

Kaplan turbines, 716
Kinematic Stirling engines, 612

L

Land-based ocean thermal power plants, 770–771
La Rance Tidal Power Plant, 786–787
  cost of power from, 791
Latent heat
  physical properties of storage materials, 749
  storage of, 747–748
Latitude, 260
Lavoisier, 541
Lead-acid batteries, 732–734
Levelized cost of energy (LCOE), 82–83
Levelized energy costs, 47–49
Life cycle cost comparison for vehicles, 819
Light-induced recombination current, 375
Lighting, terms and units, 680
Light loss factor, 687, 695
Light water reactor (LWR), example cash flow scenario, 83–87
Lignocellulosic biomass, 183–185, 209–210
Limpet OWC system, 802–803
Line focus solar thermal power system, 594
Lipids, fuels from, 219–223
Liquid-based solar-heating systems, 469–470
  design of by *f*-chart, 475–484
  fluid flow rates, 471–472
  physical configurations of active systems, 470–471
  solar collector orientation, 471
  unglazed transpired wall system, 472–474
Liquid desiccant cooling system, 508–509
Liquid fuels
  classification of biorenewable, 181
  relative land area requirement for bioresources, 811
Liquid-type flat plate collectors, 423–424; *see also* Flat-plate collectors
Lithium bromide–water system, 496
  example of, 497–505
Lithium ion batteries, 736–737
Liu–Jordan (LJ) method, 291–293
Load calculations for photovoltaics, 390–391
Load collector ratio (LCR), 631, 633
  approach to design of passive heating systems, 641–665
Load devices, 757–758
  heating load diagram for, 759
Logging residues, 184–185
Log law wind speed profile, 138–140
Long-term measured horizontal solar radiation, models based on, 290–299
Lumen, 680
Lumen methods of daylighting design, 683–697
Luminous flux, 680
Lux, 680

M

Mandates *vs.* incentives, 25
Masonry materials, conductance and resistance values for, 528
Massive electric storage (MES), 707
Mass transportation, potential of to reduce energy consumption, 813–814

McIntosh CAES plant, 724–725
Measurement of solar radiation
detectors for instrumentation, 303
instruments for, 299–302
spectral, 304
Mechanical energy storage, 708–709
compressed air energy storage (CAES), 717–726
pumped hydroelectric energy storage (PHES), 713–717
Metabolic engineering, 212
Metals, thermal properties of, 755–756
Methanation, 198
Methane
conversion of dry matter to, 193, 195
use of as transportation fuel, 225
Methanogenesis, 193
Methanol, 213–214
direct carbonylation of, 213
Methylated furans, production of transportation fuels from, 223–225
Microalgae, production of biofuels from, 220–221, 812
Microbial fermentation, synthesis of hydrocarbons from, 211
Milk processing, solar industrial process heat for, 550–552
Mixed alcohols, 212
synthesis of, 214–215
Modified internal rate of return analysis, 78–82; *see also* Internal rate of return (IRR)
Moisture content of biomass, 189
Molten carbonate fuel cells (MCFC), 740–741
Molten salt power towers, 596–603
characteristics of, 597
thermal energy storage in, 750–752
Monthly average sky temperature depression, 674
Monthly clearness index, 290
Monthly solar radiation estimation models, 287–290; *see also* Solar radiation
tilted surfaces, 290–295
Moored ocean thermal power plants, 770–771
MOUSE inputs, 89
Multijunction photovoltaic cells, 400–401
Multijunction thin films, fabrication of, 388–389
Multilayer photovoltaic technology, 400–406
Municipal solid wastes (MSWs), 182

N

Nanoparticle printing, 400
National Renewable Energy Laboratory (NREL), solar radiation data from, 304–305
National Solar Radiation Data Base (NSRDB), 305
Natural circulation systems, 464–467
Natural convection/ventilation, 670
Net building load coefficient (NLC), 631
Net cell reaction, 734
Net energy analysis (NEA), 87–88, 97
Net energy savings, 622
Net present value, role of in internal rate of return analysis, 76–82
Net value of energy saved, 66
Neutral stability, 139
Nevada Solar One, 594
Nickel metal batteries, 734–735
cost assumptions for NiMH batteries, 820
Nocturnal cooling systems, 673–675
Noise, 123
Nominal discount rate, 70
Nonmetals, thermal properties of, 754–755
North America, economic energy efficiency potentials in, 19–21
n-type semiconductors, 369
Nuclear energy, 3, 26–29
cost estimates for, 50
cost estimating example, 54
EROI estimate for, 105, 108–109
Nuclear reactors, 26
example cash flow scenario for light water reactor, 83–87
number of in operation, 27–28
Nuclear resources, availability of, 26–29

O

Ocean thermal energy conversion (OTEC), 769–771
closed-cycle, 771–775
cold-water pipe and pumping requirements for, 782–783
comparison of open-and closed-cycle systems, 780–782
economics of, 783–785
open-cycle, 775–780
Ocean wave energy, 795
wave power, 795–798
Ocean wave power, 798–800
Offshore wind farms, 164–173
Oil production, 809–810
Online resources, 60–61
Open-circuit voltage, 729
Open-cycle ocean thermal energy conversion (OC-OTEC), 775–778
capital costs for, 783–785
comparison with closed-cycle systems, 780–782
design specifications for 10 MWe plan, 779
evaporation and condensation in, 778–780
schematic, 780
Open-loop systems, 467–469
Operating and maintenance (O&M) costs
role of in internal rate of return analysis, 76–82
role of in total life cycle cost analysis, 74–75
Optical concentration ratio, 553
Optical depth, 282
Optical error, 561
Optical transmission, 421–422
Order of magnitude estimates, 46–54
Oscillating water column (OWC) systems, 802–804

P

Parabolic collectors, 553–555
compound concentrators in, 555–559
Parabolic dish-based solar thermal power plant, 605
Parabolic trough collectors (PTC)
geometric relations for, 560
single-axis tracking, 559–569
thermal energy storage in, 750–752
use of for industrial process heat, 545
Parallel hybrid-electric vehicles, 815–817
Passive cooling systems, 621
evaporative cooling, 673
natural convection/ventilation, 670–673
nocturnal and radiative, 673–675
solar control, 670
Passive heating systems, 621
design approaches, 630–632
design fundamentals, 627–628, 630
generalized approach to design of, 632–641
LCR approach to design of, 641–665
solar-load ratio design approach for, 665–670
types of, 628–629
Passive solar buildings, solar angles and the design of, 268–269
Passive solar heating, 470
Passive solar water heaters, 464–467
Passive systems
cost of, 622
defining terms, 700
distinction between energy conservation and, 621–622
elements of economic consideration, 622
general application status and costs, 622–623
net energy savings of, 622
Payback period (PP)
calculating, 66
determining using input–output analysis, 90–92
example problem for determining, 97–102
photovoltaics, 365, 398
use of internal rate of return analysis to estimate, 78–82
Peaking generators, 707
Pelamis, 804–805
Pelton turbines, 716
Pennington cycle, 507
Phase change heat storage, 747–748

Phase change materials, 748
  physical properties of, 749
Phosphoric acid fuel cells (PAFC), 740–742
Photobioreactors, 221
Photoelectric effect, 365
Photons, energy of, 250
Photovoltaic cells; *see also* Solar cells
  analysis of, 374–380
  efficiency of, 380–381, 396
  energy cap for candidate materials for, 373
  manufacture of single crystal and polycrystalline cells, 384–388
  multijunction, 381–382
Photovoltaic detectors, 303
Photovoltaic effect, 371–375
Photovoltaics
  design for remote applications, 389–390
  estimation of available solar radiation, 391–392
  estimation of loads and load profiles, 390–391
  multilayer technology, 400–406
  system sizing, 392–394
  thin-film technology, 397–400
  water pumping applications using, 394–395
Photovoltaic systems
  cost estimating example, 50–52
  design of, 382–384
  energy payback time example, 16–17
  global status and potential of, 41–42
  internal rate of return analysis example, 78–82
Physical vapor deposition (PVD), 399
Planck's law, 250–251
Plug-in hybrid electric vehicles (PHEVs), 709, 817–821
Plutonium, 28
p-n junction, 369–370
Polar jet stream, 134
Polycrystalline cells, 384–388
Population growth, 4–5
  food insecurity and, 9–10
  sustainability and, 1–2, 4–5
Power
  coefficient, 151
  definition of, 119–120
  ocean waves, 798–800
  relationship of to thrust, 148
  storage technologies to improve quality of, 709
Power generation unit (PGU), 815–817
Power tower systems, 595–596; *see also* Central receiver systems
Present discounted monies, 67
Present worth, 67
  effect of inflation on, 72
  example of calculating, 69
  factors, 68
Pressure jet dyeing, 550
Priestley, Joseph, 541
Primary energy consumption, United States, 2004, 108
Probability distribution functions, 141–143
Problems
  biomass, 237–240
  biomass conversion, 204, 236–237
  biomass feedstocks, 191–192
  economics, 109–111
  energy storage, 763–765
  photovoltaics, 406–410
  solar heating and cooling of buildings, 510–521
  solar process heat and thermal power, 613–616
  solar radiation fundamentals, 305–308
  sustainable energy, 57–59
  transportation, 829–831
  transportation fuels from biomass, 229–230
  wind energy, 174–177
Process chain analysis, 88–89
Producer gas, 195
Producer gases, composition of from various gasifiers, 200
Product gas, removal of tar from, 202–203
Proton exchange membrane fuel cells (PEMFC), 740–741
Proximate analysis of biomass, 187, 189

PTGEN, 604
p-type semiconductors, 369
Pumped hydroelectric energy storage (PHES)
  capital costs, 720
  comparison with CAES, 720–721
  turbines, 715–717
Pumping requirements for OTEC, 782–783
Pyranometers, 299–301
Pyrheliometers, 299, 301–302
Pyrolysis, 195, 231–232
  fuels from oils from, 217–219 (*see also* Bio-oil)

R

Raccoon Mt. pumped storage plant, 713–714
Radiation; *see also* Thermal radiation
  black-body, 250
  electromagnetic spectrum, 249
  intensity of and shape factor, 252, 255–257
  solar, 275 (*see also* Solar radiation)
  transmission of through a medium, 258
Radiation function tables, 250–255
Radiative cooling systems, 673–675
Radiative heat transfer, calculating for flat-plate solar collectors, 430–432
Radiative properties, 414
Ramp-assisted foil casting technique (RAFT), 386–387
Rankine cycle, 581, 583–586
  thermal energy storage and, 750–752
Rankine power plants
  choice of working fluid for, 588–591
  components of, 586–588
Rated wind speed, 146–147
  turbine power and, 155
Rayleigh probability distribution function
  determining wind speed using, 143–144
  wind speed calculation example, 145–146
Reaction turbines, 716
Real discount rate, 70, 72
Reasonably assured resources (RAR), 26
Rechargeable batteries, 732
Reflectance, 417–419
  glass-air interface, 421–422
ReflecTech Mirror Film, 569
Reflecting surfaces, 419–420
Reflectivity values, 362
Refractive index, 421–422
Refrigerant–absorbent pairs, 496
Refrigeration, 489–492
Remote photovoltaic applications
  estimation of available solar radiation, 391–392
  estimation of loads and load profiles, 390–391
  examples of, 389
  heuristic approach to system sizing, 392–393
  system operating efficiency, 393–394
Renewable energy, 2–3
  annual potential of, 43–46
  capacity factors for various options, 84
  consumption of, 15
  costs of, 47–54
  global status and potential of, 36–43
  societal costs of, 73–74
  total life cycle cost analysis example, 74–76
Reserves, 43
Resistance values
  exterior siding materials, 526
  external air surfaces, 523
  flooring materials, 531
  glass, 533
  interior finishes, 532
  internal air surfaces, 524–525
  masonry materials, 528
  roofing materials, 531
  sheathing and building paper, 527
  wall-insulation materials, 529–530
  wood, 529
Reverse saturation current, 375
Ribbon against drop (RAD) method, 386
Roofing materials, conductance and resistance values for, 531

S

Salt caverns, 718–719
  map of potential sites for in U.S., 720
Salter's Duck, 804
Saturated steam and water, 534–535
Sealed absorbed glass mat (AGM) batteries, 734
Sealed gel cell batteries, 734
Secondary cells, 731
Seed crops, production of biofuels from, 220–221
SEGS power plants, 591–594
Selective surfaces, 419
Self-discharge time (SDT), 711
Semiconductors, 367–374
Sensible heat storage, 745–746
  dual media materials for, 747
  liquid media for, 747
  physical properties of materials, 746
  solid media for, 746
Series hybrid electric vehicles, 815–817
Service hot water load calculation, 462–464
Severn barrage
  capital and annual recurring costs, 794
  cost of energy analysis for, 791–794
Severn Estuary, effects of concentration of tidal flow in, 786
Shading band correction factors, 301
Shadow-angle protractor, 271–274
Shadow flicker, 124
Shallow-water ocean thermal power plants, 770–771
Shape factor, intensity of radiation and, 252, 255–257
Sheathing, conductance and resistance values for, 527
Shelf-mounted ocean thermal power plants, 770–771
Short rotation wood crops (SRWC), 184
Shuman, F., 541
Sidelighting, 681
  lumen method of, 683–692
Significant wave height, 798
  scatter diagram of, 800
Simultaneous saccharification and fermentation (SFF), 209–210
Single-axis tracking parabolic trough collectors, 559–569
Single crystal cells, 384–388
Single-paned windows, heat loss through, 457
Single-spout evaporators, 779, 781
Sinusoidal waves, characteristics of, 795
Sky diffuse radiation, 280
  measurement of with pyranometers, 300
Skylight, 621, 681
Skylighting, lumen method of, 692–697
Sky temperature depression, monthly average, 674
Societal costs, 72–74
Softwoods, 184–185
Soil, thermal properties of (U.S.), 678
Solar altitude angle, 260, 682; *see also* Sun angles
Solar azimuth angle, 260; *see also* Sun angles
Solar cells; *see also* Photovoltaic cells
  efficiency of, 380–381
  manufacture of, 384–389
  multijunction, 381–382
Solar chimney, 672–673
Solar collectors
  determining yearly performance for, 578–580
  evacuated tube, 446–448
  experimental testing of, 448–451
  flat-plate, 423–445 (*see also* Flat-plate collectors)
  fluid flow rates, 471–472
  long-term performance of systems with storage, 758–762
  orientation of, 471
  parabolic, 553–555
  parabolic trough, 559–569
  parameters used to calculate monthly solar flux on, 574
  schematic diagram of, 429
  selection of for industrial process heat applications, 545, 549
  surfaces for, 419–420
  tilt factor for, 336
  transparent materials used in, 421–422
  unglazed transpired wall system, 472–474

Solar constant, 276
Solar control, 670
Solar cooling, 484
absorption air-conditioning, 494–505
requirements for buildings, 485–489
vapour-compression cycle, 489–494
Solar declination, 260
Solar desiccant dehumidification, 505
liquid desiccant cooling system, 508–509
solid desiccant cooling systems, 505–507
Solar energy, 13
cost estimates for, 50
global status and potential of, 41–43
Solar ephemeris, 261–262
Solar flux, parameters used to calculate, 574
Solar heating systems
methods of modeling and design for, 475–484
schematic diagram of system with storage, 753
Solar industrial process heat, 544–549
estimated date of technical readiness for systems, 546–548
long-term system performance, 569–580
milk processing, 550–552
textile industries, 549–550
Solar-load ratio, approach to design of passive heating systems, 665–670
Solar One, 603
Solar power production, options for thermal energy storage in, 751–752
Solar process heat, historical perspective, 541–544
Solar radiation, 275
atmospheric extinction of, 281–282
circumsolar or anisotropic diffuse, 295–296
on clear days, 282
clear-sky data for, 298–299, 337–361
concentration of, 553
daily models for tilted surfaces, 296–297
data and web sites, 304–305
estimation of available, 391–392
extraterrestrial, 276–279
hourly models for tilted surfaces, 297–299
measurement of, 299–305
models based on long-term measured, 290–299
monthly estimation models, 287–290
regression coefficients for Page model of, 333–335
terrestrial, 280–290
on a tilted surface, 283–287
United States daily average data, 325–332
worldwide data for, 314–324
Solar savings fraction (SSF), 632–633
relations, 635–640
sensitivity curves, 661–664
Solar thermal collection, basic principle of, 413
Solar thermal collectors
testing standards for, 452–453
thermal performance of, 427–433
types of, 413–414
Solar thermal power, 581
central receiver systems, 595–605
examples of Stirling engine systems, 609–613
generation of based on Rankine cycle, 581, 583–586
generation of with Stirling engines, 606–608
parabolic dish-based plant, 605
parabolic trough-based plant, 591–594
Solar thermal storage, 745
Solar thermosyphon water heating, 623–627
Solar time
method for computing, 308–309
sun angles and, 262–269
Solar trough thermal power system, 594
Solar Two, 596–603
Solar water-heating systems, 464
forced-circulation systems, 467–469
industrial process heat systems, 469
natural circulation systems, 464–467
Solar zenith angle, 260; *see also* Sun angles

Solid desiccant cooling systems, 505–507
Solid-fuel combustion, 231
Solid-fuel combustors, 233
Solid oxide fuel cells (SOFC), 740–742
SOLMET, 305
Solution mining, 718–719
Space-heating systems, active solar heating systems, 470–471
Specific energy, 712
Spectral solar radiation, measurement of, 304
Specular reflection, 414, 419
  values for solar reflector materials, 420
Speed of light, 248
Spinning reserves, 707
Spout evaporators, 779, 781
Sputtering, 399
Stable atmosphere, 139
Starch, 207
Steam
  saturated, 534–535
  superheated, 536–537
  use of as a wetting fluid in Rankine power plants, 588
Steam methane reforming, 742
Stefan–Boltzmann law, 250
Stirling cycle
  states and processes, 610
  thermodynamics of, 606–608
Stirling engines
  examples of use for solar thermal power generation, 609–613
  generation of solar thermal power using, 606–608
Storage; *see also* Energy storage
  energy modes of, 712
  hydrogen, 743–744
  size, 711
  thermal energy, 745–762
  thermal mass, 633
Sugar fermentation, fuels from, 210–212
Sun angles
  definitions of for a tilted surface, 283
  method for computing, 308–309
  solar time and, 262–269
Sun–earth geometric relationship, 259–274
Sunlight, 621, 680–681
Sun-path diagrams, 269–271, 310–312
Sun, physics and energy transport of, 247–248
Sunrise times, estimation of, 265
Sunset times, estimation of, 265
Sunshine duration, measurement of, 303
Sunshine switch, 303
Sunspace, 628–629
Sun–window geometry, 682
Superconducting magnetic energy storage (SMES), 727
Superheated steam, 536–537
Suspension burners, 233–234
Sustainability
  population growth and, 4–5
  U.N. report on, 1
Sustainable development, 1–3
  energy and, 10–18
  food supply and, 9–10
  water and, 5–8
Sustainable energy, economic analysis of using EROI, 91–92
Switchgrass
  calculations of enthalpies of formation for, 189–190
  use of as a biomass feedstock, 183–184
Syngas, 50, 181
  ethanol from, 212–213

T

Tank-to-wheel (TTW) efficiency, 226–228
Tapchan wave capture system, 801
Tar, removal of from biomass product gases, 202–203
Tenter frames, energy consumption and demand for, 551
Terrestrial radiation, 248
Terrestrial solar radiation, estimation of, 280–290
Textile industrial, solar process industrial heat for, 549–550
Thermal energy storage (TES), 708, 745
  applications, 748, 750
  for concentrating collector systems, 750–752

overnight storage for buildings and domestic hot water, 752–753, 756–758
phase change heat storage, 747–748
sensible heat, 745–747
thermochemical storage, 748
Thermal evaporation, 399
Thermal gasification, 195–203
process of, 199
Thermal mass storage, 633
Thermal performance, 453
Thermal radiation
black-body radiation, 250
fundamentals of, 248–250
intensity of and shape factor, 252, 255–257
radiation function tables, 250–255
transmission of through a medium, 258
Thermal storage walls, 628–629
Thermochemical storage, 748
reactions, 750
Thermodynamic limit, 439
Thermodynamics of hydrogen, 744
Thermoelectric detectors, 303
Thermo-fluid system design considerations, 625–627
Thermolysis of water, 31
Thermosyphon loop, 464
Thermosyphons
concept of, 623–625
design considerations for, 625–627
use of in solar water-heating systems, 464–467
The Whale OWC system, 803–804
Thin-film photovoltaic technology, 395
deposition techniques for, 399–400
types of, 397–398
Thin films
amorphous silicon and multijunction, 388–389
polycrystalline, 386–388
Thrust, 151
Tidal barrages
estimating power output from, 787–788
world locations for, 789
Tidal energy, 785–787
generation of power from, 790–791
Tidal power, 787–791
economics of, 791–794
Tidal power *vs.* hydropower, 785
Tilted surfaces
daily solar radiation on, 296–297
hourly solar radiation on, 297–299
monthly solar radiation estimation models for, 290–295
solar radiation on, 283–287
Time constant, 452–453
Time value of money, 66–70
Tip speed ratio (TSR), 147
calculation of, 147
Toplighting, 681
Total life cycle costs (TLCC), 74–76
Transesterification, 222
Transient thermal analysis, 442–445
Transitional acetogenic dehydrogenation, 193
Transmittance, 417
Transparent materials, 421–422
Transportation
application of energy storage for, 709
future of, 809
hydrogen for, 828–829
Transportation fuels, 204
comparison of, 205
well-to-wheel analysis of, 226–228
Transport, hydrogen, 743–744
Trickle collectors, 427
Triglycerides, conversion of into biodiesel, 219–223
TRNSYS, 581, 604
component subroutines, 584
Trombe wall, 470, 628–629
Turbine power coefficients, 129
Turbines, 117–119; *see also* Wind turbines
bird and bat kills due to, 124–125
bulb, 786, 791–792
capacity of in Severn barrage, 792
in closed-cycle OTEC systems, 782
constant speed/variable pitch (CSVP), 156–158
costs of offshore, 160
gas, 717–718
hydro, 715–717
increasing power output by changing area of, 121–122
noise concerns related to, 123

offshore, 169 (*see also* Offshore wind farms)
in open-cycle OTEC systems, 777–778
performance of, 146–147
rated power production of, 120
steam, 586–587
variable speed/variable pitch (VSVP), 156–158
wake effects of, 165
wells, 802–803
Turbulent velocity, 135
Two-tank biomass reactors, 193
Two-tank indirect systems, 752

U

Ultracapacitors, 709, 726–727
Unglazed transpired wall system for air preheating, 472–474
United States
average daily solar radiation in, 325–332
energy consumption of, 11
energy production by major source, 12–13
industrial heat usage in, 544–545
per capita energy use in, 18–19
2004 primary energy consumption, 108
size of biomass resource base in, 185–187
wind capacity potential in, 169
wind power resource map, 170
Unstable atmosphere, 139
Updraft gasifiers, 198
Uranium, 28–29
Urbanization, effect of on food availability, 10
Utility shaping, 709
Utilizability method, 484
assessing performance of SIPH systems using, 570
empirical expressions for, 573, 575
examples for, 575–577
Utilization coefficient, 695

V

Valence band, 367
Vanadium redox (VR) batteries, 737–739
Vapor compression cycle, 489–494
Variable speed/variable pitch (VSVP) turbines, 156–158
Velocity vectors for wind turbines, 133–134
Ventilation air preheating, unglazed transpired wall system for, 472–474
Vertical axis wind turbines (VAWTs), 126, 128; *see also* Wind turbines
Vertical illuminance, 685
Vibrio furnissii, hydrocarbon synthesis by, 211
Volatile matter, 189
Volta, Alexander, 727
Voltages in photovoltaic cells, 375–380

W

Wake effects, 165
Wall-insulation materials, conductance and resistance values for, 529–530
Waste materials, as biomass feedstocks, 182–183
Water
electrolysis of, 31–32
pumping using photovoltaics, 394–395
saturated, 534–535
solar systems for heating of, 464–469
solar thermosyphon heating systems, 623–627
sustainability based on, 5–8
thermolysis of, 31
Water–gas shift reaction, 198
Water-induced phase separation, 207
Wave capture systems, 801–802
Wave power, 795–798
devices, 801–804
Wave profile devices, 804–805
Wave records, 799
Wealth, energy use and, 10–11
Weibull probability distribution function, 143
Wells turbine, 802–803
Well-to-tank (WTT) efficiency, 226–228
Well-to-wheel (WTW) analysis of transportation fuels, 226–228, 812–813

Western Europe, economic energy efficiency potentials in, 23–24
Wet cell batteries, 734
Wetting fluids, 588
Whole-tree burners, 234
Wind
- aerodynamic power of, 152
- annual energy production determination, 152–154
- cost of energy, 158–163
- distribution of, 135–137
- generation of, 134–135
- power and annual energy calculation, 155
- power of, 120–121

Wind energy, 13
- calculation of EROI for DePere Wind Project, 102–105
- cost estimating example, 52–53
- environmental impact of using, 123–124
- EROI for sample systems, 106–107
- global status and potential of, 39–41
- offshore, 164–173
- resource potential, 137

Wind farms, 163–167
- offshore, 164–173

Windows
- heat loss through, 457–458
- placement of for passive cooling design, 670–671

Wind power ratio (WPR), 167
Wind power resource map (U.S.), 170
Wind speed
- increase in with height, 137–138
- log law profile, 138–140
- power law profile, 140–141
- probability of observing a given speed, 141–146

Wind turbines, 117–119, 715; *see also* Turbines
- aerodynamics of, 130–134
- bird and bat kills due to, 124–125
- capacity factor as a measure of, 155–156
- comparison of HAWTs *vs.* VAWTs, 128
- control schemes for, 156–158
- cost of energy (COE), 158–163
- floating, 172–174
- history of development of, 125–127
- increasing power output by changing area of, 121–122
- intermittency, 164
- noise concerns related to, 122–123
- performance of, 129–130, 146–147
- rated power production of, 120
- wake modeling of, 165–166

Woods, conductance and resistance values for, 529
Woody biomass, 184
Working fluid
- calculation of rate of heat transfer to, 433–437
- choice of for Rankine power plants, 588–591
- seawater as, 777

Work-plane illuminances, 687–688, 696
World Commission on Environment and Development, 1
World oil production, 809
World population, sustainable development and, 4–5
World Radiation Data Center (WRDC), solar radiation data from, 305

Z

Zenith angle, 260; *see also* Sun angles
Zero emission vehicles, 827
Zero-up-crossing period, 798–799